Techniques in
AQUATIC TOXICOLOGY

Edited by
Gary K. Ostrander
The Environmental Institute
Department of Zoology
Department of Biochemistry and Molecular Biology
Oklahoma State University
Stillwater, OK

LEWIS PUBLISHERS
Boca Raton   New York   London   Tokyo

Acquiring Editor:	Neil Levine
Project Editor:	Carole Sweatman
Marketing Manager:	Greg Daurelle
Cover Design:	Shayna Murry
Typesetting:	Pamela Morrell
Pre Press:	Kevin Luong
Manufacturing:	Sheri Schwartz

Library of Congress Cataloging-in-Publication Data

Techniques in aquatic toxicology / editor, Gary K. Ostrander.
     p.   cm.
     Includes bibliographical references and index.
     ISBN 1-56670-149-X (alk. paper)
     1. Water—Pollution—Toxicology—Measurement.   2. Water quality
bioassay—Technique.   I. Ostrander, Gary Kent.
     QH90.8.T68T43   1996
     574.2′4—dc20

96-6037
CIP

QH
90.8
.T68
T43
1996

© 1996 by CRC Press, Inc.
Lewis Publishers is an imprint of CRC Press

No claim to original U.S. Government works
International Standard Book Number 1-56670-149-X
Library of Congress Card Number 96-6037
Printed in the United States of America 1 2 3 4 5 6 7 8 9 0
Printed on acid-free paper

Dedication

Dedicated to past, present, and future graduate students, post-docs, and technicians who are largely responsible for most accomplishments that occur in our laboratories

Preface

As the second millennium draws to a close we are struck with the realization that anthropogenic perturbations of aquatic ecosystems are of relatively recent origin. Yet, the eventual remediation of these ecosystems will take us many decades and perhaps even centuries beyond the year 2000. Whether it be the development of successful remediation strategies or appropriate revision of current environmental practices, a sound knowledge base, relative to aquatic resources is essential. Acquisition of this base is best acquired through carefully constructed, concerted, and focused research efforts. Ideally, whenever possible, researchers should avail themselves to "cutting-edge" technologies to accelerate our arrival at the day when all of our aquatic ecosystems are again robust and healthy. To this end, contributors to the initial volume of this series have provided detailed protocols for techniques currently in use in the field of aquatic toxicology. It is intended that subsequent editions will follow at 2 to 3 year intervals.

This book includes a blend of established techniques, as well as recently developed novel approaches. Divided into four broad sections, the volume includes techniques for the assessment of toxicity in whole organisms, cellular and subcellular toxicity, contaminant identification, and concludes with general techniques for aquatic toxicologists. Each individual paper covers, in detail, a particular method or procedure and begins with a brief *Introduction*, highlighting the technique. The *Materials* section provides the reader with an itemized list of necessary materials, and where appropriate, their source. The *Procedures* section includes a detailed explanation of the technique such that an investigator could hand the volume to graduate students or technicians and expect them to be able to establish it in their laboratory with little or no outside consultation. This latter statement should be taken in context. For example, an author who contributed a paper detailing a new molecular biology technique or procedure for gas chromatography/mass spectrometry analysis was told to assume that the readership will already be competent on basics of the appropriate discipline. In the *Results* section the authors have been asked to provide typical results, as well as anomalous results such as false positives, artifacts, etc. In some cases the authors have provided data from their published work, in other instances new information is provided. Each paper concludes with a brief *Discussion* of the potential utility of the technique, and as necessary, other considerations. Finally, the authors have provided pertinent *References*.

Contributions to this volume were subjected to at least two external anonymous reviews. I am most grateful to the 60 plus reviewers who participated in this process. I also thank the authors for their considerable efforts and willingness to put their techniques to paper in an easily accessible format for our colleagues. Finally, I am most grateful to Ms. Melanie Storm who spent considerable effort in the processing of the manuscripts, reviews, and revisions.

Gary K. Ostrander

Editor

Gary K. Ostrander, Ph.D. is the Associate Dean of the Graduate College and Director of the Environmental Institute at Oklahoma State University, Stillwater, Oklahoma. He holds academic appointments in the Department of Zoology and the Department of Biochemistry and Molecular Biology.

Dr. Ostrander received his B.S. degree in biology from Seattle University in 1980, obtained his M.S. degree in biology at Illinois State University in 1982, and his Ph.D. degree in aquatic toxicology from the University of Washington in 1986. He was a NIH postdoctoral fellow in the Department of Pathology, School of Medicine, University of Washington from 1986 to 1989 and also served as a Staff Scientist at the Pacific Northwest Research Foundation from 1986 to 1990.

Dr. Ostrander is an editor for the journal *Aquatic Toxicology* and serves on the Board of Directors for the National Institutes for Water Resources. He is a member of the American Association for Cancer Research, Society of Environmental Toxicology and Chemistry, and Sigma Xi.

Dr. Ostrander has authored or edited over 50 technical papers, book chapters, and books. His major research interests are in the area of chemical carcinogenesis with a focus on exploitation of the novel aspects of the biology of fishes to address fundamental mechanistic questions relative to cancer in all species. Current studies in his laboratory are aimed at elucidating the functional significance of oncogenes, tumor suppressor genes, and carbohydrate antigens during chemical carcinogenesis in human and aquatic species.

Contributors

Gun Akèrman
Stockholm University Laboratory
  for Aquatic Ecotoxicology
Nykoping, Sweden

Raymond W. Alden, III
Applied Marine Research
  Laboratory
Old Dominion University
Norfolk, Virginia

James C. Allgood, Jr.
School of Pharmacy
University of Mississippi
University, Mississippi

Brian S. Anderson
Institute of Marine Sciences
University of California
Santa Cruz, California

Jack W. Anderson
Columbia Analytical Services
Carlsbad, California

Michael Arts
National Hydrology Research
  Institute
Environmental Sciences Division
Saskatoon, Saskatchewan, Canada

Gordon C. Balch
Environmental and Resource
  Studies
Trent University
Peterborough, Ontario, Canada

Lennart Balk
Stockholm University
Laboratory for Aquatic
  Ecotoxicology
Nykoping, Sweden

Joseph R. Beaman
Biomedical Department of
  Immunotoxicology
U.S. Army Biomedical Research
  and Developmental
  Laboratory
GEO-CENTERS, INC.
Fort Detrick
Frederick, Maryland

William H. Benson
School of Pharmacy
University of Mississippi
University, Mississippi

Kristen Bothner
MEC Analytical Systems
Carlsbad, California

Anthony Bulich
Microbics Corporation
Carlsbad, California

Susan E. Burbank
School of Biology
Georgia Institute of Technology
GA EPD
Atlanta, Georgia

G. Thomas Chandler
Department of ENHS
School of Public Health
University of South Carolina
Columbia, South Carolina

Peter F. Chapman
Zeneca Agrochemicals
Jealott's Hill Research Station
Bracknell, Berkshire, United
  Kingdom

Peter S. Cooper
Department of Environmental
  Sciences
School of Marine Science
Virginia Institute of Marine Science
Gloucester Point, Virginia

Marianne W. Curry
Department of Immunology
U.S. Army Biomedical Research and
  Developmental Laboratory
GEO-CENTERS, INC.
Fort Detrick
Frederick, Maryland

Geoffrey S. Ellis
U.S. Geological Survey
Water Resources Division
Arvada, Colorado

David L. Fabacher
U.S. Department of the Interior-NBS
Midwest Science Center
Columbia, Missouri

Kevin P. Feltz
National Biological Service
Midwest Science Center
Columbia, Missouri

Eliezer Flescher
NYU Medical Center
Department of Environmental
  Medicine
Tuxedo, New York

John W. Fournie
National Health and Environmental
  Effects Research Laboratory
Gulf Ecology Division
U.S. EPA
Gulf Breeze, Florida

Robert W. Gale
U.S. Department of the Interior-NBS
Midwest Science Center
Columbia, Missouri

Robert B. Gillespie
Department of Biology
Indiana University-Purdue
  University
Fort Wayne, Indiana

Denise A. Gordon
National Exposure Research
  Laboratory
U.S. EPA — Ecological Exposure
  Research Division
Cincinnati, Ohio

Kristina N. Graham
Department of Soil Science
University of Manitoba
Winnipeg, Manitoba, Canada

Andrew S. Green
Department of ENHS
USC-Columbia
Columbia, South Carolina

William E. Hawkins
Gulf Coast Research Laboratory
Ocean Springs, Mississippi

John V. Headley
National Hydrology Research
  Institute
Environment Canada
Environmental Sciences
  Division
Saskatoon, Saskatchewan,
  Canada

Michael F. Helmstetter
Brevard Teaching and Research
  Laboratories
Palm Bay, Florida

Michelle Hester
Institute of Marine Sciences
University of California
Santa Cruz, California

James P. Hickey
Great Lakes Science Center
National Biological Service
Ann Arbor, Michigan

Philip H. Howard
Environmental Sciences
  Center
Syracuse Research Corporation
Syracuse, New York

James N. Huckins
Midwest Science Center (USDI)
Chemistry Department
Columbia, Missouri

John W. Hunt
Institute of Marine Sciences
University of California
Santa Cruz, California

Huyen Huynh
Microbics Corporation
Carlsbad, California

Nazrul Islam
NYU Medical Center
Department of Environmental
  Medicine
Tuxedo, New York

Suzanne V. Jacobson
University of Maryland
School of Medicine
Department of Pathology
Baltimore, Maryland

B. Thomas Johnson
Midwest Science Center (USDI)
Environmental Toxicology
  Department
Columbia, Missouri

Andrew S. Kane
University of Maryland School of
  Medicine
Aquatic Pathobiology Center
Department of Pathology
Baltimore, Maryland

Richard M. Kocan
School of Fisheries
University of Washington
Seattle, Washington

Rena M. Krol
Gulf Coast Research Laboratory
Ocean Springs, Mississippi

David L. Lattier
Oak Ridge Institute for Science and
  Education
U.S. Department of Energy and
  NERL
U.S. EPA
Cincinnati, Ohio

J. McHugh Law
College of Veterinary Medicine
North Carolina State University
Raleigh, North Carolina

Jon A. Lebo
Midwest Science Center (USDI)
Chemistry Department
Columbia, Missouri

Edward E. Little
Midwest Science Center
Columbia, Missouri

Alexander E. Maccubbin
Grace Cancer Drug Center
Roswell Park Cancer Institute
Buffalo, New York

Gamini K. Manuweera
Pesticide Registration Office
Gatambe, Peradeniya, Sri Lanka

Stephen J. Maund
Zeneca Agrochemicals
Jealott's Hill Research Station
Bracknell, Berkshire, United
  Kingdom

Frank McCormick
National Exposure Research
  Laboratory
U.S. EPA — Ecological Exposure
  Research Division
Cincinnati, Ohio

Debra J. McMillin
Department of Veterinary
  Pharmacology and Toxicology
School of Veterinary Medicine
Louisiana State University
Baton Rouge, Louisiana

Jay C. Means
Department of Veterinary
  Pharmacology and Toxicology
School of Veterinary Medicine
Louisiana State University
Baton Rouge, Louisiana

Chris D. Metcalfe
Environmental and Resource
  Studies
Trent University
Peterborough, Ontario, Canada

William M. Meylan
Environmental Sciences Center
Syracuse Research Corporation
Syracuse, New York

Michael R. Miller
Department of Biochemistry
Health Sciences Center
West Virginia University
Morgantown, West Virginia

Joy L. Mitchell
School of Biology
Georgia Institute of Technology
Atlanta, Georgia

Michael J. Moore
Biology Department
Woods Hole Oceanographic
  Institution
Woods Hole, Massachusetts

Michael Morss
Biology Department
Woods Hole Oceanographic
  Institution
Woods Hole, Massachusetts

Michael C. Newman
The University of Georgia
Savannah River Ecology
  Laboratory
Aiken, South Carolina

James M. O'Neal
School of Pharmacy
National Center for the
  Development of Natural
  Products
University of Mississippi
University, Mississippi

Carl E. Orazio
Midwest Science Center (USDI)
Chemistry Department
Columbia, Missouri

Lisa S. Ortego
Rhône Poulenc
Department of Toxicology
Research Triangle Park,
  North Carolina

Gary K. Ostrander
The Environmental Institute
Oklahoma State University
Stillwater, Oklahoma

Kerry M. Peru
National Hydrology Research
  Institute
Environmental Sciences Division
Saskatoon, Saskatchewan, Canada

Paul H. Peterman
U.S. Department of the Interior-NBS
Midwest Science Center
Columbia, Missouri

Jimmie D. Petty
Midwest Science Center (USDI)
Chemistry Department
Columbia, Missouri

Bryn Phillips
Institute of Marine Sciences
University of California
Santa Cruz, California

Harry F. Prest
Joseph M. Long Marine Laboratory
University of California
Santa Cruz, California

Renate Reimschussel
University of Maryland School of
  Medicine
Aquatic Pathobiology Center
Department of Pathology
Baltimore, Maryland

Linda D. Rhodes
Department of Zoology
University of Maine
Orono, Maine

Brenda M. Sanders
Molecular Ecology Institute
California State University
Long Beach, California

Leonard P. Sarna
University of Manitoba
Department of Soil Sciences
Winnipeg, Manitoba, Canada

Lee R. Shugart
Environmental Science
  Division
Oak Ridge National Laboratory
Oak Ridge, Tennessee

Richard Silbiger
DynCorp
U.S. EPA
Cincinnati, Ohio

M. Kate Smith
National Exposure Research
  Laboratory
U.S. EPA — Ecological Exposure
  Division
Cincinnati, Ohio

Terry W. Snell
School of Biology
Georgia Institute of Technology
Atlanta, Georgia

David H. Swenson
School of Veterinary Medicine
Department of Veterinary
  Pharmacology and Toxicology
Louisiana State University
Baton Rouge, Louisiana

Sanjeev Thohan
University of Maryland
School of Pharmacy
Baltimore, Maryland

Donald E. Tillitt
U.S. Department of the Interior-
  NBS
Midwest Science Center
Columbia, Missouri

Robert H. Tukey
Department of Pharmacology
University of California,
  San Diego
La Jolla, California

Lorraine E. Twerdok
American Petroleum Institute
Health and Environmental
  Science Division
Washington, D.C.

Shimon Ulitzur
Department of Biotechnology
Technion-Israel Institute of
  Technology
Haifa, Israel

Rebecca J. Van Beneden
Department of Zoology
University of Maine
Orono, Maine

Peter A. Van Veld
Department of Environmental
  Sciences
School of Marine Science
Virginia Institute of Marine
  Science
Gloucester Point, Virginia

Wolfgang K. Vogelbein
Department of Environmental
  Sciences
School of Marine Science
Virginia Institute of Marine
  Science
Gloucester Point, Virginia

Tien Vu
Department of Pharmacology
University of California, San Diego
LaJolla, California

Mary K. Walker
Department of Nutritional Sciences
University of Wisconsin
Madison, Wisconsin

William W. Walker
Gulf Coast Research Laboratory
Ocean Springs, Mississippi

Weihua Wang
NYU Medical Center
Department of Environmental
  Medicine
Tuxedo, New York

G. R. Barrie Webster
Department of Soil Science
University of Manitoba
Winnipeg, Manitoba, Canada

Peggy J. Wright
National Biological Service
Midwest Science Center
Columbia, Missouri

Marilyn J. Wolfe
Experimental Pathology
  Laboratories
Herndon, Virginia

Erik W. Zabel
School of Pharmacy
Madison, Wisconsin

Judith T. Zelikoff
NYU Medical Center
Department of Environmental
  Medicine
Tuxedo, New York

Contents

III. Techniques for identification and assessment of contaminants in aquatic ecosystems

IV. Techniques for aquatic toxicologists

Section I

Techniques for assessment of toxicity in whole organisms

chapter one

The use of luminescent bacteria for measuring chronic toxicity

Anthony A. Bulich, Huyen Huynh, and Shimon Ulitzur

Introduction

Bioassays used in aquatic toxicology have taken a prominent position among analytical tests for identifying and measuring environmental hazards. In particular, chronic toxicity tests have been developed for testing effluents, surface waters, and sediment samples to estimate the safe or no effect sample concentration.

Chronic tests are intended to predict adverse biological effects that result when organisms are exposed over their entire life cycle. Such tests are difficult and expensive to perform; thus, abbreviated tests have been developed that use only the most vulnerable life stages, such as growth or reproductive capacity. Even such abbreviated tests lack sufficient standardization and are complicated and expensive to perform. Short-term tests, using chronic endpoints, are needed for applications such as routine effluent monitoring and toxicity identification evaluations.[1]

Bioassays using luminescent bacteria are routinely used to assess the acute toxicity of environmental samples. Over the last 15 years, various applications using these organisms have been validated and recognized by several standards organizations.[2,3] Luminescent bacteria possess several attributes that support their practical use for toxicity testing. Their small cell size provides a high surface-to-volume ratio which maximizes exposure potential. This structural characteristic plus (1) lack of membrane-aided compartmentalization; (2) location of most respiratory pathways (including enzymes required for bioluminescence) on or near the cell membrane; and (3) a metabolic rate 10 to 100 times mammalian cells, provide a dynamic metabolic system which can be easily quantitated by measuring the *rate* of light output. The close association of the light production pathway with the bacteria's respiratory system provides a convenient and sensitive biological system for quantitating metabolic inhibition due to the presence of toxic chemicals.[4]

1-56670-149-X/96/$0.00+$.50

3

Two additional physiological characteristics of these organisms, a relatively short division cycle and an inducible luciferase pathway, provide useful attributes for measuring chronic toxicity. The development of this chronic test method was based upon a detailed understanding of the growth requirements, physiology, and biochemistry of this organism as well as the molecular biology of luciferase induction. Appropriately prepared freeze-dried luminescent bacteria, specifically *Vibrio fischeri* (strain number NRRL B-11177), following inoculation into a defined growth medium, initiates a series of reproductive cycles while concurrently inducing numerous complex metabolic pathways including the enzymes required for light production. Toxic chemicals or samples that inhibit any aspect of the reproductive cycle, growth process, or luciferase induction process can be detected in very low concentrations by simply comparing light production in the sample cuvettes with light production in the control cuvettes.[5]

In summary, the Microtox® Chronic Toxicity Test measures inhibition of light production by luminescent bacteria when these organisms are grown in the presence of sublethal concentrations of toxic agents. The test organisms are supplied in freeze-dried form (Chronic Test Reagent), and the test is performed in a specially defined growth medium which is also supplied in freeze-dried form. After 22 h of growth, the control cells undergo several cellular divisions and induction of the luciferase system. Those cuvettes containing toxic concentrations of sample display less than normal light production. Consequently, toxicity is defined as any sample concentration with light which is at least two standard deviations lower than the average light from the control cuvettes. This biological test system was developed by Microbics Corporation to provide a sensitive, precise, and convenient method for rapidly assessing the chronic toxicity of environmental samples.

Materials required

The Microtox Chronic Test Procedure was developed to provide a sensitive and low-cost bioassay based upon the use of a standardized suspension of luminescent bacteria. Standardization of the test is enhanced by using test organisms and growth medium that have been pre-formulated to defined specifications and freeze-dried for convenient storage and shipping. The materials required to perform this test are listed below with corresponding descriptions and Microbics Corporation part numbers where appropriate.

1. Temperature-controlled luminometer: Microbics Model 500 analyzer (Part No. 50A000) with temperature-controlled read well, temperature-controlled 30-well incubator block, and 5.5°C reagent precooling well. The temperature-controlled photometer has the appropriate sensitivity and dynamic range to measure light production from the luminescent bacteria used for this and other Microtox test procedures.
2. Incubation bath or chamber able to control temperature at 27°C ± 1°C and hold at least 60 12 × 48 mm flat-bottom round cuvettes.
3. Disposable glass cuvettes (No. 686013): 12 × 48 mm, flat-bottom round cuvettes made of borosilicate glass. These cuvettes are pretested to

confirm that they do not contain any toxic materials which could interfere with the test.

4. Microtox Chronic Test Reagent (No. 989015): freeze-dried luminescent bacteria, *Vibrio fischeri* strain number NRRL B-11177. Each 6 ml vial has a 1.0 ml dry pellet containing the equivalent of 10^7 cells per ml. This reagent must be stored at $-15°$ to $-20°C$ and is used to inoculate the individual control and test vials containing the Microtox Chronic Test Medium. Each vial of the Microtox Chronic Test Reagent must be reconstituted with 3.0 ml of $5.5°C \pm 1°C$ Microtex Chronic Test Reagent Activation Solution.

5. Microtox Chronic Test Medium (No. 989220): a freeze-dried concentrate of salts and nutrients that is able to support growth and light induction of Microtox Chronic Test Reagent. This test medium must be stored at -15 to $-20°C$ and reconstituted with Microtox Reconstitution Solution.

6. Microtox Chronic Test Reagent Activation Solution (No. 989024): specially prepared nontoxic 3.5% sodium chloride (NaCl) solution used to reconstitute the Microtox Chronic Test Reagent.

7. Microbics Reconstitution Solution (No. 686028): ultrapure water used to reconstitute the Microtox Chronic Test medium.

8. Pipettes: fixed-volume and repeat pipettes able to deliver 20, 500, 1000 μl volumes. These pipettes must use disposable, nontoxic pipette tips.

9. Copper Sulfate positive control. A 10,000 ppb solution of Cu^{++} prepared using analytical grade $CuSO_4 \cdot 5H_2O$ and ultrapure water.

This bioassay is very sensitive. Care must be exercised to minimize inadvertent contamination of test materials. Test cuvettes and sample containers should be made of clean borosilicate glass. The dilution water must be equivalent to Milli Q type water which has been pretested to confirm no toxicity. The test and control cuvettes should be kept lightly covered during the 22-h incubation period to prevent toxic particulates from falling into any of the cuvettes. If the test cuvettes are to be incubated in a dry incubator, the incubator should be humidified using a pan of water. Each test vial contains 0.5 ml of test medium and under conditions of low relative humidity significant water evaporation from the test cuvettes can occur. We have observed water loss of up to 15%, which increases the salt concentration in the test medium, resulting in poor growth of the test organisms.

After inoculation and 22-h incubation of each test and control cuvette, there will be *no* visible turbidity. The inoculum concentration and the concentration of nutrients in the test medium were determined experimentally to support sufficient growth and light induction for a sensitive and reproducible test.

Procedures

The objective of this test is to inoculate a series of test cuvettes containing growth medium prepared with purified, nontoxic water alongside a series of cuvettes containing growth medium prepared with dilutions of the test sample. Each series of cuvettes is inoculated with the same concentration of luminescent bacteria. After 22 h of incubation at $27°C$, light production of the control cuvettes is compared with light production from the test sample cuvettes. The test endpoint is based upon calculating the lowest observable effect concentration (LOEC). This endpoint is defined as that sample concentration which displays an average light production which is at least two standard deviations lower than the average light production from the control cuvettes.[6]

The following procedure describes the method for testing a negative control (blank), a positive control ($CuSO_4$), and an effluent sample. The $CuSO_4$ solution is prepared to 10,000 µg/l in pure water and is used as a 100 times stock solution. Each test round requires the inclusion of multiple (ten) control cuvettes and four dilution sets of the sample. These multiple cuvettes are required to obtain statistically valid LOEC values.

1. Remove two vials of chronic test medium from the freezer and allow them to equilibrate to room temperature (20° to 25°C). Also remove the stock solution of $CuSO_4$ and effluent sample from the refrigerator and allow equilibration to room temperature.

2. Prepare a setup of test cuvettes to comply with the pattern shown in Figure 1: 10 cuvettes are to be used for the negative control; 20 cuvettes for the 5 dilutions of $CuSO_4$ to be tested in quadruplicate; 20 cuvettes for the 5 dilutions of effluent sample also to be tested in quadruplicate.

3. Hydrate one vial of test medium using 36 ml of reconstitution solution. Transfer 1.5 ml of this test medium into a cuvette and place the cuvette in the instrument reagent well. Next, pipette 1.0 ml of test medium into cuvettes A1 and B1 and 500 µl of test medium into cuvettes A2–A5 and B1–B5. These 10 cuvettes will be used as the control cuvettes.

4. Pipette 1.0 ml of test medium from step 3 (medium hydrated with reconstitution solution) into cuvettes C1, D1, E1, and F1. Pipette 500 µl of test medium into cuvettes C2–C5, D2–D5, E2–E5, and F2–F5. These four sets of cuvettes will be used to make dilutions of the $CuSO_4$.

5. Transfer 10 µl from the $CuSO_4$ stock solution into cuvettes C1, D1, E1, and F1. Each of these cuvettes now contains 100 µg/l Cu^{++}.

6. Hydrate a second vial of test medium using 36 ml of room temperature effluent sample. Pipette 1 ml of this medium/sample solution into cuvettes G1, H1, I1, and J1. Also transfer 500 µl of test medium from step 3 (medium hydrated with reconstitution water) above into cuvettes G2–G5, H2–H5, I2–I5, and J2–J5.

7. Prepare dilutions (1:2) of each of the eight sample rows (four $CuSO_4$ rows, and four effluent rows). Beginning with row C, transfer 500 µl from cuvette C1 (containing 100% concentration) to cuvette C2 (now containing 50% concentration). Continue to transfer and mix 500 µl from cuvette C2 to C3, C3 to C4, and C4 to C5 (now containing 6.25% concentration). Discard 500 µl from cuvette C5 so the remaining volume is 500 µl, as is the case with cuvettes C1–C4. Repeat for rows D–J, using a new pipette tip for each dilution set (each row).

8. In order to prepare cuvettes in rows A and B as true negative controls, prepare a false dilution set by transferring 500 µl from cuvette A1 to A2, A2 to A3, A3 to A4, and A4 to A5. Discard 500 µl from cuvette

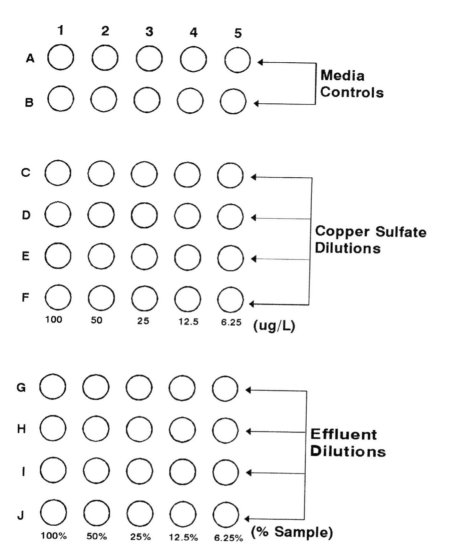

Figure 1   Cuvette setup diagram for the luminescent bacteria chronic toxicity test.

A5. Repeat the same procedure for row B. This procedure is recommended because pipette tips can contain small amounts of a mold release agent which can cause a slight toxic response with the test organisms. These false dilution sets normalize the data for any slight toxic effects imparted by the pipette tips and provide a realistic measure of light production variability between cuvettes.

9. Remove a vial of Microtox Chronic Test Reagent from the freezer, and reconstitute the dry pellet using the 1.5 ml of cold test medium from the reagent well. Mix this suspension thoroughly using the 500 μl pipette by withdrawing and expelling 500 μl of cell suspension 10 times. Proceed to step 10 immediately.

10. Using the repeat pipettor set for 20 µl delivery, inoculate each cuvette with 20 µl of test reagent. Be cautious not to touch the tip of the pipette to the liquid in the cuvette and be careful to expel the 20 µl of reagent directly into the test medium and not onto the side of the cuvette.
11. Transfer all cuvettes containing inoculated test medium to the 27°C incubator bath. Allow all control and test cuvettes to incubate for 22 h.
12. Read the light level for each cuvette.
13. Calculate the mean and standard deviation for the ten control cuvettes. Calculate the mean light value for each test sample concentration.

Results and discussion

Data from three typical experiments are summarized in Tables 1 to 3. The data presented in these tables are organized in the following manner:

1. Readings at 22 h from each of the ten control cuvettes are listed in column one along with the calculated mean value and standard deviation, respectively.
2. Column 2 contains the concentration of the sample tested.
3. Columns 3 to 6 list the actual 22-h light readings for each of the sample dilution sets.
4. Column 7 contains the average light readings for each sample concentration tested. Those values that are boldfaced represent average light levels which are at least two standard deviations lower than the mean value of the control cuvettes. The lowest concentration in this category, by definition, is the LOEC for the sample.

The data presented in Table 1 are test data from a copper sulfate test experiment. These data represent a data set with minimal variability of light production among the ten control cuvettes. A standard deviation of 57 light units for a set of control cuvettes is lower than average, but quite achievable. The calculated average light levels for the concentrations of copper tested show a good dose–response and a calculated LOEC of 25 µg/l. This LOEC value for copper is quite repeatable with LOEC values ranging from 12.5 to 50 µg/l.

Table 1   Test Data for $CuSO_4$

Control readings	Cu^{2+} (µg/l)	Dilution set 1	Dilution set 2	Dilution set 3	Dilution set 4	Calculated average values
750; 702	100	1	3	1	1	**1.5**
675; 738	50	147	105	166	140	**139**
733; 582	25	620	592	497	512	**555**
740; 799	12.5	787	704	599	780	718
699; 695	6.25	730	792	621	679	700

Note: For the control readings, the average was 711 and the standard deviation was 57. Lowest observable effect concentration for C^{2+}: 25 µg/l.

Table 2 Test Data for Effluent Sample 1

Control readings	Sample concentration (%)	Dilution set 1	Dilution set 2	Dilution set 3	Dilution set 4	Calculated average values
835; 892	100	3	1	1	1	**1.5**
964; 1052	50	1	5	1	1	**2**
997; 805	25	2	1	4	1135	**286**
794; 766	12.5	1032	830	738	998	900
985; 1045	6.25	1665	1472	1398	1482	1504

Note: For the control readings, the average was 914 and the standard deviation was 108. Lowest observable effect concentration for sample 1: 25%.

Table 2 summarizes 22-h chronic test data from an industrial effluent. These values are representative of a typical experiment. Note that the standard deviation increased to 108 light units as compared with Table 1, as did the mean light level from the control cuvettes.

It is important to note the actual light levels from each of the sample dilution sets. At sample concentrations of 100% and 50%, there is almost total inhibition of light production, because at these dilutions the sample is quite toxic to the test organisms. There is a similar trend at the 25% sample concentration, except for dilution set 4, where the cells produced a light level higher than the control. This case is not uncommon and is worth some discussion.

This whole test concept is based upon the expectation of similar bacterial growth and light induction in each test vial. We have learned from conducting hundreds of such experiments that light production can be influenced by several factors in addition to sample toxicants. For example, pipetting the cells into each test vial at the beginning of the protocol provides the inoculum for each cuvette. Although pipetting precision from a volumetric perspective is quite good, this assumes that the hydrated bacteria (inoculum) is a homogenous suspension; however, even after mixing the reconstituted bacteria, microscopic examination reveals some clumps of organisms. Consequently, there can be differences in the number of organisms actually delivered to each test cuvette.

Table 3 Test Data for Effluent Sample 2

Control readings	Sample concentration (%)	Dilution set 1	Dilution set 2	Dilution set 3	Dilution set 4	Calculated average values
1153; 945	100	1	1	1	1	**1**
1037; 1026	50	1	1	2	2	**1.5**
1045; 1030	25	156	40	427	739	**341**
641; 1004	12.5	716	770	1599	1169	1064
1002; 1087	6.25	624	900	1461	1156	1035

Note: For the control readings, the average was 997 and the standard deviation was 130. Lowest observable effect concentration for sample 2: 25%.

The data presented in Table 2 also show another interesting facet of this test. Notice that six of the eight light readings in effluent concentrations 6.25% and 12.5% exhibit light readings *higher* than the average control readings, and in most cases, even higher than the *highest* control reading. We know these increased light levels are *not* due to the presence of additional nutrients contributed by the test sample. During the development of this test, effluent and pure compound samples were tested with experimental variations of the test media, supplemented with various nutrients which could be present in samples with high biological oxygen demand (BOD). This addition of nutrients did not significantly increase the light production beyond that which was observed in the control cuvettes containing regular medium.

These examples of light increases are not uncommon and most likely are due to a phenomenon called "Hormesis." All living cells under appropriate stress conditions can alter specific physiological activities. These stress conditions include exposure to nonlethal concentrations of toxic substances. This phenomenon is well documented in the toxicology literature.[7]

Table 3 contains data from a different effluent sample. These data show a higher average light level from the control and a higher standard deviation for the control cuvettes. As with the first effluent sample, light output variability for the LOEC sample concentration shows extreme differences in light production. These types of data further show the need to perform the test using multiple replicates of each test concentration.

We have also learned that this test is quite sensitive to a broad spectrum of chemicals. LOEC values for some 30 chemicals are listed in Table 4. Residual chemicals in pipette tips and repeat pipettor syringes, or particulates that may invade some cuvettes can influence test results. The test should be performed using ten control cuvettes and four sample dilution sets because of these potential sources of vial-to-vial variability.

Table 4   Luminescent Bacteria Chronic Test Lowest Observable Effect
Concentration (LOEC) Data

Compound	LOEC (ppm)	Compound	LOEC (ppm)
Acetone	600.0	Dibromochloromethane	10.0
Aflatoxin	1.6	Dimethyl sulfoxide	800.0
Aluminum	1.6	2,4-D	0.3
Aroclor 1242	0.16	3,5-Dichlorophenol	0.20
Aroclor 1254	0.29	Ethanol	6300.0
Atrazine	0.6	Lead	1.0
2-Amino anthracine	0.098	Mercury	0.05
Beauvericin	0.78	Methoxychlor	0.18
Butanol	7.8	Nickel	0.17
Cadmium	0.06	Sodium lauryl sulfate	0.09
Chromium	0.035	Pentachlorophenate	0.06
Copper	0.02	Trichloroethylene	0.2
Diazinon	0.75	Zinc	0.1

Table 5   Lowest Observable Effect
Concentration (LOEC) Values for Effluent
Samples Tested with C. *dubia* and the
Luminescent Bacteria Chronic Test

C. *dubia* LOEC	Microtox chronic LOEC
>100	100
>100	100
>100	50
>100	50
>100	25
100	>100
100	>100
100	100
100	100
100	50
100	50
100	50
100	25
100	25
>90	100
56	100
50	>100
50	25
50	12.5
25	50
20	25
20	6.25
20	1.56
>10	25
10	25
10	6.25
10	0.39
6.25	6.25
4	6.25
4	3.13
4	0.78
3.13	12.5
2	0.39
2	0.19

One important aspect of any new test system is its comparability with other commonly used tests. We have undertaken a rigorous evaluation program whereby the sensitivity of the luminescent bacteria chronic test is being compared with the *Ceriodaphnia dubia* seven-day chronic test. Coded environmental samples from different laboratories were split and tested with each test system and the LOEC values compared. The first of these comparative data is reported in Table 5. The results from the testing of 34 samples are displayed. The comparative study continues with the ongoing efforts of active participants.

References

1. U.S. Environmental Protection Agency. 1991. Technical Support Document for Water Quality-Based Toxics Control *(EPA/505/2-90-001)*.
2. American Society for Testing and Materials. 1995. *ASTM D-5660, Test Method for Assessing the Microbial Detoxification of Chemically Contaminated Water and Soil Using a Toxicity Test with a Luminescent Marine Bacterium*, ASTM, Philadelphia, PA.
3. National Rivers Authority (U.K). 1991. *Determination of the EC50 of Test Substances to the Microtox Reagent Photobacterium Phosphoreum SOP No. 230/1.*
4. Johnson, F.H., H. Eyring and B.J. Stover. 1974. *The Theory of Rate Processes in Biology and Medicine.* John Wiley & Sons, New York.
5. Meighen, E.A. and P.V. Dunlap. 1993. Bacterial Bioluminescence. *Advances in Microbial Physiology.* 34:1–67.
6. Model 500 Microtox® Chronic Toxicity Test. 1995. Microbics Corporation. 2232 Rutherford Road, Carlsbad, CA 92008.
7. Calabrese, E.J. Ed. 1994. *Biological Effects of Low Level Exposure.* CRC Press, Boca Raton, FL, 302 pp.

chapter two

Rapid toxicity assessment with microalgae using in vivo esterase inhibition

Terry W. Snell, Joy L. Mitchell, and Susan E. Burbank

Introduction

Algae have long been used in water quality assessment as *in situ* biomonitors,[1,2] but have not been commonly used in toxicity tests.[3] A minimal database therefore exists which is not consistent with the ecological importance of algae in aquatic ecosystems. Population growth rate or photosynthetic rate have typically been used as endpoints in algal toxicity tests. In general, biomarkers for rapid toxicity assessment have not been developed for algae as they have for animals.[4]

Rapid toxicity assessment using enzyme inhibition has proven useful in microorganisms.[5,6] Esterase activity in particular has proven to be a sensitive, reliable toxicological endpoint in a variety of organisms.[7-10] Phylogenetically ubiquitous, esterases are found in plant and animal cells,[8] most bacteria, and reflect heterotrophic activity.[7] Esterase activity can be measured easily *in vivo* with the nonfluorescent fluorescein diacetate (FDA) substrate which releases a fluorescent product when cleaved by nonspecific esterases.[11] Esterase activity is a good sublethal indicator of the physiological state of a cell. As toxic stress increases, esterase activity is diminished in a dose-dependent manner. In this paper, we describe a technique for rapid toxicity assessment quantifying *in vivo* esterase activity in microalgae with a microplate fluorometer.

Materials required

Equipment

- Microplate fluorometer: model 7620 (Cambridge Technology)
- Adjustable micropipettes: 10, 100, 1000 µl ranges
- Analytical balance

- Magnetic stirrer
- Compound microscope with 400× magnification
- Centrifuge
- Vortex mixer
- Hemocytometer
- Culture chamber
- Freezer
- pH meter
- Autoclave

Supplies

- 96-Well U-bottom tissue culture plates (Corning)
- Centrifuge tubes (50 ml)
- Beakers (50, 125, 2000 ml)
- Volumetric flasks (50, 100 ml)
- Serological pipettes (5, 10, 25 ml)
- Transfer pipettes
- Microcentrifuge tubes (1.5 ml)
- Gloves
- Microscope slide cover glasses

Reagents

- Fluorescein diacetate (#F1303, Molecular Probes, 4849 Pitchford Avenue, Eugene, OR 97402-9144)
- Sea salts (Instant Ocean, Aquarium Systems)
- Sodium dodecyl sulfate (SDS) (# L-4509, Sigma)

All reagents used for algal medium preparation (Table 1) are reagent grade and were obtained from Sigma Chemical Company.

Toxicants

- Acetone (#17,997-3, Aldrich)
- Dimethyl sulfoxide (DMSO) (#D-8779, Sigma)
- Pentachlorophenol, sodium salt (#17,130-1 Aldrich)
- Phenol (#P-4161, Sigma)
- Chlorpyrifos (#PS-674, Chem Services)
- Mercuric chloride (#M1136, Sigma)
- Naphthol (#N12-100, Fisher)
- Cadmium atomic absorption standard (#C-5524, Sigma)
- Cupric chloride (#C-6641, Sigma)

Procedures

Algae cultures

The algae used in these tests were obtained from the Culture Collection of Algae at the University of Texas at Austin, Department of Botany, The Uni-

Table 1 Algal Growth Media for Freshwater
and Marine Algae

Chemical	Allen's	F (15 ppt)	BBM
$CaCl_2$	5		19
Citric acid	1.2		
K_2HPO_4	7.5		75
KH_2PO_4			175
$MgSO_4 \cdot 7H_2O$	7.5		75
$NaNO_3$	4		
Na_2CO_3	1.5 g		250
$Na_2SiO_3 \cdot H_2O$	12		
PIV metals	1 ml		
NaCl			25
KNO_3		15	
NaH_2PO_4		5	
Sea salts		15 g	
Trace metals		2 ml	2 ml
Vitamins		2 ml	2 ml

	PIV metals	Trace metals	Vitamins
$CoCl_2 \cdot 6H_2O$	2	10	
$FeCl_3 \cdot 6H_2O$	97		
$MnCl_2 \cdot 4H_2O$	41	180	
$Na_2MoO_4 \cdot 2H_2O$	4	6.4	
$ZnCl_2$	5		
Na_2EDTA	750		
$CuSO_4 \cdot 5H_2O$		10	
$ZnSO_4 \cdot 7H_2O$		22	
NaFeEDTA		5 g	
B_{12}			10
Biotin			10
Thiamine			200

Note: Table entries are mg/l unless otherwise noted.

versity of Texas at Austin, Austin, TX 78718-7640.[12] The green alga *Selenastrum capricornutum* (#1648) was cultured in Bold's Basal Medium.[13] The prasinophyte *Tetraselmis suecica* (#2286) and the diatom *Cyclotella* sp. (#1269) were grown in F-medium.[14] The cyanobacterium *Synechococcus leopoliensis* (#625) was grown in Allen's medium.[15] The composition of these algal culture media is listed in Table 1.

Unialgal cultures were propagated serially in 20-ml test tube cultures under sterile, controlled conditions of 25°C and constant light of 3000 lux. Twenty ml from a test tube culture was used to initiate a 2-l culture which was aerated in constant light at 25°C for about 3 to 4 days until log-phase growth was reached. Log-phase growth lasted 3 to 4 days and cells were collected for toxicity tests as needed. Algal cells for toxicity tests were harvested from the 2-l cultures by centrifugation in 50-ml tubes at 1150xg for 5 to 10 min. Concentrated cells were collected with a transfer pipette and resuspended in 25 ml of dilution water. Algal cells were counted with a hemocytometer at 400× and 50 to 100 μl were pipetted into a test well of a 96-well plate. These were then diluted to the appropriate final density with test solution for a final volume of 250 μl.

Enzyme substrate

The *in vivo* esterase activity of algal cells was monitored using the substrate FDA. The FDA stock was made in a microcentrifuge tube by adding 1 mg FDA to 1 ml of acetone yielding a 2.4 mM solution. Aliquots of 200 μl were frozen at –80°C for up to 2 months. Shortly before use, the final dilution was prepared by adding 3 μl of the FDA stock to 95 μl of acetone (73 μM), then 37 μl of this solution was added to each test well (2.7 μM final FDA concentration). FDA slowly hydrolyzes in water releasing the fluorescent product so it is important to prepare the final dilution immediately prior to use.

Dilution water

For freshwater algae, moderately hard synthetic fresh water[16] was prepared weekly consisting of 96 mg $NaHCO_3$, 60 mg $CaSO_4 \cdot 2H_2O$, 123 mg $MgSO_4 \cdot 7H_2O$, and 4 mg KCl in 1 l of deionized water. The pH was adjusted to 7.5 with NaOH or HCl. For *Tetraselmis*, ASPM synthetic seawater[14] was prepared which consists of 11.31 g NaCl, 0.36 g KCl, 0.54 g $CaCl_2$, 1.97 g $MgCl_2 \cdot 6H_2O$, 2.39 g $MgSO_4 \cdot 7H_2O$, 0.17 g $NaHCO_3$ in 1 liter of deionized water and has a salinity of 15 ppt. The pH was adjusted to 8.0 with HCl. For *Cyclotella*, ASPM reagent concentrations were doubled to obtain 30 ppt seawater.

Deionized water was obtained by passing tap water through a Spectra-Pure Reverse Osmosis system consisting of a 1 μm spun fiber pre-filter, a high activity carbon pre-filter and a thin film composite reverse osmosis membrane. Water was then passed through a Barnstead Nanopure II system consisting of one pretreatment, one high capacity, and two ultrapure cartridges.

Test conditions

The following environmental conditions were closely regulated during each test: temperature 25°C; exposure in darkness; 96-well plate containing 250 μl of test solution; cell density: *Selanastrum*, 7.5×10^5 cells/ml; *Synechococcus*, 5×10^6 cells/ml; *Cyclotella*, 1×10^5 cells/ml; *Tetraselmis*, 2.5×10^5 cells/ml; 1-h toxicant exposure; 5 to 15 min incubation with FDA substrate.

Toxicants

Seven chemicals representing several toxicant classes were tested including the metals: copper, mercury, and cadmium; the pesticides: pentachlorophenol (PCP) and chlorpyrifos; and the organic compounds: phenol and naphthol. Naphthol and chlorpyrifos have limited water solubility so 1 mg was dissolved first in 1 ml of dimethyl sulfoxide (DMSO), then diluted 1:50 with dilution water. For these two toxicants, a solvent control was included that

Table 2   Toxicant Concentrations Tested for the Four Algal Species (mg/l)

Toxicant	*Selanastrum*	*Tetraselmis*	*Cyclotella*	*Synechococcus*
Pentachlorophenol	0.15, 0.3, 0.5, 0.7, 0.9	10, 15, 20, 25, 30	0.15, 0.3, 0.5, 0.7, 0.9	30, 40, 50, 60, 70
Naphthol	0.3, 0.9, 1.2, 3.6, 6.0	1.5, 2.5, 4.0, 7.0, 12.0	0.5, 0.92, 2.4, 4.8, 6.0	10, 20, 30, 40, 50
Chlorpyrifos	0.07, 0.14, 0.2, 0.3, 0.5	0.5, 1.0, 1.7, 2.7, 4.5	0.5, 1.0, 2.0, 3.0, 4.0	30, 40, 50, 60, 70
Cadmium	0.01, 0.05, 0.1, 0.5, 1.0	0.1, 0.5, 1.0, 2.0, 3.0	0.1, 0.5, 1.0, 2.0, 3.0	0.5, 1.0, 1.7, 2.7, 4.5
Mercury	0.005, 0.08, 0.2, 0.4, 0.8	0.05, 0.08, 0.13, 0.2, 0.3	1.0, 1.5, 2.0, 2.5, 3.0	0.01, 0.02, 0.03, 0.05, 0.1
Copper	0.025, 0.08, 0.2, 0.4, 0.8	0.05, 0.1, 0.3, 0.5, 0.8	0.2, 0.4, 0.8, 1.0, 2.0	0.01, 0.02, 0.03, 0.05, 0.1
Phenol	0.1, 0.5, 2.5, 5.0, 20.0	—	—	30, 40, 50, 60, 70

had the same DMSO concentration as the highest concentration of test chemical. Stock solutions were prepared for each toxicant at the following concentrations (µg/ml): PCP — 5; naphthol — 20; chlorpyrifos — 36; cadmium — 20; mercury — 37; copper — 27; and phenol — 492. After each toxicant stock solution was prepared, the pH was adjusted with HCl or NaOH to 7.5 for synthetic fresh water and 8.0 for synthetic seawater. Toxicant concentrations tested are listed in Table 2.

Microplate fluorometer

Fluorescence measurements were made with a Cambridge Technology 7620 microplate fluorometer. The excitation filter was 485 nm and the emission filter was 530 nm designed for fluorescein detection.

Protocol

The algae were added to the test wells, then an appropriate volume of test solution was added and the algae were exposed to toxicant for 1 h. The final dilution of FDA was made at the end of the hour, then 37 µl of FDA was added, and the cells incubated with substrate for 5 min (*Selanastrum* and *Tetraselmis*) or 15 min (*Cyclotella* and *Synechococcus*). In the tests with cadmium, mercury, and copper, sodium dodecyl sulfate (SDS) was used to lyse the algal cells after substrate incubation, but prior to reading the fluorescence. In this case, each test well received 5 µl of 12.5 µg SDS/ml (0.43 mM), which lysed the algal cells, releasing fluorescein into the medium. SDS was added only to the metals because they had high within-treatment variance compared to the other test compounds. Cell lysis facilitated fluorescence measurements since fluorescein was distributed homogenously throughout

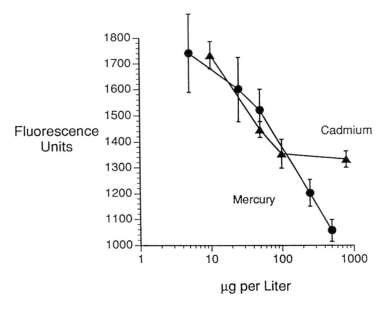

Figure 1 Sample concentration–response curves for *Selanastrum* exposed to cadmium and mercury. Fluorescence units refer to the arbitrary units of fluorescence measured by the fluorimeter and are directly proportional to esterase activity. Vertical bars indicate standard errors.

the well. Sample concentration–response curves for *Selanastrum* exposed to cadmium and mercury are presented in Figure 1.

Each experiment consisted of a control and five toxicant concentrations, each with five replicates. Mean fluorescence units and associated variances were calculated for each treatment, then a one-way analysis of variance was performed, followed by Dunnett's test. Dunnett's test indicated which of the treatments was significantly different from control ($\alpha = 0.05$) which permitted determination of the no observed effect concentration (NOEC) and the lowest observed effect concentration (LOEC).

Results and discussion

We have used this procedure to examine the toxicity of several chemicals to four species of algae (Table 3). No species was always the most sensitive to the seven toxicants; however, *Selanastrum* was most sensitive in five and *Synechococcus* in two of the seven. *Selanastrum* and *Synechococcus* had similar sensitivities to the metals, whereas the esterase inhibition NOEC for *Cyclotella* was generally at least an order of magnitude higher. *Tetraselmis* was intermediate between these two groups. The *Synechococcus* NOECs for the organics was two orders of magnitude higher than *Selanastrum*, suggesting that cyanobacteria and green algae have very different sensitivities to these xenobiotics. We were unable to estimate phenol toxicity in seawater because of the spontaneous hydrolysis of FDA. This interaction between toxicant and substrate was not observed in fresh water.

Table 3 Comparison of Toxicant Sensitivities of Four Algae Species (mg/l)

Algae		PCP	Naph	Chlor	Cd	Hg	Cu	Phenol
Selanastrum	NOEC	0.15	0.3	0.07	0.01	0.05	0.2	0.05
	LOEC	0.3	0.9	0.14	0.05	0.25	0.4	2.5
	N	2.0	2.0	4.0	3.0	4.0	3.0	3.0
Cyclotella	NOEC	0.15	4.8	2.0	0.5	1.5	0.8	—
	LOEC	0.3	6.0	3.0	1.0	2.0	1.0	—
	N	4.0	4.0	3.0	3.0	3.0	3.0	—
Synechococcus	NOEC	60.0	10.0	40.0	0.5	0.03	0.01	40.0
	LOEC	70.0	20.0	50.0	1.0	0.05	0.02	50.0
	N	3.0	3.0	3.0	4.0	3.0	3.0	3.0
Tetraselmis	NOEC	10.0	4.0	1.65	1.0	0.05	0.1	—
	LOEC	15.0	7.0	2.7	2.0	0.08	0.3	—
	N	3.0	3.0	4.0	3.0	3.0	3.0	—

Note: NOEC, no observed effect concentration; LOEC, lowest observed effect concentration; N, sample size of replicate measures and should be reported as an integer.

For biomarkers to be useful in toxicity assessments, it is essential to demonstrate their relationship to more traditional toxicological endpoints. For algae, population growth tests which measure cell reproductive rates have long been the standard. These have high ecological relevancy since toxicant-reduced algal population growth eventually would lower the primary productivity of an aquatic ecosystem, causing cascading effects at higher trophic levels. We estimated population growth rate (r) of *Selanastrum* using a 3-day test in parallel with the esterase inhibition test for 7 toxicants (Table 4). Despite the fact that the esterase test exposed algae to toxicant for only 1 h as compared to 72 h for the population growth test, the NOECs for the seven toxicants were quite similar. For copper and naphthol, the esterase and population growth NOECs were identical; for mercury, the esterase NOEC was 2× higher than the population growth NOEC and 3× higher for PCP. For cadmium, phenol, and chlorpyrifos, the esterase NOEC was con-

Table 4 Comparison of a Selanastrum 1-h Esterase Inhibition Test with a 3-Day Population Growth Test (mg/l)

Toxicant	Esterase inhibition		Population growth		Est-NOEC
	NOEC	LOEC	NOEC	LOEC	PopGr-NOEC
PCP	0.15	0.3	0.05	0.15	3.0
Naphthol	0.3	0.9	0.3	0.9	1.0
Chlorpyrif	0.07	0.14	0.2	0.3	0.35
Cd	0.01	0.05	0.5	1.0	0.02
Hg	0.05	0.25	0.025	0.05	2.0
Cu	0.2	0.4	0.2	0.4	1.0
Phenol	0.5	2.5	5.0	20.0	0.1

Note: NOEC, no observed effect concentration; LOEC, lowest observed effect concentration; Est-NOEC, esterase NOEC; PopGr-NOEC, NOEC for the population growth test.

siderably lower than that of the population growth rate test, 50× lower in the case of cadmium. These data suggest that for *Selanastrum*, an esterase biomarker with only 1-h toxicant exposure can estimate the same toxic effect threshold as a 72-h population growth test for many compounds.

The large differences in toxicant sensitivity among the four algal species that we tested is typical of what is reported in the literature.[17] Summarizing the data for nearly 500 chemicals, Lewis[3] shows that for >30% of them, there are 10- to 1000-fold differences in freshwater algal species sensitivities. No single species is always the most sensitive to all toxicants. Differential sensitivity of algal species provides the basis for marked shifts in phytoplankton assemblage composition upon toxicant exposure.[18-20] This illustrates the need for more comparative toxicology on the different groups of microalgae common in freshwater and marine environments. These large interspecific differences further suggest that a battery of algal species is required to make inferences about toxicant impacts on phytoplankton assemblages. More comparative toxicity data are required before definitive suggestions about the composition of this battery can be made. In the meantime, Swanson et al.[21] have provided a list of some possibilities. Utilization of a battery of algal test species necessitates that toxicity measurements be simple, rapid, and inexpensive. The esterase inhibition test provides a method for testing multiple algae species rapidly using a sensitive sublethal endpoint and an automated microplate reader.

It is important to know how the toxicant sensitivity of algae compares to other organisms. The best comparisons use the same endpoint, which is not so easy when comparing phylogenetically disparate organisms like bacteria, fungi, plants, and animals. We have used esterase biomarkers for such comparisions between microalgae and the bacterium *Bacillus subtilis* and the yeast *Candida tropicalis*,[22] and the rotifer *Brachionus calyciflorus*.[10] The NOECs for only *Selanastrum* are included in Table 5 since this alga was generally more sensitive than the other algae species tested. For four of the six toxicants, including PCP, naphthol, chlorpyrifos, and cadmium, *Selanastrum* had the lowest NOECs of the four test organisms. For mercury and copper, both the rotifer and bacteria were substantially more sensitive than *Selanastrum*. Freshwater plants are more sensitive to a variety of toxicants than many

Table 5 Comparison of Algae Toxicant Sensitivity with
Rotifers, Yeast, and Bacteria

Toxicant	*Selanastrum* NOEC	*Brachionus* NOEC	*Candida* NOEC	*Bacillus* NOEC
PCP	0.15	0.5	10.0	1.0
Naphthol	0.3	60.0	1.0	30.0
Chlorpyrifos	0.07	30.0	10.0	0.1
Cadmium	0.01	1.0	60.0	0.5
Mercury	0.05	0.01	1.0	0.01
Copper	0.5	0.08	—	0.001

Note: Table entries are esterase inhibition NOECs in mg/l for *Selanastrum capricornutum*, *Brachionus calyciflorus*, *Candida tropicalis*, and *Bacillus subtilus*.

animals,[3] including copper, cadmium, and phenol. The greater sensitivity of algae to many toxicants underscores the importance of including this group in toxicity assessments. Toxicity data from a phylogenetically diverse group of organisms like bacteria, fungi, algae, and animals are useful for predicting the trophic level most likely to be impacted. This should provide insight about which ecological processes are expected to be most impaired: primary or secondary production and/or decomposition and recycling.

The substrate FDA has been used for characterizing esterase activity in microalgae by a few other researchers. Dorsey et al.[8] showed that there is a relationship between esterase activity and photosynthesis as measured by ^{14}C uptake. These authors and Gala and Geisey[23] developed techniques for quantifying microalgae enzyme activity using flow cytometry. Gilbert et al.[24] used FDA to measure intracellular esterase activity in the marine microalgae *Tetraselmis suecica, Skeletonema costatum,* and *Prorocentrum lima* with a microplate fluorometer. They exposed the algae to the toxicants PCP, Cu, Zn, lindane, and phosalone for 5 h, followed by 1-h incubation with FDA. The results presented here extend their observations to freshwater microalgae and cyanobacteria and make direct comparisons of esterase inhibition with the more traditional population growth endpoint.

Ecotoxicologists should be cognizant of the major limitations of the esterase inhibition test. The 1-h toxicant exposure may be too short for some slow-acting toxicants to cause detectable enzyme inhibition. The relationship between esterase inhibition and population growth has been characterized only for *Selanastrum*. The ecological significance of laboratory toxicity tests is poorly understood. To be confident about inferring toxicant effects in natural populations, supporting field tests must be carried out in enclosures or similar exposure systems. The large variation in toxicant sensitivity among algal species indicates that one or a few species may not be representative of natural phytoplankton assemblages. Until more comparative toxicity data for algae are obtained and the relationship between lab tests and field responses is better characterized, the reliability of predicting safe levels of toxicant exposure for natural phytoplankton assemblages is unknown.

References

1. Schubert, L. E., ed. 1984. *Algae as Ecological Indicators.* Academic Press, New York.
2. Dixit, S., J. Smol, J. Kingston and D. Charles 1992. Diatoms: powerful indicators of environmental change. *Environ. Sci. Technol.* 26: 23–32.
3. Lewis, M. A. 1995. Use of freshwater plants for phytotoxicity testing: A review. *Environ. Pollut.* 87: 319–336.
4. McCarthy, J. F. and L. R. Shugart, eds. 1990. *Biomarkers of Environmental Contamination,* CRC/Lewis Publishers, Boca Raton, FL.
5. Blaise, C. 1991. Microbiotests in aquatic ecotoxicology: Characteristics, utility, and prospects. *Environ. Toxicol. Water Qual.* 6: 145–155.
6. Bitton, G. and B. Koopman 1992. Bacterial and Enzymatic Bioassays for toxicity testing in the environment. *Rev. Environ. Contam. Toxicol.* 125: 1–22.
7. Obst, U., A. Holzapfel-Pschorn and M. Wiegand-Rosinus 1988. Application of enzyme assays for toxicological water testing. *Toxicity Assessment* 3: 81–91.

8. Dorsey, J., C. M. Yentsch, S. Mayo and C. McKenna 1989. Rapid analytical technique for the assessment of cell metabolic activity in marine microalgae. *Cytometry* 10: 622–628.

9. Torslov, J. 1993. Comparison of bacterial toxicity tests based on growth, dehydrogenase activity, and esterase activity of *Pseudomonas fluorescens*. *Ecotoxicol. Environ. Safety* 25: 33–40.

10. Burbank, S. E. and T. W. Snell 1994. Rapid toxicity assessment using esterase biomarkers in *Brachionus calyciflorus* (Rotifera). *Environ. Toxicol. Water Qual.* 9: 171–178.

11. Guilbault, G. G. and D. N. Kramer 1964. Flurometric determination of lipase, acyclase, alpha- and gamma-chymotrypsin and inhibitors of these enzymes. *Anal. Chem.* 36: 409–412.

12. Starr, R. C. and J. A. Zeikus 1993. UTEX — The culture collection of algae at the University of Texas at Austin. *J. Phycol.* 29: 1–106.

13. Nichols, H. W. 1973. Growth media — freshwater. In: *Handbook of Phycological Methods*, J. R. Stein (ed.), Cambridge University Press, New York. pp. 7–24.

14. Guillard, R. R. L. 1983. Culture of phytoplankton for feeding marine invertebrates. In: *Culture of Marine Invertebrates*. C. J. Berg Jr. (ed.), Hutchinson-Ross, Stroudsberg, PA. pp. 108–132.

15. Allen, M. M. 1968. Simple conditions for the growth of blue-green algae on plates. *J. Phycol.* 4: 1–4.

16. USEPA 1993. *Methods for Measuring the Acute Toxicity of Effluents and Receiving Waters to Freshwater and Marine Organisms.* C. I. Weber, (ed.), 4th edition, EPA/600/4-90/027F, US EPA, Washington, DC.

17. Tadros, M. G., J. Philips, H. Patel and V. Pandiripally 1994. Differential response of green algal species to solvents. *Bull. Environ. Contam. Toxicol.* 52: 333–337.

18. Bringman, G. and R. Kuhn 1980. Comparison of toxicity thresholds of water pollutants to bacteria, algae, and protoza in the cell multiplication test. *Wat. Res.* 14: 231–241.

19. Brand, L. E., W. G. Sunda and R. L. Guillard 1986. Reduction of marine algae reproduction rates by copper and cadmium. *J. Exp. Mar. Biol. Ecol.* 96: 225–250.

20. Genter, R. B., D. S. Cherry, E. P. Smith and J. J. Cairns 1987. Algal periphyton population and community changes from zinc stress in stream mesocosms. *Hydrobiologia* 153: 261–275.

21. Swanson, S. M., C. P. Rickard, K. E. Freemark and P. MacQuarrie 1991. Testing for pesticide toxicity in aquatic plants: Recommendations for test species. Plants for Toxicity Assessment ASTM STP 1115. Philadelphia. PA, *Am. Soc. Test. Materials.*

22. Burbank, S. E. 1994. Development of rapid toxicity tests using enzyme inhibition as a sublethal biomarker. Georgia Institute of Technology, Master's Thesis.

23. Gala, W. R. and J. P. Geisey 1990. Flow cytometric techniques to assess toxicity to algae. Aquatic Toxicology and Risk Assessment ASTM STP 1096. Philadelphia, PA, *Am. Soc. Test. Materials.* 237–246.

24. Gilbert, F., F. Galgani and Y. Cadiou 1992. Rapid assessment of metabolic activity in marine microalgae: application in ecotoxicological tests and evaluation of water quality. *Marine Biol.* 112: 199–205.

chapter three

A 14-day harpacticoid copepod reproduction bioassay for laboratory and field contaminated muddy sediments

G. Thomas Chandler and Andrew S. Green

Introduction

Contaminated sediment effects on infaunal benthos are not usually studied beyond the acute and partial life cycle levels. Life history and multiple generation studies require either sessile field populations or protracted culture of test fauna; thus easily-cultured benthic "weed" species such as nereid and capitellid polychaetes, tube-building or sand-dwelling amphipods, and chironomid larvae have been used most often.[1-5] Unfortunately, these species may not be the best choices if they are atypical of the natural fauna in the habitat of interest, or if the same characteristics that make them hearty in the lab also make them less sensitive to contaminants from/in the field. Many meiobenthic taxa (i.e., invertebrates passing a 1-mm sieve but retaining on a 0.063 mm sieve) are ideal for "whole-sediment" and porewater bioassay of sedimented pollutants. Most spend their entire life cycles burrowing freely and feeding on/within the sediment:porewater matrix, many taxa undergo 10 to 14 generations per year, and many have easily quantifiable reproductive output. Furthermore, many meiobenthic taxa can be cultured indefinitely over multiple life cycles at high densities within sediment microcosms.[6,7] We describe here several recent technical developments exploiting meiofaunal sediment culture for rapid contaminated sediment bioassays of toxicant effects on survival, reproduction, and population growth of meiobenthic taxa. Currently benthic copepods, nematodes, foraminifera, and polychaetes are being continuously cultured to study these parameters under exposure to organochlorine, organophosphate, and pyrethroid insecticides, heavy metals, PCBs, and PAHs. Meiofauna culturing methods depend heavily on a toxicant-free muddy sediment substrate (i.e., capable of supporting ≥90% copepod survival for 14 d), and a highly "polished" recirculating seawater system. In turn, these clean culture media can easily be adulterated with toxicants to assess acute/chronic toxicant effects.

Sediment-associated toxicants are almost always the most abundant and persistent contaminants in estuarine and marine ecosystems.[8,9] The very characteristics of silty-clay sediments that make them cohesive and "muddy" also make them serve as sinks for heavy metals and nonpolar organic pollutants. Unfortunately, limited ecotoxicological methods are currently available to study higher order population:community-level effects of sediment toxicants. This is especially true in estuaries and coastal environments where muddy sediments provide an important nursery habitat for many bottom-feeding juvenile fishes and the vast numbers of meiobenthos upon which they feed. The ecological and trophic importance of meiobenthos in estuarine ecosystems is well documented,[10,11] as is the preferably oxidized nature of their sediment microhabitat. Annual production of meiobenthos is 5 to 10 times that of the more commonly studied macrobenthos,[12-14] and >95% of all meiobenthos live in the oxic zone (0 to 2-cm depth) of muddy sediments at densities of 4 to 12 million per M^2.[15,16] Few taxa can survive anoxic conditions beyond minutes to hours. This meiofaunal microhabitat preference puts this community at high risk to metals and other toxicants that are most bioavailable in oxidized sediment horizons.

While many studies have utilized macrobenthos to test for acute and chronic effects of sediment-associated contaminants,[17-22] only eight published studies[23-30] have cultured multiple single-species populations of the rapidly reproducing meiobenthos in contaminated sediments to determine toxicant effects on survivorship, fecundity, body burdens, and population growth.[31] Yet, well over 100 studies have shown meiobenthos to be sensitive indicators of environmental perturbation in the field, and typically more sensitive than the macrobenthos.[31-37] Even though meiobenthos satisfy many criteria for "ideal" bioassay organisms[38] (e.g., sensitive, rapid response, ecologically important, simple and inexpensive to collect/culture), their high potential for toxicant bioassay, microcosm, and trophic-transfer studies has been ignored because taxonomic determination and laboratory culture of this diverse assemblage can be difficult. (Recently, the logistic problems with meiofauna culture/bioassay largely have been overcome,[6,7,23,24,38-40] and many species of benthic copepods, nematodes, foraminifera, and polychaetes can now be collected from almost any muddy estuarine/coastal marine habitat and cultured indefinitely in laboratory sediments tailored to contain almost any sediment-associated toxicant(s) concentration.) Furthermore, contaminated muddy sediment from the field can also be studied with this approach and referenced against "clean site" sediments and/or laboratory culture sediments.

Even though many meiobenthic taxa can be cultured, benthic copepods have been our primary taxon for bioassay because the entire life cycle (egg to egg) can be produced in 15 to 25 days (species dependent) at 20 to 23°C.

In less than one month, enough F_1 offspring at multiple life stages can be produced to run comprehensive chronic reproductive bioassays in 14 days. The rapid life stage transitions of benthic copepods (e.g., 12 stages in <25 d) allow evaluation of such *sublethal* toxicant effects as depressed egg production, and impaired/delayed hatching and development. The methods described here provide a simple, straightforward approach to exploit a diverse but often overlooked collection of test fauna that are (1) extremely important ecologically and trophically,[11,16,41] (2) intimately associated with all compartments of the sedimentary matrix throughout their life cycles; and (3) indigenous to every sedimentary habitat studied to date.[10,16]

Materials required

The following equipment and supplies are recommended for successful execution of the 14-day harpacticoid copepod reproduction bioassays:

Material	Manufacturer	Available from	Catalog #	Estimated 1995 price
Cryo-Fridge diurnal incubator (47.2 cu.ft. capacity)	Revco, Inc.	Baxter Products	J1481-48	$8200
Finnpipette® (2–10 ml)	Labsystems, Inc.	Baxter Products	P5055-15	$300
Eppendorf® 10–100 µl Pipettor	Eppendorf-Netheler-Hinz GmbH	Baxter Products	P5061-32	$300
Stainless steel sieves (U.S. Sieve Designation Nos. 270, 230, 120, 70, 60, and 35)	W.S. Tyler, Inc	Baxter Products	S1212-270,230, 120,70,60,35	$50–$105
Kimax® crystallizing dishes	Kimble Glass, Inc.	Baxter Products	D1710-1	$60
Megafuge™ benchtop centrifuge	Heraeus, Inc.	Baxter Products	C1725-3	$3800
Intramedic® medical/surgical tubing (20 gauge)	Becton Dickenson Primary Care Diagnostics, Inc.	Baxter Products	T6065-6	$95 (100 feet)
Multiposition stirrer	PMC Industries, Inc.	Baxter Products	H2185-40	$670

Material	Manufacturer	Available from	Catalog #	Estimated 1995 price
Rose bengal stain	Sigma Diagnostics, Inc.	Baxter Products	S7538-25	$30
Fiber optic illuminator	Dolan-Jenner, Inc., Model 180	Baxter Products	M4025-15	$300
Teflon® Erlenmeyer flasks (50 ml capacity)	E.I. duPont de Nemours & Co., Inc.	Baxter Products	F4271-50	$85 (4/case)
Teflon® needle control valve	Cole Parmer, Inc.	Cole Parmer	06393-60	$45
Instant Ocean® synthetic sea salt	Aquarium Systems, Inc.	Mail Order Pet Shop	A60200	$15/bag (1 bag makes 50 gal)
Nitex® mesh	Tetko, Inc.	Tetko Inc.	3-63/30XX 3-125/45	$35/yd $20/yd

Procedures

Amphiascus tenuiremis is an ideal benthic copepod for sediment bioassay. To date, we have cultured more than a dozen benthic copepod species from estuarine, coastal, and continental shelf marine environments. Certainly, many more species are culturable and likely amenable to contaminated sediment bioassay. However, most of the test methods described here were developed using an easily-cultured, diosaccid harpacticoid copepod, *Amphiascus tenuiremis* c.f. Mielke, collected originally in 1988 from North Inlet, SC. This species is more abundant at higher latitudes than South Carolina but is amphi-Atlantic in distribution ranging from the North Sea/Baltic intertidal to the southern Gulf of Mexico.[42] Diosaccid copepods are the most abundant, diverse, and widely-distributed family of sediment-dwelling copepods.[10] Most species produce 2 egg sacs containing 12 or fewer eggs each, thus clutch size enumeration is relatively easy. *A. tenuiremis* is an epifaunal to deep burrowing (0 to 15 mm) species that ingests sediments for diatoms and bacteria, and produces burrowing larvae with no swimming ability. Sexes are dimorphic with females reaching 0.4 mm in length and males 0.25 to 0.3 mm. Males are also streamlined in shape and have a swollen geniculate segment on their first antenna to clasp the female. By using a flow-through seawater system (Figure 1), this species can be cultured under extremely tight temperature/salinity/DO (±0.3°C; ±10/00; ±1 mg-O_2/l) to >1000 individuals · 10 cm^{-2} in toxicant-free muddy sediments. However, such tight limits are not required for successful culture of this and many other species. *A. tenuiremis* does not reproduce well in the absence of sediments, but adults and juveniles (i.e., copepodites) routinely exhibit <10% mortality for 5 to 6 days in oxygenated seawater alone.[26] Previous aqueous and sediment acute

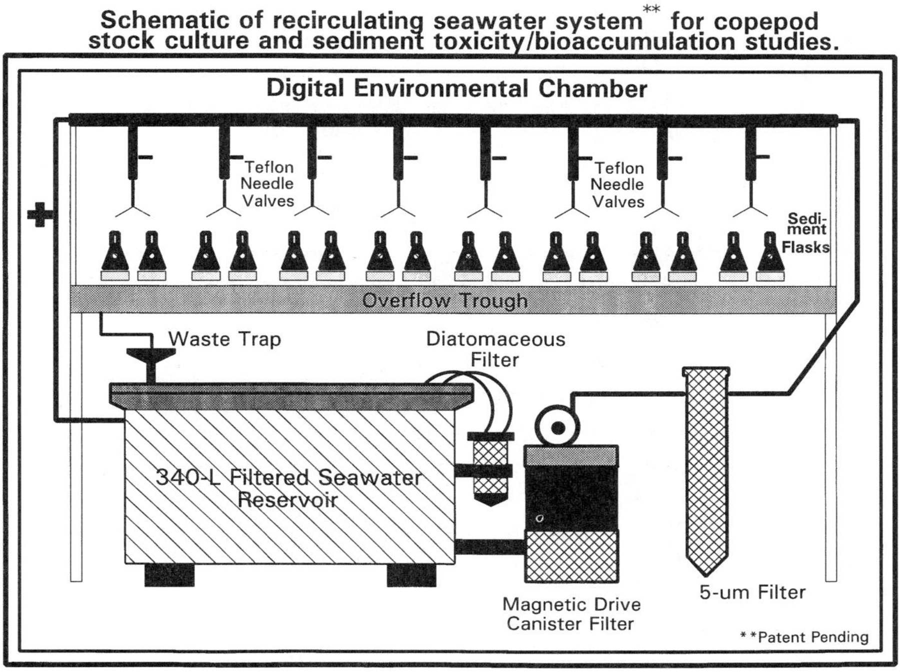

Schematic of recirculating seawater system[**] for copepod stock culture and sediment toxicity/bioaccumulation studies.

Digital Environmental Chamber

Teflon Needle Valves

Teflon Needle Valves

Sediment Flasks

Overflow Trough

Waste Trap

Diatomaceous Filter

340-L Filtered Seawater Reservoir

Magnetic Drive Canister Filter

5-um Filter

**Patent Pending

Figure 1 Micro-flow through bioassay system for multi-generation meiofaunal culture and multiple chronic tests with benthic copepods and muddy sediments. Entire recirculating system is enclosed in a refrigerated 50 ft³ environmental chamber on a 12:12 light:dark cycle. System currently is hand built from PVC and acrylic.

reference bioassays with *A. tenuiremis* have shown it to be sensitive and consistent in its mortality response. For example, for 4 replicate 24-h aqueous tests with the common reference toxicant sodium dodecyl sulfate, the mean LC_{50} and 95% confidence interval for *A. tenuiremis* was $10.78 < 12.00 < 13.37$ mg-SDS $\cdot$ l^{-1}. For cadmium, replicate 96-h bioassays[26] (30 ppt salinity) yielded aqueous LC_{50} of $0.213 < 0.224 < 0.235$ mg-Cd $\cdot$ l^{-1} and $0.183 < 0.198 < 0.216$ mg-Cd $\cdot$ l^{-1} respectively.

Amphiascus tenuiremis has a generation time of 21 days at 20°C.[43] Females are sexually mature after the 5th copepodite stage (i.e., 12th life stage). They then mate and produce their first clutch in 4 to 6 days. Nauplii hatch within 2 to 5 days of egg extrusion and reach the copepodite stage in 6 to 9 days. Thus, a sediment bioassay started with virgin or young mature females will yield endpoints for adult survival, clutch size, and naupliar production/survival in 14 days or less at 21°C. Virgin female *A. tenuiremis* (and many other harpacticoid species) can be distinguished and separated by collecting only the fifth copepodite females from stock cultures. Since insemination in most harpacticoids does not occur until immediately after the molt from the fifth stage, noninseminated females can be mated under exposure to specific toxicants. In practice, however, we have found that using virgins vs. non-gravids imparts no significant increase in test sensitivity for endpoints of clutch size or reproductive output — at least for *A. tenuiremis* which is capable of multiple clutches in short periods of time (i.e., 10 to 14 days post-insemination). Some harpacticoids may only produce one or two clutches in their lifetime; for such species virgin females may yield more sensitive bio-assays. Often these species develop large, long-term stores of ovigerous tissue soon after they have mated. Toxicant exposure after these stores have already developed would be expected to have little effect on clutch sizes (i.e., eggs are already pre-formed), but may still have effects on hatching success and offspring survival. The following bioassay method was developed for *A. tenuiremis*, but it is amenable to almost any meiobenthic taxon with a short life cycle and manageable clutch size.

Stock culture of meiobenthos in sediments

Successful meiofauna culture is keyed by three important conditions: (1) A toxicant-free muddy sediment substrate consisting of clay, silt, fine sand particles <0.100 mm in size (median grain diameter approximately 0.004 to 0.02 mm); (2) a highly-filtered carbon-polished seawater source saturated with oxygen and relatively free of ammonia (<30 μg-NH_4/l; pH = 8.0 to 8.3); (3) a varied diet of at least three easily cultured phytoplankton species (e.g., a chlorophyte, *Dunaliella tertiolecta*, a chrysophyte, *Isochrysis galbana*, and a diatom, *Phaeodactylum tricornutum*). All marine algae are cultured in nutrient-enriched seawater.[6] Preparation of clean stock sediments was described in detail previously.[6] Briefly, toxicant-free muddy sediments are collected from a pristine estuarine habitat (e.g., North Inlet, SC), passed through 0.5-mm and 0.1-mm sieves into deionized water, resuspended and washed three times after settling, and then centrifuged (3 min at 4,000 RCF) to remove 50 to 60% of the water. Sediment pellets are homogenized in an electric blender, placed into 1-liter Griffin beakers, covered and autoclaved for 15 min at 125°C and 30 psi. Autoclaved "dry" sediment stocks are stored at 4°C until use. Culture sediment substrate is prepared by taking 100 g "dry" stock, blending it for 3 min at high speed with 300-ml 5-μm-filtered seawater, and then pouring the blended stock through a 0.063-mm sieve into 3.5 l filtered seawater in a 4-l Griffin beaker. Sediments are allowed to settle for 30 min, and the supernatant water is then aspirated away. Remaining settled sediments are resuspended again in 3.5 l filtered seawater and allowed to resediment for 24 h at 4°C.

After 24 h, all supernatant water is aspirated away leaving approximately 300 ml wet sediments. For spiked sediment bioassays, these washed sediments are used for control and toxicant spiked sediments. For culture, these sediments are poured into 1-l crystallizing dishes filled with 500 ml seawater, allowed to settle, and then inoculated with the copepod taxon of interest. Typically >100 gravid copepods are required to start a viable culture. Wet sediment media has a very consistent water-to-solids percentage of 87:13 to 88:12 using this technique. This is a useful feature when computing nominal spiked-sediment toxicant concentrations based on sediment solids content (i.e., dry weight). Sediment organic carbon content (based on C:H:N analysis) consistently ranges from 3.8 to 4.2% among North Inlet, SC, sediment lots after passing them through these procedures. Raw field sediments range from 4 to 6% OC by C:H:N.

Each copepod culture dish is flushed slowly (~3 to 5 ml/min) but continuously with polished natural or synthetic seawater (Instant Ocean brand) from a recirculating or ambient 0.005-mm filtered seawater system (Figure 1). Cultures are fed a 1:1:1 mixed algal cell suspension in slight excess of what they consume (e.g., ~5 · 10^7 centrifuged cells for a mature culture) twice per week. Algal cells are cultured to exponential growth using standard f/2 enriched seawater media,[44] centrifuged 12 min at 4700 RCF, microwaved for 15 seconds (700 W) to knock out their swimming ability, cooled to 4°C, and then pipetted into each culture. One hundred gravid copepods will often yield a peak density culture in 2 to 3 months if food, sediment, and water quality are high.

14-Day harpacticoid copepod reproduction bioassay of spiked sediment toxicants

Figure 2 is a flow diagram of the general procedures used to conduct spiked toxicant and field toxicant-contaminated sediment bioassays with benthic copepods. For spiked sediment bioassay, a 300- to 400-ml wet sediment suspension is prepared as above for meiofaunal culture, placed on a magnetic stirrer, and pipetted under vortexing into 6 equal 60-ml aliquots in 100-ml beakers using Finnpipette digital pipettors (2 to 10 ml capacity). Each of the six aliquots receives a 5-mm Teflon stir bar and is placed under vortex on a multiple station magnetic stir plate. One 60-ml aliquot is randomly selected as control sediment. If a carrier solvent is used to solubilize hydrophobic toxicants, then an extra sediment aliquot must be retained and set up as an extra control for carrier solvent toxicity.[45] Five appropriate nominal concentrations of the toxicant of interest are then spiked into each of the five remaining vortexing beakers using precision volumetric pipettors (e.g., Eppendorf 10 to 100 μl capacity). All six beakers are covered immediately with parafilm and further mixed for 1 h. Each beaker is then placed in a 4°C refrigerator and allowed to "equilibrate" for 24 h. After 24 h, all beakers are again placed on magnetic stirrers and homogenized for 20 min. Sediments are now ready for adding to copepod bioassay chambers (Figure 2).

The 14-day bioassay is typically run in 36 50-ml bioassay chambers held under continuous dripping flow inside a 50 ft^3 digital environmental chamber (12:12 LD, 30 ppt, 21°C, 7 to 7.5 mg-O_2/l, 8.0 to 8.3 pH, <30 ppb NH_4). The environmental chamber encloses a complete recirculating seawater system identical to Figure 1 except that contaminated outflow water is not recycled but discarded. Bioassay chambers are made from 50-ml PVDF (Teflon) Erlenmeyer flasks with two 1 cm^2 windows cut from opposing sides. Each window is placed to allow a 30- to 35-ml test volume under flow (5 ml/h), and covered with 0.063-mm Nitex mesh to prevent copepods from escaping. Ten-ml toxicant-free or spiked sediments are pipetted into each of 5 flasks per treatment:control, and then each flask is connected to the recirculating seawater system via 20-gauge microcatheter tubes. Micro-fine flow rates are accomplished using Teflon needle control valves between the seawater system and each bioassay chamber. For each chamber, exactly 25 male

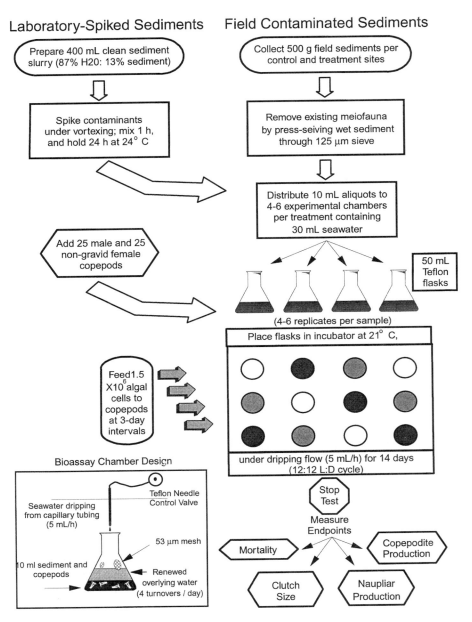

Figure 2 Procedure flow-diagram for 14-day benthic copepod bioassays with field-contaminated and laboratory spiked sediments. Lower left inset is a schematic of copepod bioassay chamber design.

and 25 nongravid female *Amphiascus tenuiremis* are sieved from stock laboratory cultures, counted twice, and pipetted into each flask. Flow rates are checked daily and adjusted as necessary. Contaminated water outflow is collected in a common drain basin and treated/discarded. For *A. tenuiremis*, each culture is fed $1.5 \cdot 10^6$ mixed algal cells (as above) every 3 days during the 14-d test. A Coulter® MultiSizer II is used to measure cell densities in food stocks and ensure consistent food additions to each bioassay chamber. Food quantities should meet, not exceed, what the copepods will eat/require in a 3-day period. Feeding consistency is very important, and the absolute cell densities required for a successful bioassay will be species-dependent. Standard ASTM[45] guidelines are followed with regard to maintenance and monitoring of pH, DO, salinity, and temperature. For recirculating seawater systems, fresh seawater must be added daily to the main seawater reservoir (Figure 1) since the bioassay slowly depletes the residual volume.

At the end of the test, each flask is removed one by one from the environmental chamber and its contents sieved onto a 0.053-mm sieve. Most of the sediments will pass the sieve leaving behind the copepod parents and their offspring. These are gently washed into 60-mm petri dishes, and all remaining parent copepods enumerated immediately for survival. Samples are then fixed in 5% buffered formalin. A 0.5% Rose Bengal red staining solution is added to mark all copepods and offspring, and thus make their enumeration easier. All copepods are counted, staged as to nauplius:copepodite:adult, and sexed if mature. Clutch sizes of all gravid females are determined by direct counts of egg sacs attached to females' bodies. Loose egg sacs are not included in clutch size determinations because clutch size is the sum of two statistically-related (i.e., nonindependent) egg sacs per female. For this reason, sieving and washing should be conducted as gently as possible to reduce egg sac disattachment and loss. Endpoints are compared statistically using ANOVA and multiple comparison tests.[46]

14-Day harpacticoid copepod reproduction bioassay of field-contaminated sediments

Bioassay of contaminated field sediments with copepods is a straightforward modification of the spiked sediment bioassay above. It should be noted, however, that this approach may not be suited to sediments characterized as coarse as, or coarser than, muddy sand. Sandy muds can be bioassayed with muddy-bottom copepod species if sufficient muds are present to pass a 0.125- to 0.225-mm pore size. The investigator should be aware, however, that he/she may be assaying a biased sediment fraction, i.e., a concentrated fine particle fraction which normally may be dispersed among larger particles. Also, since finer particles harbor most sediment toxicants, such bioassays with copepods may lead to overestimates of a bulk sandy sediment's toxicity. Figure 2 gives a simplified diagram of how the bioassay works. For our bioassays with *Amphiascus tenuiremis*, approximately 500 g bulk sediments are collected[47] from (1) each contaminated muddy field site (or replicate station) of interest, (2) a designated nonpolluted muddy control site as similar as possible to the field site of interest, and (3) a muddy "internal control" site in North Inlet, SC, from where our copepod culture stock sediments are derived. Additional sediment should be collected from each site for chemical analyses appropriate to the needs of the study. Each 500-g lot of sediments should be held at 4°C until ready for bioassay (usually <36 h after collection). Field sediments are then stirred/homogenized with stainless-steel stirring rods, and the original toxicant concentrations preserved by "press-sieving" as little as 100 g (for very fine-grained sediments) to as much as 500-g (i.e., sandy muds) sediments through a 30-cm, 0.125-mm pore size, stainless-steel sieve to remove large particulates. At no time are these sediments heated or autoclaved as before. For very sandy muds, a 0.225-mm pore size sieve is recommended. Of the <0.125-mm sediment fraction, 60 to 80 ml are scooped immediately from the bottom of the sieve, and then packed into acid cleaned 50-cm³ glass syringes for disbursement to 50-ml bioassay chambers (Figure 2). Ten ml of sieved sediments are extruded onto the bottoms of at least four replicate chambers per sediment sample. Four replicates or more are recommended for good statistical precision. Particulates >0.125 mm make recovery of copepod offspring difficult and time consuming, but not impossible. Thus muddy, fine-grained sediments are best suited to this approach. Once sediments are sieved and added to bioassay chambers, each chamber is placed under micro-flow as above, and 25-male:25-female *A. tenuiremis* are added. The bioassay is run for 14 days as above for spiked sediments. We have found in practice that sediment storage at 2 to 4°C and press sieving will kill/destroy any benthic copepods resident in field sediments; >90% of sediment nematodes will be removed as well. Endpoints for this bioassay are the same as above, but contaminated target sites can be referenced against two ecologically-relevant controls — a clean target site sediment and a clean North Inlet, SC sediment.

Bioassay of meiobenthic copepods indigenous to sediment sites of interest

In muddy coastal habitats throughout the world, only about 8 to 10 harpacticoid genera comprise >95% of the copepod fauna of these communities.[10] On smaller scales, the southeastern U.S. Gulf and Atlantic coasts for example, only 8 to 12 species in these genera make up 90 to 95% of muddy bottom coastal copepod communities.[48,49] Therefore, in almost any estuarine/coastal habitat of the southern U.S., one can collect muddy sediments and find several similar or identical harpacticoid species indigenous to a site of interest and many other sites as well. It is relatively easy to collect indigenous species from a study site, place them into culture, and in 1 to 2 months have a laboratory-acclimated, reproducing population suitable for bioassay.[24] Copepod collection/extraction from sediments is very simple. Using any flat, spatulate tool, scrape the uppermost 1 cm from a 1 to 2 m² muddy sediment surface, sieve the material through 0.5- and 0.125-mm poresize 30-cm sieves, collect and transfer the >0.125 mm fraction into translucent plastic or glass rectangular containers filled 50% with seawater, and then place the containers under intense fiber optic light. Most benthic copepods are strongly attracted to collimated light, so literally 100s to 1000s of individuals[50] will leave the sediments and congregate near the light source where they can easily be pipetted into sediment-free seawater. Once extracted from sediments, many copepods can easily be sorted live to species and placed into culture. Direct acute bioassay of such field collected copepods may also be acceptable[24] provided they are handled gently, acclimated for several days to test conditions, and not so stressed as to yield >10% control mortalities.

Results and discussion: examples from spiked and field-contaminated bioassays with *Amphiascus tenuiremis*

Our first example presents results from a 14-d reproduction bioassay of sediment-associated azinphosmethyl (APM), a moderately persistent organophosphorous insecticide (tradename Guthion), that associates >98% to sediments under the above-described conditions. The spiked-sediment procedures were followed, and two test concentration results are presented here. The 96-h sediment-associated APM LC_{50} was 504 ± 34 µg-APM/Kg-dry sediment. For the 14-d test, initial sediment concentrations were spiked at 0, 116, and 175 µg-APM/kg-dry sediment (i.e., the ~LC_{10} and ~LC_{15} 96-h exposure concentrations). There was a 58 to 60% degradation/loss of time zero concentrations by 14 days, probably due to bacterial APM degradation and/or alkaline APM hydrolysis (porewater pH = 7.8 to 8.0).

The sublethal effects of APM on *Amphiascus tenuiremis* reproductive output by the end of the 14-d exposure are presented in Figure 3. Two indices

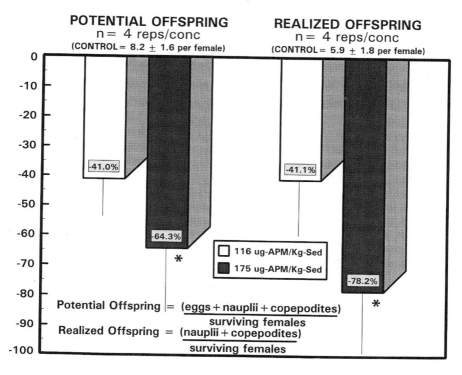

Figure 3 Reproductive effects of a 14-day azinphosmethyl spiked-sediment exposure with the benthic copepod *Amphiascus tenuiremis*. Results are presented as mean percent depression (±1 SD) relative to control reproduction.

of reproductive success are presented: *potential offspring* which is mean total reproductive output (i.e., total number of eggs per collected unhatched clutch, surviving larvae, and surviving juveniles) per surviving mother copepod per treatment, and *realized offspring* (Figure 3) which is offspring actually hatched and surviving at 14-d per surviving mother copepod. Data are denominated to the number of surviving adult mothers to eliminate bias from replicates where differential female survival may have occurred (e.g., a replicate with 30 surviving mothers producing 300 offspring would be weighted equally with a replicate having 3 surviving mothers and 30 offspring). This normalization gives a much more conservative comparison of effects than using raw production values. Figure 3 is scaled to depict percent depression of reproductive output by APM relative to APM-free controls. For both indices at the 175 µg-APM/kg-sed concentration, reproductive output was strongly and significantly (Dunnett's Test; $P < 0.05$) depressed relative to controls (–64 and –78% respectively for potential and realized offspring). At the 116 µg-APM/kg-sed concentration, both indices were depressed but not significantly so relative to controls. Realized offspring at the 175 µg-APM/kg-sed exposure were also significantly more depressed than in the 116 µg/kg exposure (Tukey's Studentized Range Test; $P < 0.05$). Copepod clutch sizes are often the most sensitive and consistent endpoint in this bioassay, but for APM, mean clutch sizes at 0, 116, and 175 µg-APM/kg-sed were not significantly different (i.e., 7.5, 6.9, and 6.5 eggs per clutch respectively; Tukey's Test, $P < 0.05$). This strongly supports that APM reproductive effects were due to naupliar and/or copepodite sensitivity.

For field-contaminated sediments, we present data from two studies where we bioassayed muddy sediments from a heavily industrialized site in the Gulf of Mexico contaminated primarily with metals (e.g., Hg and Cd), and a sewage outfall site in coastal SC contaminated with multiple PAHs, organics, and some metals. Figure 4 is a composite of *Amphiascus tenuiremis* reproductive endpoints for both sediment types relative to a site control (where available) and a North Inlet, SC, control. Both bioassays followed procedures described above for field-contaminated sediments. Mean clutch sizes (Figure 4) in the metal-contaminated industrial sediments were depressed >50% relative to North Inlet clean controls and field-site clean controls which were almost identical. No significant effects on naupliar or copepodite production were found however, suggesting that larval/juvenile survival was unaffected in the metal-contaminated sediments. Keep in mind that, for this species, clutch sizes assayed after 14 days represent a full clutch development cycle in the presence of these toxicants. The initial clutches that produced the offspring may have been unaffected on/before hatching and yielded equivalent offspring densities in control and contaminated sediments. For the sewage outfall sediments, notice that field site control sediments were considerably more supportive of reproduction than North Inlet control sediments. This is not an uncommon result if field site controls are toxicant-free and contain more diatoms, bacteria, and/or protists for food. All three reproductive endpoints — clutch size, naupliar production, and copepodite production — were significantly depressed in sewage outfall

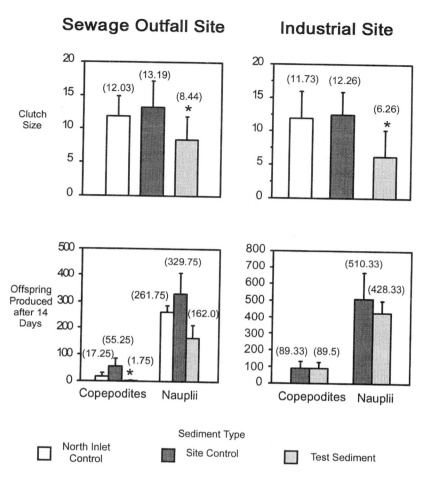

Figure 4 Results of a 14-day *Amphiascus tenuiremis* reproductive bioassay of contaminated muddy field sediments from two different land-use sites — an organics-contaminated sewage outfall site and a metals-contaminated industrial outfall site. Contaminated-site sediments are referenced against two controls — a North Inlet, SC, clean sediment and a "near test site" clean sediment. Mean clutch sizes ±1 SD and mean offspring produced per larval stage (±1 SD) are presented. Data were not normalized to number of surviving females because survival rates were almost identical across treatments and controls.

sediments relative to the site control sediments. Clutch size and naupliar production were also significantly depressed relative to North Inlet controls.

In the above examples, the sewage outfall site would be considered significantly toxic overall to *Amphiascus tenuiremis* reproduction. The industrialized site probably would not support long-term population growth/maintenance either, given the magnitude of clutch-size depression, but without longer-term tests we cannot say so with certainty. We are currently developing full to multiple generation meiofaunal bioassays which will better predict the longer-term impacts of depressed partial life cycle endpoints on meiofaunal population survival and growth. For example, a full life cycle or 2 to 3 generation bioassay of these industrial site sediments would have shown the importance of a >50% clutch-size depression on subsequent generations, even if the depression was not permanent or larval/juvenile stages became more or less sensitive to the toxicants over time. Life table/matrix models and survival analysis are also proving useful in this regard.[51,52] Although tedious and time consuming to derive for copepods and other taxa, once deciphered they can be powerful tools for predictions of long-term population maintenance. The life table for *A. tenuiremis* has been characterized. Survival analysis was used to predict life expectancies for each weekly age class cultured in toxicant-free North Inlet control sediments.[53] *A. tenuiremis* has a mean life expectancy of 5.4 weeks and a maximum life span of 22 weeks. Its mean lifetime offspring production is 48.5 offspring per female in the absence of toxicants. The next step is to apply this information to contaminated sediment culture and unravel toxicant effects on the most fundamental parameters of population growth and maintenance. Hopefully, other meiofaunal species can be similarly characterized to provide a more comprehensive multi-taxon assessment of chronic toxicant effects acting at the reproductive endpoint level.

Acknowledgments

This research was supported by grants from the U.S. NOAA Coastal Ocean Program and the U.S. EPA Office of Exploratory Research (G.T. Chandler, B.C. Coull, W.B., and F.J. Vernberg (USES), Principal Investigators). We thank Ms. Teresa Donelan for her quality assistance with bioassays, data analysis, and graphics. This research also benefitted from discussions with B.C. Coull, G.I. Scott, E.R. Long, and A.W. Decho.

References

1. Chapman, P.M., G.A. Vigers, M.A. Farrell, R.N. Dexter, E.A. Quinlan, R.M. Kocan, and M. Landot. 1982. Survey of biological effects of toxicants upon Puget Sound biota I. Broad scale toxicity survey. NOAA Tech. Mem. OMPA-25, Boulder, CO.
2. Chapman, P.M., D.R. Munday, J. Morgan, R. Frink, R.M. Kocan, M.L. Landot, and R.M. Dexter. 1983. Survey of biological effects of toxicants upon Puget Sound biota II. Tests of reproductive impairment. NOAA tech. Rep. NOS-102-OMS1, Rockville, MD.
3. Olla, B.L., A.J. Bejda, A.L. Studholme, and W.H. Peterson. 1984. Sublethal effects of oiled sediment on the sand worm, *Nereis (Neanthes) virens*: Induced changes in burrowing and emergence. *Mar. Environ. Res.* 13:121–139.
4. Olla, B.L., V.B. Estelle, R.C. Swartz, G. Braun, and A.L. Studholme. 1988. Responses of polychaetes to cadmium-contaminated sediment: Comparison of uptake and behavior. *Environ. Toxicol. Chem.* 7:587–592.
5. Pesch, C.E. and P.S. Schauer. 1988. Flow-through culture techniques for *Neanthes arenaceodentata* (Annelida:Polychaeta), including influence of diet on growth and survival. *Environ. Toxicol. Chem.* 7:961–968.

6. Chandler, G.T. 1986. High density culture of meiobenthic harpacticoid copepods within a muddy sediment substrate. *Can. J. Fish. Aquat. Sci.* 43:53–59.

7. Chandler, G.T. 1989. Foraminifera may structure meiobenthic communities. *Oecologia* 81:354–360.

8. Baker, R.A. 1980. *Contaminants and Sediments*, Ann Arbor Science, Ann Arbor, MI, 445 pp.

9. Reynoldson, T. 1987. Interactions between sediment contaminants and benthic organisms. *Hydrobiologia* 149:53–66.

10. Hicks, G.R.F. and B.C. Coull. 1983. The ecology of marine meiobenthic harpacticoid copepods. *Oceanogr. Mar. Biol. Ann. Rev.* 21:67–175.

11. Coull, B.C. 1990. Are members of the meiofauna food for higher trophic levels? *Trans. Am. Micro. Soc.* 109:233–246.

12. Gerlach, S.A. 1971. On the importance of marine meiofauna for benthos communities. *Oceologia* 6:176–190.

13. Platt, H. 1981. Meiofaunal dynamics and the origin of the metazoa. In: *The Evolving Biosphere*. (P.L. Forey, ed.), Cambridge University Press, London, pp. 207–216.

14. Coull, B.C. 1988. Ecology of the marine meiofauna. In: R.P. Higgins and H. Thiel (eds.), *Introduction to the Study of Meiofauna*, Smithsonian Institution Press, Washington, pp. 18–38.

15. Coull, B.C. and S.S. Bell. 1979. Perspectives in Marine Meiofaunal Ecology. In: *Ecological Processes in Coastal and Marine Ecosystems*, (R.J. Livingston, ed.), Plenum Press, N.Y., pp. 189–216.

16. Giere, O. 1993. *Meiobenthology*. Springer-Verlag, Berlin, 328 pp.

17. Schimmel, S.C., J.M. Patrick and A.J. Wilson. 1977. Acute toxicity to and bioconcentration of endosulfan by estuarine animals. In: *Aquatic Toxicology and Hazard Evaluation*, (F.L. Mayer and J.L. Hamelink, eds.), American Society of Testing and Materials STP Report No. 634, Washington, pp. 241–252.

18. Swartz, R.C., D.W. Schultz, G.R. Ditsworth, W.A. DeBen, and E.A. Cole. 1981. Sediment toxicity, contamination and benthic community structure near ocean disposal sites. *Estuaries*, 4:258–269.

19. Hansen, D.J. 1984. Utility of toxicity tests to measure effects of substances on marine organisms. In: *Concepts in Marine Pollution Measurements*, (Harris H. White, ed.), Maryland Sea Grant Publications, University of Maryland, College Park, pp. 33–56.

20. Swartz, R.C., G.R. Ditsworth, D.W. Schultz and J.O. Lamberson. 1986. Sediment toxicity to a marine infaunal amphipod: Cadmium and its interaction with sewage sludge. *Mar. Environ. Res.* 18:133–153.

21. Tagatz, M.E., R.S. Stanley, G.R. Plaia and C.H. Deans. 1987. Responses of estuarine macrofauna colonizing sediments contaminated with fenvalerate. *Environ. Toxicol. Chem.* 6:21–25.

22. Swartz, R.C., P.F. Kemp, D.W. Schultz and J.O. Lamberson. 1988. Effects of mixtures of sediment contaminants on the marine infaunal amphipod, *Rhepoxynius abronius*. *Environ. Toxicol. Chem.* 7:1013–1020.

23. Chandler, G.T. 1990. Effects of sediment-bound residues of the pyrethroid insecticide Fenvalerate on survival and reproduction of benthic copepods. *Mar. Environ. Res.* 29:65–76.

24. Chandler, G.T. and G.I. Scott. 1991. Effects of sediment-bound endosulfan on survival, reproduction and larval settlement of meiobenthic polychaetes and copepods. *Environ. Toxicol. Chem.* 10:375–382.

25. Strawbridge, S., B.C. Coull and G.T. Chandler. 1992. Reproductive output of a meiobenthic copepod exposed to sediment-associated fenvalerate. *Arch. Environ. Contam. Toxicol.* 23:295–300.

26. Green, A.S., G.T. Chandler and E.R. Blood. 1993. Aqueous, porewater and sediment phase cadmium: Toxicity relationships for a meiobenthic copepod. *Environ. Toxicol. Chem.* 12:1497–1506.

27. DiPinto, L.M., B.C. Coull and G.T. Chandler. 1993. Lethal and sublethal effects of the sediment-associated PCB Aroclor 1254 on a meiobenthic copepod. *Environ. Toxicol. Chem.* 12:1909–1918.

28. Green, A.S. and G.T. Chandler. 1994. Meiofaunal bioturbation effects on the partitioning of sediment-associated cadmium. *J. Exp. Mar. Biol. Ecol.* 180:59–70.

29. Chandler, G.T., B.C. Coull and J. Davis. 1994. Sediment- and aqueous-phase fenvalerate effects on meiobenthos: Implications for sediment quality criteria development. *Mar. Environ. Res.* 37:313–327.

30. Wirth, E.F., G.T. Chandler, L.M. DiPinto and T.F. Bidleman. 1994. Accumulation of PCB's from sediment by marine benthic copepods using a novel microextraction technique. *Environ. Sci. Technol.* 28:1609–1614.

31. Coull, B.C. and G.T. Chandler. 1992. Pollution and meiofauna: Field, laboratory and mesocosm studies. *Oceanogr. Mar. Biol. Ann. Rev.* 30:191–271.

32. Heip, C. 1980. Meiobenthos as a tool in the assessment of marine environmental quality. *Rapp. P.-v. Reun. Cons. Int. Explor. Mer.* 1979:182–187.

33. Warwick, R.M., M.R. Carr, J.M. Gee and R.H. Green. 1988. A mesocosm experiment on the effects of hydrocarbon and copper pollution on a sublittoral soft-sediment meiobenthic community. *Mar. Ecol. Prog. Ser.* 46:181–191.

34. Warwick, R. M. 1988. The level of taxonomic discrimination required to detect pollution effects on marine benthic communities. *Mar. Poll. Bull.* 19:259–268.

35. Herman, P.M.J. and C. Heip. 1988. On the use of meiofauna in ecological monitoring: Who needs taxonomy? *Mar. Poll. Bull.* 19:665–668.

36. Fleeger, J.W. and G.T. Chandler. 1983. Meiofauna responses to an experimental oil spill in a Louisiana salt marsh. *Mar. Ecol. Prog. Ser.* 11:257–264.

37. Moore, C. G. and B.J. Bett. 1989. The use of meiofauna in marine pollution impact assessment. *Zool. J. Linnean* Soc. 96:263–280.

38. Giesy, J.P. and R.A. Hoke. 1989. Freshwater sediment toxicity bioassessment: rationale for species selection and test design. *J. Great Lakes Res.* 15:539–569.

39. Chandler, G.T. and J.W. Fleeger. 1987. Facilitative and inhibitory interactions among estuarine meiobenthic harpacticoid copepods. *Ecology* 68:1906–1919.

40. Chandler, G.T., D.F. Williams, H.J. Spero and G. Xiaodong. Sediment microhabitats and stable isotope disequilibria phenomena in microcosm cultured benthic foraminiferal calcite. *Limnol. Oceanogr.* (In Press).

41. Gee, J.M. 1989. An ecological and economic review of meiofauna as food for fish. *Zool. J. Linn. Soc.* 96:243–261.

42. Lang, K. 1948. *Monographie der Harpacticiden*, Hakan Ohlsson, Lund, 1682 pp.

43. Woods, R. E. and B. C. Coull. 1992. Life-history responses of *Amphiascus tenuiremis* (Copepoda:Harpacticoida) to mimicked predation. *Mar. Ecol. Prog. Ser.* 14:122–130.

44. Guillard, R.R.L. 1972. Culture of phytoplankton for feeding marine invertebrates. In: *Culture of Marine Invertebrate Animals*, (W.L. Smith and M.H. Chaney, eds.), Plenum Press, NY, pp. 20–60.

45. ASTM. 1989. Standard guide for conducting acute toxicity tests with fishes, macroinvertebrates, and amphibians. American Society for Testing and Materials, E-729 88a, 20 pp.

46. Williams, D.A. 1972. A comparison of several dose levels with a zero dose control. *Biometrics* 28:519-531.

47. ASTM. 1991. Standard guide for collection, storage, characterization, and manipulation of sediments for toxicological testing. American Society for Testing and Materials, E-1391-90, 15 pp.

48. Coull, B.C. 1985. Long-term variability of estuarine meiobenthos: An 11 year study. *Mar. Ecol. Prog. Ser.* 24:205-218.

49. Chandler, G.T. and J.W. Fleeger. 1983. Meiofaunal colonization of azoic estuarine sediments in Louisiana: Mechanisms of dispersal. *J. Exp. Mar. Biol. Ecol.* 69:175-188.

50. Couch, C.A. 1988. A procedure for extracting large numbers of debris-free, living nematodes from muddy sediments. *Trans. Am. Micros. Soc.* 107:96-100.

51. Caswell, H. 1989. *Matrix Population Models.* Sinauer Associates, Inc., Sunderland MA, 328 pp.

52. Carey, J.R. 1993. *Applied Demography for Biologists with Special Emphasis on Insects.* Oxford Univ. Press, NY, 206 pp.

53. Green, A.S. and G.T. Chandler. 1995. Age-specific survival analysis of an infaunal meiobenthic harpacticoid copepod, *Amphiascus tenuiremis. Mar. Ecol. Prog. Ser.* 129:107–112.

chapter four

Fish egg injection as an alternative exposure route for early life stage toxicity studies. Description of two unique methods

Mary K. Walker, Erik W. Zabel, Gun Åkerman, Lennart Balk, Peggy Wright, and Donald E. Tillitt

Introduction

In the environment, lipophilic contaminants such as halogenated aromatic hydrocarbons (HAHs, e.g., polychlorinated biphenyls, PCBs) and polycyclic aromatic hydrocarbons (PAHs, e.g., benzo[a]pyrene) readily bioaccumulate in fish, and the bioaccumulation of these lipophilic chemicals by adult fish may have significant consequences on the development and survival of their offspring. Halogenated and polycyclic aromatic hydrocarbons translocate from adult female body stores into eggs during oocyte maturation,[1-3] and early life stages of fish are often more sensitive than adults to the toxicity of these chemicals.[4-6] Thus, the presence of persistent, bioaccumulative contaminants in the environment may pose a risk to fish early life stage survival and ultimately reduce recruitment into the adult population.

Typically, standard early life stage toxicity studies exposed embryos, larvae, and juveniles to graded concentrations of waterborne toxicants, and dose–response relationships are based on the concentrations of chemicals in the water.[7] However, use of waterborne exposure to assess the toxicity of persistent, bioaccumulative contaminants, such as HAHs and PAHs, has two significant drawbacks. First, uptake of hydrophobic chemicals, such as HAHs and PAHs, into the developing embryo from water is not a significant route of exposure in the environment since concentrations of these chemicals freely dissolved in water are extremely low.[8] Rather, maternal deposition into developing oocytes is the most significant source of these chemicals to the embryo.[1,2,8] Second, the dose received by the target tissue, in this case the developing embryo, is the most accurate predictor of the toxic response,[9] and since extrapolation from water concentrations of the chemical to egg concentrations is required, the exact dose received by the embryo can only be estimated, often with large uncertainty. Due to these drawbacks, it is important to develop an alternative exposure method that will directly expose the developing embryo without the need to chronically expose adult fish with subsequent natural deposition of hydrophobic chemicals into the oocytes.[3] Fish egg injection provides this exposure route. Embryos are exposed directly after fertilization with known doses of contaminants, the dose is delivered prior to critical developmental events, and extrapolation of the dose received by the embryo is not needed.[10-13]

We have developed two unique fish egg injection methods as alternative routes of exposure for fish early life stage toxicity studies of lipophilic environmental contaminants. With either method, individual fish eggs are injected with a known dose of chemical. The first approach, a microinjection method, originally developed to assess the developmental toxicity of HAH congeners to early life stages of salmonids,[14] utilizes micro-syringes, 30-gauge stainless steel injection needles, and micro- to nanoliter injection volume. The second approach, a nano-injection method,[15-17] utilizes glass capillary micropipettes with 2 to 10 μm tips as injection needles, and nano- to picoliter injection volume, allowing injection of nearly any size of fish egg.

Both of these egg injection methods allow an investigator to assess the toxicity of lipophilic environmental contaminants to early life stages of fish in a manner that realistically reflects environmental exposure and allows accurate quantitation of the dose to the developing embryo. These injection techniques, however, are not limited to use with only lipophilic chemicals. Since the developmental toxicity of many environmental contaminants ultimately depends on the dose received by the embryo, these egg injection methods could serve as a realistic exposure route in many fish early life stage toxicity studies.

Method I: microinjection of salmonid eggs

This procedure outlines the accurate injection of micro- to nanoliter volumes into salmonid eggs for the purpose of early life stage toxicity studies. Equipment requirements are minimal and include a rotary evaporator and sonicator for preparation of dosing solutions along with a stereomicroscope, micromanipulator, and microburette for egg injection. Vehicle carriers to inject contaminants into salmonid eggs include a variety of liquids (i.e., saline, dimethyl sulfoxide (DMSO), acetone, 1,4-*p*-dioxane, liposomes).[14] Although this procedure was developed to inject individual HAH congeners into newly fertilized rainbow trout eggs (*Oncorhynchus mykiss*), it also can be used to inject other species of salmonids,[3] more advanced stages of salmonid embryonic development,[18] mixtures of HAHs,[19,20] and other classes of environmental contaminants. The preparation of liposomes as a vehicle carrier for delivery of individual HAH congeners and the injection of newly fertilized rainbow trout eggs with those congeners will be described in detail below.

Materials required

Animals

Requests for salmonid eggs from federal or state sources should be made well in advance of the planned experiments following approval by the appropriate governing agency. We obtain rainbow trout eggs (Arlee, Eagle Lake, Erwin, Fish Lake, McCounaghy, and Shasta strains) through the Fishery Resources Program (Mr. John Leonard, U.S. Fish and Wildlife Service, Region 3, Fort Snelling, Twin Cities, MN 55111). Approximately 3,000 rainbow trout eggs are shipped from a U.S. Fish and Wildlife Service Hatchery on the day of fertilization by Federal Express. Following fertilization, eggs are allowed to water harden during which time they rapidly absorb water, forming the perivitelline fluid. Once they are water hardened they are placed in a 2-l polypropylene jar (supply list Table 2) and the jar is filled with water. The jar containing the eggs is then placed in a 38 × 34 × 30 cm styrofoam box, packed with styrofoam peanuts, and 0.1 kg of ice added.

Upon arrival, rainbow trout eggs are allowed to warm to 11°C at approximately 1.5°C/h to avoid temperature shock. Rainbow trout eggs arriving at ≥14°C or ≤4°C exhibit elevated mortality after injection; thus, eggs arriving outside of the temperature range of 5 to 12°C are assumed to have been temperature shocked and are not used for egg injection.

Equipment

The equipment needed to prepare dosing solutions of lipophilic chemicals for injection into rainbow trout eggs is detailed in Table 1. Dosing solutions are prepared using a rotary evaporator and bath sonicator as detailed under the section, Procedures. The bath sonicator must be of sufficient power to prepare liposomes.[21] Ultrasonic cleaners commonly sold by most scientific supply companies are not sufficient to properly sonicate liposomes.

All injections are performed using a binocular stereomicroscope with a lightfield illumination base (Figure 1). To clearly observe the insertion of an injection needle and delivery of the dosing solution, the stereomicroscope should provide at least 20× magnification, and the illumination base should provide light from below the egg. During injection the egg is supported by a plexiglass egg holder (Figure 1, EH). The injection needle (Figure 1, N) is mounted on a micromanipulator (Figure 1, MM) and is connected to the injection syringe (Figure 1, S) by a short piece of polyethylene tubing (Figure 1, T). The injection syringe is advanced with a microburette accurately delivering a known dose into the egg (Figure 1, MB).

Table 1 Equipment Needed for Microinjection Method

Item	Model number	Specifications	Supplier
Rotary evaporator	RE-11A	None	Brinkman Instruments, Westbury, NY
Sonicator Bath	G112SP1G	600 V capacity	Laboratory Supplies Co., Hicksville, NY
Power supply	G112Sp1T	50–60 Hz, 80 watts, 3.5 amps	
Binocular stereomicroscope	Wild M8	10–50× magnification	W. Nuhsbaum Inc., McHenry, IL
Illumination base	150-1L-PS	Light/darkfield, 0–2 amp power supply	Diagnostic Instrument Co., Sterlings Heights, MI
Micromanipulator	MM-33N	None	Narishige Medical Systems Corporation, Greenvale, NY
Microburette	SB2	Dial indicator: 0.001–1.0 inch travel range; micrometer; two fuse clips; 13 mm thick Lucite™ base	Specially ordered[a]

[a] Based on the design of a Micromatic Instruments microburette (model SB2) no longer being manufactured. This unit was fabricated by the instrument shop of the University of Wisconsin, Department of Physics. Contact a medical or scientific instrument company for a potential substitute product or special ordering.

Supplies and reagents

Supplies and reagents needed for injection of rainbow trout eggs are listed in Table 2.

Egg holder

The egg holder, which supports eggs during injection, consists of a 1.2-cm thick plexiglass disc, 10 cm in diameter. Twenty beveled holes, 4 mm in diameter at the surface of the disc and 2 mm in diameter at the bottom, are drilled into the plexiglass disc.

Injection needle and syringe

Polyethylene tubing is attached to the hub of a 30-gauge stainless steel needle, and needle and tubing are glued into a capillary tube. The other end of the polyethylene tubing is attached to a 30-gauge, removable needle on a 100-µl gas-tight injection syringe.

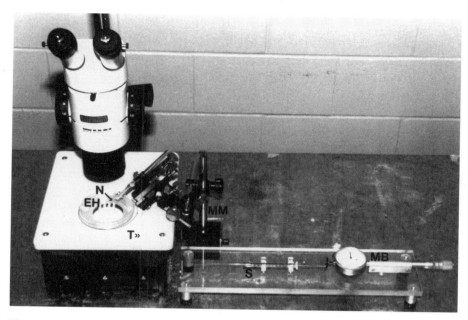

Figure 1 Microinjection apparatus consists of a 100-μl gas-tight syringe (S) mounted in the fuse clips of a microburette (MB); polyethylene tubing (T) connects a 30-gauge removable needle on the syringe to the shaft of a 30-gauge injection needle (N) mounted on a micromanipulator (MM). Injections of all rainbow trout eggs are performed using a binocular stereomicroscope, lightfield illumination, and plexiglass egg holder (EH). (Reprinted with permission from Walker, M. K. et al. 1992. *Aquat. Toxicol.* 22:15–38.[14])

Table 2 Supplies and Reagents Needed for Microinjection Method

Item	Specifications/catalog numbers	Supplier
Supplies		
Polypropylene jars	1.9 l, with screw cap, autoclavable to 121°C, Cat. #11-815-13C	Bel-Art Products
Styrofoam box	38 × 34 × 30 cm, Cat. #03-529-1	Fisher Scientific, Pittsburgh, PA
Injection needle	30-Gauge, beveled tip, 1.3-cm length Cat. #305106	Becton Dickson and Co., Rutherford, NJ
Polyethylene tubing	PE-10, 0.28 mm ID; 0.61 mm OD, Cat. #14170-12P	Becton Dickson and Co., Rutherford, NJ
Glass capillary tubing	25 μl, Cat. #21-170H	Fisher Scientific, Pittsburgh, PA
DURO Super Glue®	None	Loctite Co., Cleveland, OH

Table 2 (continued)   Supplies and Reagents Needed for Microinjection Method

Item	Specifications/catalog numbers	Supplier
Syringe needle	Removable needle, point style 3, Cat. #79630	Hamilton Co., Reno, NV
Syringe	100 μl gas-tight, model 1710RN, supplied with 22-gauge beveled needle, Cat. #81030	Hamilton Co., Reno, NV
Glass test tubes	16 × 100 ml with screw cap Tubes — Cat. #14-957-86D	
	Caps — Cat. #14-959-36A	Fisher Scientific, Pittsburgh, PA
Forceps	100 cm, blunt end, Cat. #13-812-40	Fisher Scientific, Pittsburgh, PA
Fiber filters	Rust and sediment type, S10-52, polypropylene yarn, 1-μm pore size	Raytec Watergroup Co., Milwaukee, WI
Charcoal filters	Activated charcoal, taste and odor type, S1035	Raytec Watergroup Co., Milwaukee, WI
Reagents		
Polychlorinated dibenzo-*p*-dioxins, dibenzofurans, and biphenyls	98–99% purity	Ultra Scientific, Hope, RI
2,3,7,8-Tetrachlorodibenzo-*p*-dioxin (TCDD)	1 mg, 98% purity, Cat. #ED-901-C	Cambridge Isotopes, Woburn, MA
Phosphatidylcholine	50 ml in chloroform, >99% purity, Cat. #840051	Avanti Polar Lipids, Alabaster, AL
Sodium chloride (NaCl)	Cat. #S-5886	Sigma Chemical Co., St. Louis, MO
Chloroform	HPLC grade, Cat. #36,692-7	Aldrich Chemical, Milwaukee, WI
1,4-*p*-dioxane	Anhydrous, Cat. #27,053-9	Aldrich Chemical, Milwaukee, WI
Tricaine methanesulfonate (MS-222)	100 g, Cat. #C-Finq-UE-100G	Argent Aquaculture, Redmond, WA
Soluene 350	500 ml, Cat. #6003038	Packard Instrument Co., Downers Grove, IL
Hionic	4 l, Cat. #6013431	Packard Instrument Co., Downers Grove, IL

Halogenated aromatic hydrocarbons (HAHs)

Halogenated aromatic hydrocarbons of the highest purity available (≥98% purity) were obtained as neat compound, weighed on a microelectrobalance, dissolved in anhydrous 1,4-*p*-dioxane, and stored at –20°C.

Phosphatidylcholine

Chicken egg yolk phosphatidylcholine (PC, >99% purity) in chloroform is obtained, phosphorus concentration determined,[22] diluted to 10 µmol/ml with chloroform (high pressure liquid chromatography (HPLC) grade), and stored at –20°C in 5-ml sealed glass ampules under argon. Highly pure PC can readily peroxidize when exposed to oxygen; therefore, PC should be obtained from a source that provides highly purified lipid suitably protected from oxygen.

Procedures

Liposome preparation

Incorporation of halogenated aromatic hydrocarbons (HAH) into liposomes

One to two days prior to egg injection multilamellar liposomes are prepared for control and HAH treatment groups. First, an aliquot of stock HAH, that is of sufficient quantity to prepare all dosing solutions, is evaporated to dryness under nitrogen gas. The HAH is then reconstituted in 0.5 ml chloroform for approximately 1 h, quantitatively added to 25 or 50 µmoles PC in chloroform, and transferred to a 16×100 ml glass test tube. Based on the sensitivity of rainbow trout to HAHs and the ability to incorporate HAHs into liposomes, we have typically used 25 or 50 µmoles PC so that the mole ratio HAH congener to PC does not exceed 30%.

The HAH is then incorporated into the lipid bilayer of PC liposomes using the "thin film hydration" method.[21] The HAH/PC mixture in chloroform in the 16×100 ml glass test tube is evaporated to dryness under vacuum using a rotary evaporator. The HAH/PC forms a thin, uniform film on the inside of the test tube. Multilamellar vesicles are formed by adding 0.5 ml of sterile 0.9% sodium chloride (NaCl) and vortexing. Argon is bubbled through the sterile NaCl solution for at least 1 h prior to adding it to the dried film to ensure that all dissolved oxygen is eliminated from solution. The presence of oxygen during subsequent sonication can initiate lipid peroxidation and destroy the PC. Control liposomes are prepared as described above without the addition of the HAH, and the final concentration of PC in NaCl is 50 or 100 mM. Multilamellar liposome solutions are stored at 4°C until sonicated.

Sonication and dilution of liposome dosing solutions

On the day of injection the multilamellar liposomes are sonicated to form unilamellar vesicles, and then graded doses of the HAH are prepared by dilution of the HAH/PC unilamellar liposomes with control unilamellar liposomes. Unilamellar liposomes are formed by sonication of multilamellar vesicles for approximately 1 to 2 min with a bath sonicator.[21] The 16×100 ml glass test tube can be mounted with a clamp on a ring stand above the bath sonicator or it can be held with long forceps. The sonicator should be turned on, and the test tube lowered about 1 to 1.5 cm into the center of the bath until the liposome solution forms small droplets that jump up the sides of the test tube and slide back down. The physical action of droplets jumping up and rolling back down the sides of the test tube ensures that unilamellar vesicles are being formed. If this action is not apparent, adjust the height of test tube in the bath sonicator until the liposome solution "boils" properly. Prior to sonication the multilamellar liposomes will appear viscous and somewhat clumpy, while after sonication the unilamellar liposomes will appear less viscous and smooth. The glass test tube containing the multila-mellar liposome solution should be allowed to warm to room temperature prior to sonication, because a glass test tube will shatter if sonicated cold. Safety glasses or protective eyeware should always be worn when sonicating solutions. Following sonication of both control and HAH-containing lipo-somes, control liposomes are used to serially dilute HAH-containing lipo-somes; thus, the concentration of PC remains constant for all doses.

It is imperative that all liposomes be prepared using sterile conditions and stored under argon at 4°C. Argon is preferable to nitrogen since it is heavier than air and will form a more protective barrier than nitrogen against oxidation at the liposome surface. Liposomes should be prepared freshly each week, 1 to 2 days prior to use. However, if necessary, diluted unilamellar liposomes can be stored under argon for up to 1 week at 4°C, while longer storage increases the risk of peroxidation.

Injection procedure

Injection needle

To prepare needles for injection, a 30-gauge, 1.3-cm disposable needle is removed from its plastic hub. The stainless steel shaft of the needle is scored with wire cutters as close to the hub as possible. The needle is then snapped off the hub with needle-nose pliers without crimping the shaft. Proficiency of this technique increases exponentially with practice. Approximately 30 cm of 0.28 mm ID polyethylene tubing is then attached to the shaft of the hubless needle. The needle and tubing are inserted and glued into a 2.5-cm long glass capillary tube with cyanoacrylate ester (DURO Super Glue®) such that only 1 cm of the needle tip protrudes from the end of the capillary tubing. Thus, the capillary tube provides physical support for the hubless needle. All injection needles must be examined at a minimum of 20× mag-nification prior to use to ensure sharpness, and all dull needles should be discarded. A dull injection needle significantly increases mortality above that associated with treatment. The capillary tube, with needle and polyethylene tubing attached, is then mounted on a micromanipulator.

Injection syringe

When injecting salmonid eggs with control and graded doses of an HAH, one syringe is dedicated to control liposomes and one syringe is used for the HAH doses with the lowest doses being injected first. Approximately 80 to 100 µl of the liposome solution is drawn into the 100-µl gas-tight injection syringe through the 30-gauge removable needle. The 30-gauge, removable needle is then attached to the injection needle via the 30-cm piece of polyethylene tubing (i.e., from the back of the injection needle). The syringe plunger is advanced until the liposome solution passes through the length of the poly-ethylene tubing and emanates visibly from the injection needle. The injection syringe is then placed into the fuse clips of the microburette and the microbu-rette is advanced a few turns to ensure that there is no back pressure in the tubing. Using a 100-µl syringe in the microburette and turning the micrometer by ¹⁄₁₂ of a revolution, or 5 units on the dial indicator, injects 0.2 µl of solution.

Egg injection

The stepwise procedure followed when injecting an individual rainbow trout egg is shown in Figure 2. Since salmonid embryos become more sensitive to mechanical shock during epiboly (2 to 6 days postfertilization at $10°C$[23-25]), all eggs should be injected within the first 50 h after fertilization. First, the egg is blotted on a piece of Whatman 2 filter paper to remove excess water, placed on the egg holder using a pair of blunt end forceps, and examined under the binocular stereomicroscope with lightfield illumination (Figure 2A). The egg is rejected if the chorion appears cloudy or if there are white spots in the egg indicating denaturation of protein. Once an egg is selected for injection, it is rotated gently with the forceps so the blastodisc faces downward (Figure 2B). The injection needle is slowly advanced by the micro-manipulator so that the beveled needle tip pierces the chorion of the egg (Figure 2C), and then more rapidly advanced so that the needle penetrates the yolk. The egg is injected with 0.2 µl of 50 mM PC liposomes (vehicle control) or 0.2 µl of HAH incorporated into 50 mM PC liposomes (Figure 2D). The needle is withdrawn, and the egg surface is gently wiped clear of albumin using the forceps. The injection site in the chorion is sealed with a droplet of Super Glue® applied to the egg surface with a capillary tube (Figure 2E), and allowed to dry for 6 to 10 sec (Figure 2F). It is important

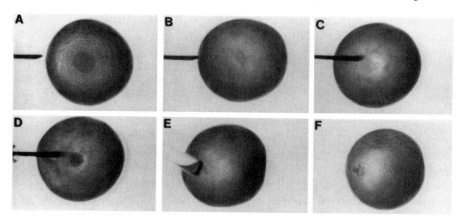

Figure 2   Sequence of injection of a rainbow trout egg: (A) egg was examined with lightfield illumination from below, (B) blastodisc rotated downward, (C) needle slow-ly advanced through the chorion, (D) more rapidly advanced into the yolk and 0.2 µl of liposomes injected, (E) needle removed and slit in chorion sealed with Super-glue® applied to the egg surface with a capillary micropipette, and (F) glue allowed to dry for 6–10 sec. (Reprinted with permission from Walker, M. K. et al. 1992. *Aquat. Toxicol.* 22:15–38.[14])

that the chorion is free of water, albumin, yolk, and perivitelline fluid before the droplet of Super Glue® is applied. Albumin or yolk on the egg surface will wick up into the capillary tubing and glue the capillary tubing to the egg, while excess water or perivitelline fluid on the egg surface will prevent the glue from sticking to the chorion. Typically, each egg remains out of water for less than 40 sec during injection.

Injection volume

In most experiments with rainbow trout eggs, we inject 0.2 µl; however, on occasion we inject up to 1.0 µl in order to deliver a sufficient mass of HAH and not exceed 30 mole percent of HAH in PC. The success of injecting this larger volume, however, depends on the size of the rainbow trout eggs. Rainbow trout eggs ≥60 mg tolerate injection of 0.5 to 1.0 µl volumes without any significant increase in injection-related mortality.

Determination of retained dose

When first validating the microinjection method, the percentage of injected dose retained by the eggs should be determined. The simplest way to determine this is to inject individual eggs with 0.2 µl of the appropriate radiolabeled chemical, and sample a minimum number of 3 eggs/sampling day at regular intervals after injection until hatching. Individual eggs are solubilized with 2 ml Soluene 350 in a scintillation vial, cap screwed on tightly, and vial placed in a 50°C oven overnight. On the following day the solubilized samples are allowed to cool to room temperature, 10 ml Hionic liquid scintillation fluid added, and low energy chemiluminescence allowed to decay. To account for quenching of radioactive disintegrations by the egg pigments and lipids, a quench curve can be made by simply solubilizing increasing numbers of noninjected eggs in scintillation vials and adding a known amount of radioactive toluene to each vial. The dpm of radiolabeled chemical/egg are calculated using cpm, counting efficiency, and the constructed quench curve. Using the microinjection method as described above and injecting 0.2 µl of ^{14}C-2,3,7,8-tetrachlorodibenzo-p-dioxin (TCDD) into individual rainbow trout eggs, 92 to 97% of the injected dose is retained through 22 days of development (2 to 4 days prior to hatching).[14]

Calculation of egg halogenated aromatic hydrocarbon

The egg dose of HAH (pg HAH/g egg) is determined by multiplying the concentration of HAH in the dosing solution (pg HAH/µl) by the injection volume (0.2 µl), and then dividing by the mean wet weight of 10 non-injected eggs (g) determined on the day of injection. Since dosing solutions are prepared before knowing the actual egg weight, the egg weight must be estimated initially and the true egg HAH concentration calculated on day of injection.

Method II: nano-injection method for any fish species

This procedure outlines the accurate injection of nano- to picoliter volumes into fish eggs for early life stage toxicity studies. Injection equipment allows small volumes of liquid to be delivered precisely using micropipettes, a regulated gas pressure system, and a digital control device. Pressure is applied pneumatically (compressed nitrogen gas) to the micropipette with digital control of dwell time which allows delivery of a range of volumes. A variety of liquids can be injected such as saline, corn oil, triglycerides, triolein, liposomes, or mixtures of lipids. In addition, eggs of various sizes can be injected with this method, such as Atlantic salmon (*Salmo salar*), fathead minnow (*Pimephales promelas*), sea trout (*Salmo trutta trutta*), and northern pike (*Esox lucius*). And although this procedure was developed to inject newly fertilized eggs prior to epiboly, it may be applied to various life stages of fish embryos. The preparation of dosing solutions in a triolein vehicle carrier and injection of newly fertilized rainbow trout eggs will be described in detail below.

This method has been developed focusing on the ability to (1) reduce high injection-related mortality reported in earlier studies,[11,13] (2) inject eggs immediately after fertilization to closely mimic exposure via maternal transfer, (3) inject both hydrophobic and hydrophilic chemicals, and (4) inject those fish species that have very small eggs. Ultimately, this methodology should facilitate the analysis of morphological, anatomical, physiological, and biochemical responses to toxic chemicals with considerable accuracy.

Materials required

Animals

Rainbow trout eggs may be obtained from numerous private, state, or federal fish hatcheries. We receive unfertilized eggs and milt, shipped on ice, during most of the year from Ennis National Fish Hatchery, U.S. Fish and Wildlife Service, Ennis, Montana. Fertilization of eggs is described in detail under the Procedures section.

Equipment

Equipment needed to prepare dosing solutions of HAHs and inject them into rainbow trout eggs is detailed in Table 3. Dosing solutions are prepared using a nitrogen evaporator and bath sonicator as detailed under the Procedures section. We use a nitrogen evaporator capable of evaporating small liquid volumes. The bath sonicator is used to prepare dosing solutions of HAHs in triolein. Fertilized rainbow trout eggs are reared in a 16-tray vertical flow incubator. An inverted microscope is used for checking sperm motility and monitoring stages of egg development. Micropipettes (injection needles) are prepared by pulling glass capillary tubing using a vertical pipette puller. The pulled tubing is then beveled with a diamond wheel grinder and filled using an evaporating unit connected to a vacuum pump. Fish eggs are injected using a pico-injector and micromanipulator. The micromanipulator holds the pulled micropipette, allowing movement in all directions which is necessary for smooth injection. Injections are performed using a zoom stereomicroscope with micrometer providing at least 50× magnification, and adequate illumination from above or below to clearly observe needle insertion and solution delivery (Figure 3).

Table 3   Equipment Needed for Nano-Injection Method

Item	Model number	Specifications	Supplier
Vertical pipette puller	P-30	None	Sutter Instrument Co., Novato, CA
Micro-pipette grinder	EG-40	None	Narishige Medical Systems Corporation, Greenvale, NY
Analytical evaporator	N-EVAP 111	100 V, 400 watts, 50–60 Hz	Organomation Associates Inc., Berlin, MA
Sonicator	B2200R-1	117 V, 1.0 amps, 50–60 Hz	Branson Ultrasonics Corp., Danbury, CT
Zoom stereomicroscope	Heerbrugg Wild M8	10–50× magnification	W. Nuhsbaum Inc., McHenry, IL
Diascope base with illumination	SMZ-U	None	Nikon, Inc., Melville, NY
Stage and adapter	GE 4 × 4/SMZ-U	None	Nikon, Inc., Melville, NY
Evaporating unit with vacuum pump and adjustable stage	18780	None	Pierce Chemical Co., Rockford, IL
16-Tray vertical flow incubator	None	None	Health Techna Corporation, WA
Inverted microscope	TMS	10–40× magnification	Nikon, Inc., Melville, NY
Pico-injector	PLI-188	Digitally regulated injection time and pressure	Nikon, Inc., Melville, NY
Micromanipulator or	MM-3	None	Stoelting Co., Wood Dale, IL
Hydraulic manipulator	MO-150	Required for injection of small fish eggs	Nikon, Inc., Melville, NY

Supplies and reagents

Supplies and reagents needed for injection of rainbow trout eggs are listed in Table 4.

Fertilization supplies

Rainbow trout eggs used for injection are fertilized on site. Eggs are placed in a plastic or glass mixing bowl, milt added, and 1% NaCl solution added to activate fertilization.

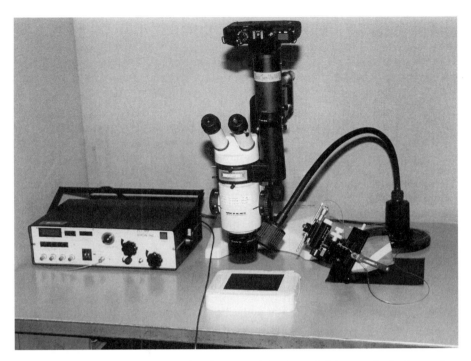

Figure 3   The basic equipment for the nano-injection of fish eggs includes a pico-injector (left), stereomicroscope (center), and micromanipulator (right).

Agarose plates
Agarose plates provide stability and support for eggs during injection. Preparation of agarose plates is detailed under the Procedures section. Briefly, hot agarose is poured into square disposable petri dishes, and wells of appropriate size (depending on the size of the fish eggs being injected) are made using stainless steel tubing (1 to 5 mm) after the agarose has gelled.

Injection micropipettes
Preparation of micropipettes as injection needles is described in detail under the Procedures section. Micropipettes are prepared by pulling aluminosilicate glass capillary tubing, applying Sigmacote® to the tips, and beveling using a diamond wheel pipette grinder. Micropipettes are filled with dosing solution made with filter-sterilized triolein.

Halogenated aromatic hydrocarbons
Halogenated aromatic hydrocarbons used for injection must be of highest purity available (98 to 99%). This injection procedure may also be used for complex mixtures extracted from environmental samples. Purity of 2,3,7,8-tetrachlorodibenzo-*p*-dioxin (TCDD) was confirmed by high resolution mass spectrometry.

Table 4   Supplies and Reagents Needed for Nano-Injection Method

Item	Specifications/catalog number	Supplier
Supplies		
Plastic mixing bowl	None	Local
Stainless steel tubing	Various size diameters (0.1 mm < egg diameter)	Local hardware store
Disposable polystyrene petri dishes with lid	15 × 100 mm, square style with numbered 13-mm grid, Falcon #1012	Becton Dickson and Co., Rutherford, NY
Aluminosilicate glass capillaries	Internal microfilament, 1.0 mm OD; 0.53-mm ID; AF100-53-10	Sutter Instrument Co., Novato, CA
Disposable sterile syringe filters	22 mm, 0.22 µm, cellulose acetate membrane, 21052-25	Corning Glass Inc., Corning, NY
5-cc Syringes	Sterile, GX23525B	Corning Glass Inc., Corning, NY
96-Well flat bottom microtiter plates	86 × 128 mm external dimensions, 269787 and 264122	Nunc Inc., Naperville, IL
Reagents		
Sodium Chloride (NaCl)	Cat. #S-5886	Sigma Chemical Co., St. Louis, MO
Agarose	Type I-A:Low EEO, Cat. #S-0169	Sigma Chemical Co., St. Louis, MO
Triglyceride ($C_{8,0}$)	Cat. #T-9001	Sigma Chemical Co., St. Louis, MO
Triolein	95% purity, Cat. #T-9275	Sigma Chemical Co., St. Louis, MO
Sigmacote®	SL-2	Sigma Chemical Co., St. Louis, MO
Glacial acetic acid	Reagent A.C.S., Cat. #A38-212	Fisher Scientific, Pittsburgh, PA
2,3,7,8-Tetrachloro-dibenzo-*p*-dioxin	99% purity	Dow Chemical Co., Midland, MI
2,3,7,8-Tetrachloro-dibenzofuran	98–99% purity	Cambridge Isotope Laboratories, Woburn, MA
Tricaine methane-sulfonate (MS-222)	100 g, Cat. #C-Finq-UE-100G	Argent Aquaculture, Redmond, WA

Triolein

We use triolein for rainbow trout egg injection, although other carriers such as liposomes, other lipids, and corn oil are alternative vehicle carriers for lipophilic contaminants. We have found that triolein greatly decreases needle clogging and carrier-related mortality of embryos and sac-fry compared with organic solvents. Sterile saline is an effective vehicle carrier for Cd and other hydrophilic chemicals. Triolein (95% purity) is obtained in sealed 10-g ampules, filtered using a 5-cc syringe and disposable sterile syringe filter, and stored in 1-ml vials at 0 to 5°C prior to use.

Procedures

Preparation of agarose plates

Eggs are supported by an agarose gel during injection and remain supported by the gel during development. A 1% agarose solution (1 g agarose in 100 ml deionized water) is prepared and heated thoroughly. For rainbow trout eggs, 55 ml of hot agarose solution is poured into the bottom of an individual petri dish and allowed to completely solidify. Each agarose plate is coded using a permanent marker with the appropriate treatment group and date of injection. Using stainless steel tubing 3.5 to 4.5 mm in diameter (0.1 mm smaller than average diameter of eggs), 1 hole per grid square (36/petri dish) is punched and the agarose plug is removed with pointed forceps. Alternatively, plastic pipette tips may be cut to the desired diameter to punch a correct size hole. Plates may be kept for several weeks if stored in the refrigerator upside-down with lids on. With minor variations, agarose plates can accommodate eggs of various sizes and species. For example, when working with fathead minnow eggs the amount of agarose per plate will only be approximately 10 ml, size of the wells will be <1.2 mm in diameter, and agarose plates should be stored at room temperature.

Preparation of micropipettes

Reliable microinjection requires certain skill in making and using micropipettes. Micropipettes are made with a vertical pipette puller from an aluminosilicate glass microcapillary tube (1.0 mm OD, 0.53 mm ID). Aluminosilicate has greater tensile strength than borosilicate glass, and thus an aluminosilicate micropipette can penetrate the chorionic membrane without breaking. A concentric heater slowly heats and melts the center of an 8-cm-long microcapillary tube and gravitational force pulls the two ends of the microcapillary tube apart, forming two micropipettes. The taper of the micropipette is a function of the speed and temperature at which the microcapillary tube is heated and pulled. Ideally, the pulled micropipette should have a gradual taper, such that the outside diameter of the micropipette is 18 to 20 μm at a distance of approximately 125 μm from the tip. This taper allows for a fine tip (8 to 10 μm) after beveling and ensures needle strength. Injection of smaller eggs, such as fathead minnow, may require a finer tip, and adjustments to puller settings should be made accordingly. We set the micropipette puller at: heater 62.5 and solenoid 55; however, heater and solenoid settings will differ when using other micropipette puller brands. Thus, exact settings needed to achieve the desired micropipette taper and tip size must be determined by trial and error.

Micropipettes are siliconized after pulling by dipping the tip into Sigmacote®. This forms a tight, microscopically thin film of silicone on the glass which retards coagulation of egg yolk and subsequent needle clogging. Micropipette tips are then beveled at a 15° angle with a diamond wheel grinder, using deionized water to keep the surface of the grinder wet. The tip of the micropipette is touched to the grinder 4 to 5 times, for a total of approximately 60 sec. When viewed under the microscope, the bevel should appear sharp with no cracks or notches, and the outside diameter of the tip at the top of the bevel should be 8 to 10 μm. Micropipettes are then thoroughly rinsed with deionized distilled water using the pico-injector fill/clear settings and dried in a drying oven before use. Micropipette quality should improve with practice.

Preparation of dosing solutions

Dosing solutions are prepared by solvent transfer from an organic solvent, in which HAHs are routinely packaged by suppliers, to the dosing vehicle, triolein. Stock solutions of pure HAHs obtained from the suppliers or made from neat compound are diluted to obtain the appropriate amount of each dose. These dilutions are made with isooctane or an appropriate solvent. The concentration needed for each is calculated from the desired dose to the egg (HAH pg/g of egg), estimated egg weight (g/egg), and dosing volume (i.e., 0.1% of egg, nl). The concentration [(dose × egg wt.)/injection volume] is converted to the mass required by multiplying by the number of eggs in each dose. The calculated mass of HAH required for each dose is prepared in isooctane and added to a small (1.8 ml or less), sealable vial. The ability to accurately measure and transfer a specific volume of isooctane or other organic solvent is the limiting factor in preparing the dosing solutions, not the injection volume. The organic solvent is evaporated under a gentle stream of nitrogen to near dryness. The required volume of sterile triolein is added and each dose is sonicated thoroughly to mix the solution. Argon (or nitrogen) is layered over triolein solutions which may then be stored, protected from light, at 4°C or at room temperature for extended periods (approximately a year).

Filling of micropipettes

It is convenient to fill micropipettes before the day of injection, and store them so that they are ready for use. Two to four micropipettes can be filled simultaneously using the evaporating unit. Micropipettes are connected to steel evaporating needles with Teflon® tubing, and the height adjusted so that the tip of the micropipettes is submerged in the liquid. Because the micropipette tips are very fragile, a flashlight can help locate the position of the tip as it is placed into the vial. Vacuum pressure (about 650 mmHg) is then applied to the evaporating unit, and triolein solution is slowly pulled up into the micropipettes. Due to the high probability of accidental breakage, micropipettes should be filled only to 15 mm from the tip. Micropipettes prepared in this way hold approximately 3 μl of dosing solution, and take 10 to 15 min to fill. The total number of micropipettes needed for each dose is dependent on injection volume and total number of eggs per dosing group. For example, to inject 75 rainbow trout eggs per dose with 50 nl, 3.75 μl of each dosing solution is needed. Although this would require only two micropipettes, it is wise to fill extra micropipettes because they break frequently. For storage, filled micropipettes are placed horizontally in petri dishes affixed to pieces of molding clay to hold them in position. Both micropipettes and vials of dosing solution should be stored under argon at 4°C to prevent oxidation and to be protected from light.

Fertilization of eggs and placement in agarose plates

Unfertilized rainbow trout eggs and milt arrive packed in ice (about 4°C). Strain, number, temperature, and comments about the eggs should be recorded immediately upon arrival. Eggs and milt are allowed to warm slowly (over several hours) until the eggs are within one centigrade degree of incubator water temperature. Sperm motility can be tested by placing a drop of milt and a drop of 1% NaCl (using water taken directly from the incubator) on a glass microscope slide and examining the slide under the microscope for sperm movement and number. Good quality milt should contain a large percentage of active sperm. Fertilization involves placement of eggs in a cool, dry mixing bowl, addition of 30 to 40 ml of 1% NaCl, followed by immediate addition of milt. Eggs and milt are then mixed gently by hand for a few minutes. Fertilization of eggs should be done quickly in order to keep the eggs from warming. Eggs are strained to dispose of excess waste water and are placed directly into the incubator. Water temperature should be between 8 and 10°C for rainbow trout. After water-hardening for about 1 h, white or discolored eggs should be discarded. Eggs are placed into agarose plates at least 1 h prior to injection, allowing dead eggs to be culled and replaced. Eggs can be transferred to agarose plates using a hand-operated automatic pipettor with glass pipette, or a valved silicone-rubber bulb pipettor. Eggs should fit easily but be held firmly in the wells in the agarose plates. Filled plates without lids are placed directly into the incubator.

Micropipette calibration

Before injection can begin, each micropipette must be calibrated. In order to determine the desired injection volume, mean egg volume must be obtained, and the percentage of egg volume to be injected determined. Average egg volume is estimated by randomly selecting a sub-sample of 25 eggs, measuring diameters, and calculating the sample mean. This value is then used in an equation that calculates the volume of a sphere (i.e., $V = \pi d^3/6$), and a percentage of this volume is then chosen as the injection volume. Selection of injection volume will depend on factors such as carrier solvent, type of study being conducted (larger volumes may be required due to solubility problems), and egg size (species and strain). For rainbow trout, injection volume typically equals 0.1% of the average egg volume. Since the average egg volume is 50 µl, the resulting injection volume would equal 50 nl.

Injection volumes can be manipulated by changing either injection pressure, injection duration, or micropipette tip diameter or taper. Micropipettes prepared according to the previously described method will all have a fairly uniform tip diameter and taper, so injection volume depends only on the injection pressure or injection duration. Injection pressure and injection duration are manipulated to increase or decrease the amount of liquid being delivered. Two methods are used to estimate the amount of liquid injected: (1) bubble pressure method and (2) measurement of the droplet diameter (Table 5). The bubble pressure method[26,27] calculates volume delivered as a

Table 5 Two Methods for Estimating Injection Volume for the
Nano-Injection Method

Technique	Explanation	Calculations
Bubble pressure[26,27]	Tip diameter is a function of surface tension of the fluid and threshold pressure	*Glass Tip Diameter:* $d_{in} = 4s/P$ where: d_{in} = inside diameter of pipette tip (μm) s = surface tension of the fluid ($H_2O = 72 \times 10^{-3}$ Newton/M) P = threshold pressure at which an air bubble forms when injected into a fluid with surface tension (s)
	Injection volume is a function of fluid viscosity, tip diameter, pipette taper, pressure and duration of injection	*Volume Delivered:* $V = (3\pi/4\eta)d_{in}^3\sigma_t\Delta PT$ where: V = volume delivered (μl) d_{in} = inside diameter of pipette tip σ = tangent of the pipette taper η = viscosity of fluid ΔP = change in pressure[a] T = duration of injection
Measurement of droplet diameter[28-30]	Diameter of a droplet of carrier solution (i.e., triolein) is measured in deionized water or preferably in fish eggs, and droplet volume computed	*Volume Delivered*[b]: $V = \pi d^3/6$ where: V = volume delivered (nl) d = droplet diameter (μm)

[a] Balance pressure up to injection pressure.

[b] For a desired injection volume, the injection pressure and/or time can be manipulated to change droplet diameter.

function of viscosity of the injection fluid, inside diameter of the micropipette, micropipette taper (taper height/taper length), pressure difference between balance and injection, and duration of injection. Inside tip diameter is measured by ejecting air through the micropipette into a liquid of known surface tension, and determining the threshold pressure where bubbles begin to form in the liquid. The inside tip diameter is then used to calculate volume delivered. Viscosity of injection fluid, inside tip diameter, tangent of the pipette taper, and change in pressure can be measured in advance. The basic equation can be used to find injection duration needed to deliver the desired injection volume. Therefore, only time of injection would need to be manipulated.

The second method for calculating volume of fluid delivered through a microcapillary is the direct measurement of the volume of a droplet of the carrier solution.[28-30] When a hydrophobic substance, such as triolein, is injected into a scintillation vial of water, the droplet formed will adhere to the micro-capillary tip. The diameter of this droplet can be measured using a stereomicroscope equipped with a micrometer, and volume of the droplet can then be calculated. To achieve a desired volume, injection pressure and/or time can be manipulated to increase or decrease the droplet diameter. For example, if an injection volume of 50 nl is desired, the corresponding droplet diameter would be 457 µm. This method negates the need to measure micropipette tip diameter and involves simpler calculations, but may result in greater error than the bubble pressure method, and the accuracy of the droplet measurement method depends on the precision of the micrometer. The most reliable method is when the droplet is measured inside the egg to account for internal egg pressure. Moreover, diameter of the droplet should be rechecked periodically because the micropipette tip diameter may change during injection of multiple eggs. In many cases the droplet diameter can be measured in each egg.

There are three methods for determination of the accuracy of the egg injection volume. The first involves the use of a radiolabeled substance in the carrier solution, followed by liquid scintillation counting (LSC). One problem which may arise with LSC is counting errors caused by egg pigment. Yellow-to-orange pigments of rainbow trout eggs may quench detection of radioactive disintegrations by the liquid scintillation counter and underestimate injection volume. A quench curve can be constructed (as described under the discussion of the microinjection method) using control eggs to account for quenching by egg pigments. Alternatively, an estimated volume can be injected straight into a liquid scintillation vial. However, this will not account for internal egg pressure. The third method to determine accuracy of injection volume is to photograph carrier droplets inside the egg and manually calculate droplet volume. This method can calculate the exact dose (injection volume) and verify the accuracy of previous calculations.

Egg injection

Once the micropipette is calibrated, the first plate of eggs can be injected. The plate is removed from the incubator, leaving a layer of water over the eggs, and is placed under the microscope. If possible, examine the eggs for signs of blastodisc formation. Although it is difficult to recognize newly fertilized eggs, time can be saved by culling unfertilized eggs. The micropipette is positioned and slowly advanced towards the egg until the needle tip can be seen under the microscope. The micropipette is carefully advanced using a hydraulic micromanipulator until the egg chorion is pierced, then further advanced until the needle tip breaches the yolk (Figure 4A). Injection is triggered either manually or with an optional footswitch, and a tone signals the end of liquid delivery. The injected droplet should be visible (Figure 4B). The micropipette is slowly moved out of the egg and the sequence repeated with the next egg. If the needle clogs or breaks, it will be necessary to recalibrate a new micropipette. With practice, an individual should be able to inject a plate of 36 eggs in 10 to 15 min. Once a plate is finished, it is returned to the incubator. Temperature control of the petri dish may be required during injection if the air temperature is vastly different

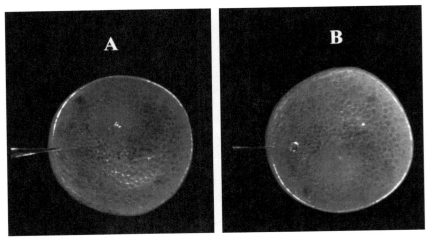

Figure 4 Glass micropipette of the nano-injection (A) pierces the chorion of a rainbow trout egg, and (B) injects approximately 50 nl triolein solution, which is visible within the egg.

than the water temperature. Rainbow trout eggs should be injected within the first 48 h after fertilization to investigate potential effects on early events in embryonic development; however, eggs can be injected until the completion of gastrulation. Between this stage and the eyed stage, eggs do not tolerate injection or excessive handling, and this generally holds true for eggs of all fish species.

Considerations which apply to both injection methods

Maintenance of eggs and sac fry after injection

Following injection, developing eggs and sac fry should be kept at a fairly constant temperature (8 to 10°C for rainbow trout) and should be protected from bright light. Once the eggs have eyed, they are less sensitive to damage from light or physical disturbances. In addition, water quality, dissolved oxygen, pH, water hardness, nitrates, and alkalinity, must be kept within acceptable limits for rainbow trout development and monitored periodically during the experiment.

Postinjection monitoring of rainbow trout eggs and sac fry

Eggs and sac fry are monitored daily for mortality, the dead removed and number recorded. Dead eggs are checked for fertilization by placing them in 10% glacial acetic acid; a developing embryo appears white while an unfertilized egg will not show any evidence of embryonic development. Mortality which occurs during hatching should be recorded. Newly hatched fry should be monitored until at least the swim-up stage of development at which time surviving fry in each group are counted.

Animal welfare

Disposal of the swim-up fry follows standard procedure, with strict attention paid to animal welfare regulations. Fry are placed in a (100 to 200 mg/l) solution of MS-222® (tricaine methanesulfonate) and euthanized. The ethical and humane treatment of experimental animals is of utmost importance in scientific investigations. It is imperative that scientists are aware of and follow guidelines and regulations for the humane treatment of all experimental animals. The U.S. Congress passed the Laboratory Animal Welfare Act in 1966, which was subsequently amended, most recently in 1985. This act stipulates conditions and practices required for animal testing in the U.S. Guidelines that more specifically address the proper handling of fishes in scientific studies can be found in "Guidelines for Use of Fishes in Field Research," a document developed by the American Society of Ichthyologists and Herpetologists, the American Fisheries Society, and the American Institute of Fisheries Research. Special regulations at your institution should be consulted prior to any testing or planning of experiments.

Safety

The proper handling of toxic chemicals is required for the safety of researchers and their co-workers, and to comply with all institutional, state, and federal regulations. The development of standard procedures for the safe handling of chemicals and waste products generated from egg injection studies are imperative. Sonication of dosing solutions should be conducted in a fume hood since aerosols can be generated. Effluent water from incubators or aquaria containing injected fish eggs must be filtered to remove any waterborne contaminants. This waste water should be pumped through a treatment system that includes particulate and carbon filtration. For small waterflow systems, an initial filter which removes particulate matter followed by two charcoal filters which remove lipophilic contaminants (see Table 2 for supplier of these filters) are sufficient to remove essentially 100% of waterborne contaminants.[31] Filters should be changed according to waterflow use and disposed of in accordance with U.S. Environmental Protection Agency regulations. Nondisposable laboratory supplies that come in contact with dosing solutions, such as needles, vials, etc., should be rinsed with a solvent (e.g., 4:1 chloroform:methanol) and the solvent collected. The solvent rinsate and any other liquid waste are placed in a quartz reaction vessel and photolyzed. The resultant solvent after photolysis may be disposed of as normal solvent waste. Disposable laboratory supplies should be disposed of in a U.S. Environmental Protection Agency-approved hazardous waste landfill, while contaminated animal waste should be incinerated at >800°C.

Statistical procedures

Mortality during the egg and sac fry stages of development is recorded and evaluated for treatment-related increases by chi-square analysis.[32] From those mortality data which exhibit treatment-related increases by chi-square analysis, a continuous dose–response curve and 95% fiducial limits are gen-

erated using a probit procedure and SAS[33] (Appendix 1). The probit procedure corrects for control mortality (option c, optc) analogous to using
Abbott's formula.[33,34] In addition, the probit procedure uses chi-square goodness of fit, estimates the slope and intercept, and is based on the assumption
that mortality is independent for fish within a dose group and among dose
groups.[33,35] Significance levels are set at $p \leq 0.05$.

Results and discussion

The data and statistical analyses presented here are based on injection of
rainbow trout eggs using the microinjection method, but the data could have
as easily been generated using the nano-injection method and any species
of fish egg. Data generated when using the nano-injection method will differ
in two ways: (1) the N per dose group can be substantially higher, and (2)
injection-related mortality in the control group is consistently $\leq 1\%$.

Mortality data

Chi-square analysis

Mortality data from a dose–response experiment where control (50 mM PC)
and 7 graded doses of 2,3,7,8-tetrachlorodibenzo-p-dioxin (TCDD) in 50 mM
PC were injected into rainbow trout eggs (Erwin strain, 30 eggs/dose, 0.6 μl
injection volume) are shown in Table 6. Previous research has shown that
TCDD manifests toxicity during salmonid early life stage development by
a dose-related increase in hatching and sac fry mortality, without an effect
on egg survival.[14,36,37] Chi-square analysis of egg mortality (based only on
dead fertilized eggs) shows that in this experiment, as in previous experiments, egg mortality is not related to TCDD exposure ($\chi^2 = 1.066$, df $= 7$, p
>0.25).

Since TCDD did not affect rainbow trout mortality during the egg stage
of development, egg mortality is disregarded and the dose-related effect of
TCDD on sac fry mortality (from hatching onset to swim-up) is analyzed
based only on those embryos surviving to hatching (adjusted N). Chi-square
analysis of sac mortality based on the adjusted N is highly related to treatment ($\chi^2 = 127$, df $= 7$, $p <0.001$).

Probit analysis

Sac fry mortality (hatching onset to swim-up) is modeled as a function of
the egg TCDD dose using a probit procedure (Appendix I) which corrects
all TCDD doses for control mortality. Input for the program (line 2) includes
the egg TCDD dose (dose), adjusted N (n), and number of dead sac fry
(response). The program models the dose–response data using a probit procedure (line 15, proc probit), the log10 dose (log10), and a threshold of control
response (optc). Abbreviated output from this statistical analysis is shown
in Appendix II. The probit procedure uses two chi-square goodness-of-fit
tests to determine if the probit model fits the data. A p value >0.05 from both
tests rejects the null hypothesis that the probit model does not fit the data.
The probit procedure also estimates the slope and intercept of the

Table 6 Mortality Data Set Generated from Dose–Response Experiment Where Control (50 mM PC) and 7 Graded Doses of TCDD in 50 mM PC Were Injected into Rainbow Trout Eggs (Erwin Strain, 30 Eggs/Dose, 0.6 µl Injection Volume)

TCDD dose (pg TCDD/g egg)	Mortality Egg Unfertilized	Mortality Egg Fertilized[a]	Sac fry[b]	Survivors at swim-up	Total eggs/dose[c]	Responders[d]	Adjusted N[e]
0	0	4	1	22	27	1	23
84	0	6	4	20	30	4	24
117	0	7	0	23	30	0	23
168	1	6	7	16	30	7	23
237	0	7	14	8	29	14	22
339	0	6	24	0	30	24	24
486	0	6	22	0	28	22	22
693	0	5	23	0	27	23	23

[a] No treatment related effect on egg mortality. $\chi^2 = 1.066$, df = 7, $p > 0.25$.

[b] Sac fry mortality includes sac fry mortality and mortality of embryos during the hatching process. Sac fry mortality was highly related to treatment. $\chi^2 = 127$, df = 7, $p < 0.001$.

[c] Ideally, total eggs/dose should be 30. However, on occasion the number of injected eggs is miscounted, or eggs and sac fry removed as dead are miscounted. Thus, only those eggs and sac fry accounted for as dead or alive are included in statistical analyses.

[d] Since chi-square analysis indicates that egg mortality is not related to treatment, then egg mortality is not included in further analyses and "responders" represent only dead sac fry.

[e] Since chi-square analysis indicates that egg mortality is not related to treatment, then those dying as eggs are removed from the total N and the adjusted N represents only those surviving to hatching.

dose–response curve, and a lower threshold of response termed "C." The threshold of response is necessary in the probit model when a zero dose (i.e., control) shows a response (i.e., sac fry mortality). In this experiment control mortality was 1 sac fry out of 23, or 4%, and the probit procedure estimated a threshold of response as 7.7 ± 3.3% (mean ± standard error) sac fry mortality. The probit procedure also determines significance of the slope and intercept. A p value <0.05 indicates that the slope and intercept of the probit model are statistically significant.

Lastly, the probit procedure generates a dose–response curve and 95% fiducial limits from 1 to 99% sac fry mortality vs. egg TCDD dose (Figure 5). The raw data points (closed circles) plotted on this graph are corrected for the threshold of response (C) calculated by the probit procedure (Table 7). This is done using Abbott's formula:

$$[(P - C)/(100 - C)] \times 100 = \text{adjusted percent mortality}^{34} \qquad (1)$$

where P is the percent mortality of a given dose group based on the raw data, and C is the threshold of response calculated by the probit procedure.

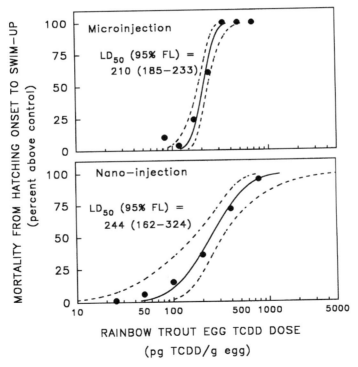

Figure 5 Probit model for mortality from hatching onset to swim-up (percent above control) for Erwin strain rainbow trout eggs injected with TCDD using the microinjection method (top panel) or using the nano-injection method (bottom panel). Solid and dashed lines represent the probit model and 95% fiducial limits, respectively, corrected for percent mortality in the vehicle control group using option C.

Table 7 Adjustment of Raw Data Percent Mortality for
Threshold of Response (C) using Abbott's Formula[a]

TCDD Dose (pg TCDD/g egg)	Percent mortality	
	Raw data	Corrected for threshold response
84	4/24 = 14	10
117	0/23 = 0	<0[b]
168	7/23 = 30	24
237	14/22 = 64	61
339	24/24 = 100	100
486	22/22 = 100	100
693	23/23 = 100	100

[a] Abbott's formula: $[(P - C)/(100 - C)] \times 100$ = adjusted percent mortality.[24]

[b] Since the percent mortality for the 117 pg TCDD/g egg treatment group is zero, the raw data cannot be corrected for threshold of response using Abbott's formula. The percent mortality can be adjusted using continuity correction as described in the text.

For the egg TCDD dose of 117 pg TCDD per g egg, percent sac fry mortality (0%) was lower than the control value (4%), and thus an adjusted value cannot be calculated. This is not uncommon, and an adjusted value can be calculated using continuity correction:[32]

$$\tfrac{1}{4} N = \text{adjusted percent mortality} \qquad (2)$$

where N is the number of dose groups (not including control, N = 7 in this example) that generate the probit model. In this case the adjusted percent mortality for 117 pg TCDD per g egg treatment group is $\tfrac{1}{28} = 3.6\%$.

Validation of egg injection

The major route of accumulation of lipophilic, environmental contaminants in salmonid eggs is from the transfer of these contaminants from adult female tissues to the eggs during oocyte maturation.[1,2] The question remains whether injection of newly fertilized fish eggs with lipophilic contaminants will reflect accurately the toxicity of these chemicals when eggs are exposed via maternal transfer. Recently, Walker et al.[3] have shown that exposure of lake trout (*Salvelinus namaycush*) eggs to the prototypical HAH, TCDD, via egg injection or via maternal transfer results in the same signs of early life stage toxicity and dose–response relationship for sac fry mortality. Thus, the comparability of early life stage toxicity following egg injection and maternal transfer validates the use of egg injection to characterize the toxicity of other HAH congeners to fish early life stage development. To strengthen the validity of using egg injection data in environmental risk assessment, it will be necessary to repeat this validation for other classes of environmental contaminants as well as other species of fish.

Other applications

The egg injection methods and statistical analyses described here were employed for injection of rainbow trout eggs with lipophilic chemicals and determining a dose–response curve for early life stage mortality. These two injection methods, however, are not limited only to rainbow trout early life stage mortality studies. The nano-injection method can be applied to fish eggs that are much smaller than rainbow trout, while both methods can be used to inject hydrophilic chemicals or to determine dose–response relationships for responses other than lethality. Two brief examples are provided where the nano-injection method was used and dose-related effects determined for nonlethal endpoints as well as for mortality: (1) injection of fathead minnow eggs,[38] and (2) injection of sea trout eggs with cadmium chloride ($CdCl_2$), a hydrophilic chemical.

Nano-injection of fathead minnow eggs

Outboard motor exhaust from two-stroke engines typically leaves "used fuel" in the wake water. Water, previously exposed to "used fuel" from a two-stroke outboard motor, was extracted with hexane and extract dissolved in corn oil. Fathead minnow eggs were then injected with graded doses of "used fuel" extract early in embryonic development using the nano-injection method. Dose-related responses observed at hatching and 3 days after hatching included pericardial edema, yolk sac precipitate, deformed vertebrae, and mortality (Figure 6A).[38]

Injection of a hydrophilic chemical

As a model substance, cadmium chloride ($CdCl_2$), dissolved in iso-osmotic saline, was injected into sea trout eggs and the incidence of deformed vertebrae and mortality recorded. The incidence of dead sac fry or sac fry exhibiting deformed vertebrae increased with increasing dose of $CdCl_2$ and increasing time after injection (Figure 6B).

Summary

Microinjection method

The microinjection method was developed to expose salmonid eggs to known, graded doses of nonradiolabeled HAHs, and to determine the dose–response relationship for early life stage mortality based on a known egg dose.[14,36] The liposome vehicle carrier is prepared from purified phosphatidylcholine, a natural constituent of fish eggs and the most common phospholipid in rainbow trout eggs;[39] and the use of liposomes significantly reduces injection-related mortality compared to use of organic solvents such as DMSO.[11,13] The microinjection method requires a modest capital investment for a stereomicroscope and micromanipulator, and minimal need to train personnel on use of injection equipment. Two investigators can inject up to 300 rainbow trout eggs in 3 to 4 h, with 93 to 97% of the injected dose retained in the eggs prior to hatching, and dose–response results for early life stage mortality obtained within 60 days. Although this method was

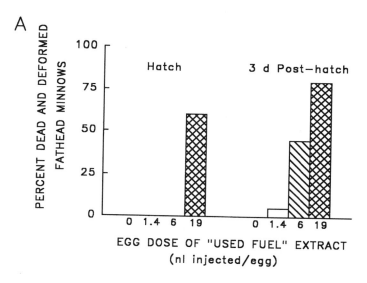

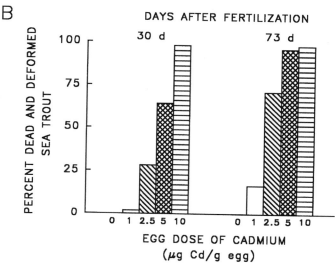

Figure 6 (A) Cumulative percent dead and deformed fathead minnows at hatch or 3 days after hatch, following injection as eggs with "used fuel" extracts from a two-stroke engine exhaust using the nano-injection method. Control injected eggs (0 dose) were injected with corn oil. Non-injected control eggs exhibit the same percent mortality and deformities (0%) as injected controls (data not shown). N = 100 eggs/dose group. (B) Cumulative percent dead and deformed sea trout 30 and 73 days after fertilization, following injection with $CdCl_2$ as eggs using the nano-injection method. Coefficient of variation ≤15%.

developed for injection of newly fertilized rainbow trout eggs, it also has been successfully applied to newly fertilized lake trout eggs[3,14] as well as lake trout eggs between eye-up and hatching, and newly hatched lake trout sac fry.[18,40]

One primary disadvantage to the microinjection method is that injection-related mortality can range from 0 to 50% depending on egg strain, egg size, shipping conditions, salmonid species injected, and experience of the investigator performing the injections. However, injection-related mortality decreases to 0 to 5% as the investigator improves his/her injection technique and if eggs are maintained at a relatively stable temperature during shipping. Furthermore, and most importantly, the statistical analyses allow an investigator to easily compare among different injection studies which exhibit different levels of control response. The probit model accounts for injection-related control mortality by calculating a threshold of response, thus allowing for comparisons among experiments with differing levels of control mortality.[36]

Nano-injection method

The nano-injection method for fish egg injection has some distinct advantages over previous methods. First, mortality of the control group (sham injected) is consistently less than 1% greater than that of noninjected control eggs. Previous methods have resulted in high mortality in the vehicle control groups and greater variability in control mortality among experiments.[11-14] Second, use of glass micropipettes as needles allows experimentation with small fish species such as the fathead minnow. Use of small fish allows for multi-generation studies to be conducted more easily. Third, the nano-injection method has shorter injection times and thus greater flexibility for experimental designs. Greater numbers of eggs can be injected than with previous methods given an equivalent amount of personnel effort. Fourth, dosing solutions using triolein, corn oil, or triglycerides as carriers may be prepared in fewer steps than liposomes, and use of these nonpolar substances reduces solubility problems associated with some HAHs in other polar carriers.

The major disadvantage of the nano-injection method is the initial capital investment required for the pico-injector, hydraulic micromanipulator and stage, diamond grinder, and pipette puller. At current prices these items cost approximately $15,000 U.S. However, the puller and grinder are common pieces of equipment that are found in many physiology departments on university campuses, and it is not uncommon for these to be available for use by others. Therefore, the cost of the pico-injector (about $6,000 U.S.) is the major piece of equipment that is not also required for the microinjection technique.

Acknowledgments

M. K. Walker and E. W. Zabel would like to thank Larry C. Hufnagle, Jr., Murray Clayton, and Linda Damos for their expert roles in helping to develop the microinjection method; Richard E. Peterson for providing perceptive advice and financial support; and Michael J. Barden and Susan M. Smith for their assistance in finalizing this manuscript. Don Tillitt and Peggy Wright would like to thank Diane K. Nicks for her critical role in helping to develop the nano-injection method.

Appendix I   SAS statistical program used to analyze dose–response data using a probit model

```
data rbttcdd;
input dose n response;
phat = response/n;
cards;
     0   23    1
    28   24    4
    39   23    0
    55   23    7
    79   22   14
   113   24   24
   162   22   22
   231   23   23
proc print:
   TITLE 'RBTTCDD';
   proc probit log10 optc
   var dose n response;
   output out=b p=prob;
   proc plot;
   plot phat*dose='x' prob*dose='p'/overlay;
   run;
```

Appendix II   Abbreviated output from probit statistical analysis

Goodness-of-Fit-Tests

Statistic	Value	DF	Prob > Chi-Sq
Pearson Chi-Square	7.1472	5	0.2099
L.R. Chi-Square	5.4102	5	0.3679

Note: Since the chi-square is small ($p > 0.10000$), fiducial limits will be calculated using a t value of 1.96.

Probit Procedure

Variable	DF	Estimate	SE	ChiSq	Pr > ChiSq
INTERCEPT	1	−17.7	3.6	24.5	0.0001
Log 10(DOSE)	1	9.6	1.9	25.1	0.0001
C	1	0.08	0.03		

Probit Procedure, Probit Analysis
of DOSE

Probability	DOSE	95% Fiducial Limits	
		Lower	Upper
0.001	119.97	79.59	145.50
.	.	.	.
.	.	.	.
.	.	.	.
0.45	203.61	178.32	226.32
0.50	209.85	185.40	233.79
0.55	216.27	192.45	241.89
0.60	223.02	199.53	250.83
.	.	.	.
.	.	.	.
.	.	.	.
0.98	343.77	294.15	472.80
0.99	367.05	309.36	524.79

References

1. Niimi, A.J. 1983. Biological and toxicological effects of environmental contaminants in fish and their eggs. *Can. J. Fish. Aquat. Sci.* 40:306–312.
2. Miller, M.A. 1993. Maternal transfer of lipophilic contaminants in Salmonines to their eggs. *Can. J. Fish. Aquat. Sci.* 49:1405–1413.
3. Walker, M.K., Cook, P.M., Batterman, A., Butterworth, B., Bernini, C., Libal, J.J., Hufnagle, Jr., L.C., Peterson, R.E. 1994. Translocation of 2,3,7,8-tetrachlorodibenzo-*p*-dioxin from adult female lake trout (*Salvelinus namaycush*) to oocytes: effects on early life stage development and sac fry survival. *Can. J. Fish. Aquat. Sci.* 51:1410–1419.
4. McKim, J.M. 1977. Evaluation of tests with early life stages of fish for predicting long-term toxicity. *J. Fish. Res. Board Can.* 34:1148–1154.
5. Walker, M.K., Peterson, R.E. 1994. Aquatic toxicity of dioxins and related chemicals. In *Dioxins and Health*, Ed. A. Schecter, New York:Plenum. pp. 347–387.
6. Walker, M.K., Peterson, R.E. 1994. Toxicity of 2,3,7,8-tetrachlorodibenzo-*p*-dioxin to brook trout (*Salvelinus fontinalis*) during early development. *Environ. Toxicol. Chem.* 13:817–820.
7. McKim, J. 1985. Early life stage toxicity tests. In *Fundamentals of Aquatic Toxicology: Methods and Applications*, Eds. G.M. Rand, S.R. Petrocelli, Washington:Hemisphere. pp. 58–95.
8. Cook, P.M., Walker, M.K., Kuehl, D.W., Peterson, R.E. 1991. Bioaccumulation and toxicity of 2,3,7,8-tetrachlorodibenzo-*p*-dioxin and related compounds in aquatic ecosystems. In *Biological Basis for Risk Assessment of Dioxins and Related Compounds*, Eds. M.A. Gallo, R.J. Scheuplein, C.A. vander Heijden, Banbury Report, Vol. 35, Cold Spring Harbor, New York:Cold Spring Harbor Laboratory Press. pp. 143–167.

9. Klaassen, C.D. 1986. Principles of toxicology. In *Casarett and Doull's Toxicology. The Basic Science of Poisons*, Eds. C.D. Klaassen, M.O. Amdur, J. Doull, 3rd edition, New York:Macmillan Publishing Company. pp. 11–32.

10. Metcalfe, C.D., Sonstegard, R.A. 1981. Microinjection of carcinogens into rainbow trout embryos: an *in vivo* carcinogenesis assay. *JNCI* 73:1125–1132.

11. Black, J.J., Maccubbin, A.E., Schiffert, M. 1985. A reliable, efficient, microinjection apparatus and methodology for the *in vivo* exposure of rainbow trout and salmon embryos to chemical carcinogens. *J. Natl. Cancer Inst.* 75:1123–1128.

12. Metcalfe, C.D., Cairns, V.W., Fitzsimons, J.D. 1988. Microinjection of rainbow trout at the sac-fry stage: a modified trout carcinogenesis assay. *Aquat. Toxicol.* 13:347–356.

13. Black, J.J., Maccubbin, A.E., Johnston, C.J. 1988. Carcinogenicity of benzo[a]pyrene in rainbow trout resulting from embryo microinjection. *Aquat. Toxicol.* 13:297–308.

14. Walker, M.K., Hufnagle, Jr., L.C., Clayton, M.K., Peterson, R.E. 1992. An egg injection method for assessing the early life stage mortality of polychlorinated dibenzo-*p*-dioxins, dibenzofurans, and biphenyls in rainbow trout, (*Oncorhynchus mykiss*). *Aquat. Toxicol.* 22:15–38.

15. Åkerman, G., Balk, L. 1992. A reliable and improved methodology to expose fish in the early embryonic stage. Seventh International Symposium on Responses of Marine Organisms to Pollutants, Abstract.

16. Åkerman, G., Balk, L. 1995. A reliable and improved methodology to expose fish in the early embryonic stage. *Marine Environ. Res.* 39:155–158.

17. Åkerman, G., Ericson, G., Westin, L., Broman, D., Näf, C., Balk, L. 1994. Potential tools to investigate reproduction disturbances in fish. In *Reproductive Disturbances in Fish*, Ed., L. Norrgren, Report from workshop held Oct. 20–23, 1993, Uppsala, Sweden, Swedish Environmental Protection Agency, Report 4346. ISBN 91-620-4346-3.

18. Zabel, E.W., Peterson, R.E. 1994. 2,3,7,8-Tetrachlorodibenzo-*p*-dioxin (TCDD) toxicity at three stages of lake trout egg development. *Soc. Environ. Toxicol. Chem.*, Fifteenth Annual Meeting Abstracts, p. 153.

19. Zabel, E.W., Walker, M.K., Clayton, M.K., Peterson, R.E. 1995. Interaction of polychlorinated dibenzo-*p*-dioxin, dibenzofuran, and biphenyl congeners for producing rainbow trout early life stage mortality. *Toxicol. Appl. Pharmacol.*, 134:204–213.

20. Walker, M.K., Cook, P.M., Butterworth, B.C., Zabel, E.W., Peterson, R.E. 1996. Potency of a complex mixture of polychlorinated dibenzo-*p*-dioxin, dibenzofuran, and biphenyl congeners compared to 2,3,7,8-tetrachlorodibenzo-*p*-dioxin in causing fish early life stage mortality. *Fundam. Appl. Toxicol.* 30: in press.

21. Johnson, S.C., Chapman, G.A, Stevens, D.G. 1983. Sensitivity of steelhead trout embryos to handling. *Prog. Fish-Cult.* 45:103–104.

22. Johnson, S.C., Chapman, G.A., Stevens, D.G. 1989. Relationships between temperature units and sensitivity to handling for coho salmon and rainbow trout embryos. *Prog. Fish-Cult.* 51:61–68.

23. Leitritz, E., Lewis, R.C. 1976. In *Trout and salmon culture*. Calif. Dept. Fish Game, Fish. Bull. 164, 197 pp.

24. Woodle, M.C., Papahadjopoulos, D. 1989. Liposome preparation and size characterization. In *Biomembranes* Part R, *Transport Theory, Cells and Model Membranes*, Eds., S. Fleischer, B. Fleischer, San Diego:Academic Press, Methods in Enzymology series, 171:193–217.

25. Bartlett, G.R. 1959. Phosphorus assay in column chromatography. *J. Biol. Chem.* 234:466–468.

26. Hagag, N., Viola, M., Lane, B., Kemp Randolph, J. 1990. Precise, easy measurement of glass pipet tips for microinjection or electrophysiology. *BioTechniques* 9:401–406.
27. Mittman, S., Flaming, D.G., Copenhagen, D.R., Belgum, J.H. 1987. Bubble pressure measurement of micropipet tip outer diameter. *J. Neurosci. Meth.* 22:161–166.
28. Minaschek, J., Bereiter-Hahn, G., Bertholdt, G. 1989. Quantitation of the volume of liquid injected into cells by means of pressure. *Exp. Cell Res.* 183:434–442.
29. Stephens, D.L., Miller, T.J., Silver, L., Zipser, D., Mertz, J.E. 1981. Easy-to-use equipment for the accurate microinjection of nanoliter volumes into the nuclei of amphibian oocytes. *Anal. Biochem.* 114:299–309.
30. Sakai, M., Swartz , B.E., Woody, C.D. 1979. Controlled micro release of pharmacological agents: measurements of volume ejected *in vitro* through fine tipped glass microelectrodes by pressure. *Neuropharmacology* 18:209–213.
31. Kleeman, J.M., Olson, J.R., Chen, S.M., Peterson, R.E. 1986. 2,3,7,8-Tetrachlorodibenzo-*p*-dioxin (TCDD) metabolism and disposition in yellow perch. *Toxicol. Appl. Pharmacol.* 83:402–411.
32. Snedecor, G.W., Cochran, W.G. 1980. *Statistical Methods*, seventh edition. Ames:Iowa State University Press, 507 pp.
33. SAS Institute Inc., 1988. SAS® Technical Report P-179, Additional SAS/STAT, Procedures, Release 6.03, Cary, NC.
34. Abbott, W.S. 1925. A method of computing the effectiveness of an insecticide. *J. Econ. Entomol.* 18:265–267.
35. Finney, D.J. 1971. *Probit Analysis*, third edition, Cambridge, England:Cambridge University Press, 333 pp.
36. Walker, M.K., Peterson, R.E. 1991. Potencies of polychlorinated dibenzo-*p*-dioxins, dibenzofurans, and biphenyls, relative to 2,3,7,8-tetrachlorodibenzo-*p*-dioxin, for producing early life stage mortality in rainbow trout (*Oncorhynchus mykiss*). *Aquat. Toxicol.* 21:219–238.
37. Walker, M.K., Spitsbergen, J.M., Olson, J.R., Peterson, R.E. 1991. 2,3,7,8-Tetrachlorodibenzo-*p*-dioxin (TCDD) toxicity during early life stage development of lake trout (*Salvelinus namaycush*). *Can. J. Fish. Aquat. Sci.* 48: 875–883.
38. Balk, L., Ericson, G., Lindesjöö, E., Petterson, I., Tjärnlund, U., Åkerman, G. 1994. Effects of exhaust from two-stroke outboard engines on fish. Studies of genotoxic, enzymatic, physiological, and histological disorders at the individual level. Tem Nord 1994:528, ISBN 92-9120-139-0.
39. Henderson, R.J., Tocher, D.R. 1987. The lipid composition and biochemistry of freshwater fish. *Prog. Lipid. Res.* 26:281–347.
40. Abnet, C. 1995. Environmental Toxicology Center, University of Wisconsin, Madison, personal communication.

chapter five

Fish embryos as in situ monitors of aquatic pollution

Richard M. Kocan

Introduction

Investigators have used *in situ* exposure of fish embryos as a method for evaluating aquatic pollution for many years. Schnack[1] studied herring mortality by exposing them *in situ* to natural conditions, and Hardy et al.[2,3] looked at the effects of sea-surface microlayer on sole eggs in Puget Sound. Likewise, the author used this technique in several pollution studies in Puget Sound and Prince William Sound following the *Exxon Valdez* oil spill of 1989.[4]

In order to maintain adequate control over experimental conditions and keep variability to a minimum, aquatic ecosystems are frequently evaluated for contamination by collecting water or sediment from suspected sites, returning these to the laboratory, and evaluating them for acute toxicity using standardized toxicity test protocols.[5] Although this procedure has been used successfully, it measures toxicity at a single point in time in a non-dynamic system, and does not accurately replicate the constantly changing conditions encountered by aquatic organisms in their natural environment. Conditions such as tide, currents, temperature cycles, light, sediment disturbance, and other natural variables must be considered when evaluating the toxicity of a site to select aquatic organisms. The passage of time inevitably results in an organism encountering change, and this may modify the effects of exposure to toxicants.

As embryo development progresses from fertilization to hatching and larval maturation, the response to toxic substances also changes.[6,7] Exposure of the earliest cleavage stages (fertilization → 64 cells) most often results in embryo mortality, while exposure of the undifferentiated cells prior to organogenesis will most often result in physical defects (teratogenesis). The later stages of development, from early organogenesis through hatching are generally resistant to overt toxic effects, but are more prone to develop neoplasms (cancer) later in life. Very little overt toxicity will be observed in the laboratory at concentrations normally found in the environment.

1-56670-149-X/96/$0.00+$.50
© 1996 by CRC Press, Inc.

To address the issue of "real world" exposure, a method for exposing organisms *in situ* was developed[8] that allowed laboratory, indigenous or native fish species to be exposed on site during the most sensitive stage of their life history — embryonic development. Following *in situ* exposure, these organisms were returned to the laboratory for final evaluation. This technique is a modification of the early life stage (ELS) toxicity test described by McKim.[9] In 83% of the cases, the ELS estimates the maximum acceptable toxicant concentration (MATC) as accurately as the more costly and time consuming complete life cycle toxicity test.

The *in situ* method permits an investigator to evaluate a potentially contaminated site under the variable conditions that normally occur over a period of time. Embryos can be continuously or pulse exposed, following which the investigator may evaluate classical toxic endpoints such as embryo mortality, hatching success, post-hatch mortality, and teratogenicity as well as genotoxic and cytotoxic damage.[10] The present report details the methodology used for *in situ* exposure, and applies these methods to recent studies in Puget Sound, WA and Prince William Sound, AK.

Materials required

Any fresh or saltwater fish with adherent eggs can be used, however, the herring will be discussed here because of the author's extensive experience with this species. Although adherent eggs are ideally suited for *in situ* monitoring, a modification of the technique will be described that enables the investigator to use pelagic eggs under some conditions. Figure 1 depicts (a)

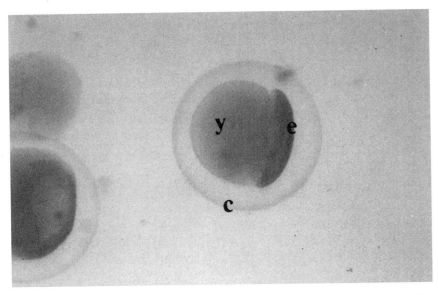

Figure 1a   Sand sole egg: (1.0 mm) Pelagic egg that incubates in the water column from 0–10 m. The exact depth depends on salinity of the surrounding water. e, embryo (blastula); y, yolk; c, chorion.

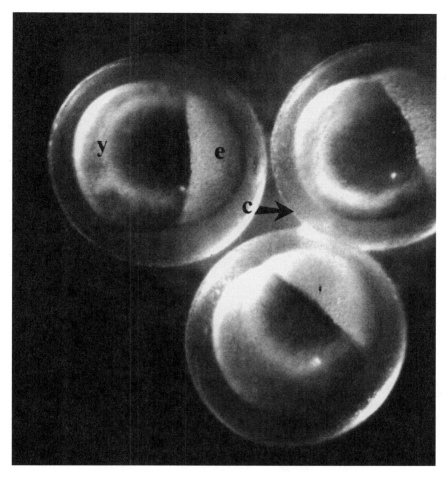

Figure 1b Herring egg (1.2–1.5 mm): Adherent demersal eggs usually found from the intertidal zone to 10 m subtidal. Eggs normally adhere to natural substrates such as kelp, rock, sunken logs, piers, and debris. They can also be made to adhere to glass, plastic, nylon net, and various other artificial substrates. e, embryo (blastula); y, yolk; c, chorion.

a pelagic non-adhesive sand sole (*Psttichthys melanostictus*) egg, (b) an adhesive demersal Pacific herring (*Clupea pallasi*) egg , and (c) a demersal Medaka (*Oryzias latipes*) egg with adhesive fibers. These represent the three basic types of fish eggs that are amenable for *in situ* studies.

1. Fish eggs/embryos: Table 1 lists several fresh and saltwater species and their egg types. There are obviously many more species that can be used for *in situ* exposure and a more comprehensive list can be found in Breder and Rosen and others.[11-13]
2. Glass slides: 2.5 × 7.5 cm (1″ × 3″) standard glass microscope slides; preferably with frosted end for labeling. Suggest cleaning with 95% alcohol prior to use.

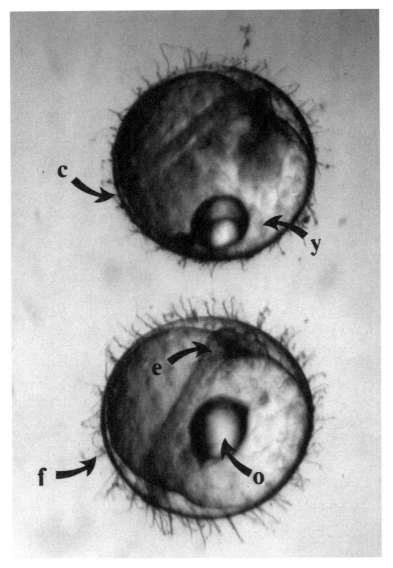

Figure 1c Medaka egg: (1.0–1.5 mm) Demersal egg with adherent fibers on the surface of the chorion. This species is euryhaline and commonly used in a variety of laboratory studies. It is amenable to *in situ* exposure and the embryos develop under a wide range of temperatures and salinities. e, embryo (early organogenesis); y, yolk; f, chorionic filaments; c, chorion; o, oil droplet.

3. Nitex net fiberglass window screen or other nontoxic netting: adequate m^2 for entire project. Generally 10 squares/cm^2 is adequate. Netting is used for both chemically adhesive eggs and eggs with adhesive fibers.
4. Heavy duty rubber bands for holding netting over end of PVC pipe.

Table 1 Some Common Easily Obtained Fresh and Saltwater Fish Species
That are Potentially Useful for *In Situ* Embryo Exposures[11]

Family	Common name	Egg type
Acipenseridae	Sturgeon	Demersal/adhesive
Polyodontidae	Paddlefish	Demersal/adhesive
Elopidae	Tarpon	Pelagic/nonadhesive
Clupeidae	Herring	Demersal/adhesive
Engraulidae	Anchovies	Pelagic/nonadhesive
Osmeridae	Smelt	Demersal/adhesive
Umbridae	Mud Minnows	Demersal/adhesive
Cyprinidae	Carp	Demersal/adhesive
Catostomidae	Suckers	Demersal/adhesive
Ictaluridae	Bullheads	Adhesive (nest builder)
Anguillidae	Eels	Pelagic/nonadhesive
Gasterosteidae	Sticklebacks	Demersal/nonadhesive
Syngnathidae	Seahorse	Nonadhesive (male pouch)
Cyprinodontidae (Adrianichthyidae)	Medaka	Demersal/adhesive threads
Gadidae	Cod	Pelagic/nonadhesive (seawater); adhesive (freshwater)
Percinae	Perch	Semi-buoyant/adhesive
Scaridae	Parrottfish	Pelagic/nonadhesive
Cottidae	Sculpin	Demersal/adhesive
Solidae	Sole	Pelagic/nonadhesive
Pleuronectidae	Flounder	Pelagic/nonadhesive

5. PVC pipe: 2 to 4 in. diameter segments cut to desired length.
6. ABS pipe: 2 to 4 in. diameter segments cut to desired length. Available from most hardware and plumbing stores.
7. Floating poly-rope: sufficient to accommodate all cassettes.
8. Anchors for holding apparatus in place.
9. Net (egg) floats for holding apparatus off bottom.
10. Velcro to fasten deployment cylinders to retrieval line.
11. Insulated coolers for transporting eggs and embryos.
12. Air pump or oxygen tank to aerate embryos during transport.
13. Hand lens or magnifying glass to determine fertilization rate in the field.
14. Plastic containers for transporting eggs to and from the field.
15. Fresh/seawater system for incubating and hatching embryos.
16. Dissecting microscope to evaluate embryos and larvae.
17. Formalin (5 to 10%) to preserve embryos and larvae for light microscopy.
18. Gluteraldehyde (5%) as alternate fixative for electron microscopy and clearing the larvae.
19. MS-222 (Tricaine Methanesulfonate) (Argent Chemical Labs, Redmond, WA 98052) for anesthetizing larvae.
20. Environmental room with air/water temperature control.

Procedures

Spawner collection

Spawning fish can be obtained from captive or wild stocks that have ripened naturally, or they can be induced to spawn by the use of hormones. The specific requirements of each species naturally differ, and can be found in the existing literature.[11] Collection and spawning techniques will be described here in detail for the Pacific herring, *Clupea pallasi*.

Ideally, ripe spawning herring should be used to ensure that each female is ready to release eggs when needed. This can be done by identifying "white-water" events (spawning) and capturing actively spawning herring with an experimental or drifting gill net. Alternatively, spawning herring can occasionally be obtained from commercial fishermen or bait dealers, the latter of which often keep large numbers of fish in floating net pens. These fish frequently begin spawning on the net mesh while in captivity and can be dipped out when needed.

If the fish are not going to be spawned on site they should be anesthetized or killed, then chilled and transported to the spawning site. It has been the author's experience that fish placed in direct contact with ice (0 to 1°C) produce a high proportion of nonviable eggs. A technique described to the author by Drs. Westernhagen and Rosenthal of the Biologische Anstalt Helgoland, FRG works very well [personal communication]. This system consists of a cooler with wet ice or artificial ice packs on the bottom, covered with several layers of paper or towels. The fish are then placed in single layers on racks above the ice and covered. Transport by this method has resulted in >90% viability of eggs after 6 h in transit. Another method, developed by the author, utilizes fish egg shipping boxes (Flex-a-Lite, Tacoma, WA) that already have stacked trays with perforations, and an ice holding rack at the top. This method allows the colder denser air to sink through the perforations and evenly chill the fish which are layered on each rack. Transport time is critical and should be kept under 6 h from the time of collection in order to obtain the highest rate of fertilization.

An alternative to shipping chilled spawners is to transport them in live tanks with aeration. This method was successfully used in a study carried out by the author and Mr. "Mike" McKay, Fishery Biologist for the Lummi Tribe of Bellingham, WA. In this study 16 live spawners were transported in two 60-l coolers aerated with pure oxygen from a 4-l oxygen bottle fitted with a two-stage regulator to control the flow of oxygen. The total time in transit of spawners was approximately 10 h, with no adverse effects to the quality of eggs or larvae.

Spawning and fertilization

Ripe females are selected and the vent area wiped with a clean gauze pad to remove fecal material, contaminating organisms, and loose scales. Sterilization of the vent is not necessary, but can be accomplished with a variety of surface disinfectants. The abdomen is then gently squeezed to expel the eggs, or the abdomen can be surgically opened and the skeins removed. The eggs can be deposited directly onto any suitable substrate, such as microscope slides, swatches of nitex netting, or directly into seawater containing some other suitable substrate. Once the eggs have attached to the substrate, sperm is expelled from the males in a similar manner, and approximately 0.5 ml added to 250 ml of clean seawater and stirred. When dispersed, the sperm can be poured over the eggs. Unlike salmonids, herring eggs can be fertilized for as long as 1 h after being placed in seawater. A complete description of the artificial spawning of herring can be found in Kocan and Landolt.[8] Other species spawning requirements are described in Breder and Rosen.[11]

Verification of fertilization

Eggs and sperm are allowed to remain in contact for at least 1 h, after which the eggs are examined with a magnifying glass, dissecting microscope or hand lens to determine whether fertilization has been successful. Unfertilized eggs have virtually no space between the chorion and yolk membrane, while fertile eggs show a distinct perivitelline space between the yolk and chorion. Fertilization techniques for each species should be practiced and mastered by the investigator prior to beginning the study.

Transport

Newly fertilized herring eggs can be transported without difficulty for several days in chilled (5 to 10°C) seawater that has been adequately aerated. Because there is very little respiratory activity during early embryogenesis (e.g., first third of incubation), it is not necessary to aerate herring eggs during transport. They should be placed in seawater at the temperature and salinity appropriate to local conditions. Temperature can range from 4°C (Alaska) to 12°C (California), and salinity may vary from 10 ppt (Baltic Sea) to 33 ppt (North Pacific).

Most adherent fish eggs are relatively resistant to trauma and can be handled without damage if they are not subjected to severe trauma, and care is taken to prevent desiccation and overheating. It is also important not to subject early cleavage embryos to rapid temperature changes (e.g., >1°C min^{-1}), because this can result in physical deformities (e.g., teratogenic effects) in some susceptible species (personal observation). A common error is to remove the embryos from the seawater and subject them to ambient air temperatures during transfer operations. Because the chorion is wet, there is a rapid temperature change in the egg when the air temperature is different than the incubation seawater. This problem can be avoided by transferring embryos from one container to another under water, or when in the laboratory, using a cold-room adjusted to appropriate temperature for egg transfers.

If an environmental room with temperature/humidity controls is not available, an insulated meat locker made of portable panels with a chilling unit can be used. A 9 m^2 unit of this type, manufactured by Uniflow Mfg. Co. (Erie, PA) has been successfully used by the author for over 10 years.

Deployment

Water column or near-bottom exposures

Embryos can be placed into deployment cassettes immediately after fertilization, or at a later time, depending on the developmental stage to be exposed. Exposure cylinders are made of schedule 40 PVC pipe which is negatively buoyant and suited for exposures near the bottom or in mid-water.

Figure 2 depicts a configuration successfully used for *in situ* deployment of herring eggs at numerous sites ranging from quiet bays to relatively high energy coastal zones. This system consists of PVC pipe with a piece of netting permanently glued to the bottom. Super Glue® or silicone are both suitable for this purpose. An all-nylon and plexiglass cassette is placed inside the cylinder. The cassette holds glass microscope slides, to which the herring eggs are attached, and prevents their breaking during exposure. The top of the cylinder is covered with a piece of nylon netting and fastened in place with an O-ring or heavy duty rubber band that fits into a groove cut into the surface about 2 cm from the top of the cylinder. This entire apparatus is fastened to a mooring line made of floating poly-rope by means of a stainless steel hose clamp or a velcro strap.

It is important that no toxic materials be used in the construction of the cassettes or holding cylinders since seawater will corrode even the most resistant materials. As a precaution against leaching of toxic material, the entire apparatus and the artificial substrate should be thoroughly washed, then soaked in deionized water for several hours, followed by a rinse with water from the deployment site. This will remove any soluble chemicals which might be present in the construction materials prior to deployment.

The cylinders are fastened to the retrieval line at the desired depth, then the line is placed on site with an anchor adequate to withstand wave and current forces. More than one cylinder can be fastened to the same line if multiple depth exposures are desired. Because the PVC/cassette system is negatively buoyant, a secondary float must be attached to the retrieving line about 0.5 m above each cylinder. This prevents the apparatus from sinking to the bottom when the tide goes out, and keeps the cylinder lined up with the direction of current, thus ensuring adequate water flow over the developing embryos.

The retrieving line can be marked with a marker buoy for location and retrieval, but it has been found that this often attracts curious boaters and fishermen. To prevent (or reduce) loss and disturbance of equipment, it is best to let the retrieval line float free or perhaps mark it with a small inconspicuous float. The author has found that a length of line that is submerged at high tide and exposed only at lower tides, lessens losses due to vandalism. Another consideration is the flow rate at the site, particularly in high current areas or rivers. The current has a tendency to pull the float and line under water, thus changing the angle of the retrieval line and bringing the apparatus closer to the bottom. This problem can be controlled by using a smaller diameter floating line and a smaller float or no float at the surface. It is critical, however, to have secondary floats above the cassettes in order to

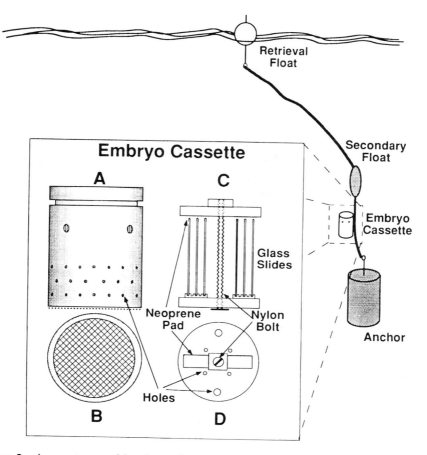

Figure 2 Apparatus used by the author to successfully expose herring embryos *in situ* in Puget Sound and Prince William Sound. (A) 7.5-cm diameter PVC cylinder; (B) top/bottom view of cylinder showing netting cover; (C) plexiglass cassette with six 2.5 × 7.5 cm glass slides, each of which can hold up to 300 eggs; (D) top/bottom view of cassette. The cassette and glass slides can be replaced with any artificial or natural substrate.

keep them off the bottom. These floats come in various sizes and can be obtained from fisherman supply stores. They should be tested in advance of deployment in water of similar specific gravity as the actual exposure site to ensure that they are sufficiently buoyant to keep the cylinder, cassette, and artificial substrate suspended in the water column.

Surface exposures

An apparatus successfully used for *in situ* surface exposures is depicted in Figure 3. The cylinder is made of ABS drain pipe which is positively buoyant and will float at the surface. In order to allow surface water (e.g., sea surface microlayer) to enter the cylinder, "windows" are cut into the pipe and covered with netting. This apparatus can accommodate both adherent eggs attached to an artificial or natural substrate, or floating (pelagic) eggs such

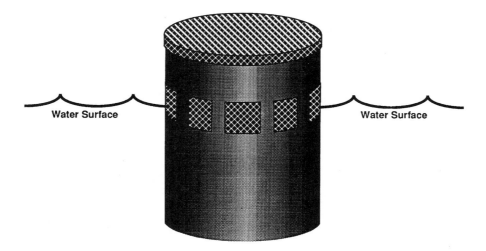

Figure 3 Buoyant apparatus used to expose fish embryos *in situ* at the water surface. The cylinder is made of ABS drain pipe which naturally floats. Windows covered with appropriate size netting allow the free exchange of water and contaminants between the inside of the cylinder and the open water. This apparatus can be anchored or allowed to float free.

as those of the flounder, cod, or sole. The eggs of these species float at or near the surface, depending on salinity, and would thus not be suited for deep water exposures. Adherent eggs, such as the herring, would be suited for both deep and surface exposures.

The ABS pipe can be anchored to the bottom in a manner similar to that described for deep water exposures, or multiple cylinders can be attached to a single mooring line. In some situations, such as small ponds or lakes, the investigator can allow the cylinders to float freely until they are retrieved.

In situ incubation

Once the embryos are on site, they can incubate for any length of time determined necessary by the investigator. Since most adherent and pelagic eggs are relatively resistant to handling damage, they can be deployed and retrieved at any time after fertilization. It is also possible to retrieve subsets of the developing embryos during the exposure period in order to obtain temporal data. Under conditions of excessive fowling by marine organisms or suspended debris, it may be necessary to retrieve and clean the cylinders and cassettes during the exposure period. This can be accomplished by vigorously raising and lowering the cylinder through the water column several times to dislodge adherent materials, or by rinsing the embryos and apparatus with a squirt bottle containing clean water.

Since the time of fertilization is known, as is the temperature at which the embryos are incubating, it is possible to determine the date of hatching (e.g., herring hatch on day 14 at 10°C and day 21 at 8°C). It is critical to retrieve the cassettes and return them to the laboratory prior to hatching so that no larvae will be lost or damaged, and an accurate evaluation of percent hatch, live hatch, and abnormal larvae can be made. If the eggs hatch *in situ*, these types of data will very likely be lost.

Retrieval

If the embryos are retrieved during the second half of incubation or near the time of hatching, it is critical that they be supplied with adequate aeration during transport to the laboratory. These late stage embryos have a much higher oxygen demand than do early undifferentiated embryos, and oxygen demand goes even higher at the time of hatching.[14] This is often the cause of heavy larval mortality at or near hatching.[7] Aeration can be supplied during transport by means of a battery operated portable air pump that plugs into the vehicle's cigarette lighter. Another method is to attach an air hose to an oxygen bottle fitted with a two-stage pressure valve and supply pure oxygen to the embryos during transport.

If pure oxygen is used and the embryos are to be shipped by air, it is important to leave adequate space for gas expansion which results from reduced pressure at altitude. This can be done by placing the embryos in an air-tight plastic bag, expelling the normal atmosphere and re-gassing with oxygen to about one half of the maximum volume. This will allow the oxygen to expand at altitude without rupturing the bag and escaping. The plastic bag can be contained within an insulated cooler or similar container to prevent puncture of the bag and maintain a constant temperature for the embryos. With adequate aeration and temperature control, herring embryos can be transported for as long as necessary during any period of their development. Due to the rapid increase in oxygen demand at hatching, it is critical that hatching occur in the laboratory where oxygen concentration can be maintained at or above $9 \text{ mg} \cdot \text{l}^{-1}$.

Hatching

Once the embryos have been returned to the laboratory, the temperature of the embryo shipping water should be equilibrated with that of the laboratory water. Once equilibrated (about 1 h) the embryos can be placed into their respective hatching vessels and supplied with adequate aeration and/or flowing water with at least 80% O_2 saturation. Lower concentrations of oxygen often result in heavy mortality during hatching for herring as well as other species.[14,15]

If only embryotoxic effects are to be considered, the newly hatched larvae can be anesthetized with methane tricane sulfonate (MS-222), a standard fish anesthetic, then placed directly into fixative and archived. A 3 to 5% Formalin® solution in fresh or seawater is adequate for fixation. If seawater-Formalin is used for fixation, this should be replaced with freshwater-Formalin within 8 h to prevent the formation of salt crystals that interfere with viewing of the larvae. The author has found that fixation of larvae in freshwater-Formalin does not adversely affect the tissues or cells.

Fixation with Formalin or Stockard's solution results in the larvae becoming opaque, thus reducing the resolution required to see minor morphologic changes. This can be alleviated by substituting 5% gluteraldehyde as a fixative. This standard electron microscopy fixative results in a transparent larva that resembles a living organism when viewed through a microscope, thus making it possible to visualize larval structures in three dimensions.

Post-hatch

Resolution or sensitivity of the system can be increased by allowing the newly hatched larvae to incubate until their yolk sac is resorbed. Teratogenic effects (physical and biochemical) often manifest themselves over a period of time after birth or hatching as critical developmental processes progress.[6] By maintaining the larvae up to the point of yolk sac resorption, it is possible to increase the sensitivity or resolution for larval mortality, developmental defects, and behavioral changes that are dependent on continued larval development.

Results and discussion

Case studies

Port Gamble Bay, Washington

In 1986 and 1987 the Washington Department of Fisheries and the Port Gamble Klallam Tribe reported unusually heavy mortality of natural herring spawn in Port Gamble Bay and Port Madison, Washington, two heavily used spawning areas for Puget Sound herring.[16,17] The mortality ranged from 20 to 100%, usually affecting the eggs during the first five days of incubation. Surveys from 1977 through 1987 indicated that the mortality was not due to predation, suffocation, desiccation, or thermal stress, all known to adversely affect herring embryo survival.[18] Although the mortality was severe at specific sites, it appeared to change from year to year. For example, eggs deposited away from affected sites, but within the same spawning area appeared to develop normally.

In order to determine whether the embryo mortalities were caused by poor environmental conditions (e.g., water or sediment) or by agents vertically transmitted from adult to embryo, a study was conducted in Port Gamble Bay in 1987 which employed *in situ* exposures of artificially spawned herring.[8] Spawning adult herring were obtained from a commercial bait dealer and artificially spawned onto glass slides. One group of slides was then deployed at three sites within Port Gamble Bay where the Washington Department of Fisheries had recently observed heavy egg mortality. A second group was deployed at Bywater Bay, a reference site within 0.5 miles of Port Gamble Bay that had similar substrate, but was not used by spawning herring. A third group was incubated in sterile seawater in the laboratory at the University of Washington School of Fisheries.

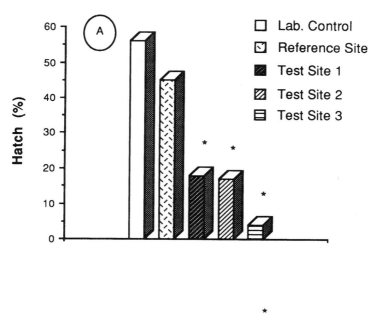

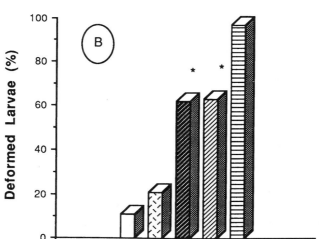

Figure 4 Data from an *in situ* herring mortality study carried out in Port Gamble Bay, Washington. Embryos exposed *in situ* at test sites had a significantly lower hatching success and higher rate of deformed larvae than did embryos incubated at a reference site or in the laboratory. Stars indicate significantly different from control and reference, p <0.05.

Embryos were exposed *in situ* for the first four days following fertilization, then returned to the laboratory where they were allowed to complete incubation and hatch. Figure 4 summarizes the results of one such exposure. There was no difference in hatching success nor larval deformities between the laboratory controls and the reference sites. However, the three sites within Port Gamble Bay showed differences in hatching success as well as in the percent of deformed larvae hatched from surviving embryos. Since all of the embryos were derived from the same spawners, it was concluded that something at the site was responsible for the observed embryo mortality. It is interesting to note here that mortality prior to hatching is rare in herring under laboratory conditions, but is frequently seen when embryos are exposed *in situ*.

A study on Port Gamble Bay sediments done in conjunction with this study by Yake and Norton[18] showed no unusual concentrations of metals, polychlorinated biphenyls (PCB), or pesticides. There was a slight elevation in the levels of polyaromatic hydrocarbons, but they concluded that these levels were too low to be responsible for the observed mortality.

This is an example of a study where observed mortalities in the field were verified by *in situ* exposures of sentinel animals, while chemical analysis of sediment could not identify the cause of the mortality.

Prince William Sound, Alaska

Three years following the *Exxon Valdez* oil spill in Prince William Sound, a study was conducted to determine whether there was any residual toxicity associated with herring spawning sites in the oil trajectory. Prince William Sound herring were artificially spawned onto artificial substrate, and their eggs deployed at previously oiled and unoiled sites.[19] The oiled sites were Naked Island (heavily oiled) and Rocky Bay on Montague Island (lightly oiled). Both of these sites had herring spawn present in 1989, but only Rocky Bay had natural spawn during the study in 1992. The unoiled site was Fairmont Bay, located northeast of the spill site and outside the oil trajectory (Figure 5).

Simultaneously with the *in situ* exposures, a laboratory study was conducted to evaluate the effects of Prudhoe Bay crude oil on developing herring embryos and larvae. In this study, artificially spawned herring embryos were exposed during incubation to varying concentrations of oil–water dispersions (OWD) ranging from 0.01 to 9.7 mg · l^{-1}.

Following the laboratory and *in situ* studies,[4] embryos and larvae were evaluated for survival, hatching, and various embryopathies. Larval dry weights were significantly lower in the oil-exposed group than in the unexposed group. A similar relationship was observed in larvae exposed *in situ* at the previously oiled and unoiled sites in Prince William Sound (Figure 6).

Evaluation of genotoxic damage to newly hatched larvae which had been exposed to a range of oil concentrations during incubation shows a clear dose–response relationship between chromosome damage and oil concentration. Likewise, chromosomes from larvae exposed *in situ* at oiled sites exhibited significantly more chromosome damage than did larvae exposed at previously oiled sites (Figure 7).[4]

Conclusions

Observations made in the field are rarely duplicated in the laboratory, and laboratory results are seldom observed under field conditions. Consequently, it becomes necessary to reconcile the different types of data, and

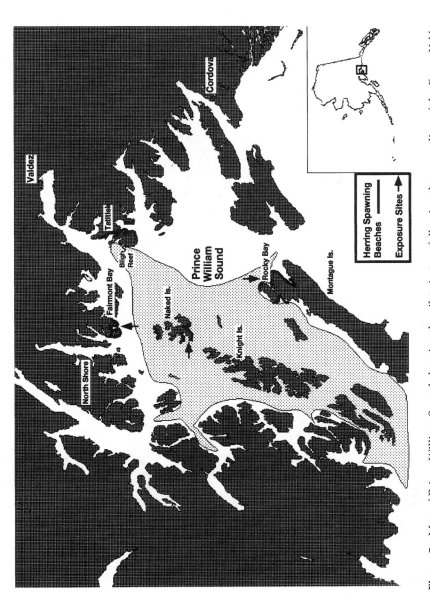

Figure 5 Map of Prince William Sound showing the oil trajectory following the grounding of the *Exxon Valdez* on Bligh Reef in 1989. Also shown are herring spawning sites active in 1989 and sites used for *in situ* exposure of herring embryos in 1992.[4,19]

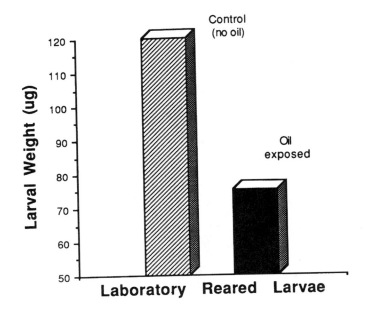

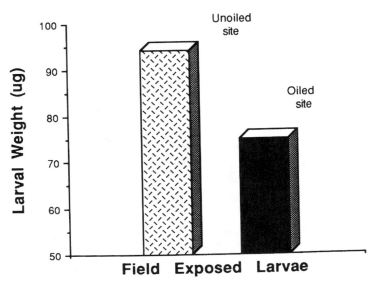

Figure 6 Dry weights of herring larvae exposed to Prudhoe Bay crude oil in the laboratory (top) and weights of larvae exposed to oiled and unoiled sites in Prince William Sound (bottom). Similar differences between exposed and unexposed embryos are observed in both instances.

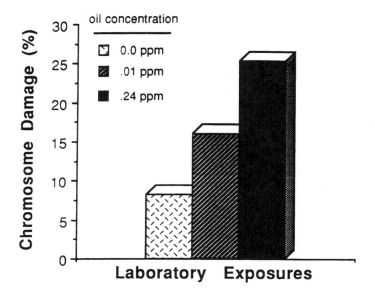

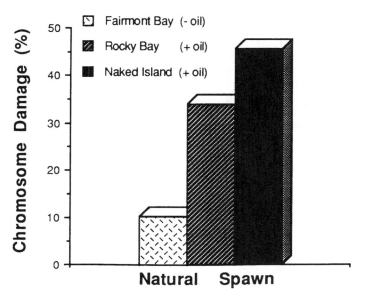

Figure 7 Chromosome damage in herring larvae exposed to increasing concentrations of Prudhoe Bay crude oil in the laboratory (top), and chromosome damage observed in herring larvae exposed *in situ* at oiled and unoiled sites in Prince William Sound (bottom). Similar patterns of damage are seen in exposed and unexposed embryos in both instances.[4,10]

the deployment of fish embryos *in situ* makes this task easier. Using controlled laboratory exposures and exposure under "natural" field conditions, one can look for trends, similar types of response and patterns that result following exposure to contaminated water and/or sediments.

By selecting the appropriate fish species and using the methods described in this chapter, it should be possible to directly evaluate the toxic potential of natural aquatic ecosystems. Having complete control over (1) time of spawning, (2) selection of brood stock, (3) number and density of eggs, (4) length of *in situ* exposure and then (5) being able to evaluate the newly hatched larvae in the laboratory, allows for more scientifically sound conclusions than if only field observations were made.

Thus far, this technique has been successfully used on several occasions to evaluate aquatic pollution, and it is felt that its expanded use in the future will add considerably to our knowledge about aquatic pollution and water quality effects on fish health.

References

1. Schnack, D. 1981. Studies on the mortality of Pacific Herring larvae during their early development, using artificial *in situ* containments. *Rapp. P.-v. Reun. Cons. Int. Explor. Mer.* 178: 135–142.

2. Hardy, J.T., Crecelius, E. and Kocan, R. 1986. Concentration and toxicity of sea-surface contaminants in Puget Sound. PNL-5834; UC-11. Natl. Technical Info. Service, U.S. Dept. Commerce, Springfield, VA. 46 pp.

3. Hardy, J., Kiesser, S., Antrim, L., Stubin, A., Kocan, R. and Strand, J. 1987. The sea-surface microlayer of Puget Sound: Part I. Toxic effects on fish eggs and larvae. *Mar. Environ. Res.* 23: 227–249.

4. Kocan, R.M., Hose, J.E., Brown, E.D., Baker, T.T. (in press). Pacific herring embryo (*Clupea pallasi*) sensitivity to Prudhoe Bay petroleum hydrocarbons: laboratory evaluation and *in situ* exposure at oiled and unoiled sites in Prince William Sound. *Can. J. Fish. Aquatic Sci.*

5. U.S. EPA. 1985. Methods for measuring the acute toxicity of effluents to freshwater and marine organisms. Toxicity data analysis, Sect. 11, pp. 70–78. EPA/600/4-85/013, Cincinnati, OH.

6. Harbison, R.D. 1908. Teratogens. In: *Toxicology,* Eds. J. Doull, C.D. Klaassen, and M.O. Amdur, (2nd ed)., 8:158–175. Macmillan Pub. Co. NY. 778 pp.

7. Kocan, R.M., Landolt, M.L. 1989. Survival and growth to reproductive maturity of Coho salmon following embryonic exposure to a model toxicant. *Mar. Environ. Res.* 27: 177–193.

8. Kocan, R.M., Landolt, M.L. 1990. Use of herring embryos for in situ and in vitro monitoring of marine pollution. In: *In situ Evaluations of Biological Hazards of Environmental Pollutants.* Eds. S.S. Sandhu, W.R. Lower, F.J. de Serres, W.A. Suk and R.R. Tice. 49–60. Plenum Press, NY. 277 pp.

9. McKim, J.M. 1985. Early life stage toxicity testing. In: *Fundamentals of Aquatic Toxicology.* Eds. G.M. Rand and S.R. Petrocelli, 58–95. Hemisphere Pub. Corp., New York/London. 666 pp.

10. Hose, J.E., McGurk, M.B., Marty, G.D., Timken, B.E., Biggs, E.D., Baker, T.T. (in press). Sublethal effects of the *Exxon Valdez* oil spill on herring embryos and larvae: Morphological, cytogenetic and histopathological assessment, 1989–1991. *Can. J. Fish. Aquat. Sci.*

11. Breder, C.M. Jr., Rosen, D.E. 1966. *Modes of Reproduction in Fishes*. T.F.H. Publications, Jersey City, NJ. 941 pp.

12. Hempe, G. 1979. *Early Live History of Marine Fish: The Egg Stage*, University of Washington Press, Seattle, WA. 70 pp.

13. Garrison, K.J., Miller B.S. 1982. *Review of the Early Life History of Puget Sound Fishes*. FRI-UW-8216. University of Washington, Seattle, WA. pp. 1–729.

14. Holliday, F.G.T, Blaxter, J.H.S., Lasker, R. 1964. Oxygen uptake of developing eggs and larvae of the herring (*Clupea harengus*). *J. Mar. Biol. Assoc. U.K.* 44:711–723.

15. McQuinn, I.H., Fitzgerald, G.J., Powles, H. 1983. Environmental effects on embryos and larvae of the Isle Verte stock of Atlantic herring (*Clupea harengus*). *Naturalaiste Can. (Rev. Ecol. Syst.)*, 110:343–355.

16. Gonyea, G., Burton, S., Pentilla, D. 1982a. Summary of 1981 Herring Recruitment Studies in Puget Sound. Wash. Dept. Fisheries Progress Report. No. 157.

17. Gonyea, G., Burton, S., Pentilla, D. 1982b. Summary of 1981 Herring Recruitment Studies in Puget Sound. Wash. Dept. Fisheries Progress Report. No. 179.

18. Yake, B., Norton, D. 1987. Port Gamble Bay: A reconnaissance survey of sediment quality. Segment No. 25-00-03; 07-15-03. Wash. State Dept. Ecol., Olympia, WA. 17 pp.

19. Brown, E. D., Norcross, B. L., Baker, T. T., Short, J. W. (in press) The distribution of oil spilled from the *Exxon Valdez*, ocean conditions affecting it, initiation of damage assessment studies in response and exposure of Pacific herring, *Clupea pallasi*, to oil in Prince William Sound. *Can. J. Fish. Aquat. Sci.*

chapter six

The medaka embryo-larval assay: an in vivo assay for toxicity, teratogenicity, and carcinogenicity

Michael F. Helmstetter, Alexander E. Maccubbin, and Raymond W. Alden III

Introduction

As the burden of pollution on aquatic environments has increased, the discipline of aquatic toxicology has seen the development of numerous assays designed to evaluate the acute and sublethal effects of environmental pollutants. Aquatic animals have become important as surrogate species for toxicological testing of materials that may have adverse biological effects in humans. Moreover, aquatic organisms can provide direct information about the impact of pollutants on the health of the aquatic ecosystem as it relates to resident organisms. Assays using whole animals as opposed to those employing cell cultures or subcellular fractions have distinct advantages. By using the whole animal, the complexity of the biological system is maintained and potential biological effects can be observed directly rather than inferred. It has been suggested that by using early life stages of fish, a broad spectrum of targets of environmental pollutants can be evaluated.[1]

Several studies have used an immersion protocol for exposing eggs to chemicals and complex mixtures.[2-5] Although these studies have provided valuable insight on the toxicity and carcinogenicity of some materials, the use of the immersion technique is limited by the solubility of the materials being tested.[4-6] Solubility problems often result in poor exposure of the eggs and uncertain dosing, thus limiting the development of true dose response data.[7] In an attempt to minimize the problems associated with the immersion exposure protocols, several methods for injecting chemicals dissolved in a suitable carrier have been developed.[8-14] Once again, these techniques have

1-56670-149-X/96/$0.00+$.50
© 1996 by CRC Press, Inc.

proved to be very useful but suffer from the drawback of relatively high mortality (34.0 to 69.9%) due to the trauma of injection.[8-14] This reduces the utility of the injection methods in studying acute toxicity of compounds.

We have developed a noninvasive exposure technique to assay the potential harmful biological effects of individual compounds or complex mixtures. The embryo–larval stages of the Japanese medaka (*Oryzias latipes*) are exposed by topical application to chemicals that have been dissolved in dimethylsufoxide. There are numerous advantages to the use of medaka in studies of normal or stressed embryology including the following:

1. The medaka is a small vertebrate species; therefore, large numbers of breeding individuals can be maintained in a limited space and at an acceptable cost. Further, sexual dimorphism is easily distinguishable through external features.
2. It is a hardy fish and easy to sustain under normal conditions in the laboratory. It is omnivorous and can be maintained on freshly hatched, dried, or synthetic foods.
3. The medaka is easy to breed under favorable laboratory conditions; therefore, large numbers of test eggs can be obtained with relative ease on a daily basis, any time of the year.
4. Time of oviposition can be predicted so that the earliest developmental stages may be employed in studies of normal or altered embryology.
5. Separated from the maternal surroundings, the egg is nearly transparent, except for oil globules in the yolk which soon coalesce at the vegetal pole away from the developing embryo. Hence, all of the internal organs of the medaka can be observed *in vivo*.
6. The various medaka life stages can tolerate and thrive in a wide range of culture temperatures and salinities.
7. The medaka has been shown to be highly susceptible to a wide range of toxic, mutagenic, teratogenic, and carcinogenic agents and has a low incidence of spontaneous, background anomalies and spontaneous neoplasia, as well as a short latency period for induction of such anomalies and/or tumors through treatment with test substances/mixtures.
8. All tissues can be assessed histologically through observation of transverse, sagittal, or serial sections of eggs, fry, or adults with comparative ease.

Originally designed as a carcinogenesis assay,[15,16] the medaka embryo–larval assay (MELA) has subsequently been adapted to measuring several other biological endpoints.[17-19] The method has been refined into a standardized 21-day protocol that can be used with a wide variety of test materials including pure chemicals, defined chemical mixtures, and complex mixtures extracted from environmental samples. In this paper we describe the details of the materials and methods of the MELA, provide some representative data demonstrating its potential, and discuss the advantages and disadvantages of the assay system.

Materials required

Japanese medaka (*Oryzias latipes*) (80 to 100 adults, Carolina Biological
    Supply, Burlington, N.C.; local supplier or other researcher stock)
10-gal Aquaria, undergravel filters, 100-watt heaters (local supplier)
4-in. Dip net (local supplier)
250-μm Mesh nitex net (Gilson Testing Equipment)
$NaHCO_3$ (J.T. Baker, #3506, >99% purity)
$CaSO_4 \cdot 2H_2O$ (J.T. Baker, #1452, >98% purity)
$MgSO_4 \cdot 7H_2O$ (J.T. Baker, #2500, >99% purity)
KCl (J.T. Baker, #3040, >99% purity)
$CaCl_2$ (J.T. Baker, #1336, >99% purity)
NaCl (J.T. Baker, #3624, >99% purity)
Dimethylsulfoxide (Burdick & Jackson, #081, >99.5% purity)
Breathing quality air (Air Products, AO1-1-00513, zero grade)
Brine shrimp cysts and TetraMin® flake food (local supplier)
Petri dishes (100 mm × 15 mm; S/P®, #D1905)
Methylene blue (optional; Mallinkrodt through S/P®, #5891, 0.05%)
Sea Salts (HW MARINEMIX, Hawaiian Marine Imports)
Stainless steel forceps (Uni-Fit, #OC9)
Glass Pasteur pipettes (S/P®, #P5202)
Dissecting microscope (Nikon 8–40×)
Incubator (25 ± 1°C; Thelco Model 6)
Balance (to 1×10^{-6} g or better)
Ultrasonic bath (Branson, Bransonic B-12)
2-ml Amber vials, caps and septa for standards (Supelco, #3-3210)
1-ml Volumetric (Kimble, #28017A1)
Crystallizing dish cover (100 mm × 10 mm; Kimble, #2306010010)
Glass fiber filter paper (S/P®, #F2320-55)
Push-button repeating dispenser (Hamilton, #PB-600-1)
5-μl Syringe (Hamilton, #1005RN)
10 cm × 0.17 mm O.D. fused silica on-column injection needle (Supelco,
    #2-1361)
0.17-mm Bore Teflon® ferrule (Supelco, #2-1362)
100-ml Beaker (Kimble, #14000100)
Pipette bulb (S/P®, P5215)
Scintillation vial and polypropylene lined cap (S/P®, #R2548-20)
Teflon® stretch tape (Scientific Specialty Services, #J79979)
Tubing (³⁄₁₆″, local supplier)
Crystallizing dish (100 mm × 50 mm; Kimble, #230009050)

Fish for freshwater studies are maintained in synthetic fresh water of
pH 7.6 ± 0.2 and medium hardness (120 ± 3 mg/l as $CaCO_3$). Synthetic fresh
water is prepared from deionized water containing 1.82 g $NaHCO_3$, 1.14 g
$CaSO_4 \cdot 2H_2O$, 1.14 g $MgSO_4 \cdot 7H_2O$, and 0.08 g KCl per 10 gal. Prior to use,
synthetic fresh water is conditioned by aeration for 24 h. In studies requiring

Table 1 Developmental Anomalies Evaluated During Daily
Observations of Treated Eggs

• Twinning (e.g., Cephalodidymus)	• Fin stunting
• Microencephaly	• Stasis of circulation
• Ophthalmic edema	• Hemorrhages
• Degeneration of eye cup	• Blood islands
• Incomplete development of bilateral symmetry of eyes	• Pericardial edema
	• Rudimentary heart/"Tube-heart"
• Unilateral or bilateral microphthalmia and anophthalmia	• Enlarged ventricles
	• Reduced or increased heart rate
• Cyclopia	• Chorionic effects
• Lordosis	• Abdominal swelling
• Kyphosis	• Deranged postural equilibrium
• Scoliosis	• Tetanic convulsions
• Rigid Coiling	• Abnormal pigmentation/"Blacktail"
• Retardation of centrum formation	• Sporadic hyperkinesis
• Centrum damage (e.g., vertebral dislocations and fractures)	• Activity level
	• Complete developmental arrest
	• Retardation of yolk absorption

seawater, sea salts are added to conditioned synthetic fresh water to yield the desired salinity.

Embryo rearing solution is prepared from stock solutions of NaCl (12.5 g/100 ml), $MgSO_4 \cdot 7H_2O$ (2.04 g/100 ml), KCl (0.375 g/100 ml), $CaCl_2$ (0.5 g/100 ml) and methylene blue (0.05 g/100 ml). For 500 ml of embryo rearing solution, 5 ml of each salt solution and 1 ml of methylene blue are added to reagent grade bottled water. For studies evaluating the effect of salinity on toxicity, sea salts are added to the embryo rearing to yield the desired salinity. The methylene blue, used for ease of identification of dead embryos and in mold growth inhibition, may be omitted from the solution if excluding this additional chemical variable is preferred.

Procedures

MELA endpoints include acute mortality (≤48 h), delayed mortality (>48 h) and a wide range of sublethal anomalies (Table 1) resulting in a simple assay delivering a wide range of information, including teratogenic characteristics. MELA may be used to study the biological effects of pure chemicals and complex mixtures of chemicals extracted from water and sediment samples collected in field studies. In studies of pure chemicals, typical statistical evaluations include probit analysis to determine ED_{50} and LD_{50} values and chi-squared analysis for control and independent variable comparisons. Studies may further be carried to 3 to 6 months, yielding data associated with delayed effects such as neoplasia and other pathological anomalies. Field studies result in data that allow quantitative ranking of the relative toxicity of study sites, assessments of relationships between land uses and MELA effects, and determination of overall spatial and temporal patterns of potential ecological effects among sites. The specific procedures associated with the short-term MELA protocol are detailed below.

Maintenance of breeding cultures and eggs

Adult Japanese medaka (*Oryzias latipes*) are sexed[20] and maintained in 10-gal aquaria containing conditioned synthetic fresh water. Due to their polygamous nature, the medaka adults are maintained at six females and four males per aquarium (this 3:2 ratio has been previously demonstrated to be ideal for optimum egg production[20-22]). In addition, this polygamous property allows for rapid genetic mixing, reducing the variations of susceptibility between eggs from differing adult pairs.[23]

The breeding sets are maintained in conditioned water at $26 \pm 2°C$ and artificial light is provided with a photoperiod of 16 h light/8 h dark. These conditions have been demonstrated to increase egg laying behavior, without increasing antagonistic behavior.[22] All adults are fed twice daily; the first feeding is coincident with the onset of light and the second feeding takes place in the early evening. The adult diet includes freshly hatched brine shrimp (*Artemia salina*; ≈1 to 3 g concentrated shrimp per aquarium per feeding event) supplemented with TetraMin® artificial flake food (≈0.3 g per aquarium per feeding event).

Eggs are stripped daily from each female by removing the female from the tank with a small dip net followed by the gentle removal of the egg cluster while she remains in the net. Slight finger pressure to the cluster results in a spontaneous response in the female to "instinctively" remove herself from the egg mass. The female is then returned to the tank. The procedure is repeated for all productive females in the multiple breeding sets. This approach to stripping appears to be less stressful on the female and more efficient than previously described techniques where the female was removed from the net and placed in an egging bowl or where the eggs were physically stripped with forceps or a pipette (see, for example, Reference 20). Eggs are placed on a 250-μm mesh nitex net and washed with the synthetic medium described previously. Each egg is then separated from the neighboring cluster by gently teasing at the long adjoining filaments with stainless steel forceps until the filaments detach. The eggs are subsequently evaluated under a dissecting microscope to ensure successful fertilization. Unfertilized eggs that lack a raised vitelline membrane (i.e., appearance of the perivitelline space) are discarded and the remaining fertilized eggs are placed in plastic petri dishes (100 mm × 15 mm) containing 20 ml of embryo rearing solution at a density of ≤20 eggs per dish. The dishes are placed in an incubator maintained at $25 \pm 1°C$. Eggs are evaluated daily for mortalities and latent development prior to use in the toxicity assays. Any eggs found to be dead (as indicated by their blue color), or underdeveloped (based on microscopic examination), are removed to avoid possible fungal or bacterial contamination of the remaining eggs in a dish. Chorions of eggs carried to term for use in future breeding sets are also removed daily, as these provide an optimal surface for rapid bacterial growth. Fry hatching from eggs to be used in future breeding sets are placed in 10-gal aquaria with reduced flow from filtration discharge. From the time of transfer, the diet of endogenously feeding fry is supplemented with freshly hatched *Artemia salina*. Exogenous feeding animals are fed a diet of *A. salina* and finely ground flake food. Following a 3- to 4-month period, the juveniles are sexed and separated into six female:four male breeding sets. As new tanks are developed for additional breeders, a portion of well-conditioned water from other breeding sets is included to induce rapid biological and algal activity in order to maintain a well conditioned environment for the breeding adults.

Carrier solvent

Dimethylsulfoxide (DMSO) is used as the carrier solvent to deliver the test chemicals across the egg chorion to the developing embryo. The goal of a chemical carrier is to provide the required physical and/or chemical properties to deliver a test substance to a test organism, but without induction of any extraneous variables, such as structural alteration of the test substance or adverse biological effects. Unlike other commonly used carriers, such as methanol or acetone, DMSO has the ability to permeate biological membranes without significant damage to the structural integrity of these "barriers."[24] In addition, numerous studies have been conducted where the primary carrier solvent was DMSO and the effects (due to the solvent itself) on developing fish embryos were similar to those seen in untreated embryos.[10-16,18,25-28]

MELA topical treatment protocol

Medaka eggs at 48-h post-fertilization are used in the topical treatment procedure of the MELA. This is the stage at which the early heart is being formed and begins to contract, the tail bud is formed, the optic lobes are forming, and the oil globules have coalesced into a single drop. Exposing eggs after the cleavage stages are completed reduces the incidence of any naturally occurring, spontaneous abnormal eggs or non-viable eggs in the experimental and control groups.

For a given test material, three replicates of ten eggs are treated. The eggs are randomly selected from all rearing dishes for a given date of collection and placed on the back of a 100-mm glass crystallizing dish cover. Each egg is subsequently blotted dry of rearing solution with a piece of glass fiber filter paper. Each egg then topically receives a 0.1-µl droplet of the DMSO/chemical mixture dispensed using a Hamilton push-button repeating dispenser (Chaney repeater) equipped with a Hamilton 5-µl syringe and a 10 cm × 0.17 mm O.D. fused silica gas chromatography on-column injection needle and a 0.17-mm bore Teflon® ferrule (Figure 1). This dosing system was calibrated to deliver 0.10 ± 0.01 µl. Recent advances in technology have resulted in a digital repeating dispenser which may potentially further improve the accuracy and precision of repetitive doses (Hamilton Digital Syringe, #7000.5).

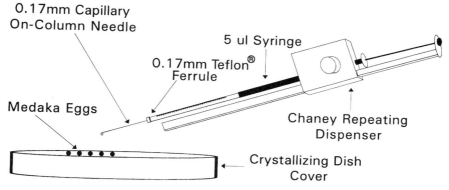

Figure 1   Topical dosing apparatus used in the MELA protocol.

After receiving the treatment, each egg is allowed to stand for 1 min to allow the dose to penetrate the chorion, washed with rearing solution, and then transferred to a 100-ml beaker containing clean rearing solution. The 1-min time period allows sufficient material to penetrate the chorion without unnecessary dehydration of the exposed eggs.[19] Once all eggs of a given replicate are transferred to this beaker, they are washed of any unabsorbed test material by vigorous swirling of the solution for 0.5 min. The eggs are then transferred to aerated scintillation vials containing 20 ml of the rearing solution (Figure 2). All treatments (and controls) are randomly distributed throughout a rearing incubator maintained at 25 ± 1°C.

Studies conducted to determine optimal rearing conditions for the treated eggs indicated that aerated scintillation vials produced the fastest

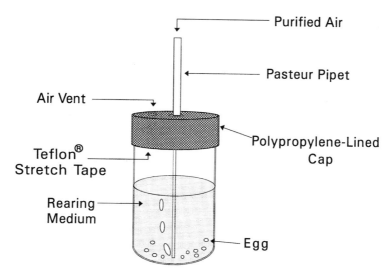

Figure 2   Schematic of the optimal rearing set-up for the MELA protocol.

hatch rates and the lowest number of unhatched eggs.[17] The variables evaluated included vessel type (scintillation vial vs. Petri dish), the presence and absence of aeration, and solution changes. It was determined that if the rearing solution in the aerated vials is changed daily, the hatch rate is significantly prolonged and the number of unhatched eggs increases. The aeration process appears to stimulate the first hatching events through increased oxygen content of the solution and/or through the agitation provided by the aeration. If the solution is frequently changed, it may be that the hatching enzyme is washed from the vial and, therefore, is not present in sufficient concentration to initiate subsequent hatching events in neighboring eggs. Therefore, after treatment, it is recommended that eggs be incubated, through hatching, in aerated scintillation vials without solution changes. The aeration should be controlled using a two-stage regulated cylinder of purified breathing-quality air (30 psi head pressure) at a rate of approximately 60 bubbles per minute for freshwater studies and 120 bubbles per minute for saline studies (differing rates of aeration are necessary to maintain a dissolved oxygen level in excess of 4 mg/l in both the freshwater and elevated salinity studies).

Each treated egg is evaluated daily under a low power stereoscopic microscope for viability, hatching events, and developmental anomalies including cardiac, skeletal, and organ-based deformities. A typical list of anomalies of interest is presented in Table 1. Eggs are considered viable if development through hatching progresses normally. Fry are considered normal when free of their chorions, they are morphologically normal compared to controls, able to maintain postural equilibrium, and swim successfully. Otherwise, they are recorded as deformed if the heart is still beating and, if not, they are recorded as dead. Dead eggs, dead fry, and egg cases of hatched fry are removed daily to avoid bacterial and fungal contamination of the remaining eggs in a given treatment. The experiment is terminated at 21 days post-treatment or when all eggs and/or fry of a given treatment are dead. Fry that hatch during the course of the experiment are transferred to 100 mm × 50 mm crystallizing dishes containing 150 ml of the conditioned water used in the breeding aquaria. The fry dishes are lightly aerated with clean, breathing-quality air. No feeding is necessary during the swim-up period, as endogenous feeding of the yolk material typically takes place for 5 to 7 days following hatching.[29] Any surviving deformed hatch fry are preserved in 10% buffered formalin, followed by exchange to 70% ethanol for subsequent examination and photographic records. If desired, surviving fry may be preserved for histological evaluation after paraffin embedding, sectioning, and staining. Fry may also be reared for 3 to 6 months followed by histological examination if carcinogenesis is considered as an endpoint.

Eggs treated with DMSO alone and untreated eggs serve as negative controls for all experiments and are evaluated concurrently with test chemicals. If more than 10% of the eggs for any given control die, the test chemical experiments evaluated in conjunction with these controls are terminated and

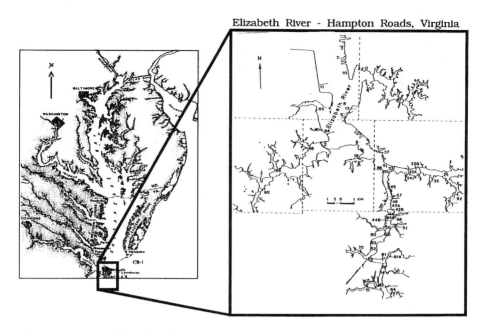

Figure 3 Map of the Elizabeth River, Virginia indicating land use activity (LUA) study sites.

repeated with new controls. Pentachlorophenol is recommended as an ideal positive control (reference toxicant) for MELA. Toxicological responses of the MELA to pentachlorophenol are discussed further in the Results and Discussion section below.

Preparation of materials for tests using MELA

Pure test chemicals are weighed to 1×10^{-6} g and diluted volumetrically with the DMSO. A stock solution of chemical is placed in an ultrasonic water bath for 1 min to ensure homogeneity of the mixture. Successively lower chemical concentrations are developed through serial dilution from the stock solutions with DMSO as the diluent. All stock solution and dilution preparation should be conducted in reduced light and in a negative pressure hood. All solutions are stored in Teflon® septa sealed 2-ml glass amber vials at $4 \pm 2°C$ in the dark until needed for the topical treatment experiments.

The procedures for collection and preparation of field samples for the MELA will vary depending on the type of field study. The following techniques were used for preparing extracts of chemicals contained in water and sediment samples collected from a number of sites in case studies conducted in southeastern Virginia (Figures 3 and 4). Water samples were initially collected from 52 sites to explore the distribution of toxicants/mutagens associated with various land use activities (LUAs) in the Port of Hampton

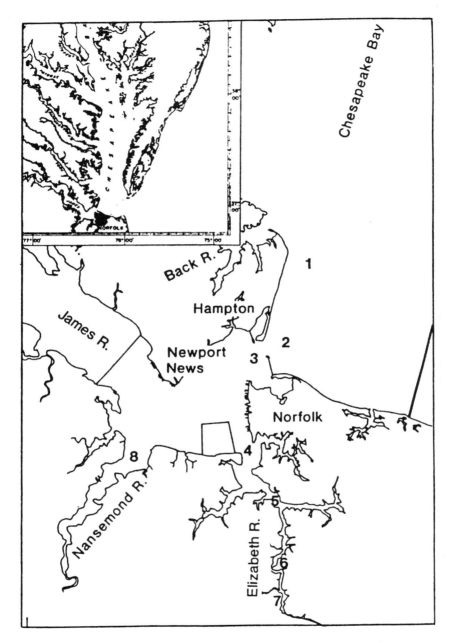

Figure 4   Map of the MELA sediment extract analysis sampling stations.

Roads, Virginia, primarily in the Elizabeth River Basin (Figure 3 and Table 2). Collection devices (termed "SCDs" for "sorbent containing devices") were deployed to concentrate organics from receiving waters in proximity to the various LUAs. Each of the three sorbent resins (cyclohexyl [CH], C_{18}, XAD-2; Alltech Associates, Inc.) was placed in one of three SCDs located on a 1-m

Table 2 LUA Categories and Corresponding Sampling Site Numbers

LUA category	Corresponding site numbers
Military installations (other than shipyards)	1, 70, 83, 95
Chemical processing industries	2, 44, 44A, 44B, 50, 54, 64, 76, 84
Oil terminals	3, 46, 48, 52, 63
General industrial and commercial sites	5, 10, 25, 43, 57, 62, 67
Publically-owned treatment works	7, 8, 9
Landfills and disposal areas	11, 12, 35
Marina and dock areas	16, 82, 86, 91
Shipyards	22A, 22B, 24, 33, 42A, 42B, 45
Urban runoff (and creek drainage basins)	37, 51, 80, 81, 81A, 87, 88, 89, 90, 92

floating sampling buoy and the entire setup was deployed for a 72-h sampling period at each of the 52 sites.

Following retrieval of the SCD buoys, the sorbents were removed, extracted with methylene chloride, solvent exchanged to DMSO, and evaluated using MELA as well as the Ames mutagenicity assay[30,31] and a toxicity test assaying the respiration responses of *Vibro alginolyticus*.[32]

A second group of water samples was collected from 12 of the "worst" sites identified in the initial study and 16 reference sites located at 1 nautical mile intervals throughout the study area and into the Chesapeake Bay. In an effort to obtain quantitative results, organic contaminants were extracted from 15-liter water samples collected from each site employing an XAD-2 resin solid phase extraction (SPE) column. The column resin was then extracted with a methylene chloride: acetone (1:1) solvent system, solvent exchanged to DMSO, and tested using MELA as well as the Ames test, toxicity to cultured cells (cytotoxicity) and analysis for organic pollutant concentrations. Grab water samples were used in a whole organism assay, and in the analysis of metals and nutrients.

The goals of the subsequent statistical analyses in this case study were threefold: (1) to determine relationships between general LUA types and potential ecological effects, (2) to characterize overall spatial patterns of potential ecological effects among sites, and (3) to quantitatively rank the LUA and ambient sites with respect to each type of potential ecological effect as well as overall effects (i.e., to identify "hot spots").

In another field study, sediment samples were collected in triplicate from eight sites in the Elizabeth, Nansemond, and James Rivers, Virginia as well as the mouth of the Chesapeake Bay (Figure 4). Samples were extracted employing U.S. Environmental Protection Agency (EPA) protocols for semi-volatile (base/neutral and acid extractable) organic compounds in solid matrices.[33] Briefly, approximately 30 g wet weight of each sample were extracted by sonication with methylene chloride:acetone (1:1). The extracts were concentrated to approximately 100 μl, 25 μl of DMSO was added, and each extract was further concentrated to 25 μl with nitrogen to remove methylene chloride and acetone. Therefore, each 0.1 μl dose represented the extract of approximately 0.12 g of wet sediment. The data were evaluated for overall acute and sublethal responses to the sediment extracts.

Results and discussion

Permeability factor

Prior to discussing the specific single chemical and field extract results, it is fitting to summarize a recent study of the permeability of several chemicals into the medaka eggs using the topical technique. A detailed study was recently conducted to assess the transchorionic permeability or "permeability factor" (PF) of several test substances using the MELA technique.[19] Numerous replicates of medaka eggs were treated in a manner analogous to that used in the toxicological assays, but were immediately evaluated for the permeability of each of seven test substances: lindane, aldrin, *n*-nitroso-diethylamine (DENA), *n*-nitosodi-*n*-propylamine (DPrNA), pentachlorophe-nol (PCP), tributyltin chloride (TBTCl), and dibutyltin dichloride (DBTCl$_2$). The findings indicated that the PF (as percent material in the eggs) for these chemicals varied significantly with values ranging from 10.5 to 82.9%. Further comparison of these data with a physico-chemical coefficient frequently used in studies of bioconcentratable ability, the *n*-octanol-water partition coefficient (K_{OW}), indicated that these chemicals are apparently passively diffusing into the eggs at rates well correlated to their K_{OW} (PF = 11.1 log K_{OW} + 3.97, R^2 = 0.96; Figure 5). These observations suggest that the amount of chemical penetrating each egg is actually based on the lipid affinity of the substance rather than active transport induced by the carrier solvent. These preliminary results indicate that it may be possible to approximate a MELA dose to represent an ambient environmental concentration employing a theoretical approach such as the following:

$$\text{Estimated MELA Dose (μg / egg)} = \frac{(K_{OW} \times AWC \times E_W)}{PF_D} \qquad (1)$$

where: K_{OW} = *n*-octanol-water partition coefficient = (μg/ml *n*-octanol) ÷ (μg/ml water), AWC = ambient water concentration of toxicant in μg/ml, E_W = weight of egg in μg and PF_D = permeability factor expressed in decimal form, determined from the regression 11.1 log K_{OW} + 3.97.

Assumptions included in this calculation are (1) 1 ml of the *n*-octanol represents 1 g of lipid-filled tissue, (2) the lipid content of each medaka egg remains constant, and (3) the environmental toxicant equilibrium is at steady state with respect to water and tissue.

Single chemicals studies

MELA has been used to evaluate a number of pure chemicals that had previously been shown to have biological effects in fish and other organisms.[17] In general, the responses observed with MELA were comparable to those observed when these pure chemicals were previously studied in other test systems. Known carcinogens (e.g., benzo[a]pyrene, *n*-nitrosodiethy-lamine) produced little acute toxicity, as was seen in previous studies with

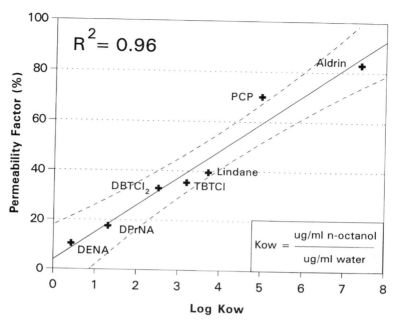

Figure 5 Relationship between permeability factors and *n*-octanol-water partition coefficients (K_{OW}). (Reproduced with permission from Helmstetter, M.F. and Alden, R.W. III. 1995. *Aquat. Toxicol.* 32:10).

medaka eggs and similar doses (see, for example, References 6, 15, 16, 21, and 34), with the primary effect being delayed hatching. These and other studies indicate that the response to these chemicals is typically detected at a later stage of development (typically 2 to 6 months) expressed as neoplasia in numerous organ systems. Naphthalene, phenanthrene, and *n*-nitrosodi-*n*-propylamine all produced delayed hatching responses analogous to those for benzo[a]pyrene and *n*-nitrosodiethylamine. Lindane exposures resulted in delayed hatching as well as elongation of the heart (i.e., tube heart as described in Reference 35), poor blood circulation, presence of blood islands, depressed heart rate, and an overall inability of hatch fry to swim successfully. The swim bladder in these fry appeared normal, suggesting a possible effect on the motor response system responsible for swimming activity. 2-Nitrofluorene produced a significant lethal response, as well as delayed hatching as a result of swelling in the abdominal region. Exposure of 10-day-old eggs (i.e., after the liver had formed) to benzo[a]pyrene, *n*-nitrosodi-*n*-propylamine and 2-nitrofluorene failed to increase the level of adverse response.

A common sublethal response detected in these studies of pure chemicals was delayed development and/or reduced embryo activity reflected in the inability of the embryo to hatch from the restrictive chorion, particularly in the lower concentrations of several substances. Embryonic movement is presumed to stir and circulate the perivitelline fluid, improving the distribution of oxygen to the embryo.[36] The reduced embryo size and/or activity may have been insufficient to circulate the oxygen and/or hatching enzymes within the eggs, resulting in the high number of unhatched eggs.

The results from more detailed case studies on pentachlorophenol (PCP) and tributyltin chloride (TBTCl) are presented below to demonstrate the utility of MELA in studying the toxicity of known environmental pollutants and factors affecting this toxicity.

Pentachlorophenol

PCP is a moderately lipophilic, relatively non-persistent environmental contaminant that is often found in industrialized estuarine seaports.[37,38] The effect of PCP on medaka eggs was evaluated using both the standardized MELA procedure and a typical immersion protocol. In addition, results from freshwater topical treatments from a range-finding study were compared to an extensive topical treatment evaluation followed by rearing in 20 ppt solution.[18] Previously, it has been suggested that the bioconcentration of PCP is lower in organisms living in saline environments when compared to those found in fresh water.[39-41] Thus, it might be expected that PCP would cause increased toxicity to medaka raised in fresh water. However, results from the MELA at salinities of 0 ppt and 20 ppt demonstrated that, over the range of PCP concentrations examined, adverse effects were similar in magnitude and type (Table 3). At the highest concentrations examined, lethal toxicity (observed as death of eggs) accounted for up to 100% of the observed effects of PCP. At the lower concentrations, sublethal effects were the dominant results of PCP exposure. Sublethal effects of the PCP treatments included unhatched eggs, dead fry, and developmental deformities. The primary deformities were swollen abdomens, which often resulted in incomplete hatching, developmental arrest of the embryo, slowed heart rate and circulation, pericardial edema, and low hatch fry activity level.

Table 3 The Effect of Salinity on Overall Adverse Effects of PCP on Medaka Eggs and Fry Using MELA Topical Exposure Protocol

	Percent adversely affected[b]	
Dose[a]	0 ppt	20 ppt
12,500	100	100 (±0.0)
1,250	100	100 (±0.0)
125	40.0	50.0 (±5.8)
12.5	30.0	13.3 (±3.3)
DMSO Control	5.0	4.4 (±1.8)
Untreated Control	5.0	7.8 (±1.5)

Note: 0 ppt data are from a single replicate rangefinding study and 20 ppt data are presented as the mean of three replicates, ± standard error of the mean.

[a] Dose in ng PCP/egg.

[b] Expressed as percentage lethal and sublethal effects observed at a given dosage.

When all adverse effects were considered, the 21-day median effective dose (ED_{50}) for PCP in medaka exposed by the MELA protocol and raised in freshwater, was 55.9 ng PCP per egg (Table 4). A similar 21-day ED_{50} of 61.8 ng PCP per egg was determined by the MELA conducted in 20 ppt water (Table 4). Statistical analysis indicated that the 21-day ED_{50} values calculated for the different salinities were not significantly different ($p > 0.05$). Moreover, the EC_{50} calculated from the results of toxicity studies of PCP using an immersion protocol (following conversion of the data to topical dose equivalents) was not significantly different from that determined using the MELA protocol (Table 4). The LD_{50} values from these two exposure modes differed by a factor of approximately two. A more detailed analysis comparing the results of the MELA topical protocol and the immersion protocol and the 0 and 20 ppt data can be found in Reference 18.

Table 4   21-Day Median Effective and Lethal Doses of PCP and TBTCl to Medaka with 95% Confidence Limits

Exposure protocol[a]	Salinity	ED_{50}[b]	LD_{50}[b]
PCP			
Topical	0 ppt	55.9 (12.0–172)	ND[c]
Topical	20 ppt	61.8 (45.7–81.7)	113 (88.4–145)
Immersion	20 ppt	59.0 (35.0–92.8)	215 (159–293)
TBTCl			
Topical	0 ppt	0.20 (5.00×10^{-3}–11.0)	47.5 (17.2–163)
Topical	20 ppt	0.10 (5.00×10^{-2}–0.2)	72.1 (12.9–1580)
Topical	35 ppt	1.00 (0.70–1.70)	112 (24.2–1100)
Immersion	20 ppt	3.34×10^{-3} (1.76×10^{-3}–5.42×10^{-3})	3.90×10^{-2} (2.77×10^{-2}–5.06×10^{-2})
Immersion	35 ppt	2.32×10^{-2} (3.26×10^{-3}–5.04×10^{-2})	0.17 (4.58×10^{-2}–0.70)

[a] Medaka were exposed either by the MELA topical exposure method or by an immersion method described in detail in Reference 18.

[b] The 21-day ED_{50} and LD_{50} values for the topical exposures were caculated from the number of adversely affected eggs observed at the doses used for the topical treatment. The 21-day ED_{50} and LD_{50} values for the immersion exposure were calculated from EC_{50} and LC_{50} values using the technique described in Reference 18. Values are in ng/egg.

[c] Value could not be determined.

This study demonstrated that for PCP the response of the medaka is a classic response (Figure 6), with acute lethality resulting only at high concentrations, and sublethal effects manifested most frequently at the intermediate and lower concentrations. The similarity in response between the numerous studies with PCP conducted over a 4-year period, suggests that PCP may be a prime candidate as a reference toxicant for MELA. The reference toxicant would be employed on a routine basis to serve as an indicator of the general health of the test population of eggs to eliminate problems associated with "weak" eggs produced from breeding adults which are unhealthy or simply suffering from "reproductive burnout". In this context,

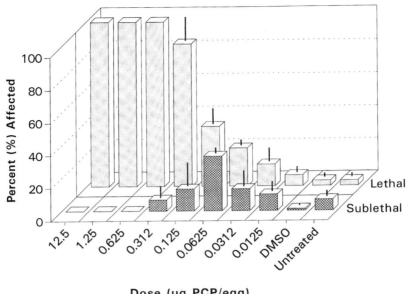

Dose (ug PCP/egg)

Figure 6 Response of medaka eggs to PCP following exposure using the MELA protocol. Values are replicate means ± standard error of each mean. (Reproduced with permission from Helmstetter, M.F. and Alden, R.W. III. 1995. *Aquat. Toxicol.* 32:21.)

PCP was used as a reference toxicant for the field case studies described below and results indicated that there were no significant differences in the field study treatments from those conducted during the single freshwater toxicant assays (mean LD_{50} for all PCP experiments was 363 ng PCP per egg ± 16.7).

Tributyltin chloride (TBTCl)

TBTCl, an organotin most commonly employed as a stabilizer in PVC and as an additive in antifoulant paints, was also extensively evaluated using MELA. MELA was used in a similar manner as that discussed for PCP. However, due to the potential for elevated open ocean concentrations of TBTCl, as a result of shipping activity, 35 ppt salinity medium was added in addition to the 20 ppt medium representing estuarine conditions. Further, as with the PCP study, immersion concentrations were developed based on an estimated TBT BCF of 4000 (Reference 42), to provide for side-by-side comparisons of the topical treatment and immersion exposure.

The results from the TBTCl exposures indicated that concentrations as low as 0.0187 ng TBTCl per egg produced a response significantly different from the laboratory controls and mean 21-day ED_{50} values of 1.0, 0.10 and 0.20 ng TBTCl per egg for the 35, 20, and 0 ppt exposures, respectively (see Table 5). Immersion exposures indicated that the no observable effects concentration (NOEC) and maximum acceptable toxicant concentration (MATC) for both the 20 and 35 ppt exposures lie below 25.0 ng TBTCl/l and, in 20

Table 5 Effect of Salinity on Overall Adverse Effects of
TBTCl on Medaka Eggs and Fry Using the MELA Topical
Exposure Protocol

Dose[a]	Percent adversely affected[b]		
	0 ppt	20 ppt	35 ppt
12,000	100	100 (±0.0)	100 (±0.0)
1,200	100	100 (±0.0)	100 (±0.0)
120	100	100 (±0.0)	100 (±0.0)
12.0	100	90.0 (±5.8)	76.7 (±6.7)
1.20	100	80.0 (±5.8)	50.0 (±10.0)
0.600	60	76.7 (±3.3)	33.3 (±6.7)
0.300	10	73.3 (±8.8)	33.3 (±8.8)
0.150	40	43.3 (±8.8)	26.7 (±8.8)
0.075	20	36.7 (±8.8)	16.7 (±8.8)
0.037	20	60.0 (±5.8)	16.7 (±3.3)
0.019	60	40.0 (±5.8)	16.7 (±3.3)
DMSO control	5.0	4.4 (±1.8)	6.7 (±2.4)
Untreated control	5.0	8.9 (±2.0)	5.6 (±2.4)

Note: 0 ppt data are from a single replicate range-finding study
and 20 and 35 ppt data are presented as the mean of three
replicates, ± standard error of each mean.

[a] Dose in ng TBTCl/egg.

[b] Expressed as percentage lethal and sublethal effects observed
at a given dosage.

ppt medium, concentrations as low as 1 ng TBTCl/l produced a response
significantly different from control groups. These values are some of the
lowest effective values reported for TBT to date; the lowest previous concentration resulting in an adverse biological response was 5 ng TBT/l, which
induced imposex, or female development of male sex organs, in the dog-
whelk (*Nucella lapillus*[43]). However, it should be noted that a previous study[44]
found that the immersion exposure of medaka eggs in a freshwater system
resulted in significantly reduced effects as compared to the present study
(e.g., Reference 44 NOEC was 320 ng TBTO/l, the present study NOEC was
determined to be less than 25 ng TBTCl/l).

A comparison of the topical and immersion data indicated that there
was a statistically significant difference between the two exposure modes,
but that the slopes of these lines for all adverse effects were not significantly
different, suggesting that the mode of action was approximately the same
for both treatments, but that there was a displacement due potentially to an
underestimated bioconcentration factor (BCF; Figure 7). The results of this
study suggest that the BCF for medaka eggs may be orders of magnitude
higher than the estimated value of 4000 (estimate from these data to be
greater than 94,000) and may possibly be one of the highest established for
TBT. This is not surprising, as the eggs contain a high concentration of lipid
material, with very little physiological mechanism for elimination of accumulated materials. These results suggest that further study is warranted to
establish the magnitude of concentration of TBT in fish eggs, particularly

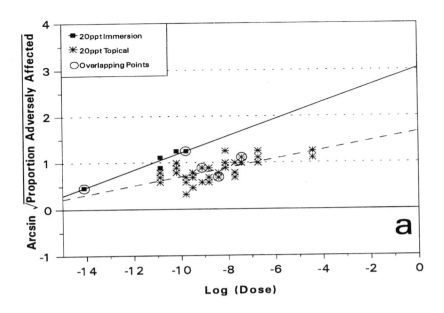

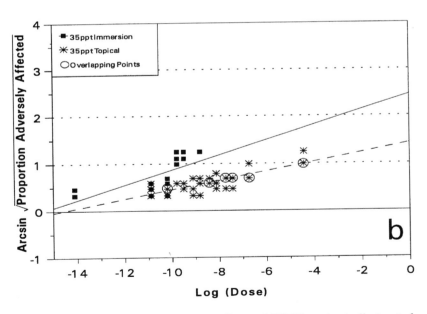

Figure 7 Regression plots of all adverse effects of TBTCl on topically treated and immersion exposed medaka eggs reared in 20 ppt (a) and 35 ppt (b) salinity media.

since most eggs (from egg laying species) reside either in the surface micro-layer or on the sediments, both of which have been shown to concentrate organotins (see, for example, References 45–50).

Numerous developmental anomalies were detected in all TBTCl exposures, the most prevalent of which were skeletal deformities (lordosis, kyphosis, pectoral stumping, etc.), cardiac malformations and gross underdevelopment (many similar to those seen in Reference 51). The appearance and extent of these anomalies was strongly concentration-dependent, with the higher concentrations resulting in full developmental arrest (subsequently followed by death) and the lower concentrations exhibiting spinal abnormalities and associated swimming problems as well as reduced hatching events. These developmental anomalies were common to all treatment and salinity exposures. Example anomalies are presented in Figure 8.

The practical implications of these results for PCP and TBTCl are enormous, as ambient field concentrations in excess of the experimental concentrations are commonly found, even to the present day, following significant legislative efforts to reduce environmental levels of these and other substances. For example, ten sampling events where water samples were collected from the Elizabeth River in Hampton Roads, Virginia, between September 1991 and September 1992, revealed the presence of TBT in all ambient samples with concentrations ranging from 0.017 to 0.228 µg TBT/l (Helmstetter and Alden, unpublished data). Further, sediment samples taken in August 1992 from a major tributary of the Chesapeake Bay contained PCP concentrations as high as 35,200 µg PCP/kg and TBT in excess of 400 µg TBT/kg (Helmstetter and Alden, unpublished data). Environments such as these study sites may very well be undergoing severe perturbations in ecological structure as a result of exposure of sensitive early life stages of finfish to these and other potentially toxic and teratogenic substances.

Field studies

The Port of Hampton Roads, Virginia is one of the largest industrial seaports in the country and is the largest military port in the world. Increased military activity as well as extensive commercial and industrial growth have resulted in the input of substantial pollution into this system which drains into the lower Chesapeake Bay via the Elizabeth and James Rivers. Previous chemical and toxicological investigations of the Elizabeth River system and its receiving waters (i.e., the lower James River and lower Chesapeake Bay) have indicated that this may be one of the most polluted areas in the country, particularly with respect to polynuclear aromatic hydrocarbons and metals contamination (see, for example, References 52–56).

Water samples

Extracts associated with 25 of the 52 LUA sites resulted in effects in over 50% of the MELA test populations, indicating that biologically active contaminants capable of producing adverse responses in medaka were fairly widespread for the study region (Figure 9). Common deformities included distended abdomens similar to those seen for the individual polynuclear aromatic hydrocarbon studies and PCP, cardiac anomalies similar to those seen for lindane, and numerous skeletal deformities similar to those seen for TBTCl.

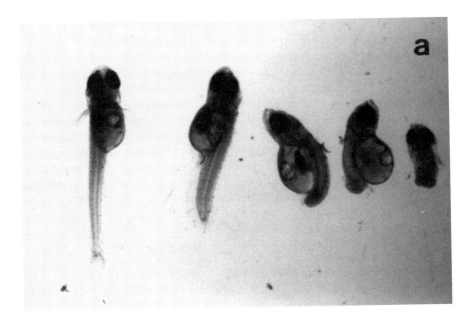

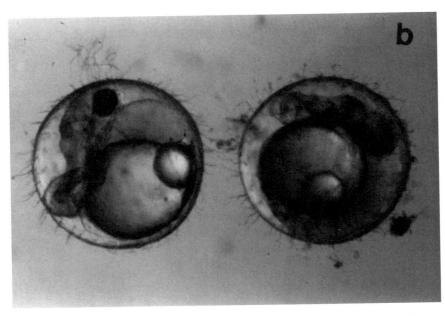

Figure 8   Example deformities following MELA topical exposure of medaka eggs to TBTCl. Variation in severity of vertebral deformities resulting from exposure to 120 TBTCl per egg is shown in (a). Severe skeletal and optical deformation as well as numerous undifferentiated cells resulting from exposure to 0.150 ng TBTCl per egg is shown in (b). All example deformities were reared in 35 ppt salinity medium.

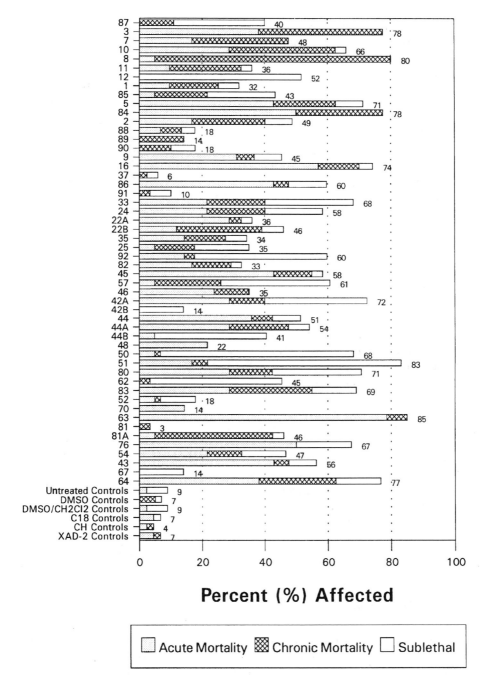

Percent (%) Affected

□ Acute Mortality ▨ Chronic Mortality □ Sublethal

Figure 9 Results of LUA study extract exposures employing MELA. Acute mortality is that occurring within 48 h of treatment, chronic mortality is egg mortality occurring after 48 h and sublethal effects include all deformities, unhatched eggs, and dead fry.

A second water sample collection was made from 12 of the "worst" sites and 16 midchannel ambient sites. Extracts from 9 of the 28 sites resulted in adverse effects in greater than 50% of MELA test organisms. Sublethal responses, primarily underdeveloped embryos and unhatched eggs, appeared to dominate the effects on the medaka. Moreover, it was apparent that the mid-channel samples appeared to induce lower effects than the nearby LUAs, indicating that there may be a dilution effect present in the ambient samples. The Chesapeake Bay sample (CB-1; see Figure 3), included to establish "background" levels of lethal and sublethal effects, showed little adverse impact on the medaka eggs, and chi-squared analysis indicated that it was not significantly ($p > 0.05$) different from any control type.

The following provides a descriptive and visual illustration of how field data generated from MELA can be used to assess general relationships and trends. Six site groups were identified from the cluster analysis of all biological effects data (Figure 10). The discriminant analysis and multiple analysis of variance (MANOVA) of the biological effects indicated that groups 4 and 6 were strongly separated from the other groups with site group 4 being associated with mutagenicity and deformed medaka fry and site group 6 being associated with acute mortality in the medaka and cytotoxicity (Figure 11a). Elimination of the variables related to mutagenicity (which overwhelmed the discriminant analysis) resulted in the ellipses shown in Figure 11b. The separation of the first discriminant function (DF1) is clearly dominated by acute mortality in medaka associated with the site group 6 samples. The second discriminant function (DF2) separated the site groups based on more chronic biological effects. Figure 12 shows the standardized, quantitative rank data for all parameters and all sites. This histogram indicates that multiple variables showed an adverse response for all sites (i.e., no site was dominated by one outstanding endpoint).

This study was designed to assess the relative potential of a very extensive series of LUAs for producing adverse biological impacts in laboratory tests with sensitive indicators of toxicity, teratogenicity, and mutagenicity. As a component of this comprehensive study of the Elizabeth River Basin, MELA proved to be a useful tool in identifying "hot spots" in the system relative to impacts on fish early life stages. However, the patterns of effects were found to be both diverse and widespread. This is not surprising when one considers the large number of potential inputs into the system and the associated diversity of chemicals introduced in close proximity to each other in this complex ecosystem.

Sediment samples

The results of the sediment extract exposures are presented in Figure 13. The pattern of response was similar to what one would expect, with the Elizabeth River stations (5, 6, and 7) eliciting the greatest impact, while the comparably pristine Nansemond site extracts resulted in a significantly reduced level of impact. Further, as one moves toward the mouth of the bay, the level of adverse impact decreases relative to the Elizabeth River sites; however, the levels of effects are still high relative to the Nansemond site (8) and the

STANDARDIZED DISTANCE

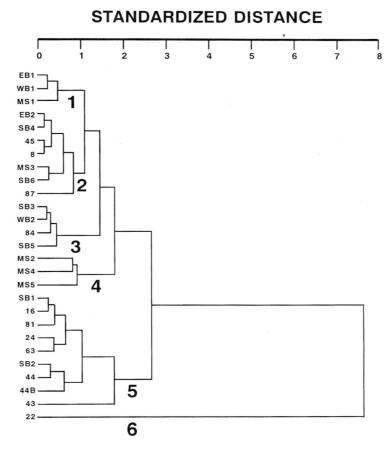

Figure 10 Standardized distance dendogram for classification of site groups with respect to biological data. The data are from the 12 priority LUA sites and the corresponding ambient stations. The numbers indicate site groups used in subsequent statistical analyses. LUA categories corresponding to the sampling site numbers can be found in Table 2.

untreated and DMSO controls. This moderate level of toxicity in sites 1 through 3 may be the result of the flushing of pollutants introduced into the estuary by the Elizabeth River and/or the James River.

These results indicate that MELA is also very useful in the screening of sediment samples from systems suspected to be contaminated with biologically active pollutants. Extracts from as little as 0.12 g of sediment resulted in a significant decrease in viability in medaka for the majority of the sites and the spatial patterns of adverse impact appeared to correlate well with other studies of the harmful biological effects of water and sediments from this region and the level of pollutants contained within the tributary sediments (Alden and Helmstetter, unpublished data). The results also appear to correlate reasonably well with the water sample exposures discussed

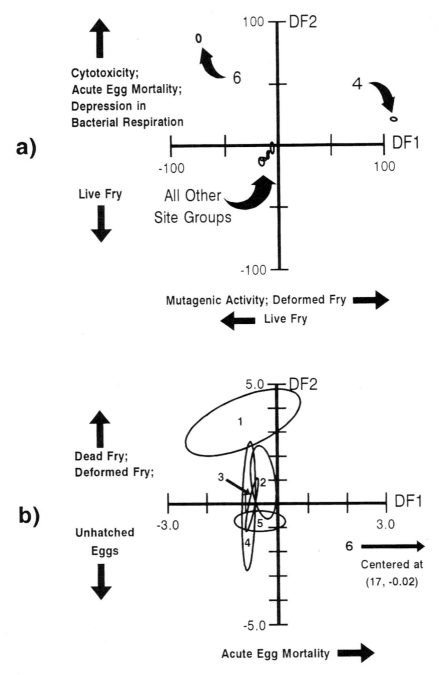

Figure 11 Confidence ellipses ($\alpha = 0.05$) for canonical scores for discriminant functions (DF1 and DF2) describing differences in biological effects data with mutagenicity (a) and without mutagenicity (b). The differences indicated are between site groups defined by cluster analysis (see Figure 10).

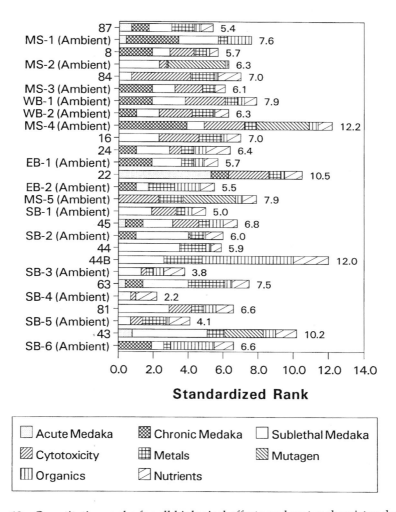

Figure 12   Quantitative ranks for all biological effects and water chemistry data.

previously, in that the Southern Branch and Main Stem sites tended to elicit the greatest effects relative to Chesapeake Bay sites and numerous laboratory controls.

Summary

The eggs of the Japanese medaka (*Oryzias latipes*) have been used to develop a standardized assay, the medaka embryo-larval assay (MELA). As described, the assay can be used to assess numerous toxicological endpoints that may result from exposure to pure chemicals or complex mixtures of chemicals such as those that might be found in samples collected from chemically contaminated environments. Numerous previous investigations with medaka eggs have indicated that short-term embryo-larval assays can be used to rapidly evaluate the toxicokinetics of single chemicals or of

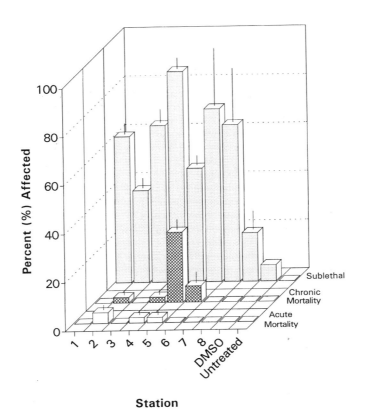

Station

Figure 13 Results of sediment study extract exposures employing MELA. Acute mortality is that occurring within 48 h of treatment, chronic mortality is egg mortality occurring after 48 h and sublethal effects include all deformities, unhatched eggs, and dead fry. Data are presented as means of field replicates ± the standard error of each mean.

complex mixtures of substances, such as those seen in field-collected samples from highly industrialized areas (see, for example, References 5, 6, 16–18, 21, 29, 34, 57–64).

Previous investigations have indicated that salinity stress patterns appear to be species specific; some may display no visible effect of salinity changes while others may show an adverse response to either increasing or decreasing sea salt concentration.[36] The medaka appear to exhibit limited stress related to variable salinity. High breeder egg production, in addition to the hatching success of control eggs and low incidence of osmotic swelling and dehydration from the various exposures, indicate that the medaka has a high tolerance for salinities up to 35 ppt and is an ideal "surrogate" test species for studies of the relative toxicity/teratogenicity of various estuarine and coastal systems and their source and receiving waters. In addition, it appears that the medaka is extremely sensitive to several chemicals commonly found in these environments. This finding is particularly important in that concentrations in coastal waters are often too low to induce effects in standard short-term acute assays, and often require a sensitive, longer-term assay which includes the observation of sublethal effects.

Our studies have also indicated that the topical technique may capably represent natural exposure to waterborne contaminants. This finding suggests that, given the proper physico-chemical data, the MELA topical exposure can be used for chemicals for which an immersion exposure regime may prove problematic or for representation of aqueous exposure. Further, it may be possible to apply the results from simultaneous topical and immersion toxicity exposures to estimate bioconcentration factors for embryonic stages. These studies would be advantageous in situations where direct bioaccumulation measurements cannot be made due to limited biomass.

Field evaluations suggest that MELA is a practical and valuable tool in the assessment of the relative biological impact of numerous sites in a large aquatic system. The ease in applying the assay to field-collected samples allows a large number of stations to be evaluated simultaneously. This assay is ideal for screening numerous sites with a sensitive test system, and at a low cost, from which several areas could be designated for more intensive, and costly biological and chemical evaluations. Further, MELA should prove very useful in evaluating temporal trends of individual sites to determine the relative impacts of these sites to the receiving system over the duration of seasonal and industrial activity fluctuations.

There are many advantages that can be realized by the investigator when using the MELA protocol including the following:

1. The use of a whole animal test organism exposed at the embryo stage, coupled with the optical clarity of the medaka eggs, allows observation of differentiation and all major organogenesis events *in vivo.*
2. The use of a simple, rapid, and uniform method of exposing test animals that needs only limited quantities of test material.
3. The use of a noninvasive exposure technique which minimizes mortality unrelated to the material being studied and may be representative of exposure in nature.
4. The ability to rapidly and inexpensively expose large numbers of individuals and to generate true dose–response data.
5. The ability to assess multiple endpoints including acute and delayed toxicity, teratogenicity, and carcinogenicity.
6. The ability to assess the effects of a wide range of chemicals which is not limited by the chemical's water solubility.
7. Unlike adult fish, which may excrete large quantities of ingested test materials (particularly if the component is highly water insoluble), little if any "excretion" occurs in treated eggs.

Although these and numerous other advantages make the MELA a practical and powerful testing system, investigators should be aware of some of the limitations of the MELA procedure which include:

1. The need for good animal husbandry practices and special attention to the quality of water being used for rearing and conducting the assay.
2. The inability to test for effects on organs or tissues that are not found in fish but may be important targets in mammals (e.g., lungs and breasts).
3. The need to determine the uptake and retention of test materials when animals are exposed via topical application in DMSO. However, in light of the results reported here and in detail in Reference 19, this may truly be an advantage rather than a disadvantage.
4. Though the topical treatment procedure solves many of the problems associated with solubility in immersion tests, the data from our studies have indicated that some level of passive permeation is occurring, resulting in low penetration of chemicals with low affinities for biological tissues.
5. If teratogenicity or carcinogenicity are endpoints, a specialist may be needed to interpret the results.
6. For compounds that require metabolic activation, external activation systems may be required if the enzyme system needed is not present in medaka.
7. The need for a balance between adequate aeration and injury to treated eggs, particularly in elevated salinity studies when increased aeration rates are required.

In spite of these limitations, the MELA has considerable utility as a valuable, practical and sensitive toxicity test system which can be used as a stand alone assay of lethal and sublethal effects or as a component of a more extensive tiered approach for toxicity testing of pure chemicals and complex mixtures.

References

1. McKim, J.M. 1985. Early life stage toxicity tests. In *Fundamentals of Aquatic Toxicology: Methods and Applications*, Eds. G.M. Rand, S.R. Petrocelli, 58–95 New York: Hemisphere.
2. Llewellyn, G.C., Stephenson, G.A., Hofman, J.W. 1977. Aflatoxin B1 induced toxicity and teratogenicity in Japanese medaka eggs (*Oryzias latipes*). *Toxicon* 15:582–587.
3. Wales, J.H., Sinnhuber, R.O., Hendricks, J.D., Nixon, J.E., Eisele, T.A. 1978. Aflatoxin B_1 induction of hepatocellular carcinoma in the embryo of rainbow trout (*Salmo gairdneri*). *J. Natl. Cancer Inst.* 60:1133–1139.
4. Hendricks, J.D., Meyers, T.R., Casteel, J.L., Nixon, J.E., Loveland, P.M., Bailey, G.S. 1984. Rainbow trout embryos: Advantages and limitations for carcinogenesis research. In *Use of Small Fish Species in Carcinogenicity Testing*, Ed. K.L. Hoover, 65:129–137. *Natl. Cancer Inst. Monogr.* Bethesda, MD:National Cancer Institute.
5. Cooper, K.R., Liu, H., Bergqvist, P.A., Rappe, C. 1991. Evaluation of baltic herring and icelandic cod liver oil embryo toxicity, using the Japanese medaka (*Oryzias latipes*) embryo larval assay. *Environ. Toxicol. Chem.* 10:707–714.

6. Hawkins, W.E., Walker, W.W., Overstreet, R.M., Lytle, J.S., Lytle, T.F. 1990. Carcinogenic effects of some polycyclic aromatic hydrocarbons on the Japanese medaka and guppy in waterborne exposures. *Sci. Total Environ.* 94:155–167.

7. Black, J.J. 1988. Carcinogenicity tests with rainbow trout embryos: A review. *Aquat. Toxicol.* 11:129–142.

8. Metcalfe, C.D., Sonstegard, R.A. 1984. Microinjection of carcinogens into rainbow trout embryos: An *in vivo* carcinogenesis assay. *J. Natl. Cancer Inst.* 73:1125–1132.

9. Metcalfe, C.D., Sonstegard, R.A. 1985. Oil refinery effluents: Evidence of co-carcinogenic activity in the trout embryo microinjection assay. *J. Natl. Cancer Inst.* 75:1091–1097.

10. Black, J.J., Maccubbin, A.E., Schiffert, M. 1985. A reliable, efficient, microinjection apparatus and methodology for the *in vivo* exposure of rainbow trout and salmon embryos to chemical carcinogens. *J. Natl. Cancer Inst.* 75:1123–1128.

11. Grizzle, J.M., Putnam, M.R. 1987. *Microinjection of fish embryos as a laboratory assay for chemical carcinogens.* U.S. EPA Environmental Research Laboratory, Florida, EPA/600/53-87-032.

12. Black, J.J., Maccubbin, A.E., Johnston, C.J. 1988. Carcinogenicity of benzo[a]pyrene in rainbow trout resulting from embryo microinjection. *Aquat. Toxicol.* 13:297–308.

13. Metcalfe, C.D., Cairns, V.W., Fitzsimons, J.D. 1988. Microinjection of rainbow trout at the sac-fry stage: A modified trout carcinogenesis assay. *Aquat. Toxicol.* 13:347–356.

14. Walker, M.K., Hufnagle, L.C., Jr., Clayton, M.K., Peterson, R.E. 1992. An egg injection method for assessing early life stage mortality of polychlorinated dibenzo-p-dioxins, dibenzofurans, and biphenyls in rainbow trout (*Oncorhynchus mykiss*). *Aquat. Toxicol.* 22:15–38.

15. Maccubbin, A.E., Black, J.J. 1986. Passive perchorionic carcinogen bioassay using rainbow trout (*Salmo gairdneri*) embryos. In *Aquatic Toxicology and Environmental Fate: Ninth Volume, ASTM STP 921*, Ed. T.M. Poston, R. Purdy, 277–286. Philadelphia: American Society for Testing and Materials.

16. Maccubbin, A.E., Ersing, N., Weinar, J., Black, J.J. 1987. *In vivo* carcinogen bioassays using rainbow trout and medaka fish embryos. In *Short-Term Bioassays in the Analysis of Complex Mixtures V*, Eds. S.S. Sandhu, D.M. DeMarini, M.J. Mass, M.M. Moore, J. L. Mumford, 209–223. New York:Plenum Press.

17. Helmstetter, M.F. 1992. Development and standardization of a short-term assay for evaluating polluted estuarine and coastal environments: The medaka embryo-larval assay. Ph.D. thesis. Old Dominion University, Norfolk, VA. 256 pp.

18. Helmstetter, M.F., Alden, R.W. III. 1995a. Toxic responses of Japanese medaka (*Oryzias latipes*) eggs following topical and immersion exposures to pentachlorophenol. *Aquat. Toxicol.* 32:15–29.

19. Helmstetter, M.F., Alden, R.W. III. 1995b. Passive trans-chorionic transport of toxicants in topically treated Japanese medaka (*Oryzias latipes*) eggs. *Aquat. Toxicol.* 32:1–13.

20. Kirchen, R.V., West, W.R. 1976. *The Japanese Medaka: Its Care and Development.* Carolina Biological Supply Company, Burlington, NC. 36 pp.

21. Klaunig, J.E., Barut, B.A., Goldblatt, P.J. 1984. Preliminary studies on the usefulness of medaka, *Oryzias latipes*, embryos in carcinogenicity testing. *Natl. Cancer Inst. Monogr.* 65:155–161.

22. Weber, D.N., Spieler, R.E. 1987. Effects of the light-dark cycle and scheduled feeding on behavioral and reproductive rhythms of the cyprinodont fish medaka, *Oryzias latipes. Experientia* 43:621–624.

23. Weis, J., Weis, P., Heber, M. 1982. Variation in response to methylmercury by killifish (*Fundulus heteroclitus*) embryos. In *Aquatic Toxicology and Hazard Assessment: Fifth Conference,* ASTM STP 766. Eds. J.G. Pearson, R.B. Foster, W.E. Bishop, 109–119. Philadelphia:American Society for Testing and Materials.

24. Rammler, D.H., Zaffaroni, A. 1967. Biological implications of DMSO based on a review of its chemical properties. *Ann. N.Y. Acad. Sci.* 141:13–23.

25. Benville, P.E. Jr., Smith, C.E, Shanks, W.E. 1968. Some toxic effects of dimethyl sulfoxide in salmon and trout. *Toxicol. Appl. Pharmacol.* 12:156–178.

26. Anderson, R.J., Prahlad, K.V. 1976. The deleterious effects of fungicides and herbicides on *Xenopus laevis* embryos. *Arch. Environ. Contam. Toxicol.* 4:312–323.

27. Arii, N., Namai, K., Gomi, F., Nakazawa, T. 1987. Cryoprotection of medaka embryos during development. *Zool. Sci.* 4:813–818.

28. Robertson, S.M., Lawrence, A.L., Neill, W.H., Arnold, C.R., McCarty, G. 1988. Toxicity of the cryoprotectants glycerol, dimethyl sulfoxide, ethylene glycol, methanol, sucrose, and sea salt solutions to the embryos of Red Drum. *Prog. Fish Cult.* 50:148–154.

29. Takimoto, Y., Hagino, S., Yamada, Y., Miyamoto, J. 1984a. The acute toxicity of fenitrothion to killifish (*Oryzias latipes*) at twelve different stages of its life history. *J. Pesticide Sci.* 9:463–470.

30. Ames, B.N., McCann, J., Yamasaki, E. 1975. Methods for detecting carcinogens and mutagens with the Salmonella/mammalian-microsome mutagenicity test. *Mut. Res.* 31:347–364.

31. Maron, D.M., Ames, B.N. 1983. Revised methods for the Salmonella mutagenicity test. *Mut. Res.* 113:173–215.

32. Gordon, Andrew. Old Dominion University. Norfolk, VA, unpublished procedure.

33. U.S. EPA. 1986. Test Methods for the Evaluation of Solid Waste. Vol. 1–2. Office of Solid Waste, Washington, D.C.

34. Marty, G.D., Nunez, J.M., Lauren, D.J., Hinton, D.E. 1990b. Age-dependent changes in toxicity of n-nitroso compounds to Japanese medaka (*Oryzias latipes*) embryos. *Aquat. Toxicol.* 17:45–62.

35. Weis, P., Weis, J.S. 1979. Congenital abnormalities in estuarine fishes produced by environmental contaminants. In *Symposium on Pathobiology of Environmental Pollutants: Animal Models and Wildlife as Monitors. Animals as Monitors of Environmental Pollutants,* 94–107. University of Connecticut:National Academy of Science.

36. Rosenthal, H., Alderdice, D.F. 1976. Sublethal effects of environmental stressors, natural and pollutional, on marine fish eggs and larvae. *J. Fish. Res. Board Can.* 33:2047–2065.

37. Cirelli, D.P. 1978. Patterns of pentachlorophenol usage in the United States of America — an overview. In *Pentachlorophenol: Chemistry, Pharmacology, and Environmental Toxicology,* Ed. K. Ranga Roa, 13–18. New York:Plenum Press.

38. Crosby, D.G. 1981. Environmental chemistry of pentachlorophenol. *Pure and Appl. Chem.* 53:1051–1080.

39. Trujillo, D.A., Ray, L.E., Murray, H.E., Giam, C.S. 1982. Bioaccumulation of pentachlorophenol by killifish (*Fundulus similis*). *Chemosphere.* 11:25–31.

40. Tachikawa, M., Sawamura, R., Okada, S., Hamada A. 1991. Differences between freshwater and seawater killifish (*Oryzias latipes*) in the accumulation and elimination of pentachlorophenol. *Arch. Environ. Contam. Toxicol.* 21:146–151.

41. Tachikawa, M., Sawamura, R. 1994. The effects of salinity on pentachlorophenol accumulation and elimination by killifish (*Oryzias latipes*). *Arch. Environ. Contam. Toxicol.* 26:304–308.

42. Cardwell, R.D., Sheldon, A.W. 1986. *Proc. Oceans '86 Conf., Organotin Symposium*, A risk assessment concerning the fate and effects of tributyltins on the aquatic environment. 4:1117–1129. Washington, D.C.

43. Gibbs, P.E., Pascoe, P.L., Burt, G.R. 1988. Sex change in the female dog-whelk, *Nucella lapillus*, induced by tributyltin from antifouling paints. *J. Mar. Biol. Assess. U.K.*, 68:715–731.

44. Wester, P.W., Canton, J.H., Van Iersel, A.A.J., Krajnc, E.I., Vaessen, H.A.M.G. 1990. The toxicity of bis(tri-n-butyltin)oxide (TBTO) and di-n-butyltindichloride (DBTC) in the small fish species *Oryzias latipes* (medaka) and *Poecilia reticulata* (guppy). *Aquat. Toxicol.* 16:53–72.

45. Salazar, M.H. 1986. *Proc. Oceans '86 Conf., Organotin Symposium*, Environmental significance and interpretation of organotin bioassays. 4:1240–1245. Washington, D.C.

46. Cleary, J.J., Stebbing, A.R.D. 1987. Organotin in the surface microlayer and subsurface waters of Southwest England. *Mar. Pollut. Bull.* 18:238–246.

47. Harris, J.R.W., Cleary, J.J. 1987. *Proc. Oceans '87 Conf.*, Particle-water partitioning and organotin dispersal in an estuary. 1370–1374. Washington, D.C.

48. Unger, M.A., MacIntyre, W.G., Huggett, R.J. 1987. *Proc. Oceans '87 Conf.*, Equilibrium sorption of tributyltin chloride by Chesapeake Bay sediments. 1381–1385. Washington, D.C.

49. Hall, L.W. Jr., Bushong, S.J., Johnson, W.E, Hall, W.S. 1988. Spatial and temporal distribution of butyltin compounds in a northern Chesapeake Bay marina and river system. *Environ. Monitor. Assess.* 10:229–244.

50. Krone, C.A., Chan, S.-L., Varanasi, U. 1991. *Proc. Oceans '91 Conf., Ocean Technologies and Opportunities in the Pacific for the 90's*, Butyltins in sediments and benthic fish tissues from the east, gulf and pacific coasts of the United States. 2:1054–1059. Washington, D.C.

51. Walker, W.W., Heard, C.S., Lotz, K., Lytle, T.F., Overstreet, R.M. 1989. *Proc. Oceans '89 Conf., Ocean Pollution*, Tumerogenic, growth, reproductive and developmental effects in medaka exposed to bis(tri-n-butyltin)oxide. 2:516–524. Washington, D.C.

52. Alden, R.W. III, Young, R.J., Jr. 1982. Open ocean disposal of materials dredged from a highly industrialized estuary: An evaluation of potential lethal effects. *Arch. Environ. Contam. Toxicol.* 11:567–576.

53. Rule, J.H. 1986. Assessment of trace element geochemistry of Hampton Roads Harbor and lower Chesapeake Bay area sediments. *Environ. Geol. Water Sci.* 8:209–219.

54. Alden, R.W. III, Butt A.J. 1987. Statistical classification of the toxicity and polynuclear aromatic hydrocarbon contamination of sediments from a highly industrialized seaport. *Environ. Toxicol. Chem.* 6:673–684.

55. Huggett, R.J., deFur, P.O., Bieri, R.H. 1988. Organic compounds in Chesapeake Bay sediments. *Mar. Pollut. Bull.* 19:454–458.

56. Long, E.R., Morgan, L.G. 1990. *NOAA Technical Memorandum NOS OMA 52.* 179 pp, Seattle, Washington.

57. Solomon, H.M. 1979. Teratogenic effects of carbaryl, malathion, and parathion on developing eggs of medaka (*Oryzias latipes*). In *Symposium on Pathobiology of Environmental Pollutants: Animal Models and Wildlife as Monitors. Animals as Monitors of Environmental Pollutants*, University of Connecticut:National Academy of Science. 393 pp.

58. Hatanaka, J., Doke, N., Harada, T., Aikawa, T., Enomoto, M. 1982. Usefulness and rapidity of screening for the toxicity and carcinogenicity of chemicals in medaka, *Oryzias latipes*. *Jap. J. Exp. Med.* 52:243–253.

59. Takimoto, Y., Ohshima, M., Yamada, H., Miyamoto, J. 1984b. Fate of fenitrothion in several developmental stages of the killifish (*Oryzias latipes*). *Arch. Environ. Contam. Toxicol.* 13:579–587.

60. Hinton, D.E., Lauren, D.H., Teh, S.J., Giam, C.S. 1988. Cellular composition and ultrastructure of hepatic neoplasms induced by diethylnitrosamine in *Oryzias latipes*. *Mar. Environ. Res.* 24:307–310.

61. Shigeoka, T., Yamagata, T,. Minoda, T., Yamauchi, F. 1988a. Toxicity test of 2,4-dichlorophenol on embryo, larval, and early-juvenile Japanese medaka (*Oryzias latipes*) by semi-static method. *Eisei Kagaku.* 34:274–278.

62. Shigeoka, T., Yamagata, T., Minoda, T., Yamauchi, F. 1988b. Acute toxicity and hatching inhibition of chlorophenols to Japanese medaka, *Oryzias latipes* and structure-activity relationships. *Eisei Kagaku.* 34:343–349.

63. Marty, G.D., Cech, J.J. Jr., Hinton, D.E. 1990a. Effect of incubation temperature on oxygen consumption and ammonia production by Japanese medaka, *Oryzias latipes*, eggs and newly hatched larvae. *Environ. Toxicol. Chem.* 9:1397–1403.

64. Wisk, J.D., Cooper, K.R. 1990. The stage specific toxicity of 2,3,7,8-tetrachlorodibenzo-p-dioxin in embryos of the Japanese medaka (*Oryzias latipes*). *Environ. Toxicol. Chem.* 9:1159–1169.

chapter seven

A multi-injection fish carcinogenesis model with rainbow trout

Gordon C. Balch and Chris D. Metcalfe

Introduction

Our current understanding of carcinogenicity is based primarily on research conducted with mammalian experimental models, but in the last two decades carcinogenicity research has expanded to include fish. Data generated with fish models indicate that teleosts respond to carcinogens in much the same way as mammals,[1-3] and in some cases, fish are more sensitive to chemical carcinogens than rodent test species.[4]

Investigations of an epizootic in the early 1960s of aflatoxin B_1-induced hepatocellular carcinomas in cultured rainbow trout (*Oncorhynchus mykiss*) led to the development of experimental carcinogenesis models with rainbow trout and other salmonid species.[5,6] In addition to being extremely sensitive to the carcinogenic effects of aflatoxin B_1 (AFB1) and other aflatoxins, rainbow trout appear to be sensitive to nitroso-compounds, polynuclear aromatic hydrocarbons (PAHs), aromatic amines, and several other classes of organic carcinogens.[1] In all carcinogenesis tests with trout, the primary site for tumor development is the liver and the most commonly observed hepatic malignancies are hepatocellular in origin.[7] Hepatic malignancies generally develop at 8 to 12 months after initial exposure to chemical carcinogens.[1] Chang et al.[8] reported that a high proportion of the liver tumors induced in trout by AFB1 showed evidence of point mutations in the Ki-*ras* oncogene.

Many different exposure protocols have been developed with trout to investigate the carcinogenic activity of various chemicals and environmental contaminants. These methods include dietary exposure and intraperitoneal (i.p.) injections,[9] stomach intubation,[10] immersion of eggs in aqueous solutions of carcinogen,[11] and microinjection of test compounds into the perivitelline fluid of eggs,[12] into the yolk of eggs,[13] or into the yolk of newly hatched sac-fry.[14]

1-56670-149-X/96/$0.00+$.50
© 1996 by CRC Press, Inc.

The sensitivity of experimental carcinogenicity models with early life stages of trout is probably due to the intrinsic promotional activity caused by the rapid rate of cell proliferation in neonates.[15] Therefore, only one dose of carcinogen at the egg or sac-fry stage is usually required to induce liver tumors in trout. However, care must be taken to dose early life stages of trout after the development of a liver bud in order to ensure metabolic activation of indirect-acting carcinogens.[16] A functioning liver develops at approximately the same time as the embryo enters the "eyed" egg stage of development. In comparison to models where adult or juvenile fish must be dosed repeatedly via the diet or by aqueous exposure, the protocols involving microinjection of test compounds into early life stages of salmonids have several advantages: (1) the dose administered to the fish is easily controlled and quantified, (2) the amount of material required for exposures is small, and (3) handling of the carcinogens by laboratory personnel and contamination of test media is kept to a minimum. A disadvantage of this carcinogenicity model is that the exposure protocol does not directly mimic the routes of contaminant exposure that would occur naturally in the environment (i.e., maternal transfer into the egg, aqueous uptake, diet).

Most carcinogenicity tests with trout and other salmonids have involved exposure to a single carcinogenic compound.[1] However, enhanced development of liver tumors in AFB1-initiated trout occurs when these fish are subsequently exposed to DDT, 17-β-estradiol, β-napthoflavone and indole-3-carbinol,[17] as well as carbon tetrachloride.[18] A few trout carcinogenicity assays have been performed with complex environmental samples, such as extracts from contaminated sediments,[19,20] and extracts from pulp mill effluents.[21] Environmental samples may contain a mixture of carcinogens, tumor promoters and agents that stimulate or inhibit cancer progression. Therefore, it may be desirable to use a fish carcinogenicity model that incorporates exposure to chemical carcinogens and to other chemicals that can modify tumor development. Such a protocol could involve initial exposure of trout to a carcinogen at an early life stage followed by repeated exposure of the developing trout with other compounds or environmental mixtures.

In this manuscript, we describe a carcinogenicity assay with trout involving microinjection at the sac-fry stage of development, with subsequent i.p. injections of developing trout over several weeks. The sensitivity of this assay is assessed by determining the incidence of liver tumors in rainbow trout repeatedly exposed to a synthetic PAH compound, 7,12-dimethylbenzanthracene (DMBA). DMBA induces DNA adducts in fish cells,[22] and is a potent carcinogen in both rodents[23] and fish.[12,24] The primary goal of this investigation was to devise a sensitive and relatively rapid assay procedure for detecting compounds with carcinogenic and promotional activity in fish.

Materials required

Trout.   The rainbow trout used in this study were of the Kamloops strain of *Oncorhynchus mykiss* and were obtained from an inbred stock at the

Rainbow Springs Trout Hatchery in Thamesford, Ontario. Previous studies with the Shasta strain of rainbow trout indicate that this strain is also a sensitive test organism.[16] A variety of sources exist for obtaining eggs of rainbow trout, including commercial trout farms and government hatcheries. There are fall-spawning and spring-spawning trout, so eggs are generally not available during the summer months or during January to February. However, many hatcheries are now employing photoperiod manipulation and hormone treatments to provide trout eggs year-round.

Commercial trout diet. The commercial trout food used in this study was sold by Martin Feed Mills, Elmira, Ontario and contained 40% crude protein, 12% crude fat, 3% crude fiber, 7,500 IU/kg of vitamin A, 3,000 IU/kg of vitamin D_3, 100 IU/kg of vitamin E, and 800 mg/kg ascorbic acid. Although any commercial diet that promotes rapid growth will be adequate for carcinogenicity tests, high protein (70%) diets have been shown to optimize the carcinogenic response in trout.[25] It is recommended that the food be stored in a freezer to avoid loss of vitamin supplements with time.

Gases. A bottle of hospital-grade CO_2 and a regulator was purchased from a local gas supplier. CO_2-saturated water was prepared by bubbling the gas through an aquarium air stone into 6 l of water for approximately 2 min.

Chemicals. DMBA was purchased from Ultra Scientific Ltd., Rhode Island. DMBA should be stored out of the light at approximately –10°C, and should be handled with extreme care, as it is a potent carcinogen. High purity DMSO was purchased from Caledon Laboratories Ltd., Georgetown, Ontario. DMSO should be stored at 4°C and warmed to room temperature prior to use. Tricaine methane sulfonate (MS-222) was purchased from Argent Chemical Laboratories, Redmond, Washington. The MS-222 should be stored at approximately –10°C, and contact with this genotoxic chemical should be avoided.

Injection supplies. The Hamilton push-button repeating dispenser (Model # PB-600-1) and Hamilton 700 Series microliter luer tip syringes with capacities of 25, 100, and 500 µl were purchased from Chromatographic Specialties, Brockville, Ontario. The 1-inch, 31-gauge syringe needles (Model # PT2) with a Teflon hub and #4 style of beveled point were also purchased from Chromatographic Specialties. The ³/₄-inch, 23-gauge standard Becton Dickinson syringe needles were purchased from Fisher Scientific, Toronto, Ontario.

Fixation supplies. Bouin's fixative consisted of a mixture of 750 ml picric acid solution (saturated aqueous), 250 ml formalin solution (40% aqueous), and 100 ml glacial acetic acid, and these chemicals were purchased from Fisher Scientific, Toronto, Ontario. Omnisette (HistoPrep) tissue cassettes were also purchased from Fisher.

Procedures

Treatment groups

In this manuscript, we describe the incidence of hepatic neoplasms in rainbow trout from three different treatment groups: (1) control, (2) single exposure, and (3) repeated exposure. The purpose of this experiment was to evaluate whether a protocol that involves repeated exposures of trout to a complete carcinogen (DMBA) over a period from the sac-fry stage to 19 weeks post-hatch increases the incidence of hepatic neoplasms in comparison to a protocol that involves a single exposure to the carcinogen at the trout sac-fry stage.

The control group received a microinjection of dimethylsulphoxide (DMSO) at the sac-fry stage of development, and subsequently received a total of 5 i.p. injections of DMSO. DMSO was the "carrier" solvent used to dilute the carcinogen. Fish from the single exposure group were injected once at the sac-fry stage with DMBA dissolved in DMSO, and were subsequently given 5 i.p. injections of DMSO only. Fish from the repeated exposure group were injected once at the sac-fry stage with DMBA dissolved in DMSO, and were subsequently given 5 i.p. injections with DMBA dissolved in DMSO. Approximately 250 sac-fry were microinjected in each treatment group at the start of each experiment to ensure that there would be approximately 100 fish remaining at the end of the experiment after cumulative mortalities.

All fish in the control, single and repeated exposure groups received a total of six injections. The first injection was administered into the yolk of the sac-fry using the microinjection methods described by Metcalfe et al.,[14] and the remaining 5 i.p. injections were administered at 7, 10, 13, 16, and 19 weeks according to the methods described by Kotsanis and Metcalfe.[18] Sac-fry of the control group were injected with volumes of 0.5 µl of DMSO, and sac-fry of the other two treatment groups were injected with 0.5 µl of DMBA dissolved in DMSO (2 mg/ml). Therefore, the dose of DMBA administered to sac-fry was 1 µg per sac-fry, and since the average weight of a sac-fry was 0.077 g, this dose corresponds to approximately 13 µg DMBA per g of sac-fry. All subsequent i.p. injections of DMSO, or DMBA dissolved in DMSO were dispensed at a volume of 6.5 µl/g of fish. The repeated exposure group was given a dose of 13 µg DMBA per g body weight for all five i.p. injections.

Fish husbandry

Rainbow trout of the Kamloops strain were obtained at the "eyed" egg stage (i.e., embryo eye pigments visible through the egg chorion). Trout eggs reach this stage of development approximately 21 days after fertilization and can be transported with minimal mortalites. Eyed eggs will hatch within approximately 10 days when held at temperatures between 10 to 12°C. The eggs, and trout fry and juveniles that hatched from the eggs, were maintained in the wet laboratory facilities at Trent University. The water supply to these facilities was Otonabee River water (pH 7.8 to 8.2; alkalinity 60 to 80 mg

$CaCO_3/l$; hardness 80 to 100 mg $CaCO_3/l$) that was sand filtered and UV sterilized. The developing embryos were kept in the dark at a water temperature of 10°C for the period from arrival at Trent to 4 days post-injection (approximately 2 weeks). A low water temperature during this period reduces the numbers of mortalities associated with the trauma of sac-fry injection. After this time, the water temperature was gradually raised over a 4 day period to 15°C, and the photoperiod was adjusted to 16 h light/8 h dark. Water temperatures were maintained at 15 ± 3°C for the remainder of the study.

Immediately after microinjection of sac-fry, the treatment groups were held in separate fish-rearing trays nested into fiberglass raceways. The fish were raised in these raceways until fish were approximately 10 g in size, and then they were transferred to circular 1,000 l fiberglass tanks. Fish were fed *ad libitum* with commercial trout chow of an appropriate pellet size. Fish were weighed and total numbers of fish remaining in each treatment group were determined during each set of i.p. injections. By the end of the 21-week experiment, the weights of the trout ranged between approximately 12 to 14 g.

In these types of experiments, the densities of fish in the holding tanks should be kept similar and should not exceed approximately 10 g of fish per liter. As the fish grow, it will be necessary to split the treatment groups into more and more holding tanks to ensure optimal growth. Trout fry should be fed at least 4 times per day, although during the swim-up period (i.e., start of exogenous feeding) the trout fry should be fed on an hourly basis in order to keep mortalities to a minimum. Larger trout should be fed at least twice per day with a total daily food ration equal to approximately 4% of the total fish biomass in the tank. Tank densities, water temperatures, aeration, and feeding rates should be regulated to maximize the growth rates of the trout, as rapid growth appears to be a promotional factor for hepatocarcinogenesis in salmonids. However, because cultured rainbow trout develop social hierarchies that limit the feeding of subdominant individuals, there is inevitably a large variation in the growth rates and sizes of the trout. This can be minimized by successively segregating the smaller fish from the larger ones throughout the course of the experiment.

Sac-fry injection

The integument surrounding the yolk of salmonid sac-fry is easily ruptured by microinjections immediately after hatch. However, it is resilient enough for an injection by 3 days post-hatch, and the yolk sac remains large enough for injections for approximately 4 days thereafter. Care must be taken to avoid direct handling of the sac-fry, as this increases the number of mortalities. It is possible to microinject 800 to 1,000 sac-fry per day using the protocol described below. However, a mortality rate of 5 to 20% should be expected over the first 48 h post-injection due to the stress and trauma of the microinjection procedure and the toxicity of the carrier solvent.[14]

The sac-fry were microinjected with test solutions at 3 to 5 days post-hatch. Approximately 10 sac-fry at a time were removed with an aquarium dip-net from a holding tank, and these animals were anesthetized by placing the dip-net with fry into a container of CO_2-saturated water for approximately 30 seconds. Sac-fry were then transferred from the dip-net to a piece of dry Whatman filter paper (15-cm diameter), where they were immobilized by adhesion to the paper. During this step, the sac-fry were evenly distributed around the filter paper and the fry were "applied" directly to the surface of the paper without handling. This was accomplished by inverting the net and lightly applying the sac-fry to the filter paper.

Each sac-fry was injected with a 0.5 µl volume of DMSO, or DMBA dissolved in DMSO. Injections were done with a 1-inch, 31-gauge needle with a Hamilton gas-tight microliter syringe (25 µl capacity) attached to a repeating dispenser. The sac-fry were injected by inserting the syringe needle into the yolk sack in an anterior-posterior direction, and pressing the dispenser button. Care was taken to avoid the highly vascularized area at the anterior of the yolk. After all of the sac-fry on the filter paper were injected, the paper and adhering sac-fry were placed into a recovery bucket containing aerated 10°C water. The anesthetized sac-fry were out of water for less than one minute during the microinjection procedure. The sac-fry in the recovery bucket were allowed to float free, or were eventually shaken off the filter paper into the water. After all sac-fry had resumed normal activity, the fry were returned to the rearing raceways.

Intraperitoneal injections

During all i.p. injections, trout were removed in groups of less than 20 from holding tanks and anesthetized individually in CO_2-saturated water. The CO_2-saturated water is used as an anesthetic to avoid use of potentially carcinogenic chemical anesthetics. Depending on the size, the fish require 2 to 5 min of exposure to CO_2-saturated water before loss of equilibrium and slowing of opercular movements indicates that they are anesthetized. Fish that completely lose opercular movements generally do not recover.

Anesthetized fish were weighed on a two-decimal place balance and the weight was used to calculate the appropriate volume of test solutions required to dose the fish (5 to 100 µl). The trout were injected with Hamilton chromatography syringes of 25, 100, or 500 µl capacity (depending on the size of the fish), which were fitted with a 3/4-inch, 23-gauge syringe needle and a repeating dispenser. After i.p. injection, fish were placed in a recovery tank filled with 15°C aerated water. When fish resumed normal swimming and ventilatory activity, they were returned to rearing tanks.

Trout should be i.p. injected using techniques that require a minimum of handling. A simple technique that can be used is to wedge the fish into a slit cut into a block of polyurethane foam, with the ventral surface of the fish exposed. The fish should be injected i.p. in an anterior to posterior direction at a point just anterior to the pectoral fins. Care must be taken to avoid puncturing of the viscera with the syringe needle. Recently, we have been sealing the needle puncture site after i.p. injections with a drop of cyanoacrylate glue to avoid leaking of injected solutions into the water.

Necropsy and histopathology

All fish were sacrificed at 21 weeks post-hatch by an overdose of tricaine methane sulfonate (MS-222) anesthetic, and each fish was weighed and grossly examined for external lesions. The ventral surface of each fish was cut open from below the gill opercula to the anal vent, and the internal organs (including the liver, kidneys, spleen) and viscera were examined for grossly visible lesions. In particular, the incidence of grossly visible hepatic lesions was noted. These lesions are generally circular, and are white to pale yellow in color; varying in size from approximately 1 mm to >10 mm in diameter. These lesions are found on both the dorsal and ventral surfaces of the liver.

The liver was removed from the gut cavity with forceps, and after removal of the gall bladder, was weighed to two decimal places. The liver tissue was placed in plastic tissue cassettes and immediately transferred to a Nalgene® container (2 l) filled with Bouin's fixative. To aid in fixation, large livers (>2 g) should be cut into 2 to 3 pieces which should be placed in separate cassettes, and there should be no more than 30 tissue cassettes per liter of fixative. To avoid bleaching of tissues by Bouin's fixative, it should be poured from the Nalgene® container within 72 h, and the tissues flushed with tap water for at least 1 h. The containers should then be filled with 70% ethanol, at which point the tissues can be held for several months before further processing. Bouin's fixative penetrates rapidly into tissues and produces excellent tinctorial contrast after staining. However, because this fixative is toxic and a strong irritant, researchers may choose to utlilize one of the nontoxic fixatives available commercially.

Liver tissues were dehydrated, embedded in paraffin, sectioned at a 5 μm thickness and stained with hematoxylin and eosin using standard histological methods.[26] A slide was prepared from a tissue section cut from the middle of each block of liver tissue. The liver sections were examined by light microscopy and examined for the presence of toxicopathic lesions, foci of altered hepatic cells, and neoplastic lesions of hepatocytic and cholangiocytic origin. Lesions were identified according to the criteria described by Hendricks et al.[7] Differences in lesion incidence were tested for statistical significance by chi-square analysis ($p \leq 0.05$) with the aid of the statistical analysis program, Minitab (Version 8.0).

In order to analyze the data from this type of histopathological survey, it is necessary to have a consistent protocol for sectioning each tissue block, and for examining the histological slides. The tissues should be oriented consistently within the paraffin block. For instance, the dorsal surface of the liver may always be oriented towards the front of the paraffin block. The number of sections prepared for histological examination should always be the same, and the location within the tissue where the sections are removed should also be consistent. Finally, the histopathologist examining the tissues should not know the treatment groups assigned to the slides while conducting the survey (i.e., "blind" scoring).

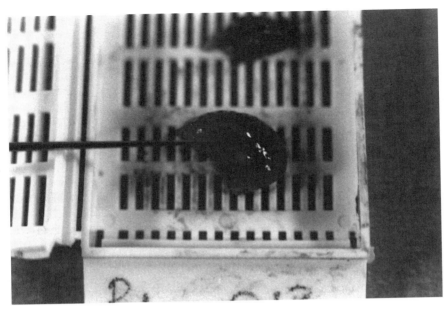

Figure 1   A photograph of a grossly visible lesion (arrow) on the surface of the liver of a rainbow trout necropsied after multiple exposures to DMBA.

Results and discussion

Lesions were grossly visible (Figure 1) on the surface of the livers of two trout in the single exposure group (2%), and of seven trout in the repeated exposure group (7%). No lesions were grossly visible on the livers of control fish. A histological survey of the livers failed to confirm the presence of the two lesions observed grossly in the single exposure group. The survey confirmed four of the seven grossly visible lesions noted in the repeated exposure group, and all four of these histologically confirmed lesions were hepatocellular carcinomas.

These data illustrate the "hit-or-miss" nature of histopathological surveys where only a small number of sections from each tissue block are examined. Serial sectioning or "step" sectioning (i.e., several sections are removed from the tissue at intervals of 100 to 200 µm) would be ideal, but these detailed surveys are not practical in carcinogenesis experiments where there are large numbers of fish from several treatment groups. Alternatively, the researcher may choose to orient the tissue during embedding to ensure that the grossly-visible lesions are sectioned. This is practical, since Hendricks et al.[7] found that lesions as small as 0.5 mm were grossly visible on trout livers after fixation with Bouin's fixative. However, we contend that any deviation from a consistent orientation of the liver tissue in the embedding medium would compromise the statistical integrity of the histological survey.

The results of the histopathological survey are summarized in Table 1. Cholangiofibrosis, and large (>1 mm) zones of necrosis (massive necrosis) and fibrotic tissue (massive scarring) were classified as "degenerative" tox-

Table 1 Hepatic Lesions in Rainbow Trout at 21 Weeks after
Beginning of Experimental Treatments

Lesion	Treatment		
	Control	Single	Repeated
Number of fish necropsied	97	99	100
(A) Degenerative			
Massive necrosis	25	30	24
Massive scarring	1	2	12
Cholangiofibrosis	0	1	1
(B) Altered foci			
Anaplastic bile ducts	0	0	5
Eosinophilic foci	0	0	2
(C) Neoplastic lesions			
Basophilic foci	0	2	3
Liver cell adenomas	0	2	1
Hepatocellular carcinomas	0	1	4
Mixed cell carcinomas	0	0	1

Note: Treatment groups received either: a single sac-fry injection of DMBA
followed by 5 i.p. injections of DMSO carrier solvent (**Single**); one
sac-fry injection followed by 5 i.p. injections of DMBA (**Repeated**);
or one sac-fry injection and 5 i.p. injections of DMSO (**Control**).

icopathic lesions, indicative of chemically induced hepatotoxicity. Eosinophilic foci of hepatocytes, and foci of anaplastic cholangiolar cells were
classified as *altered foci* of hepatic and biliary cells. These lesions could be
precursors of hepatic neoplasms, but there is not yet sufficient evidence
indicating that they can progress to malignancy in trout.[7] Basophilic foci,
liver cell adenomas, hepatocellular carcinomas, and mixed cell carcinomas
were identified as *neoplastic lesions* of the liver. The basophilic foci and liver
cell adenomas are preneoplastic hepatocellular lesions that clearly can
progress to malignancy, although they may not progress in every instance.[7]
Basophilic foci and liver adenomas are considered the first and second steps,
respectively, in the progression towards hepatocytic malignancy. The hepatocellular and mixed cell carcinomas are malignant neoplasms of hepatocytic
origin and mixed hepatocytic/cholangiocytic origin, respectively.[7]

Approximately 25% of the livers in all three treatment groups showed
signs of toxicopathic tissue necrosis (Table 1). Since the control fish showed
this response, repeated exposure to DMSO carrier solvent obviously caused
cytotoxic effects in the liver. However, there was a significantly higher proportion of fish in the repeated exposure group (12%) that showed signs of
hepatic "scarring" (fibrosis) in comparison to the control (1%) and single
exposure (2%) treatments. The incidence of cholangiofibrotic lesions was low
(1%) in both the single and repeated exposure groups, and these lesions were
absent in the control group.

Eosinophilic foci, and foci of anaplastic bile ducts were found only in
the repeated exposure group, and the incidences of these altered foci were
significantly different than in the single exposure group (Table 1). It is not
known whether eosinophilic foci are preneoplastic,[7] but this lesion was noted

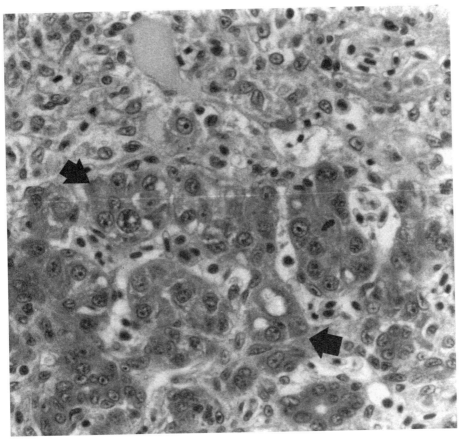

Figure 2 A photomicrograph of anaplastic bile ducts (arrows) in the liver of a rainbow trout necropsied after multiple exposures to DMBA (H&E staining, ×400 magnification).

in an earlier study in which trout eggs were microinjected with DMBA.[12] Anaplastic bile duct lesions are relatively uncommon in trout and little is known about their malignant status. This lesion (Figure 2) consists of an aggregate of non-caniculized bile duct cells that are spindle shaped, lack polarity, and contain mitotic figures. Black et al.[27] observed a similar lesion in rainbow trout injected with benzo[a]pyrene in the egg stage, and this type of lesion has been observed in brown bullheads (*Ameiurus nebulosus*) collected from the Detroit River.[28] This lesion was not categorized in this present study as a neoplasm, but the histological characteristics strongly indicate that it may be a preneoplastic lesion of bile duct cells.

No neoplastic lesions were observed histologically in the livers of fish from the control group. Neoplastic lesions were observed in the livers of a total of nine fish from the repeated exposure group in comparison to five fish in the single exposure group, but this difference in the incidence of total neoplasms was not statistically significant. However, the neoplasms observed in trout from the repeated exposure group included four fish with

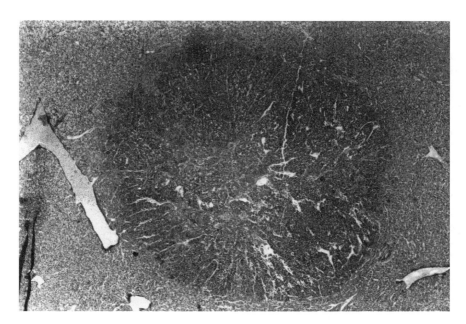

Figure 3 A photomicrograph of a hepatocellular carcinoma in the liver of a rainbow trout necropsied after multiple exposures to DMBA (H&E staining, ×40 magnification).

hepatocellular carcinomas (Figure 3) and one fish with a mixed cell carcinoma, while all lesions except one in the single exposure group were classified as preneoplastic.

The repeated exposure protocol developed in this study induced malignant neoplasms in trout within 21 weeks of first exposure. To our knowledge, this is the shortest duration reported for induction of malignant liver tumors in rainbow trout exposed to a PAH compound. Although several protocols have been developed for exposing juvenile and early life stages of trout to carcinogens,[9-14] the time between first exposure to a PAH compound and the development of malignant liver tumors is usually between 40 and 52 weeks.[1] The method developed here represents the first protocol involving repeated exposure of trout from an early life stage to several weeks of age. Since only one malignant neoplasm was observed at 21 weeks in the single exposure treatment group (1%) in comparison to 5 malignancies observed in the repeated exposure group (5%), it can be concluded that the combination of exposure at the sac-fry stage plus repeated i.p. exposure of juveniles results in a short latency period for the development of malignancies in the trout liver.

In an earlier microinjection study with trout eggs, a single exposure to DMBA at dose of 0.5 μg per egg (i.e., half of the sac-fry dose used in the present study) induced a low incidence (1/32 = 3%) of hepatocellular carcinomas when the trout were necropsied 52 weeks after exposure.[12] Therefore, it was surprising that the same type of malignant lesion was observed in this study in a trout from the single exposure group at only 21 weeks. Since there was histological evidence that DMSO was cytotoxic, the carrier solvent may have had a weak promotional effect, as well. DMSO was chosen as a solvent because (1) it can dissolve high concentrations of DMBA and other non-polar compounds, (2) it is relatively non-viscous, and therefore easily injected through a fine gauge needle, and (3) it is not as toxic to fish as many other non-polar solvents. However, the toxicity associated with the use of DMSO or other organic solvents is obviously a problem in these assays. The use of solvents could be avoided by using other technologies for delivery of injected chemicals, such as the adsorption of hydrophobic chemicals to liposomes.[29]

There are several possible mechanisms for the rapid development of tumors using the repeated exposure protocol. Trout were repeatedly exposed to DMBA with the assumption that this compound may cause cumulative genetic damage to cellular oncogenes and tumor-suppressor genes consistent with multi-step carcinogenesis.[30] Dosing was begun at the sac-fry stage since assays involving early life stages of trout are known to be sensitive carcinogenicity models.[11-14] Single doses of carcinogens to early life stages of vertebrates are often sufficient to induce liver tumors because of the rapid rate of cell proliferation in neonates.[15]

The "enzyme-altered foci" model of hepatocarcinogenesis[31] may be a mechanism that explains the results of this experiment. This model is based on the premise that exposure to chemical carcinogens alters cellular enzymes in a way which increases the risk of subsequent genetic damage. Some of the enzymatic alterations considered to be important include reduced glucose-6-phosphate, reduced canicular adenosine triphosphatase, reduced β-glucuronidase, and increased glutamyl transpeptidase activity.[31] However, the histochemical methods required to determine whether there were alterations in the activity of these enzymes in the trout liver neoplasms[32] were not used in this study.

The hepatoxicity induced by repeated exposure to DMBA may have induced regenerative proliferation of hepatocytes. The liver fibrosis observed in the repeated exposure group indicated that DMBA was hepatotoxic. There is a strong correlation between liver damage and the development of hepatic cancers in vertebrates,[33] which is likely related to regenerative cell proliferation induced in response to chronic liver damage. Exposure of trout to the hepatotoxic chemical, carbon tetrachloride following sac-fry exposure to AFB1 enhanced the carcinogenic response to AFB1.[18]

Repeated exposure to DMBA may also have created an environment which was selective for proliferation of initiated hepatocytes that were resistant to the cytotoxic effects of this chemical. The "resistant hepatocyte model" proposed by Farber and Sarma[34] is based on the fact that most hepatocarcinogens are also strong cytotoxins which suppress cell replication. According to this model, the initiated cells undergo a phenotypic change which makes this cell more resistant to the cytotoxic action of the carcinogen. Because of this resistance to cytotoxins, the initiated cells will then have a growth advantage over noninitiated cells during subsequent exposure to the carcinogen.

Electrophilic metabolites of aromatic compounds are formed during Phase I metabolism. Aryl hydrocarbon hydroxylase (AHH) is one of the cytochrome P-450IA family of enzymes involved in the metabolism of PAHs. AHH activity is increased by prior exposure to PAHs, but activity subsides to near background levels within weeks when the inducer is removed.[35] Britvic et al.[36] showed that the hepatic biotransformation activity of pre-exposed fish remains elevated for weeks beyond the return of AHH to basal activity. This indicates that fish in the repeated exposure group may have been more efficient than the single exposure group in metabolizing and metabolically activating DMBA. An increased rate of formation of genotoxic metabolites may have resulted in increased carcinogenic activity.

The repeated exposure protocol is an effective assay system for inducing malignant hepatic neoplasms in trout within a short period of time (i.e., 21 weeks). Although this carcinogenicity model is somewhat labor intensive, the amount of test material needed is relatively small, and the administered dose is more accurately regulated than exposure through the diet or water. Because of these advantages, this method is a practical and sensitive protocol for investigating the carcinogenic and promotional activity of chemicals. This protocol has already proven to be useful for determining the carcinogenic response of fish to complex environmental mixtures.[20,21] However, the single exposure protocol is still an effective carcinogenicity model with rainbow trout, provided that there is a longer latency period (e.g., 40 to 50 weeks) for the development of malignant hepatic neoplasms.

Acknowledgments

This work was supported by research grants to CDM from the Natural Sciences and Engineering Research Council (NSERC) of Canada.

References

1. Metcalfe, C.D. 1989. Tests for predicting carcinogenicity in fish. *CRC Critical Rev. Aquatic Sci.* 1:111–129.
2. Hendricks, J.D. 1982. Chemical carcinogenesis in fish. In *Aquatic Toxicology,* Ed. L.J. Weber. Raven Press, New York. 149 pp.
3. Dawe, C.J. and J.C. Harshbarger. 1975. Neoplasms in feral fishes: their significance to cancer research. In *The Pathology of Fishes,* Eds. W.E. Ribelin and E. Migaki, U. Wisconsin Press, Madison, Wisconsin. 871 pp.
4. Dawe, C.J. and J.A. Couch. 1984. Debate: mouse versus minnow: the future of fish in carcinogenicity testing. In *Use of Small Fish Species in Carcinogenicity Testing,* Ed. K.L. Hoover, National Cancer Institute Monograph 65:223–235.
5. Sinnhuber, R.O., J.D. Hendricks, J.H. Wales and G. Putnam. 1977. Neoplasms in rainbow trout, a sensitive animal model for environmental carcinogenesis. *Ann. N.Y. Acad. Sci.* 298:389–415.
6. Hendricks, J.D. 1981. The use of rainbow trout (*Salmo gairdneri*) in carcinogen bioassay, with special emphasis on embryonic exposure. In *Phyletic Approaches to Cancer,* Eds. J.C. Dawe, J.C. Harshbarger, S. Kendo, T. Sugimura and S. Takayama. Japan Scientific Soc. Press, Tokyo, pp. 227–239.

7. Hendricks, J.D., T.R. Meyers and D.W. Shelton. 1984. Histological progression of hepatic neoplasia in rainbow trout (*Salmo gairdneri*). In *Use of Small Fish Species in Carcinogenicity Testing*, Ed. K.L. Hoover, National Cancer Institute Monograph 65:321–336.

8. Chang, Y.-J., C. Mathews, K. Mangold, K. Marien, J. Hendricks and G. Bailey. 1991. Analysis of *ras* gene mutations in rainbow trout liver tumors initiated by aflatoxin B_1. *Mol. Carcinogenesis* 4:112–119.

9. Hendricks, J.D., T.R. Meyers, D.W. Shelton, J.L. Casteel and G.S. Bailey. 1985. Hepatocarcinogenicity of benzo[a]pyrene to rainbow trout by dietary expoure and intraperitoneal injection. *J. Natl. Cancer Institute* 74:839–851.

10. Kimura, I., T. Miyaki and K. Yoshizaki. 1976. Induction of tumors of the stomach, of the liver, and of the kidney in rainbow trout by intrastomach administration of N-methyl-N-nitroso-N'-nitrosoguanidine (MNNG). *Proc. Jap. Cancer Assoc.* 35:16.

11. Wales, J.H., R.O. Sinnhuber, J.D. Hendricks, J.E. Nixon and T.A. Eisele. 1978. Aflatoxin B_1 induction of hepatocellular carcinoma in the embryos of rainbow trout (*Salmo gairdneri*). *J. Natl. Cancer Institute* 60:1133.

12. Metcalfe, C.D. and R.A. Sonstegard. 1984. Microinjection of carcinogens into rainbow trout embryos: an *in vivo* carcinogenesis assay. *J. Natl. Cancer Institute* 73:1125–1132.

13. Black, J.J., A.E. Maccubbin and M. Schiffert. 1985. A reliable, efficient, micro-injection apparatus and methodology for the *in vivo* exposure of rainbow trout and salmon embryos to chemical carcinogens. *J. Natl. Cancer Institute* 75:1123–1128.

14. Metcalfe, C.D., V.W. Cairns and J.D. Fitzsimmons. 1988. Microinjection of rainbow trout at the sac-fry stage: a modified trout carcinogenesis assay. *Aquatic Toxicol.* 13:347–356.

15. Peraino, C., E.F. Staffeldt, B.A. Carnes, V.A. Ludeman, J.A. Blomquist, S.D. Vesselinovitch. 1984. Characterization of histochemically detectable altered hepatic foci and their relationship to hepatic tumorigenesis in rats treated once with diethylnitrosamine or benzo[a]pyrene within one day after birth. *Cancer Res.* 44:3340–3349.

16. Hendricks, J.D., T.R. Meyer, J.L. Casteel, J.E. Nixon, P.M. Loveland and G.S. Bailey. 1984. Rainbow trout embryos: Advantages and limitations for carcino-genesis research. In *Use of Small Fish Species in Carcinogenicity Testing*. Ed. K.L. Hoover, National Cancer Institute Monograph 65:129–137.

17. Bailey, G.S. and J.D. Hendricks. 1988. Environmental and dietary modulation of carcinogenesis in fish. *Aquatic Toxicol.* 11:69–75.

18. Kotsanis, N. and C.D. Metcalfe. 1991. Enhancement of hepatocarcinogenesis in rainbow trout with carbon tetrachloride. *Bull. Environ. Contam. Toxicol.* 46:879–886.

19. Metcalfe, C.D., G.C. Balch, V.W. Cairns, J.D. Fitzsimmons and B.P. Dunn. 1990. Carcinogenic and genotoxic activity of extracts from contaminated sediments in western Lake Ontario. *Sci. Total Environ.* 94:125–141.

20. Balch, G.C., C.D. Metcalfe and S.Y. Huestis. 1995. Characterization of fish carcinogens in contaminated sediment from Hamilton Harbour, Ontario, Canada. *Environ. Toxicol. Chem.* 14:79–91.

21. Metcalfe, C.D., M.E. Nanni and N.M. Scully. 1995. Carcinogenicity and mu-tagenicity testing of extracts from bleached kraft mill effluent. *Chemosphere*, 30:1085–1095.

22. Smolarek, T.A., S.L. Morgan, C.G. Moynihan, H. Lee, R.G. Harvey and W.M. Baird. 1987. Metabolism and DNA adduct formation of benzo[*a*]pyrene and 7,12-dimethylbenz[*a*]anthracene in fish cell lines in culture. *Carcinogenesis* 8:1501–1509.

23. Dipple, A. 1976. Polynuclear aromatic carcinogens. In *Chemical Carcinogens*, Ed. C.E. Searle, American Chemical Society Monograph 173, Washington, D.C., pp. 245–314.

24. Hawkins, W.E., W.W. Walker, J.S. Lytle, T.F. Lytle and R.M. Overstreet. 1989. Carcinogenic effects of 7,12-dimethylbenz[*a*]anthracene on the guppy (*Poecilia reticulata*). *Aquatic Toxicol.* 15:63–82.

25. Hendricks, J.D., J.E. Nixon and P.M. Loveland. 1981. High dietary protein promotes embryo-induced aflatoxin carcinogenesis in rainbow trout. *Fed. Proc.* 40:948–957.

26. Grimstone, A.V. and R.J. Skaer. 1972. *A Guidebook to Microscopical Methods*, Cambridge University Press, Cambridge, UK, 283 pp.

27. Black, J.J., A.E. Maccubbin and C.J. Johnston. 1988. Carcinogenicity of benzo[*a*]pyrene in rainbow trout resulting from embryo microinjection. *Aquatic Toxicol.* 13:297–308.

28. Leadley, T., G. Balch, C.D. Metcalfe, R. Lazar, E. Mazak, J. Habowsky and G.D. Haffner. An ecotoxicological investigation of three brown bullhead (*Ameiurus nebulosus*) populations of the Detroit River. In preparation.

29. Walker, M.K., L.C. Hufnagle, M.K. Clayton and R.E. Peterson. 1992. Development of an egg injection method for assessing the early life stage mortality of polychlorinated dibenzo-p-dioxins, dibenzofurans and biphenyls in rainbow trout (*Oncorhynchus mykiss*). *Aquatic Toxicol.* 22:15–38.

30. Barrett, J.C. 1993. Mechanisms of multistep carcinogenesis and carcinogen risk assessment. *Environ. Health Persp.* 100:9–20.

31. Pitot, H.C. 1990. Altered hepatic foci: Their role in murine hepatocarcinogenesis. *Annu. Rev. Pharmacol. Toxicol.* 30:465–500.

32. Parker, L.M., D.J. Lauren, B.D. Hammock, B. Winder and D.E. Hinton. 1993. Biochemical and histochemical properties of hepatic tumors of rainbow trout, *Oncorhynchus mykiss*. *Carcinogenesis* 14:211–217.

33. Dunsford, H.A., S. Sell and F.V. Chisari. 1990. Hepatocarcinogenesis due to chronic liver cell injury in hepatitis B virus transgenic mice. *Cancer Res.* 50:3400–3407.

34. Farber, E. and D.S.R. Sarma. 1987. Hepatocarcinogenesis: a dynamic cellular perspective. *Lab. Investigations* 56:4–22.

35. Andersson, T. and L. Forlin. 1992. Regulation of the cytochrome P450 enzyme system in fish. *Aquatic Toxicol.* 24:1–20.

36. Britvic, S., D. Lucic and B. Kurelec. 1993. Bile fluorescence and some early biological effects in fish as indicators of pollution by xenobiotics. *Environ. Toxicol. Chem.* 12:765-773.

chapter eight

Exposure of freshwater fish to simulated solar UVB radiation

Edward E. Little and David L. Fabacher

Introduction

Stratospheric ozone depletion and subsequent increases in ultraviolet-B (UVB) radiation observed on a yearly basis in Antarctica,[1,2] have recently been confirmed in the northern temperate regions of North America[3] where in Toronto, Canada, summertime UVB radiation has increased by 15% over the past 5 years. UVB radiation is highly energetic and poses significant potential for biological damage[4] including damage to DNA,[5,6] induction of other lesions,[7,8] impaired immune system function,[9,10] and reduced photosynthesis.[11,12]

Photolytic processes play an important role in the environmental persistence of agricultural chemicals as well as numerous industrial substances and products. For example, herbicides such as atrazine are readily broken down by sunlight[13] into relatively nontoxic components. However, other photoproducts may become more toxic. For example, the pesticides aldrin, dieldrin, and parathion are transformed by sunlight into more toxic compounds.[14,15] Polycyclic aromatic compounds, such as anthracene are altered to highly toxic forms following brief exposure to sunlight.[16]

Solar radiation also has a direct biological impact, inducing a range of cellular and organismal injuries. Numerous evolutionary adaptations and repair mechanisms evolved for coping with solar radiation (a likely limiting factor in the exploitation of shallow water and terrestrial habitats). Many organisms continue to exist at their limits of tolerance for solar radiation, thus changes that increase exposure may be directly harmful. For example, the loss of environmental cover, or an increase in environmental clarity will effectively increase irradiance and the risk of injury to an organism. Ozone depletion arising from the destruction of ozone by chlorofluorocarbons is expected to average around 11% annually in mid-northern latitudes through the rest of the century.[17] Depletion of the protective stratospheric ozone layer will similarly increase solar radiation exposure and UVB-induced damage.

Thus, an appreciation of the biological consequences of UVB exposure is important for understanding the role that solar radiation plays in the alteration of chemical structures, as well as to understand the influence of solar radiation on adaptive biological processes in aquatic systems.

In assessing the biological consequences of UVB exposure, a need exists for an accurate means of exposing freshwater fish to simulated solar UVB radiation. This includes an exposure system that provides consistent and repeatable irradiance and monitoring equipment that provides an accurate means of measuring UVB doses. We describe a solar simulator for exposing freshwater fish to environmentally relevant doses of UVB radiation. The simulator is characterized by a highly accurate spectroradiometer. Actual exposures from a prior study and current investigations are given, along with some of the consequent biological effects.

Materials required

Animals

We recommend using the earliest life stage that would be exposed to solar UVB in the environment.

Solar simulator

The light cap was supplied with twenty 160-watt cool white lamps, eight 100-watt incandescent lamps controlled by a 24-h recycling timer, four UVB313 lamps controlled by a second 24-h recycling timer, and specular aluminum reflective shields suspended from the light cap to enclose a 6×2 foot exposure area. [Environmental Growth Chambers Inc., 510 E. Washington Street, Chagrin Falls, Ohio 44022, (216) 247-5100].

Optional lamps

1. 160-watt UVA365 lamps to replace eight 160-watt cool white lamps. [National Biological Inc., 1632 Enterprize Parkway, Twinsburg, Ohio 44087, (800) 338-5045].
2. UVA340 lamps. [Q-Panel Co., 26200 First Street, Cleveland, Ohio 44145, (216) 835-8700].
3. 75-watt halogen lamps to replace incandescent lamps. [General Electric Co., Cleveland, Ohio].

UV (ultraviolet) filtering materials

1. 0.76-mm (30-mil) polycarbonate plastic.
2. 0.39-mm (15-mil) cellulose acetate plastic.
3. 0.13-mm (5-mil) cellulose acetate plastic.
4. 0.13-mm (5-mil) mylar B plastic.

[Cope Plastics, 6340 Knox Industrial Drive, St. Louis, Missouri 63139, (314) 644-5120].

Spectroradiometer and accessories

1. Model 752 Spectroradiometer.
2. NIST traceable calibration lamp.
3. 6.5-V calibration power supply.
4. Gain check module for voltage gain and wavelength accuracy calibration.
5. Photomultiplier tube-type optics head equipped with 4-in. diameter quartz dome integrating sphere.

[Optronic Laboratories, Inc., 4470 35th Street, Orlando, Florida 32811, (407) 648-5412].

Temperature control

Thermostatic water temperature control. [Remcor Corp., 500 Regency Dr., Glendale Heights, Illinois 60131].

Other materials from local suppliers or fabricated on site

1. Glass water bath or equivalent — fabricated on site.
2. Light blue styrofoam insulation board placed under water bath — local construction supply.
3. Airlift chambers[18] or appropriate exposure chamber provided with water and air supply — local fabrication, glass with silicon cement.
4. Air supply and air stones — local pet supply.
5. Safety glasses or face shields — general scientific catalogs.
6. Appropriate apparel to cover exposed skin — long sleeve garment and gloves.

Procedures

Exposure methods

A variety of methods have been used to determine the biological effects of solar radiation. These include approaches that supplement UVB in natural solar radiation, selectively limit UVB from solar exposures, or simulate solar UVB exposures in laboratory studies. A number of factors should be considered in the selection of approaches for solar exposures including extent of UV enhancement, spectral composition of the exposure irradiance, consistency and repeatability of exposure conditions, application in the field, and costs.

Solar supplemental methods rely largely on ambient sunlight with additional UV provided by lamps. The benefits of such an approach are the ability to apply visible radiation at natural intensities and spectral composition of visible radiation specific to the latitude of the study site. This should provide an exposure that integrates a range of variables affecting exposure such as solar angle (time of day, season), cloud cover, tropospheric pollution, and thus would require extensive monitoring to characterize the UV dose applied throughout the exposure. Caldwell et al.[19] used a grid of overhead UV lamps over field plots to provide enhancements of UVB. Bjorn and Teramura[20] used a UV sensor to monitor solar output and adjust the artificial UV intensity in order to maintain stable UVB output relative to solar irradiance. Since glass or acrylic materials of a greenhouse will also filter UVA, it is important to provide additional UVA irradiance as well.

Solar UV reduction approaches also rely on solar irradiation as the exposure source; thus the intensity and spectral composition of ambient solar radiation are applied, but UVB wavelengths are reduced. The UVB reduction can also be made through the use of filtering plastics, as utilized by Blaustein et al.[21]

We conducted pond UV exposures using styrene culture dishes which were mounted at various lengths on a PVC pipe that was suspended in the pond using weights and floats. The organisms were confined within the chambers which were transparent to UVB, and depth of placement in the water column was used to limit exposure. UVB varied with the depth at which exposure vessels were placed in the water column (Figure 1).

Solar simulation methods

The assessment of the biological impact of ultraviolet radiation often requires the use of artificial lighting sources in order to achieve consistent environmental conditions, appropriate experimental controls, and consistent intensity and duration of exposure. It is not possible to exactly replicate the spectral qualities or intensities of the sun with artificial sources. However, it is important to approximate three aspects of solar radiation in laboratory solar simulation: (1) sufficient UVB (280 to 320 nm) radiation, (2) an absence of UVC (200 to 280 nm), and (3) the presence of UVA (320 to 400 nm) and visible wavelengths (400 to 800 nm) in sufficient intensity to induce cellular repair mechanisms. Exposure to UVB radiation in the absence of photoreactive radiation of UVA and visible wavelengths will result in an overestimation of UVB effects.

Solar simulation approaches depend entirely on lamps to generate the photic exposure. Most lamp systems provide peak output over a limited range of wavelengths; therefore, the total output of lamps is usually much different than natural solar radiation. These differences can be partially compensated by using a combination of lamps with different spectral characteristics. Guiding selection of lamps for solar simulation requires consideration of target intensities of exposure within the wavelength range of interest as well as a natural balance of other wavelengths to ensure activation of cellular repair mechanisms, photosynthesis, diurnal rhythmicity and physiological cycles, and orientation.

Spectral quality of the solar simulator is attained through selection of lamps having phosphor coatings to generate different wavelength output

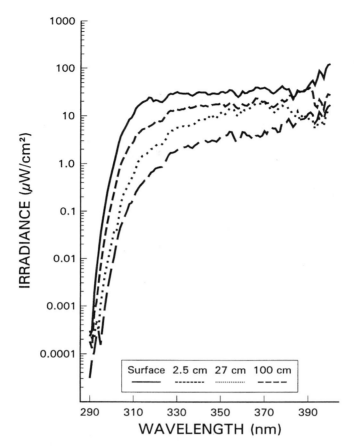

Figure 1   Semilog plot of ultraviolet spectral characteristics of sunlight at the water surface, 2.5, 27, and 100 cm beneath the water surface of a pond at 38.5° N latitude at 1430 h on July 25, 1994.

(Figure 2). Filtering materials are used to limit or block particular wave bands (Figure 3). The waveband intensities produced by the simulator are a function of the type of fluorescent lamps used, and their operational voltage. Very high output (VHO) cool white lamps, for example operate at a higher voltage (160 watts) than standard cool white lamps (60 watts). Intensity will increase with the number of lamps used, as well as the distance of the exposed surface to the lamps, since irradiant intensity varies with the square of distance from the light source. Finally, irradiance can be increased by using highly reflective materials to enclose the exposure area.

Solar simulator

The solar simulator (0.61 m wide by 1.83 m long) we used contained ten 160-watt cool white lamps (General Electric Co., East Cleveland, OH), four 160-watt UVB313 lamps (National Biological Corp., Twinsburg, OH), eight 160-watt UVA365 lamps (National Biological Corp., Twinsburg, OH), two

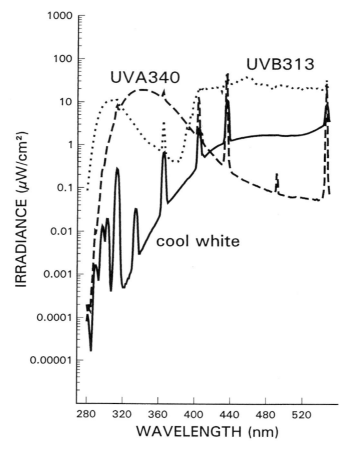

Figure 2   Semilog plot of spectral characteristics of ultraviolet and visible irradiance of cool-white, UVA-340, and UVB 313 fluorescent lamps.

20-watt cool white lamps (Osram Sylvania, Danvers, MA), two 20-watt SF20 sun lamps (Philips Lighting, Somerset, NJ), and eight 75-watt halogen incandescent flood lamps (General Electric Co., East Cleveland, OH). The simulator was suspended over a water bath of similar dimensions and was enclosed with reflective specular aluminum. The UVB lamps were controlled by a recycling 24-h timer for operating these lamps over a 5-h period during a simulated solar mid-day UVB dose. The cool white and UVA fluorescent lamps were controlled by a second timer to operate for a 16-h period, simulating a midsummer photoperiod. The UVB exposure was 5 h per day for 7 d. The testing area was air-conditioned because of the heat generated by the simulator, and the water bath was supplied with continuously flowing chilled water, or by recirculating water that was regulated by a thermostatically controlled chiller/heater. The water temperature was 18°C.

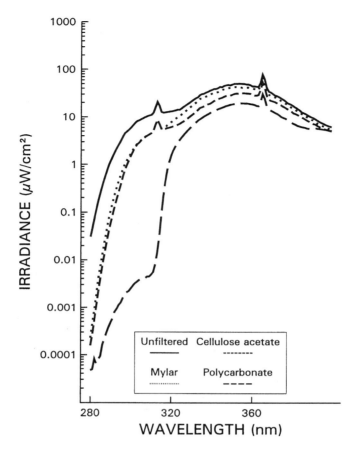

Figure 3 Semilog plot of ultraviolet spectral characteristics of unfiltered MSC solar simulation and with 0.39-mm cellulose acetate, 0.13-mm mylar B, 0.76-mm polycarbonate filters. Note convergence of intensity at 400 nm.

Calibration of the solar simulator

The intensity and spectral qualities of the solar simulator will change over time as the lamps age and as filtering materials undergo photolytic decomposition. Therefore it is important to measure the spectral output of the lamps at regular intervals (e.g., weekly) with a spectroradiometer. Frequent calibration of the solar simulator is required to ensure the spectral quality and intensity of exposure. Full calibrations, which entail measurements along a matrix of the water bath, should be performed as follows: (1) when the position of the light cap relative to the exposure vessels is changed, (2) when lamps are replaced, or (3) when filtering materials are changed. Calibration checks which include measurements at one or two locations within the water bath should also be conducted prior to, at some interval during, or at the conclusion of an exposure of aquatic organisms. A full calibration is conducted by measuring the spectral intensity of the simulator at 15.2-cm intervals along a matrix beneath the light cap.

Spectroradiometer

The Optronic Laboratories Model 752 Spectroradiometer is an analytical instrument which is used to measure the spectral qualities of radiation, and the intensity at which they occur, over a wavelength range of 250 to 800 nm, which includes wavebands for UVC, UVB, UVA, and visible radiation.

Brief description of operation and use of spectroradiometer

In our study, the output of the simulator was calibrated by measuring spectral irradiance at 15.2-cm intervals beneath the lamps using the Model 752 spectroradiometer equipped with a 15.2-cm diameter integrating sphere. The following summarizes our operation and use of the spectroradiometer from Optronic Laboratories manuals. The scanning range was 280 to 800 nm at 1-nm intervals, using a signal integration time of 1040 msec/nm. The spectroradiometer was calibrated with a U.S. National Institute of Standards and Technology traceable tungsten lamp over the range of 280 to 800 nm at 1-nm intervals. In addition, the alignment of the holographic gratings in the spectroradiometer was checked daily for wavelength accuracy at 302.2 nm using a fluorescent calibration lamp. The voltage output (gain) was also checked daily using a tungsten lamp at 550 nm, and the calibration file was adjusted for reductions or increases in the system response. Investigators should be aware that calibration files used during these measurements should always be based on the same instrumental configuration (i.e., integrating sphere size and type, lamp slit widths, PMT voltage, and operating temperature) as used for the measurement. For most measurements: (1) the spectroradiometer PMT voltage will be set at 550 volts, (2) the slit combination in the 752 optics head will be 0.25, 0.50, and 0.25 mm, and (3) the operating temperature will be 22° to 26°C. Frequent calibration checks of the 752 will be necessary if the operating conditions of the optics head increase by more than 4°C.

It is important to note that the scanning range should be 280 to 800 nm with step intervals of at least 3 nm, and preferably 1 nm steps. The system can be programmed for consecutive scans at each position to improve accuracy, and is advisable if less than the full matrix is to be measured. At each matrix point, intensity by wavelength measurements are made. The light cap is raised or lowered to increase or decrease the desired intensity. Lamps may also be turned off if the intensity is too great. In addition, filtering materials may be added or removed to achieve the target intensity. Final spectral measures will be recorded in a notebook as the sum of wavelength intensities for: (1) 280 to 320 nm, (2) from 320 to 400 nm, and (3) from 400 to 800 nm. The file for intensity by wavelength will be saved on a 3.5-inch disc in the spectroradiometer.

It may be necessary to assess filtering properties of test containers, filters, or safety glasses. This may be done by fitting the integrating sphere with the ring attachment and comparing values obtained with and without the material placed over the aperture. One might also wish to evaluate the actual experimental exposure obtained through the use of such materials, and in this case the ring is removed and a cap of the material is placed over the dome of the integrating sphere. This provides a 180° reading and measures reflected as well as direct irradiance. It is important to make the cap of material as complete as possible because gaps in the material will allow unfiltered radiation to enter the integrating sphere.

UV exposure procedures

Experimental treatments are the spectral intensities generated by the solar simulator through the filter materials applied, including the duration of spectral intensities under those conditions. The treatments include control conditions under which UVB intensities are held to a minimum. The control conditions are achieved through the use of top and bottom coverings of 0.76 mm polycarbonate, and a side wrap of 0.13 mm mylar. Cellulose acetate of varying thickness provides filtration of UVB. The filtering properties of cellulose acetate, at any thickness, are maximized by 72-h exposure to UVB. Cellulose acetate of 0.39 mm thickness and greater provides significant filtration of UVB, whereas a thickness of 0.13-mm cellulose acetate provides moderate filtration. Unaged 0.13-mm cellulose acetate provides slight filtration. High UVB treatments are direct, unfiltered exposures. Where possible, the exposure chambers can be fabricated with filtering materials such as quartz, polycarbonate, or polystyrene. A variety of commercially available bottles and boxes may be suitable for UV exposures as long as the UV filtering, and photolytic decomposition characteristics of the material are known. Filtering properties can be evaluated by comparing spectroradiographic measurements obtained when the dome of the integrating sphere is covered and uncovered with the material. Photolytic aging can be assessed by placing the materials under the highest exposure intensity for the exposure duration to be used during the test. Where photolytic aging is required to achieve optimal filtering characteristics, the material should be sufficiently aged before it is used in a study.

Treatments can be randomly distributed under the unfiltered light cap by wrapping the treatment containers with appropriate filtering materials. With randomized distributions there should be at least three replicate groups per treatment. Placement of each experimental chamber relative to the matrix should be carefully noted, so that results can be related to the unique spectroradiographic quality of each position.

The light cycle is set for a 16-h light: 8-h dark photoperiod. Five hours after the initiation of the 16-h light photoperiod the UVB lamps are illuminated for a 5-h period, then extinguished for the remainder of the 16-h photoperiod to simulate midday solar irradiance when UVB is most intense.

Exposures of freshwater fish

As an example of exposure of freshwater fish to UVB we refer to a prior study[22] and recent investigations in our laboratory. Exposures were conducted in open-top-glass $15 \times 15 \times 23$-cm airlift chambers[18] containing 2 liters of water of the same temperature and quality used during fish culture. The distance between the simulator and the water surface was 61 cm. Treatment chambers were covered with the appropriate filtering materials to generate three treatments: control, low, and high. The control conditions were created by covering the top and bottom of each chamber with 0.76-mm polycarbonate and then covering the sides with 0.13-mm mylar. A low UVB irradiance was generated by covering the top and bottom of each chamber with 0.13-mm cellulose acetate that had been aged by a 72-h exposure to UVB radiation to ensure stable UVB filtration. The cellulose acetate did not decrease the overall illumination by greater than 5%. The high UVB irradiance was a direct, unfiltered exposure. After measurement with the spectroradiometer, irradiance expressed as $\mu W/cm^2$ was converted to J(joules)/cm^2/d(day) by multiplying the number of μWatts by 3600 (seconds in an hour) to yield the number of $\mu J/cm^2/h$. This number is then multiplied by the number of hours of exposure to UVB and the decimal point moved appropriately to yield the dose in $J/cm^2/d$.

The fish used in these investigations were cultured at the Midwest Science Center, and tested at 60 to 75 d posthatch. Rainbow trout *Oncorhynchus mykiss* had a mean length [SD] of 4.0 cm [0.3] and mean weight of 0.53 g [0.1] at the time of testing. Lahontan cutthroat trout *Oncorhynchus clarki henshawi* had a mean length of 5.6 cm [0.3] and a mean weight of 1.3 g [0.4]. Apache trout *Oncorhynchus apache* had a mean length of 5.4 cm [0.9] and a mean weight of 1.4 g [0.7]. Razorback suckers *Xyrauchen texanus* had a mean length of 4.0 cm [0.1] and a mean weight of 0.58 g [0.1].

Investigators should note that stocking capacity of the chambers should follow customary practices for toxicological studies. Exposures can be conducted under static water conditions with daily renewal; in airlift chambers described in Cleveland et al.,[18] or in constant or intermittently-flowing water.[23] The studies should be conducted under the same temperature and water quality used during fish culture. However, if it is necessary to examine such variables in relation to UV exposure, then test organisms should be acclimated to the test water conditions.[24]

In our study, groups of five fish were stocked in each chamber for each of the control, low, and high treatments. Three replicate groups per species per treatment were distributed under the simulator and exposed for 7 d, and the entire experiment repeated. Prior to positioning of each chamber, a spectroradiographic scan over the range of 290 to 800 nm at 1-nm intervals was made at each treatment location using the filter material applied at that location. Each chamber was then placed in its position. The bubbling action of an air stone created a water flow of about 0.6 liters/min that entered a screened area at the bottom of one side of the chamber and drained at the top of the partially screened, partitioned corner of the chamber confining the air stone. Disturbance of the water surface by the bubbling action of the air stone was minimal.

Investigators should check the solar simulator daily for lamp function, photocycle intervals, water bath temperature, water flow, and drainage. Exposure chambers should be examined for appropriate water levels, flow, and drainage, and air stone bubble stream and screen blockage, if airlift chambers are used. The exposure chambers should be examined to ensure complete coverage by filtering materials. The UV filters should be examined for yellowing, curling, or cracking indicative of photolytic damage, and be replaced if necessary.

When feeding is necessary, it should be conducted at the onset of the white light photoperiod, or at the conclusion of the UV photoperiod. Food materials may diminish the optical clarity of the exposure chamber, and may absorb UV radiation; therefore it is necessary to minimize the amount of food materials present during the UV photoperiod and to maintain the highest optical clarity of the exposure conditions. The exposure chambers should be cleared of egg cases, food, feces, and dead organisms. In our study, fish were fed 24-h-old *Artemia* sp. and salmon starter several hours prior to and at the conclusion of the UV photoperiod each day; otherwise, the chambers were cleared of food and debris to maintain the highest clarity.

In addition to sunburn, investigators can also examine the test organisms daily for mortality, hatching, lesions, hematomas, abnormal swellings or morphological changes, fungal infections, and abnormal behavior. The time of day, water temperature, optics head operating temperature, matrix position of spectroradiometric scan, scan file name, and total UVB and UVA irradiance measured during the scan should also be recorded. At selected intervals of exposure, organisms may be temporarily removed and photographed or videotaped for an assessment of condition, morphological development, behavioral function, and growth or removed to assess biochemical or physiological responses.

Statistics

In our study, the effect of treatment level and species on the response of fish to UVB was evaluated using analysis of variance techniques performed with the Statistical Analysis System.[25] Cumulative percentages of sunburn were arcsine transformed and analyzed using a randomized complete block model to account for effects between experimental trials.[26] Mean values were compared using Fisher's protected LSD test ($p \leq 0.05$).

Results and discussion

Spectral characteristics of solar simulation

We observed the following spectral characteristics for different treatments when freshwater fish were exposed to UVB. The control conditions provided a minimal average UVB irradiance of 4.3 ± 0.3 µW/cm² (0.08 J/cm²/d), the low UVB irradiance 190 ± 17 µW/cm² (3.42 J/cm²/d), and the high UVB irradiance 357 ± 7 µW/cm² (6.43 J/cm²/d).

Ozone depletion could increase UVB irradiance in the environment of these species. The total UVB radiation applied in our study was moderate relative to ambient midsummer solar irradiance, implying that increased irradiance would be harmful. The lowest UVB irradiance generated by the

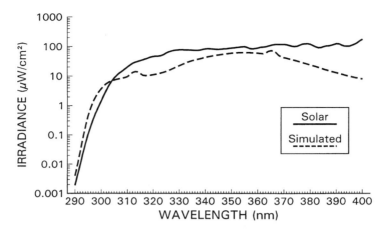

Figure 4   Semilog plot of ultraviolet spectral characteristics of MSC solar simulator and natural sunlight observed at solar noon on June 21, 1993 at 38.5°N latitude. (From Little, E.E., Fabacher, D.L. 1994. *Arch. Hydrobiol.* 43:217–226.)

solar simulator during our study was 190 µW/cm², which compares with a summertime solar maximum of 427 µW/cm² measured on June 21, 1993, at an altitude of 271 m and a latitude of 38.5°N for Columbia MO (Figure 4). In order to provide an indication of the wavelength composition and sunburning potential of the simulated solar UV irradiance, we compared the weighted irradiance values for simulated solar and solar UV using the Diffey action spectra[27] for human erythema (sunburn). Since shorter wavelengths induce erythema more readily than longer wavelengths, this weighting provides a sunburning dose relative to the spectral wavelength composition of the exposure. The Diffey erythemal weighted irradiance reflects the total of the weighted irradiances for all wavelengths necessary to produce erythema in humans. The Diffey erythemal weighted irradiance was 3.256×10^{-5} W/cm² for sunlight and 3.021×10^{-5} W/cm² for the lowest simulated solar irradiance. Thus, the sunburning potential of the simulated solar irradiance was similar to that of ambient solar irradiance. In addition, the durations of exposure were comparable to what might be expected during a normal solar photoperiod, and the levels of UVA radiation applied during the 16-h photoperiod were comparable to natural solar levels.

Simulated solar UVA levels are important to ensure sufficient photoreactivating radiation to induce cellular repair mechanisms.[28] The total visible radiation at approximately 40% of ambient solar was sufficient for the activation of photorepair mechanisms.[29]

Biological effects of exposures

Sunburn was the first effect of simulated solar UVB radiation we observed in fish.[22] The presence of sunburn was a qualitative observation which first appeared as a characteristic darkening on the dorsal surface of a fish, and usually occurred between the head and dorsal fin and ventrally to about the

Table 1   Percent Sunburn Among Juvenile Lahontan Cutthroat
Trout During a 7-Day Exposure to Ultraviolet-B Radiation

Treatment	Day of exposure						
	1	2	3	4	5	6	7
Control	0	0	0	0	0	0	6.6
	—	—	—	—	—	—	[5.1]
Low	0	16.7	53.3[a]	73.3[a]	80.0[a]	86.6[a]	90.0[a]
	—	[7.4]	[14.8]	[11.0]	[8.0]	[5.1]	[5.5]
High	0	6.7	30.0[a]	50.0[a]	70.0[a]	83.3[a]	96.7[a]
	—	[7.4]	[16.0]	[9.5]	[7.5]	[3.7]	[3.7]

Note: Data are given as mean with [standard error].
[a] Indicates treatment mean is statistically different from control.
From Little, E.E., Fabacher, D.L. 1994. *Arch. Hydrobiol.* 43:217–226.

lateral lines. After day 2 the incidence of sunburn increased significantly for Lahontan cutthroat trout in the low and high treatment groups and was apparent in 93% of the treated fish by the end of the 7-d exposure (Table 1). The onset and extent of response did not differ among fish in the low and high treatment groups. The low and high exposures caused sunburn on the dorsal skin of rainbow trout (Table 2) similar to that described for the Lahontan cutthroat trout. By day 7 of exposure 90% of the UVB-exposed rainbow trout in the low and high treatments were sunburned. In contrast to the other two species of salmonids, only 10% of UVB-exposed Apache trout in the high treatment developed sunburn (Table 3). In recently completed investigations in our laboratory, razorback suckers did not develop sunburn after 21 days of exposure to all three doses of simulated UVB, including the high dose of 6.43 J/cm²/d. Lahontan cutthroat trout developed fungal infection and began to die shortly after the appearance of sunburn. Rainbow trout developed similar patterns of pigmentation and fungal infection but survived the exposure. Apache trout developed sunburn within 5 days and did not exhibit any

Table 2   Percent Sunburn Among Juvenile Rainbow Trout
During a 7-Day Exposure to Ultraviolet-B Radiation

Treatment	Day of exposure						
	1	2	3	4	5	6	7
Control	0	0	0	0	0	0	0
	—	—	—	—	—	—	—
Low	0	3.3	26.6[a]	3.3[a]	53.3[a]	83.3[a]	93.3[a]
	—	[6.0]	[11.1]	[25.2]	[11.4]	[3.7]	[4.7]
High	0	6.6	33.3[a]	40.0[a]	66.6[a]	73.3[a]	86.7[a]
	—	[7.4]	[14.6]	[11.5]	[4.6]	[7.4]	[7.4]

Note: Data are given as mean with [standard error].
[a] Indicates treatment mean is statistically different from control.
From Little, E.E., Fabacher, D.L. 1994. *Arch. Hydrobiol.* 43:217–226.

Table 3 Percent Sunburn Among Juvenile
Apache Trout During a 7-Day Exposure to
Ultraviolet-B Radiation

Treatment	Day of exposure						
	1	2	3	4	5	6	7
Control	0	0	0	0	0	0	0
Low	0	0	0	0	0	0	0
High	0	0	0	0	3.3	3.3	10.0
					[3.6]	[3.6]	[7.5]

Note: Data are given as mean with [standard error].
From Little, E.E., Fabacher, D.L. 1994. *Arch. Hydrobiol.*
43:217–226.

subsequent fungal infection or mortality.The increased susceptibility of sun-burned fishes to fungal infection and consequent mortality could result from UVB injury to the skin, as well as to suppression of the immune system.[8]

The effects we observed in UVB-exposed fish are consistent with observations of other investigators. Bell and Hoar[30] reported a casual observation of changes in pigmentation and fungal infection in irradiated coho salmon *Oncorhynchus kisutch* fry and goldfish *Carassius auratus*. Dunbar[31] observed sunburn in rainbow trout fingerlings exposed to direct sunlight in outdoor pools and to a sunlamp in the laboratory. Fish from both exposures developed identical necrotic areas around the dorsal fin and behind the head and experienced greater mortality than unexposed fish. Bullock and Roberts[32] described skin lesions in rainbow trout exposed to direct sunlight in outdoor tanks. Similar skin damage was observed among rainbow trout exposed to solar ultraviolet radiation while held on a high-altitude fish farm in Bolivia.[33]

McArdle and Bullock[34] reported widespread dispersion of melanosomes in the dermis of Atlantic salmon *Salmo salar* that were exposed to direct solar ultraviolet radiation. Sunburn (darkening) of the dorsal areas on UVB exposed fishes in our study probably resulted from melanosome dispersion.

The differences we observed in the susceptibility of Apache trout, Lahontan cutthroat trout, rainbow trout, and razorback suckers to UVB-induced sunburn may be explained by the results of recent investigations in our laboratory.[35] When methanol extracts of surgically removed sections of dorsal skin from the four species of fish were scanned in a Beckman 5230 UV/vis recording spectrophotometer, we observed an unknown skin component with an absorption maximum around 292 nm (Figure 5). The shoulders of this peak extended over the entire UVB wavelength range, probably affording protection to the skin by absorbing any radiation in that wavelength range that would induce sunburn. We calculated semiquantitative estimates of the amount of the component using the formula $\frac{1}{2}$ baseline × height of the 292 nm peak and found significant differences in the amount of the component among the four species (Table 4).

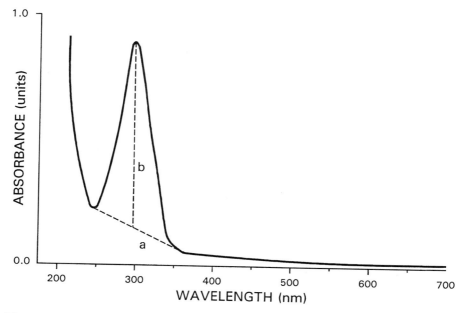

Figure 5 Representation of UV-visible light absorption spectrum of dorsal skin methanol extract from a razorback sucker. Dotted lines are baseline (a) and height (b) of peak. (From Fabacher, D.L., Little, E.E. 1995. *Environ. Sci. Pollut. Res.* 2:30–32. With permission.)

There appeared to be a direct relation between the number of days it took for each species to develop UVB-induced sunburn and the amount of this component in methanol extracts of dorsal skin. We are currently refining this technique and are in the process of identifying this component. Bullock[36] also observed considerable variability in the response of individual rainbow trout to intense simulated solar radiation and suggested that the ability of some fish to tolerate the radiation may result from elevated

Table 4 Amount of Dorsal Skin Component in Methanol Extracts from Unexposed Fish and Number of Days to Develop Sunburn in Fish Exposed to Low UVB Irradiance

Species	Amount of component[a]	Days to sunburn
Rainbow trout	23.9 [2.8]A[b]	2
Lahontan cutthroat trout	23.9 [1.2]A	2
Apache trout	49.6 [7.8]B	>7
Razorback sucker	101.2 [3.8]C	>21

[a] Values are mean [SE] area units/milligram wet weight of tissue for five fish of each species.

[b] Means with the same letter are not significantly different as determined by *t*-test and Duncan's multiple range test.

From Fabacher, D.L., Little, E.E. 1995. *Environ. Sci. Pollut. Res.* 2:30–32. With permission.

levels of a photoprotective factor of genetic origin. The skin component we observed may function as the photoprotective factor suggested by Bullock.[36]

The differences in the susceptibility to UVB-induced sunburn and the corresponding differences in the amount of this apparent photoprotective skin component among the four species of fish probably resulted from an adaptive response to different intensities of solar radiation. The effects of UVB radiation on natural populations of these salmonids depends on the clarity of their habitat, tropospheric attenuation of UVB irradiance, and behavioral habits of the fish. The salmonids occur in high-altitude, clearwater environments where solar UVB may increase to maximum levels daily, depending on the amount of substrate and canopy cover. Early life stages of these species develop within gravel redds; however, the emergence of free-swimming, exogenously feeding juveniles often occurs in spring during maximum solar UVB irradiance. Razorback suckers occur in the Colorado River basin, where the fish may intermittently inhabit clear water that lacks substrate and canopy cover and is between steep canyon walls; daily exposure to solar UVB may be sudden and intense.

In summary, we have described a solar simulator, spectroradiometer, and exposure of fish to simulated UVB. This methodology is appropriate for the laboratory screening of species of fish to determine UVB sensitive and insensitive species. Natural populations of these fish can then be ranked as to their risk of susceptibility to the adverse effects of enhanced solar UVB resulting from ozone depletion.

References

1. Smith, R.C., Prezelin, B.B., Baker, K.S., Bidigare, R.R., Boucher, N.P., Coley, T., Karentz, D., MacIntyre, S., Matlick, H.A., Menzies, D., Ondrusek, M., Wan, Z., Waters, K.J. 1992. Ozone depletion: ultraviolet radiation and phytoplankton biology in antarctic waters. *Science* 255:952–959.
2. Rowland, S. 1991. Stratospheric ozone in the 21st century: the chlorofluorocarbon problem. *Environ. Sci. Technol.* 25:622–628.
3. Kerr, J.B., McElroy, C.T. 1993. Evidence for large upward trends of ultraviolet-B radiation linked to ozone depletion. *Science* 262:1031–1034.
4. van der Leun, J.C., de Gruijl, F.R. 1993. Influence of ozone depletion on human and animal health. In *UV-B radiation and ozone depletion: effects on humans, animals, plants, microorganisms, and materials*, Ed. M. Tevini, 95–123. Boca Raton, Florida: Lewis Publishers. 248 pp.
5. Ahmed, F.E., Setlow, R.B. 1993. Ultraviolet radiation-induced DNA damage and its photorepair in the skin of the platyfish *Xiphophorus*. *Cancer Res.* 53:2249–2255.
6. Applegate, L.A., Ley, R.D. 1988. Ultraviolet radiation-induced lethality and repair of pyrimidine dimers in fish embryos. *Mut. Res.* 198:85–92.
7. Bullock, A. M. 1982. The pathological effects of ultraviolet radiation on the epidermis of teleost fish with reference to the solar radiation effect in higher animals. *Proc. R. Soc. Edinb.* 81B, 199–210.

8. Fabacher, D.L., Little, E.E., Jones, S.B., De Fabo, E.C., Webber, L.J. 1994. Ultraviolet-B radiation and the immune response of rainbow trout. In *Modulators of fish immune responses: models for environmental toxicology, biomarkers, immunostimulators,* Eds. J. Stolen, T.C. Fletcher, 1:205–217. Fairhaven, New Jersey: SOS Publications. 118 pp.

9. Fisher, M.S., Kripke, M.L. 1982. Suppressor T lymphocytes control the development of primary skin cancers in ultraviolet-irradiated mice. *Science* 216:1133–1134.

10. De Fabo, E.C., Noonan, F.P., Frederick, J.E. 1990. Biologically effective doses of sunlight for immune suppression at various latitudes and their relationship to changes in stratospheric ozone. *Photochem. Photobiol.* 52:811–817.

11. Cullen, J.J., Neale, P.J. 1993. Quantifying the effects of ultraviolet radiation on aquatic photosynthesis. In *Photosynthetic responses to the environment,* Eds. H. Yamamoto, C. Smith, 45–61. Rockville, Maryland: American Society of Plant Physiologists. 252 pp.

12. Teramura, A.H., Sullivan, J.H. 1993. Effects of UVB on plant productivity. In *Photosynthetic responses to the environment,* Eds. H. Yamamoto, C. Smith, 37–45. American Society of Plant Pathologists. Rockville, Maryland: American Society of Plant Physiologists. 252 pp.

13. Hessler, D. P., Gorenflo, V., Frimmel, F.H. 1993. UV degradation of atrazine and metazachlor in the absence and presence of hydrogen peroxide, bicarbonate and humic substances. *Aqua* 42:8–12.

14. Metcalf, R.L. 1968. The role of oxidative reactions on the mode of action of insecticides. In *Enzymatic oxidation of toxicants,* Ed. E. Hodgson, 151–174. Raleigh, North Carolina: North Carolina State University. 299 pp.

15. Brooks, G.T. 1980. Perspectives of the chemical fate and toxicity of pesticides. *J. Environ. Sci. Health* B15:755–793.

16. Oris, J.T., Geisy, J.P. 1987. The photoinduced toxicity of polycyclic aromatic hydrocarbons to larvae of the fathead minnow (*Pimephales promelas*). *Chemosphere* 16:1396–1404.

17. Madronich, S., Bjorn, L.O., Ilyas, M., Caldwell, M.M. 1991. Changes in biologically active ultraviolet radiation reaching the earth's surface. *UNEP Environmental Effects Panel Report: 1991 Update,* United Nations Environment Programme. ISBN 92 807 1309 4.

18. Cleveland, L., Little, E.E., Ingersoll, C.G., Wiedmeyer, R.H., Hunn, J.B. 1991. Sensitivity of brook trout to low pH, low calcium and elevated aluminum concentrations during laboratory pulse exposures. *Aquat. Toxicol.* 19:303–318.

19. Caldwell, M.M., Gold, W.G., Harris, G., Ashurst, C.W. 1983. A modulated lamp system for solar UV-B (280–320 nm) supplementation studies in the field *Photochem. Photobiol.* 37:479–485.

20. Bjorn, L.O., Teramura, A.H. 1993. Simulation of daylight ultraviolet radiation and effects of ozone depletion. In *Environmental UV photobiology,* Ed. A.R. Young, L.O. Bjorn, J. Moan, W. Nultsch, 41–70. New York: Plenum Press. 479 pp.

21. Blaustein, A.R., Hoffman, P.D., Hokit, D.G., Kiesecker, J.M., Walls, S.C. Hays, J.B. 1994. UV repair and resistance to solar UV-B in amphibian eggs: a link to population declines? *Proc. Natl. Acad. Sci.* 91:1791–1795.

22. Little, E.E., Fabacher, D.L. 1994. Comparative sensitivity of rainbow trout and two endangered salmonids, Apache trout and Lahontan cutthroat trout, to ultraviolet-B radiation. *Arch. Hydrobiol.* 43:217–226.

23. ASTM. 1994. Standard guide for conducting acute toxicity tests with fishes, macroinvertebrates, and amphibians. In *Annual Book of ASTM Standards. Pesticides; Resource Recovery; Hazardous Substances and Oil Spill Responses; Waste Management; Biological Effects.* 11.04: 480–499. Philadelphia, Pennsylvania: American Society for Testing and Materials. 1786 pp.

24. ASTM. 1994. Standard guide for conducting early lifestage toxicity tests with fishes. In *Annual Book of ASTM Standards. Pesticides; Resource Recovery; Hazardous Substances and Oil Spill Responses; Waste Management; Biological Effects.* 11.04: 968–995. Philadelphia, Pennsylvania: American Society for Testing and Materials. 1786 pp.

25. SAS. 1989. *SAS/STAT User's Guide.* Version 6, 4th edition, 2 volumes. Cary, North Carolina: SAS Institute, Inc. 1686 pp.

26. Snedecor, G.W., Cochran, W.G. 1980. *Statistical methods.* 7th edition. Ames, Iowa: Iowa State University Press. 507 pp.

27. McKinlay, A.F., Diffey, B.L. 1987. A reference action spectrum for ultra-violet induced erythema in human skin. In *Human exposure to ultraviolet radiation: risks and regulations,* Eds. W.R. Passchler, B.F.M. Bosnajokovic, 83–86, Elsevier, Amsterdam. 580 pp.

28. Mitchell, D.L., Scoggins, J.T., Morizot, D.C. 1993. DNA repair in the variable platyfish (*Xiphophorus variatus*) irradiated *in vivo* with ultraviolet B light. *Photochem. Photobiol.* 58:455–459.

29. Funayama, T., Mitani, H., Shima, A. 1993. Ultraviolet-induced DNA damage and its photorepair in tail fin cells of the medaka, *Oryzias latipes. Photochem. Photobiol.* 58:380–385.

30. Bell, M.G., Hoar, W.S. 1950. Some effects of ultraviolet radiation on sockeye salmon eggs and alevins. *Can. J. Res.* D28H:35–43.

31. Dunbar, C.E. 1959. Sunburn in fingerling rainbow trout. *Prog. Fish. Culturist* 21:74.

32. Bullock, A.M., Roberts, R.J. 1981. Sunburn lesions in salmonid fry: a clinical and histological report. *J. Fish Dis.* 4:271–275.

33. Bullock, A.M., Coutts, R.R. 1985. The impact of solar ultraviolet radiation upon the skin of rainbow trout, *Salmo gairdneri* Richardson, farmed at high altitude in Bolivia. *J. Fish Dis.* 8:263–272.

34. McArdle, J., Bullock, A.M. 1987. Solar ultraviolet radiation as a causal factor of "summer syndrome" in cage-reared Atlantic salmon, *Salmo salar* L.: a clinical and histopathological study. *J. Fish Dis.* 10:255–264.

35. Fabacher, D.L., Little, E.E. 1995. Skin component may protect fishes from ultraviolet-B radiation. *Environ. Sci. Pollut. Res.* 2:30–32.

36. Bullock, A. M. 1988. Solar ultraviolet radiation: a potential environmental hazard in the cultivation of farmed finfish. In *Recent advances in aquaculture,* Eds. J. E. Muir, R. J. Roberts, 3:139–224. Beckenham, Kent: Croom Helm. 420 pp.

Section II

Techniques for measurement of cellular and subcellular toxicity

chapter nine

Gene expression analysis in aquatic animals using differential display polymerase chain reaction

Linda D. Rhodes and Rebecca J. Van Beneden

Introduction

Toxic effects frequently result in changes in gene expression, and these changes are often produced by alterations in the steady state levels of messenger RNA (mRNA) for those genes. The traditional way to analyze these changes has been to use techniques such as subtractive hybridization and differential hybridization to isolate mRNA that is uniquely expressed or suppressed in treated or diseased cells. Major disadvantages of these methods are the requirement for large amounts of mRNA (e.g., 2 mg), substantial time involved in constructing and screening complementary DNA (cDNA) libraries, and the ability to compare only two populations at a time. To circumvent these drawbacks, alternative methods based on arbitrarily primed polymerase chain reaction (PCR) of total RNA have been developed, such as differential display polymerase chain reaction (ddPCR).[1] Figure 1 schematically displays an overview of this technique. First, cDNA is generated from polyadenylated RNA using a poly-T primer that is anchored at the 3′ end. The cDNA is then used as a template for radiolabeled PCR, and the amplified products (amplicons) are resolved on a polyacrylamide gel. The gel is exposed to X-ray film, and the amplicons are viewed as autoradiographic bands. If tissue or cell populations are to be compared, the amplicons from all of the comparison groups for a given pair of primers are electrophoresed on the same gel. Genes that are expressed in all of the comparison groups are expected to produce autoradiographic bands that are present in all of the reactions. In contrast, uniquely expressed genes are expected to generate bands that are present exclusively in the tissue or cell populations that express

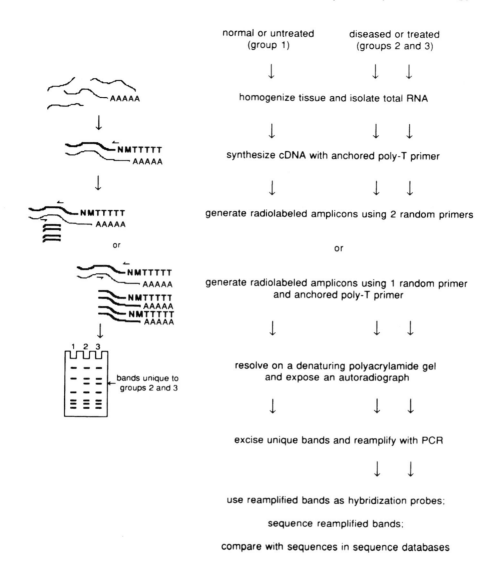

Figure 1 A schematic overview of the differential display procedure. RNA is shown as thin lines, and the polyadenylated mRNA from the total RNA is shown by a thin line adjacent to "AAAAA." DNA is shown by thick lines, and the RNA–cDNA heteroduplex is represented by both thick and thin lines. Primers are shown as small, single sided arrows. The anchored poly-T primer is indicated by "TTTTTMN" (5' → 3'), where "M" is G, A, or C and "N" is any of the four nucleotides (i.e., G, A, T, or C). The DNA amplicons are represented by pairs of thick lines.

the gene. These unique autoradiographic bands can be isolated for subsequent sequencing, cloning, and use as hybridization probes.

The application of ddPCR to a wide variety of studies has been well demonstrated since its introduction. Genes that are developmentally regulated have been identified in the early stages of murine brain development,[2] and during preimplantation embryogenesis in the mouse.[3] In the latter case, the sensitivity of ddPCR greatly facilitated the study, since the amount of biological material available at specific stages of early embryogenesis can be very small. Differential display PCR has also been applied to identify inducible genes in model systems of clinical pathology[4,5] as well as biopsy tissues collected from patients.[6] Furthermore, target genes for inducing agents such as growth factors[7,8] or ionizing radiation[9] have also been found by ddPCR. Perhaps the best known application of ddPCR has been in the investigation of genes associated with neoplasms.[6,9-11] The earliest identification of a tumor-related gene by ddPCR was of the α integrin gene as a possible tumor suppressor.[10]

Our laboratory has been using ddPCR in studies of the responses of aquatic organisms to toxicants. We have been particularly interested in the occurrence of gonadal neoplasms in feral populations of hard-shell clam (*Mercenaria* spp.) and soft-shell clam (*Mya arenaria*) and the possible molecular mechanisms of pollution response in these bivalves. For example, using affinity binding studies, we have demonstrated a cellular mechanism in *Mercenaria mercenaria* for binding halogenated aromatic hydrocarbons such as dioxin.[12] The ddPCR technique has been used both in our field surveys and controlled laboratory experiments. We have been screening RNA from *M. arenaria* for differences between normal gonad and gonadal neoplasms, and we have developed a protocol that works well for bivalves and that is similar to the protocol used for mammalian tissues. This protocol is organized into five sections:

1. Isolation of total RNA;
2. Treatment of total RNA with DNase I to remove contaminating DNA;
3. Production of cDNA by reverse transcription;
4. Generation and resolution of radiolabeled amplicons by PCR and electrophoresis; and
5. Isolation and reamplification of selected amplicons.

Preceding each protocol, there are comments about the purpose of the procedure, possible alternatives, and any caveats about "tricky" parts of the protocol.

Materials required

Chemicals can be purchased from any reputable supplier, such as Sigma or Gibco/BRL/Life Technologies. Chemicals from specific suppliers that have worked well in our laboratory are noted by the accompanying catalog number. Unless otherwise indicated to be a working solution, all of the supplies listed are stock reagents (e.g., 10×), which may be diluted in the working reagent.

Diethylpyrocarbonate is abbreviated as "depc."

Total RNA isolation

Equipment required:
- Tissue homogenizer, if preparations are from tissue samples

Reagents required:
- 4 M Acid guanidinium stock: 4 M Guanidinium isothiocyanate + 25 mM sodium citrate, pH 7.0
- 4 M Acid guanidinium working solution: add β-mercaptoethanol at a ratio of 1:140 (70 μl per 10 ml) just before use
- 10% Sarkosyl
- 2 M Sodium acetate, pH 4.1
- Water-saturated phenol, pH <5 (use depc-treated water to saturate the phenol crystals, Gibco/BRL # 15509-011)
- Chloroform
- Absolute isopropyl alcohol
- 80% Ethanol

DNase I treatment

Equipment required:
- 37°C Water bath or a programmable thermocycler

Reagents required:
- DNase I , 1 unit/μl (must be RNase-free; Boehringer Mannheim's #776785)
- 10× DNase buffer: 100 mM Tris · Cl, pH 8.3
                      500 mM KCl
                      15 mM $MgCl_2$
- RNase inhibitor, 10 units/μl (BRL # 15518-012)
- Water-saturated phenol, pH <5
- Chloroform
- 3 M Sodium acetate, pH 5.2
- Absolute ethanol
- 80% Ethanol

Reverse transcriptase reaction

Equipment required:
- 80°C Water bath or a programmable thermocycler
- 37°C Water bath or a programmable thermocycler

Reagents required:
- 5× Buffer (ready to use buffer often comes with the reverse transcriptase):   250 mM Tris · HCl, pH 8.3
                          375 mM KCl
                          15 mM $MgCl_2$
- Superscript II reverse transcriptase (BRL # 18064-014)
- 100 mM Dithiothreitol (DTT)
- 10 mM Deoxyribonucleotides (dNTPs)

- 10 µM Anchored poly-T primer
- depc-treated H$_2$O

Radiolabeling PCR reaction

Equipment required:
- Programmable thermocycler
- Standard sequencing electrophoresis apparatus (e.g., BRL model S2)
- High wattage power supply (e.g., Pharmacia ECPS 3000/150)

Reagents required:
- 10× PCR buffer (usually comes with the thermostable polymerase)
- 25 mM MgCl$_2$ (if there is none in the PCR buffer)
- AmpliTaq DNA polymerase (Perkin Elmer #N801-0060)
- 10 µM anchored poly-T primer
- 10 µM 10-mer (10 bp primer):
- 100 µM dGTP
- 100 µM dTTP
- 100 µM dCTP
- α^{35}S-dATP (Amersham # SJ1304, >1000 Ci/mmole)
- depc-treated H$_2$O

Isolation and reamplification

Equipment required:
- Programmable thermocycler
- Standard horizontal agarose gel electrophoresis apparatus

Reagents required:
- 10 mg/ml Oyster glycogen (e.g., Sigma # G8751)
- 3M Sodium acetate, pH 5.2
- Absolute ethanol
- 10× PCR buffer (usually comes with the thermostable polymerase)
- 25 mM MgCl$_2$ (if there is none in the PCR buffer)
- AmpliTaq DNA polymerase (Perkin Elmer #N801-0060)
- 10 µM poly-T primer
- 10 µM 10-mer (10 bp primer):
- 1 mM dGTP
- 1 mM dTTP
- 1 mM dATP
- 1 mM dCTP

Procedures and protocols

Figure 1 is a schematic overview of the procedures to be described here, as well as possible applications for the products to be generated by the differential display. The specific protocols to be described are total RNA isolation, DNase I treatment, reverse transcriptase reaction, radiolabeling PCR reaction, and isolation and reamplification of selected bands.

Total RNA isolation

Isolation of total RNA can be done by a variety of techniques, but the popular and relatively quick method of Chomczynski and Sacchi[13] can be successfully used for a variety of tissue types from a wide range of organisms. Briefly, cells are lysed with guanidinium isothiocyanate, which also inactivates RNA-degrading enzymes, or RNases. Extraction by acid phenol-chloroform partitions RNA from DNA and proteins. Precipitation of RNA with isopropyl alcohol is followed by a second round of dissolving in guanidinium isothiocyanate, extraction with acid phenol-chloroform, and precipitation of RNA with isopropyl alcohol. The resulting total RNA pellet is suitable for ddPCR as well as Northern hybridization and RNase protection analysis.

The tissues of some organisms, such as molluscs, have high glycoprotein or mucopolysaccharide content. While these compounds do not appear to interfere with either the reverse transcription (RT) or PCR reaction, they may prevent efficient isolation of RNA away from DNA or RNases. Groppe and Morse[14] recommend cold (4°C) guanidinium isothiocyanate for the initial lysis of tissues as an effective way of preventing mucopolysaccharide contamination. It is important to use cold (4°C), but not ice cold (0°C) guanidinium isothiocyanate, since the guanidinium isothiocyanate will precipitate out of the 4 M solution at the lower temperature. Other investigators have used progressive ethanol precipitations at 0°C to differentially precipitate RNA away from glycogen and other polysaccharides. Typically, a precipitation in a 60 to 66% ethanol solution can precipitate RNA with minimal carbohydrate contamination, but the yield of RNA is also reduced. Higher percentages (e.g., 80%) of ethanol will improve the yield of RNA, but will also precipitate carbohydrates (A. Marsh, personal communication). Alternatively, enzymes such as amyloglycosidase (Sigma # A7420) can be used to digest carbohydrates. In our laboratory, enzymatic treatment successfully "cleaned up" polyadenylated bivalve RNA that was subsequently used to synthesize a cDNA library (T. Puryear, personal communication).

A one-step cesium chloride gradient can be used to further purify total RNA, but we have found that glycoproteins and very small amounts of DNA will sediment with the RNA pellet. Furthermore, the cesium chloride gradient will remove very small RNAs, such as transfer RNA, and may also remove some of the smaller mRNAs. The expense and time involved in this procedure is only justified by its extremely efficient removal of RNases and other proteins, but it is certainly not necessary for ddPCR.

The important factor to keep in mind about total RNA isolation is minimizing RNase contamination. If the tissue is not processed immediately, it should be quickly frozen in liquid nitrogen and stored at ≤–70°C. We have recovered good quality RNA from specimens stored for 3 years in this manner. Prevention of RNase contamination during processing is facilitated by keeping tissues cold during homogenization, using RNase-free equipment and solutions, and storing RNA samples at –70°C, preferably in the presence of RNase inhibitor or under an ethanolic solution. Treatment of solutions and water with diethylpyrocarbonate (depc) to inactivate RNases is described in section 7 of Sambrook et al.[15] The following protocol is one we have successfully used to isolate total RNA from bivalve tissues.

Total RNA isolation protocol

The 4 M acid guanidinium working solution should be at 4°C.

The 2 M sodium acetate should be on ice.

The tip of the homogenizer should be chilled by placing it in a clean tube on ice.

The tissue sample should be fresh or quick frozen in liquid nitrogen and stored at –70°C.

- Drop 0.2 to 0.5 g of tissue into 3.8 ml of acid guanidinium working solution in a polypropylene tube (Falcon 2059). A smaller scale version can be done for <0.2 g of tissue using a smaller polypropylene tube (e.g., Falcon 2063) and 0.5 to 1.0 ml of guanidinium working solution.
- While keeping the tube on ice, homogenize tissue at low to moderate speed for 20 to 30 seconds. Quickly inspect the bottom of the tube for evidence of intact pieces of tissue. Repeat homogenization until there are no tissue pieces visible in the bottom of the tube. Do not allow the tube to warm up by homogenizing at room temperature for long periods (e.g., >1 minute).
- Extract once with acid phenol:chloroform.
    Add 200 µl of 10% Sarkosyl, vortex, and store on ice until all samples have been homogenized.
    Add 400 µl of cold 2 M sodium acetate and vortex.
    Add 4 ml of acid phenol and mix by inversion briefly.
    Add 800 µl of chloroform.
    Shake vigorously for 1 minute, and store on ice until all samples have been extracted.
    Centrifuge at approximately 10,000 g at 4°C for 20 minutes.
    Transfer the aqueous layer to a new polypropylene tube.
- Precipitate the RNA.
    Add 1 volume (about 4 ml) of ice cold isopropyl alcohol and mix by inversion.
    Precipitate at least 1 h at –20°C.
    Centrifuge at approximately 10,000 g at 4°C for 20 minutes.
    Wash the pellet 2 times with 80% ethanol.
- Resuspend the pellet in 475 µl of freshly prepared 4 M acid guanidinium working solution.
- Transfer to a microfuge tube.
- Add 25 µl of 10% Sarkosyl and 50 µl of 2 M sodium acetate and vortex.
- Extract once with acid phenol:chloroform.
    Add 500 µl of acid phenol and vortex.
    Add 100 µl of chloroform and shake vigorously for 1 minute.
    Store on ice for 15 minutes.
    Centrifuge at approximately 10,000 g at 4°C for 20 minutes.
    Transfer the aqueous (upper) layer to a fresh microfuge tube.

- Precipitate the RNA.
    Add 500 µl of ice cold isopropyl alcohol and vortex.
    Precipitate at least 1 h at –20°C.
    Centrifuge at approximately 10,000 g at 4°C for 20 minutes.
- Wash the pellet twice with 80% ethanol and air dry until all odor of ethanol is gone, usually about 10 minutes.
- Resuspend the pellet in 50 to 100 µl of depc-treated water, heating to 65°C for up to 15 minutes, if necessary.
- Measure absorbance at 260 nm to estimate the concentration of nucleic acid.
- It is also a good idea to run out 1 to 2 µg on a nondenaturing agarose gel to check the integrity of the RNA.

DNase I treatment

Regardless of the method of RNA preparation, DNA is still likely to be present. Because PCR can detect very small amounts of DNA, less than 1 ng of DNA can serve as a template for amplification. Furthermore, the low annealing temperature used in ddPCR (40°C) increases the chance of amplifying from any DNA present in the preparation. PCR products amplified from contaminating DNA, rather than from the RNA-cDNA hybrid, are false bands.

DNase I treatment protocol

- In a 0.5 ml microfuge tube, combine 25 µg of RNA, 10 µl of 10× DNase buffer, 10 µl (10 units) of DNase I, 10 units of RNase Inhibitor, and depc-treated H$_2$O to a total volume of 100 µl.
- Incubate 30 minutes at 37°C.
- Extract once with phenol-chloroform (this removes the DNase I enzyme):
    Add 100 µl of water-saturated phenol:chloroform (3:1; 75 µl of phenol + 25 µl of chloroform).
    Shake vigorously for 1 minute.
    Centrifuge 10 minutes at 14,000 rpm (16,000 g) at room temperature.
    Transfer the aqueous (upper) layer to a fresh tube.
- Extract once with chloroform (this removes any residual phenol, which will inhibit subsequent enzymatic reactions):
    Add 100 µl of chloroform.
    Shake vigorously for 1 minute.
    Centrifuge 5 minutes at 14,000 rpm at room temperature.
    Transfer the aqueous (upper) layer to a fresh tube.
- Precipitate the RNA:
    Add 0.1 volume (~10 µl) of 3 M sodium acetate, pH 5.2, and 3 volumes (~330 µl) of cold absolute ethanol.
    Place at –20°C for at least 30 minutes.
    Centrifuge at least 45 minutes at 14,000 rpm at 4°C.
    Carefully remove the supernatant with a pipette (you can leave a little liquid in the bottom).
    Add 400 µl of cold 80% ethanol, lightly vortex or thump the tube with your finger, and repeat the centrifugation, but spin for only 10 minutes.

Remove the supernatant with a pipette, being careful to not aspirate the pellet.

Air dry for 10 minutes.

Resuspend the pellet in 25 µl depc-treated H_2O.

- Determine the concentration of RNA by measuring the absorbance of 1 µl of RNA at 260 nm. If you don't want to sacrifice that much RNA for a reading, make a dilution with depc-treated H_2O and measure 1 µl of your dilution. Remember to factor the dilution into your concentration calculation.
- Dilute 1 µg of the DNA-free RNA to 0.1 µg/µl with depc-treated H_2O. This will be used as the template for the next step, which is reverse transcription.

Reverse transcriptase reaction

In this step, the RNA is used as a template to synthesize a cDNA strand. The enzyme used for this reaction is reverse transcriptase, which has been cloned from the Moloney murine leukemia virus. The enzyme is a DNA polymerase that can use RNA or single stranded DNA as a template when a primer is present. An anchored poly-T primer (i.e., a primer that has nucleotides other than T at the penultimate 3' position) is used to anneal to the polyadenylated region at the 3' ends of mRNA. In the presence of abundant deoxyribonucleotides, a cDNA is synthesized, resulting in a RNA–cDNA hybrid duplex.

Although the RNA has been treated with DNase I, it is a good idea to monitor for bands derived from double stranded DNA (which you don't want) rather than the cDNA (which you do want). One control we frequently use is a parallel reaction that contains all of the components except reverse transcriptase (RT⁻ reaction). Theoretically, in the RT⁻ reaction, the only double stranded PCR template that is present will be contaminating DNA. However, there is a caveat about this control: we have occasionally observed bands in the RT⁻ reaction, even after purifying the total RNA through a cesium cushion and treating with DNase I. The bands are typically weak in intensity on the autoradiograph, and the pattern differs considerably from the pattern observed in the reaction containing reverse transcriptase (RT⁺ reaction). Furthermore, these bands could be eliminated if the RNA was removed by treatment with RNase A after reverse transcription, a result that is consistent with reports of reverse transcriptase activity with *Taq* polymerase.[16,17] If the reverse transcriptase activity of *Taq* appears to contribute to the banding pattern, the RNase A treatment should remove potential RNA templates. Another control for contaminating DNA in the RT reaction reagents should be prepared for each set of RT reactions. This is simply a reaction that contains all of the ingredients (including the reverse transcriptase enzyme) except the RNA template.

Reverse transcription reaction protocol
- The total volume = 20 µl.

- Set up the 2 tubes containing the RNA and primer for each RNA sample.
  Each tube should contain:
      2.0 µl;   total RNA (0.2 µg)
      3.0 µl;   10 µM poly-T primer (final conc = 1.5 µM)
      <u>3.0 µl;   depc-treated H₂O</u>
      8.0 µl
  Label one tube as RT⁺ and the other tube as RT⁻.
  Place the tubes on ice while preparing the cocktails.
- Make up the reverse transcription reaction cocktail containing the buffer, DTT, and dNTPs.
  For each tube, the cocktail contains:
      4.0 µl;   10 mM dNTPs (final conc = 500 µM each dNTP)
      0.2 µl;   0.1 M DTT (final conc = 1 mM)
      4.0 µl;   5× buffer (the Superscript buffer from BRL)
      0.5 µl;   RNase inhibitor (5 units/reaction)
      2.3 µl;   depc-treated H₂O
      <u>1.0 µl;   reverse transcriptase (200 units/reaction)</u>
      12.0 µl
  Vortex the cocktail, briefly spin down, and store on ice until used.
- Make up the dummy reaction cocktail using the same ingredients listed for the reverse transcription cocktail, replacing the reverse transcriptase with depc-treated H₂O.
- Do the reaction:
  Heat tubes to 80°C for 5 minutes.
  Move to 37°C for 5 minutes.
  Add 12 µl of cocktail per tube. (If you add the cocktail to the bottom of the tube where the RNA and primer are located, you won't need to vortex or spin down.)
  Incubate at 37°C for 60 minutes.
  Heat to 80°C for 10 minutes. (This stops the reaction by denaturing the reverse transcriptase.)
  Store on ice (if doing the PCR reaction immediately) or at −20°C (if doing the PCR reaction later).

Radiolabeling PCR

The double stranded RNA–cDNA generated by the RT reaction is used as a template for the PCR. One of the primers can be the anchored poly-T primer used in the RT reaction, while the other is a 10-base pair primer, or a "10-mer," that is not specifically designed for a particular gene or genes. Alternatively, a PCR can be set up using two different 10-mers.

We have found that PCR by the poly-T primer and a single 10-mer has several disadvantages. First, the 3′ region of many mRNAs often contains untranslated sequences, and the possibility of matches with a Genbank sequence are reduced, especially if amplifications are being performed with tissues from nontraditional species, such as marine invertebrates. Second, the poly-T primers can often behave in a "shifty" manner during the low stringency annealing conditions, i.e., the poly-T primer may be able to anneal in slightly different positions at the same site, in spite of the 3′ anchor nucleotides. This can result in uninformative ladders on the autoradiograph, particularly with the smaller (<200 base pair) amplicons. The disadvantage of using two different 10-mers is that the position of the amplified sequences within the gene is unknown, since the primers may anneal anywhere along the mRNA.

Because the T_m of both types of primer pairs is low (<45°C), the PCR uses a low annealing temperature of 40°C. To minimize mispriming, a method of setting up polymerase chain reactions called "hotstarting" is used. This involves placing all of the ingredients into the reaction tube except the polymerase, and heating to a high temperature (e.g., 94°C) to denature the templates. The enzyme is added while the templates are still denatured, and then cycling is begun. To extend the life of the polymerase, especially if >20 reactions are being "hotstarted" simultaneously, the temperature can be lowered to 90°C while the polymerase is added. Since only a fraction of a microliter of enzyme is needed for each reaction tube, a "hotstart" cocktail is made, and 5 µl of cocktail are added to each reaction. This ensures that a uniform amount of polymerase is added to each tube and makes mixing the reaction by pipette action easier.

Since the 10-mers may be able to anneal at different positions for different mRNAs, the amplicons will be of different sizes. To visualize these products, one of the incorporated nucleotides is radiolabeled, and the reactions are electrophoresed on a polyacrylamide gel for autoradiography. If a denaturing gel is used, both strands of an amplicon may have different mobilities. Furthermore, *Taq* polymerase may add a nontemplated A at the 3′ end. Consequently, a single amplicon may be represented by up to 4 bands with slightly different mobilities. The use of nondenaturing gels to simplify the banding pattern has been reported,[18] but we have found band smearing and associated poor band resolution makes it difficult to select bands for subsequent analysis. As a result we routinely use denaturing gels in ddPCR.

SAFETY NOTE: When ^{35}S is used as the radioisotope, the thermocycler should be operated in a fume hood. There have been reports that a volatile breakdown product of the radioisotope, possibly H_2S, forms at high temperatures. This radioactive breakdown product is capable of penetrating the plastic of the thinner-walled reaction tubes, so you can expect your thermocycler block to be radioactive. So far, there are no reports of this problem with ^{32}P or ^{33}P, which can be used instead of ^{35}S.

Radiolabeling PCR protocol

Make up a cocktail containing everything except the template and the thermostable polymerase.

- If using an anchored poly-T primer, the cocktail contains:

1.5 µl	10× PCR buffer	
1.2 µl	25 mM MgCl$_2$	(final conc = 1.5 mM)
3.0 µl	100 µM dNTP (no dATP)	(final conc = 5 µM each dNTP)
0.5 µl	α^{35}S-dATP (Amersham)	(final conc = 80 µM; 5 µCi)
1.0 µl	10 µM 10-mer primer #1	(final conc = 0.5 µM)

5.0 μl   10 μM poly-T primer   (final conc = 2.5 μM)
0.8 μl   depc-treated H₂O
13.0 μl
- If using two arbitrary 10-mers, the cocktail contains:
    1.5 μl   10× PCR buffer
    1.2 μl   25 mM MgCl₂   (final conc = 1.5 mM)
    3.0 μl   100 μM dNTP (no dATP)   (final conc = 5 μM each dNTP)
    0.5 μl   α³⁵S-dATP (Amersham)   (final conc = 80 μM; 5 μCi)
    1.0 μl   10 μM 10-mer primer #1   (final conc = 0.5 μM)
    1.0 μl   10 μM 10-mer primer #2   (final conc = 0.5 μM)
    4.8 μl   depc-treated H₂O
    13.0 μl
Make up a "hotstart" mix, containing the polymerase in buffer.
- For each tube, the "hotstart" mix contains:
    0.50 μl   10× PCR buffer
    0.25 μl   AmpliTaq, 5U/μl (1.25 U/tube)
    4.25 μl   depc-treated H₂O
    5.00 μl
- Aliquot 13 μl of the cocktail into each tube.
- Add 2 μl of the reverse transcription reaction into each tube.
- Overlay with 30 μl of sterile mineral oil.
- Heat to 94°C for 5 minutes.
- Reduce to 90°C while adding 5 μl of the "hotstart" mix to each tube.
- Begin the cycling reaction:
    Cycling conditions: These cycling conditions have worked well
        with an MJ Research thermocycler equipped with a heated lid.
    The times may need to be adjusted for other types of thermocyclers.

- Store reactions at –20°C if you're not going to electrophorese the
  reactions immediately. Reaction products are stable at –20°C for
  several months.
- Pour a 5% denaturing sequencing gel in 1× TBE (see Reference 15
  for instructions on preparing and pouring this gel), but insert a 32-
  well square toothed Delrin comb, 0.4 mm (BRL # 21035-100).
    12.5 ml 40% polyacrylamide (19:1, acrylamide:bisacrylamide)
    20 ml 5× TBE
    46 g urea (qs to 100 ml with distilled H₂O)
    Degas 10 minutes.
    Add: 130 μl TEMED; 350 μl 10% ammonium persulfate.
    Polymerize at least 1 h.

- Prerun the gel at 60 watts, constant power, for 30 to 45 minutes in 1× TBE.
- Just before loading the samples, wash out the urea from each well.
- Combine 6 µl of the PCR reaction with 3.5 µl sequencing loading buffer (see Reference 15, for the recipe for sequencing loading buffer).
- Heat to 85°C for 3 to 5 minutes, place on ice for 2 minutes, then load all 9.5 µl into each well.
- Run at 60 watts for 3 to 4 h, until the xylene-cyanol dye has migrated to about 80% of the gel length.
- Separate the plates, mount the gel on a sheet of 3MM chromatography paper, and cover with plastic wrap.
- Dry at 70 to 80°C until the gel is no longer sticky (e.g., 2 h in a Savant gel dryer).
- Radioactive ink can be made by spiking 200 µl of India ink or loading buffer with 1 µl (10 µCi) of α^{35}S-dATP. Orientation stickers can be made by making random marks on adhesive-backed labels with the spiked ink.
- Attach 3 or 4 orientation stickers to the gel side of the 3MM paper before putting the gel on film.
- Exposure for 20 to 24 h is sufficient for isotope that has not decayed more than one half-life.

Isolation and reamplification of amplicons

Once differentially expressed bands have been found, the banding pattern should be confirmed by repeating the radiolabeling PCR. An even better test for reproducibility would be a repeat of both the reverse transcription and the radiolabeling PCR. The next step involves recovering the PCR product from the ddPCR gel and using it to generate more DNA (with PCR, naturally). Remember that what appears to be a big band on the autoradiograph is usually less than what is required for double-stranded sequencing or for cloning, because the sensitivity of radioactive labeling allows visualization of less than picogram (10^{-12} g) amounts of nucleic acid. Since the reamplification PCR has only the excised band as a template, the amount of dNTPs can be reduced. Also, the annealing temperature will be raised slightly and the amount of enzyme will be reduced to discourage nonspecific priming.

Isolation and reamplification of amplicons protocol
Recovery of the DNA from the polyacrylamide gel.

- Superimpose the autoradiograph on the dried gel, using the orientation stickers, and tape the gel to the autoradiograph.
- Cut through the autoradiograph and the gel (and the 3MM backing paper) with a clean scalpel or razor blade.
- Place the gel slice (and the backing paper) into a 1.7 ml microfuge tube.
- Add 100 µl of distilled H_2O, and heat to 100°C for 15 minutes.

- Spin down any condensation (brief spin), and transfer the liquid to another 1.7 ml microfuge tube.

  Add: 5 μl   10 mg/ml oyster glycogen
             10 μl   3 M sodium acetate, pH 5.2
            350 μl   absolute ethanol.
- Precipitate for at least 30 minutes at –20°C.
- Spin at for at least 30 minutes at 4°C at 14,000 rpm (16,000 g).
- Wash the pellet 1× with 80% ethanol.
- Resuspend in 10 to 20 μl of distilled H_2O; heat to 65°C for 5 minutes to help the pellet dissolve.

Reamplification PCR: This reaction will be in a total volume of 50 μl.

- For each reaction, the cocktail contains:

5.0 μl	10× PCR buffer	
3.0 μl	25 mM MgCl$_2$	(final conc = 1.5 mM)
4.0 μl	1 mM dNTPs	(final conc = 20 μM of each dNTP)
2.0 μl	10 μM 10-mer primer #1	(final conc = 400 nM)
2.0 μl	10 μM poly-T primer or 10-mer #2	(final conc = 400 nM)
23.9 μl	distilled H_2O	
0.1 μl	AmpliTaq	(0.5 U/reaction)
50.0 μl		

- Into each reaction tube, aliquot 2 to 4 μl of DNA recovered from the acrylamide gel, and qs to 10 μl with distilled H_2O.
- Add 40 μl of cocktail to each tube, and overlay with 30 μl of sterile mineral oil.

The cycling conditions for reamplification are:

$$
\left.\begin{array}{ll}
94°C & 30\ seconds \\
42°C & 2\ minutes \\
72°C & 1\ minute
\end{array}\right\} — 40\ cycles
$$

72°C   5 minutes
4°C   hold

Electrophorese 10 to 20% of the reaction (5 to 10 μl) on a 2% agarose gel to check for the product.

If you are getting lots of product but multiple bands, try reducing the amount of template (a 10-fold dilution is a good starting point). Raising the annealing temperature a couple of degrees might work, but often that results in no products whatsoever. Titrating (i.e., reducing) the MgCl$_2$ is another option.

If there is no reamplification product, increase the amount of template (up to 10 μl). Also, the annealing temperature may be reduced to 40°C.

Results and discussion

Anticipated results

The desired outcome of differential display PCR is isolation of a fragment of a transcript that is differentially expressed. The protocol presented here will take the investigator to the point of isolating and reamplifying a transcript fragment that is at least differentially amplified. Since the autoradiograph that displays the amplified bands is the principal product, its interpretation is an important step. An example from our work with clam gonadal neoplasms can help to illustrate some of the typical features of a ddPCR autoradiograph. Figure 2 displays part of an autoradiograph of ddPCR using two different pairs of PCR primer with gonadal RNA from six *M. arenaria* (A through F), showing bands that are approximately 100 to 350 bp in length. Clams A and B were a normal female and a normal male, respectively. Clams C, D, E, and F had extensive gonadal neoplasms. Clam F was a female, but the gender of clams C, D, and E could not determined due to the abundance of neoplastic cells. This figure demonstrates several commonly observed features in a ddPCR autoradiograph. First, although both sets of reactions were performed with the same population of cDNA (synthesized from a T12AC primer) and only one of the two 10-mers in each primer pair was different, the banding pattern is quite different between the primer pairs, showing that changing just one primer can dramatically affect the banding pattern. Second, the efficiency of amplification can vary for a given specimen, depending upon the primer pair. In reaction #1, the intensity of the bands that are common to all six individuals appears to be equivalent among all six reactions. However, in reaction #2, the intensity of the bands for animal B appears reduced, relative to the other five specimens. This kind of variation occurs frequently and it may be due to the efficiency of the primer interaction with the particular cDNA. Third, occasionally very strong differences can be observed among the samples which are not correlated with a phenotype or characteristic such as gender. An example of this can be seen in the band marked with an asterisk in reaction #1, where a strong band appears only for animals C and D. Fourth, an example of "laddering" that is frequently observed in the smaller amplicons can be seen near the bottom of the lanes for reaction #2. This is typically observed among amplicons that are <200 bp in size. In thermocycling reactions that are performed with the anchored primer, the laddering can appear among larger amplicons. As previously discussed, some laddering may be due to complementary strands that are migrating at different mobilities or the slight "shiftiness" of the anchored poly-T primer in the amplification reaction. Alternatively, laddering can result if there are regions of dinucleotide repeats, resulting in shortening or lengthening of the repeat region. Finally, there are bands that appear to be differentially amplified from the RNA of neoplastic animals, and these bands are marked by arrowheads. The marked band in reaction #1 actually is a doublet, but it is likely to be complementary strands of the same amplicon.

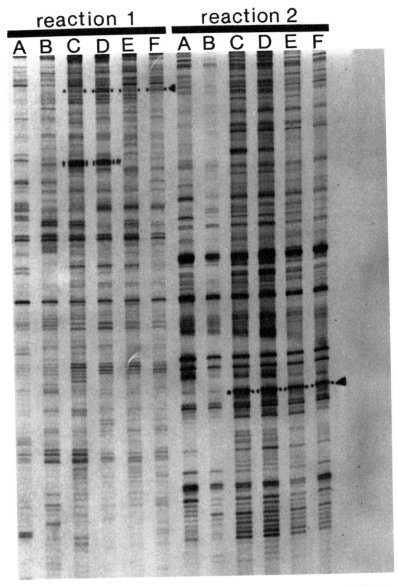

Figure 2 A representative autoradiogram from differential display PCR of gonadal RNA from normal or neoplastic *Mya arenaria*. Total RNA was isolated from gonads of six individuals (A through F), and 200 ng of RNA was reverse transcribed with the primer T12AC. One tenth of the reverse transcription reaction was placed into a PCR reaction containing α^{35}S-dATP and either primer pair #1 (5' GGCTAACCGA 3' and 5' GGAGTGGACA 3') or primer pair #2 (5' AGGCTGGGTG 3' and 5' GGAGTGGACA 3'). Clams A and F are female, clam B is male, and the gender of clams C, D, and E could not be determined histologically due to the extent of the neoplasm. Bands that are present in all of the clams bearing gonadal neoplasms (C through F), but absent in the normal clams (A and B) are indicated by arrowheads. An asterisk indicates bands that are present in only two of the clams with neoplasms (C and D).

The marked band in reaction #2 is faintly visible in the lanes for the normal animals, but the large difference in amplification suggests that the expression may be much higher or that the efficiency of amplification may be greater in the neoplastic animals, due perhaps to alterations at the primer binding site. These bands warrant further investigation.

Once a band is isolated and reamplified, it may be cloned into a plasmid vector or the PCR product may be directly sequenced. There are vectors that are specifically designed for cloning PCR products (e.g., the pCRII vector from Clontech, Palo Alto, CA). If a standard, and less expensive, vector is desired, the ends of the PCR product may be modified for cloning into more common plasmids. For example, the ends of the PCR product may be filled in using Klenow enzyme, and the resulting blunt-ended PCR product may be ligated into the Sma I site of a pUC-derived plasmid. Direct sequencing of the PCR product typically is a less reliable reaction than sequencing a plasmid, but it does eliminate the time and costs of cloning. Since the primers used to generate the amplicon are typically poor sequencing primers, it may be necessary to use modified primers for the reamplification reaction. For example, the addition of bases at the 5′ end of the arbitrary 10-mers, such as an EcoR I sequence, will increase the T_m of the primer as well as adding the option of cloning the reamplified product into the EcoR I site of a plasmid. The obvious disadvantage is the cost of purchasing the modified primer. A variation to direct sequencing is ligation linked PCR.[17a] In this method, the ddPCR band that is eluted from the polyacrylamide gel is ligated into a cloning vector, then immediately placed into PCR using one primer that anneals to the insert (e.g., the anchored poly-T primer) and one primer that anneals to the cloning vector (e.g., the M13 reverse primer). The sensitivity of PCR permits detection of low abundance, transiently ligated products, and the vector primer sequence allows direct sequencing of the PCR product. Given the variety of options for further characterizing ddPCR amplicons, the goals of the investigator will most likely determine which option is selected.

Primer selection

Aside from the quality of the RNA, the choice of primers will have the greatest effect on the observed banding patterns. Selection of the 10-mer primers has few rules, and optimal primers vary, depending on the species under analysis. In general, the criteria typically used for designing PCR primers should be followed: (1) G+C content of approximately 50%; (2) lack of self complementarity or hairpin secondary structure; and (3) G/C clamp at the 3′ terminus. Bauer et al.[18] compared 26 primers, of which 13 contained GATC at the 5′ end. The mean number of bands per PCR reaction did not vary significantly among all 26 of the primers, suggesting that all of the tested primers worked equally well with human and mouse RNA. Mou et al.[19] observed that primers containing a G or C at the 3′ end worked better than primers lacking these nucleotides at this position.

We have made a small scale comparison of 14 arbitrary primers with cDNA generated from 3 anchored poly-T primers. Surprisingly, one pair of primers consistently failed to produce more than ten bands, although each primer generated many bands when used with other primers. Aside from this anomalous pair, the average number of bands per PCR reaction was 59 (standard deviation = 17.8). We do not know the degree of overlap among bands or the rate of redundant display for any given message in our PCR reactions. If two to four closely spaced bands had equivalent densities in our autoradiographs, we counted them as a single band, since these are likely to be the complementary strands of a single PCR product, with and/or without the Taq polymerase-added A at the 3′ terminus.[1,18] Thus, our counts may underestimate the true number of different amplification products actually produced.

Time and effort considerations

In spite of the potential speed of ddPCR in isolating differentially expressed genes, the number of PCR amplifications required to screen the entire mRNA population can be intimidating. Originally, Liang and Pardee[1] estimated that 240 PCR reactions (20 upstream 10 bp primers and 12 anchored poly-T primers) would statistically screen the entire mRNA population. This estimate was based on the degenerate annealing behavior of the 10-mer primers and the use of nondegenerate anchored poly-T primers. Subsequently, Liang et al.[20] reported that the penultimate 3′ position of the poly-T primer annealed in a degenerate manner, reducing the number of required poly-T primers to 4, and the total number of PCR reactions to 80. Assuming that the 10 bp primers annealed as 6 bp primers and that there were approximately 1000 cDNA templates generated by an anchored poly-T primer, Bauer et al.[18] have estimated that at least 25 upstream 10 bp primers are needed to achieve a 95% confidence level that the entire mRNA population has been screened. Thus, if 4 degenerate anchored poly-T primers are used, at least 100 PCR reactions would be needed. Since comparisons would be made among two or more individuals or populations of cells and since replicate individuals or populations reduce false positives (see below), the required number of PCR reactions could range to more than several hundred.

Handling a large number of reactions can be eased by performing them in 96 well plates with a multichannel pipette. Some investigators have attempted to speed the screening of banding patterns by using an automated DNA sequencer to acquire banding pattern data, allowing more than one set of samples to be run consecutively on the same gel.[21] A slightly different approach uses a different fluorescent dye for each RNA sample to be compared. This permits four different samples (e.g., a control plus three test samples) to be electrophoresed simultaneously in a single lane in the gel.[18] However, both of these techniques require expensive fluorescein- or rhodamine-modified poly-T primers and do not allow bands to be recovered for subsequent analysis. Since software to quantitatively analyze the separate dyes is not yet available, inspection of the banding patterns is still performed manually.

False positives

Although ddPCR is designed to select for messages that are uniquely or preferentially expressed, a significant number of the selected messages are not. The production of false positives is a well-recognized feature of ddPCR. In a comparison of normal mammary and breast cancer cell lines, Liang et al.[20] reported that 33% of the isolated bands were true positives, 27% were false positives, and 40% failed to yield a signal by Northern analysis. Aiello et al.[4] examined altered gene expression in retinal pericytes upon exposure to glucose by ddPCR. Of the 25 differentially displayed bands they examined by total RNA Northern analysis, 28% were true positives, 8% were false positives, and 64% failed to produce a signal. When those investigators used polyadenylated RNA for Northern analysis, the percentage of true positives and false positives increased to 40% and 12%, respectively, suggesting that the sensitivity of the ddPCR method certainly exceeds that of Northern analysis, especially when total RNA is used for the Northern.

Several strategies have been proposed to minimize false positives. Since the annealing temperatures used in ddPCR are very low, some of the spurious bands are simply a result of the low stringency conditions. Identification of these uninformative bands can usually be made by repeating the procedure, preferably with a replicate cDNA preparation. Further improvements can be made by using multiple samples for each tissue group to be compared, since it is less likely that spurious bands will arise in independent samples. In a comparison of gonadal neoplasms with normal gonads of soft-shell clam (*M. arenaria*), we found the percentage of false positives was reduced by nearly 80% simply by increasing the number of individuals per comparison group from two to four.

Another source of false positives is the presence of multiple products in the excised gel piece. This may result from slight inaccuracies in aligning the autoradiograph and dried gel or from multiple PCR products superimposed in the same location. Misalignment can be easily minimized by using at least three radiolabeled markers, and by keeping the gel flattened until alignment. The gel and its 3MM paper backing slowly distort with time, and if it is not kept in a flattened condition (as in a film cassette), it becomes difficult to make an accurate alignment.

Multiple PCR products with similar mobilities in the polyacrylamide gel is a more difficult problem. The reamplification reaction contains a limited number of templates, compared to the initial amplification reaction, and the low annealing temperature and high number of cycles increases the likelihood that most or all of the templates present will be amplified. When we sequenced 3 to 4 clones each of 7 cloned ddPCR bands, identical sequence in all clones was found for only one band. For the remaining six bands, an average of 62.5% of the clones contained the same insert. We have attempted to avoid amplification of minor contaminating bands by cloning directly from the eluted gel slice, but the quantity of DNA is often too low for consistently efficient cloning.

One method for eliminating false positives due to minor contaminating amplicons uses the original ddPCR band to screen plasmid DNA from bacterial clones containing inserts from the reamplification reaction.[22] The ddPCR is performed using ^{32}P-dCTP, and the original ddPCR band is eluted from the polyacrylamide gel slice. Half of the eluate is used for cloning, and the cloned DNA is dotted onto a membrane. The other half of the eluate, which is already radiolabeled, is used as a probe against the membrane. The plasmid matching the most abundant amplicon in the original ddPCR band will be detected, while low abundance contaminants will have either a very low signal or no signal.

Screening false positives before cloning would be a desirable option. Probes made directly from the reamplified band by random priming have been used to screen Northern blots or RNA dot/slot blots for evidence of differential expression.[4,19] We have generated probes from the cloned insert or from the reamplified band, but often the low abundance of reamplified material may limit the amount of probe synthesized. The relatively short length of many ddPCR bands can be problematic if the random primed labeling method is used, since many of the probe fragments will tend to be short. For example, for bands that are 300 to 350 bp, the lengths of the dominant probe fragments can be <100 bp. An obvious drawback of screening Northern or RNA dot/slot blots is that a separate hybridization is required for each ddPCR band. An alternative method uses dot/slot blots of the reamplified bands (up to ~250 ng per dot/slot) which are then hybridized with random primed probes generated from cDNA of the tissues being compared.[19] Care must be taken to remove all DNA from the RNA that is used to make the probes, since probes will be produced from contaminating DNA.

Alternatives to ddPCR

A technique related to ddPCR has been developed as an extension of arbitrarily primed PCR (AP-PCR) analysis of DNA. This method, called "RNA fingerprinting by arbitrarily primed PCR" or RAP, relies on a single primer, approximately 20 bp in length, to generate both the cDNA and the PCR product.[23] Although RAP is intuitively less systematic than ddPCR, which divides the mRNA into subpopulations at the reverse transcription stage, the RAP technique can be converted to use on an automated DNA sequencer with minimal modification. Furthermore, RAP produces fewer bands per PCR reaction, implying that a smaller portion of the mRNA population is screened by each reaction.

We have been employing a slight modification of the ddPCR technique, which uses the degenerate anchored poly-T primers for cDNA synthesis. However, for the PCR amplification we have been using two different 10-bp primers. The rationale for this change was an attempt to generate amplification products away from the polyadenylated region of the message, which is likely to contain untranslated sequences. When we compared the number of GenBank Blast searches[24,25] that could find no matches at either the nucleotide or amino acid level, the original ddPCR protocol produced

A.

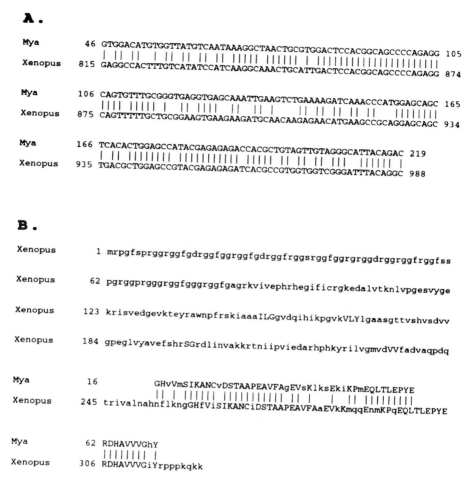

B.

Figure 3   Nucleic acid and amino acid comparison of a differential display PCR product isolated from a neoplastic gonad of *Mya arenaria* with *Xenopus* fibrillarin. Total RNA was isolated, reverse transcribed with a T12CG primer, and amplicons were generated by PCR with primers 5' GGAGTGGACA 3' and T12CG. Amplicons were resolved on a 6% polyacrylamide-urea gel, and bands specific to neoplastic gonads were cut from the gel and reamplified by PCR. Reamplified products were cloned into a TA cloning vector (pCRII, Clontech, Palo Alto, CA) and sequenced by Sanger dideoxynucleotide sequencing (Sequenase 2.0, US Biochemical, Cleveland, OH). Nucleic acid and predicted amino acid sequences were compared with sequences in GenBank by the Blast algorithm.[23,24] The best matches at the nucleotide level were with *Xenopus* fibrillarin ($p = 5.9 \times 10^{-32}$), *Leishmania* fibrillarin ($p = 3.8 \times 10^{-19}$) and yeast fibrillarin ($p = 2.2 \times 10^{-12}$). The best matches at the amino acid level were with *Xenopus* fibrillarin ($p = 2.9 \times 10^{-26}$), human fibrillarin ($p = 3.8 \times 10^{-25}$) and mouse fibrillarin ($p = 3.9 \times 10^{-25}$). Part A shows the nucleic acid match between the *Mya* amplicon and the region of the *Xenopus* fibrillarin cDNA with the best match. The length of the *Xenopus* cDNA is 1172 bp. Part B shows the entire *Xenopus* fibrillarin amino acid sequence and the best match with the *Mya* amplicon. Sequence identities are shown with upper case letters connected by a vertical line. The numbers at the beginning of each line is the number of the nucleic acid (A) or amino acid (B) at the start of the line.

20% (6 "no matches" out of 30 searches), while the two 10-mer primer method resulted in 0% (0 "no matches" out of 20 searches). The number of bands per PCR reaction was not significantly different between these methods (p >0.1, 2 sample t-test, 2-tailed, df = 17), suggesting that the two approaches may screen the mRNA (cDNA) subpopulations equally well. An example of the 3' bias when using the anchored poly-T primer in the amplification reaction is shown in Figure 3. Figure 3 displays the nucleotide and amino acid matches between amplicons from *M. arenaria* and the reported sequences for *Xenopus* fibrillarin. It can be clearly seen that the clam amplicon has high sequence identity with the 3' end of the polypeptide.

Summary

In summary, ddPCR is a screening method that can be used to compare multiple samples simultaneously with the straightforward techniques of reverse transcription, PCR amplification, and gel electrophoresis. With ddPCR, it is possible to identify and sequence partial messages without resorting to cloning. The major drawbacks of ddPCR are the large number of reactions needed to screen an entire mRNA population and the substantial number of partial messages that are false positives or produce no signals in Northern analysis. As the application of the automated DNA sequencer to ddPCR becomes more prevalent, the first drawback will be largely overcome. The second drawback may simply be more of a reflection of the sensitivity of PCR rather than a true disadvantage. As more investigators employ this technique in their laboratory and field studies, refinements and improvements are likely to make ddPCR a regular part of the molecular biologist's toolbag.

References

1. Liang, P., Pardee, A. B. 1992. Differential display of eukaryotic messenger RNA by means of the polymerase chain reaction. *Science* 257: 967–71.
2. Joseph, R., Dou, D., Tsang, W. 1994. Molecular cloning of a novel mRNA (neuronatin) that is highly expressed in neonatal mammalian brain. *Biochem Biophys Res Commun* 201: 1227–34.
3. Zimmermann, J. W., Schultz, R. M. 1994. Analysis of gene expression in the preimplantation mouse embryo: use of mRNA differential display. *Proc Natl Acad Sci U S A* 91: 5456–60.
4. Aiello, L. P., Robinson, G. S., Lin, Y.-W., Nishio, Y., King, G. L. 1994. Identification of multiple genes in bovine retinal pericytes altered by exposure to elevated levels of glucose by using mRNA differential display. *Proc Natl Acad Sci U S A* 91: 6231–5.
5. Utans, U., Liang, P., Wyner, L. R., Karnovsky, M. J., Russell, M. E. 1994. Chronic cardiac rejection: identification of five upregulated genes in transplanted hearts by differential mRNA display. *Proc Natl Acad Sci U S A* 91: 6463–7.
6. Watson, M. A., Fleming, T. P. 1994. Isolation of differentially expressed sequence tags from human breast cancer. *Cancer Res* 54: 4598–602.
7. Hsu, D. K., Donohue, P. J., Alberts, G. F., Winkles, J. A. 1993. Fibroblast growth factor-1 induces phosphofructokinase, fatty acid synthase and Ca(2+)-ATPase mRNA expression in NIH 3T3 cells. *Biochem Biophys Res Commun* 197: 1483–91.

8. Burn, T. C., Petrovick, M. S., Hohaus, S., Rollins, B. J., Tenen, D. G. 1994. Monocyte chemoattractant protein-1 gene is expressed in activated neutrophils and retinoic acid-induced human myeloid cell lines. *Blood* 84: 2776–83.

9. Jung, M., Kondratyev, A. D., Dritschilo, A. 1994. Elongation factor 1 delta is enhanced following exposure to ionizing radiation. *Cancer Res* 54: 2541–3.

10. Sager, R., Anisowicz, A., Neveu, M., Liang, P., Sotiropoulou, G. 1993. Identification by differential display of alpha 6 integrin as a candidate tumor suppressor gene. *FASEB J* 7: 964–70.

11. Zhang, L., Medina, D. 1993. Gene expression screening for specific genes associated with mouse mammary tumor development. *Mol Carcinogen* 8: 123–6.

12. Brown, D. J., Van Beneden, R. J., Clark, G. C. 1996. Identification of two binding proteins for halogenated aromatic hydrocarbons in the hard-shell clam, *Mercenaria mercenaria*. *Arch Biochem Biophys* 319: 217–24.

13. Chomczynski, P., Sacchi, N. 1987. Single-step method of RNA isolation by acid guanidinium thiocyanate-phenol-chloroform extraction. *Anal Biochem* 162: 156–9.

14. Groppe, J. C., Morse, D. E. 1993. Isolation of full-length RNA templates for reverse transcription from tissues rich in RNase and proteoglycans. *Anal Biochem* 210: 337–43.

15. Sambrook, J., Fritsch, E. F., Maniatis, T. 1989. *Molecular Cloning: A Laboratory Manual 2nd ed.* Cold Spring Harbor Laboratory, Cold Spring Harbor, N.Y.

16. Jones, M. D., Foulkes, N. S. 1989. Reverse transcription of mRNA by *Thermus aquaticus* DNA polymerase. *Nucl Acids Res* 17: 8387–8.

17. Tse, W., Forget, B. 1990. Reverse transcription and direct amplification of cellular RNA transcripts by *Taq* polymerase. *Gene* 88: 293–6.

17a. Reeves, S. A., Rubio, M.-R., Louis, D. N. 1995. General method for PCR amplification and direct sequencing of mRNA differential display products. *Biotechniques* 18: 18–20.

18. Bauer, D., Müller, H., Reich, J., Riedel, H., Ahrenkiel, V., Warthoe, P., Strauss, M. 1993. Identification of differentially expressed mRNA species by an improved display technique (DDRT-PCR). *Nucl Acids Res* 21: 4272–80.

19. Mou, L., Miller, H., Li, J., Wang, E., Chalifour, L. 1994. Improvements to the differential display method for gene analysis. *Biochem Biophys Res Commun* 199: 564–9.

20. Liang, P., Averboukh, L., Pardee, A. B. 1993. Distribution and cloning of eukaryotic mRNAs by means of differential display: refinements and optimization. *Nucl Acids Res* 21: 3269–75.

21. Ito, T., Kito, K., Adati, N., Mitsui, Y., Hagiwara, H., Sakaki, Y. 1994. Fluorescent differential display: arbitrarily primed RT-PCR fingerprinting on an automated DNA sequencer. *Febs Lett* 351: 231–6.

22. Callard, D., Lescure, B., Mazzolini, L. 1994. A method for the elimination of false positives generated by the mRNA differential display technique. *Biotechniques* 16: 1096–1103.

23. Welsh, J., Chada, K., Dalal, S. S., Cheng, R., Ralph, D., McClelland, M. 1992. Arbitrarily primed PCR fingerprinting of RNA. *Nucl Acids Res* 20: 4965–70.

24. Altschul, S. F., Gish, W., Miller, W., Myers, E. W., Lipman, D. J. 1990. Basic local alignment search tool. *J Mol Biol* 215: 403–10.

25. Gish, W., States, D. J. 1993. Identification of protein coding regions by database similarity search. *Nat. Genet.* 3: 266–72.

chapter ten

DNA adduct analysis in small fish species using GC/mass spectrometry*

J. McHugh Law, Debra J. McMillin, David H. Swenson, and Jay C. Means

Introduction

One of the emerging fields of investigation in aquatic toxicology is the use of aquatic organisms to investigate environmental risks associated with exposure to mutagenic and carcinogenic substances. Field observations of tumors in populations of wild fish collected near known sources of chemical contamination led to controlled laboratory investigations of the potential for fish and other aquatic organisms to respond to exposures to genotoxic agents.[1-3] Observations that fish have metabolic capabilities similar to mammals in terms of critical biochemical pathways which are involved in xenobiotic chemical activation, conjugation/detoxification, and DNA repair have led to an increasing acceptance of fish as viable test models for the investigation of mutagen and carcinogen impacts. For example, fish from a wide spectrum of families have been shown to have P450 isoenzymes analogous to those found in rodent models and humans.[4,5] These enzymes have been found to be inducible using agents which elicit a similar response in mammalian models. Recently, the presence of the inducible form of P450-1A1 in embryos and larval fish has been reported.[6]

Studies in fish have documented the responsiveness of numerous species to *classical* alkylating mutagens such as ethylmethane sulfonate, methyl-N-nitroso-guanidine, and ethylnitrosourea, as well as to mutagens of environmental origin such as aflatoxin B1, benzo[a]pyrene, and 2-amino-fluorene.

* This work was presented, in part, at the "Fifth Symposium on Environmental Toxicology and Risk Assessment: Biomarkers and Risk Assessment," American Society for Testing and Materials, Denver, Colorado, April 3–5, 1995. Figures 1, 2, 3, and 5 were used in the report.

Adaptation of methods originally developed in mammalian systems to evaluate mutagenic potential, such as sister chromatid exchange assays, micronucleus tests, chromosomal aberration assays, and other endpoints, have all been evaluated in fish models with varying degrees of success. Chromosomal aberration assays have been shown to be useful in assessing the impacts of alkylating agents[7] as well as complex environmental mixtures of polycyclic aromatic hydrocarbons.[8,9] Biochemical methods to measure DNA breakage through DNA unwinding have been successfully used in fish models to assess DNA damage subsequent to both laboratory and field exposures to mutagens and carcinogens.[10]

DNA adduct formation is considered to be one of the earliest steps in the multi-stage process of carcinogenesis. Assays for DNA adduct formation following exposure to both direct and indirect acting mutagens and carcinogens have been shown to be adaptable to aquatic organisms including fish and various invertebrates (see recent review by Maccubbin[11]). Among the methods that have been applied to DNA adduct detection in aquatic organism models are ^{32}P-postlabeling[12] combined with multi-dimensional thin layer chromatography or high performance liquid chromatography (HPLC), fluorescence detection combined with HPLC,[13] radio- or fluorescence-immunoassays using monoclonal antibodies for specific adduct forms, and GC/MS analysis of adducted purine and pyrimidine bases. The relative sensitivities of these methods range from 1 adduct in 10^6 bases for fluorescence methods to 1 adduct in 10^9 bases for ^{32}P-postlabeling or GC/MS. However, only the GC/MS method has the ability to specifically identify the adduct structure, the base involved in the adduct, and the position of base substitution at which the adduct has formed. Mass spectrometry as an analytical tool has the advantage that isotopically modified forms of the adducts of interest may be synthesized and used as internal standards in an isotope dilution experiment. This approach allows for discrimination of positional isomers of adducts as well as detection of the bases involved in adduct formation, all at femtomole (10^{-15} moles) levels.

In this chapter, a method for the detection of two positional isomeric forms of both methyl and ethyl adducts of guanine is presented using the stable isotope internal standard approach. The method may be expanded to include analyses of adducts at other positional isomers or bases by the addition of other stable isotope standards. Further, the method represents a procedural template that may be adapted to other classes of mutagens and carcinogens with the preparation of appropriate standards and modification of chromatographic parameters. The following describes the test fish, reagents, and instrumentation used in development of this method in our laboratories. Initial experiments were limited to a single, model carcinogen, methylazoxymethanol acetate, that has been shown to induce a high rate of liver neoplasia in several small fish species, including *Gambusia affinis*.[14,15] These experiments are presented and discussed along with typical calculations and data reduction strategies.

Materials required

Test animals

The western mosquitofish (*Gambusia affinis*) is a small, native, freshwater fish that is distributed from New Jersey to northern Mexico and has been introduced worldwide as a mosquito control agent.[16] Mosquitofish are easily cultured using techniques developed for other small fish species such as the guppy (*Poecilia reticulata*). They breed prolifically and appear resistant to many infectious diseases. They have been shown in previous studies to express measurable changes in a variety of biomarkers following contaminant exposure and, thus, may prove useful for direct validation of field studies by enabling substances to be tested under both laboratory and field conditions.[17] In the context of this chapter, these fish represent one of several possible small fish models that may provide tissue samples for the investigation of DNA damaging agents in laboratory or field studies.

All specimens used in this study were the progeny of laboratory-maintained western mosquitofish which were originally caught in Ocean Springs, Mississippi. However, feral organisms may be utilized in other experimental designs. In both variants of the present experimental protocol, adult fish ranging from 9 to 14 months old were used. The fish were acclimated for a minimum of 2 weeks in 40-l aquaria containing dechlorinated, activated carbon-purified, organic-free water maintained at $27 \pm 1°C$ by a recirculating water bath. The aquarium room was maintained on a 12-h light/12-h dark cycle with incandescent lighting. The fish were fed commercial flake food (Prime Tropical Flakes-Yellow®, Zeigler Brothers, Inc., Gardners, PA) two times daily. Water quality was monitored and tank maintenance performed once weekly.

Reagents

A list of reagents and equipment needed for this procedure is provided in Table 1. All reagents used in this procedure are of the highest possible purity to avoid interferences from extraneous substances in this highly sensitive system. For example, DNA extraction was accomplished using high purity, biotechnology grade phenol:chloroform:isoamyl alcohol (25:24:1, pH 8.0, Amresco®, Solon, Ohio) to avoid artifactual adduct formation by phenol oxidation products. Conical bottom, glass vials (Alltech, Deerfield, IL) fitted with PFTE-sealed Mininert Valves® (Alltech) were used for all heating steps. These vials concentrate small samples in the center of the container to assure more consistent sample heating. The PFTE-sealed valves allow for creation of negative pressure drawn within the head space of vials using a 22-gauge hypodermic needle prior to performing the heating step.

It should be noted that reagents which are stored *cold*, such as the phenol:chloroform mixture or the derivatization reagent (BSTFA + 1% TMCS), should always be allowed to equilibrate to room temperature before opening in order to minimize absorption of water. This is especially important in the derivatization step, since silylating reagents are extremely water

Table 1   Reagents and Equipment Needed

Reagents, glassware, and equipment	Purpose, model/stock no.	Source
Methylazoxymethanol acetate	Methylating carcinogen	Sigma Chemical Co., St. Louis, MO
Microtube pellet pestles, polypropylene	Maceration of small, frozen tissue samples within 2 ml microtubes; stock # 95050-99	KIMBLE DELTAWARE, Division of Owens-Illinois
100 mM NaCl (AR)	Digestion buffer	Mallinckrodt, Paris, KY
10 mM Tris Cl, pH 8 (AR)	Digestion buffer	AMRESCO, Solon, OH
25 mM EDTA, pH 8 (AR)	Digestion buffer	Sigma
0.5% Sodium dodecyl sulfate (AR)	Digestion buffer	Pierce Chemical Co., Rockford, IL
Proteinase K, 20 mg/ml solution, biotechnology grade	Digestion buffer	AMRESCO
Blue M Magni Whirl	Constant temperature water bath	Blue M Electric Co., Blue Island, IL
Phenol:chloroform:isoamyl alcohol 25:24:1, biotechnology grade, pH 8	Organic extraction of DNA; stock # 0883	AMRESCO
7.5 M ammonium acetate	Precipitation of DNA	Mallinckrodt
Iso-propyl alcohol	Precipitation of DNA	EM Science, Gibbstown, NJ
Ethyl alcohol USP, absolute	Pellet rinse, to remove residual salt and phenol	Midwest Solvents Co., Pekin, IL
60% Formic acid (AR)	DNA hydrolysis	Mallinckrodt
Reactivials, conical bottom, 5 ml	Hydrolysis and derivatization steps	Alltech, Deerfield, IL
Mininert valves, with PFTE seals	Allow evacuation of reaction vials	Alltech
Manifold freeze dryer	Model B66	New Brunswick Scientific Co., Inc., Edison, NJ
Bis(trimethylsilyl)-trifluoroacetamide/1% trimethylsilylchlorosilane (BSTFA + 1% TMCS)	Stock # 18089; derivatization	Alltech
Acetonitrile (AR)	Derivatization	Mallinckrodt

reactive. Care should always be taken to minimize exposure to room air and to seal the vials under nitrogen, argon, or other inert gas. Since these experiments are performed on a micro scale, even when a 100 to 1000 molar excess is used for derivatization, a small amount of water can consume a large portion of the active derivatization reagent and result in incomplete or highly variable yields of derivatized bases. Further, all containers/glassware and other implements should be meticulously cleaned as for all analytical work, i.e., solvent-rinsed and oven-dried. Reagent blanks should be run with each set of samples to allow detection of interfering peaks or sample cross-contamination.

Methylazoxymethanol acetate (MAM-Ac), a model alkylating carcinogen used in this study, was purchased from Sigma Chemical Company (St. Louis, Missouri). MAM-Ac forms methyl-DNA adducts by release of a highly reactive carbonium ion.[18] The compound is carcinogenic in several mammalian animal models, and has been shown to cause liver neoplasia in several fish species.[19,20] For example, a 2-h pulse exposure to 10 mg/l (76 μM) MAM-Ac in ambient water resulted in an approximately 33% incidence of liver tumors in western mosquitofish after 25 weeks and greater than a 50% incidence after 40 weeks.[15] In the laboratory, any potential alkylating agent or ionizing radiation could be tested for adduct formation and repair with this method, provided that suitable internal standards were available.

Stable isotope-labeled standards

7-Methylguanine (7-MG) was purchased from Sigma. 7-Ethylguanine (7-EG), O⁶-methylguanine (O-MG), and O⁶-ethylguanine (O-EG) were prepared using published procedures.[21,22] Procedures specific for the synthesis of methyl derivatives were modified to yield ethyl derivatives by selecting the corresponding ethyl donating reagent. Deuterium-labeled analogs of the four analytes were prepared from the corresponding fully deuterated methyl or ethyl donating reagents. Briefly, for the synthesis of uniformly deuterium-labeled 7-alkylguanines, 2'-deoxyguanosine 5'-monophosphate (Aldrich Chemical Company, Milwaukee, Wisconsin) was dissolved in DMSO and stirred overnight with iodomethane-d_3 (Sigma, 99 + atom % D) or iodo-ethane-d_5 (Aldrich, 99 atom % D). The resulting product was precipitated in ethyl acetate and purified by recrystalization. For the synthesis of the deuterated O⁶-alkylguanines, either methyl-d_3 alcohol-d (Aldrich) or ethyl-d_5 alcohol-d (Aldrich) was first reacted with sodium metal to form Na⁺O⁻CD₃ or Na⁺O⁻C₂D₅. These products were heated with 2-amino-6-chloropurine (Aldrich) at 70°C overnight in sealed vials. The solutions were neutralized with 3.5 mM glacial acetic acid, solvents removed by evaporation, and the resulting products purified by recrystalization.

The identity and purity of the methylated and ethylated guanine bases and their stable isotope-labeled analogs were confirmed by GC/MS (see below), calculation of extinction coefficients,[23] and HPLC with diode array detection.[21] The identity of the derivatized compounds was verified on the basis of reference mass spectra (NIST/EPA/NIH Spectral Library, U.S. Department of Commerce).

Instrumentation

In the development of these methods, emphasis was placed upon utilizing instrumentation commonly available in analytical and toxicological laboratories. In this study, DNA adducts were separated and detected using a Hewlett-Packard (HP) GC/MS system. A HP5890 gas chromatograph equipped with a 30 m by 0.25 mm ID DB-5 column with a liquid film of 0.25 μm was directly interfaced with a HP 5970 mass selective detector. Samples and standards were introduced onto the GC using a HP 7673 liquid autosampler. Instrument parameters/settings are given below.

Procedures

Experimental design

Experiments were performed to investigate the extent and kinetics of DNA alkylation and subsequent DNA repair following single, short, pulse exposures to MAM-Ac. The exposure designs and tissue sample preparation sequence are shown schematically in Figure 1.

Exposure methods

A variety of controlled laboratory exposure methods exist for testing mutagens and/or carcinogens in small fish species (see References 24 to 28). These include: (1) static, single-pulse exposures; (2) intermittent, multiple-pulse exposures; and (3) flow-through exposures (see Chapter 32). Because of its sensitivity, GC/MS DNA adduct quantification can potentially be adapted for use with feral fish that may be caught in polluted environments. Aquatic organisms with such "unknown" exposures could be assessed for a variety of DNA-carcinogen adducts following appropriate laboratory synthesis of internal standards and verification of sample preparation procedures to optimize recovery.

In development of this method, single-pulse exposures were used. Two basic protocols, differing only in post-exposure sampling times (Figure 1) and numbers of specimens used, were followed in these experiments. Test fish were randomly distributed into 4-l Pyrex® beakers containing dechlorinated, organic-free water (pH 8.7, dissolved oxygen 9.0 mg/l) maintained at 27 ± 1°C in a heated recirculating waterbath contained within a ventilated laboratory hood. Five randomly selected adult fish ranging from 30 to 35 mm in length were used for each treatment. Ten fish of similar size served as unexposed controls during each experiment. Specimens were exposed for 2 h to MAM-Ac at 10 mg/l (76 mM solution) in test water, then rinsed and transferred to clean water until sampled for DNA adduct analysis. This protocol mimicked that used in a previous histopathology study in which liver tumors were diagnosed in mosquitofish as early as 25 weeks after exposure to 10 mg/l MAM-Ac.[15] Exposed and control fish were held in separate 40-l aquaria maintained at 27 ± 1°C. At each designated sampling time, five fish were randomly selected and were euthanatized by stunning in ice water followed by decapitation.

DNA extraction and purification

Livers from five fish were immediately excised, pooled, weighed, and flash-frozen in liquid nitrogen. DNA was extracted and purified according to

DNA ADDUCT ANALYSIS
Experimental Design:

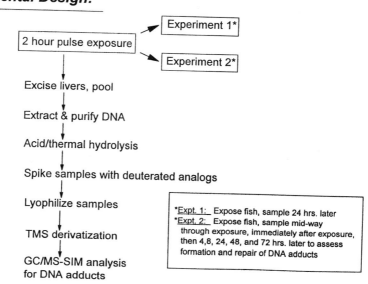

Figure 1 Experimental design. TMS = trimethylsilyl.

standard methods,[29] with some modifications.* Briefly, the frozen liver tissue was pulverized in 2 ml polypropylene microtubes, then suspended in digestion buffer containing proteinase-K and incubated with shaking at 50°C for 14 to 16 h. A reagent blank which contained all reagents and materials except liver tissue was also prepared. DNA extraction was accomplished using phenol:chloroform:isoamyl alcohol (25:24:1, pH 8.0). The extracted DNA was then precipitated by addition of isopropanol and 7.5 M ammonium acetate (to reduce residual RNA) followed by centrifugation at 13,000 × g. The DNA pellet was then rinsed in 70% ethanol and briefly air-dried.

Hydrolysis and derivatization

Isolated DNA samples were hydrolyzed in 0.5 ml of 60% formic acid in evacuated, PFTE-sealed and mininert-capped, conical bottom, glass vials at 75°C for 30 min. After cooling, all samples were spiked with a mixture of the four deuterium-labeled adduct standards. Standards were also prepared at this time by adding both the deuterated and a non-deuterated guanine adduct standard mixtures to clean vials. Nanopure grade water (2 ml) was

* Note: A number of methods and "rapid DNA prep" kits are currently available for preparation of genomic DNA from animal tissues. Regardless of the method chosen, care should be taken that the method results in virtually complete removal of proteins and quantitative recovery of nucleic acids in pellet form. Purity is important. However, size of the DNA product is not a major concern with this procedure, since the DNA will later be hydrolyzed into individual bases (some procedures, such as construction of phage libraries, require high- molecular-weight DNA; with such methods, care must be taken to minimize shearing forces). With the current method, some shearing of the DNA strands is acceptable.

added to each tube except the standards.* Samples and blanks were lyophilized and trimethylsilyl derivatives were produced using a mixture of 80
µl of bis(trimethylsilyl)-trifluoroacetamide (BSTFA)/1% trimethylsilylchlo-
rosilane and 20 µl of acetonitrile under nitrogen at 75°C for 30 min. After
cooling, derivatized mixtures were injected directly onto the gas chromatograph (GC). Lyophilized mixtures of standard methylated and ethylated
guanine bases and their deuterium-labeled analogs were silylated in the
same fashion and used to create standard response curves.

Quantitative analysis

The following parameters were used for the GC/MS, after the method of
Dizdaroglu.[30] The column was initially held at 100°C for 3 min and then
temperature programmed to 300°C at a rate of 7°C/min. The transfer line was
held at 300°C. The autosampler was set to deliver 2 µl aliquots into a
split/splitless injection liner packed with deactivated glass wool (Restek
Corp., Bellefonte, Pennsylvania) and held at 250°C. The electron multiplier
was operated at 400 volts above the standard instrument perfluorotributy-
lamine (PFTBA) tune value. Initial mass spectral identifications and confir-
mations were carried out using the scanning mode (50 to 550 amu, at 1.3
scan/sec). Three intense, characteristic ions were selected for each of the 8
analytes which also avoided mass spectral interferences such as column bleed.
A selected ion monitoring (SIM) method (dwell times = 30 msec) was then
developed based on retention times and characteristic ions of each component.

Results and discussion

Adduct standards

Table 2 shows the chromatographic and mass spectral parameters for each
analyte. Trimethylsilyl-guanine adducts and their deuterated analogs were
initially analyzed separately by GC/MS. Examples of extracted ion chromatograms within selected retention time windows for the non-deuterated
adduct standards are given in Figure 2. Their corresponding spectra are
shown in Figure 3. Gas chromatographic retention times of the derivatized
adduct standards were not significantly different (<0.08 min) from their
respective deuterium-labeled analogs.

Standard curves

Calibration plots (standard curves) are used to quantify sample levels and
to determine the precision of the measurements by GC/MS-SIM across the
expected range of sample concentrations.[31] For the experiments described

---

* Note: This addition of water dilutes the formic acid, thereby raising the freezing point, to
allow for more consistent freeze drying. The refrigeration system on the freeze dryer should
cool the condenser to about 20°C lower than the eutectic point of the samples, generally around
–65 to –80°C. Care should be taken that samples remain frozen solid while freeze drying and
that all water is removed. A dry powder should remain in the bottom of the vial.

Table 2 Retention Times and Characteristic Mass Fragments of Guanine Adducts

Adduct	Retention time	m/z	Relative abundance[a]
O[6]-methylguanine	21.86	294	100%
		309	35%
		295	24%
d$_3$ O[6]-methylguanine	21.83	297	100%
		312	33%
		298	24%
7-Methylguanine	22.56	294	100%
		309	22%
		295	21%
d$_3$ 7-Methylguanine	22.54	297	100%
		312	21%
		298	25%
O[6]-ethylguanine	22.38	308	67%[b]
		323	79%
		324	20%
d$_5$ O[6]-ethylguanine	22.31	313	93%[b]
		328	93%
		329	25%
7-Ethylguanine	22.87	308	100%
		323	20%
		324	52%
d$_5$ 7-Ethylguanine	22.84	313	100%
		328	22%
		329	13%

[a] Spectra obtained under standard tune conditions.

[b] The most abundant mass fragments for the deuterated and nondeuterated O[6] ethylguanines were 280 and 281, respectively, the same as seen in liquid phase background bleed of the GC column. Therefore, these ions were not used for SIM analyses.

above, standard curves were prepared by plotting amount ratios (non-deuterated vs. deuterated) over a range representing an average of 30 fmol to 6.1 nmol of alkylguanines on column (Figure 4). Example raw data used in generating these curves are shown in Table 3. Duplicate dilutions at nine levels of the nondeuterated compound containing a constant amount of its deuterated analog were prepared by lyophilization and derivatization. Regression statistics were generated by plotting amount ratios against response ratios. The standard curves resulted in linear responses ($R^2 = 0.998$ or greater) across all data points and indicated detection limits as low as femtomoles (10^{-15} moles) of DNA adduct molecules.

Data from a typical experiment

Examples of DNA adduct data generated from a typical small fish exposure are shown in Table 4 and Figure 5. In this experiment, western mosquitofish

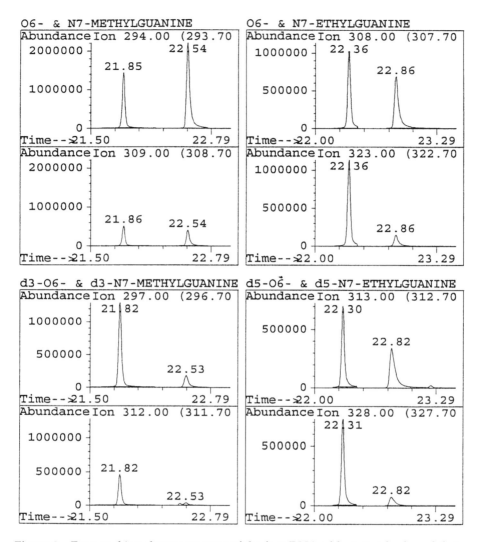

Figure 2   Extracted ion chromatograms of the four DNA adduct standards and their deuterated analogs.

(*Gambusia affinis*) were exposed to 10 ppm (76 μM) methylazoxymethanol acetate for 2 h, as described above, then sampled at several intervals after exposure and analyzed for the presence of methylguanine adducts in the liver. Note that the concentration levels of O^6-methylguanine (O-MG) detected averaged approximately 12% (molar ratio 0.12) of the levels of 7-methylguanine (7-MG). This is similar to the O-MG:7-MG ratios reported *in vitro* for another methylating agent, *N*-methyl-*N*-nitrosourea, by Lawley.[32] Although substitution at the N-7 position of guanine is a preferential site of attack for most alkylating agents, substitution at the O^6 position is thought to be of substantially greater biological consequence.[21,32-34] Methylation at O^6 fixes guanine in an unfavorable tautomeric configuration that tends to base-

```
File        : C:\HPCHEM\1\DATA\4291\4291B01.D
Operator    : DJ
Acquired    : 19 Oct 94  12:17 pm using AcqMethod CBLSME
Instrument  :   5970 - In
Sample Name : CBL STD @ 0.5 PPM W/ 2FBP (5PPM)
Misc Info   : 2UL ALS EMV=+200  5/12/94 TR
Vial Number : 1
```

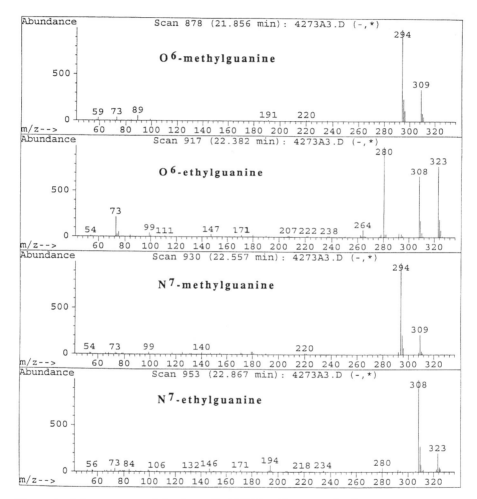

Figure 3 Mass spectra of the four alkyl DNA adduct standards.

pair with thymine rather than cytosine and could, therefore, lead to a transition mutation.[35] Mutagens and carcinogens that form simple base modifications, such as methylating agents (e.g., MAM-Ac) or ethylating agents (e.g., diethylnitrosamine), differ from bulkier adduct-forming agents in that they do not significantly distort the DNA helix. Thus, although simple base modifications may block or alter the fidelity of DNA replication, more often they directly produce promutagenic adducts in DNA.[36,37]

Also note in Table 4 and Figure 5 that, once formed, O-MG adducts do not appear to undergo significant repair by 72-h postexposure. This is in

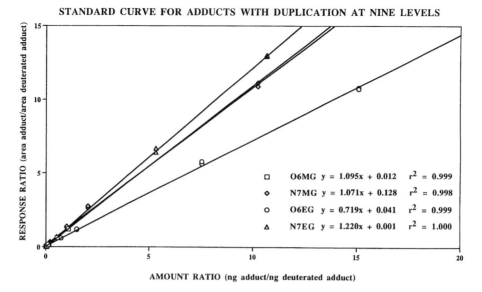

STANDARD CURVE FOR ADDUCTS WITH DUPLICATION AT NINE LEVELS

Figure 4   Standard curves for methylguanine and ethylguanine adducts. Standard curves were prepared by plotting amount ratios (nine levels of nondeuterated compound vs. constant amount of its deuterated analog) over a range representing an average of 30 fmol to 6.1 nmol of alkylguanines on column.

agreement with other workers who have concluded that fish have relatively slow rates of repair of alkylguanine adducts at the O^6 position.[11] Although O^6-alkylguanine-DNA alkyltransferase activity has been shown to be present, this repair enzyme appears to be depleted rapidly and is probably not inducible in fish.[13]

Quantitative analysis

With mass spectrometry, deuterium-labeled internal standards can be added to facilitate quantitative estimation that minimizes problems associated with variability in isolation techniques and variations in instrument performance. Since stable isotope-labeled analogs have virtually the same chemical and physical characteristics as the analytes, corrections are made to compensate for losses due to chemical and physical characteristics of the analytes.[31] Losses due to these parameters should be proportionally the same for the internal standards and the analytes. Thus, the *ratio* of each deuterium-labeled standard to its corresponding analog should be the same in the final step of analysis as when the standards are added, even though the absolute levels of both compounds may be lower in the final step.[31]

Examples of equations used to determine final values or adduct concentrations with this method are shown in Figure 6. Sample concentrations were determined by multiplying the sample peak area by a response factor which is dependent upon the area of the standard peak and the amount of standard injected onto the column. Sample concentrations were then corrected for

Table 3   Standard Response Curve Raw Data for O-MG and dO-MG

Level	O-MG ng on-col	Target m/z 294 peak area	Q1 m/z 309 peak area	Q2 m/z 295 peak area	dO-MG ng on-col	Target m/z 297 peak area	Q1 m/z 312 peak area	Q2 m/z 298 peak area	Ratios Amount	Ratios Response
1	0.0088	22370	ND	ND	1.6	1079773	378218	ND	0.0055	0.0207
1	0.0088	19855	ND	ND	1.6	1140163	398575	ND	0.0055	0.0174
2	0.018	21439	ND	ND	1.6	996947	351353	ND	0.011	0.0215
2	0.018	21533	ND	ND	1.6	1076620	375103	257665	0.011	0.0200
3	0.044	38365	ND	ND	1.6	1091366	389216	262247	0.028	0.0352
3	0.044	37881	ND	ND	1.6	1109003	393333	264166	0.028	0.0342
4	0.088	75299	ND	ND	1.6	1279383	448411	300465	0.055	0.0589
4	0.088	80684	ND	ND	1.6	1295207	455451	309537	0.055	0.0623
5	0.18	223391	77184	61662	1.6	1490311	522610	351355	0.11	0.1499
5	0.18	229798	76065	68569	1.6	1492975	514950	355636	0.11	0.1539
6	0.88	1157509	390276	281331	1.6	1854722	656190	440108	0.55	0.6241
6	0.88	1106338	372521	279523	1.6	1793102	631158	429635	0.55	0.6170
7	1.8	1959031	658652	473150	1.6	1622535	576925	383437	1.1	1.2074
7	1.8	1981477	660330	487049	1.6	1628194	567884	380194	1.1	1.2170
8	8.8	ND	ND	ND	1.6	ND	ND	ND	5.5	ND
8	8.8	ND	ND	ND	1.6	ND	ND	ND	5.5	ND
9	18	ND	ND	ND	1.6	ND	ND	ND	11	ND
9	18	ND	ND	ND	1.6	ND	ND	ND	11	ND

Note: ND, no data was obtained for O-MG at the top 2 levels due to detector electronics shutdown at these high levels.

Table 4 DNA Adduct Formation and Repair after Exposure to 10 ppm MAM-Acetate

Liver Tissue Concentrations (ng/mg), Internal Standard Calculation Method

Sample	Tiss wt mg	OMG ng/mg	OMG Q1 ratio*	7-MG ng/mg	7-MG Q1 ratio	O-EG ng/mg	O-EG Q1 ratio	7-EG ng/mg	7-EG Q1 ratio
Reag blk	79	0.051	44%	0.227	0%	0.050	85%	0.165	0%
C72	87	ND	NA	0.397	80%	ND	NA	0.017	0%
E-1B	83	0.146	55%	1.170	88%	0.032	68%	0.090	67%
E0B	74	0.248	46%	1.538	90%	ND	NA	0.050	0%
E4B	95	0.157	109%	2.697	108%	ND	NA	0.025	0%
EG8B	70	0.296	75%	3.450	90%	ND	NA	0.051	70%
E24B	65	0.364	57%	4.115	103%	ND	NA	0.048	47%
EG48B	60	0.296	85%	2.705	88%	0.246	94%	0.460	99%
E72B	99	0.279	20%	1.558	71%	ND	NA	0.038	0%

* Q1 response ratio expressed as percent of Standard response ratio, values <50% in italics.

Recovery of Surrogate Standards

Sample	dOMG	d7-MG	dO-EG	d7-EG
Reag blk	40%	3%	31%	7%
C72	17%	27%	17%	43%
E-1B	20%	31%	17%	60%
E0B	17%	23%	14%	38%
E4B	16%	14%	17%	14%
EG8B	21%	45%	17%	65%
E24B	15%	36%	15%	46%
EG48B	40%	36%	29%	61%
E72B	9%	16%	9%	28%

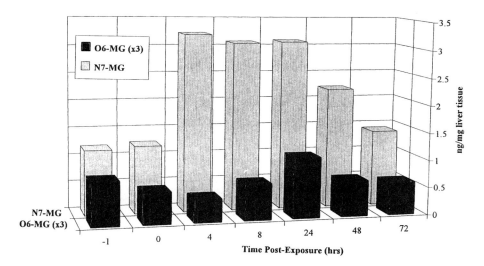

Figure 5 Methylguanine adduct concentrations in liver DNA of western mosqui-tofish before, mid-way through (–1), and after 2-h pulse exposure to 10 ppm MAM-Ac. The 0 time point denotes sampling immediately following removal of fish specimens to carcinogen-free water.

recovery of corresponding internal standards and tissue wet weight. Final values were expressed as nanograms of adducted guanine base per mg of liver tissue. Quality ratios are discussed below.

Figure 7 shows a total ion chromatogram (TIC) obtained in full scanning mode from a typical control sample. This illustrates that each sample contains an abundance of information that could be used in detection and quantification of DNA adducts, provided that suitable standards were obtained. In fact, the presence and corresponding concentrations of numerous DNA adducts, even in tissues from fish with uncertain exposure histories, could potentially be monitored simultaneously using this method.

Data reduction and quality assurance

Quality assurance procedures included the preparation of a reagent blank with each set of samples, repetitive analysis of standards and samples, and evaluation of internal standard recoveries and quality ratios (QR). QR for each analyte were determined by comparing the confirming ion (CI) intensity ratios for standards and samples, as follows: The response of the first CI, e.g., m/z 309 for O-MG, was divided by the response of the target ion (e.g., 294 for O-MG). Next the QR was calculated as a ratio of the sample CI ratio to standard CI ratio, expressed as a percentage. A result of 100% would be a perfect match, while values either less than or greater than 100% would indicate that there is less confidence in the identification of the analyte or that either the target ion response or CI response contained a co-eluting interference. A second confirming ion (m/z 295 for methylguanines, m/z 324 for ethylguanines) was also monitored; however, this signal was often

EQUATIONS USED:
Example: d7-MG and 7-MG for sample E24B (24-hr Exposed fish liver)

RECOVERY OF DEUTERATED INTERNAL STANDARD (d7-MG)

Given:	Measure:		Calculate:	
Conc. of Std.: 1.6ng/µl	Std. Peak Area:	=339,067	Response Factor:	=Amount Standard Injected/Std. Peak Area
Injection Vol.: 2µl				=(1.6*2)/3.39e05
				=9.44e-06
	Sample Peak Area: =120,769		Amount detected:	=Sample Peak Area*Response Factor
				=1.21e05*9.44e-06
				=1.14ng
			Recovery (%)	=Amount detected/Amount expected
				=1.14ng/3.2ng
				=36%

QUALITY (Q1) RATIO (7-MG)

Measure Peak Areas:			Calculate:	
Standard:	Target Ion:	=3,052,718	Standard Ion Ratio:	=5.19e05/3.05e06
	Confirming Ion	=518,962		=0.17
Sample (E24B):	Target Ion:	=1,038,554	Sample Ion Ratio:	=1.82e5/1.04e6
	Confirming Ion	=181,812		=0.18
			Quality Ratio:	=Sample Ion Ratio/Standard Ion Ratio
				=0.18/0.17
				=103%

SAMPLE CONCENTRATION

Given:	Measure Peak Areas:		Calculate:	
Conc. of Std.: =2.8ng/µl	Standard:	=3,052,718	Response Factor:	=Amount Standard Injected/Std. Peak Area
Injection Vol.: =2µl	Sample (E24B)	=1,038,554		=(2.8*2)/3.05e6
				=1.84e-06
			Sample Amount:	=Sample Peak Area*Response Factor
				=1.04e6*1.84e-06
				=1.91ng

LIVER TISSUE CONCENTRATION, Corrected for Internal Standard Recovery

Given:		Calculate:	
Total ng detected:	=1.91ng	Conc. Factor:	=Final Vol./(Tissue Wt.*Injection Volume)
Recovery of Internal Standard:	=36%		=100/(65*2)
Amount of Tissue:	=65mg		=0.77
Final Volume of Extract:	=100µl	Tissue Conc.:	=(Total ng detected/Recovery)*Conc. Factor
Injection Vol.:	=2µl		=(1.91/.36)*0.77
			=4.11ng/mg

Figure 6 Equations used in quantitative analysis and data reduction.

too noisy to provide useful QR data, and alternate ions are being evaluated for future inclusion in the QR calculation. A QR of 50% was arbitrarily defined as an operational lower limit for identification of analytes. Quality ratios for all samples from one experiment are shown in Table 4.

During development of this methodology, some samples expected to have similar DNA adduct concentrations have shown considerable variation. This variation may be attributed to methodological error from sample preparation and extraction, sample matrix effects, and instrument variability, as well as to biological variability (i.e., individual fish will have different rates of xenobiotic metabolism). Wet tissue weights, which are inherently variable, were used in initial experiments. In future work, adduct concentrations will be expressed on an "amount of adduct per unit of *guanine*" basis as measured by GC/MS and, thus, may be compared among samples and between tissue types.

Recovery of the deuterated internal standards has also shown considerable variation, and has been generally low. Much effort went into eliminating moisture from the samples prior to addition of the derivatizing reagents, and it has become apparent that moisture alone is not the sole factor in low recoveries. We observed sample matrix effects and nonspecific binding of

File   : C:\HPCHEM\1\DATA\ADDUCTS\ADDT4178\4178CG2S.D
Operator   : DJ
Acquired   : 27 Jun 94   5:28 pm using AcqMethod ADDCTSCN
Instrument :   5970 - In
Sample Name: MAC LAW:CG2 SCAN
Misc Info : EMV=+400 2UL ALS   5UL STDS 6/27/94
Vial Number: 8

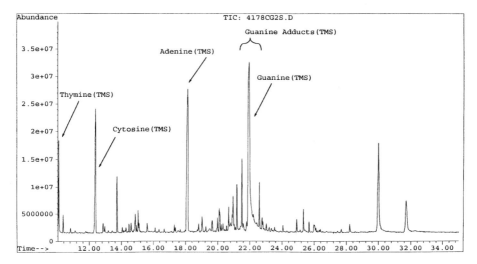

Figure 7 Total ion chromatogram (TIC) obtained in full scanning mode from a typical control sample. Note that multiple purine and pyrimidine bases, both modified and unmodified, could be monitored simultaneously from a single sample, provided that suitable standards were obtained.

sample components in the GC injection port which resulted in a mostly reversible loss of response. Injections of samples were followed by injections of the standards previously used for calibration of the instrument, and loss of response was noted. This response then returned to original levels upon repetitive injections of the same standards. Presumably, injections of the derivatizing reagents alone would result in an improved response. We are currently investigating additional sample handling and cleanup techniques, such as filtration, activated carbon adsorption, and ion exchange steps prior to derivatization in order to eliminate this problem and improve recovery of the internal standards.[38]

Conclusions

DNA adducts resulting from exposure to mutagenic and/or carcinogenic agents are detectable at femtomole levels in small fish species using the isotope dilution GC/mass spectrometry method described above. In method development experiments, levels of O^6-methyl guanine ranging from approximately 330 to 1100 fmol (55 to 185 pg) per microgram of liver DNA in *Gambusia affinis*, measured in the first 72 h following short-term methylating agent exposure, were correlated with a 33% liver tumor incidence after 25 weeks in parallel experiments. Although further studies are needed with a broader range of mutagenic and carcinogenic agents representing various mechanistic classes and with other small fish species, this method appears to be useful for studies of DNA adduct formation and repair in aquatic species. These findings, along with additional mechanistic studies, should also aid in validation of small fish species both as environmental sentinels and as useful models for mechanistic carcinogenicity studies in the laboratory.

References

1. Couch, J.A. and J.C. Harshbarger. 1985. Effects of carcinogenic agents on aquatic animals: an environmental and experimental overview. *Environ. Carcinogen. Revs.* 3:63–105.
2. Harshbarger, J.C. and J.B. Clark. 1990. Epizootiology of neoplasms in bony fish of North America. *Sci. Total Environ.* 94:1–32.
3. Harshbarger, J.C., P.M. Spero, and N.M. Wolcott. 1993. Neoplasms in wild fish from the marine ecosystem emphasizing environmental interactions. In *Pathobiology of Marine and Estuarine Organisms*, Eds. J.A. Couch and J.W. Fournie, 157–176. Boca Raton, Florida: CRC Press.
4. Stegeman, J.J. and J.J. Lech. 1991. Cytochrome P-450 monooxygenase systems in aquatic species: carcinogen metabolism and biomarkers for carcinogen and pollutant exposure. *Environ. Health Perspec.* 90:101–109.
5. Stegeman, J.J. and M.E. Hahn. 1994. Biochemistry and molecular biology of monooxygenases: current perspectives on forms, functions, and regulation of cytochrome P450 in aquatic species. In *Aquatic Toxicology: Molecular, Biochemical, and Cellular Perspectives*, Eds. D.C. Malins and G.K. Ostrander, 87–206. Boca Raton, Florida: CRC/Lewis Publishers.
6. Means, J.C., G.W. Winston, and Z.-M. Yang. 1995. Induction of P450-1A1 in embryo and larval fish by selected mutagens. In: *Bioavailability and Genotoxicity of Produced Water Discharges Associated with Offshore Production Operations*, Eds. G.W. Winston and J.C. Means, 185 pp. U.S. Department of the Interior, Minerals Management Service Final Report.
7. Baksi, S.M. and J.C. Means. 1987. Preparation of chromosomes from fish eggs and larvae for cytogenetic analysis. *J. Fish. Biol.* 32:321–325.
8. Daniels, C.B. and J.C. Means. 1990. Assessment of the genotoxicity of produced water discharges associated with oil and gas production using a fish egg and larval test. *Marine Environ. Res.* 28:303–307.
9. Daniels, C.B., C.B. Henry, and J.C. Means. 1990. Coastal oil drilling produced waters: Chemical characterization and assessment of genotoxicity using chromosomal aberrations in *Cyprinodon variegatus*. In *Aquatic Toxicology and Risk Assessment: Thirteenth Volume*, ASTM STP 1096, Eds. W.G. Landis and W.H. van der Schalie, 356–371. Philadelphia: ASTM.
10. Shugart, L., J. Bickham, G. Jackim, et al. 1992. DNA alterations. In *Biomarkers: Biochemical, Physiological, and Histological Markers of Anthropogenic Stress*, Eds. R.J. Huggett, R.A. Kimerle, P.M. Mehrle, Jr., and H.L. Bergman, 125–154. Boca Raton: CRC/Lewis Publishers.
11. Maccubbin, A.E. 1994. DNA adduct analysis in fish: laboratory and field studies. In *Aquatic Toxicology: Molecular, Biochemical, and Cellular Perspectives*, Eds. D.C. Malins and G.K. Ostrander, 267–294. Boca Raton: CRC/Lewis Publishers.
12. Varanasi, U., W.L. Reichert, and J.E. Stein. 1989. ^{32}P-postlabeling analysis of DNA adducts in liver of wild English sole (*Parophrys vetulus*) and winter flounder (*Pseudopleuronectes americanus*). *Cancer Res.* 49:1171–1177.

13. Fong, A.T., J.D. Hendricks, R.H. Dashwood, S. Van Winkle, and G.S. Bailey. 1988. Formation and persistence of ethylguanine in liver DNA of rainbow trout (*Salmo gairdneri*) treated with diethylnitrosamine by water exposure. *Food Chem. Toxicol.* 26:699–704.

14. Hawkins, W.E., R.M. Overstreet, and W.W. Walker. 1988. Carcinogenicity tests with small fish species. *Aquatic Toxicol.* 11:113–128.

15. Law, J.M., W.E. Hawkins, R.M. Overstreet, and W.W. Walker. 1994. Hepatocarcinogenesis in western mosquitofish (*Gambusia affinis*) exposed to methylazoxymethanol acetate. *J. Comp. Pathol.* 110:117–127.

16. Dees, L.T., 1961. The mosquitofish, *Gambusia affinis*. Bureau of Commercial Fisheries, Fish and Wildlife Service, U.S. Department of the Interior.

17. Law, J.M. 1995. *Mechanisms of chemically-induced hepatocarcinogenesis in western mosquitofish (Gambusia affinis)*. Ph.D. dissertation. Louisiana State University, Baton Rouge. 195 pp.

18. Feinberg, A. and M.S. Zedeck. 1980. Production of a highly reactive alkylating agent from the organospecific carcinogen methylazoxymethanol by alcohol dehydrogenase. *Cancer Res.* 40:4446–4450.

19. Stanton, M.F. 1966. Hepatic neoplasms of aquarium fish exposed to *Cycas circinalis*. *Fed. Proc.* 25:661.

20. Hawkins, W.E., R.M. Overstreet, J.W. Fournie, and W.W. Walker. 1985. Development of aquarium fish models for environmental carcinogenesis: tumor induction in seven species. *J. Appl. Toxicol.* 5:261–264.

21. Beranek, D.T., C.C. Weis, and D.H. Swenson. 1980. A comprehensive quantitative analysis of methylated and ethylated DNA using high pressure liquid chromatography. *Carcinogenesis* 1:595–606.

21a. Swenson, D.H. 1995. unpublished data.

22. Farmer, P.B., A.B. Foster, M. Jarman, and M.J. Tisdale. 1973. The alkylation of 2′-deoxyguanosine and of thymidine with diazoalkanes. *Biochem. J.* 135:203–213.

23. Balsiger, R.W. and J.A. Montgomery. 1960. Synthesis of potential anticancer agents. XXV. Preparation of 6-alkoxy-2-aminopurines. *J. Am. Chem. Soc.* 25:1573–1575.

24. Hawkins, W.E., W.W. Walker, and R.M. Overstreet. 1995. Practical carcinogenicity tests with small fish species. In *Fundamentals of Aquatic Toxicology*. Ed. G.M. Rand. 421–446. London: Taylor and Francis.

25. Powers, D.A. 1989. Fish as model systems. *Science* 246:352–358.

26. Walker, W.W., C.S. Manning, R.M. Overstreet, and W.E. Hawkins. 1985. Development of aquarium fish models for environmental carcinogenesis: an intermittent-flow exposure system for volatile, hydrophobic chemicals. *J. Appl. Toxicol.* 5:255–260.

27. Metcalfe, C.D. 1989. Tests for predicting carcinogenicity in fish. *CRC Rev. Aquatic Sci.* 1:111–129.

28. Malins, D.C. and G.K. Ostrander, Eds. 1994. *Aquatic Toxicology: Molecular, Biochemical, and Cellular Perspectives*. Boca Raton: CRC/Lewis Publishers. 539 pp.

29. Strauss, W.M. 1987. Preparation of genomic DNA from mammalian tissue. In *Current Protocols in Molecular Biology*, Eds. F.M. Ausubel, et al., 2.2.1–2.2.3. Media, Pennsylvania: John Wiley & Sons, Inc.

30. Dizdaroglu, M. 1993. Quantitative determination of oxidative base damage in DNA by stable isotope-dilution mass sprectrometry. *FEBS Lett.* 315:1–6.

31. Watson, J.T. 1990. Selected-ion measurements. *Meth. Enzymol.* 193:86–106.

32. Lawley, P.D. 1984. Carcinogenesis by alkylating agents. In *Chemical Carcinogens*, Ed. C.E. Searle, 325–484. Washington, D.C.: American Chemical Society Monographs.

33. Loveless, A. 1969. Possible relevance of O-6 alkylation of deoxyguanosine to the mutagenicity and carcinogenicity of nitrosamines and nitrosamides. *Nature* 223:206–207.

34. Swenson, D.H. and P.D. Lawley. 1978. Alkylation of deoxyribonucleic acid by carcinogens dimethyl sulphate, ethyl methanesulphonate, *N*-ethyl*N*-nitrosourea, and *N*-methyl-*N*-nitrosourea. *Biochem. J.* 171:575–587.

35. Swenson, D.H. 1983. Significance of electrophilic reactivity and especially DNA alkylation in carcinogenesis and mutagenesis. In *Developments in the Science and Practice of Toxicology*, Eds. A.W. Hayes, R.C. Schnell, and T.S. Miya, 247–254. London: Elsevier Science Publishers B.V.

36. Dresler, S.L. 1989. DNA repair mechanisms and carcinogenesis. In *The Pathobiology of Neoplasia*, Ed., A.E. Sirica, 173–197. New York: Plenum Press.

37. Singer, B. and J.T. Kusmierek. 1982. Chemical mutagenesis. *Annu. Rev. Biochem.* 52:655–693.

38. Lakings, D.B., C.W. Gehrke, and T.P. Waalkes. 1976. Determinations of trimethylsilyl methylated nucleic acid bases in urine by gas-liquid chromatography. *J. Chromatog.* 116:69–81.

chapter eleven

Application of the alkaline unwinding assay to detect DNA strand breaks in aquatic species

Lee R. Shugart

Introduction

DNA integrity

Maintaining the integrity of an organism's DNA is of utmost importance to its survival, and to ensure that DNA damage is corrected, an elaborate monitoring/repair system[1] has evolved. Shugart et al.[2] have detailed the various types of structural changes that may occur to DNA under normal cellular conditions as well as after exposure to genotoxicants. It should be noted that the persistence of DNA damage, for whatever reasons, makes it possible to detect changes in the integrity of the DNA that may be indicative of exposure to genotoxic agents.

Chemical or physical agents are considered genotoxicants if they modify the structure of DNA and adversely affect the integrity (structure and/or function) of DNA. When an organism is exposed to genotoxicants in its environment, determination of the effects on the integrity of its DNA can be assessed by observing changes at many different levels of biological complexity.[2] Approaches for detecting changes to DNA integrity include assays that demonstrate nucleotide base modifications,[3-7] DNA strand breakage,[8] sister chromatid exchange or other chromosomal aberrations,[9,10] unscheduled DNA synthesis,[11] micronuclei formation,[12,13] point mutations,[14,15] and variations in nuclear DNA content.[16] Of these approaches, the determination of DNA strand breakage is probably the least technically complex and/or time consuming because it does not involve extensive chemical or cytological manipulation of the samples.

DNA strand breakage

DNA strand breakage is not an uncommon occurrence in a cell. Heat energy causes thousands of abasic sites per cell per day which, however, are rapidly repaired.[17] This is an example of an insult to DNA that indirectly results in strand breakage (i.e., the initial damage is a loss of a base from the DNA chain, the repair of this damage results in a temporary gap in the DNA molecule). Ionizing radiation can cause strand breakage directly, whereas other physical agents such as UV light and chemical agents that are genotoxic, potentiate alterations to the DNA molecule that are candidates for repair (e.g., photoproducts, adducts, etc.) and thus for the occurrence of strand breaks.[2]

Alkaline unwinding assay

From the previous discussion it is obvious that the detection and quantitation of transient strand breakage in DNA can provide useful information on the integrity of the genetic material of an organism and how that organism may be responding to any of a number of genotoxic agents in its environment.

Most methodologies for performing a particular DNA strand break assay are based on the general principle that, under *in vitro* conditions, the rate at which single stranded DNA is released from the duplex DNA at high pH is proportional to the number of strand breaks in the DNA molecule.[18] It should be noted that under these assay conditions alkaline labile sites in the DNA molecule will also be detected because they are chemically converted to single strand breaks.

The alkaline unwinding assay described in detail below[8,19] is a modification of several existing techniques and methods previously described in the scientific literature.[18,20,21] The assay was modified to aid in the rapid and facile isolation of highly polymerized, double-stranded DNA from the liver and other organs of fish, as well as intact small aquatic species, and to minimize the effort needed to estimate the amount of single-stranded molecules that occur subsequent to strand separation under alkaline conditions. DNA isolation is accomplished by homogenizing the intact organism, or selected organs, in 1 N NH_4OH/0.2% Triton X-100.[22] Further purification is achieved by differential extraction with phenol/chloroform/isoamyl alcohol (25/24/1 v/v) and passage through a molecular sieve column (Sephadex G50). The DNA is then subjected to strand separation under defined conditions of pH and temperature. Periodically thereafter, aliquots of the preparation are neutralized and mechanically sheared in the presence of detergent to separate single-stranded from double-stranded DNA and to prevent renaturation. The amounts of these two types of DNA that are present after neutralization are then quantified by measuring the fluorescence that results when bisbenzamidazole (Hoechst dye 33258) is added.[23-25] Since strand unwinding takes place at strand breaks within the DNA molecule, the amount of double-stranded DNA remaining after a given period of alkaline unwinding will be inversely proportional to the number of strand breaks present at the initiation of the process, provided renaturation is prevented.

Materials required

Equipment and supplies

Fluorometer: Perkin-Elmer LS-5 fluorescence spectrophotometer
Centrifuge: Beckman J-21B, refrigerated
Constant temperature bath: Sybron thermolyne, DR1 bath
Pipette: Gilson pipetman (1 to 20 μl)

Chemicals

Reagent grade chemicals: HCL, NaOH, NH_4OH, NaCl, $MgCl_2$, KH_2PO_4, Na_2HPO_4
Bisbenzamidazole dye: Polyscience Inc., Hoechst dye 33258, product #9460
DNA: Sigma Chemical Co., calf thymus "highly polymerized," product #D-1501
Molecular sieve gel: Pharmacia Fine Chemicals, Inc., Sephadex-G50, dry powder
Ethylenediaminetetraacetic acid: Sigma Chem. Co., disodium salt, product #E-4884
Tris base: Sigma Trizma pH 7.4, product #T-4003
Surfactant: Sigma Chemical Co., Triton X-100, product #X-100 and sodium dodecyl sulfate, product #L-4505
Phenol/Chloroform/Isoamyl Alcohol: Gibco BRL (25/24/1 v/v), product #15593
Water: J.T. Baker, Inc., product #4218

Reagents

The following reagents are prepared according to standard laboratory procedures from the chemicals listed above.

1 N NH_4OH in 0.2% Triton X-100 (v/v)
G-50 buffer (150 mM NaCl, 10 mM Tris, pH 7.4, 1 mM $MgCl_2$, 0.5 mM EDTA)
25 mM NaCl
2 mM EDTA in 0.2% Triton X-100 (v/v)
0.2 M potassium phosphate buffer, pH 6.9
0.05 N HCl
0.05 N NaOH
Hoechst dye 33258 in water (1 mg/ml)
DNA standard in G-50 buffer (50 μg/ml)

Procedures

The protocol outlined below describes the facile isolation of DNA from the soft tissue of aquatic species (or intact small species) in a form suitable for the estimation of strand breakage by the alkaline unwinding method. DNA in solution is forced to unwinding (i.e., pass from the double-stranded to the single-stranded moiety) by raising the temperature and pH of the solution. At the end of the unwinding period the fraction of DNA remaining in the double-stranded form is estimated with Hoechst dye 33258 which demonstrates a reduced fluorescence when bound to the single-stranded as compared to the double-stranded form of DNA.[23-25]

DNA sample preparation

A modification of the procedure of Downs and Wilfinger[22] was used. Note: all steps are performed at 4°C except the G-50 step which can be done at room temperature. The sample is placed in a ground glass homogenizing tube (2-ml), and 1 ml of 1N NH_4OH in 0.2% Triton X-100 is added for each 200 to 400 mg of sample. Complete homogenization occurs after 6 to 10 strokes. The homogenate is transferred to a centrifuge tube with 2 ml of water and 6 ml of phenol/chloroform/isoamyl alcohol (25/24/1 v/v) are added, the contents mixed by inversion, and the mixture allowed to stand for 10 min. The phases are separated by centrifugation at $10,000 \times g$ for 20 min, and 1.25 ml of the aqueous phase are placed on a Sephadex G-50 column (1 cm i.d.; 3.5 ml settled bed volume) previously equilibrated in G-50 buffer. The sample is allowed to flow into the column and the eluate is discarded. This is followed by 1.25 ml of G-50 buffer, and the eluate, which contains the DNA, is recovered.

DNA determination

A modification of the procedure of Kanter and Schwartz[24] is used. To a test tube are added 100 μl of 25 mM NaCl, 5 μl of 0.2% SDS in 2 mM EDTA, 3 ml of 0.2 M potassium phosphate buffer, pH 6.9, and 3 μl of Hoechst dye 33258 (1 mg/ml). The contents are mixed, 100 μl of DNA sample or G-50 buffer (blank) are added, and the contents mixed again. After 15 min in the dark, the fluorescence of the sample is measured (λ_{ex}: 360 nm, λ_{em}: 450 nm).[23] A standard DNA solution is prepared using cal thymus DNA dissolved in G-50 buffer (1 A_{260} = 50 μg DNA).

Alkaline unwinding

A modification of the procedures of Kanter and Schwartz,[24] and Daniel et al.[25] were used. To a test tube containing 100 μl of DNA sample in G-50 buffer, 50 μl of 0.05 N NaOH is added with rapid mixing (3 sec total time) and the contents incubated at a specified temperature (see discussion below). Incubation continues for a defined period of time at the end of which the sample is neutralized by the rapid addition (3 sec as above) of 50 μl of 0.05 N HCl. This is followed immediately by the addition of 5 μl of 0.2% SDS in 2 mM EDTA and the forceful passage of the sample through a 20-gauge needle several times (15 sec total time). Fluorescence of the sample is measured as described above upon the addition of 3 ml of 0.2 *M* potassium phosphate buffer, pH 6.9, and 3 µl of Hoechst dye 33258 (1 mg/ml).

Results and discussion

Theoretical background

In dilute alkali, DNA undergoes a time-dependent transformation to the single-stranded form, a process that is accelerated by the presence of strand breaks. The theoretical background for estimating strand breaks in DNA by alkaline unwinding was established by Rydberg,[18] and is summarized by the equation:

$$\ln F = -(K/M)(t^{\beta}) \tag{1}$$

where F is the fraction of double-stranded DNA (referred to as the F value) that is present in the mixture of double- and single-stranded DNA at the end of the alkaline unwinding assay, K is a constant, t is time, M is the number average molecular weight between two breaks, and β is a constant less than 1 that is influenced by the conditions for alkaline unwinding.

lnF values are used to determine the relative number of strand breaks (N value) between two DNA samples and is expressed as follows:

$$N = (\ln F_s / \ln F_r) - 1 \tag{2}$$

where F_s and F_r are the mean F values of DNA from a sampled and reference preparation respectively. N values greater than zero indicate that the sampled DNA has more strand breaks than the reference DNA. Calibration of number of strand breaks per unit of molecular weight of DNA will require comparison of alkaline unwinding data with physical measurements of the molecular weight of DNA that lies between DNA strand breaks. However, the reader is referred to the scientific literature for a more detailed discussion on this subject, References 18, 21, 24, and 25.

F value calculations

Hoechst dye 33258 binds with double-stranded DNA in solution (G-50 buffer) to form a stable, fluorescent complex. When the DNA is in the single-stranded form, the intensity of fluorescence of the bound dye is reduced (Figure 1) to about one-half of that observed with the double-stranded moiety. This phenomenon constitutes the basis for determining the amount of double- and single-stranded DNA that is present in an aliquot of DNA during the alkaline unwinding assay[8,19,24] and thus the F value.

An F value for a particular DNA preparation is obtained as follows (see Procedures section for details). DNA is isolated through the Sephadex G-50 column step and divided into three aliquots. A DNA determination is performed on aliquot No. 1, and the other two are subjected to alkaline unwind-

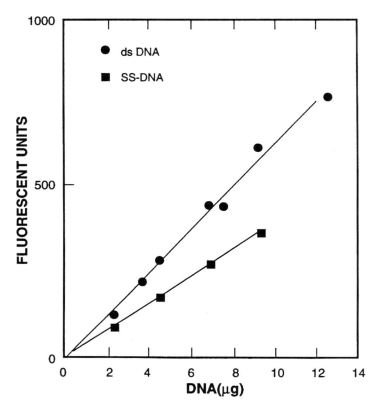

Figure 1 Fluorescence of double- and single-stranded calf thymus DNA exposed to Hoechst dye 33258. Double-stranded DNA was converted to single-stranded DNA by heating at an alkaline pH for 60 min at 80°C. (From Shugart, L.R. 1988. *Aquat. Toxicol.* 13:43–52. With permission.)

ing. Of these two, aliquot No. 2 is allowed to denature until no double-stranded DNA remains, usually by incubating the sample at an elevated temperature (80°, see Figure 2). Aliquot No. 3 is allowed to only partially denature (i.e., incubation is at a temperature of 38° and the reaction is stopped while there is still double- and single-stranded DNA present). The F value is obtained using the following equation:

$$F = (f \text{ No. } 3 - f \text{ No. } 2)/(f \text{ No. } 1 - f \text{ No. } 2) \tag{3}$$

where f is the measured fluorescence of the various aliquots at the end of the incubation period. The data presented in Figure 2a illustrates the rate of change of fluorescence under alkaline unwinding conditions for DNA obtained from the liver of a bluegill sunfish (*Lepomis macrochiris*) as a function of the temperature at which the alkaline unwinding was performed. At zero time when the DNA in the duplex form, f is approximately 230. Because denaturation is a chemical reaction involving the breaking of hydrogen bonds, temperature will affect the rate of unwinding. Thus, after 20 min

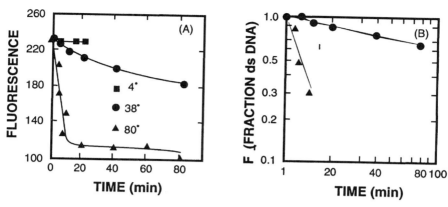

Figure 2 Effect of temperature on the rate of alkaline unwinding of DNA isolated from the liver of a bluegill sunfish. Each data point represents the average of 3 separate determinations. Panel A: fluorescence of Hoechst dye 33258 with 4 µg of DNA plotted on linear scale vs. time of incubation at temperature specified. Panel B: replot of data from Panel A according to Eq. 1 as the log–log plot of fraction of double-stranded DNA (F) value vs. time of incubation. (From Shugart, L.R. 1988. *Aquat. Toxicol.* 13:43–52. With permission.)

incubation at 80°C all the double-stranded DNA present at time zero has unwound to the single-stranded form and the measured fluorescence has decreased by 50% ($f = 115$). Conversely, at 4°C during the same time interval no change in fluorescence is detectable, while at 38°C the fluorescence has decreased by about 15%. F values are arbitrary, but become fixed when the time and temperature for alkaline unwinding are specified. For example from Figure 2a, after 40 min at 38°C, F is approximately 0.74. From equation (2): $F = (200 - 115)/(230 - 115) = 0.74$.

The experimental data from Figure 2a was replotted according to Equation 1 (log–log plot of F value vs. time of incubation). The fit of a straight line shown in Figure 2b indicates that the Rydberg equation (Eq. 1) can be applied to the data generated by this technique.

It should be noted that the decrease in fluorescence observed when DNA is denatured in the alkaline unwinding assay is an intrinsic property of DNA[23,24,26] and therefore the magnitude of this change should not always be assumed to be 50%, but should be determined for the DNA of each organism investigated.

Examples

Laboratory investigations

Polycyclic aromatic hydrocarbons, of which benzo[a]pyrene (BaP) is a representative compound, are an important class of toxicologically relevant chemicals because of their prevalence in the environment and the indication that they are the etiological factors associated with the high frequency of certain tumors in aquatic species.[27,28] The DNA obtained from the liver of

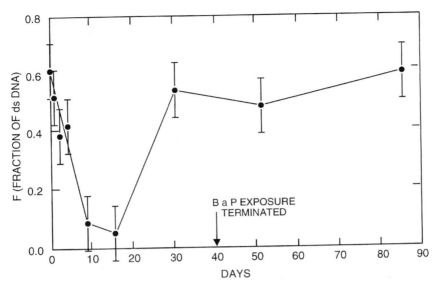

Figure 3 Occurrence of DNA strand breaks in liver of DNA of bluegill sunfish during chronic exposure to benzo[a]pyrene. Data expressed as *F* values (fraction of double-stranded DNA present in fish examined). Each point is the average of data obtained from four separate fish (± s.e.m.). (From Shugart, L.R. 1988. *Aquat. Toxicol.* 13:43–52. With permission.)

bluegill sunfish exposed chronically in the laboratory to a concentration of 1 µg/l of BaP in their water for a period of 40 days was examined by the alkaline unwinding technique outlined above.[19] The *F* data obtained are shown in Figure 3 and indicate that upon exposure to BaP strand breaks occur in the organism's liver DNA in a temporal manner. It should be noted that under the experimental conditions imposed, strand breakage in the DNA is a transitory state with the level reaching a maximum around day 16 and returning to a normal value by day 32. This later response is attributed to the organism's ability to repair such damage and therefore makes it difficult to correlate strand breakage with carcinogenic potencies of deleterious chemicals. Furthermore, other variables associated with a specific chemical such as absorption, metabolism, and excretion will ultimately affect the formation of this type of lesion and thus adds another layer of complexity to the dose response relationship. By substituting ionizing radiation as the genotoxicant, the phenomenon of DNA repair can be examined in the absence of chemical metabolism. The alkaline unwinding assay has been used to assess repair of DNA strand breakage in fathead minnows (*Pimephales promelas*) exposed to X-rays and gamma radiation.[29]

The exposure of bluegill sunfish to a mixture of chemicals via contaminated sediment under laboratory conditions[30] was performed to in order to examine the expression of a number of biological responses in the organism. Analysis of liver DNA for the occurrence of strand breaks showed kinetics initially similar to that reported above from exposure to a single chemical. With time, however, DNA strand breakage reappeared.

Field studies

Early in 1987, the detection of excessive strand breakage in the DNA of several aquatic species was implemented as a biological monitor for environmental genotoxicity as a part of the Biological Monitoring and Abatement Program for the U.S. Department of Energy (USDOE) Reservation in Oak Ridge, Tennessee. DNA strand breakage as an endpoint of genotoxicant insult was used for two important reasons. First, it is compatible with routine monitoring as the analysis (alkaline unwinding assay) for this type of damage is easy to perform and cost effective; and second, the assay provides a measure of DNA strand breaks arising from several contaminant-mediated processes.[2] Examples with two different aquatic species will suffice to demonstrate the suitability of the approach under field conditions.

Two species of turtles, the common snapping turtle (*Chelydra serpentina*) and the pond slider (*Trachemys scripta*) were compared for their usefulness as biological sentinels for environmental genotoxicants in White Oak Lake on the USDOE Reservation.[31] White Oak Lake is a settling basin for low-level radioactive and nonradioactive wastes generated at ORNL since 1943 and supports a high diversity of turtle species with *T. scripta* the most abundant and *C. serpentina* as the second most abundant. Cesium-137, cobalt-60, strontium-90, and tritium contribute most of the radioactivity to the lake. Species-specific data collected on DNA strand breakage in turtles captured in White Oak Lake were compared to Bearden Creek embayment, a reference site with similar biota but with no known history of contamination by hazardous chemicals. Over the entire course of the study, genotoxic stress was evident in both species taken from White Oak Lake. This is graphically represented in Figure 4, in which individual F values are plotted in relation to when and where the turtles were captured. The F values for both species of turtles reveal a significant ($p < 0.001$) site effect and indicate that the DNA in these species have higher levels of strand breaks than the same species from the reference site. It should be noted that Lamb et al.[32] also detected DNA damage by flow cytometric analysis in turtles occupying seepage basins containing radioactive contaminants.

Analyzing for strand breaks in the DNA of sunfish has been employed as a biological marker for environmental genotoxicity as part of the Biological Monitoring and Abatement Program at East Fork Poplar Creek.[33] This creek is the receiving stream for industrial effluent from the USDOE reservation in Oak Ridge, TN. Water and sediments downstream contain metals, organic chemicals, and radionuclides discharged over many years of operation.[33]

DNA strand break data (F values), measured in sunfish from the head waters of the creek (near the USDOE reservation) and at Hinds Creek (reference stream) over a period of four years are presented in Figure 5. Two points are clear: (a) DNA structural integrity of the sunfish from the reference stream is high and relatively constant (large F value); and (b) DNA structural integrity of the sunfish from East Fork Poplar Creek improved during the study period to reach levels similar to those for Hinds Creek. In all probability, the large genotoxic response observed in sunfish from East Fork Poplar Creek during the years 1987 and 1988 (small F value) was related to the

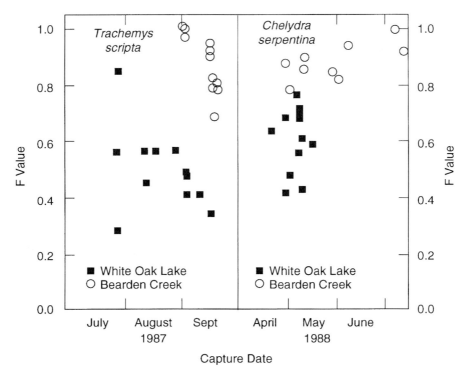

Figure 4   Fraction of double-stranded DNA (*F* value) in liver samples of *Trachemys scripta* and *Chelydra serpentina* collected from the Oak Ridge Reservation in TN. (From Meyers-Schöne, L.R. et al. 1993. *Environ. Tox. Chem.* 12:1487–1496. With permission; copyright SETAC.)

release of chemicals from the USDOE reservation. Diminution of this response in subsequent years may be due to the remedial activities that occurred on the USDOE reservation to attenuate the release of pollutants. Included in these activities were the capping of existing settling basins, the creation of a new settling basin, and the treatment of wastewater before discharge. However, the possibility that there has been an adaptive response over time by the resident population of sunfish to their environment cannot be excluded.

New directions

Advances in any scientific field, such as aquatic toxicology, will depend upon the implementation of new methods and techniques that possess the required selectivity and sensitivity. Current methodologies must be constantly refined or discarded if they fail to answer questions posed by the scientists in the field. For example, chemical genotoxicants primarily produce single-strand breakage, ionizing radiation may produce both single- and

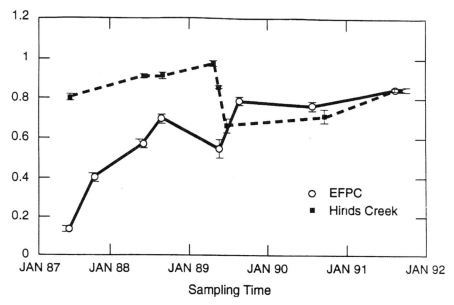

Figure 5 Temporal status of double-stranded DNA (*F* value) in liver samples of sunfish from East Fork Poplar Creek (contaminated stream) and Hinds Creek (reference stream) over a 4-year period.[35]

double-stranded breaks. Because the alkaline unwinding assay cannot differentiate between these two types of strand breaks, another method, electrophoresis of DNA on agarose, was introduced as an ancillary analytical technique to overcome this difficulty.[34] Migration of the DNA within the gel matrix is size-dependent, and detection is easily accomplished after staining with ethidium bromide. By conducting the assay under alkaline (pH 12) and neutral (pH 7) conditions, both double- and single-strand breaks can be detected. Under alkaline electrophoretic conditions the duplex structure of the DNA molecule is completely disrupted, resulting in various single strands of DNA, the lengths of which will depend on the type (both single- and double-stranded breaks) and numbers of both initially present in the DNA molecule. Conversely under neutral electrophoretic conditions the duplex structure of DNA is not disrupted, and migration of DNA within the gel will depend on small duplex structures produced by double-stranded breaks only.

Initial observations using the alkaline unwinding assay documented strand breaks in the DNA of a resident population of mosquitofish (*Gambusia affinis*) inhabiting retention ponds heavily contaminated with radionuclides. Gel electrophoresis to detect both types of strand breaks have helped expand upon our initial observations as to the ecological consequences of this exposure by demonstrating a correlation between reduced fecundity and double-strand breaks.[35]

Acknowledgment

The author is a senior research staff member in the Environmental Sciences Division of the Oak Ridge National Laboratory which is managed by Martin Marietta Energy Systems, Inc. for the U.S. Department of Energy under contract number DE-AC05-84OR21400. This is Environmental Sciences Division publication No. 4426.

References

1. Sancar, A., and Sancar, G. B. 1988. DNA repair enzymes. *Ann. Rev. Biochem.*, 57:29–67.
2. Shugart, L. R., Bickham, J., Jackim, G., McMahon, G., Ridley, W., Stein, J., and Steiner, S., DNA alterations, in *Biomarkers: Biochemical, Physiological, and Histological Markers of Anthropogenic Stress*, Huggett, R. J., Kimerle, R. A., Mehrle, P. M., and Bergman, H. L., Eds., Lewis Publishers Inc., Boca Raton, FL, 1992, pp. 125–153.
3. Shugart, L.R., J.F. McCarthy, B.D. Jimenez, and J. Daniel. 1987. Analysis of adduct formation in the bluegill sunfish (*Lepomis macrochirus*) between benzo[a]pyrene and DNA of the liver and hemoglobin of the erythrocyte. *Aquat. Toxicol.* 9:319–327.
4. Shugart, L.R. 1990. 5-Methyl deoxycytidine content of DNA from bluegill sunfish (*Lepomis macrochirus*) exposed to benzo[a]pyrene. *Environ. Toxicol. Chem.* 9:205–208.
5. Dunn, B., J. Black, and A. Maccubbin. 1987. ^{32}P-Postlabeling analysis of aromatic DNA adducts in fish from polluted areas. *Cancer Res.* 47:6543–6548.
6. Malins, D.C., and R. Haimanot. 1991. The etiology of cancer: Hydroxyl radical-induced DNA lesions in histologically normal livers of fish from a population with liver tumors. *Aquat. Toxicol.* 20:123–130.
7. McCarthy, J.F., H. Gardner., M.J. Wolfe and L.R. Shugart. 1991. DNA alterations and enzyme activities in Japanese medaka (*Oryzias latipes*) exposed to diethylnitrosamine. *Neurosci. Biobehav. Rev.* 15:99–102.
8. Shugart, L.R. 1988. An alkaline unwinding assay for the detection of DNA damage in aquatic organisms. *Marine Environ. Res.*, 24:321–325.
9. Dixon, D.R. 1982. Aneuploidy in mussel embryos (*Mytilus edulis* L.) originates from a polluted dock. *Mar. Biol. Lett.* 3:155–161.
10. Maddock, M.B., H. Northrup, and T.J. Ellingham. 1986. Induction of sister-chromatid exchanges and chromosomal aberrations in hematopoietic tissue of a marine fish following *in vivo* exposure to genotoxic carcinogens. *Mutat. Res.* 172:165–175.
11. West, W.R., P.A. Smith, G.M. Booth and M.L. Lee. 1988. Isolation and detection of genotoxic components in a black river sediment. *Environ. Sci. Technol.* 22:224–228.
12. Das, R.K. and N.K. Nanda. 1986. Induction of micronuclei in peripheral erythrocytes of fish *Heteropneustes fossilis* by mitomycin C and paper mill effluent. *Mutat. Res.* 175:67–71.
13. Hose, J.E., J.N. Cross, S.C. Smith, and D. Deihl. 1987. Elevated circulating erythrocyte micronuclei in fishes from contaminated sites off southern California. *Mar. Environ. Res.* 22:167–176.

14. Yung-Jin, C., C. Mathews, K. Mangold, K. Marien, J. Hendricks, and G. Bailey. 1991. Analysis of *ras* gene mutations in rainbow trout liver tumors initiated by aflatoxin B1. *Mol. Carcinog.* 4:112–119.

15. McMahon, G., L.J. Huber, M.J. Moore, J.J. Stegeman and G.N. Wogan. 1990. Mutations in c-K-ras oncogenes in diseased livers of winter flounder from Boston Harbor. *Proc. Natl. Acad. Sci. U.S.A.* 87:841–845.

16. McBee, K. and J.W. Bickham. 1988. Petrochemical-related DNA damage in wild rodents detected by flow cytometry. *Bull. Environ. Contam. Toxicol.* 40:343–349.

17. Alberts, B., D. Bray, L. Lewis, M. Raff, K. Roberts, and J. D. Watson, *Molecular Biology of the Cell*, 2nd. ed., Garland Publishing Company Inc., New York, 1989, 220–227.

18. Rydberg, B. 1975. The rate of strand separation in alkali of DNA of irradiated mammalian cells. *Radiat. Res.*, 61:274–287.

19. Shugart, L.R. 1988. Quantification of chemically induced damage to DNA of aquatic organisms by the alkaline unwinding assay. *Aquat. Toxicol.* 13:43–52.

20. Kanter, P.M., and H.S. Schwartz. 1979. A hydroxylapatite batch assay for quantitation of cellular DNA damage. *Anal. Biochem.* 97:77–84.

21. Ahnstrom, G., and K. Erixon. 1980. Measurement of strand breaks by alkaline denaturation and hydroxyapatite chromatography, In: *DNA Repair*, Vol. 1, part A, E.C. Friedberg and P.C. Hanaweldt, Eds., Marcel Dekker, NY, pp. 403–419.

22. Downs, T.R., and W.W. Wilfinger. 1983. Fluorometric quantification of DNA in cells and tissue. *Anal. Biochem.* 131:538–547.

23. Cesarone, D.F., C. Bologenesi, and L. Santi. 1979. Improved microfluorometric DNA determination in biological material using 33258 Hoechst. *Anal. Biochem.* 100:188–197.

24. Kanter, P.M. and H.S. Schwartz. 1982. A fluorescence enhancement assay for cellular DNA damage. *Mol. Pharmacol.* 22:145–151.

25. Daniel, F.B., D.L. Haas, and S.M. Pyle. 1985. Quantitation of chemically induced DNA strand breaks in human cells via an alkaline unwinding assay. *Anal. Biochem.* 144:390–402.

26. Latt, S.A., and G. Stetten. 1976. Spectral studies on 33258 Hoechst and related bisbenzimidazole dyes useful for fluorescent detection of deoxyribonucleic acid synthesis. *J. Histochem. Cytochem.* 24:24–33.

27. Kraybill, H.F., O.J. Dawe, J.C. Harshbarger, and R.C. Tardigg. 1977. Aquatic pollutants and biological effects with emphasis on neoplasia. *Ann. N.Y. Acad. Sci.* 298:1–604.

28. Malins, D.C., M.M. Krahn, M.S. Myers, L.D. Rhodes, D.W. Brown, C.A. Krone, B.B. McCain, and S.L. Chan. 1985. Toxic chemicals in sediments and biota from creosote-polluted harbor; relationships with hepatic neoplasms and other hepatic lesions in English sole (*Parophrys vetulus*). *Carcinogenesis* 6:1463–1469.

29. Shugart, L.R., M.K. Gustin, D.M. Laird, and D.A. Dean. 1989. Susceptibility of DNA in aquatic organisms to strand breakage: effect of X-rays and gamma radiation. *Marine Environ. Res.* 28:339–343.

30. Theodorakis, C.W., S.J. D'Surney, J.W. Bickham, T.B. Lyne, B.P. Bradley, W.E. Hawkins, W.L. Farkas, J.F. McCarthy, and L.R. Shugart. 1992. Sequential expression of biomarkers in bluegill sunfish exposed to contaminated sediment. *Ecotoxicology* 1:45–73.

31. Meyers-Schone, L., L.R. Shugart, J.J. Beauchamp, and B.T. Walton. 1993. Comparison of two freshwater turtle species as monitors of radionuclide and chemical contamination: DNA damage and residue analysis. *Environ. Tox. Chem.* 12:1487–1496.

32. Lamb, T., J.W. Bickham, J.W. Gibbons, M.J. Smolen, and S. McDowell. 1991. Genetic damage in a population of slider turtles (*Trachemys scripta*) inhabiting a radioactive reservoir. *Arch Environ. Contam. Tox.* 20:138–142.

33. Shugart, L.R. DNA damage as an indicator of pollutant-induced genotoxicity. In: 13th Symposium on Aquatic Toxicology and Risk Assessment: Sublethal Indicators of Toxic Stress, Landis, W.G., and W.H. van der Schalie, Eds., Philadelphia, PA:ASTM, 1990;348–355.

34. Theodorakis, C.W., S.J. D'Surney, and L.R. Shugart. 1994. Detection of genotoxic insult as DNA strand breaks in fish blood cells by agarose gel electrophoresis. *Environ. Tox. Chem.* 13:1023–1031.

35. Shugart, L.R., and C.W. Theodorakis. 1994. Environmental genotoxicity: probing the underlying mechanisms. *Environ. Health Perspec.* 102:13–17.

chapter twelve

Autoradiographic detection of DNA replication and DNA repair in cells from aquatic species

Michael R. Miller

Introduction

Radiolabeled thymidine (dT) exposure followed by autoradiography provides a sensitive, reliable method of assessing both DNA repair synthesis, or unscheduled DNA synthesis (UDS), which is induced by genotoxic agents, and DNA replication, which indicates the proliferation state of cells or tissues. In this procedure, cultured cells, or animals, are treated with [^{3}H]dT, which is incorporated into newly synthesized DNA; specimens are fixed and coated with a thin film of photographic emulsion; after exposure in the dark for a suitable period of time, samples are developed and examined under a light microscope. Emulsion above nuclei in growing cells in the S phase of the cell cycle contains a high density of grains, due to the extensive incorporation of [^{3}H]dT into replicating DNA, whereas cells involved in UDS display limited grains over nuclei, due to the relatively small amount of DNA synthesis associated with UDS. This autoradiographic procedure is used in aquatic toxicology research to determine the genotoxic potential of test samples which induce UDS, as well as to characterize the proliferation state of cells or tissues and determine if test samples stimulate or inhibit cell proliferation. Autoradiographic analysis is easily applied to cultured cells from aquatic animals and can also be adapted to animal studies. Detecting DNA replication and UDS using [^{3}H]dT/autoradiography is directly compared with BrdU/immunocytochemistry procedures; advantages and limitations of these procedures are discussed.

1-56670-149-X/96/$0.00+$.50
© 1996 by CRC Press, Inc.

Materials required

Chemicals

[³H]dT, 5-10 Ci/mmol, from any radiochemical supplier. For long-term storage, [³H]dT should be obtained in a 70% ethanol solution, and this must be removed, by blowing N_2 over an aliquot, before adding to cells or animals. Kodak® NTB3 photographic emulsion (cat. no. 1654441) or comparable emulsion is supplied in a light-tight box and must only be opened in a dark room (with a safe light). The emulsion should be stored in the refrigerator and must be warmed to 50°C before attempting to pipette. The emulsion bottle should be carefully wrapped with foil and emersed in a water bath in the dark room. Kodak D-19 (cat. no. 1464593) and Kodak fixative (cat. no. 1971746) are used to develop and fix autoradiographic samples. These reagents are prepared and used exactly as described by the manufacturer.

Dark room facilities

A dark room is required for heating the emulsion, for coating samples with emulsion and for developing and fixing samples following exposure. A light-tight dark box is constructed from wood or plexiglass in any size or shape, which is large enough to accommodate emulsion-coated samples and a drying compound (Drierite). We use a wood box ~12 in.(l) × 12 in.(w) × 8 in.(h) with a lid which overlaps the sides ~1.5"; felt material is glued to the top edge of the box, box sides, inside top and sides of lid to make a light-tight seal. The interior of the box is painted black; several coats of flat paint are often required to achieve a totally black box. Alternatively, emulsion-coated cells can be exposed in a dark room in a container with Drierite, as long as the only light in the dark room is a safe light.

Sample fixation and analysis

Tissue culture cells are fixed in methanol:acetic acid (3:1) (1), prepared fresh and used ice cold; other fixatives, including 4% paraformaldehyde or ½ strength Karnofsky's solution, can also be used. Animal tissues embedded in glycol methacrylate (1) or other polymers can be utilized following sectioning. Any microscope with high quality resolution at 200 to 400× can be used to examine and analyze the samples. Various types of image analysis equipment and software can be used as an alternative to manually scoring samples.

Procedures

Overview of data to be collected

With this procedure, it is possible to determine which cells in a population were synthesizing DNA when the cells were exposed to [³H]dT. By light microscopic examination, cells which were not undergoing DNA replication or UDS during exposure to [³H]dT display a low density of grains over nuclei, which is no greater than the "background" density of grains over non-nuclear regions. Cells that were in the S phase of the cell cycle appear to have very dark nuclei, due to a high density of silver grains over these nuclei. The percent cells in the observed population with dark nuclei, or labeled cells, is determined. By comparing the % labeled cells in control and treatment groups, it is possible to determine if the treatment protocol stimulated, inhibited or had no effect on cell growth. On the other hand, cells which were undergoing DNA repair synthesis (UDS) have a much lower density of silver grains over their nuclei (typically 5 to 50 grains per nucleus) than actively growing S phase cells, but a higher density of nuclear grains than the "background" density of silver grains in non-nuclear regions. Often the genotoxic potential of different compounds is indicated by the amount of induced UDS (2) — the greater the number of UDS grains per nucleus, the greater the genotoxicity of the compound.

Considerations for [³H]dT exposure

The following considerations pertain to cultured cells. Cells are treated with different compounds (treatment protocols) and control cells are mock treated. All cells are then exposed to medium containing 5 to 10 µCi/ml [³H]dT for 1 to 8 h. Although higher specific activity [³H]dT can be purchased, due to metabolism of dT high specific activity [³H]dT (with lower amounts of dT) are not necessarily advantageous. The length of [³H]dT exposure depends on the goals of the experiment and may need to be optimized for specific cells and treatment protocols. For example, to determine if a treatment protocol inhibits cell proliferation in a growing population, [³H]dT is added for 2 to 4 h after the inhibitory effect is likely to be exerted, which may be shortly after beginning the treatment protocol. A somewhat different approach is used to determine if a protocol stimulates cell proliferation, because quiescent cells generally enter S phase 18 to 24 h following addition of mitogens. In this case, 1- to 2-h exposures to [³H]dT at 2 to 4 h intervals following addition of mitogen will best determine the percentage of cells replicating DNA and the kinetics of cell cycle traverse. In addition, rates of DNA synthesis and cell cycle traverse may be considerably slower in cells from cold water species as opposed to cells from warm water species. Different considerations apply to studies examining the potential of a treatment protocol to induce UDS. Most direct-acting compounds damage DNA quickly, and often DNA is rapidly repaired.[2-6] Other compounds must first be metabolized[7] before reacting with DNA and inducing DNA repair synthesis. Usually the major concern in UDS studies is to determine the extent of UDS. For these reasons, [³H]dT is frequently added to cells coincident with compounds being assessed for UDS potential, or soon (within h) after the addition of the test compounds.[1,2] In UDS studies, especially initial or range-finding studies, the [³H]dT can be incubated with cells for 12 to 24 h.

Several modifications can be employed to enhance [³H]dT incorporation into DNA. First, cells can be exposed to both [³H]dT and 2 µM FdU (5-fluoro,2'-deoxyuridine) (1); FdU inhibits thymidylate synthetase, blocking endogenous dT synthesis and thereby increasing the specific activity of

[³H]dT incorporated into DNA. It is also possible to preferentially inhibit DNA replication without adversely affecting UDS.[8] In this case, cells can be incubated in medium containing 1 to 2 mM hydroxyurea for approximately 1 h prior to addition of [³H]dT. Hydroxyurea inhibits ribonucleotide reductase, depleting intracellular pools of dNTPs below the level required for DNA replication, but apparently not below the level required for UDS.

Once [³H]dT has been incorporated into cellular DNA, the cells are prepared for fixation and autoradiography. Medium containing [³H]dT is removed and cells are washed with radioactive-free medium, then briefly incubated in radioactive-free medium for 5 to 10 min to allow removal of unincorporated [³H]dT.

Fixing cells

After medium is removed, cells are fixed in fresh, ice-cold 75% methanol, 25% acetic acid for 30 min then air dried.[1] Samples can be stored in this dry state for an indefinite period of time before further processing. Various other fixatives can be used, or fixed tissue sections can be embedded, sectioned, and processed as described.[1] Cells or sections can be stained with hematoxylin/eosin or Giemsa;[9] however, nuclear staining should be limited, to permit visualization of silver grains above nuclei.

Coating with emulsion and exposure

NTB3 emulsion should be stored at 0 to 4°C; however, at this temperature the emulsion is too viscous to pour or pipette. All manipulations with emulsion must be performed in a dark room, under a safe light. The stock emulsion bottle is wrapped in foil (to prevent exposure to light) and emersed in a 50°C water bath, in the dark room. A working solution of emulsion is prepared by diluting an aliquot of the stock emulsion with an equal volume of 50°C distilled water. The stock emulsion is immediately returned to its dark box and stored in the refrigerator, and the working emulsion solution is kept in the 50°C bath. Cells are covered with a thin film of emulsion — slides are dipped in the working emulsion and set on end (angled in a test tube rack) to drain in a dark box or the emulsion is added to culture vessels, swirled to coat the cells, and vessels are set on an angle to drain. Excess emulsion is removed with a Pasteur pipette. Emulsion-coated cells are then stored in a dark box with a desiccating material, such as Drierite, for 2 to 4 days (DNA replication) or 6 to 20 days (UDS). When using autoradiography for the first time, it may be advisable to set aside several samples to determine optimum time for emulsion exposure. Following exposure in the dark box, samples are developed with Kodak D-19, rinsed with water and treated with Kodak Fixative, as recommended by the manufacturer.

Scoring autoradiography samples

Light microscopy is used to score cells prepared for autoradiography. Although manual observation and scoring is routinely used, user-friendly image analysis equipment and software greatly facilitates scoring and quantitating autoradiographic analysis.[10] For DNA replication studies where the fraction of cells in S phase is determined, cells are scored as either positive (many grains above nuclei) or negative (nuclear grain density similar to that of non-nuclear regions), and the percentage of labeled cells is determined. Usually 300 to 500 total cells must be scored for reliable statistic data. In UDS studies, cells can be scored as positive or negative, to determine the percentage of cells in which DNA repair synthesis was induced. However, often all cells in a population exhibit induced UDS, and the number of silver grains/nucleus[2,11] or the density of nuclear silver grains[10] is instead determined. The latter measurements provide a quantitative estimation of the genotoxicity of many samples.[2]

Results and discussion

Figure 1 shows autoradiographic analysis of rapidly growing cells exposed to 10 µCi/ml [³H]dT for 30 min prior to processing for autoradiography. Panel A shows control cells and Panel B is cells treated with a test compound, a DNA polymerase inhibitor.[9] There is a high density of grains above nuclei of replicating, control cells (panel A); the three cells indicated by arrows are not in the S phase (DNA replication) of the cell cycle. The density of nuclear grains is greatly diminished in cells exposed to the test compound (panel B), indicating inhibition of cell proliferation. Quantitative analysis of these autoradiographic studies is presented in Table 1,[9] where the "fraction of labeled cells" indicates the ratio of labeled, treated cells to labeled, control cells. Treatment with SJK-287 (antibody against DNA polymerase α) significantly reduced the fraction of labeled cells, whereas other antibodies (PAb122, anti-p53; PM-8, anti-c-myc; IgG, preimmune IgG) did not significantly alter proliferation. These studies clearly demonstrate that one of the treatments (SJK-287) inhibited cell proliferation and others had no effect on cell proliferation. This type of analysis can be used to determine if various endogenous or xenobiotic compounds exert either a positive or negative effect on cell proliferation.

Table 2[11] shows that autoradiography can be utilized in subcellular or nuclear preparations capable of DNA synthesis. Permeabilized cells were allowed to synthesize DNA in the presence of [³H]dTTP, to label DNA, then cells were processed for autoradiography. The effect of a DNA polymerase inhibitor on DNA replication in permeable cells is shown in Table 2. In the control population (treatment = "none") 24.7% of the cells were synthesizing DNA, with an average of 62.3 grains/nucleus. The addition of an anti-DNA polymerase α antibody (SJK 287-38), did not significantly reduce the percentage of cells synthesizing DNA, but reduced the average number of grains/nucleus to 19.5. These autoradiographic results show that in the permeable cells the DNA polymerase inhibitor reduced, but did not abolish, DNA synthesis. In a similar fashion, the effects of other compounds on DNA replication[11] or on UDS[10] can be investigated in subcellular preparations from aquatic organisms.[12]

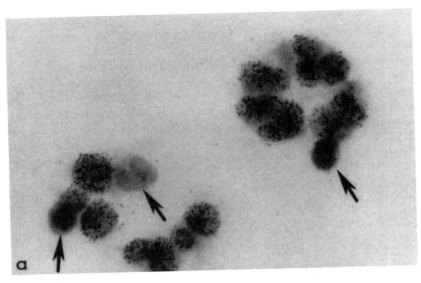

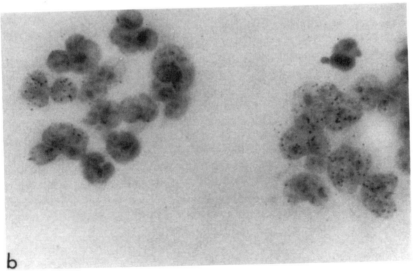

Figure 1 Autoradiography of growing cells and effect of a DNA polymerase inhibitor on nuclear labeling. Rapidly growing cultured cells were labeled with 10 μCi/ml [³H]dT for 30 min in the absence (control cells, panel a) or presence (treated cells, panel b) of a DNA polymerase inhibitor, then processed for autoradiography after Giemsa staining.[9] Arrows indicate cells without grains above nuclei, indicating the cells were not synthesizing DNA during the [³H]dT labeling period. (Reprinted with permission from Kaczmarek, L., et al. 1986. *J. Biol. Chem.* 261:10802–10807.)

Autoradiographic detection of UDS induced by the genotoxic compound N-methyl-N'-nitro-N-nitrosoguanidine (MNNG) is shown in Figure 2A.[1] Cultured human fibroblasts were grown to confluency in 8-chambered slides, exposed to 20 μg MNNG/ml for 1 h then placed in MNNG-free medium containing 2 μM FdU and 5 μCi/ml [³H]-dT for 1 h and processed for

Table 1 Effect of Different Compounds on
DNA Synthesis in Cultured Cells Determined
by Autoradiography

Treatment	Fraction of labeled cells
SJK-287	0.36
Pab122	0.98
PM-8	0.92
IgGs	0.95

Note: Cultured cells were labeled with [^{3}H]dT and
processed for autoradiography as described
in Figure 1, following treatment with SJK-287
(antibody against DNA polymerase α),
PAb122 (antibody against p53), PM-8 (antibody against c-myc) or IgGs (preimmune
IgG). The "fraction of labeled cells" indicates
the ratio of labeled, treated cells to labeled,
control (not treated) cells.

Reprinted with permission from Kaczmarek, L. et al.
1986. *J. Biol. Chem.* 261:10802–10807.

autoradiography. Cells replicating DNA have such a high density of grains
that their nuclei appear black (Figure 2A, arrows). All other cells exhibit
UDS, which is seen as lower density silver grains over nuclei than replicating
cells but higher density grains than cytoplasmic regions. In cells not exposed
to MNNG (not shown), only high density silver grains were observed over
nuclei of replicating cells; all other cells were not labeled (density of grains
over nuclei was no greater than that of cytoplasmic regions). Automated
image analysis procedures have been used to determine grain density over
nuclei,[10] permitting rapid quantitation of autoradiographic data.

Another method of assessing the effect of test compounds on DNA
synthesis utilizes incorporation of BrdU (5-bromodeoxyuridine) into DNA
(in place of [^{3}H]dT) and detection of BrdU in nuclear DNA with anti-BrdU

Table 2 Autoradiographic Analysis of Effect of
a DNA Polymerase Inhibitor on DNA Synthesis
in Permeabilized Cells

Treatment	% Labeled nuclei	Average grains/nucleus
None	24.7	62.3
SJK 287-38	22.5	19.5

Note: Permeable cells capable of DNA synthesis were incubated in
a solution containing [^{3}H]dTTP to label replicating DNA, as
described (Hammond et al.), in the absence (none) or presence of an antibody directed against DNA polymerase (SJK
287-38). The samples were then processed for autoradiography, and the average number of grains/nucleus was determined by manually counting nuclear grains for ~50 cells in
each group. The "% labeled nuclei" indicates % nuclei in each
population with more than 5 grains above background levels.

Reprinted with permission from Miller, M.R., et al. 1985. *J. Biol.
Chem.* 260:134–138.

antibodies.[13] This procedure is presented in detail in the following chapter. We directly compared these procedures[1] in assessing effects of a genotoxic agent, MNNG, on DNA replication and on UDS (Figure 2). As discussed above, Figure 2A demonstrates that autoradiography following [³H]dT labeling successfully detects both "scheduled" DNA synthesis (DNA replication) and UDS. Figure 2, B and C, shows fluorescent staining of control cells and MNNG-treated cells, respectively, after cells are labeled with a medium containing 25 μ*M* BrdU and 2 μ*M* FdU for 1 h, then processed for immunocytochemical detection of fluorescent labeling, as described in Chapter 13. In the absence of MNNG, intense nuclear staining is exhibited by some of the control cells (Figure 2B), indicating incorporation of BrdU into nuclear DNA during DNA replication in these growing cells. Following treatment with MNNG (Figure 2C), nuclear staining is greatly diminished, fewer cells exhibit nuclear staining, and low level nuclear staining or labeling in all other cells is not observed. This is in sharp contrast with the results obtained using [³H]dT labeling and autoradiographic detection (Figure 2A). These results are interpreted to indicate that the BrdU labeling and immunocytochemical detection procedure detected DNA replication, which was inhibited by MNNG-treatment (compare panels B and C). However, the BrdU labeling and immunocytochemical detection procedure was not sensitive enough to detect the smaller amount of BrdU incorporated into DNA due to MNNG-induced UDS (compare panels A and C).

The primary advantage of the [³H]dT labeling and autoradiography detection procedure, relative to the BrdU labeling and immunocytochemical procedure is that the former procedure is sensitive enough to detect both DNA replication and UDS. A limitation of the autoradiographic procedure is that a long half-live radioisotope is used, which can generate a sizeable volume of radioactive solution when aquatic organisms are dosed with [³H]dT, incurring significant expense associated with disposal. The BrdU labeling approach is inherently more rapid and is sensitive enough to detect DNA replication without generating waste disposal problems; however, this approach is limited by not being sensitive enough to detect at least some types or amounts of UDS, precluding this approach in genotoxicity studies.

The most commonly encountered problems associated with the [³H]dT/autoradiographic procedure are due to a high "background" of grains which interferes with analysis of specific nuclear DNA labeling. Usually these problems are attributed to light leaks in the dark room or in the "dark box" or to contamination of stock NTB3 emulsion with radioisotope or with a light source. Often these possibilities can be differentiated by coating slides alone (without cells or [³H]dT) with the original and with fresh emulsion, maintaining the slides in the dark box for the same amount of time and developing the slides. A high background only in the original emulsion indicates contamination of this stock; whereas a high background with both emulsions indicates some sort of light leak during the dark exposure period.

An assumption of the autographic procedure is that various treatment protocols do not alter the intracellular phosphorylation of dT to dTTP or the

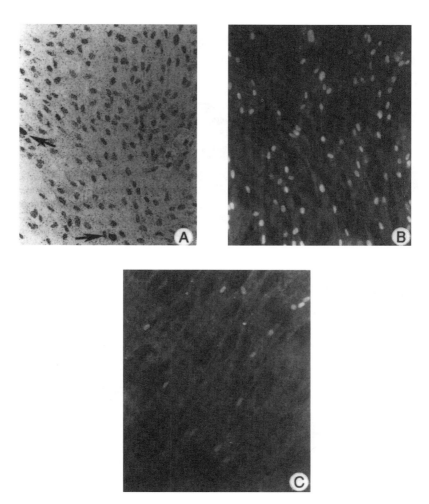

Figure 2 Autoradiography of UDS in cultured cells and comparison with BrdU/immunocytochemical labeling. Cultured cells were grown to confluence in 8-chambered slides with medium containing 10% fetal calf serum, then maintained in medium with 0.1% fetal calf serum to arrest the majority of cells in the G_o phase of the cell cycle minimizing interference by DNA replication. UDS was induced by the addition of 20 µg MNNG/ml medium for 1 h, then cells were incubated in MNNG-free medium containing 2 µM FdU and 5 µCi/ml [³H]dT for 1 h. Cells in panel A (1) were processed for autoradiography as described, and arrows indicate heavily labeled nuclei, indicative of replicating cells. All other cells exhibit low level nuclear labeling, indicative of UDS. Cells in panels B and C were labeled for 1 h with 25 µM BrdU in place of [³H]dT and processed for immunocytochemical detection of nuclear BrdU-labeled DNA as described in Chapter 13. Panel B shows cells incubated without MNNG (control cells), and panel C shows cells that had been treated with MNNG as described in panel A. (Reprinted with permission from Droy, B.F. et al. 1988. *Aquat. Toxicol.* 13:155–166.)

catabolism of dT. To test this possibility, it is possible to measure intracellular concentrations of dTTP; however, the procedures are complex and this potential problem appears to be relatively rare.

References

1. Droy, B.F., Miller, M.R., Freeland, T.M. and Hinton, D.E. 1988. Immunohistochemical detection of CCl_4-induced, mitosis-related DNA synthesis in livers of trout and rat. *Aquatic Toxicol.* 13:155–166.
2. Walton, D.G., Acton, A.B. and Stich, H.F. 1983. DNA repair synthesis in cultured mammalian and fish cells following exposure to chemical mutagens. *Mutat. Res.* 124:153–161.
3. Fong, A.T., Hendricks, J.D., Dashwood, R.H., Van Winkle, S. and Bailey, G.S. 1988. Formation and persistence of ethylguanines in liver DNA of rainbow trout (*salmo gairdneri*) treated with diethylnitrosamine by water exposure. *Fd Chem. Toxicol.* 26:699–704.
4. Vos, J.-M. H. and Wauthier, E.L. 1991. Differential introduction of DNA damage and repair in mammalian genes transcribed by RNA polymerases I and II. *Mol. Cell. Biol.* 11:2245–2252.
5. Ball, S.S., Neshat, M.S., Mickey, M.R. and Walford, R.L. 1989. DNA damage and repair in female C57BL/10 mice of different ages injected with the carcinogen benzo[a]pyrene-trans-7,8-diol. *Mutat. Res.* 219:241–246.
6. Singh, N.P., Tice, R.R., Stephens, R.E. and Schneider, E.L. 1991. A microgel electrophoresis technique for the direct quantitation of DNA damage and repair in individual fibroblasts cultured on microscope slides. *Mutat. Res.* 252:289–296.
7. Harris, C.C. 1989. Interindividual variation among humans in carcinogen metabolism, DNA adduct formation and DNA repair. Commentary, *Carcinogenesis* 10:1563–1566.
8. Erixon, K. and Ahnstrom, G. 1979. Single-strand breaks in DNA during repair of ultraviolet-induced damage in normal human and xeroderma pigmentosum cells as determined by alkaline DNA unwinding and hydroxylapatite chromatography: effects of hydroxyurea, 5-fluorouracil and 1-β-D-arabinofuranosyl cytosine on the kinetics of UV repair. *Mutat. Res.* 59:257–271.
9. Kaczmarek, L., Miller, M.R., Hammond R.A. and Mercer, W.E. 1986. A microinjected monoclonal antibody against human DNA polymerase-ainhibits DNA replication in human, hamster, and mouse cell lines. *J. Biol. Chem.* 261:10802–10807.
10. Hammond R.A., McClung, J.K. and Miller, M.R. 1990. Effect of DNA polymerase inhibitors on DNA repair in intact and permeable human fibroblasts: Evidence that DNA polymerases α and β are involved in DNA repair synthesis induced by N-Methyl-N'-nitro-N-nitrosoguanidine. *Biochemistry* 29:286–291.
11. Miller, M.R., Ulrich, R.G., Wang, T.S.-F. and Korn, D. 1985. Monoclonal antibodies against human DNA polymerase-α inhibit DNA replication in permeabilized human cells. *J. Biol. Chem.* 260:134–138.
12. Miller, M.R., Blair, J.B. and Hinton, D.E. 1989. DNA repair synthesis in isolated rainbow trout liver cells. *Carcinogenesis* 10:995–1001.
13. Miller, M.R., Heyneman, C., Walker, S. and Ulrich, R.G. 1986. Interaction of monoclonal antibodies directed against bromodeoxyuridine with pyrimidine bases, nucleosides and DNA. *J. Immunol.* 136:1791–1795.

chapter thirteen

Bromodeoxyuridine as a marker of cell proliferation in histological sections of aquatic organisms

Michael J. Moore and Michael S. Morss

Introduction

Metazoan organisms grow by cellular enlargement, the deposition of extra-cellular structures and most especially cell division. An orderly, integrated progression and control of cell population growth is central to all physiological and pathological processes. Responses to infectious, chemical, and physical trauma often include altered rates of cell proliferation. Therefore, a broadly applicable tool to measure cell proliferation is important in the study of many diverse processes in all metazoa including aquatic vertebrates and invertebrates.

The study of cell proliferation is important to aquatic toxicology because many xenobiotic compounds exert sublethal, chronic toxic effects at least in part by subtle alterations in the expression of genes that control rates of cell proliferation and death in selected target organs and tissues. Examples of such compounds include many carcinogens especially those classed as tumor promoters such as dioxins, polychlorinated biphenyls, and chlorinated pesticides such as DDT and its metabolites.

Available methods to study cell proliferation are listed in Table 1. The methods all detect cells undergoing cell division. Such cells leave the resting phase, termed G_0, and progress through three stages of preparation for mitosis, G_1, S and G_2. DNA synthesis occurs during S phase. The classic method to assess proliferation, other than counting mitoses in histological sections, involves autoradiographic detection of ^{3}H-thymidine incorporated in place of endogenous thymidine. ^{3}H-thymidine techniques involve the use and disposal of radioisotopes. A more recent approach has been to label organ-

Table 1 A Comparison of Methods for Measuring DNA Synthetic Activity
in Histological Sections

	Marker		
	³H Thymidine	Bromodeoxyuridine	Proliferating cell nuclear antigen
Label	Radioactive	Carcinogenic	None
Detects DNA repair?	Yes	No	Yes
Detects DNA replication?	Yes	Yes	Yes
Phases of cell cycle detected	S	S	G1,S,G2
Pulse labeling possible?	Yes	Yes	No
Time to conduct assay	Long	Moderate	Moderate
Detection	Autoradiography	Immunohistochemistry	Immunohistochemistry
Toxic to test organism?	Yes	Yes	No

isms with a thymidine analog, bromodeoxyuridine (BrdU — Figure 1). BrdU incorporated into the new nucleotide chain is visualized immunohistochemically by incubation with one of a variety of commercially available monoclonal antibodies, with subsequent routine chromogen or fluorochrome tagging. A third assay for cell proliferation, described elsewhere in this volume involves detection of the expression of an endogenous protein associated with cell cycling: proliferating cell nuclear antigen (Chapter 8). The advantages and disadvantages of these approaches are summarized in Table 1.

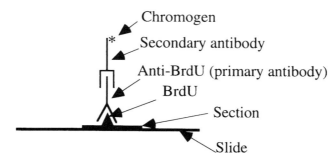

Chromogen
Secondary antibody
Anti-BrdU (primary antibody)
BrdU
Section
Slide

Figure 1 Immunohistochemical detection of bromodeoxyuridine incorporated into nuclear DNA. The organism is exposed to BrdU which is incorporated into genomic DNA in cells that are actively synthesising DNA. Sections of paraffin-embedded, formalin-fixed BrdU-labeled tissues are incubated on glass slides with an anti-BrdU monoclonal antibody. A secondary antibody then binds the primary antibody. The secondary antibody bears a peroxidase moiety that, on exposure to appropriate substrates, generates a polymerized chromogen that is visible with light microscopy. With appropriate washing the chromogen is limited to nuclei that have incorporated BrdU. In this way nuclei undergoing DNA synthesis are highlighted.

This chapter describes the methods used for detection of BrdU incorporation. Briefly, tissues are labeled *in vivo* or *in vitro* with BrdU, sampled at some time thereafter, fixed in formalin, embedded in paraffin, sectioned onto glass slides, rehydrated, processed to expose the antigen, and incubated with anti-BrdU monoclonal antibody followed by other chromogenic reagents to visualize bound anti-BrdU. Nuclei incorporating BrdU are counted microscopically and a labeling index is generated as the proportion of total nuclei of each cell type that have incorporated immunohistochemically detectable BrdU. Typically the labeling index is expressed as the number of labeled nuclei per 1000 nuclei examined. Evaluation of 1000 cells is generally accepted as a representative sample.

This study focuses on specific details of BrdU methodology. The protocols described are not necessarily optimally efficient in time and materials, but they work reliably and repeatably for us. The protocol described is often one of a number of alternatives that could be used. Where possible we have referred to alternative methods. We will also touch on a number of issues that are relevant to immunohistochemical methods in general, many of which can be found in general immunohistochemical texts,[1] but are also directly relevant to the subject of this chapter. One specific publication that concerns the use of ^{3}H-thymidine and BrdU in rodents[2] is recommended as are other important papers concerning BrdU.[3-8] The most important factor in this protocol is to minimize between sample variation in technique. Because differences in processing can easily introduce artifactual differences between data from different samples, fastidious attention to the detailed replication of methods between samples is critical.

Exposure to BrdU has been shown to reduce life-span, and increase cancer risk in rodents.[9] This information should be considered both in the design of experimental protocols and in the appropriate protection of investigators handling the chemical.

Materials required

In addition to providing the sources for the materials to be used, this section includes comments on the advantages and disadvantages of alternates that can be used at various stages in the protocol.

Animals

Enough organisms should be available to provide adequate sample sizes at each time point, and in each treatment group. Ideally 1000 cells of each cell type of interest should be in the plane of stained section(s) for each data point. Sufficient organisms should also be included for the controls described below. For samples where the cell(s) or tissue(s) of interest are within a larger tissue mass, it is important to allow for wastage of a proportion of samples where the target cells are outside of the plane of the stained section(s). There is a high inherent variability in proliferation indices within treatment or lesion types, thus sample sizes should be as large as possible to ensure statistical adequacy.

Nucleotide analogs

Bromodeoxyuridine and fluorodeoxyuridine can be obtained from Sigma Chemical (B-5002 5 Bromo-2′-deoxyuridine 1 g, and F-0503 Fluoro-2′-deoxyuridine 100 mg). BrdU crystals should be stored frozen and desiccated, and FdU should be stored at room temperature. Both should be kept in the dark at all times. BrdU:FdU stock solution should be formulated at 300:30 mg/100 ml distilled water not more than 1 week prior to use. It should be refrigerated. Other protocols use buffered saline, but BrdU and FdU are less soluble in saline than in water, and we have had good success using an aqueous solution. FdU is used in addition to BrdU as it inhibits the endogenous thymidylate synthetase preventing the conversion of dUMP to dTMP, thus enhancing BrdU substitution for thymidine and maximizing signal strength.[10]

Microscope slides

Adhesion of sections to microscope slides through the staining procedure can be a major problem, especially with small specimens and when substantial antigen-retrieval is required prior to immuno-staining. Proteolysis and acid hydrolysis antigen retrieval techniques are commonly used for BrdU studies. For these purposes, sialinized slides are best: commercially available sialinized slides include Superfrost Plus® (Baxter or Fisher). However, we have found better adhesion with slides sialinized with 3-aminopropyltriethyoxysilane (APTS) Sigma #A3648. APTS should be refrigerated and parafilm sealed between use. Slides are dipped five times each in three changes of 95% ethanol, air dried, dipped five times in freshly made 0.2% APTS in acetone, drained quickly, dipped five times in each of three changes of distilled water (slides should be well drained between each change) and then air dried and boxed. Slides can be stored for up to 12 months. In contrast, protocols that use microwave antigen-retrieval methods such as prior to proliferating cell nuclear antigen (PCNA) detection (see Ortego et al., Chapter 18) are best used with poly-L-lysine coated slides.

Immunohistochemical detection of BrdU incorporation

The choice of immunohistochemical reagents depends on availability, affordability, function in each investigator's hands, and primary antibody.

Anti-BrdU monoclonal antibody

Anti-BrdU antibodies are available from a number of sources (Amersham, Arlington Heights, IL; Becton Dickinson, Palo Alto, CA; Caltag, San Francisco, CA; Dako, Carpinteria, CA; Bioclone Australia, Marrickville, NSW, Australia; Sera Labs, Crawley, UK). We have mainly used the product supplied with the cell proliferation kit from Amersham (RPN20). When we have had to use more extensive antigen retrieval methods, and in particular have had to adjust the concentration of DNAase in the primary antibody solution, we have used an antibody from Becton Dickinson (Cat. # 347580).

Immunohistochemical detection system

The temptation to establish detection systems by buying individual reagents is substantial given the initial cost savings, however these savings can be outweighed by the substantial time required to optimize the numerous steps involved. The high cost of a kit is usually offset by the likelihood that the secondary reagent protocols will be close to optimal, and the only variables to be optimized will be antigen retrieval and primary antibody concentration. For these reasons we have primarily used the Amersham kit listed above.

Staining system

Commonly, humidified chambers are used to incubate slides with small reagent droplets overlying the tissue section. A major limitation to consistent, reproducible results is variation in processing resulting from incomplete coverage of the section by reagents at one or more stages of the protocol. This can result from collapse of the droplet of reagent with partial dehydration of the tissue section. The Coverplate™ system from Shandon, Pittsburgh, PA (Figure 2) overcomes this problem. It has an 80 µl chamber that overlies the entire slide, giving total hydrated coverage of the section(s) on the slide. Reagents are pipetted through a funnel to the chamber. Each

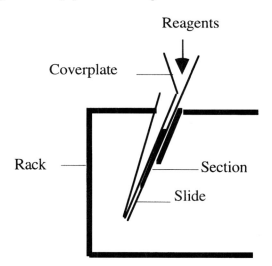

Figure 2 A sideview of the Coverplate™ (Shandon, Pittsburgh, PA) staining system. Glass slides bearing 5 µm sections are overlain by a plastic molded plate that has a chamber overlying the section. The chamber has a volume of 80 µl. Successive reagents or buffers are added to the funnel in volumes of at least 100 µl. This displaces previous reagents into the base of the rack, avoiding the risk of incomplete reagent exposure, a feature common to protocols relying on a humidified atmosphere.

incubation is then displaced by addition of the next reagent or wash. The coverplate/slide combination is held in a rack over a trough where waste reagents collect. This system can be purchased with or without the micro-processor driven staining system, as "single use" coverplates (Cat. # 72110013) with holding racks (Cat. # 73310015). Each rack holds ten slide/coverplate combinations. Ten racks and a case of 250 coverplates is an appropriate purchase. The expense of the coverplates encourages reuse. We routinely reuse coverplates up to ten times. After use coverplates are immersed in cold water, washed in detergent and rinsed in deionized water. Excessive use can lead to antibody adsorption onto the coverplate with loss of staining intensity, a phenomenon seen first at the downhill end of the slide. A cheaper but more time consuming option is to use a humidified chamber, and cover each reagent droplet with a piece of polyethylene film trimmed to cover the tissue section. This achieves the same end of ensuring full hydrated coverage of the section. This need only be done for steps where reagent cost precludes flooding the whole slide.

Buffers

Two buffers must be made fresh, to avoid bacterial contaminant growth which may affect staining quality: (1) Phosphate Buffered Saline (PBS): 27.6 g Na_2HPO_4 and 11.7 g NaCl per 2 l water (pH 7.5), and (2) Phosphate Buffer (PB): 6.88 g Na_2HPO_4 per 500 ml water (pH 7.5).

Chromogens and counterstains

A critical factor is the contrast between the chromagen used to highlight the specifically bound primary antibody, and the counterstain used to highlight background morphology. Choice of stains can only be made for specific applications after some experimentation, as each tissue and processing protocol gives subtly different end results that are unique in the final product. Optimization of the end product is essential to allow relatively easy, confident, and reproducible scoring of positive and negative nuclei. The ability to recognize negative nuclei of the cell type of interest is as important as the ability to highlight positive nuclei. The maximization of contrast is also of great importance for the production of good micrographs, especially if they are to be monochrome. Two useful combinations of chromogen and counterstain are (1) aminoethylcarbazole (red) or diaminobenzidine (brown) with hematoxylin (blue) counterstain, and (2) nickel chloride intensified diaminobenzidine (black) with either methyl green or eosin (pink) counterstain. The commonly used combination of nickel chloride intensified diaminobenzidine with hematoxylin should be avoided, as the distinction between a blue hematoxylin stained nucleus and a nucleus with a light black to dark gray nickel chloride intensified diaminobenzidine precipitate is hard to differentiate unless the hematoxylin counterstain is very light. All of these products can be used from or in conjunction with the various kits available. Suppliers include: Vector, Burlingame, CA; Signet, Dedham, MA; Dako, Carpinteria, CA; Sigma, St Louis, MO.

Procedures

BrdU incorporation

Cells, tissues, or organisms can be exposed to BrdU in a number of ways: *in vivo* by injection, osmotic minipump (Alza, 950 Page Mill Road, PO Box 10950, Palo Alto, CA — Phone 415-494-5517) or bath, or *in vitro*[11] for tissue slices or cultured cells. A limitation of the *in vitro* method with tissue slices is the limited penetration (1 to 2 mm) of the label into the tissue block. The stock BrdU/FdU label should be used at 10 ml/kg of fish by intraperitoneal injection or per liter of aquarium tank and at 1 ml/l of tissue culture medium. Small fish and invertebrates (especially filter feeders) are best labeled by addition of the BrdU/FdU to a bath.[12] Duration of label exposure has to be controlled. It should be increased if the target cell type is proliferating slowly, such as in liver tissue of adults. Routine rodent liver studies use a 3-day osmotic pump exposure. Single injections give a brief exposure to BrdU, as it is apparently rapidly metabolized.[13] In this case animals are usually sampled 3 h after exposure. Prolonged exposure can be achieved either by bath exposure, or by implantation of osmotic minipumps. References and protocols for minipump use are available from Alza. We have no experience with their use.

Bath concentrations of BrdU:FdU can be maintained by spectrophotometrically monitoring the absorption of the exposure medium for absorbance at a wavelength of 277 nm, a peak absorbance defined in previous studies.[12] Total uridine concentration in μg/ml can be estimated after generation of a standard curve with dilutions of the BrdU:FdU stock solution.

Fixation

Fix tissues with standard 10% neutral-buffered formalin, pH 7.2. This is the fixative of choice for the majority of immunohistochemical applications. Over-fixation is the commonest cause of false-negative results in these protocols; therefore, never fix for more than 24 h, and preferably not more than 1 h for each mm of tissue thickness. Otherwise, excessive crosslinking by the fixative reduces or eliminates access of the reagents to the bound BrdU. If tissue processing cannot follow after a few hours of fixation, transfer to 70% ethanol has been a suggested substitute, but in our hands this has met limited success. It is very important to handle all tissues equivalently at all stages in this protocol but especially with respect to fixation. Therefore, samples from experiments with multiple time points should be processed into paraffin with the same delay for each batch and not stock piled until the end of the experiment.

Tissue processing

Routine histopathological processing can be used for these samples as described elsewhere in this volume. Times and temperature of each stage should remain constant. Once samples are embedded in paraffin, the label is stable, and the samples can be analyzed further at any time. Tissue blocks, preferably sealed with a layer of paraffin wax, should be stored at room temperature.

Sectioning

Routine sectioning onto coated slides should be followed by drying at room temperature for a minimum of 48 h. Cover the slide rack to avoid dust accumulation. Do not oven dry, as this adds another uncontrolled variable to signal strength. Less than 48 h drying will result in poor tissue adhesion.

Controls

Careful use of controls, especially during method development is crucial to appropriate data interpretation. The following controls should be used initially:

1. Samples not labeled with BrdU.
2. BrdU-labeled samples stained with the full protocol, but with the anti-BrdU step replaced with a non-specific monoclonal antibody of the same isotype, or failing this, PBS. A typical image would appear as in Figure 3.

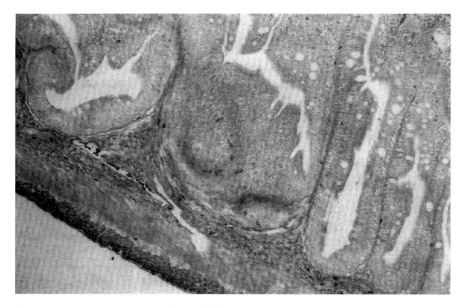

Figure 3   Transverse section through the intestine of a winter flounder injected with bromodeoxyuridine 3 h before sampling. The sample was processed and stained using the protocol described in the text, as a method control. For this the anti-BrdU step was replaced by incubation with phosphate buffered saline. No positive (black) staining is seen. A serial section stained with anti-BrdU is shown in Figure 4. Counterstained with eosin.

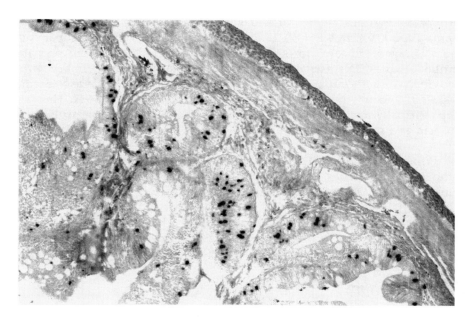

Figure 4 Serial section of that in Figure 3, stained in the same manner, except anti-BrdU was used as the primary antibody. Black stain highlights epithelial nuclei that have incorporated BrdU, as evidence of active DNA synthesis. These nuclei are overlain by nickel chloride intensified diaminobenzidine that is specifically bound to BrdU incorporated into the nuclear DNA.

   3. Samples known to have a high proliferative rate. In fish such tissues include intestinal epithelia, and renal or splenic hematopoietic tissues. Figure 4 illustrates such a positive control, for which Figure 3 is the negative control.
   4. Samples known to have a low proliferative rate. In fish, such tissues include skeletal or cardiac muscle.

   The positive and negative tissue controls (Nos. 3 and 4 above) can often be included as small tissue fragments in the same histological cassette as the tissue(s) of interest. The advantage of doing this is the avoidance of inter-slide variability in comparisons between control and test tissues.

Visualization of incorporated BrdU

Three discrete steps are involved: (1) antigen retrieval, (2) primary antibody incubation, (3) visualization of bound primary antibody.

   Antigen retrieval. The extent to which the antigen (nuclear BrdU) has to be exposed prior to primary monoclonal incubation depends on the tissue and duration of fixation. This must be established for each experiment. One or more of the following methods can be used: (1) acid hydrolysis (1 M HCl), with or without heat (40°C) for 5 to 60 min, (2) proteolysis (0.05% protease, Sigma Type XIV Cat. # P5147), in PBS prewarmed to 40°C but incubated at room temperature for 5 to 60 min, and (3) coincubation of the primary antibody with DNAase either by using the Amersham primary antibody/DNAase formulation as supplied, or by adding DNAase (Sigma D4527) at 3 to 100 units per ml to the Becton Dickinson antibody. This last step is usually always used, and can be augmented with either or both of the acid and protease steps.

Primary antibody incubation.   This step should initially be run for 1 h, although some tissues, such as fish liver require overnight incubation to allow adequate DNAase penetration.

Visualization of bound primary antibody.   These steps should follow the protocol relevant to the reagents selected. General guidelines for such immunohistochemical methods are available.[1]

Counterstaining

Use one of the three following options depending on the chromogen used, as discussed above.

Hematoxylin.   5 min in Mayers modified hematoxylin (Newcomer Supply, Oak Park, IL). After 5 min in this stain, rinse slides by immersion into many changes of tap water. Slides are then slowly dipped five times in ammonia water, washed as above, dehydrated, cleared, and mounted.

Eosin.   Slides are immersed in 1% alcoholic eosin Y diluted 1:3 in 80% ethanol for 5 min and then dehydrated in 1 change of 95% and 2 of 100% ethanol, then cleared and mounted.

Methyl green.   Slides are immersed in 0.5% (w:v) methyl green (Sigma, M6776) in 0.1 M sodium acetate, pH 4.0 (adjust pH with acetic acid), for 10 min. Dip slides 10 times in each of two changes of 100% butanol, and then hold in a third change of butanol for 30 seconds. Dehydrate in 3 changes of clearant and mount.

Counting labeling indices

Data can be generated in one of three ways: (1) Semiquantitatively by examining a set number of fields of view and estimating visually the occurrence (proportion of each cell type staining on a scale of 0 = none to 3 = all) and the intensity (0 = none to 4 = all) of staining observed. This is particularly useful for rapid assessment of results during method development. (2) Quantitatively by counting the number of stained nuclei per 1000 nuclei for each cell type of interest. Counting the number of unstained nuclei can be exceedingly time consuming. Eye piece 10 × 10 grid graticules aid in keeping track of the nuclei counted. In homogeneous tissues, the total number of nuclei present in each grid can be counted for a subset and extrapolated for the remainder. Alternatively, the relationship between the number of cells transected by a line through the sampling area, and the total number of cells in the area can be calculated and extrapolated between areas examined. (3) Quantitatively by using one of the many computer image analysis systems, some of which have software options dedicated to the calculation of proliferation indices. A major concern in the use of image analysis is in the security with which positive and negative nuclei are segregated. Most, if not all image analysis paradigms depend on the ability to threshold images into binary information for processing. The inclusion or exclusion of individual nuclei is often equivocal even when observed by the human eye and its cerebral processor. No video camera and software package comes close to the human level of visual color discrimination. The attractiveness of computer objectivity is obvious, but the limitations of digital signal processing are nonetheless significant. Results between these latter two methods have been compared.[14]

Step by step summary

As discussed in detail above, the procedures use the Amersham kit RPN20, with protease antigen retrieval and eosin counterstaining.

1. Expose animals/tissues/cells to BrdU/FdU stock solution by injection or bath at 10 ml/kg or *in vitro* at 1 ml/kg.
2. Fix experimental and control samples at appropriate time(s) after labeling.
3. Embed in paraffin, at 24 h or less of fixation.
4. Cut 5 μm sections onto sialinized slides. Dry at room temperature for 48 h.
5. Take slides through staining dishes or coplin jars in a staining hood, in the following sequence: clearant (such as Baxter Americlear, or other xylene substitute), clearant, clearant, 100% ethanol, 100% ethanol, 70% ethanol, PBS, PBS, PBS for 2 min each.
6. Hold each slide face up, flood with PBS, and lay a coverplate upon it, ensuring no bubbles are trapped and that the slide rests against the plastic lugs on the base of the coverplate. Load each slide and coverplate into a Shandon staining rack. Put 2 ml of PBS in the funnel between plate and slide, and ensure slow flow through. Excess speed of flow indicates that the plate is not resting flat on the slide. No flow indicates air bubbles in the staining chamber. Ensure that all reagents are at room temperature before adding to coverplates or bubbles will form as the reagent warms to room temperature.
7. Prepare protease just before use: 5 mg vortexed in 10 ml PBS warmed to 37°C. Incubate slides at room temperature. Time of digestion (0 to 60 min) may have to be titrated to balance signal strength with loss of morphology and of section adhesion. Additional HCl incubation may be employed (see Table 2). Stop incubation by adding 2 ml PBS to the funnel.

Table 2   Mean BrdU Incorporation, on a Scale of 0 to 3, in the Brain of
Winter Flounder Larvae Bathed in BrdU for 3 h Before Sampling

Time in protease (min)	HCl treatment			
	None	1 h @ 20°C	1 h @ 40°C	2 h @ 40°C
0	0	0.7	nd	nd
5	0	0.8	2.6	nd
20	nd	nd	2.8	nd
30	nd	nd	2.8	nd
40	nd	nd	3	0
60	0	2.6	nd	nd

Note: Sample size for each data point ranged from 5 to 14; nd = not determined.

8. Add 100 µl primary antibody to each slide for between 1 and 16 h. Stop incubation by adding 2 ml PBS to the funnel.
9. Add 100 µl secondary antibody for 1 h. Stop incubation by adding 2 ml PBS to the funnel.
10. Prepare diaminobenzidine (DAB) — 50 ml per 10 slides. Take a 1 ml aliquot of DAB stock (dissolve 500 mg DAB in 20 ml of phosphate buffer and frozen at –20°C). Thaw in cold water. Add to 50 ml phosphate buffer. Stir well. Add 5 drops of substrate/intensifier from the kit. On thawing, DAB solution should be clear to light brown. If it is dark brown do not use. Add 2 ml DAB per slide. Repeat at 5 min. Stop reaction at 10 min with 2 ml water. Overdevelopment can lead to non-specific staining of the entire section.
11. Place slides in a staining dish of distilled water.
12. Counterstain in eosin for 5 min.
13. Dehydrate through 2 min each of 95% alcohol, 100% alcohol twice, and clearant twice.
14. Mount in accumount, or permount.

Results and discussion

A specimen experiment

We present a study that highlights both the method and results. In a study of organ development in winter flounder larvae, we exposed larvae to a 3-h BrdU bath at a variety of times before fixation. The effects of using protease and HCl as antigen retrieval agents are shown in Table 2. A section of the the liver from a winter flounder larvae stained with BrdU is shown in Figure 5. To follow DNA synthetic activity in liver cells through time, we sampled individuals at a series of time points after bath labeling. Figure 6 shows the percentage of hepatocyte and non-hepatocyte populations labeled in the liver. The interpretation of these observations is unclear without further data about the relative half lives of each population. Such data could be generated with a parallel analysis of cell death by apoptosis. Figures 7 and 8 illustrate other applications of the BrdU technique.

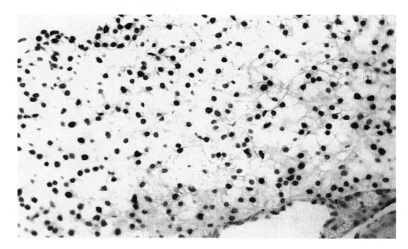

Figure 5   Section of larval winter flounder liver. This fish was bath exposed to BrdU for 3 h before sampling. A large proportion of the liver cells are undergoing replicative DNA synthesis, as seen by the black hepatocyte and non-hepatocyte nuclei. The nuclei that appear gray in this monochrome image are negative, counterstained with methyl green. The image illustrates the need to counterstain adequately but lightly. In color, the positive black nuclei were readily distinguishable from the negative green nuclei, whereas once printed in monochrome, the difference is less obvious.

Bromodeoxyuridine pulse labeling, by bath exposure, of winter flounder larvae

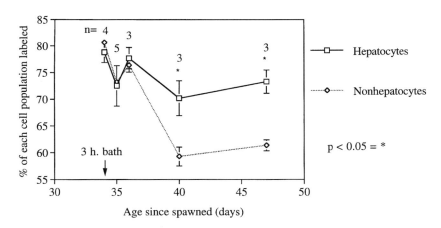

Mean labeling indices (BrdU positive nuclei/1000) ± SE

Figure 6   Labeling index in larval winter flounder liver cells at various times after bath exposure to BrdU. There is a significant reduction in the proportion of cells labeled with increasing time, although this was more marked for the non-hepatocytes, than the hepatocytes. This could suggest that some non-hepatocytes serve as a stem cell pool for the longer lived hepatocytes. Differences between cell types were tested by ANOVA.

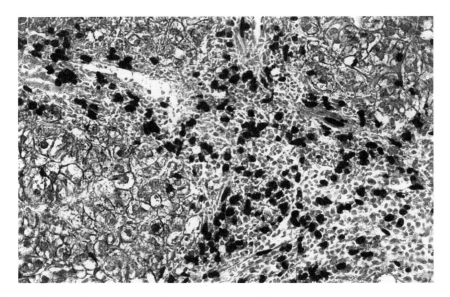

Figure 7   Adult winter flounder liver sampled 3 h after an intraperitoneal injection of BrdU. This animal had been fed a diet contaminated with technical grade chlordane for 1 year prior to this. The intense BrdU labeling of poorly differentiated cells, which are probably biliary epithelia, contrast with the quiescent hepatocytes. The low level of activity in the hepatocytes is typical of adult teleost liver, in comparison to that for rapidly growing larvae seen in Figure 5. Eosin counterstain.

The above illustrations all show positive staining, or controlled negative staining. There are many instances where results may be falsely negative. Reasons for false negative data include: failure to adequately expose tissues to bromodeoxyuridine, over-fixation or inadequate antigen retrieval, inactivation or microbial contamination of one or more reagents, omission of one or more reagents from the protocol, and inadequate duration of incubation, especially of primary antibody/DNAase mix.

Likewise, false positive data can arise, although far less commonly, for two reasons; (1) excessive duration of the final color development step, which can be controlled by periodic brief microscopic monitoring of developing slides; (2) non-specific binding of anti-BrdU, which can be confirmed by the observation of positive staining in tissue that was not labeled with BrdU.

Comparability of BrdU with other techniques

The general comparability of ^{3}H-thymidine and BrdU was described by Lanier et al.[5] It has been shown by Droy et al. in fish,[15,16] that ^{3}H-thymidine autoradiography detects both reparative and replicative DNA synthesis, whereas BrdU immunohistochemistry is not sensitive enough to detect reparative nucleotide addition. Thus the comparability of these two techniques will depend in part on the specific context. Specifically, if there is substantial DNA repair present, the indices may differ significantly.

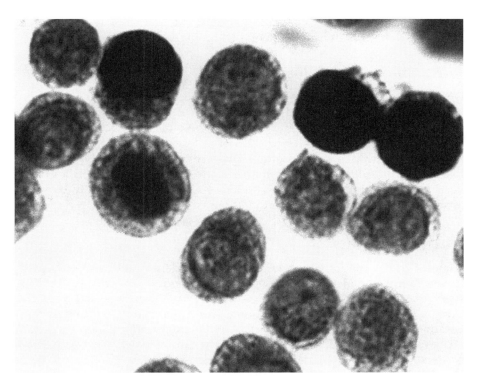

Figure 8 Neoplastic hemocytes from a soft shell clam bath exposed for 12 h to BrdU. Hemolymph was allowed to settle on an APTS coated slide, fixed in acid ethanol, and stained with the Amersahm kit, using the primary antibody for 1 h. No antigen retrieval was employed. Postive nuclei stain black. Eosin counterstain.

Applications of BrdU incorporation

A major use of BrdU in mammals has been in the study of chemically induced cell proliferation in rodent carcinogenicity studies.[2,3,17] It has also been used as part of various double-labeling techniques to characterize the kinetics of the cell cycle, and S-phase in particular.[6,18-21] These techniques use monoclonal antibodies that differentially bind two different nucleotide analogs. Specimens are labeled at different times with the two analogs, allowing determination of actual rates, in contrast to the purely relative indices that are generated with the methods that simply use exposure to one analog. These relative indices only allow between-sample comparison of the proportion of cells that are incorporating BrdU. A further use of BrdU is to expose organisms to a pulse of BrdU and then sample at serial time points thereafter to trace the maturation of particular cell type(s).[22] Other applications in mammals include aspects of repair and host response in disease.[23-26]

BrdU has been used in aquatic organisms to document the effects of cytotoxic and hyperplastic agents on DNA synthesis,[15,16,27,28] development in *Hydra*,[29] whole mounts,[8] and neurobiology.[7] We are currently using this technique to index short-term changes in growth in larval fish and copepods as affected by changes in environmental temperature and food supply.

Interpretation of labeling indices

DNA can be replicated for a number of purposes: reparative regeneration, the development of polyploidy, and for actual hyperplasia. Full interpretation of BrdU labeling indices is often only possible in the context of other cell turnover parameters such as rates of cell death by necrosis, and apoptosis, and alterations in cell volume (hypertrophy). One study[30] has combined data on cell proliferation markers and cell death. There is a greater need to use the BrdU technique as part of a battery of assays to establish actual cell turnover in particular situations.

Acknowledgments

Some of the methodology described here was developed by Dale Leavitt. Coverplate is a registered trademark with Shandon, Pittsburgh, PA. This work was partially supported by Grant # NA-90-AA-D-SG480 from the Coastal Ocean Program of the National Atmospheric and Oceanic Administration (NOAA) to the Woods Hole Oceanographic Institution Sea Grant Program and by the U.S. GLOBEC program, funded jointly by NOAA and NSF (contribution 9194 of Woods Hole Oceanographic Institution and 53 of U.S. GLOBEC). The views expressed herein are those of the authors and do not necessarily reflect the views of NOAA or any of its sub-agencies.

References

1. Beltz, B.S., Burd, G.D. 1989. *Immunocytochemical techniques: principles and practice.* Blackwell, Cambridge, MA, 182 pp.
2. Goldsworthy, T., Morgan, K., Popp, J., Butterworth, B. 1989. *Measurement of chemically-induced cell proliferation in specific rodent cell target organs.* Chemical Industry Institute of Toxicology, Box 12137, Research Triangle Park, N.C. 27709, 98 pp.
3. Eldridge, S.R., Tilbury, L.T., Goldsworthy, T.M., Butterworth, B.E. 1990. Measurement of chemically induced cell proliferation in rodent liver and kidney: a comparison of 5-bromo-2'-deoxyuridine and [^{3}H]thymidine administered by injection or osmotic pump. *Carcinogenesis* 11: 2245–2251.
4. Gratzner, H.G. 1982. Monoclonal antibody to 5-bromo and 5-iododeoxyuridine: a new reagent for detection of DNA replication. *Science* 218: 474–475.
5. Lanier, T.L., Berger, E.K., Eacho, P.I. 1989. Comparison of 5-bromo-2-deoxyuridine and [^{3}H]thymidine for studies of hepatocellular proliferation in rodents. *Carcinogenesis.* 10: 1341–1343.
6. Casas, F.C., Jefferies, A.R. 1992. Estimation of S-Phase Duration in Goat Epidermis by an *in vivo* Intradermal Double Labelling Technique Using Bromodeoxyuridine and Tritiated Thymidine. *Res Vet Sci* 52: 5–9.
7. Bollner, T., Meinertzhagen, I.A. 1993. The Patterns of Bromodeoxyuridine Incorporation in the Nervous System of a Larval Ascidian, Ciona intestinalis. *Biol Bull* 184: 277–285.

8. Plickert, G., Kroiher, M. 1988. Proliferation kinetics and cell lineages can be studied in whole mounts and macerates by means of BrdU/anti-BrdU technique. *Development* 103: 791–794.

9. Anisimov, V.N., Osipova, G.Y. 1993. Life Span Reduction and Carcinogenesis in the Progeny of Rats Exposed Neonatally to 5-bromo-2'-deoxyuridine. *Mutat Res* 295: 113–123.

10. Ellwart, J., Dormer, P. 1985. Effect of 5-fluoro-2'-deoxyuridine (FdUrd) on 5-bromo-2'-deoxyuridine (BrdUrd) incorporation into DNA measured with a monoclonal BrdUrd antibody and by the BrdU/Hoechst quenching effect. *Cytometry* 6: 513–520.

11. Prasad, K.V., Wheeler, J., Robertson, H., Mawhinney, W.H.B., Mchugh, M.I., Morley, A.R. 1994. In vitro bromodeoxyuridine labelling of renal biopsy specimens: Correlation between labelling indices and tubular damage. *J Clin Pathol* 47: 1085–1089.

12. Moore, M.J., Leavitt, D.F., Shumate, A.M., Alatalo, P., Stegeman, J.J. 1994. A cell proliferation assay for small fish and aquatic invertebrates using bath exposure to bromodeoxyuridine. *Aquat Toxicol* 30: 183–188.

13. Kriss, J.P., Revesz, L. 1962. The distribution and fate of bromodeoxyuridine and bromodeoxycytidine in the mouse and rat. *Cancer Research* 22: 254–265.

14. Rostagno, P., Birtwisle, I., Ettore, F., Courdi, A., Gioanni, J., Namer, M., Caldani, C. 1994. Immunohistochemical determination of nuclear antigens by colour image analysis: Application for labelling index, estrogen and progesterone receptor status in breast cancer. *Anal Cell Pathol* 7: 275–287.

15. Droy, B.F., Miller, M.R., Freeland, T.M., Hinton, D.E. 1988. Immunohistochemical detection of CCl$_4$-induced, mitosis-related DNA synthesis in livers of trout and rat. *Aquat Toxicol* 13: 155–166.

16. Miller, M.R., Blair, J.B., Hinton, D.E. 1989. DNA repair synthesis in isolated rainbow trout liver cells. *Carcinogenesis* 10: 995–1001.

17. Popp, J.A., Marsman, D.S. 1991. Chemically induced cell proliferation in liver carcinogenesis, In *Chemically induced cell proliferation implications for risk assessment,* Ed. Butterworth, B.B., Slaga, T.J., Farland, W., and McClain, M., Wiley-Liss, New York, 389–395 pp.

18. Casas, F.C., Jefferies, A.R. 1992. Comparison of Intradermal Injections of Bromodeoxyuridine and Tritiated Thymidine in the *in vivo* Measurement of Epidermal Cell Turnover Time in Goats. *Res Vet Sci* 52: 10–14.

19. Aten, J.A., Bakker, P.J.M., Stap, J., Boschman, G.A., Veenhof, C.H.N. 1992. DNA Double Labelling with IdUrd and CldUrd for Spatial and Temporal Analysis of Cell Proliferation and DNA Replication. *Histochem J* 24: 251–259.

20. Lin, P., Allison, D.C. 1993. Measurement of DNA Content and of Tritiated Thymidine and Bromodeoxyuridine Incorporation by the Same Cells. *J Histochem Cytochem* 41: 1435–1439.

20. Lin, P., Allison, D.C. 1993. Measurement of DNA Content and of Tritiated Thymidine and Bromodeoxyuridine Incorporation by the Same Cells. *J Histochem Cytochem* 41: 1435–1439.

21. Ritter, M.A., Fowler, J.F., Kim, Y.M.J., Gilchrist, K.W., Morrissey, L.W., Kinsella, T.J. 1994. Tumor Cell Kinetics Using Two Labels and Flow Cytometry. *Cytometry* 16: 49–58.

22. Arber, N., Zajicek, G., Schapiro, J.M., Rattan, J., Nordenberg, J., Rozen, P., Sidi, Y. 1994. In vivo bromodeoxyuridine incorporation in normal rat liver: Immunohistochemical detection of streaming cells and measurement of labelling indices. *Acta Histochem Cytochem* 27: 45–49.

246   *Techniques in aquatic toxicology*

23. Watanabe, S., Hirose, M., Yasuda, T., Miyazaki, A., Sato, N. 1994. Role of actin and calmodulin in migration and proliferation of rabbit gastric mucosal cells in culture. *J Gastroenterol Hepatol* 9: 325–333.

24. Ito, T., Ikemi, Y., Ohmori, K., Kitamura, H., Kanisawa, M. 1994. Airway epithelial cell changes in rats exposed to 0.25 ppm ozone for 20 months. *Exp Toxicol Pathol* 46: 1–6.

25. Harkema, J., Hotchkiss, J. 1993. In vivo effects of endotoxin on DNA synthesis in rat nasal epithelium. *Microsc Res Tech* 26: 457–465.

26. Miyaki, K., Murakami, K., Segami, N., Iizuka, T. 1994. Histological and immunohistochemical studies on the articular cartilage after experimental discectomy of the temporomandibular joint in rabbits. *J Oral Rehabil* 21: 299–310.

27. Moore, M.J., Stegeman, J.J. 1992. Bromodeoxyuridine uptake in hydropic vacuolation and neoplasms in winter flounder liver. *Marine Environ. Res.* 34: 13–18.

28. Koza, R.A., Moore, M.J., Stegeman, J.J. 1993. Elevated ornithine decarboxylase activity and cell proliferation in neoplastic and vacuolated liver cells of winter flounder (*Pleuronectes americanus*). *Carcinogenesis* 14: 399–405.

29. Duebel, S., Schaller, H. 1990. Terminal differentiation of ectodermal epithelial stem cells of Hydra can occur in G2 without requiring mitosis or S phase. *J Cell Biol* 110: 939–945.

30. Mundle, S., Iftikhar, A., Shetty, V., Dameron, S., Wright, V., Marcus, B., Loew, J., Gregory, S., Raza, A. 1994. Novel in situ double labeling for simultaneous detection of proliferation and apoptosis. *J Histochem Cytochem* 42: 1533–1537.

chapter fourteen

Allozyme frequency variation as an indicator of contaminant-induced impacts in aquatic populations

Robert B. Gillespie

Introduction

For nearly 30 years, electrophoretic analysis has been used to estimate genetic variation in populations and to estimate the genetic relatedness of species and sub-species.[1] The application of electrophoresis to a number of taxa has resulted in the identification of allelic variants (allozymes) for many enzyme-coding loci. Allele and genotype frequencies can be estimated for several enzyme loci in populations and used to test hypotheses about genetic structure, gene flow, and natural selection. Measures, such as interlocus heterozygosity, percent polymorphic loci and mean numbers of alleles per locus integrate allozyme frequencies into single measures of genetic variation. Electrophoretically-detected genetic variation can be monitored through time or compared among populations to determine the role of environmental influences on genetic variation in aquatic species.

In a typical analysis, either whole organisms or tissues of organisms are collected and frozen in the field. With larger organisms, protein extracts of selected tissues are created through homogenization and centrifugation and frozen until analyzed. With very small organisms, whole bodies can be ground in a small amount of buffer or water and analyzed immediately. Protein homogenates are applied to a gel medium and subjected to an electric current. Allelic variants of proteins migrate differentially based on molecular weight and net electrical charge and are visualized through specific staining techniques. Determination of allelic variants and genotypes can be learned quickly and analysis of allozyme frequency data can be accomplished with simple computer programs.

1-56670-149-X/96/$0.00+$.50
© 1996 by CRC Press, Inc.

Electrophoretic analysis of allozymes can be successful with a modest amount of equipment and supplies. The traditional standard for the analysis of allozymes has been starch-gel electrophoresis. Both vertical and horizontal starch-gel techniques have produced the majority of allozyme data to date.[1] Recently, cellulose acetate gel electrophoresis, developed for clinical use, has been adapted for population genetic analysis in plants and animals. In my lab, we have adapted cellulose acetate gel electrophoresis for use on fish populations based on the protocols published by Hebert and Beaton[2] for Helena Laboratories. This paper describes a cellulose acetate technique for quantifying electrophoretically-detected genetic variation in aquatic populations. The cellulose acetate technique is easy to do and can be initiated with modest start-up monies.

Increasing interest in the effects of environmental contaminants on genetic variation in aquatic populations is evidenced by recent publications that recommend allozyme frequency analysis as a biological monitoring tool.[3-6] Researchers have correlated allozyme frequencies in different aquatic species with varying water quality and exposure to contaminants (Table 1). Additionally, laboratory studies have associated differences in sensitivity to the toxic effects of contaminants with individual allozyme genotypes.[7-16] In whole, these studies suggest that contaminant-induced selection against sensitive allozyme genotypes significantly changes allozyme frequencies in impacted populations and may reduce genetic variation. Alternatively, significant mortality may produce a "genetic bottleneck" through an extreme reduction in population size. Since genetic variation is essential for a population to adapt to changing environmental selection pressures,[17] understanding the effects of poor water quality on genetic variation in natural populations is important.

To fully understand the implications of contaminant-induced selection on genetic variation in aquatic populations, research is needed in three areas. (1) Field studies are needed that can confidently link changes in genetic variation with exposure to contaminants. (2) Laboratory research is needed to determine the links between genes and toxic effects and to determine the heritability of resistance to toxicity. (3) Finally, we need to understand if changes in allozyme frequencies and genetic variation negatively affect the survival and reproduction (reduce fitness) of a population. Because of these needs, electrophoretically-detected analysis of allozyme variation in natural populations will play an important role in ecotoxicology.[4]

Materials

The lists of materials needed to perform adequate cellulose acetate allozyme analysis are described by groups of activities. The entire process of allozyme analysis includes (1) tissue collection and preparation, (2) electrophoretic separation, (3) biochemical staining, and (4) data collection and analysis. Because there are many reagents necessary for allozyme analysis, I have placed these lists in the Appendix (I–IV). The list below describes only equipment and supplies necessary for cellulose acetate gel electrophoresis. The researcher who adopts electrophoretic analysis should consult Hebert and Beaton[2] and Harris and Hopkinson[25] for further information about starting

Table 1 Summary of Results from Field Studies that Assessed Effects of Environmental Stressors on Allozyme Frequencies in Populations of Aquatic Organisms

Stressor	Taxon	Species	Enzyme[a]	Ref.
Thermal discharge	Largemouth bass	*Micropterus salmoides*	MDH	18
	Mummichog	*Fundulus heteroclitus*	MDH	
			EST	19
	Barnacle[b]	*Balanus amphritrite*	MDH	
			EST	20
Mercury	Shrimp	*Palaemon elegans*	PGM	21
	Gastropods	*Monodonta turbinata*	GPI	21
Heavy metals	Freshwater snail	*Heliosoma trivolvis*	GPI	22
	Mosquitofish[c]	*Gambusia holbrooki*	GPI	
			IDH	22
Complex effluent[d]	Stoneroller minnow	*Campostoma anomalum*	PGM	12
Complex effluent[e]	Spotfin shiner	*Notropis spilopterus*	GPI	23
Acidic waters	Mudminnow[f]	*Umbra limi*	IDH	24
			PGD	
DDT, DDE, Lindane	Mosquitofish	*Gambusia affinis*	GPI	8

[a] To interpret enzyme codes, see Appendix II.

[b] Three other enzymes exhibited effects (ADA, MPI, MDH).

[c] Three other enzymes exhibited effects (ACP, ME).

[d] Point source/nonpoint source from uranium enrichment facility.

[e] Point sources from steel production facility.

[f] Three enzymes exhibited slight effects (GPI, MDH, MPI).

From Gillespie, R.B. and Guttman, S.I. 1993. *Environmental Toxicology and Risk Assessment:* 2nd Vol., J. Gorsuch et al., Eds., American Society for Testing and Materials, Philadelphia. With permission; copyright ASTM.

an electrophoretic lab. The list of equipment below can be purchased without great expense, unless the researcher does not have certain items, such as an ultracold freezer, centrifuge, balance, pH meter, incubator, and drying oven. If these equipment items need to be purchased, then start-up costs will be considerably greater. Certain items, such as the drying oven and incubator may be optional, however all other equipment is necessary. Because our research focuses on fishes, the list below will reflect that bias. The researcher is advised to consult literature appropriate for the species of interest.

Tissue collection and preparation

Field preservation equipment and supplies

Ultracold Freezer — tissues/organisms are stored at –70°C.

Option 1: usually for collecting large numbers of whole organisms.

1. Large coolers
2. Dry ice
3. Ziploc® plastic bags

Option 2: usually for collecting small whole organisms or tissues that are dissected in the field.

1. Dewar flask
2. Liquid nitrogen
3. Tissue vials for Dewar

Tissue preparation

1. Homogenizer: we use a hand-held Biomizer
2. Tubes for homogenization: (15 to 50 ml)
3. Centrifuge tubes: (15 to 50 ml)
4. High-speed, refrigerated centrifuge
5. Vials for supernatant: small (3 to 5 ml) screw-cap vials

Electrophoretic separation

Items with catalog numbers refer to Helena Laboratories, P.O. Box 752, Beaumont, TX 77704-0752. Phone: (800) 231-5663.

1. Zip Zone Electrophoresis Chamber (#1283)
2. Titan III Cellulose Acetate Gels (#3033)
3. Zip Zone Applicator Kit (#4093)
4. Zip Zone Chamber Wicks (#5081)
5. Acetate Gel Rack (#5110)
6. Glass microscope slides
7. Vessel to soak gels: e.g., plastic food storage container
8. Power supply (at least 200 volts)

Biochemical staining

See Appendices I to IV for buffers, reagents, and biochemicals.

1. Balance: accurate in milligram range
2. Pippetors (20 to 1,000 μl)
3. Glass Pasteur pipettes
4. Microwave oven
5. Drying oven (60°C)
6. Plastic or plexiglass plates to hold gels for staining
7. Incubator 37°C
8. Glass beakers (600 to 1,000 ml)
9. Glass pipettes (2 to 5 ml) for agar
10. Glass vials for stains: we use glass scintillation vials

Data collection and analysis

1. Camera: automatic 35 mm attached to a copy stand
2. Light box: for transmitted light through gels

3. BIOSYS-1 genetic analysis program: see Procedures section
4. Statistics program: e.g., Statview for Macintosh
5. Spreadsheet program: e.g., Microsoft Excel for Macintosh

Procedures

The process from organism in the field to data analysis involves five different steps: (1) field collections; (2) tissue preparation; (3) electrophoretic separation; (4) biochemical staining; and (5) scoring and data analysis.

Field collections

The techniques for collecting organisms or tissues is as varied as the species themselves. In our research, we usually collect 20 to 50 fish at one site. If we are close to the laboratory, we will collect animals and quickly freeze them alive on wet ice in a cooler. If there will be an extended period of time between collection and deep freezing, then dry ice is recommended. Many researchers who collect tissues or small organisms freeze specimens with liquid nitrogen. This method is probably the best technique for preserving tissues for electrophoresis. However, there is some expense (e.g., Dewar) and logistics (e.g., liquid nitrogen) involved with this method. No matter what the species or type of sample, the tissues must be frozen until stored in an ultracold freezer at –70°C. In our work, individual fish are usually separated in individual Ziploc plastic bags, however, we have collected many small fishes from a single site in a one bag. For the later method, individuals should be rinsed thoroughly with distilled water before homogenization.

Tissue preparation

Whole organisms or tissues (1 to 5 g) can either be ground or homogenized. Small whole organisms or tissue sections should be sliced or minced cold. Homogenization buffer (see Appendix A) or water is added at about a 1:1 ratio (tissue:buffer) by volume and the sample is homogenized for about 30 seconds. We homogenize tissues on ice with low speed for about 10 seconds and on high speed for and additional 20 seconds. Our samples are usually homogenized in centrifuge tubes (15 ml, 50 ml) appropriate for the volumes of tissue and buffer. Tubes with samples are kept cold on crushed ice in an ice bucket until centrifuged. The homogenates are spun at about 20,000 to 30,000 × g for 30 to 45 min. The supernatant is pipetted into glass vials and stored at –70°C until analyzed. For small whole organisms, individuals can be thawed and ground in a grinding plate (drop plate) with a small amount of buffer just prior to electrophoresis. For repeated analyses, the plates can be frozen for later use.

Electrophoretic separation

The electrophoretic separation of proteins comprises three separate steps: (1) soaking gels in buffer; (2) loading the gel with protein extract; and (3) applying current to the gel. The cellulose acetate protocols we use are slight modifications of those published by Hebert and Beaton.[2] Prior to electrophoretic runs, the number of gels and appropriate gel buffer types (see Appendix I) are determined. We use the TG-8.5 buffer for most enzymes; however, other buffers are presented that have been successful. The researcher should search for literature pertinent to the species and/or enzyme of interest.[2] If no prior studies are available, then a screen should be used to determine the best buffer for a particular enzyme. This can be done quickly and cheaply. Cellulose acetate gels are soaked in buffer for at least 20 min prior to electrophoresis and can stay in buffer for as long as 24 h if refrigerated. The gels must be moved slowly in and out of the buffer or bubbles will form in the acetate. If bubbling in the acetate is serious, discard the gel. In an effort not to waste gels, it is best to soak only the number that is going to be used in the short term.

While gels are soaking, add electrode buffer (same as gel buffer) to the electrophoretic chamber, soak and position wicks on the bridge (Figure 1). Shortly before electrophoresis, pipette protein extracts into the well plates. The wells we use hold about 10 μl of solution and care should be taken so that solution does not "leak" from one well to another. Usually, 10 μl is enough extract to load as many as 6 gels. If samples must set in the wells between runs, they should be covered (plastic wrap) and refrigerated. The zip zone system we use is designed to run 12 individual samples on each gel; however, we have found that 10 individuals is more appropriate because samples too close to the edge of the gel often run off and are lost.

The samples are applied to the gel with an automatic applicator that loads an equal amount of solution (about 2 μl) from each well. The acetate gels have a plastic back that can be labeled with an indelible ink pen. Once the gel is labeled, it is placed on a loading plate for alignment and excess buffer is dabbed from the load zone. The applicator loads protein to the acetate side and the gel is then placed on the wick with the acetate side down on the electrophoretic chamber bridge (Figure 1). Because most proteins have a net negative charge, the gels are usually loaded near the cathode side. However, if it is known or expected that a protein might be positively charged, then the gel should be loaded near the center (see Reference 2).

The gel must be placed so that the load zone is on the cathode (–) side of the chamber, but the protein samples must not directly contact the wick. Care must be taken so that the gel does not dry before it is placed on the wick. If the gel dries, it should be discarded. Once on the wick, the gel will stay wet with buffer.

The electrophoretic chambers used for the 76 × 76 mm gels can run three gels at a time; however, we have found that running only two gels at a time (positioned centrally) gives better results. Once the gels are positioned on the chamber, glass slides are placed on the edges of the gel over the bridge to provide maximum contact with the wick. Once the electrodes are plugged into the power unit and the top is on the chamber, the power is set at 200 volts. Depending upon the number of gels and buffer type, the current will

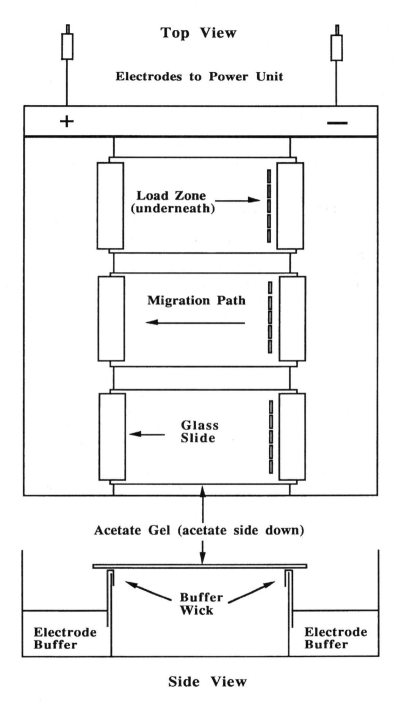

Figure 1 Cellulose acetate electrophoresis chamber. The model presented represents the zip zone electrophoresis chamber by Helena Laboratories (see Materials section).

range from 2 to 5 mA per gel. We typically run the gels for about 20 min; however, it is necessary to run for shorter periods of time for very fast migrating enzymes.

Biochemical staining

Choosing appropriate enzymes to analyze is accomplished either by using pertinent literature or by performing screens to detect variable enzymes. For fishes, many variable enzyme loci have been identified[26,27] and our lab is supplied to screen 26 enzymes (see Appendix II). We usually screen 10 individuals among all possible enzymes and use only variable loci for further analysis.

While the gels are running, stain mixtures can be prepared. All stain recipes (see Appendix II) are formulated for a total of about 2 ml solution. We mix stain ingredients in 20-ml glass scintillation vials in the order listed on the recipe. All chemicals, except stain dyes, can be added ahead of time. Because they are photosensitive, the stain ingredients (e.g., MTT, PMS) must be added just prior to staining the gel. Once the gels are ready, the final dyes are added to the stain mixture and 2 ml of hot agar (60°C) are added with a hot glass pipette. The vial is swirled once or twice and the stain/agar mixture is poured onto the gel. It is necessary to work quickly because of the photosensitive chemicals and because the agar gels quickly. Care must also be taken so that the acetate gel does not dry before being stained.

Once the stain mixture is applied, the agar should be allowed to congeal slightly (30 seconds) before moving the gel to a dark 37°C incubator. For some enzymes, (e.g., glucose phosphate isomerase, phosphoglucomutase, malate dehydrogenase, lactate dehydrogenase), banding appears very quickly, however, for other enzymes the gel may need to stay in the incubator for a longer time. The stain should not be allowed to darken too much that resolution is lost from "bleeding." The allozyme bands should be dark enough to photograph, but distinct enough to detect allelic variation. Once the resolution is good, the agar can be removed by simply placing a paper towel on the gel and applying light pressure. If soft enough, the agar will lift off of the acetate. Once the agar is removed, the gel should be placed in a beaker of cold water. This stops the stain development and prepares the gel for photographing. If the gel is overstained, then the background of the gel will become dark. To avoid this, simply change the gel to new water after about 10 min.

To keep a permanent record, the gel should be labeled and photographed. We use a 35-mm camera with tungsten ektachrome slide film. The camera is on a copy stand above a light box that transmits light through the gel. Although most investigators use prints, we have found that the slides produce good quality reproductions and are cheaper than producing prints. We photograph two gels at a time and take two shots of each gel at slightly different aperture settings. This gives us a backup and ensures at least one good slide. The gels must be photographed wet or light will not transmit through them. Once photographed the gels can be dried in a 60°C oven and they will retain their stain intensity for several days. We usually score the gels just after staining so that we have data recorded based upon the best resolution.

Scoring and data analysis

Scoring genotypes

A detailed explanation for recognizing allelic variants and assigning genotypes for allozymes is beyond the scope of this paper. Harris and Hopkinson[25] and Hebert and Beaton[2] provide good instructions for scoring alleles. Because different species produce varying patterns for the same enzyme, the researcher should also seek literature for a particular enzyme and species. A photograph of a typical gel is presented in Figure 2. The upper gel is stained for phosphoglucomutase (PGM), which is a monomeric enzyme. Therefore, homozygotes are single bands, while heterozygotes are

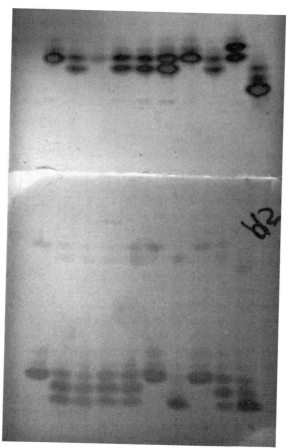

Figure 2 Photograph of two cellulose acetate gels stained for PGM (upper) and TPI (lower). For PGM, a monomeric enzyme, homozygotes are single banded (lanes 1, 3, 7, 8, 10). Heterozygotes for PGM are double banded (lanes 2, 4–6, 9). Lanes 8 and 10 appear to be heterozygotes because of sub banding. For TPI, a dimeric enzyme, homozygotes are single banded (lanes 1, 6, 7, 8, 10). Heterozygotes for TPI are three-banded (lanes 2–5, 9). Lane 10 appears to be a heterozygote because of "bleed" over from lane 9.

double bands. The lower gel is stained for triose phosphate isomerase (TPI), a dimeric enzyme. For dimeric enzymes, homozygotes are single bands, but heterozygotes are triple banded. The middle band of the dimeric heterozygote is an allelic heterodimer formed from the subunits of the two allelic forms (see References 2 and 25 for more detail).

Figure 3 shows allelic variation for two hypothetical monomeric and dimeric enzymes. Although there is no universal consensus for assigning allele and genotype identities, we assign the fastest moving allele an "A" and the next fastest a "B" and so on. In the same way, if there were two loci (isozymes) that coded for the same protein (e.g., glucose phosphate isomerase; GPI), then the faster moving locus would be GPI-1 and the next fastest locus GPI-2, etc. Individual genotypes are recorded (Figure 3) and entered into a spreadsheet program, mainly for a permanent record.

Data analysis

The simplest analysis for allozyme frequency data includes a statistical test for differences in allele and genotype frequencies (chi-square test), measures of genetic variation, and deviation from Hardy-Weinberg equilibrium (chi-square test). We use a combination of the genetic analysis program BIOSYS-1[28] and the statistics program Statview for Macintosh to analyze allozyme frequency data.

The most commonly-used statistical analysis for allozyme data is the chi-square contingency test. This statistic simply tests the hypothesis that frequencies of allozymes (alleles or genotypes) are similar (homogeneous) between two populations or for the same population between time intervals. If no significant differences in frequencies are found, then it is unlikely that other analyses are necessary. However, the researcher should not necessarily stop here. Significant differences in allozyme frequencies between populations or between time intervals is used as a first approximation of effects, but must be followed with more detailed analyses. Most statistical packages written for personal computers include a chi-square contingency analysis.

Measures of genetic variation include interlocus heterozygosity, mean heterozygosity per locus, mean number of alleles per locus, and percent loci polymorphic. These analyses are valuable; however, they must be interpreted carefully because their significance is dependent upon the number of loci analyzed. These measures are generic (not locus specific) and assume that a minimum number of loci were analyzed. In our work, we often focus specifically on a few variable enzyme loci for information and thus, we use these generic measures carefully so as not to overinterpret their significance.

The chi-square test for deviation from Hardy-Weinberg equilibrium tests the hypothesis that observed genotype frequencies of allozymes are similar to that expected by the assumptions of the Hardy-Weinberg equation. If there is random mating, large effective population size, no gene flow, and no selection acting on the population, then the genotype frequencies should be a function of the Mendelian probabilities for random segregation and assortment of alleles and genes (see Reference 1). If genotype frequencies differed significantly from Hardy-Weinberg expectations, then there may be some "force" acting on the population to cause the disequilibrium from the pre-

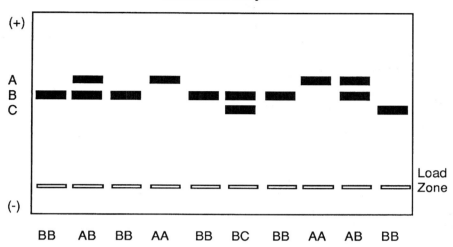

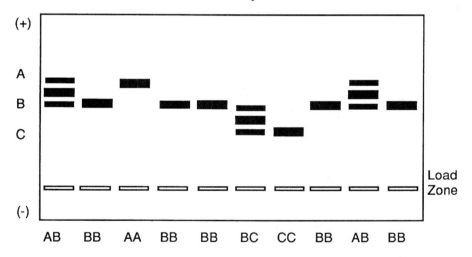

Figure 3 Graphic representation of a cellulose acetate gel. Allelic variants (left side of gel) and genotype designations (bottom of gel) show how we score allozymes for individuals. Capital letters are used for all alleles to designate codominance.

dicted frequencies. One possible explanation for deviation from Hardy-Weinberg equilibrium could be contaminant-induced selection against individuals with certain allozyme genotypes.

   Although basic statistic packages usually include chi-square analysis, the program developed by Swofford and Selander[28] analyzes many aspects of genetic variation, including those mentioned above. The BIOSYS-1 program was written in FORTRAN and is available for IBM and Macintosh computers. This program calculates allele frequencies, measures of genetic variation and performs chi-square analyses. A brief example of the simplest type of data analysis for allozyme variation is presented in the Results and Discussion section. I recommend that the researcher become familiar with other types of genetic analyses (see Reference 29) to seek methods that may be more powerful or more appropriate for a specific study.

Results and discussion

To illustrate a "typical" allozyme analysis, some unpublished data are presented from a study that is in its preliminary stages. This research project, when completed, will attempt to associate differences in water quality with allozyme variation in fishes from tributaries of the Chesapeake Bay. Although data on water quality are not included here, allozyme analysis is presented for the Atlantic silverside (*Menidia menidia*) from two tributaries (Rock Creek, Curtis Creek). The following results represent only a simple analysis of electrophoretically-detected genetic variation.

Of about 15 loci screened, three enzymes showed allelic variation in populations (n = 40 each) of silversides. Four alleles were identified for PGM-1, GPI-1, and GPI-2 in both populations of *Menidia* (Table 2). Although frequencies of alleles for all three loci varied between Rock Creek and Curtis Creek, GPI-2 showed the greatest differences between sites for both alleles and genotypes (Table 3). Allozyme frequencies of GPI-2 for fish from Curtis Creek are dominated by a single allele (GPI-2 B) and genotype (GPI-2 BB), whereas frequencies at Rock Creek are more evenly distributed (Table 3, Figure 4). Allele frequencies of GPI-2 differed significantly (chi-square; p = 0.03) between sites. Genotype frequencies of GPI-2, however, did not differ significantly (chi-square; p 0.07) between sites, although the probability approached significance (p 0.05).

Although the mean heterozygosity per locus (0.283, 0.285) for all three loci was similar for both sites (Table 2), the interlocus heterozygosity was greater in silversides from Rock Creek than those in Curtis Creek (Table 3). Whereas mean heterozygosity per locus is an average of three individual-locus heterozygosities, interlocus heterozygosity is the total number of heterozygous individuals among all loci divided by the number of individuals sampled times three (see Table 3). It is difficult to generalize about overall heterozygosity in these populations with only three variable loci; however, being somewhat conservative, we may conclude that heterozygosity was similar in both populations.

When using a few number of loci, it is probably best to use the average heterozygosity per locus to make comparisons of genetic variation. The ability of a single locus to highly influence the interlocus heterozygosity, when only a few loci are included, makes this measure less reliable than the average heterozygosity among a few loci. It should also be noted that of 20 loci examined, only these three showed variability with the techniques used. Thus, the heterozygosity presented here represents only a few loci and the overall heterozygosity of the genome would be much smaller.

Table 2 Output Files from BIOSYS 1 Analysis of Allele Frequencies of Three Enzymes from Atlantic Silversides (*Menidia menidia*) Collected in Two Tributaries of Chesapeake Bay (Data comprise partial output of complete BIOSYS 1 output)

	Population	
Locus	Rock Creek	Curtis Creek
PGM-1		
(N)	30.0	40.0
A	0.300	0.350
B	0.617	0.600
C	0.033	0.038
D	0.050	0.013
GPI-1		
(N)	32.0	40.0
A	0.047	0.125
B	0.828	0.813
C	0.109	0.063
D	0.016	0.000
GPI-2		
(N)	38.0	40.0
A	0.013	0.025
B	0.829	0.950
C	0.066	0.000
D	0.092	0.025

Allele Frequencies and Genetic Variability Measures

Population: ROCK CREEK, 93

Allele	PGM-1	GPI-1	GPI-2
A	0.300	0.047	0.013
B	0.617	0.828	0.829
C	0.033	0.109	0.066
D	0.050	0.016	0.092
H(D.C.)	0.367	0.250	0.237

Note: Mean heterozygosity per locus = .285 (S.E. .041)

Population: CURTIS CREEK, 93

Allele	PGM-1	GPI-1	GPI-2
A	0.350	0.125	0.025
B	0.600	0.813	0.950
C	0.038	0.063	0.000
D	0.013	0.000	0.025
H(D.C.)	0.475	0.275	0.100

Note: Mean heterozygosity per locus = .283 (S.E. .108)

Table 2 (continued)   Output Files from BIOSYS 1 Analysis of Allele Frequencies of Three Enzymes from Atlantic Silversides (*Menidia menidia*) Collected in Two Tributaries of Chesapeake Bay (Data comprise partial output of complete BIOSYS 1 output.)

Chi-Square Test for Deviation from Hardy-Weinberg Equilibrium with Pooling
Population: ROCK CREEK, 93

Class	Observed freq.	Expected freq.	Chi-square	DF	P
PGM-1					
Homozygotes for most common allele	13	11.4			
Common/rare heterozygotes	11	14.2			
Rare homozygotes and other heterozygotes	6	4.4	1.5	1	0.219
GPI-1					
Homozygotes for most common allele	23	21.9			
Common/rare heterozygotes	7	9.1			
Rare homozygotes and other heterozygotes	2	0.9	1.7	1	0.190
GPI-2					
Homozygotes for most common allele	27	26.1			
Common/rare heterozygotes	9	10.8			
Rare homozygotes and other heterozygotes	2	1.1	1.0	1	0.310

Chi-Square Test for Deviation from Hardy-Weinberg Equilibrium with Pooling
Population: CURTIS CREEK, 93

Class	Observed freq.	Expected freq.	Chi-square	DF	P
PGM-1					
Homozygotes for most common allele	15	14.4			
Common/rare heterozygotes	18	19.2			
Rare homozygotes and other heterozygotes	7	6.4	0.1	1	0.693
GPI-1					
Homozygotes for most common allele	28	26.4			
Common/rare heterozygotes	9	12.2			
Rare homozygotes and other heterozygotes	3	1.4	2.7	1	0.098
GPI-2					
Homozygotes for most common allele	36	36.1			
Common/rare heterozygotes	4	3.8			
Rare homozygotes and other heterozygotes	0	0.1	0.1	1	0.739

Contingency Chi-Square Table

Locus: GPI-2		Allele			
Population		A	B	C	D
ROCK CREEK, 93	Obs (O)	1.000	63.000	5.000	7.000
	Exp (E)	1.462	67.718	2.436	4.385
	(O-E)**2/E	0.146	0.329	2.699	1.560
CURTIS CREEK, 93	Obs (O)	2.000	76.000	0.000	2.000
	Exp (E)	1.538	71.282	2.564	4.615
	(O-E)**2/E	0.138	0.312	2.564	1.482

Note: Chi-square, 9.230; D.F., 3; P, 0.02638.

The genotype frequencies of all enzyme loci did not differ significantly from Hardy-Weinberg equilibrium in either population. Although BIOSYS-1 calculates a chi-square value for individual loci, when more than two alleles are detected at a locus, it is recommended that alleles are pooled for analysis.[28] Therefore, Table 2 presents the results of the chi-square deviation test with pooling.

The results of the allozyme analysis applied to this limited data set show only that the frequencies of alleles at a single locus (GPI-2) differ significantly between two populations of silversides. The genotype frequencies of GPI-2 deviated between populations, but the chi-square analysis was insignificant using the standard critical value ($p = 0.05$). The allele and genotype frequencies of GPI-2 in Curtis Creek were more homogeneous (dominated by single allele, genotype) than those in Rock Creek. Although demanding more study,

Table 3 Frequencies of Allozymes and Interlocus Heterozygosity of GPI-2 in Atlantic Silversides (*Menidia menidia*) from Two Tributaries of Chesapeake Bay

Allozyme	Rock Creek	Curtis Creek
Alleles[a]		
GPI-2 A	0.01	0.03
GPI-2 B	0.83	0.95
GPI-2 C	0.07	0.00
GPI-2 D	0.09	0.02
Genotypes[b]		
GPI-2 AB	0.07	0.05
GPI-2 BB	0.68	0.90
GPI-2 BC	0.13	0.00
GPI-2 BD	0.07	0.05
GPI-2 DD	0.05	0.00
Interlocus Heterozygosity[c]	0.343	0.267

[a] Chi-square p value = 0.03, from BIOSYS-1.

[b] Chi-square p value = 0.07, from Statview for Macintosh.

[c] Number heterozygotes among three loci/number of individuals among three loci (N × 3).

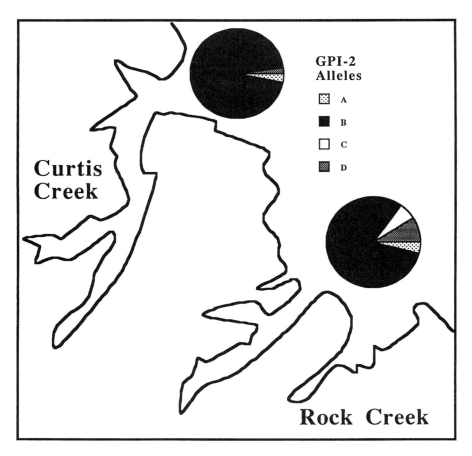

Figure 4   Allele frequencies (proportions) of GPI-2 from two populations of Atlantic silversides. Curtis Creek and Rock Creek are tributaries to the Chesapeake Bay.

the lesser heterogeneity at Curtis Creek could be hypothesized to result from selection against individuals with certain "sensitive" genotypes or from a genetic bottleneck resulting from high mortality.

Although the mean heterozygosity per locus was similar, the interlocus heterozygosity was higher in the Rock Creek population than in the Curtis Creek. However, with limited analyses (three loci), it is difficult to compare genetic variation between populations. Although the lack of deviation from Hardy-Weinberg equilibrium indicates no selection against any genotype within a single population, the allele frequencies of GPI-2 differed significantly between populations, indicating that something may be causing these differences.

In a more comprehensive study (e.g., more loci, more species, more sites), differences in frequencies and genetic variation could be correlated with water quality variables to test the hypothesis that contaminant-induced selection caused changes in allozyme frequencies. However, the researcher should keep in mind that good correlations demand good environmental exposure data, such as contaminant residues in fish tissues, water, and possibly sediments. If correlations, based upon contaminant analysis, link allozyme variation with exposure, then laboratory studies could help determine cause and effect mechanisms for the genetic differences observed in the field.

To adequately test the hypothesis that contaminants and reduced water quality significantly affect genetic variation in aquatic populations, we need to conduct more field studies that adequately link exposure/water quality with genetic variation. The use of allozyme analysis can be a very effective technique to measure genetic variation and should play a major role in this effort.

Appendix I

Tissue grinding solution and gel buffers for cellulose acetate electrophoretic analysis.

Grinding solution

We have used this solution successfully for fishes. Many others exist and the investigator should consult pertinent literature to identify the most appropriate buffer for any species analyzed.

2%, 2-phenoxyethanol

20 ml 2-phenoxyethanol
85 g sucrose (0.25 M)
To 1 liter of water
NADP may be added (100 µl) just prior to grinding (see G6PD and ACON).

Gel and electrode buffers

For cellulose acetate electrophoresis, gel and electrode buffers should be the same type. We use primarily TG 8.5[2], but for some enzymes, CT 6.1 and TC 8.0 give better results. Many other gel buffers exist and the investigator should review pertinent literature for a particular species. A screen for each enzyme can be run with several buffers to identify the best one for resolution. The recipes below are one-liter stock solutions in distilled-deionized water. Stock buffers keep well for long periods of time in a refrigerator. For electrophoretic runs, buffers are diluted 1:9 (stock:dd H_2O) and pH is adjusted (except CT 6.1) with 1 M HCL or 1 M NaOH.

Gel and running buffers (1 liter stock solution)

Tris Glycine: pH 8.5 (TG 8.5)

30 g trizma base
144 g glycine

Clayton and Tritiak: pH 6.1 (CT 6.1)

8.4 g citric acid (monohydrate)
10 ml N-3 (3-aminopropyl)-morpholine
Adjust pH with N-3 (3-aminopropyl)-morpholine

Tris Borate: pH 8.0 (TB 8.0)

60.6 g trizma
40 g boric acid
6 g EDTA

Tris Maleate: pH 7.4 (TM 7.4)

12.1 g trizma
11.6 g maleic acid
3.7 g EDTA
2.03 g $MgCl_2$
10 ml 10 M NaOH

Tris Citrate: pH 8.0 (TC 8.0)

83.2 g trizma
33 g citric acid (monohydrate)

Appendix II

Recipes for stains of enzymes used successfully for cellulose acetate electrophoresis. Although all or some of these may work well with all species, the
investigator should use pertinent literature to find variable enzymes for other
taxa. Unless indicated, all tris buffers are 0.2 M. When called for, drops are
from a glass Pasteur pipette.

Enzyme stain recipes

Acid Phosphatase (ACP)
2 ml α-naphthyl acid PO_4
5 drops garnet GBC salt or 2 ml 4-MUB
Expose to fluorescent light

Aconitase (ACON)*
16 drops cis-aconitic acid
1.5 ml NADP
6 drops $MgCl_2$
50 µl IDH (7 units/gel)
5 drops MTT
5 drops PMS

Adenosine Deaminase (ADA)
1.0 ml tris HCl-8.0
10 drops adenosine
5 drops Na_2HAsO_4
250 µl xanthine oxidase
10 µl nucleoside
 phosphorylase
5 drops PMS
5 drops NBT

Adenylate Kinase (AK)
0.6 ml tris HCL-7.0
1.0 ml NADP
6 drops MgCl$_2$
1.5 ml ADP + D-glucose
15 µl hexokinase
3 drops G6PDH
5 drops MTT
5 drops PMS

Alcohol Dehydrogenase (ADH)
0.6 ml tris HCl-7.0
1.5 ml NAD
3 drops ethanol or isopropanol
5 drops MTT
5 drops PMS

Alkaline Phosphatase (ALP)
2 ml α-naphthyl acid PO$_4$
5 drops MgCl$_2$
5 drops fast blue BB salt

Creatine Kinase (CK)
0.6 ml 0.1 M tris HCl-7.0
1.0 ml NAD
6 drops MgCl$_2$
1.0 ml ADP + d-glucose
0.5 ml phosphocreatine
15 µl hexokinase
3 drops G6PDH
5 drops MTT
5 drops PMS

Esterase (EST)
50 mg β-naphthyl acetate
15 mg fast blue BB salt
2.5 ml acetone
1.0 ml deionized water
1.0 ml tris maleate-5.3
Mix acetate in 2 ml acetone
Mix water and buffer
Slowly add water/buffer
Add fast blue just prior to staining.
Add 2 ml agar to 3 ml stain solution.
If acetate precipitates, add 1 or 2
  drops acetone to redissolve.

Fumarate Hydratase (FH)
1.0 ml tris HCl-7.0
1.5 ml NAD
5 drops fumaric acid
50 µl malic dehydrogenase
5 drops MTT
5 drops PMS

Glucose Dehydrogenase (GDH)
0.6 ml 0.5 M phosphate-7.0
1.5 ml NAD
1 ml D-glucose
5 drops NBT
5 drops PMS

Glucose-6-Phosphate Dehydrogenase (G6PD)*
0.6 ml tris HCl-8.0
1.0 ml NADP
6 drops MgCl$_2$
12 drops G-6-P
5 drops MTT
5 drops PMS

Glucose Phosphate Isomerase (GPI)
1.0 ml tris HCl-8.0
1.0 ml NADP
5 drops F-6-P
5 drops G6PDH
5 drops MTT
5 drops PMS

Glutamate Dehydrogenase (GLUD)
0.6 ml 0.5 M phosphate-7.0
1.5 ml NAD
1.0 ml L-glutamic acid
5 drops NBT
5 drops PMS

Glutamic Oxaloacetic Transaminase (GOT)

10 ml 0.1 M phosphate-7.0
(A) 1 mg pyridoxal-5-phosphate
(B) 50 mg L-aspartic acid
(C) 30 mg α-ketoglutaric acid
(D) 20 mg fast blue BB salt
Mix A to C and adjust pH to 7.4.
Add Fast Blue (D) just prior to staining. Add 2 ml agar to 3 ml stain solution. Develop at room temperature in the dark.
If quantity is needed, make batches and freeze (mixture degrades in a few days in refrigerator.

Glyceraldehyde-3-Phosphate Dehydrogenase (GAPD)

1.5 ml NAD
10 drops D-F-1, 6-diphosphate
5 drops Na_2HAsO_4
50 μl aldolase
5 drops MTT
5 drops PMS

Glycerol-3-Phosphate Dehydrogenase (GPD)

0.6 ml tris HCl-8.0
1.5 ml NAD
22 drops DL-α-glycerophosphate
5 drops MTT
5 drops PMS

Hexokinase (HK)

0.6 ml tris HCl-7.0
1.5 ml NADP (2.5 mg/ml)
16 drops ATP + D-glucose
5 drops $MgCl_2$
10 μl G6PDH
5 drops MTT
5 drops PMS

Isocitrate Dehydrogenase (IDH)

1.0 ml tris HCL-7.0
1.0 ml NADP
8 drops $MgCl_2$
15 drops DL-isocitric acid
5 drops MTT
5 drops PMS

Lactate Dehydrogenase (LDH)

1.0 ml tris HCL-7.0
1.5 ml NAD
10 drops DL-lactic acid
5 drops MTT
5 drops PMS

Leucine Amino Peptidase (LAP)

2.0 ml L-leucine naphthylamide
1 drop saturated fast black salt

Malic Enzyme (ME)

0.6 ml tris HCl-8.0
1.0 ml NAD
2 drops $MgCl_2$
6 drops malic substrate
5 drops MTT
5 drops PMS

Malate Dehydrogenase (MDH)

1.0 ml tris HCl-8.0
1.0 ml NADP
10 drops malic substrate
5 drops MTT
5 drops PMS

Mannose-1-Phosphate Isomerase (MPI)

1.0 ml tris HCl-8.0
1.0 ml NAD
5 drops d-mannose-6-phosphate
3 drops G6PDH
5 μl GPI
5 drops MTT
5 drops PMS

Peptidase (PEP)
2.0 ml 0.02 M Na_2HPO_4-pH 7.5
2 drops $MnCl_2$
8 drops o-dianisidine
8 drops peptide
4 drops peroxidase
4 drops L-amino acid oxidase

Phosphoglucomutase (PGM)
1.0 ml tris HCl-8.0
1.0 ml NADP
5 drops $MgCl_2$
5 drops G-1-P
5 drops G6PDH
5 drops MTT
5 drops PMS

Phosphogluconate Dehydrogenase (PGD)
0.6 ml tris HCl-8.0
1.0 ml NADP
6 drops $MgCl_2$
6 drops 6-phosphogluconic acid
5 drops MTT
5 drops PMS

Superoxide Dismutase (SOD)**
1.0 ml tris HCL-8.0
5 drops MTT
5 drops PMS
Expose to light

Sorbitol Dehydrogenase (SORD)
0.6 ml tris HCl-8.0
1.5 ml NAD
10 drops D-sorbitol
5 drops pyruvic acid
5 drops pyrazole
5 drops MTT
5 drops PMS

Triose Phosphate Isomerase (TPI)
1.5 ml drops TPI substrate
1.5 ml NAD
5 drops Na_2HAsO_4
15 μl G3PDH
5 drops MTT
5 drops PMS

Xanthine Dehydrogenase (XDH)
1.0 ml tris HCl-8.0
1.0 ml NAD
1.0 ml hypoxanthine (heat to dissolve
 precipitate)
5 drops MTT
5 drops PMS

* According to Hebert and Beaton,[2] G6PD and ACON are very labile and may be difficult to resolve unless NADP is added to homogenization buffer.
** SOD does not work well with cellulose acetate.[2]

Appendix III

Solutions for buffers and stains used for cellulose acetate electrophoresis.

Agar solution
4.0 g bact. grade agar
250 ml water
Microwave 30% 10–15 min
Store covered 60°C

1 M HCl (Sigma 920-1)
or 85.2 ml conc HCl to 1 liter

$MgCl_2$ Solution
1 g $MgCl_2$
To 50 ml water

Sodium azide*
2.5 g to 10 ml water

Na_2HASO_4
4.0 g to 25 ml water

1 M NaOH solution
40 g NaOH
To 1 liter water

$MnCl_2$ Solution
4.9 g $MnCl_2$
To 100 ml water

0.2 M Tris HCl-8.0
2.46 g trizma base
10.4 ml 1 M HCl
To 100 ml

0.1 M Tris HCl-8.0
1.23 g trizma
6.2 ml 1 M HCl
To 100 ml

0.2 M Tris HCl-9.0
2.46 g trizma
3.0 ml 1 M HCl
To 100 ml

0.1 M Tris Maleate-5.3
1.2 g trizma
1.2 g maleic acid
2.4 ml 1 M NaOH
To 100 ml

0.2 M Tris HCl-7.0
2.46 g trizma base
18 ml 1 M HCl
To 100 ml

0.1 M Tris HCl-7.0
1.23 g trizma
8.75 ml 1 M HCl
To 100 ml

0.02 M Tris HCl-8.0
0.246 g trizma
1 ml 1 M HCl
To 100 ml

0.1 M Phosphate-7.0
30.5 ml disodium PO_4
19.5 ml monosodium PO_4
Dilute to 100 ml with water
Adjust pH (20 KOH pellets in 10 ml
 water)

Monosodium PO_4
2.76 g NaH_2PO_4
To 100 ml
pH 4.4

* For cofactors (NAD, NADP) and certain substrates that are made in batches and sored for days, use sodium azide (1 μl/ml) to preserve their activity.[2]

Disodium PO$_4$
5.36 g Na$_2$HPO$_4$
To 100 ml
pH 8.7

0.02 M Disodium PO$_4$-7.5
0.284 g Na$_2$HPO$_4$
To 100 ml

0.05 M Citrate PO$_4$-4.0
0.710 g Na$_2$HPO$_4$
1.0 g citric acid monohydrate
To 100 ml

NAD
1 g NAD
To 50 ml water

NADP
500 mg NADP
To 50 ml water

MTT
0.200 g MTT
To 25 ml water (room temp)

PMS
0.5 g PMS
To 25 ml water

4-MUB Phosphate
20 mg MUB
10 ml 0.05 M citrate/PO$_4$-4.0
Solubility is low (stir)
Need 2 ml/gel (ACP)

Fast Garnet GBC Salt
Saturate in water
Need 5 drops/gel (ACP)

Fast Blue BB salt
Dissolve in water (saturate)
Need 10 drops/gel (ALP, EST)

NBT
20 mg NBT
To 20 ml water
5 drops/gel for ADA, GDH, GLUD

Fast Black Salt
Saturate in water
Need 1 drop/gel (LAP)

ADP + D-glucose
200 mg ADP
300 mg D-glucose
To 20 ml deionized water

G-6-P
1 g G-6-P
To 50 ml water

G-1-P
250 mg grade III (67000)
250 mg grade VI (G 1259)
To 10 ml water

D-Fructose-1,6-diphosphate
500 mg D-F-1,6-diphosphate
To 10 ml water

DL-Isocitric acid
1 g isocitric acid
100 mg D-glucose
To 50 ml water

Peptides
100 mg to 5 ml water
Use in combination/single
Phe-Pro, Phe-Leu, Leu-Gly

6-Phosphogluconic acid
200 mg 6-P acid
To 10 ml water

α-Naphthyl acid phosphate (ACP)
10 mg α-naphthyl acid PO$_4$
20 ml water

1-Leucine-β-naphthylamide
20 ml 0.1 *M* tris malate-5.3
10 mg L-leucine naphthylamide

Phosphocreatine
40 mg phosphocreatine
10 ml water

D-Sorbitol
400 mg sorbitol
20 ml water

Pyrazole
100 mg pyrazole
20 ml water

Malic substrate-8.0
45 ml water
5 ml tris HCl-9.0
2.0 g L-malic acid
Adjust pH to 8.0

Aconitic Acid-8.0
200 mg cis aconitic acid
20 ml tris HCl-8.0
Adjust to pH 8.0

Pyruvic Acid
100 mg pyruvic acid
20 ml water

ATP + D-glucose
200 mg ATP
300 mg D-glucose
To 20 ml deionized water

F-6-P
0.5 g F-6-P
To 25 ml water

D-Mannose-6-Phosphate
100 mg M-6-P
To 10 ml water (watch solubility!)

DL-α-Glycerophosphate
1 g DL-1-glycerophosphate
To 25 ml water

Hypoxanthine
250 mg hypoxanthine
To 25 ml water-boil/dissolve
Mix just prior to use and store at
   60°C

Adenosine
40 mg adenosine
20 ml water

o-Dianisidine
100 mg dianisidine
To 25 ml water

α-Napthyl acid phosphate (ALP)
50 ml tris HCl-9.0
200 mg NaCl
100 mg polyvinylpyrrolidone
10 mg α-naphthyl acid PO$_4$
Make in batches and freeze

Fumaric acid
1 g fumaric acid
20 ml tris HCl-8.0
Adjust to pH 8.0

L-Glutamic acid
1 g glutamic acid
20 ml water

TPI Substrate
20 ml 0.02 *M* tris HCl-8.0
650 mg DL-α glycerophosphate
220 mg pyruvic acid

20 mg NAD
20 µl glycerophosphate
  dehydrogenase
20 µl lactic dehydrogenase
Incubate TPI substrate at 37°C for
2 h. Stop reaction by quickly
  adjusting pH to 2 with conc HCl.
  Quickly readjust pH to 80 with conc
  NaOH.
Substrate is stable for 1 month

D-*Glucose Solution*

1 g D-glucose
20 ml water

DL-*Lactic acid*

Dilute 1:3 w/water

Enzymes — abbreviations refer to stain recipes that use these

Hexokinase (ADK)

2500 units per ml
Use as is (15 µl/gel)

G6PDH* (ADK, CK, GPI, MPI, PGM)

Dilute to 50 units per ml

Aldolase (GAPD)

Use as is (50 µl/gel)

Glycerophosphate dehydrogenase (TPI)

Use as is (20 µl/TPI stock)

Isocitric dehydrogenase (ACON)

100 mg to 500 µl water
(140 units/ml)

Nucleoside phosphorylase (ADA)

100 units to 250 µl water

Peroxidase (PEP)

50 mg to 5 ml
Small amounts only!

GPI (MPI)

Use as is (5 µl/gel)

G3PDH (TPI)

Use as is (15 µl/gel)

L-*Amino acid oxidase (PEP)*

100 mg to 10 ml water

Malic dehydrogenase (FH)

Use as is (50 µl/gel)

Xanthine oxidase (ADA)

25 units to 5 ml water

* A cheaper alternative uses G6PDH (XXIV) with NAD as a cofactor. All recipes with G6PDH would use NAD instead of NADP.

Appendix IV   Chemical list for cellulose acetate electrophoretic analysis

Chemical/item	Catalog no. (Sigma Chem. Co.)	Amount or conc
Bacterial grade agar	A 6674	500 g
Sodium azide	S 2002	25 g
Magnesium chloride	M 0250	100 g
Manganese chloride	M 3634	100 g
2-Phenoxyethanol	P 1126	250 ml
Sodium hydroxide	S 5881	500 g
Hydrochloric acid (1 N)	920-1	1 gal
Disodium phosphate	S 0876	500 g
Monosodium phosphate	S 0751	500 g
Potassium hydroxide	P 1767	500 g
Sucrose	S 8501	500 g
α D-Glucose	G 8270	1 kg
Polyvinylpyrrolidone	PVP-40	50 g
Trizma base	T 1503	1 kg
Citric acid (monohydrate)	C 7129	1 kg
Glycine	G 7126	1 kg
N-3(3-aminopropyl)-morpholine	A 9028	100 ml
Boric acid	B 0252	500 g
Maleic acid	M 0375	1 kg
EDTA ($Na_2EDTA \cdot 2H_2O$)	ED 4SS	500 g
NAD	N 0632	10 g
NADP	N 0505	1 g
ADP (III)	A 2754	5 g
ATP	A 5394	1 g
Adenosine	A 9251	5 g
NBT nitroblue tetrazolium	N 6876	1 g
MTT tetrazolium	M 2128	5 g
PMS phenazine methosulfate	P 9625	25 g
Fast blue salt BB	F 0250	100 g
Fast garnet GBC salt	F 0875	25 g
Fast black salt	F 7253	25 g
4-MUB phosphate	M 1508	10 g
G-6-P	G 7879	1 g
G-1-P (III)	G 7000	10 g
G-1-P (VI)	G 1259	10 g
F-6-P	F 3627	5 g
6-Phosphogluconic acid	P 7877	500 mg
D-Mannose-6-phosphate	M 8754	500 mg
D-Mannose-6-phosphate	M 6876	500 mg
β-Naphthyl acetate	N 6875	25 g
DL-Lactic acid	L 1375	500 ml

Chemical/item	Catalog no. (Sigma Chem. Co.)	Amount or conc
DL-Isocitric acid	I 1252	5 g
L-Malic acid	M 9138	25 g
Pyridoxal-5-phosphate	P 9255	5 g
α Ketoglutaric acid	K 1875	25 g
L-Aspartic acid	A 9256	25 g
Hypoxanthine	H 9377	25 g
Cis aconitic acid	A 3412	1 g
α-Naphthyl acid phosphate	N 7000	1 g
L-Leucine-β naphthylamide HCl	L 0376	1 g
Fumaric acid	F 1506	25 g
Phosphocreatine	P 7936	1 g
D-Sorbitol	S 1876	100 g
Pyrazole	P 2646	5 g
L-Glutamic acid (Na salt)	G 1626	100 g
Peptide (PHE-PRO)	P 6258	1 g
Peptide (PHE-LEU)	P 3876	1 g
Peptide (LEU-GLY)	L 9625	1 g
o-Dianisidine (di HCl) salt	D 3252	5 g
DL-α-Glycerophosphate	G 2138	25 g
Na_2HAsO_4	A 6756	50 g
Pyruvic acid	P 2256	100 g
D-Fructose-1,6 diphosphate	752-1	10 g
Peroxidase	P 6782	100 mg
L-Amino acid oxidase	A 5147	500 mg
Isocitrate dehydrogenase	I 1877	100 mg
Glyceraldehyde-3-PDH	G 0763	5,000 units
α-Glycerophosphate dehydrogenase	G 6751	1,000 units
L-Lactic dehydrogenase	L 2500	5,000 units
Aldolase	A 1893	1,000 units
Malate dehydrogenase	M 7383	5,000 units
Nucleoside phosphorylase	N 8264	100 units
Xanthine oxidase	X 4376	25 units
G-6-PDH (XI-NADP)	G 8878	500 units
G-6-PDH (XXIV-NAD)	G 5885	500 units
Glucose phosphate isomerase	P 5381	1,000 units
Hexokinase	H 5500	2,500 units

References

1. Ayala, F. J. 1982. *Population and Evolutionary Genetics.* Benjamin/Cummings Publishing Co., Menlo Park, California.
2. Hebert, P. D. N., and Beaton, M. J. 1989. *Methodologies for allozyme analysis using cellulose acetate electrophoresis.* Helena Laboratories, Beaumont, TX.

3. Anderson, S., Sadinski, W., Shugart, L., Brussard, P., Depledge, M., Ford, T., Hose, J., Stegeman, J., Suk, W., Wirgin, I. and Wogan, G. 1994. Genetic and molecular ecotoxicology: A research framework. *Environ. Health Perspect.* 102 (Suppl 12):3–8.

4. Guttman, S. I. 1994. Population genetic structure and ecotoxicology. *Environ. Health Perspect.* 102 (Suppl 12):97–100.

5. Gillespie, R. B. and Guttman, S. I. 1993. Allozyme frequency analysis of aquatic populations as an indicator of contaminant-induced impacts. *Environmental Toxicology and Risk Assessment:* 2nd Volume, ASTM STP 1173, J. Gorsuch, F. J. Dwyer, C.G. Ingersoll, and T.W. La Point, Eds., American Society for Testing and Materials, Philadelphia.

6. Mulvey, M. and Diamond, S. A. 1991. Genetic factors and tolerance acquisition in populations exposed to metals and metalloids. *Metal Ecotoxicology: Concepts and Applications*, Lewis Publishers, Chelsea, Michigan. pp. 301–321.

7. Benton, M. J. and Guttman, S. I. 1992. Allozyme genotype and differential resistance to mercury pollution in the caddisfly, *Nectopsyche albida*. I. Single-locus genotypes. *Can. J. Fish. and Aquat. Sci.* 49:142–146.

8. Hughes, J. M., Harrison, D. A. and Arthur, J. M. 1991. Genetic variation at the PGI Locus in the mosquito fish *Gambusia affinis* (Poecilidae) and a possible effect on susceptibility to an insecticide. *Biol. J. Linn. Soc.* 44:153–167.

9. Diamond, S. A., Newman, M. C., Mulvey, M., Dixon, P. M. and Martinson, D. 1989. Allozyme genotypes and time to death of mosquitofish, *Gambusia affinis* (Baird and Girard), during acute exposure to inorganic mercury. *Environ. Toxicol. Chem.* 8:613–622.

10. Newman, M. C., Diamond, S. A., Mulvey, M. and Dixon, P. 1989. Allozyme genotype and time to death of mosquitofish, *Gambusia affinis* (Baird and Girard), during acute toxicant exposure: A comparison of arsenate and inorganic mercury. *Aquat. Toxicol.* 15:141–156.

11. Chagnon, N. L. and Guttman, S. I. 1989. Differential survivorship of allozyme genotypes in mosquitofish populations exposed to copper or cadmium. *Environ. Toxicol. Chem.* 8:319–326.

12. Gillespie, R. B. and Guttman, S. I. 1989. Effects of contaminants on the frequencies of allozymes in populations of the central stoneroller. *Environ. Toxicol. Chem.* 8:309–317.

13. Lavie, B. and Nevo, E. 1986. The interactive effects of cadmium and mercury pollution on allozyme polymorphisms in the marine gastropod *Cerithium scabridum*. *Mar. Pollut. Bull.* 17:21–23.

14. Lavie, B., Nevo, E. and Zoller, Y. 1984. Differential viability of phosphoglucose isomerase allozyme genotypes of marine snails in nonionic detergent and crude oil-surfactant mixtures. *Environ. Res.* 35:270–276.

15. Lavie, B. and Nevo, E. 1982. Heavy metal selection of phosphoglucose isomerase allozymes in marine gastropods. *Mar. Biol.* 71:17–22.

16. Nevo, E., Perl, T., Beiles, A. and Wood, D. 1981. Mercury selection of allozyme genotypes in shrimps. *Experientia* 37:1152–1154.

17. Allendorf, F. W. and Leary, R. F. 1986. Heterozygosity and fitness in natural populations of animals. *Conservation Biology: The Science of Scarcity and Diversity*, M.E. Soule, Ed. Sinauer Associates, Inc., Sunderland, Massachusetts. pp. 57–76.

18. Smith, M. H., Smith, M. W., Scott, S. L., Liu, E. H. and Jones, J.C. 1983. Rapid evolution in a post-thermal environment. *Copeia* 1983:193–197.

19. Mitton, J. B. and Koehn, R. K. 1975. Genetic organization and adaptive response of allozymes to ecological variables in *Fundulus heteroclitus*. *Genetics* 79:97–111.

20. Nevo, E., Shimony, T. and Libni, M. 1977. Thermal selection of allozyme polymorphisms in barnacles. *Nature* 267:699–701.

21. Nevo, E., Ben-Shlomo, R. and Lavie, B. 1984. Mercury selection of allozymes in marine organisms: Predictions and verification in nature. *Proc. Nat. Acad. Sci.* 81:1258–1259.

22. Benton, M. J., Diamond, S. A. and Guttman, S. I. 1994. A genetic and morphometric comparison of *Helisoma trivolvis* and *Gambusia holbrooki* from clean and contaminated habitats. *Ecotox. Environ. Saf.* 29:20–37.

23. Gillespie, R. B. and Guttman, S. I. 1993. Correlations between water quality and frequencies of allozyme genotypes in spotfin shiner (*Notropis spilopterus*) populations. *Environ. Pollut.* 81:147–150.

24. Kopp, R. L., Guttman, S. I. and Wissing, T. E. 1992. Genetic indicators of environmental stress in central mudminnow (*Umbra limi*) populations exposed to acid deposition in the Adirondack Mountains. *Environ. Toxicol. Chem.* 11:665–676.

25. Harris, H. and Hopkinson, D. A. 1976. *Handbook of Enzyme Electrophoresis in Human Genetics*, North Holland Publishing Company, Amsterdam, Holland.

26. Whitmore, D. H., Ed. 1990. *Electrophoretic and Isoelectric Focusing Techniques in Fisheries Management*. CRC Press, Inc., Boca Raton, FL.

27. Ryman, N. and Utter, F., Eds. 1987. *Population Genetics and Fishery Management*. University of Washington Press, Seattle, WA.

28. Swofford, D. L. and Selander, R. K. 1981. BIOSYS-1: A FORTRAN program for the comprehensive analysis of electrophoretic data in population genetics and systematics. *J. Heredity* 72:281–283.

29. Weir, B. S. 1990. *Genetic Data Analysis*. Sinauer Associates, Inc., Sunderland, MA.

chapter fifteen

Using a biomarker (P450 RGS) test method on environmental samples

Jack W. Anderson, Kristen Bothner, Tien Vu,
and Robert H. Tukey

Introduction

This chapter provides the recommended guidelines for performing a bio-marker test for toxicity and carcinogenicity, using a transgenic cell line (101L) derived from the human hepatoma cell line, HepG2. Under appropriate test conditions, induction of the enzyme P450 1A1 is evidence that the cells have been exposed to one or more xenobiotic organic compounds, including dioxins, furans, coplanar PCBs, and several polycyclic aromatic hydrocarbons (PAHs). Detection of induction has been made simple and rapid, by the stable integration of the firefly plasmid, such that Ah-receptor binding results in the production of luciferase instead of P450 1A1. This Reporter Gene System (RGS) has shown concentration–response relationships using 2,3,7,8-TCDD, 5 coplanar PCBs, and 8 PAHs. This paper describes test conditions under which solvent extracts of environmental samples (water, tissue, soil, or sediments) may be tested for the presence of toxic and carcinogenic organic compounds. Those samples found to most strongly induce the P450 RGS should then be chemically characterized. Examples of the test sensitivity are shown in Table 1.

The organic compounds that induce P450 1A1 are toxic, often carcinogenic, and they have been shown to bioconcentrate and biomagnify. Testing with birds, mammals, and fish species has shown that exposure to these compounds can produce physiological, reproductive, and histopathological effects.[1-3] Concern for the possible contamination of water, food, wildlife, soil, and aquatic sediment from these compounds has led to the requirement for analytical chemical analyses of a great many environmental samples. Use of a screening tool such as P450 RGS will allow selection of the most contaminated samples, and also those that do not require further expensive chemical characterization.

Table 1   Concentrations of Organics Producing a
10-Fold Induction in P450 RGS[5]

Chemical	Conc. in well (ng/ml)
2,3,7,8-TCDD	0.003
2,3,7,8-TCDF	0.025
PCB #126*	2.5
PCB #169	200.0
PCB #77	300.0
PCB #114	300.0
PCB #156	800.0
Dibenzo(a,b)anthracene	2.5
Benzo[k]fluoranthene	10.0
Benzo[b]fluoranthene	10.0
Indeno(1,2,3-cd)pyrene	12.5
Benzo[a]pyrene	50.0
Benzo[a]anthracene	50.0
Benzofluorene	500.0
Chrysene	3000.0

* PCB numbering system from Ballschmiter and Zell.[11]

Summary of procedure

The test system is based on a transgenic cell line developed and tested previously.[2-5] When this human liver cancer cell (HepG2) is induced at the CYP 1A1 gene, the stably integrated firefly plasmid (reporter gene) produces luciferase, instead of P450 1A1. Solvent (DMSO or dichloromethane) extracts of environmental samples are added to individual wells (6-well plates), containing approximately one million cells, and the exposure time is from 6 to 18 h (generally the latter). Volumes of solvent successfully tested are 2 to 20 µl, which produce a low background (blank) induction when applied to the 2 ml of culture medium. In each assay, the luminescence (in relative light units, RUL) of the combined cytoplasm from each of three replicate wells is determined for each sample, the solvent control, and two reference toxicants (3-methylcholanthrene and 2,3,7,8-dioxin), using a 96-well luminometer. The mean RLUs of the control wells is set equal to unity. Mean RLUs of samples are converted to fold induction by dividing by the mean RLUs of the solvent (control). This biochemical response represents the integrated induction from all planar organic compounds present in the extract, that bind to the Ah-receptor in the same manner as dioxin.[6] Final results may be expressed as fold induction, or by use of the reference toxicant, results can be expressed as dioxin equivalents. For best comparisons between samples, the initial dry weight (determined on a separate subsample) of the extracted sample, the final volume the solvent containing the extracted material (1 or 2 ml), and the amount applied to the cells are all recorded. Induction may then be expressed as nanograms of dioxin (or another reference inducer; e.g., benzo[a]pyrene) equivalents per gram dry weight or per liter.

Materials required

Instruments

Sonic Probe, as Braun-Sonic 1510
Microcentrifuge
Dynatech ML2251 or ML3000 96-well Luminometer
Laminar Flow Hood
Incubator with CO_2 regulation

Chemicals

B[a]P   Benzo[a]pyrene
DMSO   Dimethylsulfoxide
DCM   Dichloromethane (methylene chloride)
TCDD   2,3,7,8-Tetrachlorodibenzo-p-dioxin
3MC   3-Methylcholanthene

Reagents and supplies

Lysis Buffer
Luciferase Assay Kit (Substrate A and B, with Luciferin)
Luciferase Standard for Calibration
  The solutions above are available from Analytical Luminescence, Inc.
  in Ann Arbor, MI.
Centrifuge tubes
Sterile 6-well culture plates, with covers
Tissue culture flasks, 75 cm^2
96 Microwell plate
Cell culture media
  Dulbecco's modified Eagle medium, with 10% fetal calf serum, and
  in 0.4 mg/ml G418
Cell scraper

Sample extraction

While other unique methods may be developed, tested, and used to extract organic contaminants from water, tissue, sediment, or soil samples, the preferred procedures are the EPA methods 3540 (Soxhlet) and 3550 (sonication), both using dichloromethane (DCM) in the extraction. Samples of approximately 20 g (dry weight) of tissue, sediment or soil, and 1 liter of water are appropriate. Tissue samples are generally about 80% water and some aquatic sediments may be as much as 70% water, so care must be taken to estimate the dry grams of sample. Extraction and evaporation will result in a small volume of DCM (1 ml) in a tightly sealed vial. These samples may readily be shipped from an extraction laboratory to the testing laboratory, if these are not the same. If the test sediments are highly contaminated, it may be possible to test the porewater for P450 induction. Sediments should be centrifuged to produce about 1 ml of porewater, which may be applied in volumes up to 200 µl (after 0.45 µm filtration for sterilization) to the cells in the wells (in 2 ml of medium).

Volume selection

Experience has shown that the darkness of the extract is an indication of the quantity of such compounds as PAHs. Therefore, a lower volume of a dark extract (2 to 5 µl) is often appropriate, avoiding the possibility of saturating the cells. Fold induction as high as 1000 times control has been observed, so the range of response is quite broad. Light colored samples are often tested with a volume of 10 or 20 µl of extract. Volumes up to 200 µl of interstitial or porewater may be applied, after 0.45 of 0.22 µm filtration to remove bacteria.

Controls

If more than one volume of test sample is added to the test wells, then it is necessary to use the same volumes of blank solvent, for measuring control luminescence (in RLUs). The control of the same volume as the test sample should be used to determine fold induction, by dividing by the RLUs exhibited by the solvent (blank). A second (background) control is the use of cells and medium alone, without the addition of any solvent or test sample.

Reference toxicants

Reference toxicants and concentrations that have been used are 10 nM 2,3,7,8-TCDD, 10 µM 3-methylcholanthrene, and 100 ng/ml benzo[a]pyrene. These concentrations all represent the final concentration in the exposure wells (in 2 ml of medium). The use of one or two reference toxicants with each set of samples provides a quality control check on the performance of the cells, and also allows conversion of the data to equivalents of the reference toxicant.

Procedures

Figure 1 provides an overview of the protocol used in testing the P450 RGS response to either organic extracts or water (including porewater) samples. This assay requires proper training on cell culture (sterile techniques), the correct use of all instruments, and spiking techniques. Any personnel with adequate training in the techniques mentioned above will be able to perform the assay. Sterile techniques should be employed when handling cells. It is very important that all equipment is handled properly so that no contaminants are introduced. Proper record keeping is required so that all reagents are labeled, all samples logged in, and all cell culture material is properly stored.

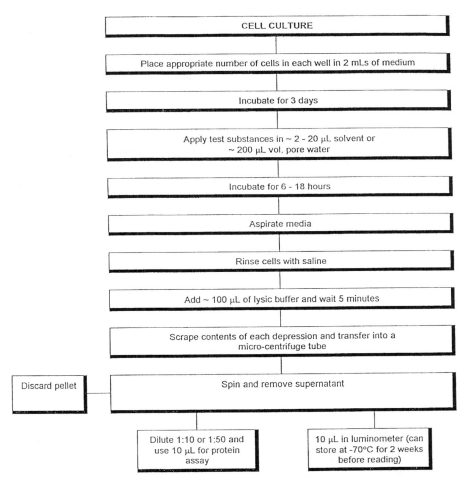

Figure 1 Outline of P450 RGS testing protocol. (Reprinted with permission from Anderson, J.W. et al. 1995. Environ. Toxicol. Chem. 14(7): 1159–1169. Copyright 1995 SETAC.)

Culture maintenance

The maintenance of the cell line requires biweekly changes of media. The cells are incubated at 37°C in an atmosphere of 5% CO_2. The disulfate salt antibiotic Geneticin (G418) is added to media at a concentration of 0.4 mg/ml with each change of medium.

Instrument calibration

The luminometer should be calibrated within two days of the analysis of a set of samples. The calibration curve produced should be compared to previous curves to demonstrate suitable sensitivity of the instrument. Using the Luciferase Standardization Kit (Analytical Luminescence, Inc., La Jolla, CA), it should be possible to detect 1 pg of luciferase.

Safety

All laboratory health and safety procedures should be followed. This includes the use of glasses, gloves, and other protective clothing when handling the reagents. Information on toxicity, handling procedures, and waste procedures should be reviewed prior to use of all chemicals.

Test protocol

The test procedures presented in this paper have been published previously.[4-7] These references should be consulted to obtain details regarding the cell line and the response of the cells to various organic substances.

1. Transfer 0.25×10^6 cells to each well in 2 ml media (1.5×10^6 cells per plate).
2. Incubate for 3 days to allow for cell increase and assure adhesion to plastic.
3. Apply 2 µl of 10 mM 3MC to 2 ml in 3 wells, producing a 10 µM exposure.
4. Apply 2 µl of 10 µM 2,3,7,8-TCDD to 3 wells, producing a 10 nM exposure.
5. Apply each sample with microsyringe or positive displacement micropipettor to 3 replicate wells (2 to 20 µl solvent or 100 to 200 µl of water) and record the volumes.
6. Incubate for 18 h in an incubator at 37°C and 5% CO_2.
7. After 18 h, aspirate media and rinse cells with a saline solution (PBS w/o Mg^{++}).
8. Add 200 µl of lysis buffer to each well and incubate for 15 min at 4°C.
9. Scrape contents of each well and transfer the suspension to a microcentrifuge tube.
10. Spin cells for 10 s at about 10,000 RPM to separate the cellular debris, remove supernatant and discard pellet.
11. Add 50 µl of cell extract from each test well (1 to 3 replicates) microwells of the 96-well plate, designed for the specific luminometer. A single well luminometer may also be used.
12. Inject 100 µl of Substrate A (from Luciferase assay kit) into each well and then inject 100 µl of Substrate B (Luciferin from kit) within 10 min. The time between adding Substrate B and measurement of luminescence should be as short as possible (no longer than 5 min) and consistent from plate to plate.
13. Measure luminescence in each well of 96-well plate in the luminometer.
14. Record all information on the weights of samples, the volume of extract produced from the extraction, any dilution, the volume applied to the cells in the 2 ml of media, and the duration of exposure. This information should be recorded on a bench sheet and later en-

tered onto the final spreadsheet. Luminometer readings should either
be captured on print-out or on disk, if available on the instrument.

15. To normalize the fold induction values to protein content, a subsample
of the supernatant of each well may be analyzed for protein.[8]

Data analysis

Data from the luminometer should be entered on a standard spreadsheet
(example attached, Table 2). From the spreadsheet (formulas embedded in
sheet) the mean RLUs for each control, reference toxicant, and sample may
be determined. Setting the control response equal to unity, it is then possible
to calculate the mean fold induction (with standard deviations) for each
reference toxicant and test sample. If protein is determined for each sample,
the data may be expressed as fold induction per mg of protein.

Responses to solvent blanks and reference toxicants should be compared
to previous data by use of control charts to determine if the test was valid.
If the response to reference toxicants was greater than 2 standard deviations
from the mean, the test may not be acceptable. Replicate plates may be used
to access the viability of cells, by adding the vital stain trypan blue (5%).
Normal viability is 70 to 90%, and a decrease to less than 60% is unacceptable.
When protein is measured, a reduction in protein is also a measure of cell
viability. Dead cells do not adhere to the plates, and are washed away during
rinsing, before the centrifugation step which produces the supernatant tested
for both luminescence and protein content.

Findings should be reported as the fold induction produced by the
samples, with the dry weight of each sample, the volume of the final extract,
and the volume applied to the cells. If the reference toxicant results were
within acceptable limits, this should be noted. The responses of the cells may
also be reported on a basis of equivalents of the reference toxicant (2,3,7,8-
dioxin or benzo[a]pyrene) per g dry weight of sample. In other words, the
listing of the sample response could include the nanograms of dioxin or the
micrograms of B(a)P that would have been present in a dry g of sample to
produce the induction observed.

The general nature of contamination in environmental samples is a mixture of organics, possibly including PAHs, chlorinated hydrocarbons (pesticides, PCBs, dioxins, and furans), and metals. The total response from the
P450 RGS gives an integrated response to the mixture of planar organic
compounds. The response is often additive (multiple PAHs and PAHs plus
a coplanar PCB), but not all combinations have as yet been tested. Since
organic extracts of the samples are applied, there is little chance for metals
to interfere with the response. Tests with extracts of sediment, highly contaminated with a range of toxic divalent metals (cadmium, copper, chromium,
lead, nickel, and zinc) have shown strong induction and thus no interference
from those metals. Tributyltins (TBT) have been shown to denature P450, both
in vitro and *in vivo*.[9,10] To our knowledge environmental samples containing
a significant amount of TBT have not been tested with this method.

Table 2. Sample Spreadsheet for P450 RGS Testing

FileName:LT040194.WK1,DH32(3M),03/26/94. Macro at AY8000.Range A13..R61.

FileName:LT040194.XLS

TV101L P2, 1.5x10^6/plate.

CAS SPIKE #2, 18h.   HARVEST 03/30/94

PROTEIN ASSAY 04/04/94   LUCIFERASE ASSAY 03/30/94

SAMPLES Triplicate	INDUCER (FINAL CONC.)	VOL. USED	Protein Conc(1) ug/uL	Conc(2) ug/uL	ConcAv ug/uL	VOL. USED	ug USED	RLU(1) U	RLU(2) U	AVG U	AVG-BGd U	RLU/ug U/ug	AVG U/ug protein	Fold Induct. XDMSO	Indiv. F.I.	Std. Dev.
BkGd								53	49	51						
1	DMSO	2uL	3.32	3.37	3.34	10	33.4	1114	1137	1126	1075	32				
2	DMSO	2uL	3.67	3.55	3.61	10	36.1	1279	1253	1266	1215	34				
3	DMSO	2uL	3.61	3.81	3.71	10	37.1	1285	1290	1288	1237	33	33.0	1.00		0.6
4	DMSO	20uL	3.39	3.29	3.34	10	33.4	1547	1587	1567	1516	45.39				
5	DMSO	20uL	3.24	3.43	3.33	10	33.32	1637	1546	1591.5	1540.5	46.23				
6	DMSO	20uL	3.29	3.29	3.29	10	32.9	1552	1548	1550	1499	45.56	45.73			0.36
7	3MC 10uM(final)	2uL	3.28	3.31	3.29	10	32.93	69895	72020	70957.5	70906.5	2153.55			65.19	
8	3MC 10uM(final)	2uL	3.21	3.19	3.20	10	31.98	71975	73775	72875	72824	2277.03			68.93	
9	3MC 10uM(final)	2uL	3.39	3.61	3.50	10	34.98	80208	79770	79989	79938	2284.92	2238.5	67.76	69.17	1.82
10	TCDD 10nM(final)	2uL	3.95	4.26	4.11	10	41.05	230918	232136	231527	231476	5638.87			170.71	
11	TCDD 10nM(final)	2uL	3.74	3.98	3.86	10	38.58	216539	218486	217513	217461.5	5636.54			170.64	
12	TCDD 10nM(final)	2uL	3.75	3.95	3.85	10	38.5	217182	219488	218335	218284	5670.29	5648.57	171.01	171.66	0.46

References

1. Safe, S. 1994. Polychlorinated biphenyls (PCBs): Environmental impact, biochemical and toxic responses, and implications for risk assessment. *Crit. Rev. Toxicol.* 24: 87–149.

2. Hahn, M.E., A. Poland, E. Glover and J.J. Stegeman. 1992. The Ah receptor in marine animals: phylogenetic distribution and relationship to cytochrome P450 A inducibility. *Mar. Environ. Res.* 34: 87–92.

3. Stegeman, J., M. Brouwer, R.T. Di Giulio, L. Forlin, B.A. Fowler, B.M. Sanders and P.A. Van Veld. 1992. Molecular responses to environmental contamination: enzyme and proteins systems as indicators of chemical exposure and effect. In *Biomarkers: Biochemical, Physiological, and Histological Markers of Anthropogenic Stress*, R.J. Huggett, R.A. Kimerle, P.M. Mehrle, Jr., and H.L. Bergman, Eds., CRC/Lewis Publishers, Boca Raton, FL, p. 235.

4. Anderson, J. W., S.S. Rossi, R.H. Tukey, T. Vu, and L.C. Quattrochi. 1995. A biomarker, P450 RGS, for assessing the induction potential of environmental samples. *Environ. Toxicol. Chem.* 14(7): 1159–1169.

5. Anderson, J.W., K. Bothner, D. Edelman, S. Vincent, T. Vu, and R.H. Tukey. (In press). A biomarker, P450 RGS, for assessing the potential risk of environmental samples. In *Field Applications of Biomarkers for Agrochemicals and Toxic Substances*, J. Blancato, R. Brown, C. Dary, and M. Saleh, Eds., American Chemical Society, Washington, D.C.

6. Postlind, H., T.P. Vu, R.H. Tukey and L.C. Quattrochi. 1993. Response of human CYP1-luciferase plasmids to 2,3,7,8-tetrachlorodibenzo-p-dioxin and polycyclic aromatic hydrocarbons. *Toxicol. Appl. Pharmacol.* 118: 255–262.

7. Quattrochi, L.C. and R.H. Tukey. 1993. Nuclear uptake of the Ah (dioxin) receptor in response to omeprazole: Transcriptional activation of the human CYP1A1 gene. *Mol. Pharmacol.* 43: 504–508.

8. Bradford, M.M. 1976. A rapid and sensitive method for the quantitation of microgram quantities of protein utilizing the principle of protein-dye binding. *Anal. Biochem.* 72: 248–254.

9. Yao, C., B. Panigrahy, and S. Safe. 1990. Utilization of cultured chick embryo hepatocytes as *in vitro* bioassays for polychlorinated biphenyls (PCBs): quantitative structure-induced relationships. *Chemosphere* 21: 1007.

10. Fent, K. and J.J. Stegeman. 1993. Effects of tributyltin *in vivo* on hepatic cytochrome P450 forms in marine fish. *Aquat. Toxicol.* 24: 219.

11. Ballschmiter, K. and M. Zell. 1980. Analysis of poylchlorinated biphenyls (PCB) by glass capillary gas chromatography: Composition of technical Aroclor- and Clophen-PCB mixtures. *Fresenius Z. Anal. Chem.* 20: 302.

chapter sixteen

Assays of reactive oxygen intermediates and antioxidant enzymes: potential biomarkers for predicting the effects of environmental pollution

**Judith T. Zelikoff, Weihua Wang, Nazrul Islam,
Lorraine E. Twerdok, Marianne Curry, Joseph Beaman,
and Eliezer Flescher**

Introduction

The study of biochemical responses in aquatic animals comprises an area of vigorous investigation within ecotoxicology for a number of reasons, including the need for highly sensitive biomarkers useful for biomonitoring in aquatic settings, and the observations of elevated rates of neoplasia and infectious diseases in some aquatic systems. Because of the role of reactive oxygen intermediates (ROI) in carcinogenesis and as cytotoxic mediators, the generation of free radicals and subsequent oxidative stress in biological systems is of particular concern to environmental toxicologists.

Molecular oxygen (O_2) first appeared on our planet as a contaminant in a highly reducing environment. Virtually all aerobic life is dependent upon a direct transfer of electrons through a series of specially-adapted molecules to O_2 as a part of the energy-producing process. Molecular O_2 requires four electrons for its complete reduction to water. Since this reduction occurs sequentially, one-, two-, and three-electron-reduced intermediates are possible. The one electron reduction of molecular oxygen results in the formation of the superoxide anion radical (O_2^-) which is produced in all respiring cells by a number of biological reactions including the autoxidation of hydroxyquinones, thiols, and reduced ferredoxins.[1] Reduction by a second electron results in the formation of hydrogen peroxide (H_2O_2) and the third electron reduction results in the production of the potent, highly reactive hydroxyl radical (OH·). All of the aforementioned oxygen species can be toxic and destructive not only to ingested parasites/infectious agents, but to living host cells, producing inflammation, tissue damage, and/or mutations.[2]

Superoxide anion, produced both inside and outside mitochondria, is the first radical produced in the reduction of oxygen to water.[3,4] Mammalian[5-7] and fish[1,8-10] phagocytes, when activated by microbes or other stimulants, quickly produce O_2^- and hydrogen ions as part of their antimicrobial defense. These potent cytotoxic free radicals then participate in the ultimate destruction of internalized infectious agents. Endogenous, non-mitochondrial sources of O_2^- include NADPH oxidase from activated phagocytes, prostaglandin synthesis, and the reaction of xanthine oxidase with hypoxanthine or xanthine. Within mitochondria, electrons are ultimately derived from the oxidation of carbohydrates, proteins, and lipids. In a series of energy conserving reactions, these electrons reduce O_2 to water with the concomitant production of adenosine triphosphate. One to four percent of the oxygen, however, is only partially reduced and forms potentially toxic oxyradicals.

Hydrogen peroxide is a powerful oxidant with cytotoxic properties.[3] Like O_2^-, H_2O_2 is produced by activated phagocytes, as well as by mitochondrial NADH and ubiquinone. Probably the most important reaction contributing to the toxic effects of H_2O_2 is its reduction to OH· radical in the presence of ferrous iron.

The biological potential for oxygen radical production is great. Complex multi-enzyme systems such as mitochondrial and microsomal electron transport, as well as leukocyte phagocytosis, can generate large quantities of oxygen radicals.[1-4] Certain classes of xenobiotics such as metals, quinones, aromatic hydroxylamines, and bipyridyls represent particularly prolific sources of oxygen radicals because of their abilities to be reduced to their corresponding radicals via NAD(P)H-dependent reductases (NADPH-cytochrome P450 reductase, xanthine oxidase, ferredoxin reductase, NADH-ubiquinone oxidoreductase) prior to subsequent redox cycling.[1,3] In addition to the ability of certain xenobiotics to generate ROI based on their chemical reactivities, these toxicants may also affect cellular components (i.e., receptors, protein kinases, and phosphatases), thereby modulating the response of phagocytes to stimulators of the oxidative burst. The analysis of data is even more complex when fish are exposed *in vivo* and the observed effects on immune cells may be secondary to other xenobiotic-induced effects such as tissue damage.

An important aspect of free radical-mediated toxicity is that it is moderated in the host by several antioxidant cellular defense mechanisms including enzymatic [i.e., superoxide dismutase (SOD), catalase (CAT), glutathione peroxidase] as well as nonenzymatic (e.g., vitamins E and C) systems.[1,11-13] Antioxidant enzymes allow the organism to protect itself against xenobiotic oxidants. In addition, these enzymes enable ROI-producing cells and adjacent tissues to withstand the oxidative stress exerted by endogenously-generated active oxygen species.[3,11,12] Therefore, efficient elimination of ROI constitutes an essential component of the defense systems used by impacted species exposed to environmental toxicants. While mammalian and bacterial antioxidant enzymes have been widely studied,[2,3,11,14-17] much less is known about these activities in fish in general,[12,13,18,19] and in fish immune cells in particular.

Procedures

This chapter will focus upon three microassay methods that can be utilized easily and economically to measure the production of superoxide dismutase-inhibitable extracellular and intracellular O_2^-, as well as the extracellular production of H_2O_2 by immune cells recovered from the kidney (bone marrow equivalent in fish) of the small, warm-water fish *Oryzias latipes* (Japanese medaka). In addition, biochemical assays used to quantitate SOD, quinone reductase (QR), and CAT activity in cells recovered from the same fish species will be described. While the assay protocols described in this paper have been used in medaka, they can easily be applied to other aquatic species.

Fish sacrifice and cell collection

Materials

Reagents and prepared solutions

1. Supplemented L-15 medium
   a. Leibovitz Medium (Sigma Catalog No. L-5520)
   b. 0.5% Medaka plasma (collected in the investigator's laboratory)

Plasma is prepared by collecting blood from an individual medaka into a heparanized microhematocrit capillary tube (Fisher Catalog No. 02-668-86). The capillary tube containing the recovered blood is then sealed at one end with Cha-seal (Chase Instruments Corp. Catalog No. 510) and centrifuged in a microhematocrit centrifuge (at room temperature) for 4 min at 1100 rpm. Following centrifugation, the tube is cut with a fine-point glass marking pencil 1 mm above the "leukocyte band" and the plasma extruded into a 1.5 ml Eppendorf microcentrifuge tube (Fisher Catalog No. 05-402-25). Plasma from 16 fish (total plasma per fish = ~3 µl) are then pooled into a single microcentrifuge tube and the recovered plasma diluted 30 × with supplemented FPS. The entire contents from the Eppendorf tube is then transferred for sterilization into a Centrex Centrifugal Microfilter unit (PGC Catalog No. 33-8670-55). The entire unit is then spun for 10 min at 1350 rpm (~300 g) and the "receiver portion" of the filter containing the sterile medaka plasma stored at –20°C until needed.

   c. 1% Penicillin/streptomycin (Sigma Catalog No. P-0781)
   d. 1% L-Glutamine, 200 mM (GIBCO Catalog No. 320-5030)

2. Fish physiological saline (FPS) + 1% glucose (Fisher Catalog No D16-1) = supplemented FPS
   a. NaCl   6.44 g
   b. KCL   11 mg
   c. CaCl$_2$ 22 mg

    d. $MgSO_4$ 12 mg
    e. KH_2PO_4 7 mg
    f. $NaHCO_3$ 10 mg

Fish physiological saline is prepared by dissolving the aforementioned salts in 1 l of double-distilled water. The solution is then autoclaved and 1% (10 g) glucose is added prior to filtration; pH of the final solution is adjusted to 7.2.

    3. 3-Aminobenzoic acid ethylester, phosphate buffered at pH 7.2 (MS-222, 200 mg/l; Sigma Catalog No. A-5040)

Equipment

1. Illuminated magnifier (VWR Catalog No. 36935-100)
2. Very fine pointed forceps (VWR Catalog No. 25607-865)
3. Delicate scissors, 4 1/2 inch (VWR Catalog No. 25608-203)
4. Seven-ml glass/glass tissue grinder (Fisher Catalog No. 06-435A)
5. 3-ml syringes (Fisher Catalog No. 14-823-40) and glass wool (Fisher Catalog No. 11-390)
6. Paper towels and diapers
7. Petri dishes (35 × 60 mm) (Falcon Catalog No. 3001)
8. 10-gal glass aquaria equipped with charcoal filter and aerator stones

Methods

1. Medaka (6 to 12 months of age) are selected randomly from 10-gal aquaria maintained at a temperature of 25 ± 1°C with a 16/8 h light/dark cycle. Fish are maintained on a diet of flake food (Tetramin) and brine shrimp (shrimp are prepared in sea salt water by adding 17.5 g sea salt to 500 ml of tap water). Water quality (i.e., ammonia, alkalinity, chlorine, oxygen, hardness, and pH) is checked twice daily (morning and evening) when fish are kept under static conditions.
2. Fish are anesthetized by placing medaka into 500 ml deionized water containing phosphate-buffered MS-222 (100 mg MS-222/500 ml) for approximately 1 min. Immediately following their removal, weights and lengths of the individual fish are determined.
3. The fish are then sacrificed by decapitation (the head is cut just posterior to the gills). Following sacrifice, select body organs (e.g., anterior kidney, spleen, or liver) are removed and weighed; organs collected from each of 5 fish are weighed in combination. Because of their extremely small size, organ weights are determined as a group to provide for more accurate weight determinations. The dissected organs are then homogenized immediately or stored overnight at 4°C in 3-ml of supplemented FPS and used the next day as a source of cells. The potential for overnight organ storage (for any given fish

species) should be determined in pilot studies comparing oxyradical production in freshly collected cells vs. that measured in cells recovered from organs stored overnight.

4. Each specific organ from the 5 fish are pooled and ground using glass/glass homogenizers. This is accomplished by stroking the tissue with two loosely fitting pestles; the tissue is ground 20 times with pestle A and 15 times with pestle B. The resultant homogenates are then passed through a loosely-packed glass wool-syringe column (3-ml syringe barrel) to remove tissue/cellular debris/red blood cells. The recovered cells are carefully collected in a 15-ml sterile polypropylene centrifuge tube.

5. The single cell suspensions are then centrifuged twice at room temperature for 15 min at $400 \times g$. After the final wash, the supernatant is removed, the pelleted cells resuspended in 1 ml of supplemented FPS, and the cell number and viability determined by hemocytometer counting and trypan blue exclusion, respectively.

Average cell yields and purity of Japanese medaka anterior kidney cell preparations are shown in Table 1. In all cases, cell viability following tissue disruption is ≥96% and most of the kidney cells are mononuclear cells. Initial experiments evaluating oxyradical production utilized both whole kidney cell homogenates and kidney phagocytes isolated by Percoll separation. Since O_2^- production by both populations is comparable, whole kidney cell homogenates are used routinely for measuring oxyradical production.

Table 1   Characterization of Japanese Medaka Kidney Cell Preparations and Superoxide Production

	Total no. of cells (×10⁶)	Cell viability	Esterase positive cells[a] (%)	Polymorphonuclear leukocytes[b,c] (%)	Superoxide production[d]
Whole cell homogenates	2.2 ± 0.5	96 ± 0.9	60.3 ± 0.9	8 ± 2	1.3 ± 0.1
Isolated cells[e]	0.8 ± 0.01	98 ± 1	90 ± 1.0	Not done	1.5 ± 0.2

Note: Mean ± SEM (n = 6).

[a] Cells that are positive by nonspecific esterase staining are considered to be monocytes/macrophages.

[b] PMNs identified by positive myeloperoxidase staining.

[c] Those cell types which are nonesterase positive or PMN-like are classified as "others."

[d] nmoles PMA-stimulated O_2^- /4×10^5/30 min.

[e] Monocytic cells were isolated by Percoll separation.

Assays of reactive oxygen intermediates (ROI)

Medium without phenol red should be used for all of the assays described in this section.

Superoxide anion production

Materials needed for extracellular production

Reagents and prepared solutions

1. Ferricytochrome C (stock solution = 4 mg/ml prepared in supplemented FPS; Sigma Catalog No. C-7752)
2. Bovine superoxide dismutase [(SOD); stock solution = 300 µg/ml prepared in Hanks balanced salt solution (HBSS); Sigma Catalog No. S-2515]
3. Phorbol 12-myristate 13-acetate [(PMA); stock solution = 1 mg/ml prepared in dimethylsulfoxide (DMSO), working solution = 10 µg/ml prepared in HBSS; Sigma Catalog No. P-8139].
4. DMSO (Sigma Catalog No. D-5879)
5. Supplemented L-15 medium
6. Supplemented FPS
7. HBSS (GIBCO Catalog No. 310-4025)

Equipment

1. 96-Well flat-bottom microtiter plates (Falcon Catalog No. 3072)
2. Microtiter platereader equipped with 550 nm filter
3. Humidified incubator set at 30°C
4. Eight channel multipipettor (Fisher Catalog No. 21-233)
5. 35-mm Sterile Petri dishes (Falcon Catalog No. 3001)

Methods

1. Medaka, selected at random, are anesthetized and sacrificed as described above. The kidneys are removed from the fish and placed in a 35-mm Petri dish containing 3-ml of cold supplemented FPS.
2. Following tissue disruption and passage of the cells through a loosely-packed glass wool column, the single cell suspensions are adjusted to a cell concentration of 4×10^6 kidney cells/ml.
3. The reaction mixtures used to measure extracellular O_2^- production are prepared by adding 10^6 kidney cells (in a total volume of 250 µl) to each of 4 previously labeled (1 to 4) sterile polypropylene tubes containing 500 µl ferricytochrome C (final concentration = 2 mg/ml). The second and fourth tubes receive 125 µl of SOD (final concentration = 37.5 µg/ml) and the third and fourth tubes receive 50 µl of PMA at a final concentration of 0.5 µg/ml. Fish physiological saline is then added to each tube to bring the final volume up to 1 ml. An additional tube ("B") that contains all of the reagents, but without cells, serves as the reaction blank.
4. After vortexing each tube for approximately 30 seconds, 200 µl aliquots (2×10^5 cells/well) are placed into the individual wells of a 96-well microtiter plate and the absorbance measured at 550 nm for up

to 2 h. Timepoints suggested for measurement include: 0, 10, 20, 30, 60, 90, and 120 minutes; plates should be incubated at 30°C (in a humidified environment) between readings. The rate of O_2^- production can be determined from measurements taken over time, while OD readings at a single timepoint (time of peak O_2^- production = 60 min) can also be used to make comparisons between different exposure groups.

Time to peak O_2^- production by medaka kidney phagocytes appears to be dependent upon fish age; production by kidney cells peaks after 135 min in fish 7 to 9 months of age and after 60 min in fish ≥10 months.

5. Change in absorbance is calculated by subtracting the mean of the "blank" wells and the wells containing SOD from the absorbance measured in the non-SOD-containing wells. By multiplying the change in absorbance by 15.87 (20), the nmol concentration of SOD-inhibitable O_2^- can be computed. Data is expressed as nmol O_2^- /2 × 10^5 cells/unit time.

Materials needed for intracellular production

Reagents and prepared solutions

1. Nitroblue tetrazolium [(NBT); 1 mg/ml stock prepared in HBSS; Sigma Catalog No. N-6876]
2. Bovine SOD prepared in HBSS (300 µg/ml)
3. PMA (1 mg/ml stock prepared in DMSO; working concentration = 10 µg/ml prepared in HBSS)
4. Supplemented L-15 medium
5. DMSO
6. 70% Methanol (J. B. Baker Catalog No. 9093-03)
7. 2 *M* KOH
8. Supplemented FPS
9. HBSS

Equipment

1. 96-Well flat-bottom microtiter plates
2. Microtiter platereader equipped with 630-nm filter
3. Humidified incubator set at (30°C)
4. Eight channel multipipettor

Methods

1. Medaka are selected randomly and sacrificed as described above.
2. The kidney cells recovered following tissue homogenization are counted using a hemocytometer and viability determined by trypan blue exclusion.

3. The cell concentration is adjusted to 6×10^6 cells/ml in supplemented L-15 medium.

4. One-hundred µl of the cell suspension (6×10^5 cells) is then added to 12 wells of the microtiter dish and the plate incubated in a humidified environment for 90 min (at 30°C) to allow for cell attachment.

5. During incubation, reaction solutions are prepared in each of 4-sterile polystyrene tubes. Tubes 1 and 2 will be used to measure unstimulated (background) intracellular O_2^- production, while tubes 3 and 4 will contain PMA and be used to measure stimulated production. To all 4 tubes, 1.6 ml of NBT at a final concentration of 0.8 mg/ml is added. Two hundred µl of SOD is then added to tubes No. 2 and 4; tubes 3 and 4 receive 100 µl of PMA. Final concentrations of SOD and PMA are 30 µg/ml and 0.5 µg/ml, respectively. To evaluate NBT reduction at four timepoints (15, 30, 60, and 90 min), the reaction mixture in each tube is brought up to a final volume of 2.0 ml.

Nitroblue tetrazolium solution should be made up fresh for each experiment, and PMA is light sensitive and should be wrapped in aluminum foil during storage at –20°C. A 10-µg/ml PMA solution should be prepared by mixing 50 µl of the stock PMA solution (made up in DMSO) with 4.95 ml HBSS.

6. Following incubation of the cells, the supernatant is removed and the cells washed once with warm (30°C) HBSS. The covering medium (CM) and HBSS wash is saved to determine the total numbers of detached cells; these are used for calculating the actual numbers of attached cells.

Detached cell numbers can be determined up to 24 h following collection if kept in polypropylene tubes and stored at 4°C.

7. 100 µl of the previously prepared, well-mixed reaction mixtures are then added to the wells in quadruplicate.

8. The cells are incubated at 30°C for the designated time period i.e., 15, 30, 60, and 90 min (time of peak intracellular O_2^- production = 60 min), and then washed three times with 70% methanol, and air dried at room temperature.

9. Aliquots of 2 M KOH (120 µl) and DMSO (140 µl) is then added to each well and mixed thoroughly to dissolve the dried formazan (the final reaction product). The same amounts of KOH and DMSO are also added to wells without cells that serve as the reaction blanks.

10. The absorbance in each well is measured at 630 nm in a microtiter platereader. The OD values measured in both the "blank" wells and those wells containing SOD are subtracted from the OD values measured in the wells with cells, but without SOD. The final nmole concentration of intracellular O_2^- is calculated by multiplying the final OD values by 15.87.[20]

The measurements of extracellular and intracellular O_2^- are in fact measurements of ferricytochrome C and NBT reduction, respectively. These reductions can occur with a variety of electron donors and it is therefore essential to use SOD as a means to determine O_2^- as the specific reductant. If O_2^- cannot be detected in this system, a positive control of xanthine and xanthine oxidase can be included.

Hydrogen peroxide production

Hydrogen peroxide production is assayed using a method based upon the horseradish peroxidase-dependent oxidation of phenol red by H_2O_2.

Materials

Reagents and prepared solutions

1. Phenol red [(PR); Sigma Catalog No. P-4758; stock solution = 8 mg/ml and working solution prepared in HBSS = 0.3125 mg/ml]
2. Horseradish peroxidase [(HRPO); Sigma Catalog No. P-8250; stock solution = 4 mg/ml and working solution prepared in HBSS = 0.156 mg/ml]
3. PMA (Stock solution prepared in DMSO = 1 mg/ml; working concentration prepared in HBSS = 10 µg/ml]
4. Supplemented L-15 medium
5. Supplemented FPS
6. DMSO
7. 1 N Sodium hydroxide [(NaOH); Sigma Catalog No. 930-65]
8. HBSS
9. 30% H_2O_2 (Sigma Catalog No. H-1009) — stored at 4°C

Equipment

1. 96-Well flat bottom microtiter plates
2. Microtiter platereader equipped with a 620-nm filter
3. Humidified incubator set at 30°C
4. Eight-channel multipipettor
5. Repeating pipette capable of dispensing 10 µl quantities (Fisher Catalog No. 21-380-8)

Methods

1. Medaka are sacrificed and their kidney cells recovered as described above.
2. Kidney cells are counted using a hemocytometer and the final cell concentration adjusted to 6×10^6 cells/ml in supplemented L-15 medium.
3. One-hundred µl of the cell suspension (6×10^5 cells/well) is then added to each of 8 wells of the microtiter dish and the plates incubated for 90 min at 30°C to allow for cell attachment.

4. Following incubation, the CM is removed and the cells rinsed one time with warm HBSS. Both the CM and rinsing medium are saved for determining the total numbers of detached cells.
5. For each timepoint to be evaluated (i.e., 0, 30, 60, and 90 min), 100 μl of the PR/HRPO solution (PR/HRPO = 2.4 ml; HBSS = 0.6 ml) is added into each of 3 wells containing the attached medaka kidney cells. The remaining 4 wells are used to measure stimulated production of H_2O_2 and receive 100 μl of the same solution supplemented with PMA (PR/HRPO = 2.4 ml; 0.3 ml PMA; HBSS = 0.3 ml). Three wells without cells also receive 100 μl of the reaction mixture (with and without PMA) and serve as the reaction blanks.

Phenol red is light sensitive and should be wrapped in aluminum foil and stored at 4°C. Phorbol myristate acetate should be stored as a 1 mg/ml stock solution in DMSO. The working solution of PMA is prepared by diluting 50 μl of PMA with 4.95 ml HBSS to yield a final PMA concentration of 10 μg/ml.

6. Immediately following addition of the reaction mixture to the individual wells, the plates are incubated for 60 min (this timepoint represents peak H_2O_2 production by the recovered kidney cells). Following incubation, 10 μl of 1 N NaOH is added to every well and the plates reincubated at 30°C for an additional 3 min.
7. Absorbance in each well is determined in a microtiter platereader using a 620-nm filter.
8. The final nmole concentration of H_2O_2 is determined against a freshly prepared standard curve using commercial H_2O_2. The recommended range for the H_2O_2 standard curve is 1 to 60 μ*M*.

Assays of antioxidant enzymes

Catalase, an antioxidant enzyme localized in the cytoplasmic peroxisomes, directly catalyzes the decomposition of H_2O_2, thus, protecting against its potential cytotoxic effects. The assay described in this paper is based on the ability of catalase to metabolize H_2O_2 into O_2 and water. Reduction in the absorbance of the H_2O_2 solution is followed over time and translated into the level of enzymatic activity.

Catalase (CAT) activity

Materials

Reagents and prepared solutions

1. 50 m*M* phosphate buffered saline (PBS; pH 7.2; Sigma Catalog No. 1000-3) stored at 4°C
2. 30% H_2O_2 — stored at 4°C

Equipment

1. Sonicator: 550 Sonic Dismembrator
2. Ultracentrifuge: L8-M80
3. Spectrophotometer: DU 640 spectrophotometer with kinetic studies capability

Methods

1. Five million cells from medaka liver, kidney, or spleen are suspended in 0.5 ml 50 mM PBS and sonicated for 20 sec (two 10-sec pulses with a 10-sec interval between pulses).
2. The sonicate is then centrifuged at 20,000 × g for 10 min (at 4°C).
3. Fifty μl of the supernatant are added to a solution of 10 mM H_2O_2 in PBS and the decrease in absorption (due to H_2O_2 decomposition by CAT) is measured at 240 nm over a 3-min period.

It is not possible to define CAT units according to international conventions due to aberrant kinetics. However, enzymatic activity can be expressed as the rate constant of a first-order reaction (k). A_1 and A_2 refer to the absorbance before and after a given time interval of measurement, respectively. Eq. 1 is an example for a 15-sec time interval:[15]

$$k = (2.3/15)(\log A_1/A_2) = 0.153(\log A_1/A_2)(sec^{-1}) \qquad (1)$$

Superoxide dismutase (SOD)

Superoxide dismutase is found in mitochondria where it protects cells from oxidation damage by catalyzing the dismutation of O_2^- to H_2O_2. The assay method described in this chapter is based upon the inhibition by SOD of O_2^--dependent auto-oxidation of pyrogallol.[16] When pyrogallol is incubated in air-equilibrated buffer it autoxidizes, yielding a yellow-brown solution whose absorbance can be measured at 420 nm.

Materials

Reagents and prepared solutions

1. Lysing solution
   a. 0.08% Digitonin (Sigma Catalog No. S-1407)
   b. 2 mM ethylenediamine tetraacetic acid [(EDTA); Sigma Catalog No. E-9884, pH 7.8]
2. Supplemented FPS
3. Solution A (completely saturated by bubbling air through the solution)
   a. 50 mM Tris acetate buffer (Sigma Catalog No. T-1258) with 1 mM EDTA (pH 8.2)

4. Solution B (deoxygenated by bubbling nitrogen through the solution)
   a. 0.2 mM pyrogallol (Sigma Catalog No. P-2923) in 50 mM Tris acetate buffer

Both solutions A and B should be used within 15 to 30 min of enrichment by or depletion of air, respectively.

5. Bovine SOD (stock solution = 1 mg/ml prepared in sterile double-distilled water to a final working concentration between 2.5 to 15 µg/ml)

Equipment

1. Refrigerated centrifuge: GS-6KR
2. Spectrophotometer: DU 640 spectrophotometer with kinetic studies capability

Methods

1. 0.2×10^6 to 1×10^6 medaka cells are suspended in 0.1 ml lysing solution and incubated for 20 min at 30°C. Cell lysis is visually confirmed microscopically.
2. The cell lysate is centrifuged at $250 \times g$ for 10 min at 4°C.
3. The supernatant is then collected and 0.1 ml supplemented FPS is added.
4. Enzymatic activity in the supernatant is measured by adding to solution A (9.8 ml) and to 0.1 ml of solution B either: (a) 0.1 ml water to determine the maximal rate of autoxidation of pyrogallol, (b) 0.1 ml of a known SOD concentration to generate the standard curve, or (c) 0.1 ml of an unknown sample.
5. Samples from these mixtures are then transferred to a cuvette and their absorbance measured for 3 min at 240 nm in a spectrophotometer. This particular time point was selected because of the linear correlation between decreased OD and time.
6. A standard curve is generated from a set of known SOD concentrations and is used to derive the concentration of SOD in the test sample. A typical slope of 0.02 OD/min is observed when no SOD is added. Superoxide dismutase used for generating the standard curve is prepared by diluting a 1 mg/ml stock solution in sterile, double-distilled water to yield a final working concentration between 2.5 to 15 µg/ml.

NAD(P)H: quinone reductase (QR)

NAD(P)H:QR is a phase II enzyme that catalyzes the reduction of quinones and quinoneimines. Since QR promotes a two-electron reduction of quinones, it inhibits the generation of $O_2^{\bar{\ }}$ and, therefore, functions as an antioxidant enzyme.

Materials

Reagents and prepared solutions
All the necessary reagents can be purchased from Sigma:

1. Digitonin (S-1407)
2. EDTA (E-9884)
3. Dicoumarol (M-1390)
4. DMSO (D-5879)
5. Tris chloride (T-6666)
6. Bovine serum albumin [(BSA); A-4503]
7. Menadione (No. 5625)
8. Tween-20 (P-1379)
9. Flavin adenine dinucleotide [(FAD); F-6625]
10. Glucose-6-phosphate (G-7879)
11. NADP (N-0505)
12. Crystalline yeast glucose-6-phosphate dehydrogenase (G-6378)
13. 3-(4,5-dimethylthiazol-2-yl)-2,5-diphenyltetrazolium bromide [(MTT), M-2128]

Equipment

1. Refrigerated centrifuge with adaptors for plates
2. Microtiter platereader equipped with a 610-nm filter

Methods
The QR activity in medaka cells is measured in duplicate wells and the average (after subtraction of the OD of wells with the same components, but in the presence of a QR specific inhibitor — dicoumarol) is used to calculate the activity of the sample assayed.

1. Medaka cells are plated at $2.5 \times 10^6/0.2$ ml/well in a 96-welled microtiter plate.
2. The cells are then lysed with 0.08% digitonin in 2 mM EDTA (pH 7.8) for 10 min.
3. The plate is spun at $250 \times g$ and the supernatants collected and added to the wells of a second microtiter dish. The final assay is performed in the second plate in a final volume of 230 µl/well.
4. Each well in the second plate contains 40 µl cell lysate supernatant, 40 µl of 0.3 mM dicoumarol [NAD(P)H:QR inhibitor], or its solvent (0.5% DMSO/5 mM potassium phosphate, pH 7.4), and 150 µl of the assay solution. The assay solution consists of: 0.025 M Tris-HCl, pH 7.4; 0.67 mg/ml BSA; 50 mM menadione in acetonitrile; 0.01% Tween-20; 5 mM FAD; 1 mM glucose-6-phosphate; 0.12 mM NADP; 2 U/ml crystalline yeast glucose-6-phosphate dehydrogenase and 0.3 mg/ml MTT.
5. After a 1-h incubation at 37°C, the absorbance in each well is measured at 610 nm in a microtiter platereader.

6. Specific activity is expressed as nmol of MTT reduced (calculated on the basis of extinction coefficient = 11,300 M^{-1} cm^{-1}) per minute per mg of protein. Non-specific quinone reduction, in the presence of dicoumarol, is subtracted from the enzymatic measurements. Typical specific activity of QR in the medaka liver cells is 13 nmol/min/mg protein.

Protein determination

The activities of all three antioxidant enzymes are expressed on a per mg of cellular protein basis to allow distinctions to be made between effects of environmental toxicants on enzymatic activities and general effects on cell protein levels. Protein concentration is measured in the cytosolic preparation in which the corresponding enzymatic activity is also measured. The Bio-Rad assay used to measure protein is based on a modification of the Lowry reaction.[21] In this assay, color development takes only 15 min and the product is stable for 1 to 2 h (only 5 to 10% change).

Materials

Reagents and solutions

1. BSA as a protein standard (Sigma No. A-4503)
2. Protein assay kit (detergent compatible - DC), Bio-Rad VWR Catalog No. 500-0112)

Equipment

1. Vortex Genie mixer
2. Spectrophotometer: DU 640 spectrophotometer

Methods

A standard curve is prepared with BSA and readings from each sample are inserted into the curve's formula to generate the final protein concentrations.

One fifth of the volume of each of the kit's components can be used successfully, thus, significantly reducing cost.

1. Prepare BSA solutions at 0.1 to 1.6 mg/ml in the same buffer as the enzyme preparation is being assayed.

Since each enzymatic assay uses a different buffering system, the buffer in which the corresponding enzyme is being prepared should also be used for dilution of the protein sample.

2. Prepare 2- to 5-fold dilutions of your samples.
3. To 20 µl of diluted sample or standard add 100 µl of reagent A (which contains 20 µg/ml of reagent S) and vortex for 10 to 20 sec.

4. To this solution, add 0.8 ml of reagent B into each tube and vortex immediately.
5. Incubate the tubes for 15 min at room temperature.
6. The absorbance is measured on a spectrophotometer at 750 nm.
7. The protein concentrations are determined from a linear curve generated by plotting OD vs. BSA concentrations. Data are then analyzed by using a variety of possible programs (e.g., Macintosh-Cricket).

Results and discussion

Toxicity from O_2 arises from the formation of oxygen-derived free radicals. These radicals arise within mitochondria as oxygen is reduced to water, as well as by activated phagocytes (i.e., macrophages and neutrophils) as part of the immune response.[2,4,5-7] Although it is slow to react with most target molecules, O_2^- is the radical most commonly produced. The primary reaction of O_2^- is its spontaneous dismutation to H_2O_2, which is able to diffuse through lipid membranes. Hydrogen peroxide reacts with transition metals to produce the highly reactive OH· which can initiate chain reactions of lipid peroxidation leading to cell rupture. Oxygen radical scavengers such as SOD and CAT, present in a variety of tissues and cells, protect the host against normal levels of oxyradicals.[1,3,11-14,17-19]

Like their mammalian counterparts, phagocytic cells from different fish species generate large quantities of oxyradicals[8-10,12,13,23,24,26] that appear to be produced by similar mechanisms.[8] In addition, the enzymatic antioxidant activities of SOD, CAT, glutathione peroxidase, and a QR-like enzyme have also been demonstrated in the organs/tissues of some fish species.[12,13,18,19,22] In fact, Rodriguez-Ariza et al.[25] have used the level of some detoxifying enzymes in fish as an early warning indicator of marine pollution. Recently, antioxidant enzymes in fish, specifically SOD and glutathione transferases, have also been proposed as bioindicators for assessing the environmental impact of metals and certain organic xenobiotics that generate oxidative stress.[18,19,22,25] Few studies, however, have examined these same parameters in warm, freshwater fish, or in those species exposed to toxicants under controlled laboratory conditions.

Studies in our laboratory have demonstrated that liver cells recovered from medaka and processed as described above, produce substantial amounts of CAT constitutively, while the activity in spleen and kidney cells is considerably lower; QR is produced constitutively in large quantities by both the liver and spleen cells (Figure 1). To provide the opportunity to make comparisons between the tissues, enzymatic activity is expressed in arbitrary units.

Exposure to the aquatic metal pollutant cadmium (Cd) at 60 ppb for 5 days produced a significant change in QR activity by both kidney and liver cells (Figure 2). Interestingly, while QR activity in the liver cells decreased, activity in the kidney cells increased significantly. Cadmium exposure significantly increased SOD activity in both the liver and spleen, but had no significant effect on kidney-associated SOD activity. It is critical to ensure that the same number of viable cells be used in each experimental condition,

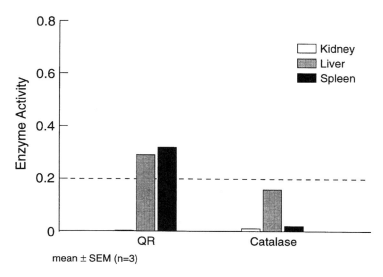

Figure 1 Constitutive levels of catalase and quinone reductase (QR) activity in the kidney, liver, and spleen of Japanese medaka. QR activity in the kidney is not clearly visible in the figure; it is .008 units of enzyme activity.

in order to assure that observed effects on enzyme activity are due to a specific toxicant rather than simply to cell death.

Production of O_2^- by both fish and mammalian phagocytes has been shown to be altered by exposure to different environmental pollutants.[23,24,26] As shown in Figure 3, waterborne exposure for five days to Cd, at near environmental levels and above, increases the intracellular and extracellular production of O_2^- by medaka whole kidney cell homogenates. While extra-

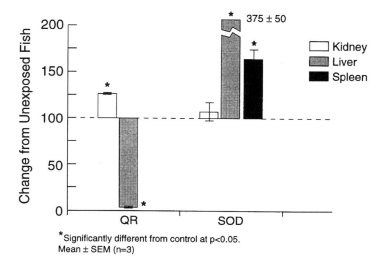

Figure 2 Activity of quinone reductase (QR) in the kidney, liver, and spleen of Japanese medaka following waterborne exposure to cadmium (Cd) at 60 ppb for 5 days.

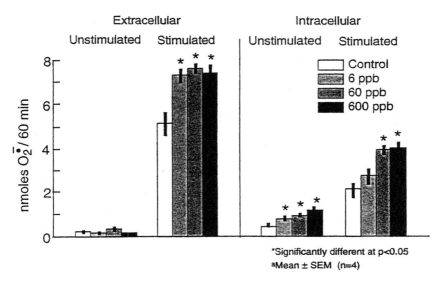

Figure 3   Effects of waterborne cadmium (Cd) exposure at concentrations ranging from 6 to 600 ppb for five days on unstimulated and PMA-stimulated extracellular and intracellular superoxide anion (O_2^-) production by medaka whole kidney cell homogenates.

cellular PMA-stimulated production of O_2^- was increased ~30% by all exposure concentrations, unstimulated and PMA-stimulated intracellular production appeared to increase with increasing Cd concentration. Although, at first glance, potentiation of this response may appear beneficial for the host, increased O_2^- production can lead to localized tissue damage, as well as increased quantities of related toxic mediators such as the highly reactive hydroxyl radical (OH·) and H_2O_2, a powerful oxidant with cytotoxic and possible tumor-promoting properties.

Hydrogen peroxide production by activated phagocytes, like that of O_2^-, is also altered by exposure to different classes of environmental xenobiotics including metals.[23,24,26] Figure 4 demonstrates the stimulatory effects of waterborne Cd exposure for 5 days on H_2O_2 production by unstimulated and PMA-stimulated medaka kidney cells; stimulated production is significantly increased (2.5-fold) by Cd concentrations as low as 6 ppb.

Many of the procedures used in this chapter are microassays and, therefore, offer the advantage of requiring relatively few cells. This is an important consideration since the number of phagocytes available from the anterior kidney of medaka (length = 20 to 30 mm; weight 200 to 500 mg) is quite low (5 to 19 × 10^5). Actual cell numbers are dependent upon the size/age of the fish.[27] Because of the low cell yield associated with the use of small aquarium fish, the immunotoxicological assays described in this paper must be performed using pooled cells from a number of fish. The exact number of fish to be used in a single experiment will depend upon the age of the fish (i.e., older fish yield larger numbers of cells), as well the number of variables to

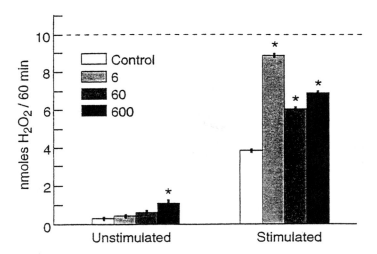

Figure 4 Effects of waterborne cadmium (Cd) exposure at concentrations ranging from 6 to 600 ppb for five days on unstimulated and PMA-stimulated hydrogen peroxide (H_2O_2) production by medaka whole kidney cells homogenates.

be examined in the experiment. To determine statistical differences between exposure groups, experiments should be repeated 4 to 6 times.

Acknowledgments

This work was supported in part by Contract Numbers DAMD-17-93-C-3059 and DAMD-17-93-C-3006, from the U.S. Army Medical Research and Materiel Command.

The views, opinions, and/or findings contained in this report are those of the authors and should not be construed as official Department of the Army position, policy, or decision, unless so designated by other official documentation. Research was conducted in compliance with Animal Welfare Act, and other federal statutes and regulations relating to animals and experiments involving animals and adheres to principles stated in the *Guide for the Care and Use of Laboratory Animals*, NIH Publication 86-23, 1985 or most recent edition.

References

1. DiGiulio, R.T., Washburn, P.C., and Wenning, R.J. 1989. Biochemical responses in aquatic animals: A review of determinants of oxidative stress. *Environ. Toxicol. Chem.* 8:1103–1123.
2. Freeman, B.A., and Crapo, J.D. 1982. Biology of disease. Free radicals and tissue injury. *Lab. Invest.* 47:412–426.
3. Bostek, C.C. 1989. Oxygen toxicity: An introduction. *J. Am. Assoc. Nurse Anesth.* 57:231–237.
4. Drath, D.B., and Karnovsky, M.L. 1975. Superoxide production by phagocytic leukocytes. *J. Exp. Med.* 141:257–262.

5. Zelikoff, J.T., Sisco, M.P., Yang, Z., Cohen, M.D., and Schlesinger, R.B. 1994. Immunotoxicity of sulfuric acid aerosol: Effects on pulmonary macrophage effector and functional activities critical for maintaining host resistance against infectious diseases. *Toxicology* 92:269–286.

6. Nathan, C.F., Brukner, L.H., Silverstein, S.C., and Cohn, Z.A. 1979. Extracellular cytolysis by activated macrophages and granulocytes. I. Pharmacologic triggering of effector cells and the release of hydrogen peroxide. *J. Exp. Med.* 149:84–100.

7. Johnston, R.B. 1978. Oxygen metabolism and the microbicidal activity of macrophages. *Fed. Proc.* 37:2759–2764.

8. Secombes, C.J. 1990. Isolation of salmonid macrophages and analysis of their killing activity. In *Techniques in Fish Immunology–1*, Eds. J. Stolen, T. C. Fletcher, D.P. Anderson, B.S. Roberson, and W.B. vanMuiswinkel, SOS Publications, Fair Haven, NJ., 1:137–155.

9. Zelikoff, J.T., and Enane, N.A. 1991. Assays used to assess the activation state of rainbow trout peritoneal macrophages. In *Techniques in Fish Immunology–2*, Eds. J.S. Stolen, T.C. Fletcher, D.P. Anderson, S.L. Kaattari, and A.F. Rowley, SOS Publications, Fair Haven, NJ., 2:107–123.

10. Zelikoff, J.T., Enane, N.A., Bowser, D., Squibb, K.S., and Frenkel, K. 1991. Development of fish peritoneal macrophages as a model for higher vertebrates in immunotoxicological studies. I. Characterization of trout macrophage morphological, functional and biochemical properties. *Fund. Appl. Toxicol.* 16:576–589.

11. Jaiswal, A.L. Antioxidant response element. 1994. *Biochem. Pharmacol.* 48:439–444.

12. Filho, D.W., Giulvi, C., and Boveris, A. 1992. Antioxidant defenses in marine fish-I. Teleosts. *Comp. Biochem Physiol.* 106C:409–413.

13. Izokum-Etiobhio, B.O., Oraedu, A.C.I., and Ugochukwu, E.N. 1990. A comparative study of superoxide dismutase in various animal species. *Comp. Biochem. Physiol.* 95B:521–523.

14. Halliwell, B. 1994. Free radicals, antioxidants, and human disease: Curiosity, cause, or consequence? *Lancet* 344:721–724.

15. Aebi, A. 1984. Catalase *in vitro*. *Meth. Enzymol.* 105:121–126.

16 Marklund, S., and Marklund, G. 1974. Involvement of the superoxide anion radical in the autoxidation of pyrogallol and a convenient assay for superoxide dismutase. *Eur. J. Biochem.* 47:469–474.

17. Prochaska, H.J., and Santamaria, A.B. 1988. Direct measurement of NAD(P)H:Quinone reductase from cells cultured in microtiter wells: A screening assay for anticarcinogenic enzyme inducers. *Anal. Biochem.* 169: 328–336.

18. Hasspieler, B.M., and DiGiulio, R.T. 1994. Dicoumarol-sensitive NADPH: Phenanthrenequinone oxidoreductase in channel catfish (*Ictalurus punctatus*). *Toxicol. Appl. Pharmacol.* 125:184–191.

19. Gabryelak, T., and Tawfek, N.S. 1991. Seasonal changes of superoxide dismutase activity in erythrocytes of *Abramis brama* from two different types of Poland lakes. *Bull. Environ. Contam. Toxicol.* 47:912–917.

20. Pick, E., and Mizel, D. 1981. Rapid microassays for the measurement of superoxide and hydrogen peroxide production by macrophages in culture using an automatic enzyme immunoassay reader. *J. Immunol. Meth.* 46:211–216.

21. Lowry, O.H., Rosebrough, N.J., Farr, A.L., and Randall, R.J. 1951. Protein measurement with the Folin phenol reagent. *J. Biol. Chem.* 193:265–275.

22. Pedrajas, J.R., Peinado, J., and Lopez-Barea, J. 1993. Purification of Cu, Zn-superoxide dismutase isoenzymes from fish liver: Appearance of new iso-forms as a consequence of pollution. *Free Rad. Res. Commun.* 19:29–41.

23. Zelikoff, J.T. 1993. Metal pollution-induced immunomodulation in fish. *Ann. Rev. Fish Diseases.* 3:305–325.

24. Bowser, D.H., Frenkel, K., and Zelikoff, J.T. 1994. Effects of *in vitro* nickel exposure on the macrophage-mediated immune functions of rainbow trout (*Oncorhynchus mykiss*). *Bull. Environ. Contam. Toxicol.* 52:367–373.

25. Rodriguez-Ariza, A., Dorado, G., Peinado, J., Pueyo, C., and Lopez-Barea, J. 1991. Biochemical effects of environmental pollution in fishes from the Spanish South-Atlantic littoral. *Biochem. Soc. Trans.* 19:301S.

26. Zelikoff, J.T. 1994. Immunological alterations as indicators of environmental metal exposure. In *Modulators of fish Immune Responses. Models for Environmental Toxicology/Biomarkers, Immunostimulators,* Eds. J.S. Stolen, J.T. Zelikoff, L.E. Twerdok, S. Kaattari, D.P. Anderson, C. Baynes, and T. C. Fletcher. SOS Publications, Fair Haven, NJ 1:101–110.

27. Twerdok, L.E., Curry, M.W., Beaman, J.R., and Zelikoff, J.T. 1995. Development of routine health monitoring methods for use with an aquatic model (*Oryzias latipes*) used in immunotoxicological testing. *Toxicologist* 19:306.

chapter seventeen

Immunohistochemical and immunoblot detection of P-glycoprotein in normal and neoplastic fish liver

Peter S. Cooper, Wolfgang K. Vogelbein, and Peter A. Van Veld

Introduction

P-Glycoprotein expression in mammals

P-glycoprotein (Pgp) is a membrane transport ATPase that plays a role in efflux of cytotoxic agents from multidrug-resistant (MDR) mammalian cell lines.[1-3] The MDR phenotype arises in cell lines selected for resistance to one cytotoxic agent. These cells are cross-resistant to a variety of other toxicants with differing structures and modes of action. A consistent biochemical change in these cell lines is increased expression of P-glycoprotein (reviewed in References 2 and 3). Studies of accumulation and efflux of labeled Pgp substrates in MDR cell lines indicate that Pgp expression is associated with decreased accumulation and increased ATP dependent efflux of drugs in these cells.[4,5] Although drug-resistant cell lines often have many biochemical alterations that distinguish them from sensitive cells, transfection of sensitive cells with full length cDNA clones of Pgp transcripts has been found to be sufficient to confer drug resistance.[6,7] This is convincing evidence that over-expression of Pgp alone is sufficient to confer the MDR phenotype.

Since the initial discovery of Pgp in MDR cell lines, normal tissues have been found to express Pgp. The apical surface of many excretory epithelia including those of the intestine, liver and kidney show Pgp expression.[8-10] This pattern of expression along with the role of Pgp in efflux of cytotoxins from cell lines suggests that the normal role of Pgp may be the excretion of endogenous or exogenous toxic compounds or their metabolites. Pgp may therefore represent a line of defense against toxic compounds. Upregulation of Pgp expression could be an adaptive response of organisms to toxicants.

In the liver of the rat, Pgp mRNA levels increase following exposure to many xenobiotics including some of the classic P450 inducers.[11] Thus it is possible that Pgp expression in mammals may be a part of the inducible xenobiotic response pathway that includes the phase I and phase II xenobiotic metabolizing enzymes.[11,12] Although there are many hints that the normal function of Pgp is elimination of xenobiotics or toxic metabolites, the normal substrates for Pgps are unknown, and the relevance of these proteins to toxicology remains uncertain.

In addition to expression in normal tissues, Pgp expression and overexpression are seen in human and animal neoplasia. Overexpression of Pgp occurs in a number of human neoplasms derived from a variety of tissue types.[13-18] In many human tumors Pgp expression is associated with resistance of the tumors to chemotherapy and poor prognosis.[13,16,17] Some cases of overexpression are associated with tumors that recur following chemotherapy.[16,18] Chemotherapy may select for tumor cells that are resistant because of Pgp overexpression.[13] Pgp overexpression has also been observed in several different laboratory models of rodent hepatocarcinogenesis.[19-22] Evidence from one model suggests that Pgp overexpression occurs early in carcinogenesis and is related to promotion of xenobiotic resistant preneoplastic liver lesions.[19,20] Other evidence suggests that Pgp overexpression in hepatocellular carcinoma may arise without promotion by xenobiotics.[21,22]

P-Glycoprotein expression in aquatic animals

Because of the importance of Pgp in conferring MDR, there is interest in the role that Pgp expression may play in tolerance or resistance of aquatic organisms to toxicants. Pgp expression and activity have been detected in tissues of a number of aquatic organisms.[23-28] Kurelec and co-workers have hypothesized that certain strains or species of aquatic organisms that are tolerant of pollutants may be so because of the presence and activity of Pgp homologs (reviewed in Reference 28). They suggest that chemical pollutants or natural toxins may be excluded from these organisms by the externally directed pumping activity of Pgp homologs in epithelial tissues. They have termed such a mechanism "multixenobiotic resistance" (MXR) since, by analogy to the MDR phenotype, such organisms may show cross-resistance to a variety of structurally unrelated toxicants.

Although the above studies suggest a role for Pgp in resistance or tolerance to toxicants, the significance of Pgp expression in aquatic organisms is not clear. Additional studies are needed to establish whether differences in relative levels of Pgp expression or activity are associated with sensitivity of aquatic organisms to toxicants. There is also a need for the development of methods that would allow measurement of Pgp expression in aquatic organisms.

We are currently investigating expression of Pgp in the liver of the mummichog, *Fundulus heteroclitus*, a small estuarine killifish. Our study site in the Elizabeth River, Virginia is heavily contaminated with polycyclic aromatic hydrocarbons (PAHs) derived from creosote.[29] High prevalences of hepatic[30] and extra-hepatic neoplasms[31,32] have been reported in mummichog from this site. This situation gives us a unique opportunity to study both the response of Pgp expression to environmental exposure to PAHs in fish liver and to examine alterations in Pgp expression occurring in neoplastic and putative preneoplastic lesions during environmental hepatocarcinogenesis in fish.

The purpose of this paper is to describe protocols that we have developed for the immunochemical detection of Pgp expression in fish tissues, primarily liver and liver tumors. These methods should be applicable to other fish tissues and possibly tissues of other aquatic organisms as well. Detailed protocols are given for both immunohistochemical staining and immunoblotting procedures. Whenever possible we use both techniques because of the complementary information that can be derived from each.

Materials required

Where manufacturers are not given, materials are readily available from many sources.

Pgp immunohistochemistry

1. Staining rack and jars: (Shandon, Tissue Tek II).
2. PAP pen: (Research Products International, Mount Prospect IL, cat. no. 195500).
3. Ethanol: 100%, 95%, and 70%.
4. Xylene.
5. Elite Avidin-Biotin Complex (ABC) immunohistochemical staining kit, murine: (Signet Laboratories, Dedham MA, cat. no. 2158).
6. Normal Horse Serum: (Sigma, cat. no. H0146).
7. Bovine Serum Albumin (BSA), Fraction V: (Sigma, cat. no. A9647).
8. 3,3'-Diaminobenzidine Substrate Tablets: (Sigma, cat. no. D4168).
9. Permanent Mounting Medium: (Preservaslide, EM Science, or equivalent).
10. Phosphate Buffered Saline (PBS): (10 mM sodium phosphate, pH 7.4, 150 mM NaCl). This is prepared as a 10× stock. 14.2 g of anhydrous dibasic sodium phosphate and 87.7 g of NaCl are added to 900 ml of deionized water. The pH is adjusted to 7.4 with 6 M NaOH. The volume is then made up to 1 l with deionized water. The solution is sterilized by filtration (0.2 µm) and stored at room temperature. The 10× stock is diluted with 9 parts deionized water immediately before use.
11. Blocking Solution: (10% Normal horse serum, 0.02% sodium azide in 1× PBS). This solution can be divided into aliquots and stored at –20°C.
12. 3% Hydrogen Peroxide: This solution must be prepared fresh daily by dilution of 1 part 30% hydrogen peroxide into 9 parts deionized water.
13. Antibody Buffer: (1% w/v BSA, 0.02% w/v sodium azide in 1× PBS). This solution is stored at 4°C. It is stable for at least 2 weeks.

14. C219 mAb: (Signet, cat. no. 2158). This reagent is provided lyophilized (100 μg) by the manufacturer. It is diluted with 1 ml of antibody buffer and stored frozen (–70°C) in 20 μl aliquots. For immunohistochemistry, aliquots are diluted immediately before use with 1.3 ml of antibody buffer (1% BSA in PBS).

15. Mouse Myeloma (UPC 10) Protein (IgG 2a,κ): (Sigma, cat. no. M 9144). The lyophilized protein is diluted to 1.5 μg/ml with antibody buffer and stored in 1 ml aliquots at –20°C.

16. Harris Hematoxylin: 5 g of hematoxylin powder (Fisher Scientific, cat. no. H345) is dissolved in 50 ml of 95% ethanol. 100 g of potassium alum is dissolved in 1000 ml of deionized water in a 2-l Erlenmeyer flask on a heating stirplate. The hematoxylin solution is added to the alum solution, and the mixture is brought to a boil while stirring. 2.5 g of mercuric oxide are carefully added to the hot solution. The solution is heated until a deep purple color is obtained and then is cooled by plunging the flask into cold water. It then is filtered and stored at room temperature. The solution is filtered before each use. The stain is stable for several months at room temperature.

Immunoblotting

1. Power Homogenizer: We use an electric drill (Sears Roebuck Co., model 15 315.10042) mounted in a drill press stand (Sears model 25921) to drive a 2-ml Potter-Elvehjem tissue grinder (Wheaton, no. 358003).

2. Vertical Slab Gel Electrophoresis Apparatus, MiniPROTEAN II: (Bio-Rad, cat. no. 165-2940).

3. Electrophoretic Transfer Apparatus, Mini Trans-Blot: (Bio-Rad, cat. no. 170-3935).

4. Electrophoresis Power Supply: (Bio-Rad, Model 200/2.0, cat. no. 165-4761).

5. Nitrocellulose Membranes: (Bio-Rad, cat. no. 162-0147).

6. 40% Acrylamide Stock (37.5:1, Acrylamide:Bis-Acrylamide): (Fisher Scientific, cat. no. BP1410-1).

7. C219 mAb: (Signet, cat. no. 2158). For immunoblotting, the lyophilized mAb (100 μg) is dissolved in 40 ml immunoblot antibody dilution buffer (see below), divided into two aliquots and stored at –20°C. Each aliquot can be saved after use, frozen, and used again at least ten times.

8. Goat Anti-Mouse IgG Alkaline Phosphatase Conjugate: (Bio-Rad, cat. no. 170-6520). The undiluted conjugate is stored at 4°C. Immediately before use the conjugate is diluted 1:3000 with immunoblot antibody dilution buffer.

9. Lysis Buffer Stock: 6.06 g tris(hydroxymethyl)aminomethane (Tris) (Sigma, cat. no. T-8524), 8.77 g NaCl, 1.86 g disodium ethylenediaminetetraacetate (EDTA) (Sigma, cat. no. E-4884), 10.0 g sodium deoxycholate (Sigma, cat. no. D-6750), 1.0 g sodium dodecyl sulfate

(SDS)(Sigma, cat. no. L-4390) and 10 ml Triton X-100 (Sigma, cat. no. X-100) are dissolved in 800 ml deionized water. The solution is adjusted to pH 7.4 with 6 M HCl. The volume is made up to 960 ml. The buffer is stored at 4°C. Just before use, the volume of lysis buffer required is prepared in a tube on ice by adding the protease inhibitors from stock solutions.

10. Protease inhibitor stocks:

N-p-tosyl-L-arginine methyl ester HCl (TAME): (Sigma, cat. no. T 4626). A stock solution of 38 mg/ml in deionized water is prepared fresh before use.

Penylmethylsulfonyl fluoride (PMSF): (Sigma, cat. no. P 7626). PMSF is dissolved at a concentration of 10 mg/ml in 100% 2-propanol. The solution is stable for several months at room temperature.

Aprotinin solution (10 units/ml): (Sigma, cat. no. A 6279). This stock is used as is from the producer.

Leupeptin: (Sigma, cat. no. L 2023). Leupeptin is dissolved in deionized water at 5 mg/ml. The solution is divided into 20 µl aliquots and is stored at –20°C.

11. 1× Lysis Buffer: (50 mM Tris HCl, pH 7.4, 150 mM NaCl, 1% v/v Triton X-100, 1% w/v sodium deoxycholate, 0.1% w/v SDS, 5 mM EDTA, containing freshly added 2% v/v aprotinin solution, 380 µg/ml N-tosylamido-L-arginine methyl ester, 100 µg/ml phenylmethylsulfonyl fluoride, 10 µg/ml leupeptin). The following are mixed in a tube on ice immediately before use.

Aprotinin solution (10 U/ml)	0.200 ml
TAME (38 mg/ml)	0.100 ml
PMSF (10 mg/ml in 2-propanol)	0.100 ml
Leupeptin (5 mg/ml)	0.020 ml
Lysis buffer stock	9.6 ml

12. Electrode Buffer: (25 mM Tris, 192 mM glycine, 0.1% w/v SDS). This is made as a 5× stock containing 15 g Tris base, 72 g glycine and 5 g SDS per liter of solution. The pH of the 5× buffer is near 8.5. The pH should not be adjusted with acid or base as this will increase the electrical conductivity. 5× electrode buffer is diluted with 4 parts water before use.

13. 1 M Tris HCl, pH 6.8: This buffer is stable for several months at 4°C.

14. 1.5 M Tris HCl, pH 8.8: This buffer is stable for several months at 4°C.

15. 10% w/v SDS: This is stored at room temperature.

16. 1 M Dithiothreitol (DTT): This is divided into aliquots and stored at –20°C.

17. 10% w/v Ammonium Persulfate (10% APS): 100 mg of APS is added to 1 ml of deionized water. This solution is prepared fresh daily.

18. N,N,N',N'-Tetramethylethylenediamine(TEMED): This is used undiluted from the bottle.

19. SDS Sample Buffer: (60 mM Tris-Cl, pH 6.8, 2% w/v SDS, 10% v/v glycerol, 0.025% bromphenol blue, 100 mM dithiothreitol). This is prepared as a stock without the reducing agent dithiothreitol (DTT). The stock is stable at room temperature.

Water	24.5 ml
1.0 M Tris, pH 6.8	3.0 ml
10% SDS	10.0 ml
Glycerol	5.0 ml
0.5% Bromphenol blue	2.5 ml

Just before use, 100 µl of 1 M DTT is added to every 0.9 ml of sample buffer needed for the experiment.

20. Resolving Gel (5.6% total acrylamide): This is made in a 15-ml disposable screwcap tube. This recipe makes enough for two mini-gels (0.075 cm × 8.0 cm × 7.3 cm).

Water	5.9 ml
40% Acrylamide stock	1.4 ml
1.5 M Tris, pH 8.8	2.5 ml
10% SDS	0.1 ml

Just before pouring, 100 µl of 10% APS and 8 µl of TEMED are added. The tube is mixed by inversion, and the gels are poured immediately.

21. Stacking Gel (3% total acrylamide): This is made in a 15-ml disposable tube. This recipe makes more than enough for two 1-cm stacking gels for the mini-gels.

Water	2.34 ml
40% Acrylamide stock	0.225 ml
1 M Tris, pH 6.8	0.375 ml
10% SDS	0.030 ml

Just before pouring, 30 µl of 10% APS and 3 µl TEMED are added. The tube is mixed by inversion, and the gels are poured immediately.

22. Tris Buffered Saline (TBS): (20 mM Tris HCl, pH 7.5, 0.5 M NaCl). This buffer is prepared as a 10× stock. 24.2 g of Tris and 292.2 g of NaCl are dissolved in 800 ml of deionized water. The pH is adjusted to 7.5 with 6 M HCl. The volume is made up to 1 l with deionized water. The buffer is sterilized by filtration (0.2 µm) and is stored at room temperature. Before use, the 10× buffer is diluted with 9 volumes of deionized water.

23. Tris Buffered Saline with Tween 20 (TTBS): (1× TBS + 0.05% v/v Tween-20 (Sigma, cat. no. P-9416)). This buffer is usually made up fresh for each run. About 1 l is required.

24. Transfer Buffer: (25 mM Tris, 192 mM glycine, 20% v/v methanol). This buffer is made up for each run (1 l) and is chilled before use. 14.4 g glycine (Sigma, cat. no. G-7032) and 3.0 g Tris are mixed with

200 ml methanol. The volume is made up to 1 l with deionized water. As with SDS-PAGE electrode buffer, the pH should be around 8.5 and should not be adjusted with acid or base.

25. Blocking Buffer: (5% w/v nonfat dry milk, 0.02% sodium azide in TTBS). This buffer is stable for at least two weeks at 4°C.

26. Immunoblot Antibody Dilution Buffer: (1% nonfat dry milk, 0.02% sodium azide in TTBS). This buffer is stable for at least two weeks at 4°C.

27. Alkaline Phosphatase Buffer (AP Buffer): (100 mM Tris HCl, pH 9.5, 100 mM NaCl, 50 mM MgCl$_2$). 12.1 g of Tris and 5.8 g of NaCl are dissolved in 900 ml of deionized water. The pH is adjusted to 9.5 with 6 M HCl before adding 10.0 g of MgCl$_2$ · 6H$_2$O. The volume is then adjusted to 1 l with deionized water. This buffer is filter sterilized (0.2 μm) and stored at room temperature.

28. 5-Bromo-4-chloro-3-indolyl-phosphate, toluidine salt (BCIP): (50 mg/ml in N,N-dimethylformamide, Boehringer Mannheim, cat. number 1383 221). This solution is stored at –20°C.

29. 4-Nitro blue tetrazolium chloride (NBT): (100 mg/ml in 70% v/v N,N-dimethylformamide, Boehringer Mannheim, cat. no. 1383 213). This solution is stored at –20°C.

30. Alkaline Phosphatase Substrate Solution: 45 μl NBT stock solution and 35 μl BCIP are added to 10 ml of AP buffer. This solution is unstable and light sensitive. It should be prepared immediately before use.

31. TE: (10 mM Tris HCl, pH 8.0, 1 mM EDTA). This buffer is filter sterilized (0.2 μm) and stored at room temperature.

Methods

For both immunoblotting and immunohistochemical staining we have used the monoclonal antibody (mAb) C219. This mAb was originally raised against membrane fractions from MDR Chinese hamster ovary cells.[33] C219 recognizes a highly conserved linear epitope which is found in all Pgp deduced amino acid sequences including the only published fish sequences.[34,35] This epitope is denaturation resistant which makes the mAb useful both in immunoblots and in immunohistochemical staining on formalin-fixed paraffin-embedded tissues.

Method I: Pgp immunohistochemistry

Immunohistochemical staining provides information on Pgp expression at the cellular and subcellular levels. Relative levels of expression in different cell types can be compared on the same slide within the same section. However, variability from slide to slide and from block to block makes this technique difficult to use for quantitative comparisons of expression between individuals. More quantitative information can be obtained from immunoblots (see below).

Several controls are necessary to avoid artifacts in this technique. It is standard practice to include both a positive and a negative control on each slide in the form of a tissue that is known to express Pgp and one that does not. It is also important to incubate sections with an irrelevant antibody to control for nonspecific binding of IgG. For our studies mummichog liver sections serve as both positive and negative tissue controls since these contain cell types that express Pgp (hepatocytes) and cells that apparently do not such as red blood cells. For an antibody control we use a matched isotype mouse myeloma protein (IgG2a, kappa) at the same dilution as C219 (1.5 µg/ml). The irrelevant antibody control can be performed on the adjacent section on the same slide if the sections are circled with a PAP pen (Research Products International). This pen deposits a hydrophobic film on the slide which will retain the antibody solution on the section and prevent spill-over between sections.

We have developed the immunohistochemical staining protocol for use with Bouin's fixed paraffin-imbedded tissues. This has allowed us to examine much of the archival material from past histopathological studies when we have routinely fixed specimens in Bouin's fluid. Other fixatives, such as neutral buffered formalin, should give similar results.

Whenever fresh material is used, sub-samples of normal liver and tumors are taken and frozen immediately in liquid nitrogen for analysis by immunoblot (see below). The remaining tissue is fixed and rinsed as follows. Tissues are fixed in Bouin's fluid[36] for 18 to 24 h. The fixed specimens are rinsed in running tap water overnight and are then soaked in several changes of Li_2CO_3 saturated 50% ethanol to remove soluble picrates. Specimens are then stored in 70% ethanol prior to processing.

The Signet ABC reagents in conjunction with mAb C219 are used for detection of antigen. Biotinylated horse anti-mouse IgG is used as a secondary antibody. Bound secondary antibody is detected with an avidin-biotinylated horseradish peroxidase complex (ABC). 3,3'Diaminobenzidine and hydrogen peroxide are used as chromogenic substrates.

The procedure given below begins with 5-µm thick paraffin embedded tissue sections adhered to slides with gelatin. A section from each specimen is also stained with hematoxylin and eosin (H&E) to allow histopathologic diagnosis of any lesions present. For details on histopathological processing of fish tissues see Chapter 31 by Fournie et al. in this volume.

Procedure

This procedure is most easily carried out with the aid of a hand staining rack (Shandon, TissueTek II) to hold the various baths. A humidity chamber is required for the antibody incubations. Plastic food storage boxes with tight-fitting lids work well for this purpose. The bottom of the box is covered with several layers of wet paper towels or other bibulous paper (Whatman 3MM or similar).

Tissue sections on slides are deparaffinized and rehydrated by the following series of baths: xylene, 2 changes, 5 min each; 100% EtOH, 2 changes, 3 min each; 95% EtOH, 2 changes, 3 min each, deionized water, 2 changes, 2 min each followed by a 5-min bath in deionized water. Then slides are transferred to PBS for 5 min. Endogenous peroxidase activity in the sections is then blocked by a bath in 3% hydrogen peroxide for 5 min. The slides are returned to the PBS bath for 5 more minutes. Slides are removed from the PBS, and the area surrounding the sections is wiped dry. The section is circled with a PAP pen. This deposits a ring of hydrophobic material around the specimen and serves as a well to retain the antibody solutions on the specimens.

The slides are placed in the humidity chamber and a solution of 10% (w/v) normal horse serum in PBS is applied to each section using a transfer pipette. In this and all subsequent steps, enough solution is applied to completely fill the ring made by the PAP pen and form a bead of solution over the section. The sections are blocked with the horse serum for 1 h at room temperature. The slides are then held upright and tapped to remove excess blocking solution. The primary antibody solution (C219, 1.5 µg/ml in PBS containing 1% BSA) is applied and the slides are incubated in the humidity chamber overnight (18 h) at 4°C.

The following morning the ABC reagent is prepared according to the manufacturers instructions. This solution must be prepared fresh daily and requires a 30-min incubation at room temperature before use to ensure complete complex formation. The slides are tapped to remove the primary antibody solution and then are rinsed with PBS by gently flooding the surface of the slide several times using a transfer pipette. The biotinylated secondary antibody (biotinylated horse anti-mouse, used at the kit concentration) is then applied to the sections, and the sections are incubated in the humidity chamber at room temperature for 20 min. The secondary antibody solution is tapped off, and the slides are rinsed with PBS. The sections are then incubated with the ABC reagent for 20 min at room temperature. During this incubation the DAB/H_2O_2 (*Sigma*Fast, Sigma) solution is prepared. After rinsing the slides with PBS, the DAB solution is applied to the sections, and the reaction is allowed to proceed for 3 to 5 min. The reaction is quenched by rinsing the slides with deionized water. The slides are then given a 5-min tap water rinse, and are counterstained by dipping them for 3 to 5 sec in Harris' hematoxylin. The sections are then differentiated in saturated, aqueous $NaHCO_3$ for 2 min. The slides are rinsed in tap water and dehydrated through graded alcohols to xylene. The sections are coverslipped using synthetic mounting media (Preservaslide, EM Science).

Method II: immunoblot analysis of tissue extracts

Immunoblotting provides both a means of quantifying Pgp expression and a check on specificity of detection. Immunoblots have been used frequently to assay expression of Pgp in mammalian organs and tumors.[37-39] In these studies crude membrane fractions or microsomes are prepared to generate a Pgp enriched sample. The small size of tissues and especially tumors from small fishes makes this approach of limited use for our application. We routinely use detergent lysates (modified RIPA lysis buffer) of unfractionated organ and tumor samples in immunoblots. This greatly simplifies sample preparation and, in our system, results in sufficient Pgp signal on the blots.

Immunoblots can provide supporting evidence that the signal seen in immunohistochemistry results from Pgp expression and not from an unrelated cross-reactive protein. The expected molecular weight for Pgps is in the vicinity of 170 kDa.[1,2] In addition, known Pgps are heavily glycosylated and are expected to produce a somewhat diffuse band in an immunoblot due to microheterogeneity in glycosylation.[40] A single band in immunoblots in this molecular weight range and of this characteristic appearance is consistent with specific detection of Pgp.

We use pre-stained, calibrated molecular weight standards for SDS-PAGE (Bio-Rad, cat. no. 161-0309) in our immunoblots. These give visible molecular weight markers on the membrane following transfer. They also provide a check on transfer efficiency, thus eliminating the need to stain the nitrocellulose membrane for total protein following transfer.

Usually a Pgp positive control sample is loaded with each run. We use extracts of the colchicine-resistant CH^{R}C5 cell line (a generous gift from Victor Ling, Ontario Cancer Institute) for this purpose. Extracts of any convenient mammalian tissue known to have a high level of expression of Pgp should also be useful for this (e.g., adrenal gland).

Densitometric scans of immunoblots can be used to quantify relative levels of expression between individuals with different exposure histories or between tumor and normal tissue. For this purpose it is necessary to load several different amounts of a Pgp standard. This can be prepared from the tissues of interest (e.g., pooled mummichog liver extract) or can be made from any convenient source of Pgp. The different loadings of the Pgp standard are used to form a standard curve of densitometric peak area vs. total protein. It is important that standards and samples fall within the linear range of this assay. In our blots using mummichog liver extracts as a standard, this range is usually between 1 and 20 µg of total protein per lane.

The following is the procedure used in our laboratory for extraction of Pgp from mummichog liver and liver tumors and subsequent analysis and quantitation of expression by immunoblotting. We use the traditional Laemli discontinuous sodium dodecyl sulfate-polyacrylamide gel electrophoresis (SDS-PAGE) system[41] followed by "wet" electrophoretic transfer to nitrocellulose membranes.[42] Pgp is detected on the membrane using an indirect technique. The mAb C219 is incubated with the membranes. Bound C219 is detected with alkaline phosphatase conjugated goat anti-mouse IgG followed by reaction with NBT/BCIP.

Procedure

Preparation of tissue extracts

Fish are anesthestized by placing them in a bath of tricane methanesulfonate (200 mg/l in seawater). Fish are then killed by cervical transection. Samples of liver or other tissues are excised as quickly as possible and either used immediately or snap frozen in liquid nitrogen and stored at –70°C. Homogenization of soft tissues such as liver is accomplished in a Teflon® glass homogenizer powered by a electric drill motor. The sample is allowed to just thaw and is placed in the prechilled (on ice) Teflon glass homogenizer

(Potter-Elvejehm type) containing five volumes of cold lysis buffer. The tissue is homogenized by three up and down strokes at full speed.

The resulting crude homogenate is transferred to a 1.5-ml polypropylene microcentrifuge tube and is spun at full speed for 5 min in a microcentrifuge (Eppendorf, model 5415) in the cold. The homogenate separates into three layers during centrifugation; a floating layer of lipid material, a layer of clear supernatant and a pellet of insoluble tissue debris plus nuclear DNA. The lipid layer is carefully removed with a cotton tipped applicator. The supernatant is then transferred to a fresh tube on ice. The clear lysate can be stored frozen at –70°C at this stage or can be used immediately in immunoblots.

Protein concentrations are determined using a modified Lowry assay that is compatible with the detergents in the lysis buffer (Bio-Rad, *Dc* Protein Assay, cat. no. 500) using BSA as a standard.

SDS-PAGE
Samples are resolved on discontinuous SDS-PAGE Laemli gels using the Bio-Rad Miniprotean II apparatus. We use resolving gel concentration of 5.6% total acrylamide and a 3% stacking gel with the standard buffer. We find it unnecessary to degas the gels if the concentration of ammonium persulfate in the gel recipes is increased to 1%. Gels are prepared from a 40% acrylamide stock solution (acrylamide:bismethylene acrylamide ratio 36.5:1). The electrophoresis apparatus is assembled according to the manufacturer's instructions. The resolving gel is prepared and is poured between the plates using a 10-ml pipette to within 1 cm of the bottom of the gel comb. The gel is then immediately overlain with 0.5 ml of deionized water using a syringe needle. The resolving gel is allowed to polymerize for 30 min. The water overlay is poured off, and the residual liquid is blotted out with filter paper. The stacking gel is prepared and poured with the comb in place. After another 30 min the combs are removed, and the gel sandwiches are set up in the electrophoresis box according to the manufacturer's instructions.

While the gels are polymerizing, the samples are prepared for electrophoresis. Samples are thawed and kept on ice at all times. Each sample is diluted with Laemli sample buffer to a final protein concentration of 0.5 to 2 µg/µl. Samples are heat denatured in a 65°C water bath for 4 min and then placed on ice until they are loaded onto the gel. (Note: Higher denaturation temperatures result in polymerization of Pgp.) Usually between 1 and 20 µg of total protein per lane is loaded in a 10-µl volume of sample buffer for mummichog liver samples. Samples are loaded with a Hamilton syringe. Using the small well combs of the minigel apparatus, up to 15 µl of sample can be loaded in each well. Normally a lane of molecular weight standards and a lane containing a Pgp positive control sample are also loaded.

The gels are run at 200 V (constant voltage) at room temperature until the dye front just reaches the bottom of the gel. This takes about 30 min.

Electrophoretic transfer to nitrocellulose membranes
At the end of the run, the gel sandwiches are disassembled. The somewhat fragile gels are best manipulated while still stuck to the glass plate. Each

sandwich is held horizontally with the large plate down, and the small plate is gently pried off using one of the spacers. Usually the gel remains stuck to the large plate. The stacking gel is cut off of each gel and is discarded. This is easily accomplished using one of the gel spacers to gently pull this stacking gel away. The first lane of each gel is marked by cutting off the corner of the gel over the first lane. If two gels are run at the same time they may be identified by cutting different sized pieces from each one.

Each gel is gently floated off of its plate by dipping the plate with the attached gel in a plastic dish containing 100 ml of transfer buffer until the gel slides into the dish. The gel is washed in transfer buffer for 20 min at room temperature with gentle agitation.

The blotting materials are prepared while the gels are equilibrating. The nitrocellulose membranes are cut to a size slightly larger than the gels. These are then soaked in transfer buffer. Four sheets (per gel) of chromatography paper (Whatman 3MM) cut to the size of the transfer cassette are also soaked in cold transfer buffer. After 20 min the transfer buffer is carefully poured off of the gels and replaced with 100 ml of fresh transfer buffer. The gels are now ready to be placed in the transfer cassette.

The cassette is placed cathode side (black side) down in a baking dish containing a few cm of cold transfer buffer. One of the porous plastic pads provided by the manufacturer (a rectangular plastic scouring pad also works) is placed in the dish and any bubbles are forced out. The saturated pad is placed on top of the cathode side of the transfer cassette. The gel is then moved onto a stack of two sheets of saturated paper by submerging the papers in the dish containing the gel and floating the gel onto the center of the paper stack. The papers and the gel are then placed on top of the plastic pad in the transfer cassette. All air bubbles are carefully removed from beneath the gel and the blotting paper by rolling a test tube over the surface of the gel. The pre-wetted nitrocellulose membrane is then placed on the gel, and all air bubbles are removed as before. The transfer sandwich is completed by placing two more sheets of buffer-saturated paper on top of the nitrocellulose followed by another saturated plastic pad. The cassette is then placed in the slot in the transblot device. The ice pack is placed next to it, and the buffer chamber is filled with cold transfer buffer. About 1 l of buffer is required for each run. The assembled transfer chamber is placed in a chromatographic refrigerator (4°C). The gels are transferred for 16 to 18 h at 34 V (constant voltage). Under these conditions, the apparatus draws about 50 mA of current at the beginning of the run. By the next day the ice in the cooling pack has melted, and the current has increased to about 80 mA.

Detection of immune complexes

After transfer, the cassettes are disassembled and each membrane is cut to the exact size of its gel. The corresponding corner is cut off to provide unambiguous orientation and the membrane is labeled on the back with a pencil to identify it.

It is possible to stain the membrane for total protein at this point using Ponceau S to verify transfer efficiency. Membranes are placed directly into Ponceau S solution (0.1% w/v Ponceau S in 5% acetic acid, Sigma, cat. no.

P7170) for 10 min. This reversibly stains all protein bands on the blot. If molecular weight standards are included, the position of these is marked at this time with a pencil. If no bands appear, it is not worth continuing, since transfer did not occur. A common cause of this is failure to orient the transfer cassette correctly in the transfer apparatus (cathode(black)-gel-nitrocellulose-anode(red)).

An easier alternative to check transfer efficiency and provide size standards is to use pre-stained molecular weight standards as mentioned previously. These are readily visible on the membrane. All but the highest molecular weight standard should completely transfer from the gel. This lane unambiguously shows the orientation of the blot and makes cutting a corner of the blot unnecessary.

After transfer is verified, the membranes are placed in a covered plastic dish containing 100 ml of blocking buffer for 1 h at room temperature. During all incubation and washing steps, constant gentle agitation is provided by an orbital shaker platform. The membranes are transferred to another plastic dish containing 20 ml of diluted C219 at a concentration of 2.5 µg/ml (1:40 dilution of reconstituted C219) in antibody buffer. Membranes are incubated in the primary Ab for 1 h at room temperature. (Note: The primary antibody solution can be saved at −20°C and reused at least 10 times before loss in sensitivity becomes noticeable.) After 1 h, the membranes are transferred to a container with 100 ml of TTBS and washed for 5 min. The wash step is repeated before transferring the membranes to the secondary Ab conjugate solution. The membranes are incubated for 1 h at room temperature with 25 ml of freshly diluted goat anti-mouse alkaline phosphatase conjugate (1:3000 in antibody dilution buffer). The membranes are then washed with two changes of TTBS as before, followed by another 5 min wash in TBS without Tween 20. The membranes are then given a final wash for 2 min in AP buffer. The membranes are then placed in a plastic bag containing 10 ml of BCIP/NBT substrate solution. All air bubbles are removed from the membrane surfaces, and the membranes are incubated in the dark until the lowest concentration bands become visible. This usually takes about 15 min. The color development reaction is stopped by washing the membrane in deionized water and placing it in TE. Membranes are stored at 4°C in TE.

Blots are then photographed (Polaroid® Pos./Neg. film type 665) and are subjected to densitometry if quantitation is desired. We use a Shimadzu® 930 TLC scanner set at 550 nm for quantitation.

Results and discussion

Immunohistochemistry

Immunohistochemical staining has been used to detect Pgp expression in normal mammalian tissues. This technique has shown Pgp expression on the canalicular surface of hepatocytes in normal liver.[8-10] We have found that immunohistochemical staining of sections of normal mummichog liver showed specific staining that corresponds to the mammalian pattern of expression with allowances for the differing architecture of fish liver.[43] The immunoperoxidase stain clearly showed the structure of the bile canaliculi within the section (Figure 1A). There was no stain in the cytoplasm or other organelles in normal hepatocytes. No other hepatocellular elements were stained in these preparations. This pattern of subcellular localization is consistent with the hypothesis that P-glycoprotein may have an excretory function in fish liver as has been suggested for mammals.[12,44]

Increased expression of Pgp has been detected by C219 immunohistochemical staining in human neoplasms.[17] Immunohistochemical staining has shown Pgp overexpression in experimentally induced rat hepatocellular carcinoma and advanced preneoplastic lesions.[21] Using mummichog from a creosote contaminated site with a high prevalence of hepatic neoplasms, we were able to examine Pgp expression in hepatic tumors of fish. Many tumors showed increased C219 immunoperoxidase staining (overexpression) as well as alterations in the pattern of expression (Figure 1B). Often there was a loss of polarity with the entire plasma membrane staining. Cytoplasmic staining and some paranuclear (Golgi) staining was also evident. The alterations in pattern and degree of immunohistochemical staining observed in these fish liver tumors were similar to those reported for rat liver hyperplastic nodules and hepatocellular carcinoma.[21]

In both tumor and normal liver preparations we observed no staining when mouse myeloma protein was used in place of C219 (Figure 1C).

Immunoblots

Immunoblots have been used to identify and quantify Pgp expression in drug resistant mammalian cell lines[33] and in mammalian tumors and normal tissues.[37-39] Immunoblots have also been used to identify Pgp family members in some aquatic organisms.[25-27] We have identified a single main immunoreactive band in immunoblots of mummichog liver and tumor extracts (Figure 1D). The mummichog Pgp band migrated at a relative molecular weight of 177 kDa. This is a higher molecular weight than that of the immunoreactive band in extracts of the multidrug-resistant CH^{R}C5 cell line in our system (165 kDa) but is within the molecular weight range reported for P-glycoproteins. The mummichog liver and liver tumor Pgp bands also had the same characteristic diffuse appearance as the mammalian Pgp. This is consistent with glycosylation of the putative mummichog Pgp. These results help confirm that the antigen labeled in immunohistochemical staining is a Pgp family member.

Densitometric analysis of the immunoblot shown in Figure 1D indicated approximately a threefold increase in expression of Pgp in a mummichog liver adenoma compared to the adjacent normal tissue. These results are consistent with the overexpression seen in immunohistochemical staining of this adenoma and allow a quantitative estimate of overexpression.

Advantages of combined immunochemical methods

We have found that using immunohistochemical staining and immunoblotting together provides information that neither one alone could give. Immu-

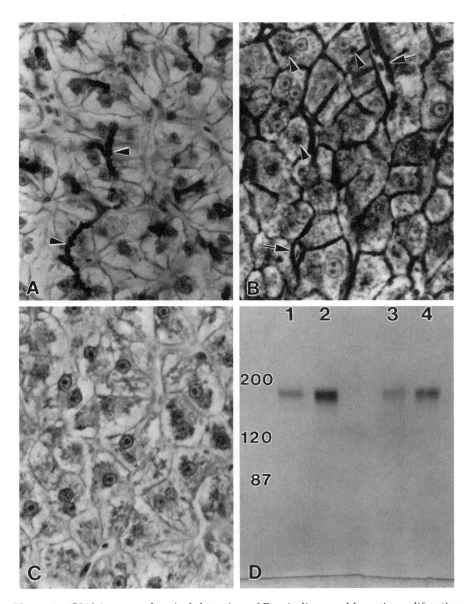

Figure 1 C219 immunochemical detection of Pgp in liver and hepatic proliferative lesions of mummichog. **A**, immunochemical localization of Pgp in normal mummichog liver on the bile canalicular surface (arrowheads) of hepatocytes. **B**, immunochemical localization of Pgp in a mummichog liver adenoma showing overexpression, paranuclear (Golgi) staining (arrowheads) and degeneration of polar membrane expression including staining of the sinusoidal surface (arrows) of hepatocytes. **C**, matched isotype mAb negative control of section adjacent to that shown in B. **D**, immunoblot of paired samples of grossly visible lesions (lanes 2 and 4) and adjacent uninvolved liver (lanes 1 and 3); the pair in lanes 1 and 2 are subsamples of the liver and lesion shown in **A** and **B**; positions of standards are shown with their molecular weight (kDa) at the left.

noblots combined with densitometric analysis have allowed comparison of relative levels of expression between individuals and between discrete proliferative lesions and surrounding non-neoplastic liver.

Immunohistochemical staining has proved less quantitative but has provided important information on the relative level of expression at the cellular and subcellular levels which is destroyed in whole organ homogenates. In our system the two techniques have supported the assumption that the protein labeled is a P-glycoprotein. Immunoblots show specific staining of a single band of the correct molecular weight and characteristics. Immunohistochemical staining shows that in normal fish liver, as in mammalian liver, Pgp is present only on the canalicular surface of hepatocytes, whereas in hepatic tumors there is often overexpression as well as loss of polar expression of this antigen.

Further applications

We are continuing to use these techniques in our studies of Pgp expression in the mummichog. We are investigating induction of Pgp in mummichog liver in response to environmental and laboratory exposures to PAHs. We are also studying relationships between relative Pgp expression and hepatic lesion classification in fish from the PAH contaminated site.

The highly conserved nature of the C219 epitope and the high specificity demonstrated with mummichog tissues should make the immunochemical techniques described here useful in detecting Pgp expression in many other aquatic organisms.

Acknowledgments

We are grateful to Pat Blake, Donna Westbrook, and Patrice Mason for expert technical assistance. This is contribution number 2001 from the Virginia Institute of Marine Science.

References

1. Kartner, N., Riordan, J.R., Ling, V. 1983. Cell surface P-glycoprotein associated with multidrug resistance in mammalian cell lines. *Science* 221:1285–1287.
2. Gottesman, M.M., Pastan, I. 1993. Biochemistry of multidrug resistance mediated by the multidrug transporter. *Annu. Rev. Biochem.* 62:385–427.
3. van der Bliek, A.M., Borst, P. 1989. Multidrug resistance. *Adv. Canc. Res.* 52:165–203.
4. Horio, M., Gottesman, M.M., Pastan, I. 1988. ATP dependent transport of vinblastine in vesicles from human multidrug-resistant cells. *Proc. Natl. Acad. Sci. USA* 85:3580–3584.
5. Hamada, H., Tsuruo, T. 1988. Purification of the 170- to 180-kilodalton membrane glycoprotein associated with multidrug resistance: 170- to 180-kilodalton membrane glycoprotein is an ATPase. *J. Biol. Chem.* 263:1454–1458.
6. Ueda, K., Cardarelli, C., Gottesman, M.M., Pastan, I. 1987. Expression of the full-length cDNA for the human *MDR1* gene confers resistance to colchicine, doxorubicin, and vinblastine. *Proc. Natl. Acad. Sci. USA* 84:3004–3008.

7. DeVaut, A., Gros, P. 1990. Two members of the mouse *mdr* gene family confer multidrug resistance with overlapping specificities. *Mol. Cell. Biol.* 9:1346–1350.

8. Thiebaut, F., Tsuruo, T., Hamada, H., Gottesman, M.M., Pastan, I., Willingham, M.C. 1987. Cellular localization of the multidrug resistance gene product P-glycoprotein in normal human tissues. *Proc. Natl. Acad. Sci. USA* 84: 7735–7738.

9. Thiebaut, F., Tsuruo, T., Hamada, H., Gottesman, M.M., Pastan, I., Willingham, M.C. 1989. Immunohistochemical localization in normal tissue of different epitopes in the multidrug transport protein, P170: evidence for localization in brain capillaries and cross reactivity of one antibody with a muscle protein. *J. Histochem. Cytochem.* 37:159–164.

10. Bradley, G., Georges, E., Ling, V. 1990. Sex-dependent and independent expression of the P-glycoprotein isoforms in chinese hamster. *J. Cell. Phys.* 145:398–408.

11. Burt, R.K., Thorgeirsson, S.S. 1988. Coinduction of MDR-1 multidrug-resistance and cytochrome P-450 genes in rat liver by xenobiotics. *J. Natl. Cancer Inst.* 80:1383–1386.

12. Thorgeirsson, S.S., Silverman, J.A., Gant, T.W. 1991. Multidrug resistance gene family and chemical carcinogens. *Pharmac. Ther.* 49:283–292.

13. Gottesman, M.M. 1993. How cancer cells evade chemotherapy: Sixteenth Richard and Hinda Rosenthal Foundation award lecture. *Cancer Res.* 53:747–754.

14. Goldstein, L.J., Galaki, H., Fojo, A., Willingham, M., Lai, S., Gazdar, A., Pirker, R., Green, A., Crist, W., Brodeur, G., Lieber, M., Cossman, J., Gottesman, M.M., Pastan, I. 1990. Expression of a multidrug resistance gene in human tumors. *J. Natl. Cancer Inst.* 82:1133–1140.

15. Fojo, A.T., Ueda, K., Slamon, D.J., Poplack, D.G., Gottesman, M.M., Pastan, I. 1987. Expression of a multidrug-resistance gene in human tumors and tissues. *Proc. Natl. Acad. Sci. USA* 84:265–269.

16. Chan, H., Thorner, P.S., Haddad, G., Ling, V. 1990. Immunohistochemical detection of P-glycoprotein: prognostic correlation in soft tissue sarcoma of childhood. *J. Clin. Oncol.* 8:689–704.

17. Weinstein, R.S., Jakate, S.M., Dominguez, J.M., Lebovitz, M.D., Koukoulis, G.K., Kuszak, J.R., Klusens, L.F., Grogan, T.M., Saclarides, T.J., Roninson, I.B., Coon, J.S. 1991. Relationship of the expression of the multidrug gene product (P-glycoprotein) in human colon carcinoma to local tumor aggressiveness and lymph node metastasis. *Cancer Res.* 51:2720–2726.

18. Miller, T.P., Grogan, T.M., Dalton, W.S., Spier, C.M., Scheper, R.J., Salmon, S.E. 1991. P-glycoprotein expression in malignant lymphoma and reversal of clinical drug resistance with chemotherapy plus high dose verapamil. *J. Clin. Oncol.* 9:17–24.

19. Fairchild, C.R., Ivy, S.P., Rushmore, T., Lee, G., Koo, P., Goldsmith, M.E., Myers, C.E., Farber, E., Cowan, K.H. 1987. Carcinogen-induced *mdr* overexpression is associated with xenobiotic resistance in rat preneoplastic liver nodules and hepatocellular carcinomas. *Proc. Natl. Acad. Sci. USA* 84:7701–7705.

20. Thorgeirsson, S.S., Huber, B.E., Sorrell, S., Fojo, A., Pastan, I., Gottesman, M.M. 1987. Expression of the multidrug-resistant gene in hepatocarcinogenesis and regenerating rat liver. *Science* 236:1120–1122.

21. Bradley, G., Sharma, R., Rajalakshmi, S., Ling, V. 1992. P-glycoprotein expression during tumor progression in the rat liver. *Cancer Res.*, 52:5154–5161.

22. Teeter, L.D., Becker, F.F., Chisari, F.V., Li, D., Kuo, M.T. 1990. Overexpression of the multidrug resistance gene *mdr3* in spontaneous and chemically induced mouse hepatocellular carcinomas. *Mol. Cell. Biol.* 10:5728–5735.

23. Kurelec, B., Pivcevic, B. 1991. Evidence for a multixenobiotic resistance mechanism in the mussel *Mytilus galloprovincialis. Aquat. Toxicol.* 19:291–302.

24. Kurelec, B., Pivcevic, B. 1992. The multidrug resistance-like mechanism in the marine sponge *Tethya aurantium. Marine Environ. Res.* 34:249–253.

25. Kurelec, B., Krca, S. Pivcevic, B., Ugarcovic, D., Bachmann, M., Imsiecke, G., Muller, W.E.G. 1992. Expression of P-glycoprotein gene in marine sponges. Identification of the 125 kDa drug-binding glycoprotein. *Carcinogenesis*, 13:69–76.

26. Toomey, B. H., Eppel, D. 1993. Multixenobiotic resistance in *Urechis caupo* embryos: Protection from environmental toxins. *Biol. Bull.* 185:355–364.

27. Minier, C., Akcha, F., Galgani, F. 1993. P-Glycoprotein expression in *Crassostrea gigas* and *Mytilus edulis* in polluted seawater. *Comp. Biochem. Physiol.* 106B:1029–1036.

28. Kurelec, B. 1992. The multixenobiotic resistance mechanism in aquatic organisms. *Crit. Rev. Toxicol.* 22:23–43.

29. Beiri, R.H., Hein, C., Huggett, R.J., Shou, P., Su, C.-W. 1986. Polycyclic aromatic hydrocarbons in surface sediments from the Elizabeth River subestuary. *Int. J. Environ. Anal. Chem.* 26:97–113.

30. Vogelbein, W.K., Fournie, J.W., Van Veld, P.A., Huggett, R.J. 1990. Hepatic neoplasms in the mummichog *Fundulus heteroclitus* from a creosote-contaminated site. *Cancer Res.* 50:5978–5986.

31. Fournie, J.W., Vogelbein, W.K. 1994. Exocrine pancreatic neoplasms in the mummichog (*Fundulus heteroclitus*) from a creosote-contaminated site. *Tox. Path.* 22:237–247.

32. Vogelbein, W.K., Fournie, J.W. 1994. Ultrastructure of normal and neoplastic exocrine pancreas in the mummichog, *Fundulus heteroclitus. Tox. Path.* 22:248–260.

33. Kartner, N., Evernden-Porelle, D., Bradley, G., Ling, V. 1985. Detection of P-glycoprotein in multidrug-resistant cell lines by monoclonal antibodies. *Nature* 316:820–823.

34. Georges, E., Bradley, G., Gariepy, J., Ling, V. 1990. Detection of P-glycoprotein isoforms by gene specific monoclonal antibodies. *Proc. Nat. Acad. Sci. USA* 87:152–156.

35. Chan, K.M., Davies, P.L., Childs, S., Veinot, L., Ling, V. 1992. P-glycoprotein genes in the winter flounder, *Pleuronectes americanus*: isolation of two types of genomic clones carrying 3' terminal exons. *Biochem. Biophys. Acta* 1171:65–72.

36. Luna, L.G. 1968. *Manual of Histologic Staining Methods of the Armed Forces Institute of Pathology,* Ed. 3. New York: McGraw-Hill. 258 pp.

37. Hitchins, R.N., Harman, D.H., Davey, R.A., Bell, D.R. 1988. Identification of a multidrug resistance associated antigen (P-glycoprotein) in normal human tissues. *Eur. J. Cancer Clin. Oncol.* 24:449–454.

38. Lieberman, D.M., Reithmeier, R.A.F., Ling, V., Charuk, J.H.M., Goldberg, H., Skorecki, K.L. 1989. Identification of P-glycoprotein in renal brush border membranes. *Biochem. Biophys. Acta* 162:244–252.

39. Fredericks, W.J., Yangfeng, C., Baker, R.M. 1991. Immunoblot detection of P-glycoprotein in human tumors and cell lines. In *Molecular and Clinical Advances in Anticancer Drug Resistance*, Ed. R.F. Ozols 6:121–148. Boston. Kluwer.

40. Richert, N.D., Aldwin, L., Nitecki, D., Gottesman, M.M., Pastan, I. 1988. Stability and covalent modification of P-glycoprotein in multidrug-resistant KB cells. *Biochemistry* 27:7607–7613.

41. Laemmli, U.K. 1970. Cleavage of structural proteins during the assembly of the head of bacteriophage T4. *Nature* 227:680–685.

42. Towbin, H., Staehelin, T., Gordon, J. 1979. Electrophoretic transfer of proteins from polyacrylamide gels to nitrocellulose sheets: Procedures and some applications. *Proc. Natl. Acad. Sci. USA* 76:4350–4354.

43. Hampton, J.A., McKuskey, P.A., McCuskey, R.S., Hinton, D.E. 1985. Functional units in rainbow trout (*Salmo gairdneri*) liver: 1. Arrangement and histochemical properties of hepatocytes. *Anat. Rec.* 213:166–175.

44. Arias, I.M. 1990. Multidrug resistance genes, p-glycoprotein and the liver. *Hepatology* 12:159–165.

chapter eighteen

Immunohistochemical assay for proliferating cell nuclear antigen (PCNA) in aquatic organisms

Lisa S. Ortego, William E. Hawkins, Rena M. Krol, and William W. Walker

Introduction

Cell proliferation is critical to repair processes following damage caused by toxic and pathogenic agents and is important in all stages of carcinogenesis.[1,2] Methods used to detect proliferating cells include directly counting mitotic figures, tritiated thymidine autoradiography (TTA), bromodeoxyuridine immunohistochemistry (BrdU), and proliferating cell nuclear antigen (PCNA) immunohistochemistry (see Reference 3). This chapter describes the adaptation of the PCNA technique for use on aquatic organisms.

PCNA is an endogenous marker of cell proliferation first described in systemic lupus erythematosus.[4] Later identified as DNA polymerase δ auxiliary factor,[5] PCNA is a evolutionarily highly conserved[6] 36-kD nuclear protein[7] directly involved in DNA synthesis.[8] PCNA is stable, can be detected in cells of all tissues[9] and is detectable in all active stages of the cell cycle.[9-11] PCNA can be detected in rodent tissues fixed and stored in formalin for over 24 months.[12]

PCNA immunohistochemistry was developed for mammalian applications and has been used to study only a few non-mammalian subjects[6] such as embryonic development in the fruitfly,[13] cell dynamics in goldfish retina,[14] a tumor in a mirror carp[15] and cell proliferation in plants.[16] Recently, this technique was adapted to several small fish species commonly used in toxicity studies[17] and to three species of penaeid shrimps[18] in preparation. PCNA staining correlates well with BrdU staining in three feral fish species: winter flounder (*Pleuronectes americanus*), mummichog (*Fundulus heteroclitus*) and English sole (*Pleuronectes vetulus*).[19]

The process and rationale for the methodology used to detect PCNA is described below. The method is a biotin-streptavidin/horseradish peroxidase procedure modified from Greenwell et al.[12] In general terms, an antibody probe for PCNA is applied to tissue sections on glass slides. After application of the anti-PCNA, a combination of reagents is used to visualize PCNA-tissue binding. Sites of endogenous peroxidase in the paraffin-embedded tissue are blocked because the assay depends on the addition of horseradish peroxidase to visualize the PCNA. For blocking endogenous peroxidase, several solutions are available including 3% hydrogen peroxide (H_2O_2) which exhausts the peroxidase naturally occurring in the tissue section. Fixation sometimes can obscure the PCNA signal. For tissues that are overfixed or preserved in formalin-based fixatives, antigen retrieval is usually required to reveal epitopes masked during the fixation process. The antigen retrieval process relies on treating sections with a dilute heavy metal solution which is believed to precipitate the protein cross-linked by the aldehyde, improving antigen preservation.[20,21] Mild microwave heating of the heavy metal solution is usually employed.[22]

A protein solution such as gelatin, animal sera or a combination of bovine serum albumin (BSA) and powdered milk is used to block charged sites that may react non-specifically with the antibody in the tissue. The PCNA antibody theoretically binds PCNA at the tissue sites where the antigen is expressed. To detect PCNA visually, a secondary antibody that is conjugated with biotin is applied and binds the species-specific Fc portion of the PCNA antibody. Streptavidin conjugated with horseradish peroxidase has a high affinity for biotin and binds the biotin sites on the secondary antibody. The chromogen diaminobenzidine (DAB) is added to the tissue along with hydrogen peroxide. The horseradish peroxidase reduces the peroxide to water and forms two brown oxidation products from the DAB where the PCNA originally bound to the tissue.

Materials required

Item	Source	Order No.
Sodium phosphate, dibasic, anhydrous (ACS)	Mallinckrodt Specialty Chemicals, Paris, KY	7917
Potassium phosphate, monobasic, anhydrous (ACS)	Mallinckrodt Specialty Chemicals	7100
Sodium chloride (ACS)	Fisher, Pittsburgh, PA	S271-500
Tween 20	Mallinckrodt Specialty Chemicals	H285-01
30% H_2O_2 (ACS)	Fisher	H325-100

Methanol (histologic grade)	Fisher	A433-4
Sodium azide	Fisher	S227-25
Citric acid monohydrate (ACS)	Fisher	A104-500
Trisodium citrate dihydrate (99%)	Aldrich Chemical Co.	85,578-2
Bovine serum albumin, Fraction V, 98–99% albumin	Sigma Chemical Co., St. Louis, MO	A-7906
Nonfat dry milk	grocery store	
Fetal calf serum (after receipt, pipet 200 µl aliquots into cryotubes and store at −20°C)	Gemini Bioproducts, Calabasas, CA	100-106
PCNA clone PC10	Dako Corp., Carpienteria, CA	M879
Concentrated Stravigen kit	BioGenex, San Ramon, CA	ZPOOO-UM
3,3′-Diaminobenzidine tetrahydrochloride (97%)	Sigma Chemical Co.	D-5637
Xylene substitute	Shandon, Pittsburgh, PA	99900502
100% ethanol	McCormick Distilling, Weston, MO	—
Poly-L-lysine slides	Newcomer Supply, Middleton, WI	5010
Sigma mouse IgG2a kappa purified immunoglobin (after receipt, dilute and store at −20°C)	Sigma Chemical Co.	M-9144
Nochromix	Fisher	04-345-20
Concentrated sulfuric acid (technical grade)	Fisher	A298-212
Harris hematoxylin	PolyScientific, Bay Shore, NY	S212
Staining equipment	Shandon	
Staining trays		7331017
Coverplates		7219950
Mounting medium	Richard Allen, Richland, MI	4111
Coverslips (1½, 24 × 50)	Richard Allen	—

Solutions

Phosphate-buffered saline with Tween 20 (PBSt) (10×)

To prepare PBSt, add the following salts one at a time to 1 l distilled, deionized water (ddH$_2$O): 29.6 g sodium phosphate (dibasic, anhydrous), 8.6 g potassium phosphate (monobasic, anhydrous), 144 g sodium chloride, and 10 ml Tween 20. Dissolve and adjust pH to 7.7 with 1 N NaOH in ddH$_2$O. Add ddH$_2$O to make 2 l of solution. Store at room temperature because the concentrated solution will precipitate if refrigerated. For the working solution, add 100 ml of 10× PBSt to 900 ml ddH$_2$O. Seal and store at room temperature. The stock buffer lasts at least 3 months but the working solution should be used within 2 weeks. Discard if a flocculent precipitate develops in the buffer.

Peroxidase blocking reagent

This solution contains 3% H$_2$O$_2$, 25% methanol and 0.1% sodium azide. To prepare, dilute 30% H$_2$O$_2$ with 10 parts ddH$_2$O containing enough methanol so that the resulting solution contains 25% methanol. For example, to make 200 ml of solution use 130 ml ddH$_2$O and 50 ml methanol and 20 ml 30% H$_2$O$_2$. Add 0.1% sodium azide (0.2 g/200 ml of 3% H$_2$O$_2$). Prepare immediately before using.

Antigen retrieval solution

The antigen retrieval solution should be prepared fresh daily. It consists of 12.75% citric acid monohydrate and 11.75% trisodium citrate dihydrate in ddH$_2$O. To prepare 500 ml, dissolve 63.75 g citric acid monohydrate in 250 ml ddH$_2$O. Dissolve and add 58.75 g trisodium citrate dihydrate. Add ddH$_2$O to 500 ml. Discard remaining solution after use.

Protein blocking reagent

Prepare 1% bovine serum albumin (BSA) in PBSt and 1% nonfat dry milk in ddH$_2$O (0.1 g in 10 ml). Prepare daily. Immediately prior to use, mix equal parts of the BSA and milk solutions. Dissolution is facilitated by mechanical stirring for 15 min. Discard remaining solution after use.

Negative control

Fetal calf serum is prepared immediately prior to use. The concentration used is equivalent to the total protein concentration in the 1:100 dilution of the primary antibody (Dako PC10 PCNA). Dilute with PBSt. Discard remaining solution after use.

Primary antibody

PCNA clone PC10 (Dako Corp.) is diluted 1:100 with PBSt immediately prior to use. For a test of up to 50 slides, dilute 0.05 ml of antibody with 4.95 ml PBSt. Discard remaining solution after use.

Secondary antibody and conjugated streptavidin

BioGenex Concentrated Stravigen kit is used at 1:65 dilution. Dilute with PBSt immediately prior to use. For a test of up to 50 slides, dilute 0.1 ml of reagent with 6.4 ml of PBSt. Discard remaining solution after use.

Chromogen

Allow DAB to reach room temperature then prepare 400 ml of a 0.05% DAB solution in PBSt (200 mg DAB/400 ml PBSt). Protect from light. Immediately prior to use, add 0.6% 30%H_2O_2 (240 µl/400 ml DAB solution). Prepare 400 ml of solution for a 50-slide test. Discard remaining solution after use. DAB is considered carcinogenic and must be properly disposed.

Isotype negative control antibody

Sigma mouse IgG2a kappa purified immunoglobin is the appropriate isotype control for PC10. Dilute to the same specific protein concentration as the primary antibody (anti-PCNA) with PBSt and test each new species or tissue to determine if the chemical species of the antibody reacts non-specifically with the tissue of interest.

Glassware cleaning

Nochromix is used as a metal-free oxidizing solution. This powder is mixed with concentrated sulfuric acid. After a detergent wash, glassware is soaked overnight in this mixture. Rinse glassware seven times in tap water followed by twice in distilled water.

Procedures

The method described here results in data used to establish PCNA-labeling indices (PCNA-positive cells/total cells) for a particular organ. The labeling indices represent the proliferative activity of a given tissue or organ.

PCNA immunohistochemistry

1. Cut paraffin sections at 4.5 µm and float them on distilled water. Mount the sections on poly-L-lysine-coated slides (Newcomer Supply) and air-dry overnight. Longitudinal sections of whole small fish include positive control tissues such as the intestine, gill, and kidney. For larger fish, a positive control tissue should be included in the paraffin block along with the study tissue.
2. Prepare 3 l PBSt, 10 ml BSA solution, 10 ml milk solution, 500 ml antigen retrieval solution, and 400 ml peroxidase blocking reagent. Add H_2O_2 immediately prior to use. Use all prepared solutions in this assay at room temperature.
3. Deparaffinize slides following routine procedures, for example, xylene or xylene substitute 2×, 3 min each; 100% ethanol 2×, 1 min each; 95% ethanol 2×, 1 min.
4. Rinse 5 min in PBSt. Note that during PBSt steps the assay can be suspended but only if the slides are immersed in PBSt, not in the cover plates.
5. Add H_2O_2 to peroxidase blocking solution and immerse slides. Incubate for 15 min at room temperature in the dark.
6. Rinse 5 min in ddH_2O.

7. To perform antigen retrieval processing use two containers with 25 slides each. Remove metal handles, pour 250 ml of antigen retrieval solution into each container and cover. Place containers in another shallow container with approximately 1.5 cm tap water. Heat containers in a 700-watt microwave at full power for 2 min. Allow to cool in the microwave for 1 min. Repeat 2 min heating and 1 min cooling twice then cool for 15 min.

8. Rinse in two changes of ddH₂O and then PBSt for 5 min each.

9. Place Shandon cover plates on slides and insert into staining trays. To accomplish this, flood a slide with PBSt and gently slide cover plate over tissue section using the raised guides on the cover plate. Press the slide against the cover plate and check for air bubbles under the cover plate. If air bubbles are evident, remount. Insert the properly mounted slide into staining tray. Steps 10–18 are performed in the dark. Also, all prepared solutions must be at room temperature before application to cover plates.

10. Mix equal parts of milk and BSA to prepare protein blocking solution. To apply to the sections, pipet 100 μl milk/BSA solution into cover plate well and incubate for 20 min. During this time, thaw the fetal calf serum.

11. Prepare negative control (fetal calf serum) and primary antibody solutions. Pipet 100 μl primary antibody or appropriate controls into well and incubate 1 h.

12. Apply PBSt for 5 min by addition to well.

13. Prepare biotinylated secondary antibody solution. Pipet 100 μl into well and incubate for 30 min.

14. Repeat buffer rinse.

15. Prepare peroxidase-conjugated streptavidin solution. Pipet 100 μl into well and incubate 30 min. During incubation, prepare DAB solution (wait to add H₂O₂ until use).

16. Repeat buffer rinse.

17. Flood well with PBSt and gently slip off cover plates.

18. Place slides in staining rack. Add H₂O₂ to DAB and place staining rack in 200 ml DAB solution. Incubate slides in darkness for 10 min.

19. Rinse in running tap water for 3–5 min.

20. Counterstain about 5 sec with Harris hematoxylin.

21. Rinse in tap water.

22. Place in PBSt for about 1 min to blue the hematoxylin.

23. Dehydrate slides following routine procedures, for example, 95% ethanol 1×, 1 min, 100% ethanol 1×, 1 min, and xylene or xylene substitute 1×, 3 min; coverslip.

Positive staining with PCNA results in a brown nucleus and occasionally in brown cytoplasm. In PCNA-negative cells, nuclei are blue from the hematoxylin counterstain.

Label quantitation

Procedures described here for determining PCNA labeling indices in livers of small fish species are based on those of Foley et al.[23] For each liver, 1000 hepatocyte nuclei are counted at 400× magnification in randomly-selected lesion-free fields and the labeled nuclei are tallied. Cells with stained cytoplasm are counted only if the nucleus is stained as well and if examination of the negative control slide reveals that nonspecific background staining was not responsible for the cytoplasmic staining. Labeling indices (LIs) are expressed as labeled hepatocyte nuclei divided by total nuclei counted.

For determining LIs of hepatic lesions, the lesions are identified on PCNA and on an adjacent hematoxylin and eosin (H&E) stained slide. For lesions, LIs are expressed as labeled nuclei per total nuclei counted. The PCNA slide containing the lesion is examined at 400× and the LI determined by one of three methods based on the size and histological characteristics of the lesion:

1. If the lesion is homogeneous in cellular structure and composition, all hepatocyte nuclei are counted in three random fields on PCNA-stained slides to determine the number of fields necessary to estimate 1000. Labeled cells are then counted on PCNA-stained slides in the necessary number of random fields.
2. For lesions with a high degree of cellular pleomorphism or that contain cells other than hepatocytes, 1000 total nuclei are counted on the PCNA-stained slides and the labeled nuclei are tallied.
3. For lesions smaller than three 400× fields, total nuclei in the lesion are counted on the PCNA-stained slides and total labeled nuclei are tallied.

Linden et al.[24] reviewed other methods for label quantitation. In addition to the manual counting method described above, there have been advances in the use of image analysis to quantitate tissue PCNA LIs[25] including fish tissues.[26]

Proliferation in liver cell types other than hepatocytes such as biliary or perisinusoidal cells or cells from other organs such as kidney, gill or intestine, also can be assessed by PCNA labeling. In cases where morphometric analyses are available, assessments are based on labeled cells per length of basement membrane, volume or area. A major consideration in the development of a technique to quantitate LI in other tissues is selecting the appropriate denominator (e.g., total hepatocytes counted) for calculation of the LI.

Results and discussion

Optimal PCNA staining in tissues of aquatic animals is distinct, intense and limited to the nucleus. Figure 1 shows PCNA staining in mosquitofish (*Gambusia affinis*) intestine. The majority of PCNA-positive cells are those lining

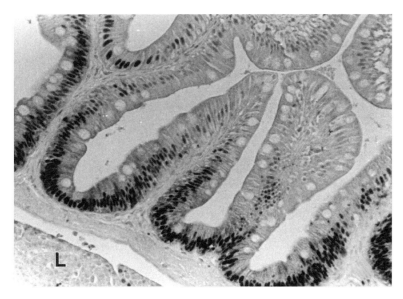

Figure 1   PCNA staining of crypt cells in intestine of a mosquitofish (*Gambusia affinis*). Liver (L); magnification ×310.

the crypts, the location of the most mitotically active cells in the intestine. This specimen, fixed whole in 10% neutral buffered formalin, illustrates an advantage of using whole specimens of small fish species which have positive control tissues such as intestine, kidney, testes and gill usually present on each slide.

In our laboratory, successful PCNA staining has been applied to several other aquatic organisms including the guppy (*Poecilia reticulata*), medaka (*Oryzias latipes*) and several species of penaeid shrimps, the blue shrimp (*Penaeus stylirostris*), the Pacific white shrimp (*P. vannamei*), and the white shrimp (*P. setiferus*).[18] In addition, PCNA immunohistochemistry can be used in cell culture studies as shown in Figure 2 in which isolated hepatocytes from rainbow trout (*Onchorynchus mykiss*) were enrobed in agar and embedded in paraffin. After sectioning and standard PCNA staining, proliferative activity was evaluated in two cell types and correlated with age of the culture and degree of confluence.[27]

Although the PCNA technique appears to be broadly adaptable to biologic specimens, care must be taken with several processing and staining steps to ensure successful results. The quality of PCNA staining varies with the type of fixative used, the protein blocking reagent, the endogenous peroxidase blocking reagent and the source of the primary antibody.[17,18] Additionally, there appear to be species-dependent effects with regard to intensity of PCNA staining.

Fixative-dependent PCNA staining effects have been widely reported.[17,28-34] We evaluated for PCNA immunohistochemistry in tissues fixed in a variety of fixatives including 10% neutral buffered formalin, zinc-buffered formalin, 4% paraformaldehyde, acetone, and Lillie's, Bouin's,

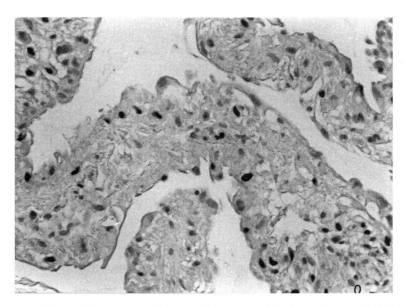

Figure 2 Cultured hepatocytes from rainbow trout (*Onchorynchus mykiss*). Magnification ×280.

Davidson's, and Karnovsky's fixatives.[17] The best fixatives for PCNA staining were zinc-formalin and 10% neutral buffered formalin. The other fixatives that provided adequate PCNA staining were Bouin's, Lillie's, Davidson's and 4% paraformaldehyde, in that order. We did not achieve good results with PCNA immunohistochemistry with either Karnovsky's or acetone. Regardless of fixative used, isolated organs typically stained more intensely for PCNA than did specimens fixed whole, perhaps due to the quality of fixation.

With regard to protein blocking, the use of unpurified animal sera such as normal goat serum resulted in heavy background staining that interfered with staining interpretation. We recommend the combination of BSA and powdered milk or casein to block nonspecific binding in aquatic animals.

Endogenous peroxidase, which must be removed or it will interfere with PCNA detection, was difficult to quench from aquatic animal tissues, particularly those in penaeid shrimp. However, the peroxidase blocking protocol reported here (3% hydrogen peroxide, 25% methanol, 0.1% sodium azide for 15 min) seemed to work sufficiently well for the aquatic species examined.

Three types of primary antibodies are presently commercially available for PCNA immunohistochemistry. Mouse anti-human PCNA 19A2 is an IgM monoclonal antibody (Mab) that has been applied to rodent tissues[12,23] and goldfish retina.[14] PC10 is a mouse anti-rat IgG2a Mab that has proven useful in human cell lines[31] and in a carp tumor.[15] A third clone, Mab 19F4, is an IgG$_1$ to rabbit PCNA. We evaluated 19A2, PC10, and 19F4 on aquatic specimens.[17,18] Clone 19A2 did not adequately label positive control cells whereas 19F4 caused excessive background staining. Further studies with nonspecific blocking and appropriate concentration of 19F4, however, may lead to acceptable staining. Of the three clones, PC10 gave the most satisfactory results in aquatic species. Other investigators have reported that PC10 is sensitive to aldehyde fixation.[35,36] In our study, PC10 stained intensely with aldehyde fixatives (neutral buffered formalin, Bouin's and Lillie's), but the acidity of fixatives such as the latter two that contain picric and acetic or formic acids seemed to decrease reactivity as did the presence of paraformaldehyde or glutaraldehyde as in Karnovsky's fixative.

All species do not react uniformly to the PCNA probe. Of the aquatic species examined, mosquitofish stained most intensely followed by guppy, medaka, and then the penaeid shrimps. Investigators therefore should be aware that modification of this technique may be required to adapt it to a particular species.

Antigen retrieval may not be required for some tissues such as isolated organs fixed in 10% neutral buffered formalin for less than 24 h. However, the retrieval procedure is required following many other treatments and can be modified to yield adequate staining from almost any tissue with nearly any fixative used. In the event that the antigen retrieval method described above does not provide adequate staining, a more rigorous antigen retrieval method may be necessary. This procedure involves boiling the slide-mounted tissue in 1% zinc (as zinc sulfate) twice for 2 min each time. We have used this protocol to successfully stain some difficult tissues. However, this method often damages the tissues, reducing their histological quality.

Although antigen retrieval can be used to restore PCNA immunogenicity to tissues preserved in a wide variety of fixatives, LIs obtained from tissues preserved with different fixation methods should be compared cautiously. Too little is known at this time to be confident that retrieval steps always restore maximal staining efficacy.

Best results with isolated tissues such as cell cultures were obtained with those preserved in 10% neutral buffered formalin for 24 h and stained with Mab PC10. However, because of the variability in staining with the parameters described above, modification at the antigen retrieval and primary antibody steps may be required for optimal staining in a particular tissue.

Another issue in which care must be taken with regard to PCNA staining interpretation is that of LIs obtained from different cell types. Different cell types have been demonstrated to express PCNA differently.[37] Also, comparing LI data gathered from different lesion types is complicated by the possibility that deregulation of PCNA synthesis uncouples PCNA measurements from the estimation of cell replication.[31]

In unpublished studies from our laboratory, PCNA staining has been adapted to several aquatic species and used to investigate dose–response relationships between N-nitrosodiethylamine exposure and cell proliferation, the induction of cell proliferation after toxicant exposure, the role of cell proliferation in neoplastic lesion progression, and cell proliferation as an indicator of environmental contamination. Overall, PCNA immunohistochemistry is an extremely versatile technique that will serve aquatic toxicology in furthering the study of cell proliferation in toxicity and carcinogenesis in aquatic organisms.

References

1. Butterworth, B.E. 1990. Consideration of both genotoxic and nongenotoxic mechanisms in predicting carcinogenic potential. *Mutat. Res.* 239:117–132.
2. Butterworth, B.E. and Goldsworthy, T.L. 1991. The role of cell proliferation in multistage carcinogenesis. *Proc. Soc. Exp. Biol. Med.* 198:683–688.
3. Goldsworthy, T.L., Butterworth, B.E. and Maronpot, R.R. 1993. Concepts, labeling procedures, and design of cell proliferation studies relating to carcinogenesis. *Environ. Health Perspect.* 101(Suppl. 5):59–66.
4. Miyachi, K., Fritzler, M.J. and Tan, E.M. 1978. Autoantibody to a nuclear antigen in proliferating cells. *J. Immunol.* 121:2228–2234.
5. Bravo, R., Frank, R., Blundell, P.A. and Macdonald, B.H. 1987. Cyclin/PCNA is the auxiliary protein of DNA polymerase-delta. *Nature* 326:515–517.
6. Suzuka, I., Daidoji, H., Matsuoka, M., Kadowaki, K., Takasaki, Y., Nakane, P.K. and Moriuchi, T. 1989. Gene for proliferating-cell nuclear antigen (DNA polymerase δ auxiliary protein) is present in both mammalian and higher plant genomes. *Proc. Natl. Acad. Sci. USA* 86:3189–3193.
7. Prelich, G., Tan, C.-K., Kostura, M., Mathews, M.B., So, A.G., Downey, K.M. and Stillman, B. 1987. Functional identity of proliferating cell nuclear antigen and a DNA polymerase-δ auxiliary protein. *Nature* 326:517–520.
8. Mathews, M.B., Bernstein, R.M., Franza, B.R. and Garrels, J.I. 1984. Identity of the proliferating cell nuclear antigen and cyclin. *Nature* 303:374–376.
9. Kurki, P., Vanderlaan, M., Dolbeare, F., Gray, J. and Tan, E.M. 1986. Expression of proliferating cell nuclear antigen (PCNA/cyclin) during the cell cycle. *Exp. Cell Res.* 166:209–219.
10. Morris, G.F. and Mathews, M.B. 1989. Regulation of proliferating cell nuclear antigen during the cell cycle. *J. Biol. Chem.* 264:13856–13864.
11. Bravo, R. and Celis, E. 1980. A search for differential polypeptide synthesis throughout the cell cycle of Hela cells. *J. Cell Biol.* 84:795–802.
12. Greenwell, A., Foley, J.F. and Maronpot, R.R. 1991. An enhancement method for immunohistochemical staining of proliferating cell nuclear antigen in archival rodent tissues. *Cancer Lett.* 59:251–256.
13. Yamaguchi, M., Date, T. and Matsukage, A. 1991. Distribution of PCNA in *Drosophila* embryo during nuclear division cycles. *J. Cell Sci.* 100:729–733.
14. Negishi, K., Stell, W.K., Teranishi, T., Karkhanis, A., Owusu-Yaw, V. and Takasaki, Y. 1991. Induction of proliferating cell nuclear antigen (PCNA)-immunoreactive cells in goldfish retina following intravitreal injection with 6-hydroxydopamine. *Cell. Mol. Neurobiol.* 11:639–659.
15. Manera, M. and Bivati, S. 1994. An immuno-histochemical technique used to demonstrate the transition form of a squamous cell carcinoma in a mirror carp, *Cyprinus carpio* L. *J. Fish Dis.* 17:93–96.
16. Daidoji, H., Takasaki, Y. and Nakane, P.K. 1992. Proliferating cell nuclear antigen (PCNA/cyclin) in plant proliferating cells: Immunohistochemical and quantitative analysis using autoantibody and murine monoclonal antibodies to PCNA. *Cell Biochem. Funct.* 10:123-132.
17. Ortego, L.S., Hawkins, W.E., Walker, W.W., Krol, R.M. and Benson, W.H. 1994. Detection of proliferating cell nuclear antigen (PCNA) in tissues of three small fish species. *Biotechnic Histochem.* 69:317–329.
18. Ortego, L.S., Murata, S.A., Krol, R.M., Gardner, H.S. and Hawkins, W.E. Improved detection of proliferating cell nuclear antigen (PCNA) in aquatic organisms. In preparation.

19. Ortego, L.S. Moore, M.J., Myers, M.S., Vogelbein, W.K., Stegeman, J.J., Varanasi, U. and Hawkins, W.E. Cell proliferation in normal and neoplastic tissues of three feral fish species: comparison of bromodeoxyuridine and proliferating cell nuclear antigen labelling indices. In preparation.

20. Jones, M.D., Banks, P.M. and Caron, B.L. 1981. Transition metal salts as adjuncts to formalin for tissue fixation. *Lab. Invest.* 44:32A.

21. Herman, G.E., Chlipapa, E., Bochenski, G., Sabin, L., and Elfont, E. 1988. Zinc formalin fixative for automated tissue processing. *J. Histotechnol.* 11:85–89.

22. Shih, S.-R., Key, M.E. and Kalra, K.L. 1991. Antigen retrieval in formalin-fixed, paraffin-embedded tissues: An enhancement method for immunohistochemical staining based on microwave oven heating of tissue sections. *J. Histochem. Cytochem.* 39:741–748.

23. Foley, J.E., Dietrich, D.R., Swenberg, J.A. and Maronpot, R.R. 1991. Detection and evaluation of proliferating cell nuclear antigen (PCNA) in rat tissue by an improved immunohistochemical procedure. *J. Histotechnol.* 14:237–241.

24. Linden, M.D., Torres, F.X., Kubus, J. and Zarbo, R.J. 1992. Clinical application of morphologic and immunocytochemical assessments of cell proliferation. *Am. J. Clin. Pathol.* 97 (Suppl. 1):S4–13.

25. Knuechel, R., Burgau, M., Rueschoff, J. and Hofstaedter, F. 1993. Proliferating cell nuclear antigen in normal urothelium and urothelial lesions of the urinary bladder: A quantitative assessment using a true color image analysis system. *Virchows Archiv. B Cell. Pathol.* 64:137–144.

26. Willis, M.L., Myers, M.S. and Gardner, H.S. 1995. Personal communication.

27. Ostrander, G.K., Blair, J.B., Stark, B.A., Marley, G.M., Bales, W.E., Veltri, R.W., Hinton, D.E., Okihiro, M., Ortego, L.S. and Hawkins, W.E. 1995. Long-term primary culture of epithelial cells from rainbow trout (*Onchorynchus mykiss*) liver. *In Vitro Cell. Dev. Biol.* 31:367–378.

28. Bravo, R. and Macdonald-Bravo, H. 1987. Existence of two populations of cyclin/proliferating cell nuclear antigen during the cell cycle: Association with DNA replication sites. *J. Cell. Biol.* 105:1549–1554.

29. Garcia, R.L., Coltera, M.D. and Gown, A.M. 1989. Analysis of proliferative grade using anti PCNA/cyclin monoclonal antibodies in fixed, embedded tissues. *Am. J. Path.* 134:733–739.

30. Golick, M.L. and Rice, M. 1992. Optimum staining of PCNA in paraffin section is dependent on fixation, drying, and intensification. *J. Histotechnol.* 15:39–41.

31. Hall, P.A., Levison, D.A., Woods, A.L., Yu, C.C.-W., Kellock, D.B., Watkins, J.A., Barnes, D.M., Gillett, C.E., Camplejohn, R., Dover, R., Waseem, N.H. and Lane, D.P. 1990. Proliferating cell nuclear antigen (PCNA) immunolocalization in paraffin sections: An index of cell proliferation with evidence of deregulated expression in some neoplasms. *J. Pathol.* 162:285–294.

32. Robbins, B.A., Vega, D., Ogata, K., Tan, E.M. and Nakamura, R.M. 1987. Immunohistochemical detection of proliferating cell nuclear antigen in solid human malignancies. *Arch. Pathol. Lab. Med.* 11:841–845.

33. Suzuki, K., Katoh, R. and Kawaoi, A. 1992. Immunohistochemical demonstration of proliferating cell nuclear antigen (PCNA) in formalin-fixed, paraffin-embedded sections from rat and human tissues. *Acta Histochem. Cytochem.* 25:13–21.

34. Takahashi, H., Oishi, Y., Chuang, S.-S. and Sotokichi, M. 1992. Effects of tissue fixation and processing on proliferating cell nuclear antigen (PCNA) immunohistochemistry. *Acta Pathol. Jpn.* 42:621–623.

35. Casasco, A., Giordano, M., Danova, M., Casasco, M., Icaro Cornaglia, A. and Calligaro, A. 1993. PC10 monoclonal antibody to proliferating cell nuclear antigen as probe for cycling cell detection in developing tissues. A combined immunocytochemical and flow cytometric study. *Histochemistry* 99:191–199.
36. Jones, H.B., Clarke, N.A. and Barrass, N.C. 1993. Phenobarbital-induced hepatocellular proliferation: anti-bromodeoxyuridine and anti-proliferating cell nuclear antigen immunocytochemistry. *J. Histochem. Cytochem.* 41:21–27.
37. Coltrera, M.D. and Gown, A.M. 1991. PCNA/cyclin expression and BrdU uptake define different subpopulations in different cell lines. *J. Histochem. Cytochem.* 39:23–30.

chapter nineteen

Characterization of the cellular stress response in aquatic organisms

Leslie S. Martin, Sylvia R. Nieto, and Brenda M. Sanders

Introduction

The cellular stress response entails the rapid synthesis of a suite of proteins, called stress proteins or heat-shock proteins, in response to environmentally adverse conditions.[1,2] The response is highly conserved and stress proteins have been found in diverse organisms including bacteria, invertebrates, fishes, plants, and mammals. Members of the stress90, stress70, and chaperonin60 (cpn60) stress protein families are molecular chaperones that are present in cells under normal conditions where they function to facilitate protein folding, assembly, and distribution.[3] Under stressful conditions, these and other stress proteins protect and repair cellular proteins that are common targets of environmentally induced damage.[4]

A number of studies have suggested that the induced synthesis of stress proteins may reflect subcellular damage to protein targets, a condition which has been called proteotoxicity.[5] Further, the accumulation of stress proteins correlates with acquired tolerance, in which exposure to a mild stress regime confers the ability to survive a subsequent but more severe stress that otherwise would be lethal to the organism. This protection appears to involve universal targets of environmentally induced damage because tolerance is enhanced as long as stress proteins are elevated and is independent of the specific chemical or physical properties of the stressor. Therefore, elevations in stress protein levels may be particularly useful for addressing specific questions of interest in aquatic toxicology including understanding the integrated effects of multiple stressors. It may also provide a mechanistic basis for developing sensitive measures of stress in organisms exposed to contaminants in the environment. In addition, the stress response may increase our understanding of the mechanisms of toxicity of various contaminants by helping to identify target organs, tissues and subcellular compartments which are particularly vulnerable to contaminant induced damage.[6]

The interest in the stress response in environmental toxicology has provided both the need and interest to better understand this ubiquitous line of defense against environmentally-induced damage. Further, the high degree of conservation of stress proteins across phyla provides the opportunity to use stress protein antibodies raised against one species as probes to detect and measure stress proteins in another species. While such heterologous systems can be quite useful, their strengths and limitations need to be completely understood when employing these techniques. In this chapter we discuss a number of techniques that are commonly used to examine the cellular stress response with a particular focus on those with the greatest utility in aquatic toxicology. While several of the techniques discussed are methods applicable to the measurement and detection of many proteins (i.e., one- and two-dimensional SDS-PAGE and western blotting), we will emphasize their applications to examination of the cellular stress response. Throughout this chapter we discuss, within the context of the cellular stress response, the specific nature of the information which is obtained, the limitations of the technique, and precautions one must take in interpreting the data.

Materials required

Equipment

> Fume hood
> Incubator (with orbital platform shaker)
> Tissue homogenizer (Omni Mixer, Dounce, dismembrator, or sonicator)
> Centrifuge (Eppendorf microfuge)
> Pipettor (1 to 20 µl volume)
> Hamilton syringe (1 to 25 µl)
> Vacuum pump with cold trap
> Power supply
> Minigel vertical gel unit (Hoeffer SE250/280 w/glass plates, clamps, buffer chambers)
> Teflon spacers and gel comb (0.75 mm)
> Minigel 2-D tube gel adaptor kit (Hoeffer SE220 w/glass tubes, tube gel caster, adaptor, tube gel extractor)
> Tank electroblotting apparatus
> Gel dryer
> Densitometer

Reagents and solutions

Note: Milli-Q water (or double-distilled water) should be used to prepare all reagents unless stated otherwise.

Metabolic labeling with radiolabeled amino acids

Radiolabeled amino acid. ^{35}S-methionine/cysteine translabel (ICN #51006 or Dupont-NEN #NEG-072). *The Dupont product contains tricine buffer to minimize the level of volatile radioactivity.*

5× Marine physiological saline.   (1) **Fish**: 700 mM NaCl, 37.5 mM Na$_2$SO$_4$, 5.0 mM NaHCO$_3$, 7.5 mM CaCl$_2$ · H$_2$O, 5.0 mM MgCl$_2$, 15.0 mM KCl, 2.5 mM NaH$_2$PO$_4$, 2.5 mM Tris-HCl (pH 7.8). (2) **Molluscan**: 2.3 M NaCl, 0.05 M KCl, 0.055 M CaCl$_2$ · 2H$_2$O, 0.25 MgCl$_2$ · 6H$_2$O, 0.25 M Tris-HCl (pH 7.8).[7] For other classes of aquatic organisms, the appropriate physiological saline must be determined.

5× Glucose.   Prepare a 0.05 M stock solution. (*May be aliquoted and stored at –20°C.*)

500× Methionine-cysteine.   Prepare a stock solution containing 0.01 M cysteine and 0.005 M methionine. Mix, then add 1 volume of the 500× solution to 9 volumes of H$_2$O to obtain a 50× stock. (*May be aliquoted and stored at –20°C.*)

Incubation media (for 10 ml)

- 2 ml 5× physiological saline (for the appropriate species)
- 2 ml 5× glucose
- 200 µl 50× methionine/cysteine
- 5.8 ml H$_2$O

(*Make fresh; keep on ice.*)

Sample processing

3× Homogenization buffer stock (pH 7.2 at room temperature).   200 mM Tris-base (Sigma #T-1503), 3% Nonidet P-40 (Sigma #N6507), pH to 7.2 with HCl. (*May be aliquoted and stored at –20°C. Note: The pH of tris buffer is sensitive to changes in temperature and should be checked prior to each use.*)

1000× Phenylmethylsulfonylfluoride (PMSF) stock.   (Sigma #P7626). Prepare a 0.1 M stock solution in isopropanol. (*May be aliquoted and stored at –20°C.*)

1× Homogenization Buffer.   66 mM Tris (dilute 1 volume of 3× homogenization buffer stock with 2 volumes of H$_2$O). 0.1 mM PMSF (add 1000× PMSF stock to a final concentration of 0.1 mM). (*Make fresh. Add PMSF immediately prior to homogenization: PMSF becomes unstable in H$_2$O. Keep homogenization buffer on ice.*)

Tris Buffer (500 mM).   Prepare a 500 mM Tris-Base solution, pH to 7.4 with HCl. *(Store at 4 °C.)*

One-dimensional SDS-polyacrylamide electrophoresis
Recipes for the following stock solutions (except where noted) can be obtained by referring to *Current Protocols, Section 10*[9] or the vertical gel unit manufacturer's instructions.

1.5 M Tris-Cl/SDS stock (pH 8.8 at room temperature).   To pH this solution, use a meter that is compatible with tris buffers. *(Store at 4 °C for up to 2 months.)* (SDS: Bio-Rad #161-0302)

0.5 M Tris-Cl/SDS stock (pH 6.8 at room temperature).   To pH this solution, use a meter that is compatible with Tris buffers. *(Store at 4 °C for up to 2 months.)*

30% Acrylamide/0.8% bis-acrylamide stock.   Bio-Rad #161-0107/#161-0201). It is very important to filter poor grades of acrylamide but less important if ultrapure acrylamide is used. Be sure to filter if the solution is not clear, not colorless or has particles in it. *(Store in a dark bottle at 4 °C for up to 30 days.)*

CAUTION!   Wear a face mask when weighing acrylamide (and bisacrylamide). Acrylamides are neurotoxins in their powder and monomer form. Once polymerized they are no longer considered neurotoxins, but should be treated as such because monomer may still be present.

10% Ammonium persulfate (AMPS).   (Bio-Rad #161-0700). Prepare a 10% (w/v) solution. AMPS begins to break down almost immediately when dissolved in H_2O. Therefore, AMPS solutions should be prepared immediately prior to use.

N,N,N'N'-tetramethylethylenediamine (TEMED).   Purchased as a solution (from Bio-Rad or Sigma) and used undiluted. *(Store at 4 °C.)*

12.5% Separating gel solution.   Separating gels containing 12.5% polyacrylamide are most suitable for separating stress proteins because they provide good resolution of proteins ranging from 20 to 100 kDa. The total amount of gel solution required will depend on the type of electrophoresis unit being used (i.e., full size or mini).

5% Stacking gel solution.   Stacking gel polymerizes rapidly and must be used immediately. Failure to form a firm gel usually indicates a problem with the ammonium persulfate, TEMED or both. The total amount of stacking gel solution required will depend on the number of gels and the type of electrophoresis unit being used (i.e., full size or mini).

Hydrated isobutanol. Shake H_2O and reagent grade isobutanol in a separatory funnel. Remove the aqueous (lower) phase. Repeat this process several times. The final upper phase is hydrated isobutanol. *(Store at room temperature.)*

10× SDS/Electrophoresis running buffer. Dilute to 1× for a working solution. The pH is 8.3 when diluted and should not need adjusting. If necessary, pH solution to 8.3 with 1 N HCl. Make fresh and store at 0° to 4°C until use. *(10× solution can be stored at room temperature.)*

RNase/DNase solution. Mix RNase [1 mg/ml], DNase [1mg/ml] and 50 mM $MgCl_2 \cdot 6H_2O$ in 500 mM Tris buffer.

2× and 5× SDS/Laemmli sample buffer.

Coomassie blue staining solution. (Sigma #B-0149)

Destaining solution.

Autoradiography

Film. Kodak X-Omat AR (#165-1454) (8 × 10 inch sheets) or Kodak Bio-Max.

Cassettes. Kodak X-OMATIC, regular 8 × 10 inch screen (#153-5335).

Chemicals. Kodak GBX Developer and Fixer pack (#190-1859); Kodak Indicator Stop Bath (#146-4247).

Two-dimensional SDS-polyacrylamide gel electrophoresis
Recipes for the following stock solutions (except where noted) can be obtained by referring to *Current Protocols, Section 10*[9] or the 2-D gel unit manufacturer's instructions.

30% Acrylamide/1.8% bis-acrylamide stock.

Urea. (Bio-Rad #161-0730)

Nonidet P-40.

Ampholytes. A pH range of 6 to 8 separates most stress proteins but the ampholytes can be adjusted to suit the proteins being separated as long as the total volume is not changed. It is important however to maintain a continuous pH range; otherwise, the gel could spark and burn.

Isoelectric focusing (IEF) gel solution. The total amount of IEF gel solution required will depend on the type of electrophoresis unit being used (i.e., full size or mini). With the Hoeffer SE220 system, 15 ml of IEF gel solution is enough for approximately 15 mini-tube gels.

IEF running buffers.

- 20 mM NaOH (Upper chamber buffer)
- 0.085% H$_3$PO$_4$ (Lower chamber buffer)

IEF sample buffer.
Triton-X may be substituted for Nonidet P-40. Similarly, BME may be substituted for DTT. Solution should be warmed at 37°C, or with the warmth of your hands, until the urea is dissolved. Do not use excessive heat to dissolve the urea as evaporation will cause the urea to precipitate when the solution cools. *(May be aliquoted and stored at −20°C.)*

Equilibration buffer.

1% Agarose in equilibration buffer.
(For sealing IEF tube gel to slab gel.) Prepare equilibration buffer without BME. Melt 1 g agarose in 100 ml of this buffer. Add BME to 5%. Prepare a 0.05% stock solution of bromophenol blue and add 1.25 ml per 50 ml of equilibration buffer. Bromophenol blue is added as a tracking dye.

Protein molecular weight (MW) markers for second dimension.
Solubilize protein molecular weight markers (Bio-Rad #161-0303/#161-0304) in SDS/Laemmli sample buffer. Boil for 5 min, then dilute 1:1 with hot 1% agarose solution. Draw up the solution in a 2-D capillary tube the same diameter as the first-dimension tube gel. After the gel has cooled and solidified, extrude the gel onto parafilm. Cut the gel into equal pieces.

Western blot

Primary antibodies.
(Aliquot and freeze the Ab concentrate. Before using, dilute the Ab in 1× TBS that contains 5% BSA.) The following antibodies are broadly cross-reactive.

- *For stress70*
  Monoclonal Ab clones 3a3 (# MA3-006), 5a5 (#MA3-007), 7.10 (#MA3-001)
  (Affinity Bioreagents, Neshanic Station, NJ)
  Monoclonal Ab N27 (#N27F3-4) (StressGen, Victoria, B.C., Canada)
  Monoclonal Ab BRM-22 (#H5147) (Sigma)

- *For cpn60*
  Monoclonal Ab clone LK-2 (SPA-807) (StressGen)

Secondary antibodies.
Anti-mouse (Bio-Rad #170-6520) OR anti-rat (Sigma# A9654) antibody with an alkaline phosphatase conjugate. *(Freeze Ab in aliquots and dilute in 1× TBS containing 5% BSA before using.)*

Prestained protein molecular weight standards. Low molecular weight (LMW) protein standards (Bio-Rad #161-0305) are suitable for most stress proteins.

Western transfer buffer. See *Current Protocols, Section 10.*[9] (*Make fresh and do not reuse.*)

10× Tris buffered saline (TBS) stock. 150 mM Tris-base, 5 M NaCl. (*Store at room temperature.*)

1× TBS. Dilute 1 volume of 10× TBS stock in 9 volumes of H_2O, pH to 7.5. (*Make fresh.*)

TTBS. (TBS with Tween-20.) Prepare 1× TBS with 0.5% Tween-20 (Sigma #P 7949) (5 ml Tween-20/1 liter 1× TBS). (*Make fresh.*)

Blocking solution. 10% (w/v) non-fat dry milk in 1× TBS. (*Make fresh.*)

5% Bovine serum albumin (BSA). Prepare a 5% BSA solution in 1× TBS. (*Make fresh.*)

Bicarbonate buffer. 0.1 M $NaHCO_3$, 1.0 mM $MgCl_2$, pH to 9.8 with NaOH. (*Store at 4°C.*)

BCIP/NBT visualization solution. Just prior to use, mix 100 μl NBT stock (300 mg NBT in 7 ml 70% DMF, stored at 4°C in a dark bottle for up to 6 months) and 100 μl BCIP stock (15 mg BCIP in 10 ml 100% DMF, stored at 4°C in a dark bottle for up to 6 months) with 10 ml of bicarbonate buffer. (DMF = Sigma #D4254; NBT = Bio-Rad #170-6532; BCIP = Bio-Rad#170-6539.)

CAUTION! Wear gloves and protective clothing when preparing BCIP/NBT Visualization Solution. DMF and BCIP can cause skin and eye irritation, and contact should be avoided. Avoid inhalation of vapors.

Nitrocellulose membrane. Nitrocellulose which contains a plastic mesh (Amersham, Hybond-C extra #RPN 303E) is the easiest type to use because it can be handled repeatedly without damage and does not become brittle after drying. *Always wear gloves when handling nitrocellulose because proteins from the skin will bind to the membrane.*

Dot blot fixer. 10% acetic acid, 25% isopropanol, 65% H_2O. (*Store at 4°C. Can be reused.*)

Procedures

Collection and maintenance of organisms

Care must be taken not to expose animals collected in the field to temperatures greater than that of their immediate environment and to minimize stress to the organisms. If possible, the animals should be dissected at the collection site. Otherwise, the animals should be kept on ice and dissected immediately upon returning to the lab.

Animals collected for research in the lab should be maintained under the best possible conditions for that species and acclimated under optimal conditions for at least one week prior to experimentation. For example, mussels and other molluscs should be maintained in an aerated seawater tank at 17°C and fed a plankton mixture twice a week prior to and during experiments. Some animals, such as limpets, do not survive acclimation and should be processed as soon as possible after collection.

Heat shock treatments

Although it is possible to dissect the tissue of interest and subject it to heat shock (HS) treatment *in vitro*, it is more physiologically relevant to subject organisms to HS *in vivo*. Regardless of the nature of the stressor being examined (i.e., heavy metals, anoxia, xenobiotics), a positive control (in which organisms are subjected to HS treatment) is needed when developing an assay to characterize the stress response with a new species. A good rule is to HS at approximately 10°C higher than the acclimation temperature but it is often necessary to determine, through preliminary experiments, the specific heat shock temperature at which heat shock proteins are induced. It is also important that the exposure is of sufficient duration to allow time for the organisms to reach the treatment temperature and then remain at that temperature for at least 30 min.

In vivo heat shock to organisms such as mussels is accomplished by placing the animals in an aerated seawater tank that is regulated to the temperature of interest. Cells grown in culture and embryonic animals, such as sea urchin embryos, are heat shocked by submerging the flask in which they are growing into a thermoregulated water bath. With sea urchin embryos, great care must be taken to keep the embryos aerated and to maintain the pH and salinity of the culture media.

After HS treatment, organisms should be returned to their acclimation temperatures and stress proteins should be allowed to accumulate 1 to 3 h for metabolic labeling studies and 8 to 24 h for detection by immunoblotting.

Metabolic labeling with radiolabeled amino acids

Metabolic labeling of proteins is a common method for examining the stress response. It entails the incorporation of radiolabeled amino acids into proteins. The pattern of protein synthesis during the incubation period can be determined by separating the metabolically-labeled proteins by electrophoresis and detecting labeled proteins by autoradiography. This measure is not equivalent to measuring protein synthesis because the free amino acid pool in the tissue is usually not known.

Labeling is straightforward for cells in culture, sea urchin embryos, algae, and dissected tissue but poses some complications when tissues from the whole organism are to be examined. Labeling can be carried out in organisms *in vivo* by injection of radioisotope into the peritoneal cavity.[8] However, a number of problems arise. Leakage at the injection site is a common problem. Further, *in vivo* labeling does not allow for the comparison of incorporation of label between tissues because the efficiency of uptake and distribution for each tissue is not known. *In vivo* labeling of multicellular animals by immersing the organism into radiolabeled media is usually inefficient and expensive due to the large quantities of radioisotope required. We have found that *in vitro* techniques allow more efficient incorporation of the radiolabel and less variability between samples. Precautions must be taken, however, due to the risk of anoxia and tissue breakdown. In most cases vigorous shaking, or aeration, is sufficient to assure that the tissue is not subjected to anaerobic conditions.

For *in vitro* metabolic labeling the tissue is dissected immediately following the stress regiment and placed in the appropriate incubation media along with the ^{35}S-methionine/cysteine or ^{3}H-amino acid label. It has been reported that some mammalian low molecular weight stress proteins do not contain methionine and in those cases ^{3}H-leucine is recommended. However for most purposes ^{35}S-methionine/cysteine is an effective and inexpensive radiolabel for aquatic organisms. For cells in culture and organisms labeled *in vivo* such as sea urchin embryos and algae, concentrations in the range of 10 to 100 µCi ^{35}S-methionine/cysteine per ml of culture media or water allows ample incorporation of label in 30 to 60 min.

For *in vitro* labeling higher concentrations of label in the range of 1 to 2 mCi/ml of ^{35}S-methionine/cysteine is required to radiolabel whole tissue in a 1 to 4 h time frame. In developing the incubation media the appropriate physiological saline and pH should be determined for the tissue and species under study. Glucose is added to the media at a final concentration of 0.01 *M* and non-radioactive methionine-cysteine is added as a carrier to ensure that the amino acids are present at high enough concentrations to be transported into the cell.

The following procedure is used to metabolically label tissues *in vitro*:

1. Adjust the incubation media (minus the radiolabel) to the acclimation temperature of the organism to avoid heat shock (or cold shock) conditions during incubation.
2. Thermoregulate an incubator (with orbital shaker) to the appropriate acclimation temperature.
3. Dissect, slice, and if appropriate, weigh the tissue of interest. Be sure to keep the tissue moist and on ice during this process.
4. Place the dissected tissue into the bottom of a culture tube and cover it with the non-radioactive incubation media. The amount of incubation media to add will depend on the type of tissue being labeled, its thickness, mucous content, etc. As a general rule, tissue and incubation media should be mixed at a 1:1 ratio (w/v). For example, 200 mg of tissue should be covered with 200 µl of incubation media.

CAUTION! The following step should be carried out under a fume hood.

5. Carefully add the ^{35}S-methionine at a concentration of 1 to 2 mCi/ml for whole tissue (or 10 to 50 µCi/ml for cultured cells).
6. Tightly cap the culture tube and transfer it to the incubator placing it securely on the shaking platform. Incubate the tissue at the acclimation temperature for approximately 6 h (2 h minimum) with constant agitation to keep the tissue aerated. An activated charcoal trap should be placed in the incubator to minimize volatile radioactivity.
7. After the incubation is completed, discard the radioactive aqueous waste and rinse the tissue with appropriate physiological saline once. Freeze at −80°C until analysis.

NOTE: Radioactive aqueous waste should *not* be discarded at the sink! Rather, radioactive waste should be collected in appropriate containers and disposed of according to local radiation safety practices.

Cultured cells are more easily metabolically labeled. To metabolically label cultured cells, the media in the flasks containing the cells must be changed to methionine-free media 3 h prior to experimentation. With some cell lines, methionine deprivation may not be necessary; preliminary experiments should be conducted to determine if this treatment is necessary. After the stressor treatment, ^{35}S-methionine/cysteine label is added to the media of each flask at a concentration of 10 to 50 µCi/ml. After 1 to 2 h, the cells are washed with PBS (Dulbecco's, Sigma D-5652), harvested using trypsin-EDTA (M Life Technologies 25300-054), and collected by centrifugation. The resultant pellets are washed again with PBS and stored at −80°C until analysis.

Sample processing

A rigorous homogenization of tissues and cells is critical in order to detect all the stress proteins in the various cellular compartments. For many aquatic organisms, tissues such as gill and hepatic organs require specific modifications to the homogenization procedure because of the presence of cartilage, mucus or digestive enzymes. Plants also require rigorous homogenization because of the presence of the cell wall. The type of homogenizer used will depend on the tissue, but a handheld homogenizer such as the Omni mixer (Omni International) works well with many tissue types. Very fibrous tissues require the use of a glass Dounce homogenizer, sonicator or dismembrator. Plant cells may also require enzymes to break down the cell wall. In contrast, cultured cells are lysed in homogenization buffer by passing the cells repeatedly through a 200-µl pipette tip.

1. Rinse tissue samples with 500 m*M* Tris-buffer.
2. Place samples in a small plastic test tube with the 1× homogenization buffer covering the tissue (a 1:4 w/v ratio often provides an appropriate dilution). Too much homogenization buffer will result in a very dilute protein sample. Keep the test tube submerged in ice while homogenizing tissue to prevent protein degradation.
3. Homogenize the tissue at 20,000 rpm for 30 sec with a motor driven homogenizer or several minutes with a glass Dounce homogenizer,

sonicator or dismembrator. Homogenization efficiency should be checked microscopically and the duration of homogenization adjusted if necessary.

4. Remove cellular debris in a single centrifugation step at 10 to 15 × g for 10 min at 4°C. Tissues, such as gill, that contain a lot of mucus may require an additional centrifugation step of 150,000 × g for 1.5 h at 4°C.

5. Freeze samples at –80°C until analysis.

One-dimensional SDS polyacrylamide gel electrophoresis

Samples are usually loaded on gels using either equal protein or equal counts (i.e., counts per minute for metabolically labeled samples). A protein assay is performed subsequent to homogenization to determine the amount of total protein in the sample.[9] Since SDS can interfere with many photometric assays, the protein assay must be done before the sample is suspended in SDS/Laemmli sample buffer. A simple way to minimize the effect of interfering substances is to analyze a very small volume of sample so that pigments and other substances which might cause interference are highly diluted.

SDS-PAGE is an essential tool for the study of proteins. It involves the separation of proteins by size in polyacrylamide gels of varying concentrations.[10] We will discuss the electrophoresis system of Blattler[10] which is a modification of Laemmli.[11] SDS polyacrylamide gels are actually composed of two gels; the smaller top section is called the stacking gel and the larger bottom section is called the resolving or separating gel. Separating gels which contain 12.5% polyacrylamide are good for separating most stress proteins because they provide good resolution of proteins from approximately 20 to 100 kDa. By adjusting the acrylamide concentration, one can resolve proteins in other molecular weight ranges.

Once the proteins are separated in the gel, the stacking gel is discarded and the resolving gel can be stained to visualize the proteins within the gel. If the proteins that have been separated are metabolically labeled then the dried gel can be placed in a cassette with X-ray film. This procedure, called autoradiography, allows the radiolabeled proteins to be visualized on the developed film.[9] In addition, SDS polyacrylamide gels can be blotted onto nitrocellulose and the proteins can be visualized using western blotting techniques as described in a later section.

Sample preparation. Combine 1 volume of 5× Laemmli sample buffer and 4 volumes of protein sample in a microfuge tube. Similarly, more concentrated protein samples can be diluted 1:1 (v/v) with 2× SDS/Laemmli sample buffer. Boil for 3 min on a heating block. Let the sample cool to room temperature and then add RNase/DNase to the tube (0.5 µl RNase/DNase per 60 µl sample). Let the sample sit at room temperature for 5 to 10 min and then pass it through a 28-gauge needle 5 times. Boil the sample again for 5 min to inactivate the RNase/DNase and then centrifuge at room temperature. The sample is ready to load onto the gel. You must take into account the dilution of the protein sample by the 2× or 5× SDS/Laemmli sample buffer when calculating the volume of protein to load onto the gel. Therefore, remember to multiply the calculated volume to load by "2/1" or "5/4", the dilution factors for 2× and 5× sample buffer, respectively. If the samples are frozen or put into the refrigerator, they must be reboiled for 5 min and spun down before loading onto the gel.

Casting and assembling. Cast the separating and resolving gels and assemble the gel unit according to *Current Protocols, Section 10*[9] or the gel unit manufacturer's instructions.

Loading protein samples. Carefully pipet 5 μl of low molecular weight protein standards (prepare as directed by manufacturer) into the first well using a gel-loading tip attached to a micropipettor. Carefully load the protein samples into the remaining wells. For gels to be stained with Coomassie brilliant blue, use a standard mixture of MW protein standards. If gels are to be subjected to autoradiography, radioactive MW standards may be used but are not necessary. For gels that are to be subjected to western blotting, use prestained MW markers.

Running the gel. Attach the lid to the electrophoretic unit and connect the red lead into the positive terminal and the black lead into the negative terminal of the power supply (it is helpful to remember "run to red"). Gels may be run under constant current or constant voltage. The amount of current or voltage to use will vary with the type of electrophoresis unit being used. Run the gels according to the manufacturer's suggestion. Allow the gels to run until the dye front reaches the bottom of the plates. Most mini-gel systems take 45 min to 1 h to complete a run. When electrophoresis is complete, turn off power supply, disconnect the power leads and disassemble the gel sandwich. When separating radiolabeled proteins, unincorporated radiolabel may run into the lower buffer. Therefore, radioactive buffer should be disposed of in radioactive waste containers. The gel can be subjected to autoradiography or western blotting.

Staining and drying gels. Place the gel in Coomassie stain for 0.5 to 1 h (until a deep uniform blue stain is achieved). Transfer the gel into the destaining solution and let it destain overnight. Gels are dried using a commercial gel dryer (i.e., Hoeffer) attached to a vacuum system or pump. Be sure that there is a vapor trap attached to the vacuum system to collect the water and that the trap is cooled with dry ice so that none of the moisture gets into the pump. Place the gel on a piece of cellophane (Bio-Rad# 165-0963) or blotting paper (Bio-Rad# 165-0962) cut slightly larger than the gel. Wet the gel and gently press it to remove any air bubbles that may be trapped. Cover the gel with a 0.05 μm Mylar film (AMBIS #3800-0008). Mylar is used if the gel is to be subjected to autoradiography. It is thin enough to allow the passage of ^{35}S particles and will prevent the gel from sticking to the X-ray film. If the proteins are not radioactive, then cellophane can be used to cover the gel. Do not use commercially available plastic wrap because it is not porous and the gel will not dry. Place the blotting paper/gel/Mylar assemblage on the dryer unit, lower the dryer cover and turn on the vacuum/pump. After a tight seal has formed around the gel, turn on the heating element for approximately 45 min. When the gels are dry they will be hard and barely visible under the cover. Turn off the heating element and the vacuum/pump. Break the vacuum seal around the gel by gently pulling up a corner of the dryer cover. When the gels are dry, proteins are easily visualized as dark blue bands. The stress70 protein is often found in great abundance and easily detected on a stained gel.

Autoradiography

After a gel is dry, it can be subjected to autoradiography if the protein samples were radiolabeled. This is accomplished by placing the dry gel in an X-ray cassette (wear gloves while handling radioactive gels). The next step is carried out in a darkroom under dim red safelights. Place a piece of X-ray film on top of the gel (cut slightly larger than the gel), close the cassette and wrap the cassette in foil. Place the cassette in a –80°C freezer and allow the X-ray image to develop (1 to 7 days depending on the amount of radioactivity in the protein bands). Develop the film according to the manufacturer's instructions. If radioisotope other than ^{35}S is used (e.g., ^{3}H), commercial fluors are available that enhance the isotopic signal and reduce the length of exposure time needed for the image to develop.[9]

Two-dimensional SDS-PAGE

Two-dimensional electrophoresis allows the separation of proteins by both charge and size. The first dimension consists of an isoelectric focusing (IEF) tube gel which separates proteins by isoelectric point (pI). The proteins are applied to a gel with a continuous pH range; a pH range of 6 to 8 is optimal for stress proteins. An electrical current applied to the gel causes the proteins to move through the gel due to their intrinsic charge until they reach their pI and have a net charge of zero (Figure 1).[10] The second dimension is an SDS polyacrylamide gel, as described earlier, which separates the proteins by molecular weight. Two-dimensional SDS-PAGE is useful for separating the multiple isoforms of stress proteins from large multigene families such as stress70 (Figure 2, bottom left). We have found that mini-IEF gel systems (Hoeffer SE220) are the easiest and quickest systems for separating proteins by pI.

For procedures and recipes on preparing first-dimension tube gels and apparatus setup, refer to *Current Protocols, Section 10*[9] or the gel unit manufacturer's instructions. When preparing first-dimension tube gels, it is important to use a sterile pipette tip or syringe to remove ampholytes from their container to ensure that they do not become contaminated by bacteria which break down the ampholytes very rapidly. It is also helpful to bring the IEF running buffers (upper and lower chamber buffers) to room temperature; chilled buffers cause the urea to crystallize in the gel which will interfere with protein separation. Some manufacturers recommend that you

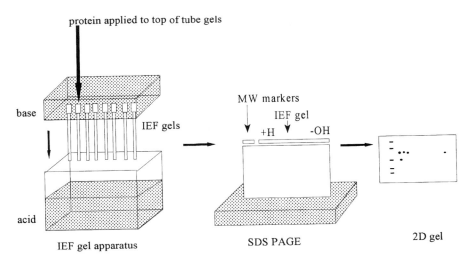

Figure 1 Two-dimensional gel electrophoresis procedure. The proteins are applied to an isoelectric focusing (IEF) tube gel and separated by isoelectric point. Then the tube gel is placed onto an SDS-polyacrylamide gel and the proteins are separated by size. This yields a 2-dimensional gel which can be stained or subjected to western blotting.

prefocus the tube gels prior to loading protein sample. Others do not think it is necessary. We suggest that you refer to the manufacturer's instructions and experiment with prefocusing to determine if it is required for your samples.

Loading protein sample.   Better resolution is obtained if small concentrations of protein are loaded on the tube gel. Depending on the purity of the protein sample, 25 to 50 μg of total protein is sufficient for radiolabeled proteins that will be visualized by autoradiography, whereas 50 to 100 μg of total protein is often necessary to detect stress proteins that will be subjected to western blotting. Disconnect the power supply from the IEF unit before loading sample. Place the protein in a microcentrifuge tube and add an equal amount of IEF sample buffer. Let the sample incubate for a few minutes at room temperature, then deliver it to the top of the tube gel using a Hamilton syringe. Do not introduce bubbles to the tube when loading the sample. If bubbles form, gently dislodge them with the syringe. If the tube gels were prefocused, place a small amount of overlay buffer on top of the protein sample. 2-D SDS-PAGE standards (Bio-Rad #161-0320) can also be run on a separate tube gel to use as pI markers. Attach the lid and turn on the power supply.

First dimension.   Focus the tube gels as instructed in the apparatus manual. Most mini-gel systems will focus protein in 4 h, however the length of focusing required to achieve good protein separation should be determined through preliminary trials. After focusing, turn off the power supply and disassemble the tank.

BRM22

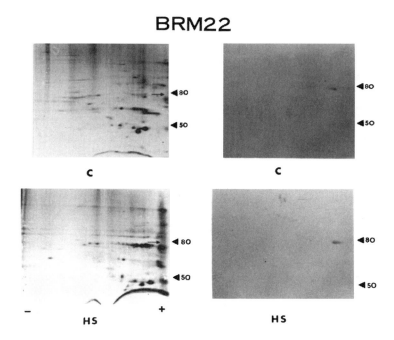

Figure 2   Autoradiograph of two-dimensional western blots. Two-dimensional western blot analysis and autoradiograph of the western blot of proteins from *S. purpuratus* embryos maintained at 17°C (C) and subjected to a heat shock of 25°C/30 min (HS). The proteins were separated by 2D SDS-PAGE, subjected to western blot analysis and then the western blot was exposed to Kodak X-AR film for autoradiography. The western blot was probed with BRM22 monoclonal antibody. Autoradiograph is on the left and western blot is on the right. Arrows on the right indicate molecular weight markers in kDa. Smaller arrows indicate the proteins that were detected by the Abs.

Second dimension.   Prepare a 12.5% separating gel as described in *Current Protocols, Section 10*[9] and cast a stacking gel (without the comb) almost to the top of the glass plate leaving a narrow space deep enough to accommodate a tube gel. When the stacking gel has polymerized, extrude the tube gel and place it in equilibration buffer. For many samples, the equilibration step may not be necessary. Place the tube gel onto the surface of the stacking gel. Be consistent with the orientation of the tube gel (e.g., acid end always pointed left). Place a piece of the MW protein standard in 1% agarose on the left side of the slab gel next to the tube gel. Make sure there are no air bubbles between the gels. Seal the noodle into place with 1% agarose in equilibration buffer. The bromophenol blue in the equilibration buffer will provide a dye front to monitor the run. Run the gel as described in 1-D SDS-PAGE section (Figure 1).

Immunological techniques

Western blot analysis

Regular western blotting techniques are, at best, semi-quantitative since there is not a linear relationship between the antigen and primary antibody, or the primary antibody and the secondary antibody. As a consequence, the conditions of the assay should be such that the antigen (i.e., the protein being measured), not the antibodies, is limiting. In addition, the efficiency of the transfer step can be highly variable. These limitations should be kept in mind during data interpretation.

1. Run either a one- or two-dimensional SDS polyacrylamide gel as described in previous sections. On each gel, include an internal standard that contains a predetermined amount of stress protein to monitor differences in transfer efficiency. The internal standard can be comprised of a large quantity of heat-shocked tissue that has been aliquoted and frozen for this purpose. Also, use prestained MW protein standards on each gel.
2. Prepare the western transfer buffer and chill to 4°C.
3. Assemble the western tank apparatus and the gel "sandwich" according to *Current Protocols, Section 10*[9] or the manufacturer's instructions. The sandwich often consists of a cassette containing a "sponge-filter paper-nitrocellulose-gel-filter paper-sponge" arrangement (Figure 3). *It is important to eliminate any bubbles that might form between the gel and nitrocellulose; otherwise the proteins will not transfer completely.* Insert the cassette into the transfer unit facing the nitrocellulose toward the positive (red) anode so that the proteins, which are negatively charged in the presence of SDS, run toward the positive electrode. Run the transfer according to the manufacturer's suggestion. Depending on the type of unit, many western transfers can be accomplished in a few hours. The same transfer can also be done in a dry blotting system.[9]
4. After the transfer, dismantle the cassette. Keep track of the side of the nitrocellulose to which the proteins have adhered (it can be marked with pencil or notched). Keep this side *up* during all subsequent steps. To determine whether the transfer was complete, visually check that the prestain markers have been transferred onto the nitrocellulose. In addition, the gel can be stained and destained to determine whether the transfer of proteins from the gel was complete.
5. Place the nitrocellulose membrane into a glass dish or heat sealable plastic bag (glass is favored over plastic because protein can adhere to plastic). Block and incubate the nitrocellulose as detailed in *Current Protocols, Section 10.*[9] Do not allow the nitrocellulose to dry out at any time. If necessary, use 1× TBS to keep the blots moist between transfer steps.

   Preliminary western blots should be made to determine the correct dilution of primary antibody to use for protein samples. Typically a 1:1000 dilution will be sufficient but some samples may require more or less antibody. Similarly, the correct dilution of secondary antibody to use must be determined; again, a 1:1000 dilution is usually sufficient. Remember to use the appropriate secondary antibody (*i.e., if a monoclonal made in a mouse was used as the primary Ab, then goat anti-*

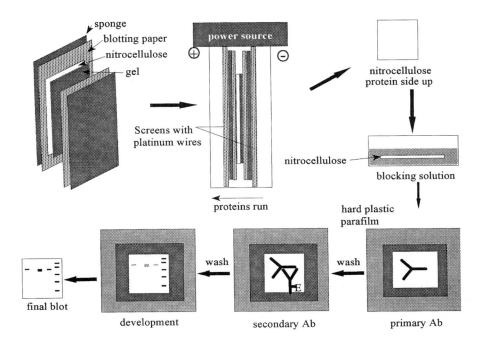

Figure 3 Western blot analysis. A western blot sandwich containing an SDS-Poly-acrylamide gel and nitrocellulose is prepared and fitted into a western transfer tank through which a current is applied. The sandwich is removed and the nitrocellulose is suspended in blocking solution. The nitrocellulose is then probed with a primary antibody to a specific stress protein. A secondary antibody with an enzyme conjugate is then used as a probe. The blot is developed by the addition of substrate and the resulting color product (catalyzed by the enzyme) is used to visualize the stress protein bands.

> *mouse must be used as secondary Ab; if a monoclonal made in a rat was used as the primary Ab, then goat anti-rat must be used as secondary Ab).*

6. The final two rinses of the nitrocellulose should be in TBS (tris-buff-ered saline *without* Tween-20). It is important to be thorough in the last two rinses because Tween-20 interferes with the visualization developer.

7. Return the blot (face up) to a clean container. Add developer to each blot and develop until bands are visible. Do not stop the development process until a band is visible in the control lane (internal standard). The time can vary considerably (from minutes to hours) depending on the assay. To stop the development process, wash the blot with double-distilled water and store in the refrigerator until it can be scanned on a densitometer and dried (Figure 2 illustrates a typical western blot). If the bands on the blot are barely visible and cannot be developed any further without heavy background staining, use a commercially available kit to amplify the signal (Bio-Rad). These kits provide anti-alkaline phosphate Ab which binds the ALP on the sec-

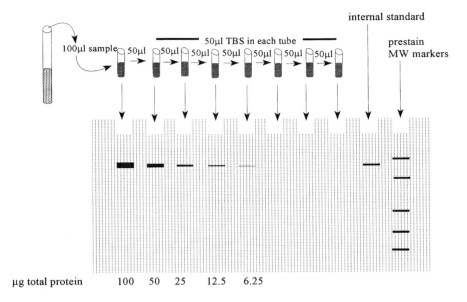

Figure 4 Quantitative western blot analysis. A protein solution is serially diluted and applied to an SDS-polyacrylamide gel for separation. The gel is subjected to western blot analysis to visualize the stress proteins. After the blots are developed, they are scanned on a densitometer to determine the dilution at which the color intensity from bound antibody reaches background. The total protein concentration in the last serial dilution in which the antibody is detected is then calculated.

ondary Ab so that more of the enzyme is produced. Another alternative is to use chemiluminescent secondary Ab probes (Bio-Rad) or radioactive probes (Bio-Rad).

Quantitative western blot techniques

Method 1. Classical quantitative western blotting techniques are used to quantify the concentrations of stress proteins. This procedure requires the serial dilution of a known amount of total protein from the sample (Figure 4). One gel is required for each sample; therefore, it can be very time-consuming and expensive if there are many samples to be analyzed. If possible, a "standard curve" of pure antigen should also be run on the same assay so that stress protein concentration can be expressed as ng/μg total protein. It should be kept in mind that since the binding affinity of the antibody and antigen are not known, these data are relative and cannot be compared to similar data in other species or in which another antibody is used. If a standard curve is not determined, the data can be expressed as the inverse of the total protein calculated in the last detectable band. Despite the disadvantages, quantitative western analysis is a very powerful technique for determining the concentration of stress proteins.

1. The sample (typically 200 μg total protein) is placed in an Eppendorf tube and the volume in the tube is brought up to 100 μl with TBS and

mixed. A series of 7 tubes each containing 50 µl of TBS are lined up in an Eppendorf test tube rack and 50 µl from the tube containing the sample is pipetted into the first tube containing 50 µl of TBS. The sample is mixed by *gently* pipetting up and down. This process is repeated until there are 8 tubes each containing a serial dilution of the protein: 100, 50, 25, 12.5, 6.2, 3.1, 1.6 and 0.8 µg of protein.

2. The samples are digested in SDS/Laemmli sample buffer and loaded onto a 12.5% SDS-polyacrylamide gel, along with an internal standard sample and prestained MW protein standards. After electrophoretic separation, the gel is processed as described for western blotting.

3. After the blots are developed, they are scanned on a densitometer to determine the detection limit of the assay. The dilution at which the color intensity from bound antibody reaches background defines the detection limit of the assay. The total protein concentration in the last detectable serial dilution is then determined.

4. Ideally, a known quantity of *pure* stress protein, belonging to the same stress protein family being studied, should also be analyzed by the same quantitative western blotting technique to determine the specific detection limit of the assay for that stress protein.

5. To determine the stress protein concentration for your sample, the concentration of *pure* stress protein (in ng) determined at the detection limit *(see step 4)* is divided by the total protein concentration (in µg) determined for your sample by the detection limit assay *(see step 3)*. If the detection limit for purified antigen is not characterized for the assay, the stress protein concentration is expressed as the inverse of the total protein calculated (µg) in the sample at the detection limit.[12] It is important to note that since the data obtained is a reciprocal, the geometric mean, not the arithmetic mean, is used for data analysis.

Method 2. We have developed an abbreviated method to quantify stress proteins which has been determined to be more efficient and provides data as reliable as the classical technique. This method also involves separating samples of known total protein concentration by SDS-PAGE but does not require serial dilution of the protein sample. Two conditions must be met when using this method: (1) the samples must be within the linear range of the assay, and (2) the transfer step, which is the most variable step in the blotting procedure, must be controlled for with the use of an internal standard.

1. A standard curve, to establish the linear range of the assay, is determined using protein sample from the tissue being studied and the appropriate primary antibody. A range of total protein concentrations (5 to 100 µg) is separated by SDS-PAGE and subjected to western blot analysis. The bands are visualized and then scanned on a densitometer. The total protein concentrations are plotted against their absorbance values and the linear range of the assay is determined.

2. The protein samples are separated by SDS-PAGE and subjected to western blot analysis. The concentration of protein sample loaded onto the gel must fall within the linear range of the standard curve.

If possible, all samples from an experiment should be run at the same time. An internal standard (of known protein concentration) must be included on each gel.

3. After visualization, the western blots are scanned on a densitometer to determine the absorbance of each protein band; the internal standard is also scanned and its absorbance is determined. The density of the internal standard band should fall within 20% of the predetermined concentration; if it does not, the blot should be rejected. The absorbance of each protein band is then normalized to the internal standard:

$$A^{normalized} = (A^{unknown} / A^{internal\ std}) / A^{mean\ internal\ std}$$

where A = absorbance units, normalized = normalized for differences in transfer efficiency, internal std = internal standard that is on same blot as unknown, and mean internal std = mean of internal standards on all the blots that were run on this set of samples.

The normalized absorbance is divided by the µg of protein applied to the SDS-polyacrylamide gel to yield a value for the stress protein concentration in units stress protein/µg protein. These units can be calibrated to a standard curve developed using the same procedure for a purified antigen. If the antigen is for a different species, it is important to remember that the concentration obtained (in ng) is a relative number and cannot be compared across species.

Dot blot immunobinding assay

While dot blot immunobinding assays (DIA) can be used for stress protein quantification, the results are not as consistent as found with a western analysis.[13] Its major strength is that a number of samples can be ran quickly. However, we have found the procedure difficult to replicate, and the potential for nonspecific binding and false positive readings to be very high in heterologous assays. Further, samples that contain mucus or lipids such as gill tissue will clog the wells and prevent the sample from passing through the membrane. The dot blotting protocol described here is a modification of that described in Sanders et al.[12] and was developed by Dr. Steve Miller.

1. Cut a piece of nitrocellulose to the size of the dot-blot unit and place it between the two plates of the dot-blotter. Tighten down the screws under a vacuum.
2. Rinse the dot-blot unit with 90°C water and let it dry.
3. Pipet 150 µl of TBS into each of the wells and pull it through with gentle suction (two times). *Do not let the wells dry out completely.*
4. The samples being analyzed cannot be in SDS/Laemmli sample buffer and must be free of debris. Place a volume of protein sample containing 200 µg of total protein into a microcentrifuge tube and bring the total volume up to 150 µl by adding 1× TBS. Heat the tube at 65°C for 5 min to denature protein, add another 150 µl of TBS and then

place the samples on ice. Put 150 µl of 1× TBS into each of the dot blot wells starting with the second column of rows. Pipette the samples into the first well of each row (1 well per sample). Serially dilute each sample by removing 150 µl from the first well and placing it in the second well (in the same row). Repeat this process across the rows until you come to the last well in the row. Remove 150 µl from the last well of each row and discard it. Pull the samples through the nitrocellulose filter with gentle vacuum until the membrane is dry. *Be sure that the sample does not leak between wells and contaminate the whole membrane.*

5. Dismantle the dot blot unit and allow the nitrocellulose to air dry. Soak the membrane in a 10% acetic acid/25% isopropanol solution for 10 min and then rinse twice with water.

6. Incubate the nitrocellulose with an Ab raised against the stress protein of interest, as described in the western blotting procedure. The bound antibody is detected as described in quantitative western blotting technique (Method 1). Figure 5 shows an immunoblot (inset) and data obtained from western blot procedures. Note, in this example, hsp60 (cpn60) abundance in samples exposed to copper is expressed relative to controls.

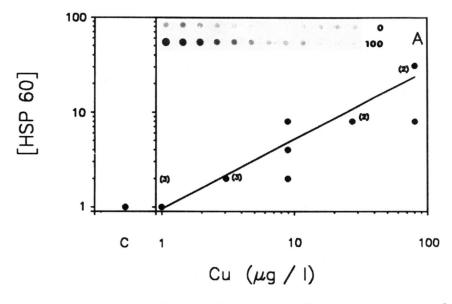

Figure 5 Dot blots. Scatter diagram of the relative hsp60 concentrations in mantle tissue vs. copper concentrations to which *Mytilus* were exposed for seven days. There were three replicate organisms per treatment. Numbers in parentheses indicate the number of data points of the same value represented by a single solid circle. Relative hsp60 for controls is on left. Inset: dot blots of serial dilutions of tissue samples from mantle tissues of a control (0) mussel and a mussel exposed to 100 µg/l copper. (Reprinted with permission from Sanders, B.M., et al. 1991. *Mar. Environ. Res.* 31:81–97.)

7. If possible, a known amount of pure stress protein should also be analyzed to determine the detection limit of the dot blot assay for that stress protein. To express the concentration of stress proteins, use the same procedures described under the quantitative western blotting technique (Method 1).

ELISA

Another standard technique for determining the concentration of specific proteins is the enzyme-linked immunosorbent assay (ELISA). An ELISA has not yet been developed for stress proteins other than for those found in mammals[14] and the development of such an assay is not a trivial task. A number of complications arise with aquatic species including nonspecific binding of the antibody to other proteins and lack of consistent adherence of samples to the plastic wells of the ELISA plate. Interfering substances such as mucus and glycoproteins can also cause problems. Therefore, at this time, the western analysis is the recommended technique to examine and quantify stress proteins in aquatic species.

Results and discussion

Characterizing the cellular stress response by metabolic labeling with radiolabeled amino acids

Initial studies to characterize the stress response often involve examining changes in the pattern of protein synthesis upon exposure to a stressor. This entails monitoring the net incorporation of radiolabeled amino acids into specific proteins during the incubation period.

One- or two-dimensional electrophoresis is followed by autoradiography to visualize the proteins being synthesized and the level of their induction during the period of incubation. This technique does not require antibodies and provides information for all proteins within the molecular weight range of the gel system used. Since the basal levels of stress proteins present in the cells are not labeled, they cannot be measured by this technique. Moreover, the technique cannot give quantitative information on protein synthesis unless the size of the free amino acid pool has been determined. Figure 6 shows an autoradiograph of proteins metabolically labeled in fathead minnow cells heat-shock for 30 min over a range of temperatures. Note the increase in synthesis of stress70 following heat-shock treatments of 32 to 40°C and the complete translational arrest observed at 44°C.

Immunological techniques

The most specific method for detecting stress proteins involve immunological techniques that use antibodies (specific for the stress protein of interest) as probes. In contrast to metabolic labeling, which detects only the stress proteins whose synthesis is induced during incubation, immunological techniques detect the total amount of the stress protein present. Several types of immunological techniques are commonly employed to measure protein con-

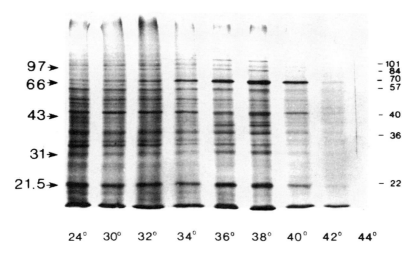

Figure 6 Autoradiograph. Autoradiograph of fathead minnow cultured cell homogenate acclimated at 24°C and subjected to a 30-min heat-shock treatment over a range of temperatures. Cells were exposed to the indicated temperatures, returned to 24°C, and metabolically labeled with [³⁵S] methionine/cysteine for 2 h. For each sample, 25 μg of total protein was separated by SDS-PAGE, autoradiographed, and exposed to film for 24 h. Numbers at the bottom indicate the temperature of the heat-shock treatment in degrees centigrade. Numbers on the left indicate molecular weight markers in kDa. Numbers on the right indicate the approximate molecular weights of heat-inducible proteins in kDa. (Reprinted from Nguyen, J.T. 1994. M.S. thesis. California State University, Long Beach.)

centrations including radioimmunoassay, ELISA, dot blot analysis and western blotting.[9] Of these, only western blotting analysis allows the determination of the molecular weights of the proteins which react with the antibody, an important check against nonspecific binding and sample breakdown. Figure 7 shows a western blot of the gill tissue of mussels exposed to a range of copper concentrations during heat shock. Gill tissue was probed with a polyclonal antibody against human stress70. Note that with increasing concentrations of copper there is an increase in abundance of stress70 and an increase in the number of isoforms of the protein that can be resolved on a one-dimensional gel. This induced synthesis of isoforms of slightly different sizes for stress70 appears to be a common feature of this large multigene family. The number of isoforms and their size class is dependent upon the species and tend to be more variable in invertebrates than in vertebrates.

Commercially available stress protein antibodies

In the last few years many antibodies (Abs) to stress proteins have become commercially available. Two companies, StressGen and Affinity Bioreagents, specialize in providing stress protein reagents, antibodies, antigen(s) and probes to researchers. Most of the Abs commercially available are monoclonal antibodies (mAbs). The cross-reactivity of these mAbs for diverse aquatic species is quite variable and is presented in Table 1. Several mAbs

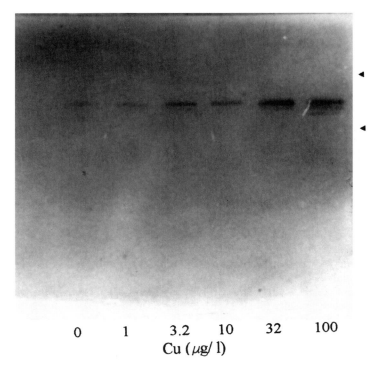

0   1   3.2   10   32   100

Cu (μg/ l)

Figure 7   Western blot. Western blot probed with anti-stress70 antibodies of *Mytilus* gill from mussels continuously exposed to a range of copper concentrations for 7 days prior to sampling. Numbers at the bottom of the blot indicate nominal copper concentrations in μg/l copper. Sixty μg of total protein were loaded on the gel for each sample. Arrows denote prestain molecular weight standards of 84 (top arrow) and 48 (bottom arrow) kDa. (Reprinted with permission from Sanders, B.M. et al. 1994. *Toxicol. Appl. Pharmacol.* 125:206–213. Copyright 1994 Elsevier Science Ltd, The Boulevard, Kidington DX5 1GB, UK.)

have proven to be particularly broadly reactive. The mAb clone 3a3 (Affinity Bioreagents) recognizes a region of the peptide involved in stress-induced nucleolar localization and cross reacts with hsp70, hsc70, hsp72, and p75 in humans.[15-18] The mAb 5a5 (raised against the ATP binding region of human stress70) and clone 7.10 also appear to be broadly cross-reactive.

The mAb N27 (StressGen) raised against both the cytoplasmic cognate and inducible forms of hsp70 in HeLa cells, reacts with these same proteins in human tissue.[18-20] While N27 mAb cross-reacts with stress70 for most vertebrate species, it reacts with fewer invertebrates examined in our lab (Table 1). The mAb C92F3A-5 (StressGen) is specific for the "inducible" form of mammalian stress70 and does not cross-react with any of the non-mammalian tissues we have tested.[18,21-22]

The mAb BRM-22 (Sigma), raised against the cognate form of stress70 from bovine brain, cross-reacts with both the constitutive and inducible forms (70, 72 kDa) of that protein in mammals. It is interesting that its cross-reactivity is notably different in other species where it appears to have a

Table 1 Molecular Weight (kDa) of Heat Inducible Proteins from Tissues of Various Species Cross-Reacting with Five Monoclonal Antibodies Raised Against the Mammalian Stress70 Protein

	Monoclonal antibody			
Organisms	3a3	7.10	BRM-22	N27
Alga				
Cyanidium caldarium	70,72		—	60,70
Crustacea				
Platoichestia platensis (amphipod)	70	nt	—	—
Carcinus maenas (green crab)	70,72	nt	70,72	72
Palaemon adspersus (shrimp)	70	nt	—	—
Astacus astacus (crayfish)	68,70	nt	70,72	—
Crangon crangon (shrimp)	70	nt	70	—
Homarus americanus (lobster)	72	nt	nt	nt
Echinodermata				
Strongylocentrotus purpuratus (urchin)	72	70, 72	78	—
Fish				
Pimephales promelas (fathead minnow)	70	nt	70,72,78	70
Cyprinodon nevadensis amargosae (pupfish)	70	nt	70	70,72
Cyprinodon nevadensis nevadensis (pupfish)	70	nt	70	70
Oncorhynchus mykiss (rainbow trout)	nt	nt	nt	70
Mollusca				
Mytilus edulis (blue mussel)	72	—	78	72
Mytilus californianus (mussel)	72	—	78	72
Crassostrea virginica (oyster)	—	70, 72	nt	—
Macoma nasuta (clam)	72	nt	nt	72

Note: Cross-reactivity was determined by Western blot analysis (see section on western blotting for details on each antibody); (–) = no reaction; (nt) = not tested. (Modified from Reference 23.)

high affinity for a 78-kDa protein that may be homologous to the mammalian grp78 stress protein.[23] Another mAb, clone LK-2 (StressGen), cross-reacts with cpn60 in many vertebrate and invertebrates species. While this monoclonal Ab was raised against human cpn60, it appears to be broadly cross-reactive with mammalian, non-mammalian and bacterial cpn60 proteins.

Characterization of antibody cross-reactivity

The highly conserved nature of the major stress protein families, stress70 and cpn60, can allow for broad cross-reactivity of the antibodies in species from diverse phyla.[23] Other stress proteins, such as the low molecular weight heat-shock proteins, are not as highly conserved and cross reactivity tends to be more species specific. The extent to which a particular antibody might cross react across species is a function of the extent to which the epitope to which it interacts is conserved. In general, polyclonal antibodies often provide the broadest reactivity (Table 2). However, some monoclonals can also be quite broad in reactivity (Table 1). Some antibodies cross react to only

Table 2   Cross-Reactivity of Heat Inducible Proteins from Tissues of
Various Species with Two Polyclonal Antibodies Raised Against
Chaperonin and Stress70

Organisms	Polyclonal antibody	
	cpn 60	Stress70
Alga		
Cyanidium caldarium	+	+
Annelida		
Capitella (polychaete)	−	−
Neanthes (polychaete)	−	−
Crustacea		
Homarus americanus (lobster)	+	+
Carcinus maenas (green crab)	+[c]	+[b]
Astacus astacus (crayfish)	+[d]	+[b]
Mysidopsis (shrimp)	+	+
Palaemon adspersus (shrimp)	+[e]	+
Crangon crangon (shrimp)	+[e]	+
Platorchestia platensis (amphipod)	+	+
Orchestoidea californiana (amphipod)	+	−
Corophium acherusicum (amphipod)	−	−
Echinodermata		
Arbacia punctulata (Atlantic sea urchin)	+	+
Strongylocentrotus purpuratus (Pacific sea urchin)	+	+
Fish		
Genyonemus lineatus (white croaker)	+	+
Micropogonias undulatus (Atlantic croaker)	+	+
Pleuronectes vetulus (English sole)	+	+
Leiostomus xanthurus (spot)	+	+
Pleuronectes americanus (winter flounder)	+	+
Platichthys stellatus (starry flounder)	+	+
Pleuronichthys verticalis (hornyhead turbot)	+	+
Oncorhynchus mykiss (rainbow trout)	+	nt
Pimephales promelas (fathead minnow)	+	+
Cyprinodon nevadensis amargosae (pupfish)	+	+
Cyprinodon nevadensis nevadensis (pupfish)	+	+
Mollusca		
Collisella scabra (limpet)	+	+
Collisella pelta (limpet)	+	+
Collisella digitalis (limpet)	nt	+[b]
Collisella limatula (limpet)	nt	+
Mytilus edulis (blue mussel)	+	+[b]
Mytilus californianus (mussel)	+	+
Macoma nasuta (clam)	+	+
Crassostrea virginica (oyster)	+	+
Nematoda		
Anisakis	+[a]	+

Note: Cross-reactivity was determined by western blot analysis (see section on
Western blotting for details on each antibody) and indicated by (−) = no cross-
reactivity; (+) = detected either a 60-kDa protein (chaperonin) or a protein in
the low 70-kDa range (stress70); (nt) = not tested. Superscripts indicate that
a doublet of (a) 58,60; (b) 70, 72; (c) 61,62; (d) 57,58; (e) 58,62 kDa was detected.
(Modified from Reference 23.)

closely related species, while others can be specific for only one isoform of a large multigene family. For these reasons, cross-reactivity must be thoroughly characterized for each new species to be examined in a heterologous immunoassay.

The most rigorous screening entails using metabolically labeled heat-shocked and control tissues. These samples are subjected to one-dimensional (and sometimes two-dimensional) SDS-PAGE, transblotted onto nitrocellulose and probed with the antibody. After this western blotting procedure is completed, the blot is dried and subjected to film to obtain an autoradiograph. The autoradiograph and blot can then be superimposed and the isoforms that are recognized by the antibody can be compared with the heat-inducible isoforms. If western blotting reveals the same amount stress protein in the heat-shocked sample and in the control sample, yet autoradiography clearly demonstrates induction of the stress response, it is likely that the antibody recognizes only constitutive, and not the heat-inducible, isoform(s) of the protein. This scenario is not uncommon for stress70. Further, if the protein recognized by the Ab is not in the 68 to 72 kDa range it is likely that it is one of the noncytosolic isoforms.

A common mistake made when developing a heterologous assay to examine stress70 for a new species is the assumption that the antibody used as a probe recognizes all of the major isoforms of this multigene family. For example, we carried out a rigorous antibody screening of the reactivity of monoclonal antibody BRM22 (which was raised against bovine brain stress70) for sea urchin stress70. First we carried out a heat-shock experiment in which we subjected embryos to a 10°C increase in temperature for 1 h. These and control embryos were then metabolically labeled for 4 h with ^{35}S-methionine/cysteine. The samples were then run on a one-dimensional SDS-PAGE gel, transblotted and probed with the BRM22 antibody (Figure 8). The blot was then dried and autoradiographed. Note that the western blot revealed that the control and HS samples appeared to have approximately equivalent amounts of the reactive stress70. When the blot and autoradiograph were superimposed, it was clear that the protein which reacted with the antibody was larger than the heat-inducible isoform. The autoradiograph also demonstrated a clear induction of the stress response. These results immediately cast doubt on the use of this antibody as a probe for this species of sea urchin. To further clarify which isoform of stress70 the antibody recognized, we ran the same samples on two-dimensional gels and carried out western blot analyses and autoradiography (Figure 2). As with the previous results, the autoradiograph of the 2D gel demonstrated that the stress response was clearly induced by the heat-shock treatment and that a number of different stress70 isoforms were synthesized in the heat-shocked embryos. However, western blotting showed only one isoform recognized by the antibody. This isoform appeared not to be heat-inducible. Further, when the blot and the autoradiograph were superimposed, the protein which reacted with the antibody was not radiolabeled. These data and the estimated 78-kDa size of the protein lead us to conclude that this commercially available antibody cross reacted with grp78, a member of the stress70 family found in the endoplasmic reticulum which is not heat inducible. From this example

BRM22

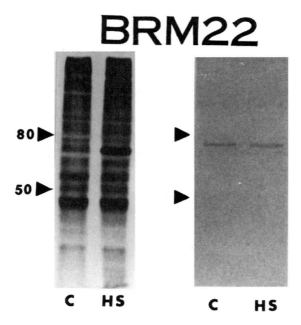

Figure 8 Autoradiograph of one-dimensional western blots. One-dimensional western blot analysis and autoradiograph of the western blot of proteins from *S. purpuratus* embryos maintained at 17°C (C) and exposed to a heat-shock of 25°C/30 min (HS). The proteins were separated by 1D SDS-PAGE, subjected to western blot analysis and the western blot was exposed to Kodak X-AR film for autoradiography. The western blot was probed with BRM22 monoclonal antibody. Autoradiograph is on the left and western blot is on the right. Arrows on the left indicate molecular weight markers in kDa.

it is easy to see how improper antibody screening could result in mistakingly using this antibody to examine the induction of the major cytosolic isoforms of stress70.

This example illustrates how, through a combination of metabolic labeling, electrophoresis and immunological techniques, it is possible to thoroughly characterize the stress response. Together these techniques are powerful tools for addressing specific questions of interest involving the stress protein response in aquatic organisms.

Acknowledgments

The authors wish to thank Chloe Taylor and Anne Slaughter for their assistance in preparing this manuscript. This research was supported by a Grant from the Air Force (F49620-94-1-0364).

References

1. Nover, L. 1991. *The Heat Shock Response.* CRC Press: Boca Raton, FL.
2. Morimoto, R.I., Tissieres, A., Georgopoulos, C. 1994. Progress and perspectives on the biology of heat shock proteins and molecular chaperones. In *The Biology of Heat Shock Proteins and Molecular Chaperones.* Eds. Morimoto, R.I., Tissieres, A., Georgopoulos, C. Cold Spring Harbor Laboratory Press: Cold Spring Harbor, NY. 31 pp.
3. Gething, M.J., Sambrook, J. 1992. Protein folding in the cell. *Nature* 355:33–45.
4. Jaenicke, R., Creighton, T.E. 1993. Junior chaperones. *Curr. Biol.* 3:234–235.
5. Hightower, L.E. 1991. Heat shock, stress proteins, chaperones, and proteotoxicity (Meeting review). *Cell* 66:1–20.
6. Sanders, B.M. 1993. Stress proteins in aquatic organisms: An environmental perspective. *Crit. Rev. Toxicol.* 23:49–75.
7. Cornell-Bell, A.H., Munro, D.R. 1982. Isolation and separation of cells from the digestive gland epithelium of *Busycon canaliculatum. J. Exp. Zool.* 219:293–300.
8. Dietz, T.J., Somero, G.N. 1993. Species- and tissue-specific synthesis patterns for heat-shock proteins hsp70 and hsp90 in several marine teleost fishes. *Physiol. Zool.* 66:863–880.
9. *Current Protocols in Molecular Biology, Section 10.* 1995. John Wiley & Sons, NY.
10. Blattler, D.P., Garner, F., Van Slyke, K., Bradley, A. 1972. Quantitative electrophoresis in polyacrylamide gels of 2–40%. *J. Chromatogr.* 64:147–155.
11. Laemmli, U.K. 1970. Cleavage of structural proteins during the assembly of the head of bacteriophage T4. *Nature* 22:680–685.
12. Sanders, B.M., Martin, L.S., Nelson, W.G., Phelps, D.K., Welch, W. 1991. Relationships between accumulation of a 60-kDa stress protein and scope-for-growth in *Mytilus edulis* exposed to a range of copper concentrations. *Mar. Environ. Res.* 31:81–97.
13. Hawkes, R., Niday, E., Gordon, J. 1982. A dot-immunoblot assay for monoclonal and other antibodies. *Anal. Biochem.* 119:142–147.
14. Anderson, R.L., Wang, C.Y., Vankersen, I., Lee, K.J., Welch, W.J., Lavagnini, P., Hahn, G.M. 1993. An immunoassay for heat shock protein 73/72 — use of the assay to correlate hsp73/72 levels in mammalian cells with heat response. *Int. J. Hyperthermia* 9(4):539–552.
15. Amici, C., Sistonen, L., Santoro, M.G., Morimoto, R.I. 1992. Antiproliferative prostaglandins activate heat shock transcription factor. *Proc. Natl. Acad. Sci. USA* 89:6227–6231.
16. Lewis, V.A., Hynes, G.M., Zheng, D., Saibil, H., Wilison, K. 1992. T-complex polypeptide-1 is a subunit of a heteromeric particle in the eukaryotic cytosol. *Nature* 358:249–252.
17. Milarski, K.L., Welch, W.J., Morimoto, R.I. 1989. Cell-cycle dependent association of HSP70 with specific cellular proteins. *J. Cell Biol.* 108:413–423.
18. Nadler, S.G., Tepper, M.A., Schacter, B., Mazzucco, C.E., 1992. Interaction of the immunosuppressant deoxyspergualin with a member of the hsp70 family of heat shock proteins. *Science* 258:484–485.
19. Minota, S., Cameron, B., Welch, W.J., Winfield, J.B., 1988. Autoantibodies to the constitutive 73-kD member of the hsp70 family of heat shock proteins in systemic lupus erythematosus. *J. Exp. Med.* 10:1475–1480.
20. Vass, K., Welch, W.J., Nowak, T.S., 1988. Localization of 70-kDa stress protein induction in gerbil brain after ischemia. *Acta Neuropathol.* 77:128–135.

21. Welch, W.J., Suhan, J.P., 1986. Cellular and biochemical events in mammalian cells during and after recovery from physiological stress. *J. Cell Biol.* 103:2035–2053.

22. Welch, W.J., Mizzen, L.A. 1988. Characterization of the thermotolerant cell. II. Effects on the intracellular distribution of Heat Shock Protein 70, intermediate filaments and small ribonucleoprotein complexes. *J. Cell Biol.* 106:1117–1130.

23. Sanders, B.M., Martin, L.S., Nakagawa, P.A., Hunter, D.A., Miller, S., Ullrich, S.J. 1994. Specific cross-reactivity of antibodies raised against two major stress proteins, stress 70 and chaperonin 60, in diverse species. *Environ. Toxicol. Chem.* 13(8):1241–1249.

24. Nguyen, J.T. 1994. The Role of Stress Proteins in Conferring Heat-Resistance in a Fathead Minnow, *Pimephales Promelas*, Cell Line. M.S. thesis. Calif. State Univ., Long Beach. 104 pp.

25. Sanders, B.M., Martin, L.S., Howe, S.R., Nelson, W.G., Hegre, E.S., Phelps, D.K. 1994. Tissue-specific differences in accumulation of stress proteins in *Mytilus edulis* exposed to a range of copper concentrations. *Toxicol. Appl. Pharmacol.* 125:206–213.

chapter twenty

Dynamic culture of fish hepatic tissue slices to assess phase I and phase II biotransformation

Andrew S. Kane and Sanjeev Thohan

Introduction

The use of fish as sentinels to detect effects of, or exposure to, environmental contamination has been well established over the last 20 years. In this time, there have been notable advances in techniques and the application of bio-markers in environmental and aquatic toxicology. These biomarkers include a variety of *in vivo* (changes in growth, reproduction, behavior, histopathology, and death) and *in vitro* (alteration of xenobiotic metabolizing enzymes, changes in stress protein expression, etc.) endpoints. *In vitro* methodologies have been favored in the recent past because they tend to be less costly and offer enhanced sensitivity relative to many *in vivo* models. *In vitro* methodologies also offer more reproducible results due to the ability to control independent variables in the laboratory and reduce the numbers of experimental animals. These attributes are needed in order to decrease inter-individual variability and demonstrate subtle yet significant differences between control and experimental groups.

Induction of xenobiotic metabolizing enzyme (XME) activity in fish has been used as an indicator of exposure to petroleum hydrocarbons since the mid-1970s.[1] Since then, molecular techniques have enabled the identification of over 221 plant and animal cytochrome P-450-related genes.[3] Of these, 71 mammalian[2,3] and 5 piscine[4,5] constitutive cytochrome P450 isozymes have been discerned, with attempts being made to examine specific induction due to environmental exposure.[6] Induction of phase I, particularly cytochrome P4501A (CYP1A), and phase II XMEs, has been widely studied and shown to be an indicator of exposure to polyaromatic hydrocarbons and other environmental contaminants in fish.[7-15]

XMEs have classically been studied using a variety of *in vitro* methodologies. In fish, as in mammals, liver tissue tends to exhibit greater constitutive XME activities as compared to other tissues.[16-19] Cell cultures, crude tissue homogenates, and subcellular components (i.e., microsomes and cytosol S-9 fractions) have all been used to evaluate XMEs. Microsomes, prepared by differential centrifugation,[20] represent a concentrated form of membrane-bound enzyme proteins (cytochrome P-450s and glucuronosyl-transferases) and offers good sensitivity when metabolic activity is relatively low, as is often seen in fish. However, microsomal preparations, when used for measuring XME activity, represent a non-cellular, optimized (with regard to cofactors necessary for metabolite formation) environment which is not necessarily representative of metabolism as seen at the tissue level or *in vivo*.

Use of freshly isolated hepatocytes in short-term (primary) culture offers a more realistic representation of oxidative (phase I) metabolism as well as conjugation (phase II) based on intact cellular machinery.[21-24] Hepatocyte harvesting techniques offer good cell yields with high viability (>90%). Cells may either remain in suspension or be plated. Cell isolation techniques utilize collagenase-based digestion, and cells are maintained for short periods of time (usually < 48 h). However, methodologies for longer-term (i.e., up to 70 d) primary cultures of fish hepatocytes have recently been reported.[58] Harvested cells may be suspended in culture, or be plated to confluency, where cell-to-cell contact may be established.[25-27] Immortalized (transformed) cell lines offer reproducibility and convenience, but tend to have altered (often diminished) metabolic activity and may not accurately represent *in vivo* systems.

Use of tissue slices to examine metabolism was proposed over 60 years ago.[28] Since then, there have been vast strides in tissue slice technology, both in laboratory hardware and biochemical applications.[29-34] Hepatic tissue slice technology allows investigation of metabolic activity while maintaining cells intact within their three-dimentional tissue architecture. Tissue slices may be produced rapidly for simultaneous, integrated phase I and phase II metabolism measurement, and remnant tissue may be used for additional or comparative subcellular fraction studies.

Historically, early tissue slice methodology suffered from technical difficulties. Slices were produced manually or with the aid of keratotome-like instruments. These slices were thick (>500 µm) and varied in uniformity. The relatively high metabolism of liver tissue coupled with inadequate diffusion of oxygen and nutrients through the thick slices led to rapid depletion of high energy intermediates and tissue necrosis.[29,35-37] Krumdieck et al.[38] introduced a tissue slicer that was capable of rapidly generating thin (<250 µm) reproducible slices with minimal trauma to the tissue. Uniformly thin slices allow greater exchange of gasses and nutrients, resulting in increased metabolic and physiologic viability. To further the utility of this technology a dynamic culture system was developed by Smith et al.[37] This enhancement allows for both sides of the tissue slice to have exposure to the incubation medium and a gas environment by supporting the slice on a screen within a rotating incubation vial. Other submersion and suspension incubation systems have also been used with success.[34,39]

This chapter will outline the use of hepatic tissue slices *in vitro* to discern phase I and phase II XME metabolism using the prototypic assay substrates 7-ethoxycoumarin (7-EC) and 7-hydroxycoumarin (7-HC). The assay substrate 7-EC measures direct phase I metabolism via ethoxycoumarin O-deethylase (ECOD). Several cytochrome P-450 isozymes in fish may play a role in the metabolism of 7-EC,[40] although CYP1A1 is possibly the predominant isoform. The phase I metabolite of 7-EC, 7-HC, when used as a parent assay substrate, measures phase II metabolism (glucuronidation and sulfation) via glucuronosyltransferase- and sulfotransferase-mediated pathways, respectively. Since intact tissue slices endogenously produce cofactors needed for metabolism, coupled phase I and phase II metabolism can be measured in slices incubated in physiological tissue culture media.

Materials

Animals. Fish should be obtained from known sources with data on any previous exposures, water quality and general history.

Chemicals and buffers. Krebs Henseleit buffer (K3753) (supplemented with 0.37 g $CaCl_2 \cdot 2H_2O$ and 2.1 g $NaHCO_3$ (as per package insert) and 0.025 M HEPES), 7-EC (E3179), 7-HC (U7626), D-saccharic acid 1,4 lactone (S0375), aryl sulfatase (S9626), β-glucuronidase (G-4882), and ATP (kit 366), LDH (228-50) and ALT (DG159-K) kits may be purchased from Sigma Chemical Co., St. Louis, MO. ^3H leucine may be obtained from New England Nuclear. All other reagents should be of the highest grade commercially available: 0.2 M sodium acetate buffer (adjusted to pH 5.1 with acetic acid); 0.2 M glycine:NaOH (1:1) buffer, neutral buffered formalin (100 ml 37% formaldehyde, 900 ml deionized H_2O, 4.0 g NaH_2PO_4, and 5.5 g Na_2HPO_4), ether:isoamyl alcohol (1:0.014). Stock for the sulfatase preparation contains 20 mM saccharic acid 1,4 lactone in 0.2 M sodium acetate buffer with 100 units of aryl sulfatase/ml.

7-EC stock (10 mM) should be made up in ethanol (may be aliquoted and frozen); use 10 μl/ml incubation medium for final concentration of 100 μM. 7-HC stock (10 mM should be made up in glycine:NaOH buffer); make fresh; use 10 μl/ml incubation medium for final concentration of 100 μM.

Apparatus and hardware. The Stadie-Riggs microtome (6727-C10) is available through Thomas Scientific Co., Philadelphia, PA. The Krumdieck precision slicer is available from Alabama Research and Development, Munford, Alabama. The Brendel/Vitron precision slicer is available from Vitron, Inc., Tucson, AZ.

Coring tools are available from Alabama Research and Development and Vitron, Inc., or may be manufactured from sharpened precision-I.D. stainless tubing. Wire mesh inserts are available from Alabama Research and Development and Vitron, Inc., or made from 80 mesh 316 stainless steel. Note that commercially obtained inserts will rotate at different rates relative to the stationary insert described herein.

Roller culture apparatus may be devised by adapting a commercial hot dog cooker with the heating supply disconnected. Depending on the model, modification to the stepper motor may be necessary to optimize vial RPMs on the roller; in our laboratory 6 rpm is routinely used. Similar equipment is commercially available from Alabama Research and Development or Vitron, Inc. The apparatus may be placed inside of a temperature-controlled incubator (or cold room) to maintain appropriate temperatures throughout preincubation and incubation periods.

Procedures

Fish acquisition and acclimation

The use of tissue slices is ideally suited for piscine laboratory studies, since animals may be acclimated under similar group conditions (with regard to variables such as diet, water quality, and exogenous stressors) and sacrificed immediately prior to tissue harvest. The laboratories at the Aquatic Pathobiology Center typically acclimate animals for at least four weeks (or longer until animals are stable). The source of animals, as with any experimental protocol, is important with regard to prior animal history, capture techniques and methods of transport and shipping (including use of drugs and anesthetics). It is best for the investigator to take charge of the transport and handling of fish derived from local sources. Minimizing stress to experimental animals is often not a consideration for many suppliers. Suppliers may not necessarily attempt to reduce transport time and exercise appropriate use of clean ice, salt (without sodium yellow prussate, an anticaking agent), oxygen, drugs, insulated containers, etc. For shipment of animals from non-local sources, proper boxing and overnight Federal Express (Priority Next Day) works effectively. Animals can be packaged at the end of the business day, shipped, and arrive at the lab door before 10:30 a.m. the next day. Shipping costs may be an order of magnitude greater than the cost for the animals, but this is a worthwhile investment to receive animals in good condition. During acclimation and holding, stress should be kept to a minimum, and water quality should be monitored and maintained appropriately.[41-45]

When experimentally assessing *in situ* metabolism from feral or wild fish, animals should be captured, transported under optimal conditions to the laboratory, sacrificed and tissues harvested within several hours from the time of capture (metabolism may be altered by most stressors, including transport). Alternately, animals may be field-sacrificed using laboratory techniques, and livers transported back to the laboratory in ice-cold buffer for immediate use. Use of frozen fish livers for generating physiologically viable tissue slices is not recommended.

Sacrifice and organ harvest

After experimental treatments fish should be humanely sacrificed using buffered methane tricanesulfonate (MS-222)[46] to overdose. Use of MS-222 is recommended for sacrifice of fish by the American Medical Veterinary Association (NIH Guidelines for Animal Care and Use). However, earlier studies[47-49] have indicated that unbuffered, prolonged or high tricane exposures may have marginal to minimal effect on enzyme metabolism when used prior to the time of sacrifice. Therefore, investigators may want to use either a low, buffered dose of tricane or an ice bath to sedate animals in combination with cervical transection (with a sharp serrated knife or guillotine).

After the fish is sacrificed, work quickly to obtain morphometric data (whole animal length and weight), remove the liver, and get it into ice cold buffer. Tools used for necropsy should be kept clean. A separate set of tools, including appropriately-sized Metzenbaum scissors and tissue forceps, should be assigned for clean necropsy purposes only. Use bone cutters and heavier scissors for cutting through bone, skin and muscle. Reserve the use of finer instruments for soft visceral tissues. Open the abdomen from the pectoral girdle and cut posteriorly to just anterior of the insertion of the anal fin. Reflect the liver away from the viscera by grasping its anterior connection to the sinus venosus with a fine pair of rat-toothed forceps. While holding this vasculature, cut anterior to the forceps using fine scissors, and gently begin reflecting the liver posteriorly away from the esophagus (cutting mesenteric connections as the liver is pulled back). Take extreme care not to cut or puncture the gall bladder. Remove the gall bladder by securing its connection to the liver with fine, blunt end forceps; cut the connection to the gall bladder "upstream" of the forceps, close to the liver. One alternative is to use a piece of surgical thread to ligate the common bile duct connection to the liver prior to removing the gall bladder; this helps to prevent puncture and leakage of bile. Another alternative would be to drain the bile from the gall bladder using a syringe with a fine needle. Bile contains proteolytic enzymes and cholic acid which may cause tissue damage. Trim away any non-hepatic tissue, determine whole liver weight, and place the liver in ice-cold Krebs Henseleit buffer.

Slice preparation

This section describes the preparation of liver slices using a Stadie-Riggs manual slicer and a Krumdieck motorized precision slicer. The authors point out that there is another mechanized precision slicer manufactured by Vitron, Inc. Comparison between the three commercially-available slicers is illustrated in Table 1. Slices produced by manual and motorized devices provide viable tissue sections for short-term culture methods.

Coring.   Coring is not necessary when generating slices on a Stadie-Riggs slicer. However, making cylindrical cores of liver tissue for use in precision (Krumdieck and Vitron) slicers helps to attain uniform slices. Keep corers in ice cold buffer prior to use. Place liver on sterile filter paper on top of a supportive cutting surface (dental wax works well) and make cores from the thickest portion of the organ (Figure 1). Corer size should be matched with tissue holder size on the slicer apparatus. Keep all liver tissue in cold buffer between processing steps. Remnant tissue may be further processed into microsomes using 0.15 M KCl.

Table 1   Comparison of Commercially Available Instruments to Generate
Liver Tissue Slices for Metabolism and Toxicity Studies

	Stadie-Riggs	Krumdieck	Brendel/Vitron
Slice uniformity	Variable	Consistent	Consistent
Ease of setup	Easy	Moderate	Moderate
Maintenance/mech. adjustments	None	Relatively labor intensive	Relatively labor intensive
Tissue slice operation	Manual	Manual or automatic	Manual
Materials contacting tissue	Methacrylate	Stainless steel	Lucite
List price	$288	$ 8,500 (excl. chiller) $13,500 (incl. chiller)	$4,500 (incl. chiller)

Note: Information is based on product literature available from the manufacturers.
The authors did not do a direct comparison using the Brendel/Vitron slicer.

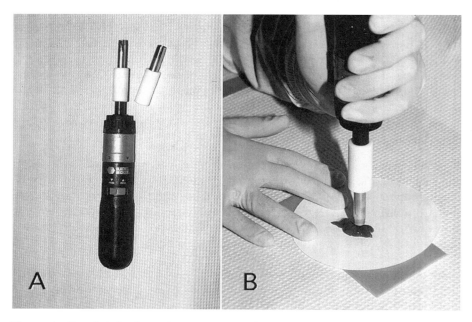

Figure 1   Panel A depicts a battery-operated screwdriver with coring tools of different inside diameters. Coring tools may also be used manually. Panel B shows fish liver (approximately 0.8 g) placed on filter paper and dental wax for coring. At least two cores (and 6–10 good slices/core) may be produced from the central portion of this liver.

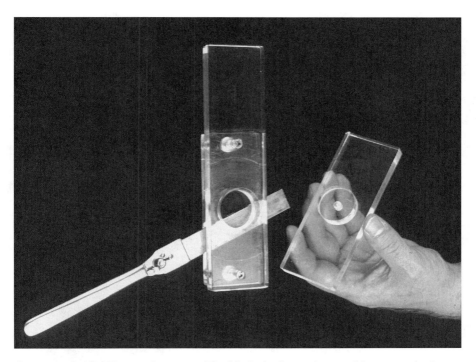

Figure 2   Stadie-Riggs microtome. The blade is shown inserted between the lower base plate and the tissue holder block.

Stadie-Riggs slicer.   The Stadie-Riggs manual slicer consists of a plexi-glass base, a tissue holder block, a weighted top, and a keratotome-like blade (Figure 2). The base has a shallow (~0.5 mm) machined depression. The tissue holder block is attached to the base by two thumbscrews and has a cylindrical hole which aligns with the depression in the base to form a cutting well. Keep the slicer and knife refrigerated until just prior to use. Place the liver in the cutting well and place the weighted top over the specimen. Use a sharp blade which has been wiped with ethanol-saturated gauze; this will clean the blade and remove any oils from new blades. Insert the blade between the base and the tissue holder block and use smooth even back-and-forth cutting motions to pass through the tissue. Remove the cover and carefully lift up the remaining liver to get to the slice. Remove the slice using a sterile swab wetted with buffer. Use a gentle rolling motion with the swab to pick up the slice at its edge and transfer it to a beaker of cold Krebs Henseleit buffer. Replace the liver in the cutting well and repeat the process until a sufficient number of slices have been collected.

Krumdieck slicer.   The Krumdieck precision tissue slicer consists of three main parts: the microtome, the buffer reservoir with slice trap, and the base containing the motor and electronics (Figure 3). The microtome and buffer reservoir may be autoclaved and kept in a cold room until ready to use. Assemble the reservoir onto the base, place a new blade that has been

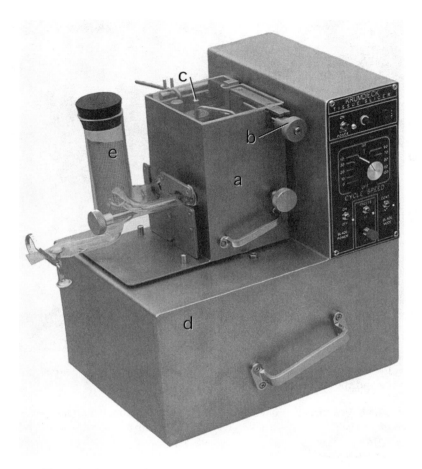

Figure 3 Krumdieck tissue slicer: (a) microtome and buffer reservoir; (b) base plate adjustment; (c) tissue holder piston weight; (d) motorized base; and (e) collection trap to contain slices ejected by a flow of buffer from the microtome assembly. Chiller attachment not shown.

wiped with ethanol in the microtome, and adjust the cutting platform to set slice thickness. Assemble the microtome into the reservoir and add ice-cold buffer. Briefly turn on the motor to insure that the blade is reciprocating and the media is circulating. Load liver tissue cores into the tissue holder of the microtome and gently place weighted piston on top. Gently tamp or twirl the weighted piston over the specimen (without applying any downward pressure) to help remove any trapped air bubbles. Generate a few slices to set thickness as necessary; they will be retained in the collection trap.

One method to gage slice thickness is to place a slice (either intact or cut in half with a sharp blade) onto a platform of known thickness such as a glass coverslip or an automotive gapping tool. Slice thickness may then be measured directly using a dissecting microscope fitted with an ocular micrometer. Adjustment for slice thickness on the motorized slicer is accom-

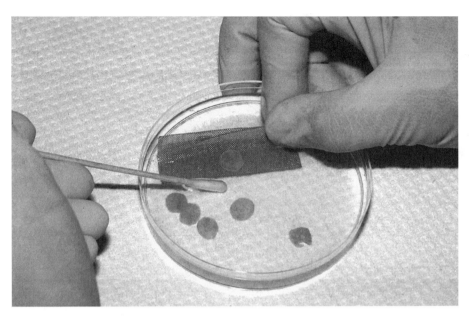

Figure 4 Slices are gently teased under buffer onto stainless steel screens using a cotton swab.

plished by manipulating the adjustable base plate, and varying the weight placed on top of the tissue holder.

Remove cut slices from the collection trap by opening up the tube clamp and allowing slices and buffer to flow into a small beaker. Decant most of the buffer out of the beaker; then pour the slices into a 160-mm petri dish. Individual slices may then be gently floated from the petri dish (under buffer) onto a screen insert using a sterile cotton swab or forceps (Figure 4). Optimal screen size for use in a 20-ml scintillation vial is 20 × 55 mm. After the slice is centrally placed on the screen, blot the underside of the screen to remove excess buffer; this also helps to ensure that the slice adheres to the screen during incubation. Place the loaded screen into a prelabeled incubation vial with incubation medium (without assay substrate) and cover tightly. Use scintillation vial covers which are lined with polypropylene (not aluminum) to minimize risk of contamination. Have incubation vials labeled and clean titanium or stainless steel inserts ready to accept slices once they are generated. The atmosphere inside the vials is sufficiently humid to prevent tissue desiccation in the short time it takes to load an individual batch of vials.

Buffers and alternate isotonic incubation media. Krebs Henseleit buffer is available commercially and can be made up in advance and stored refrigerated for up to a month. Alternatively, one recipe for isotonic Krebs Henseleit contains (per liter) 6.9 g NaCl, 0.35 g KCl, 0.14 g $MgSO_4$, 0.16 g NaH_2PO_4, 2.1 g $NaHCO_3$, 2.0 g glucose, 0.37 g $CaCl_2 \cdot 2H_2O$ and 5.96 g HEPES.

Adjust the pH to 7.4 with NaOH. Filter through a 0.2 µM filter before storage. If any cloudiness develops upon storage, discard and make new buffer. If incubation times are to be longer than several hours, necropsy tools, media, microtome and reservoir, vials, screens, etc. should be sterilized and the use of antimicrobials in the media should considered. An antimicrobial mixture which has been used for long-term (>24 h) dynamic organ culture may include gentamicin (84 µg/ml) plus 10 ml/l Fungi-Bact solution (penicillin G — 10,000 units/ml, Streptomycin sulfate — 10,000 µg/ml and fungizone — 25 µg/ml) (K. Brendel, personal communication). In mammalian slice culture systems, more sophisticated buffer solutions, such as Waymouth's media (without phenol red) supplemented with L-glutamine (3.5 mg/l) and fetal calf serum (10%)[29,31,50] have been utilized. The effect of the different media on metabolism, and use of antimicrobials, with fish tissue slice metabolism has not been thoroughly examined. Use of serum for metabolism studies may be contraindicated since it can bind substrates and metabolites.

Preincubation

Place the incubation vials containing 5 ml Krebs Henseleit buffer and slice onto the roller culture apparatus (Figure 5). Preincubate the slices for 30 min to equilibrate the tissue with the media. Transfer the loaded screens into new vials with fresh media containing assay substrate (7-EC or 7-HC).

Incubation and metabolite determination

Integrated phase I–phase II metabolism using the 7-EC assay substrate. The O-deethylation of 7-EC results in the formation of 7-HC. 7-HC is then subject to conjugation with glucuronic acid or a sulfate moiety via glucuronosyltransferases and sulfotransferases, respectively. The procedure outlined below will quantify all three metabolites of 7-EC. The summation of free 7-HC, and the glucuronide and sulfate conjugates provides a measure of phase I-mediated metabolism. The summation of the glucuronide and sulfates, using the described assay substrates, delineates total phase II capacity of the liver slice. Results are generally normalized to total protein content per slice or per milligram protein. This procedure allows for the quantification of alterations to the patterns of phase I, phase II, and integrated phase I–phase II metabolism.

Begin the batch assay by placing screen-loaded vials horizontally on the dynamic (roller) culture incubator at room temperature (20°C). Incubate for 4 h and terminate assay by placing vials vertically on a counter such that the tissue (adhering to the screen) is no longer in contact with the media. Shorter or longer incubation times may be appropriate depending on preliminary results obtained with different species or experimental protocols.

Determination of unconjugated phase I metabolites of 7-EC (7-HC) and phase II conjugates (glucuronides and sulfates). All metabolites are measured as 7-HC as outlined in Figure 6.

Figure 5 Roller culture apparatus (modified hot dog cooker) inside an environmental chamber.

Pipet 1 ml aliquots of media into two replicate 13×100 mm screw cap extraction tubes uniquely labeled glucuronide or sulfate. Add 5 ml of ether:isoamyl alcohol (1:0.014) and shake horizontally for 15 min on a reciprocating shaker. Transfer 1 ml of the organic phase to a new tube and add 5 ml of a glycine:NaOH buffer (pH > 10). The presence of free 7-HC is determined spectrofluorimetrically at an excitation wavelength of 370 nm and emission wavelength of 450 nm as suggested by Greenlee and Poland.[51] Discard the remaining organic layer from the above extraction. This may be facilitated by freezing the aqueous layer and decanting the organic layer or vacuum-pipetting off the organic layer. The remaining aqueous phase will contain the glucuronide and sulfate conjugates of 7-HC. Acidify the aqueous phase using 1 ml of 0.2 M acetate buffer (pH 5.1). Glucuronide conjugates are hydrolyzed by adding 500 units β-glucuronidase. Sulfate conjugates are hydrolyzed using a sulfatase preparation containing 20 mM D-saccharic acid 1,4 lactone (10 units).

Incubate overnight (approximately 16 h) at 37°C. Add 5 ml of ether:isoamyl alcohol (1:0.0014) and shake for 15 min. Transfer 1 ml of this organic phase which now contains the hydrolyzed conjugates to a new tube and add 5 ml of the glycine:NaOH buffer. The presence of free 7-HC from each of the conjugates is determined spectrofluorometrically. An authentic standard (2.5 to 1000 pmol/ml) is used for quantification.

Phase II metabolism using the 7-HC assay substrate. 7-HC is directly subjected to conjugation by glucuronic acid or sulfate. This bypasses the

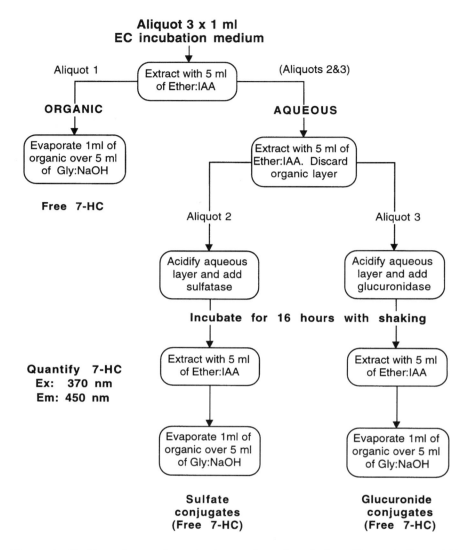

Figure 6 Outline of procedures to detect phase I and phase II metabolites of 7-ethoxycoumarin by measuring free 7-hydroxycoumarin.

phase I conversion of 7-EC to 7-HC and allows for direct quantification of phase II metabolism. The procedure for quantifying the phase II glucuronidation/sulfation capacity of the liver slice is identical to that described above. Metabolism is normalized to total protein content per slice per milligram protein.

Protein content. Slice protein content may be determined by the methods of Lowry[52] or using the kit methodology of BioRad[53] or Pierce.[54]

Tissue viability. A variety of techniques have been applied to organ slices in culture to evaluate tissue viability.[29,30,34,49] These techniques include biochemical endpoints (ATP content, lactate dehydrogenase [LDH] and alanine aminotransferase [ALT] release, intracellular potassium concentration, protein synthesis) and histopathology. ATP content can be determined using a Sigma kit or the luciferin-luciferase bioluminescent assay.[55] Intracellular K⁺ concentration is measured by flame photometry. Protein synthesis may be determined by measuring incorporation of [³H] leucine into acid precipitable protein. Results are typically normalized to mg wet weight, protein content or DNA content. Liver enzymes, LDH and ALT, can be measured using Sigma kits, and results expressed as percent of total ALT or LDH released. Tissue slices may also be evaluated histologically. Extra control and experimental tissues may be preserved in neutral buffered formalin, embedded on edge, stained with hematoxylin and eosin, and evaluated for alterations (Figure 7). Signs of cellular injury include eosinophilia, cell swelling, pyknosis, karyolysis and alterations in glycogen content.[56] However, morphologic determination of tissue viability after short-term culture involves electron microscopic evaluation.

Discussion

Use of liver tissue slices to examine metabolism from fish provides a source of XMEs maintained within the functional, compartmentalized architecture

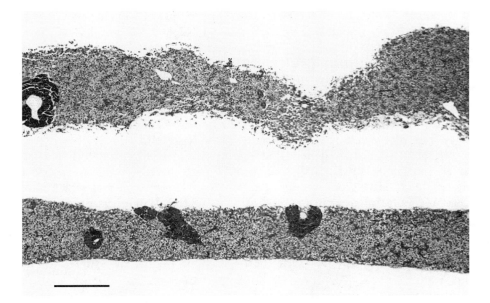

Figure 7 Histological cross-sections through control channel catfish liver slices. Top specimen was cut using a Stadie-Riggs microtome and shows lack of uniform thickness and loss of border definition. Thicker portions of this slice might become relatively anoxic in extended culture. Bottom specimen was cut using a Krumdieck slicer and shows greater uniformity. Hematoxylin and eosin. Bar = 250 μm.

of the liver tissue. Advantages and comparison of using liver slices relative to microsomes and freshly isolated hepatocytes is summarized in Table 2. It should be noted that fish species such as carp and goldfish would be inappropriate hepatic slice donors because of the nondiscrete (diffuse) morphology of their livers. Suitable donor fish should be of sufficient size such that their livers are at least 0.75 g. (Therefore it would be difficult to obtain slices from some juvenile fish or adults of such genera as *Fundulus* or *Cyprinodon*, where the liver size is <0.5 g.)

Figure 7 indicates the obvious advantage of slice uniformity using a mechanized slicer. Production of thin, uniform slices is critical to reproducable metabolism and longer term slice culturing. On the other hand, the maunal Stadie-Riggs slicer does have its merits: in addition to being significantly less expensive than the mechanized slicers (Table 2), users may, with practice, produce relatively uniform slices adequate for most preliminary studies. However, on a protein weight basis, slices produced on a Stadie-Riggs microtome will appear to have lower, less reproducible metabolism due to a higher (and variable) percentage of fragmented cells.

As previously discussed, a variety of model substrates may be used to measure metabolism from liver slices. An operational definition of a model substrate might include stable compounds whose metabolism is well-characterized, sensitive and rapidly/easily analyzed. Preliminary studies in our laboratory have demonstrated varied rates of time-dependent metabolism using two alkoxycoumarin derivatives, 7-methoxycoumarin (MC) and 7-EC. These substrates were used for quantification of integrated phase I–phase II metabolism. Rates of metabolism were approximately linear over a 4 h incubation period with average rates being 0.2 and 0.5 nmol 7-HC produced/h/mg slice protein, for 7-MC and 7-EC respectively (Figure 8). It has been previously demonstrated in the mammalian literature that different CYP 450 isoforms have varying affinities for different substrates. Qualitative results from our study with channel catfish are consistent with previous data collected using rainbow trout microsomes.[57] It has been well documented in mammalian literature that different cytochrome P450 isoforms may have both substrate-specific or overlapping substrate affinities.

Total metabolism may be further subdivided into individual components of an integrated phase I–phase II profile, e.g., the unconjugated phase I metabolite (7-HC), glucuronides, and sulfates. Therefore, factors which may alter metabolic integration can be studied using this dynamic approach (e.g., specific induction or inhibition along an integrated pathway).

Our data indicate that glucuronidation is the predominant pathway in reactions where substrate is limiting as is the case with the metabolism of 7-MC or 7-EC to 7-HC (Figure 8). However, with the use of 7-HC as parent substrate, the phase II profile demonstrated that rates of glucuronidation and sulfation were similar (Figure 9). Results shown in Figure 10 help to underscore the difference in capacity between total phase I and phase II metabolism in channel catfish liver slices; up to 20-fold greater phase II metabolism was observed as compared with phase I metabolism of alkoxycoumarins.

Table 2 Comparison of Microsomes, Hepatocytes and Liver Tissue Slices
for Metabolism Studies

	Microsomes	Hepatocytes	Slices
Maintenance of three-dimensional tissue architecture	No	No	Yes
Compartmentalized drug (substrate) distribution	No	Yes	Yes
Reaction rates	High (optimized)	Intermediate (physiological)	Low (physiological)
Phase I metabolism	Yes	Yes	Yes
Phase II metabolism	Glucuronidation only	Complete	Complete
Integrated phase I-II metabolism	No	Yes	Yes
Requires cofactor addition	Yes	No	No
Use in toxicity assessment	Limited	Yes	Yes
Storage stability	Can freeze at –80°C	Must be used fresh	Must be used fresh
Preparation considerations	Labor intensive; several centri-fugation steps	Moderately labor intensive; collagenase perfusion;	Not very labor intensive; can make micro-somes from remnant tissue

In addition to metabolism mediated by monooxygenase activity and glucuronosyl- and sulfotransferases, additional biotransformations such as amino and mercapturic acid conjugation may be assessed using adapted slice technology. Other organ systems including skin, gills, and intestines may also be assessed for their XME capacity or alteration due to environmental exposure. Combined biotransformation data derived from multiple organ systems, using a variety of substrates with overlapping enzyme specificity may benefit the modeling of *in vitro* to *in vivo* risk assessment. Additional assay substrates which potentially have relatively high specificity for induced XME isoforms in fish requires further investigation. These substrates include (but are certainly not limited to) the fluorescent substrates 7-methoxy-, 7-ethoxy-, 7-propoxy-, 7-butoxy-, 7-pentoxy- and 7-benzyloxyphenoxazones, coumarins and quinolines.[11,56]

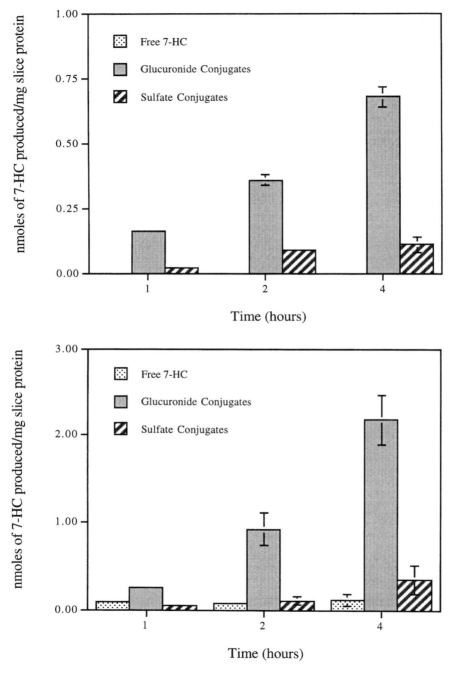

Figure 8 Integrated phase I–phase II metabolism of 7-MC (top) and 7-EC (bottom) by channel catfish liver slices using 100 µ*M* concentrations of each substrate (± S.E., n = 5).

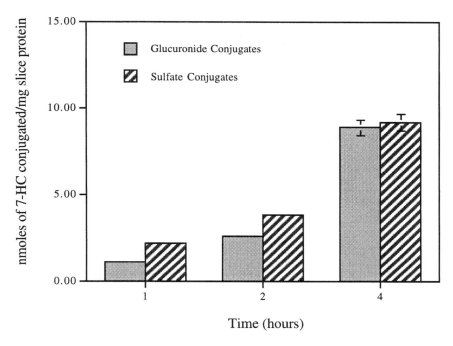

Figure 9   Phase II metabolism of 7-HC by channel catfish liver slices using 100 μ*M* concentration of substrate (± S.E., n = 5).

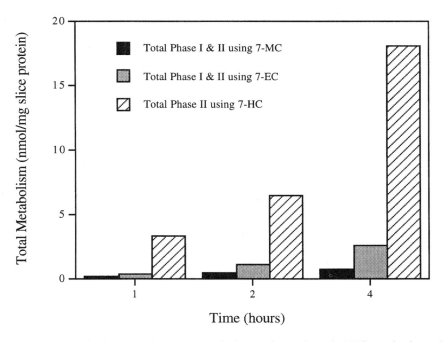

Figure 10   Total phase I–phase II metabolism of 7-MC and 7-EC, and phase II metabolism of 7-HC (integrated data same as presented in Figures 8 and 9).

Tissue slice technology as applied to aquatic models is still in its infancy. Many technical aspects of experimentally measuring baseline activities or XME alteration(s) due to exposure have not been fully addressed. Slice incubation temperature, for example, as is true for microsomal incubation temperature, in the opinion of these authors, should be within the appropriate range for the species and protocol under consideration. Therefore, slices generated from warmwater fishes such as bluegill and channel catfish might be incubated at room temperature (20 to 22°C), whereas cold water fishes such as trout might require that their liver slices be incubated at 8 to 15°C if the data is intended to link *in vitro* to *in vivo* metabolism. Other variables pertaining to the incubation environment may also need to be considered. These include the type of incubator platform used to keep the slices bathed in media (roller vs. gyratory apparatus) and exposure of the tissue by means of alternate devices (i.e., 6 well plates, vertical incubation vials, horizontal incubation vials) and different gaseous environments.[50]

Acknowledgments

We thank Charles Ruegg for his helpful insight and review of this manuscript. We also thank Kim Moore, Vallen Emory, Christina Råbergh, Michael Lipsky, Mary Haasch, Victor Melendez Jr., Myron Weiner, and Renate Reimschuessel for their assistance. This work was supported in part by an Institutional Research Grant from the Maryland Cancer Program/American Cancer Society, IRG-147L, and Special Research Initiative Support from the University of Maryland School of Medicine.

References

1. Payne, J.F. 1976. Field evaluation of benzopyrene hydroxylase induction as a monitor for marine pollution. *Science*. 191: 945–946.
2. Nebert, D.W., Adesnik, M., Coon, M.J., Estabrook, R.W., Gonzalez, F.J., Guengerich, F.P. , Gunsalus, I.C., Johnson, E.F., Kemper, B., Levin, W., Phillips, I.R., Sato, R., and Waterman, M.R. 1989. The P450 gene superfamily: updated listing of all genes and recommended nomenclature for the chromosomal loci. *DNA*. 8(1):1–13.
3. Nelson, D.R., Kamataki, T., Waxman, D.J., Guengerich, F.P., Estabrook, R.W., Feyereisen, R., Gonzalez, F.J., Coon, M.J., Gunsalus, I.C., Gotoh, O., Okuda, K., and Nebert, D.W. 1993. The P450 superfamily — update on new sequences, gene mapping, accession numbers, early trivial names of enzymes and nomenclature. *DNA Cell Biol*. 12:1–51.
4. Williams, D.E. and Buhler, D.R. 1982. Purification of cytochromes P-448 from β-naphthoflavone-treated rainbow trout. *Biochim. Biophys. Acta*. 717:398–404.
5. Miranda, C.L., Wang, J.-L., Henderson, M.C., and Buhler, M.C. 1990. Immunological characterization of constitutive isozymes of cytochrome P-450 from rainbow trout. Evidence for homology with phenobarbital-induced rat P-450s. *Biochim. Biophys. Acta*. 1037:155–160.
6. Haasch, M.L., Prince, R., Wejksnora, J.P., Cooper, K.R., and Lech, J.J. 1993. Induction of hepatic cytochrome P450 (CYP1A1) as an environmental biomonitor. *Environ. Toxicol. Chem*. 12:885–895.

7. Chambers, J.E. and Yarbrough, J.D. 1976. Xenobiotic transformation systems in fishes. *Comp. Biochem. Physiol.* 55C:77–84.

8. Kleinow, K.M., Melancon, M.J., and Lech, J.J. 1987. Biotransformation and induction: implications for toxicity, bioaccumulation and monitoring of environmental xenobiotics in fish. *Environ. Health Perspec.* 71:105–119.

9. Payne, J.F., Fancey, L.L., Rahimtula, A.D., and Porter, E.L. 1987. Review and perspective on the use of mixed-function oxygenase enzymes in biological monitoring. *Comp. Biochem. Physiol.* 86C(2):233–245.

10. Varanasi, U. 1987. *Metabolism of polyaromatic hydrocarbons in the aquatic environment.* CRC Press, Boca Raton, FL.

11. Lubet, R.A., Guengerich, F.P., and Nims, R.W. 1990. The induction of alkoxyresorufin metabolism: a potential indicator of environmental contamination. *Arch. Environ. Contam. Toxicol.* 19:157–163.

12. Goksøyr, A. and Forlin, L. 1992. The cytochrome P-450 system in fish, aquatic toxicology and environmental monitoring. *Aquat. Toxicol.* 22:287–311.

13. Stegeman, J.J., Brouwer, M., Di Giulio, R.T., Förlin, L., Fowler, B.A., Sanders, B.M., and Van Veld, P.A. 1992. Molecular responses to environmental contamination: enzyme and protein systems as indicators of chemical exposure and effect. in: *Biomarkers: biochemical, physiological, and histological markers of anthropogenic stress.* C.H. Ward, B.T. Walton, and T.W. LaPoint, editors. pp. 235–335. CRC Press/Lewis Publishers, Boca Raton, FL.

14. George, S.G. 1994. Enzymology and molecular biology of Phase II Xenobiotic-conjugating enzymes in fish. in: *Aquatic toxicology — molecular, biochemical and cellular perspectives.* D.C. Malins, and G.K. Ostrander, editors. pp. 37–86. CRC Press/Lewis Publishers, Boca Raton, FL.

15. Stegeman, J.J. and Hahn, M.E. 1994. Biochemistry and molecular biology of monooxygenases: current perspectives on forms, functions, and regulation of cytochrome P450 in aquatic species. in: *Aquatic toxicology — molecular, biochemical and cellular perspectives.* D.C. Malins, and G.K. Ostrander, editors. pp. 87–206. CRC Press/Lewis Publishers, Boca Raton, FL.

16. DeWaide, J.H. 1971. *Metabolism of xenobiotics; comparative and kinetic studies as a basis for environmental pharmacology.* 164 pp. Drukkerij, The Netherlands.

17. Kasper, C.B., and Henton, D. 1980. Glucuronidation. in: *Enzymatic basis of detoxification, vol. 2.* W.B. Jakoby, editor. Academic Press, New York.

18. James, M.O. 1986. Xenobiotic conjugation in fish and other aquatic species. in: *Xenobiotic conjugation chemistry,* G.D. Paulson, J. Caldwell, D.H. Hutson and J.J. Menn, editors. pp. 29–47. American Chemical Society, Washington, DC.

19. James, M.O. 1987. Conjugation of organic pollutants in aquatic species. *Environ. Health Perspec.* 71:97–103.

20. Vodicnik, M.J., Elcombe, C.R., and Lech, J.J. 1981. The effect of various types of inducing agents on hepatic microsomal monooxygenase activity in rainbow trout. *Toxicol. Appl. Pharmacol.* 59:364–374.

21. Moon, T.W., Walsh, P.J., and Mommsen, T.P. 1985. Fish Hepatocytes: a model metabolic system. *Can. J. Fish. Aquat. Sci.* 42:1772–1782.

22. Andersson, T., and Koivusaari, U. 1986. Oxidative and conjugative metabolism of xenobiotics in isolated liver cells from thermally acclimated rainbow trout. *Aquat. Toxicol.* 8:85–92.

23. Buhler, D.R. and Williams, D.E. 1988. The role of biotransformation in the toxicity of chemicals. *Aquat. Toxicol.* 11:19–28.

24. Baksi, S.M. and Frazier, J.M. 1990. Isolated fish hepatocytes — model systems for toxicology research. *Aquat. Toxicol.* 16:229–256.

25. Klauning, J.E., Ruch, R.J., and Goldblatt, P.J. 1985. Trout hepatocyte cultures: isolation and primary culture. *In Vitro Cell. Dev. Biol.* 21:221–228.
26. Kocal, T., Quinn, B.A., Smith, I.R., Ferguson, H.W., and Hayes, M.A. 1988. Use of trout serum to prepare primary attached monolayer cultures of hepatocytes from rainbow trout (*Salmo gairdneri*). *In Vitro Cell. Dev. Biol.* 24:304–308.
27. Råbergh, C., Kane, A.S., Reimschuessel, R., and M.M. Lipsky. 1995. Viability and induction of tyrosine aminotransferase in rainbow trout hepatocytes cultured on laminin and polylysine in serum-free media. *Methods Cell Sci.* 17(3):207–215.
28. Warburg, O., Franz, W., and Negelein, E. 1926. Über den stoffwechsel von tumoren im körper (Metabolism of tumors in the body). *Klin. Wochenschr.* 5(19):829–832.
29. Smith, P.F., Krack, G., McKee, R.L., Johnson, D.G., Gandolfi, A.J., Hruby, V.J., Krumdieck, C.L., and Brendel, K. 1986. Maintenance of adult rat liver slices in dynamic organ culture. *In Vitro Cell. Dev. Biol.* 22(12):706–712.
30. Ruegg, C.E., Gandolfi, A.J., Nagle, R.B., and Brendel, K. 1987. Differential pattern of injury to the proximal tubule of renal cortical slices following in vitro exposure to mercuric chloride, potassium dichromate, or anoxic conditions. *Toxicol. Appl. Pharmacol.* 90:261–273.
31. Sipes, I.G., Fisher, R.L., Smith, P.F., Stine, E.R., Gandolfi, A.J., and Brendel, K. 1987. A dynamic liver culture system: a tool for studying chemical biotransformation and toxicity. *Arch. Toxicol., Suppl.* 11:20–33.
32. Azri, S., Gandolfi, A.J., and Brendel, K. 1990. Precision-cut liver slices: an in vitro system for profiling potential hepatotoxicants. *In Vitro Toxicol.* 3(4):309–320.
33. Connors, S., Rankin, D.R., Gandolfi, A.J., Krumdieck, C.L., Koep, L.J., and Brendel, K. 1990. Cocaine hepatotoxicity in cultured liver slices: a species comparison. *Toxicology.* 61:171–183.
34. Dogterom, P. 1993. Development of a simple incubation system for metabolism studies with precision-cut liver slices. *Drug Metab. Dispos.* 21(4):699–704.
35. Campbell, A.K. and Hales, C.N. 1971. Maintenance of viable cells in an organ culture of mature rat liver. *Exp. Cell Res.* 68:33–42.
36. Grisham, J.W., Charlton, R.K., and Kaufman, D.J. 1978. In vitro assay of cytotoxicity with cultured liver: accomplishments and possibilities. *Environ. Health Perspect.* 25:161–171.
37. Smith, P.F., Gandolfi, A.J., Krumdieck, C.L., Putnam, C.W., Zukoski, C.F., Davis, W.M., and Brendel, K. 1985. Dynamic organ culture of precision liver slices for in vitro toxicology. *Life Sci.* 36:1367–1375.
38. Krumdieck, C.L., Dos Santos, J.E., and Ho, K.J. 1980. A new instrument for the rapid preparation of tissue slices. *Anal. Biochem.* 104:118–123.
39. Smith, P.F., Fisher, R.F., Shubat, P.,Gandolfi, A.J., Krumdieck, C.L. and Brendel, K. 1985. *In vitro* cytotoxicity of allyl alcohol and bromobenzenein a novel organ culture system. *Toxicol. Appl. Pharmacol.* 87:509–522.
40. Klotz, A.V., Stegeman, J.J., and Walsh, C. 1983. An aryl hydrocarbon hydroxylating hepatic cytochrome P-450 from the marine fish *Stenomus chrysops. Arch. Biochem. Biophys.* 226:578–592.
41. Spotte, S. 1979. *Fish and invertebrate culture.* Wiley-Interscience, New York.
42. Moe, M.A., Jr. 1992. *The marine aquarium reference.* Green Turtle Publications, Plantation, FL.
43. Munro, A.L.S. and Roberts, R.J. 1989. The aquatic environment. in: *Fish pathology.* R.J. Roberts, editor. pp. 1–12. Baillière Tindall, Philadelphia, PA.

44. U.S. EPA (U.S. Environmental Protection Agency). 1989. Short-term methods for estimating the chronic toxicity of effluents and receiving waters to freshwater organisms. 2nd edition. EPA/600/4-89-001, March 1989.

45. Stoskopf, M.K., editor. 1993. *Fish medicine.* W.B. Saunders Company. Philadelphia, PA.

46. Brown, L.A. 1993. Anesthesia and restraint. in: *Fish medicine.* M. Stoskopf, editor. pp. 79–90. W.B. Saunders Co. Philadelphia, PA.

47. Kleinow, K.M., Haasch, M.L., and Lech, J.J. 1986. The effect of tricane anesthesia upon induction of select P-450 dependent monooxygenase activities in rainbow trout (*Salmo gairdneri*). *Aquat. Toxicol.* 8:231–241.

48. Laitinen, M., Neiminen, M., Pasanen, P., and Hietanen, E. 1981. Tricane (MS-222) induced modification on the metabolism of foreign compounds in the liver and duodenal mucosa of the splake (*Salvelinus fontinalis* × *Salvelinas namaycush*). *Acta Pharmacol. Toxicol.* 49:92–97.

49. Fabacher, D.L. 1982. Hepatic microsomes from freshwater fish — II. Reduction of benzo(a)pyrene metabolites by the fish anesthetics quinaldine sulfate and tricane. *Comp. Biochem. Physiol.* 73c:285–288.

50. Fisher, R.L., Shaughnessy, R.P., Jenkins, P.L., Austin, M.L., and Roth, G.L. 1995. Dynamic organ culture is superior to multiwell plate culturing systems for maintaining liver slice viability. Presented at the Society of Toxicology 1995 Annual Meeting, Baltimore, MD. (Abstract).

51. Greenlee, W.F., and Poland, A. 1977. An improved assay of 7-ethoxycoumarin O-deethylase activity. *J. Pharmacol. Exp. Ther.* 205: 596–605.

52. Lowry, O.H., Rosebrough, N.J., Farr, AL., and Randall, R.J. 1951. Protein measurement with the Folin phenol reagent. *J. Biol. Chem.* 153:265–275.

53. Bradford, M.M. 1976. A rapid and sensitive method for the quantitation of microgram quantities of protein utilizing the principle of protein-dye binding. *Anal. Biochem.* 72:248–254.

54. Smith, P.K., Khron, R.I., Hermanson, G.T., Mallia, A.K., Gartner, F.H., Provenzano, M.D., Fujimoto, E.K., Goeke, N.M., Olson, B.J., and Klenk, D.C. 1985. Measurement of protein using bicinchoninic acid. *Anal. Biochem.* 150:76–85.

55. Kricka, L.J. 1988. Clinical and biochemical applications of luciferases and luciferins. *Anal. Biochem.* 175:14–21.

56. Rubin, E. and Farber, J.L. 1994. Cell injury. in: *Pathology, 2nd edition.* Rubin, E. and Farber, J.L., editors. J.B. Lippincott, Philadelphia, PA.

57. Haasch, M.L., Graf, W.K., Quardokus, E.M., Mayer, R.T., and Lech, J.J. 1994. Use of 7-alkoxyphenoxazones, 7-alkoxycoumarins and 7-alkoxyquinolines as fluorescent substrates for rainbow trout hepatic microsomes after treatment with various inducers. *Biochem. Pharmacol.* 47:893–903.

58. Ostrander, G.K., Blair, J.B., Stark, B.A., Marley, G.M., Bales, W.D., Veltri, R.W., Hinton, D.E., Okihiro, M., Ortego, L.S., and Hawkins, W.E. 1995. Long-term primary culture of epithelial cells from rainbow trout (*Oncorhynchus mykiss*) liver. *In Vitro Cell. Dev. Biol.* 31:367–378.

Section III

Techniques for identification and assessment of contaminants in aquatic ecosystems

chapter twenty-one

Sources and estimations of octanol–water partition coefficients and water solubilities

William M. Meylan and Philip H. Howard

Introduction

The octanol–water partition coefficient is a physical property that describes a chemical's lipophilic or hydrophobic properties. It is the ratio of a chemical's concentration in the octanol phase to its concentration in the aqueous phase of a two-phase system at equilibrium.[1] Since measured values range from <10^{-4} to >10^{+8} (at least 12 orders of magnitude), the logarithm (log P) is commonly used to characterize its value.[2] Log P is a valuable parameter in many quantitative structure-activity relationships (QSAR) developed for the toxicological, pharmaceutical, environmental, biochemical sciences, and aquatic bioaccumulation and toxicity.[1-5] For example, log P is used by the U.S. EPA's Office of Pollution Prevention and Toxics to develop 49 different QSARs for predicting aquatic toxicity and to estimate bioconcentration factors in aquatic organisms.[4,5] Applications in aquatic toxicity QSARs require either experimental or estimated log P values. This article provides information concerning sources of experimental log P values and common methods for estimating them.

Water solubility is another important physical property in aquatic toxicology. For example, water solubility is critical in determining the concentration to be used in an aquatic toxicity test. A great deal of literature is available concerning the estimation of water solubility, especially using regression-derived equations involving log P. Yalkowsky and Banerjee[6] have reviewed most of the recent literature in a book devoted entirely to the subject of aqueous solubility estimation. They concluded that, at present, the most practical means of estimating water solubility involves regression-

1-56670-149-X/96/$0.00+$.50
© 1996 by CRC Press, Inc.

derived correlations using log P.[6] This article provides information concerning sources of experimental water solubility values and common methods for estimating them including an improved method for estimating water solubility from log P.[7]

Materials required

Experimental log P values

Several extensive compilations of experimental log P values are available and are the most convenient source of this data. The following compilations are recommended:

1. Sangster LOGKOW DataBank:[8] Contains data for more than 15,000 compounds. All available values are presented for each compound; for example, more than 20 experimental values are listed for benzene. The source of the data, analytical methods, pH, and temperature of measurements, and recommended values are also listed.
2. MEDCHEM Compilations:[9] Also lists multiple values for the same chemical. The recommended value is "star-listed" with an asterisk to indicate that it is the preferred value.
3. KOWWIN/LOGKOW Programs:[10] Estimation program that contains a database of more than 9,500 recommended values for different compounds.
4. MacLogP™ Program:[11] Estimation program that contains a database of 9,000 carefully-reviewed measured values. The CLOGP™ Program[12] has a similar database.

For ionizable compounds, log P measurements can vary greatly with pH. For example, log P of the herbicide chlorsulfuron varies from 1.09 to –0.69 over a pH range of only 4.5 to 7.1.[13] In general, log P values of a compound are lower (sometimes significantly) when it exists predominantly in the ionized form as compared to existing primarily in the non-ionized form. In order to compare log P values of different, ionizable compounds on a relative basis, it is common to report log P values as "corrected for ionization." When an experimental measurement is made at a pH where a compound is primarily in the non-ionized form, the resulting log P value is already "corrected for ionization." For many compounds, however, log P measurements are made at physiologically important pHs near 7.4, and at pH 7.4, many of these compounds are predominantly ionized. As an example, consider the drug timolol; at pH 7.4, timolol has a measured log P value of 0.11[14] and is primarily ionized as it has a pKa of 9.21. The following formula is used to "correct" measured log P values at pH 7.4 for ionization:

$$\log P_{(corrected)} = \log P_{(at\ pH\ 7.4)} + \log(1 + 10^{pKa-7.4})$$

For timolol, the corrected log P would be 1.93 using the formula. When measured at pH 13 where it is predominantly non-ionized, timolol has an experimental value of 1.98[15] which confirms the "corrected" value. The recommended values from the compilation sources above pertain almost exclusively to log P that are "corrected for ionization."

Estimated log P values

The literature contains many methods for estimating log P. The most common are classified as "fragment constant" methods in which a structure is divided into fragments (atom or larger functional groups) and values of each group are summed together (sometimes with structural correction factors) to yield the log P estimate.[2] Examples of the fragment constant approach include the Hansch and Leo method,[3,9] the Meylan and Howard method,[2] the Rekker method,[16] the Broto method,[17] the Klopman method,[18] the Ghose method,[19] and the Suzuki and Kudo method.[20]

General estimation methods based upon molecular connectivity indices[21] and UNIFAC-derived activity coefficients[22] have been developed, but they lack the accuracy of the better fragment constant methods and the UNIFAC derived methods have limited structural fragments. Recent methods have been proposed that utilize properties of the entire solute molecule (charge densities, molecular surface area, volume, weight, shape, and electrostatic potential) to estimate log P.[23-25] These methods attempt to overcome various inefficiencies of the fragment constant approach (i.e., oversimplification of steric and conformational effects of complex structures, the need for correctional factors, and the inability to estimate log P for uncorrelated or unknown fragments). Although the "entire solute" methods show promise for various structures, they have not been validated sufficiently to prove their merit as a general estimation method.[2] Also, hypervalent sulfur and phosphorus are estimated poorly.[24]

The estimation methods discussed above can be classified as comprehensive methods since they are capable of estimating log P for a wide variety of chemical structures. In contrast, methods have been developed for very specific types of structures. For example, excellent methods are available for polychlorinated biphenyls[26] and acetanilides/benzamides with specific substituents;[27] however, methods for specific structures have little or no utility outside their specificity.

Two prime considerations when using a log P estimation method are its accuracy and its ease of use. Table 1 compares various comprehensive log P estimation methods. Statistically, the Meylan and Howard method[2] is better than the other methods, especially when validation datasets are considered. To be effective, an estimation method must be capable of making accurate predictions for chemicals not included in the training sets. The Meylan and Howard method[2] has been tested on a validation dataset of 7167 chemicals, and it yields excellent statistics. A newer version of the Hansch and Leo method[3] (e.g., UNIX version of CLOGP™) is also very good; however, its statistics were achieved after excluding nearly 10% of their reliable experimental database as having "deviant" (poorly) predictable values.[29] The Rekker method[16] has an excellent correlation for a training set of only 1054 compounds, but little is available in terms of validation.

Table 1 Comparison of Comprehensive Log P Estimation Methods

Method [Ref.]	Methodology	Statistical results[a]
Meylan and Howard[2,10]	125 Fragments 230 Correction factors	Total: $n = 9518$; $r^2 = 0.956$; sd $= 0.376$; me $= 0.280$ Training: $n = 2351$; $r^2 = 0.982$; sd $= 0.216$; me $= 0.161$ Validation: $n = 7167$; $r^2 = 0.949$; sd $= 0.416$; me $= 0.319$
Hansch and Leo[3] (PC-CLOGP3[28])	Fragments + correction factors	Total: $n = 8265$[b]; $r^2 = 0.899$; sd $= 0.607$; me $= 0.402$
Hansch and Leo[3] (CLOGP™UNIX[29])	Fragments + correction factors	Total: $n = 7250$; $r^2 = 0.96$; sd $= 0.3$ (using equation: Log P $= 0.914$ CLOGP $+ 0.184$)[c]
Rekker and de Kort[16]	Fragments + correction factors	Training: $n = 1054$; $r^2 = 0.99$ Validation[d]: $n = 20$; $r^2 = 0.89$; sd $= 0.53$; me $= 0.40$
Niemi et al.[21]	MCI[e]	Training: $n = 2039$; $r^2 = 0.77$ Validation: $n = 2037$; $r^2 = 0.49$
Klopman et al.[18]	98 Fragments and correction factors	Training: $n = 1663$; $r^2 = 0.928$; sd $= 0.3817$
Suzuki and Kudo[20]	424 Fragments	Total: $n = 1686$; me $= 0.35$ Validation: $n = 221$; me $= 0.49$
Ghose et al.[19]	110 Fragments	Training: $n = 830$; $r^2 = 0.93$; sd $= 0.47$ Validation: $n = 125$; $r^2 = 0.87$; sd $= 0.52$
Bodor and Huang[24]	Molecular orbital	Training: $n = 302$; $r^2 = 0.96$; sd $= 0.31$; me $= 0.24$ Validation: $n = 128$; sd $= 0.38$
Broto et al.[17]	110 Fragments	Training: $n = 1868$; me ≈ 0.4

[a] Abbreviations are: n = number of compounds, sd = standard deviation, me = absolute mean error.

[b] Taken from the present method database; the difference between the entire database (9518) and the number used (8265) is primarily due to "missing fragments" in the PC-CLOGP program.

[c] These statistics were determined after removing large systemic deviant compounds and other large deviant structures where the underlying difficulty is conformational (Leo[29]).

[d] Tabulation of 20 drug chemicals from Rekker et al.[30]

[e] Molecular Connectivity Indices + algorithmically-derived variables.

Table 2 Computer Software for Estimating Log P

Software [Ref.]	Estimation method [Ref.]	Comments
LOGKOW©[10] KOWWIN©[10]	Meylan and Howard[2]	MS-DOS and MS-Windows operating systems; database of 9500+ experimental log P values; structure entry via SMILES[a] notations; database of 60,000 SMILES notations indexed by CAS number; MS-Windows version intergrates with other commercial drawing programs and SMILES depiction programs
PC-CLOGP[28]	Hansch and Leo[3]	MS-DOS operating system; structure entry via SMILES[a] notations; database of 19,900 SMILES notations indexed by CAS number
CLOGP™[12]	Hansch and Leo[3,9]	UNIX operating system; older versions available for the VAX operating system; database of experimental log P values; structure entry via SMILES[a] notations
MacLogP®[11]	Hansch and Leo[3,9]	Macintosh® operating system; database of 9,000 experimental log P values; structure entry via SMILES[a] notations
ProLogP[31]	Broto et al.[17] Rekker and de Kort[16]	MS-DOS operating system; structure entry via graphic drawing module, MolFile, or MolNote
ATOMLOGP[32]	Ghose et al.[19]	MS-DOS operating system; database of 4,500 experimental log P values; structure entry via SMILES[a] notations

[a] SMILES (Simplified Molecular Input Line Entry System), see Reference 33.

Ease of use is important because most users will not remember the intricacies or details of applying an estimation method correctly unless they use it on a very frequent basis. All of the estimation methods discussed here involve some complicated operations in order to yield estimates for most complex structures. While it is possible to "hand-calculate" log P values for most methods, only users extremely familiar with the method should even attempt it, and even then, hand-calculation mistakes will occasionally occur. Therefore, "ease of use" almost certainly requires computer software as a computational aid. Table 2 lists computer programs that are available for various log P estimation methods.

Experimental water solubility values

Several extensive compilations of experimental water solubility values are available and are the most convenient source of this data, although they are not as comprehensive as the sources for log P. The following compilations are recommended:

1. AQUASOL dATAbASE™ of the University of Arizona:[36] Contains data for more than 3,000 chemicals. All available values are presented for each compound. The source of the data, analytical methods, units in the orginal article, converted units, temperature, and an evaluation of the quality of the documentation which would allow selection of a best value.
2. SRC's PHYSPROP© Database:[37] Contains data in a Chembase® or ISIS Base® format that allows substructure searching and value range searching for single recommended values for over 900 chemicals. The source of the data and temperature are given.
3. SRC's Environmental Fate Data Base (EFDB) system:[34,35] SRC's CHEMFATE© file has single recommended experimental or estimated water solubilities for approximately 1700 chemicals. The DATALOG© has pointers to references that have water solubility values for approximately 6000 chemicals. Indexed in the latter file are journal articles as well as book compilations, such as Horvath,[38] Riddick et al.[39] and Shiu et al.[40] and others, which have large listings of selected water solubility values.

As discussed for log P, ionizable compounds can have water solubility values that vary greatly with pH. As with log P, water solubility values can be "corrected for ionization" as long as the pKa is known.

Estimated water solubility values

Estimation methods for water solubilities values are primarily constrained to the 20 to 25°C temperature range and generally are for the unionized form. As mentioned earlier, Yalkowsky and Banerjee[6] have reviewed most of the recent literature on aqueous solubility estimation. They concluded that, at present, the most practical means of estimating water solubility involves regression-derived correlations using log P.[6] In general, most equations that correlate log P and water solubility are in a format similar to the following:[6]

$$\log S = a \log P + b (T_m - 25) + c$$

where a, b, and c are coefficients. Some equations use the melting point (T_m) parameter and some do not. When included, T_m applies only to solids; for liquids, the value of T_m is always set equal to 25°C which zeroes out the melting point parameter. In the following, we will only review the equations used by two computer programs for calculating water solubility: PCCHEM[41] and WS/KOW.[42] The other estimation equations that are available have been reviewed by Yalkowsky and Banerjee.[6]

PCCHEM is a fairly accurate water solubility estimation method that is available in the U.S. EPA's Graphical Exposure Modeling System (GEMS).[41] This method involves application of three separate log P equations; one equation if the log P is less than 0.5,

$$\log S = 0.686 - 1.123 \log K_{ow} - 0.0099 \, (T_m - 25)$$

another for log P greater than 0.5,

$$\log S = 0.455 - 1.034 \log K_{ow} - 0.0099 \, (T_m - 25)$$

and a third for organic acids with log P between –0.5 and 3.2.

$$\log S = 0.279 - 0.650 \log K_{ow} - 0.0099 \, (T_m - 25)$$

The PCGEMS method performs better than most equations available in literature sources[6] because it uses three equations as opposed to a single equation. Meylan et al.[7] evaluated a dataset of 1450 organic compounds which had measured log P and water solubility values (941 solids and 509 liquids) and a validation set of 817 compounds (482 solids and 335 liquids) which had measured water solubility values, but no measured log P value. Using the PCCHEM method they found the following statistics: $r^2 = 0.915$, sd = 0.664, and mean error = 0.506 for the 1450 compound training dataset[7] and $r^2 = 0.846$, sd = 0.763, and mean error = 0.573 for the 817 compound validation dataset[42] (CLOGP program[28] was used for calculating the log P value).

Meylan et al.[7] developed an expanded equation for estimating water solubility from log K_{ow} that has a melting point term for solids, but also includes terms for molecular weight (MW) and fifteen correction factors, three of which are used with no melting point term:

$$\log S = a \log P + b \, (T_m - 25) + cMW + d\Sigma f_i$$

Using this new method they found the following statistics: $r^2 = 0.970$, sd = 0.409, and mean error = 0.313 for the 1450 training dataset and $r^2 = 0.902$, sd = 0.615, and mean error = 0.480 for the 817 validation dataset (LOGKOW program[10] was used for calculating the log P values). The importance of melting point has been demonstated by a number of investigators as well as Yalkowsky and Banerjee.[6] Meylan et al.[7] found that the molecular weight, although not as important as the melting point, is an meaningful variable in improving the correlation between log P and water solubility. When measured melting points are available, they should be used in calculating the water solubility. However, Meylan et al.[7] demonstrated that if no measured melting point is available, one would get a better estimation if one used an equation with MW and correction factors without the melting point then to use an estimated melting point.

The calculation of water solubility from a known log P value is not as complicated as calculating log P from structure. Nevertheless, it is more convenient to use programs to calculate the value, especially if the correction factors of Meylan et al.[7] are used. PCCHEM[40] allows the input of experimental log P, melting point, and SMILES notation and will calculate the log P (CLOGP[28]) if not entered, calculate the melting point if not entered, and then calculate the estimation of water solubility using the three equations discussed above. Similarly, Syracuse Research Corporation (SRC) has a program that allows input of experimental log P, melting point, and SMILES notation and will calculate the log P (LOGKOW[10]) if not entered, and then calculate the estimation of water solubility using the log P, melting point (only if value is enter), molecular weight, and correction factors. SRC also has a program that calculates an estimated melting point (MPBPVP[43]), but it is not recommended that the estimated value be used in the water solubility estimation.

References

1. Lyman, W.J. 1990. Octanol/water partition coefficient. In *Handbook of Chemical Property Estimation Methods: Environmental Behavior of Organic Compounds*, Ed. W.J. Lyman, R. Rosenblatt. Washington, D.C.: American Chemical Society, pp. 1-1 to 1-51.
2. Meylan, W.M., Howard, P.H. 1995. Atom/fragment contribution method for estimating octanol–water partition coefficients. *J. Pharm. Sci.* 84:83–92.
3. Hansch, C., Leo, A.J. 1979. *Substituent Constants for Correlation Analysis in Chemistry and Biology.* New York: John Wiley & Sons.
4. Zeeman, M.G. 1995. Ecotoxicity testing and estimation methods developed under section 5 of the Toxic Substances Control Act (TSCA). Chapter 23. In *Fundamentals of Aquatic Toxicology: Effects, Environmental Fate, and Risk Assessment.* Ed. G. Rand, Washington, D.C.: Taylor & Francis, pp. 703–715.
5. Bysshe, S.E. 1990. Bioconcentration factor in aquatic organisms. In *Handbook of Chemical Property Estimation Methods: Environmental Behavior of Organic Compounds*, Ed. W.J. Lyman, R. Rosenblatt. Washington, D.C.: American Chemical Society, p. 5.
6. Yalkowsky, S.H., Banerjee, S. 1992. *Aqueous Solubility Methods of Estimation for Organic Compounds.* New York: Marcel Dekker, Inc. 262 pp.
7. Meylan, W.M., Howard, P.H., Boethling, R.S. 1996. Improved Method for Estimating Water Solubility from Octanol/Water Partition Coefficient. *Environ. Toxicol. Chem* 15:100–106.
8. Sangster, J. 1994. LOGKOW DataBank (a databank of evaluated octanol–water partition coefficients on microcomputer diskette). Sangster Research Laboratories, Montreal, Quebec, Canada.
9. Hansch, C., Leo, A.J., and Hockman. 1995. *Exploring QSAR: Hydrophobic, Electronic, and Steric Constants*, Washington, D.C.: American Chemical Society.
10. Syracuse Research Corp. 1995. *LOGKOW & KOWWIN Programs. Estimation of Log Octanol–Water Partition Coefficient*, computer software for MS-DOS and MS-Windows 3.1 versions 1.35. W.M. Meylan, P.H. Howard, Syracuse Research Corporation, Environmental Science Center, Merrill Lane, Syracuse, NY 13210.
11. Hansch, C., Leo, A., Leo, D. 1994. BioByte Corp Newsletter, Fall 1994. MacLogP™ computer software for the Macintosh®. BioByte Corporation, P.O. Box 517, Claremont, CA 91711-0517.
12. Daylight Chemical Information Systems. CLOGP™ Program. 111 Rue Iberville, No. 610, New Orleans, LA 70130.
13. Rib, J.M. 1988. The octanol/water partition coefficient of the herbicide chlorsulfuron as a function of pH. *Chemosphere* 17: 709–715.
14. Nieder, M., Stroesser, W., Kappler, J. 1987. Octanol/buffer partition coefficients of different beta-blockers, *Arzneim.- Forsch.*, 37: 549–550.

15. Barbato, F., Caliendo, G., LaRotonda, M.I., Morrica, P., Silipo, C., Vittoria, A. 1990. Relationships between octanol–water partition data, chromatographic indexes and their dependence on pH in a set of beta-adrenoceptor blocking agents. *Farmaco*, 45: 647–663.

16. Rekker, R.F., de Kort, H.M. 1979. The hydrophobic fragmental constant; an extension to a 1000 data point set. *Eur. J. Med. Chem.* 14: 479–488.

17. Broto, P., Moreau, G., Vandycke, C. 1984. Molecular structures: perception, autocorrelation descriptor and SAR studies. *Eur. J. Med. Chem.* 19: 71–78.

18. Klopman, G., Li, JY, Wang, S., Dimayuga, M. 1994. Computer automated log P calculations based on an extended group contribution approach. *J. Chem. Inf. Comput. Sci.* 34: 752–781.

19. Ghose, A.K., Pritchett, A., Crippen, G.M. 1988. Atomic physicochemical parameters for three dimensional structure directed quantitative structure-activity relationships III: modeling hydrophobic interactions. *J. Computational Chem.* 9: 80–90.

20. Suzuki, T., Kudo, Y. 1990. Automatic log P estimation based on combined additive modeling methods. *J. Computer-Aided Mol. Design* 4: 155–198.

21. Niemi, G.J., Basak, S.C.,Veith, G.D., Grunwald, G. 1992. Prediction of octanol–water partition coefficient (K_{ow}) with algorithmically derived variables. *Environ. Toxicol. Chem.* 11: 893–900.

22. Banerjee, S., Howard, P.H. 1988. Improved estimation of solubility and partitioning through correction of UNIFAC-derived activity coefficients. *Environ. Sci. Technol.* 22: 839–841.

23. Bodor, N., Gabanyi, Z., Wong, C.K. 1989. A new method for the estimation of partition coefficient. *J. Am. Chem. Soc.* 111: 3783–3786.

24. Bodor, N., Huang, M.J. 1992. An extended version of a novel method for the estimation of partition coefficients. *J. Pharm. Sci.* 81: 272–281.

25. Sasaki, Y., Kubodera, H., Matuszaki, T., Umeyama, H. 1991. Prediction of octanol/water partition coefficients using parameters derived from molecular structures. *J. Pharmacobio.-Dyn.* 14: 207–214.

26. Sabljic, A., Guesten, H., Hermens, J., Opperhuizen, A. 1993. Modeling octanol/water partition coefficients by molecular topology: chlorinated benzenes and biphenyls. *Environ. Sci. Technol.* 27: 1394–1402.

27. Nakagawa, Y., Izumi, K., Oikawa, N., Sotomatsu, T., Shigemura, M., Fujita, T. 1992. Analysis and prediction of hydrophobicity parameters of substituted acetanilides, benzamides and related aromatic compounds. *Environ. Toxicol. Chem.* 11: 901–916.

28. Pomona Medicinal Chemistry Project. 1987. *PC-CLOGP Version 3.32*, U.S. EPA Version 1.2 computer software. Claremont, CA: Pomona College.

29. Leo, A.J. 1992. *30 years of calculating Log P_{oct}*. QSAR Meeting, Duluth, MN. July 23, 1992.

30. Rekker, R.F., ter Laak, A.M., Mannhold, R. 1993. *Quant. Struct.-Act. Relat.* 12: 152–157.

31. CompuDrug. 1993. *ProLogP, Expert System for the Calculation of logP*, computer software version 4.2. CompuDrug NA, Inc., P.O. Box 23196, Rochester, NY 14692.

32. GS Corp. 1994. *ATOMLOGP*, computer software. Produced by General Sciences Corp., available from American Chemical Society, Distribution Office, Dept. 190, P.O. Box 57136, West End Station, Washington, DC.

33. Weininger, D. 1988. SMILES, a chemical language and information system. 1. introduction to methodology and encoding rules. *J. Chem. Info. Sci.* 28:31–36.

34. Howard, P.H., Sage, G.W., LaMacchia, A., Colb, A. 1982. The development of an environmental fate data base. *J. Chem. Inf. Comput. Sci.* 22: 38.

35. Howard, P.H., Hueber, A.E., Mulesky, B.C., Crisman, J.C., Meylan, W.M., Crosbie, E., Gray, D.A., Sage, G.W., Howard, K., LaMacchia, A., Boethling, R. and Troast, R. 1986. BIOLOG, BIODEG, and fate/expos: new files on microbial degradation and toxicity as well as environmental fate/exposure of chemicals. *Environ. Toxicol. Chem* 5: 977–980.

36. Yalkowsky, S.H., Dannenfelser, R.M. 1992. AQUASOL dATAbASE of Aqueous Solubility. Fifth edition. Tucson, AZ: University of Arizona, College of Pharmacy.

37. SRC. 1994. PHYSPROP© Database. Physical/chemical property database. Syracuse, NY: Syracuse Research Corp, Environmental Science Center (P.H. Howard).

38. Horvath, A.L. 1982. *Halogenated Hydrocarbons: Solubility — Miscibility with Water.* New York: Marcel Dekker. 889 pp.

39. Riddick, J.A., Bunger, W.B., Sakano, T.K. 1986. *Techniques of Chemistry. Vol II. Organic Solvents.* 4th edition, New York: Wiley-Interscience.

40. Shiu, W.Y., Ma, K.C., Mackay, D., Seiber, J.N., Wauchope, R.D. 1990. Solubilities of pesticides chemicals in water. II. Data compilation. *Rev. Environ. Contam. Toxicol.* 116: 15–187.

41. U.S. Environmental Protection Agency. 1987. PCGEMS User's Guide. Report prepared under task 3-10 of EPA Contract No. 68-02-3970. General Sciences Corporation, Laurel, MD.

42. Meylan, W.M., Howard, P.H. 1994. *Upgrade of PCGEMS Water Solubility Estimation Method.* SRC-TR-94-009. Syracuse Research Corp., prepared for R.S. Boethling, USEPA, OPPT, Washington, DC.

43. Syracuse Research Corp. 1995. *MPBPVP Programs. Estimation of Melting Point, Boiling Point, and Vapor Pressure,* computer software for MS-DOS and MS-Windows 3.1 versions 1.35. W.M. Meylan, P.H. Howard, Syracuse Research Corporation, Environmental Science Center, Merrill Lane, Syracuse, NY 13210.

44. Syracuse Research Corp. 1995. *WS/KOW Programs. Estimation of Water Solubility from Log Octanol–Water Partition Coefficient,* computer software for MS-DOS and MS-Windows 3.1 versions 1.35. W.M. Meylan, P.H. Howard, Syracuse Research Corporation, Environmental Science Center, Merrill Lane, Syracuse, NY 13210.

chapter twenty-two

Quantum chemical parameters in QSAR: what do I use when?

James P. Hickey

Introduction

The use of indices derived from quantum mechanical methods[1] and molecular mechanics[2] in quantitative structure activity relationships (QSAR) has become increasingly popular, primarily due to the development of reliable computational methods, the powerful desktop personal computer, and the availability of computer programs from which these indices can be obtained.

The two techniques are not competitive techniques; each has an appropriate use in molecular modeling. The concept of transferability of atoms and bonds from one related molecule to the next (relying on the analogy of the spring), which can be verified quantum mechanically, is the basis of molecular mechanics. And, while many useful properties can be predicted from quantum mechanics, fundamentally it involves computing the electron distribution of an atom.

The *ab initio* calculated parameters, such as LUMO, HOMO, and electron population densities at certain groups or atoms (e.g., amine, nitro, hydroxy) and molecular mechanical parameters such as ΔH, geometry, and total energy are increasingly being used to correlate effects of polar, reactive compounds with physical properties such as toxicity. Their power of prediction appears to combine varying modes of action.

The quantum mechanical indices described in this paper have been compiled from many sources and are generally obtained from classical molecular mechanical methods,[2] as well as from calculated molecular wavefunctions.[1] The quality of the model or wavefunction and, consequently, of the indices depend entirely on the formalism and the level of approximation used. Because the molecular waveform is often described as a linear combination of atomic orbitals (LCAO), one can easily obtain some of the indices. These include the atomic net charges, sigma and pi charges, frontier electron densities, E_{HOMO} and E_{LUMO}, and the superdelocalizability parameters. Within the LCAO approximation one can assign the reactivity indices to specific atoms or bonds within the molecule. These indices reflect the stationary reactivity properties of the atoms and bonds in the molecule as described by charges or orbital energies and can therefore serve only as indicators of the reactivity.

This chapter will provide a brief overview of the numerous quantum chemical parameters that have been/are currently being used in QSAR, along with a representative bibliography. The parameters will be grouped according to their mechanistic interpretations, and representative biological and physical chemical applications will be mentioned. Parameter computation methods and the appropriate software are highlighted, as are sources for software.

Materials required

A cornucopia of software exists that will compute the quantum parameters by any number of methods. Several good compilations of software sources include:

> *Directory of Chemistry Software 1992.* W. Warr, P. Willett, and G. Downs, Eds., Cherwell Scientific Publishing, Oxford, U.K., and American Chemical Society, Washington, D.C.[3]
> Boyd, D.R. 1990. Appendix: Compendium of software for molecular modelling. In: *Reviews in Computational Chemistry.* K.B. Lipkowitz and D.B. Boyd, Eds., VCH Publishers, Inc., New York. pp. 383–392.[4]
> Boyd, D.R. 1991. Appendix: Compendium of software for molecular modelling. In: *Reviews in Computational Chemistry II.* K.B. Lipkowitz and D.B. Boyd, Eds., VCH Publishers, Inc., New York. pp. 481–498.[5]

A good general software source is

> *QCPE Bulletin.* Quantum Chemistry Program Exchange (QCPE), Creative Arts Building 181, Indiana University, Bloomington, IN 47405. QCPE is an excellent source for a multitude of inexpensive computational software, including molecular mechanics and quantum mechanics.

Some recommended software for a basic toolbox include:

> Rogers, D.W. 1990. *Computational Chemistry Using the PC.* New York, NY: VCH Publishers, Inc.[6] This is actually a hands-on workbook with software that guides the user through many, many aspects of computational quantum chemistry, and can serve as a very beneficial reference.
> Stewart, J.J.P. 1990. MOPAC: a semiempirical molecular orbital program. *J. Computer-Aided Design.* 4:1–105.[7] This is an extensive catalog of programs for quantum mechanics and molecular mechanics, and is available through both QCPE and Serena Software, PO Box 3076, Bloomington, IN, 47402.

Gilbert, K. 1993. *PCMODEL*. Excellent stand-alone molecular mechanics software including MMX. This is an excellent companion program "kit" to MOPAC, and is available through Serena Software, PO Box 3076, Bloomington, IN, 47402.

Procedures

Computational methods

For citations on computational methods, see References 2, 6, and 8 to 10.

The two major methodologies considered in this paper are molecular mechanics and advanced molecular orbital methods. These are based on quantum chemical or semiclassical (e.g., molecular mechanical) models of molecular structure. Many useful parameters can be determined using either method, but there are fundamental differences between them. Molecular mechanics (MM) regard the geometry and energy of a molecule as the result of classical spring mechanics operating through the medium of chemical bonds replaced by mathematical functions that describe the various interactions of the connected masses (atoms) constituting the molecule. Advanced molecular orbital (MO) methods provide a mathematical description of molecular structure in terms of atomic nuclei and the electron distribution in a molecule; the different approaches address numerous approximations to the Schrödinger equation. Figure 1 (modified from Reference 10) gives an overview of the approximate MO methods.

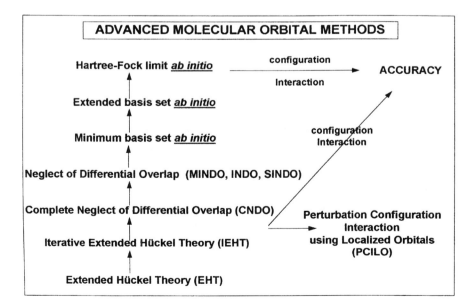

Figure 1 Summary of the various approximate orbital methods. (From Richards, W.G. 1977. *Quantum Pharmacology*. Chap. 15, p. 115. With permission.)

Quantum chemical models are further divided into two categories: *ab initio* and semiempirical.

Ab initio quantum chemical methods exactly evaluate all integrals of the theory, whether variational or perturbative. The "level" of theory then refers to the type of theory used. Common levels of theory would include Hartree-Fock, or molecular orbital theory, configuration interaction (CI) theory, perturbation theory (PT), coupled-cluster theory (CC, or coupled-perturbed many-electron theory, CPMET), etc. "Model" refers to approximations to the Hamiltonian. The distinction between *ab initio* and semiempirical is often not clear.

Semiempirical quantum chemical computation methods are much quicker due to the neglect of many difficult integrals suggested by truly first-principles application of quantum mechanics to molecular problems. The error introduced using semiempirical models is compensated through the use of parameters determined through comparison of calculation with experiment. This procedure can often produce a model of greater accuracy when compared to experiment than a similar *ab initio* calculation.

Ab initio methods that include correlations can have an accuracy comparable with experiment for many purposes. The principle drawbacks are the cost in terms of computer resources, and large basis set requirements. However, MM methods, based on the classical spring concept, are extremely fast, and capable of handling very large systems, such as entire enzymes, with ease. Some MM methods are also as accurate as the best *ab initio* methods, particularly for hydrocarbons. Most MM methods are parameterized for common ground-state bonding situations, and are incapable of handling excited state bonding situations. The semiempirical quantum methods are somewhere in between *ab initio* and MM methods. They use experimentally determined parameters for accuracy (like MM) and are quantum-mechanical in nature (like *ab initio*). In practice, with today's technology, MM methods can be applied to thousands of atoms, semiempirical quantum chemistry to hundreds, and *ab initio* quantum chemistry to tens of atoms. Figure 2 summarizes accuracies and capabilities of the various methods.[11]

Some computational modeling problems are best addressed by quantum mechanics (QM) and others by MM. QM has the advantage of generality of parameterization, applicability to short-lived species, treatment of intermediate/unknown hybridizations, determination of electronic properties, and theoretical rigor. MM is favored for concept simplicity, computation speed, capability to treat large molecules, inclusion of solvent effects, conformational analysis of "normal" molecules, and determination of geometries and relative energies.

Computation strategy

Computation time is much less for molecular mechanics than for advanced MO methods. A strategy often adopted in contemporary computational chemistry would be (a) calculate the energy and geometry of the molecule by MM; (b) stop, if sufficient information is obtained; or (c) if need be, use the MM geometry to construct an input file for an MO or an *ab initio* com-

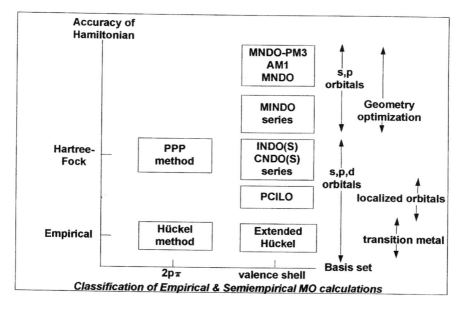

Figure 2 Schematic classification of molecular orbital calculations. (Adapted from Katagi, T. 1992. *J. Pesticide Sci.* 17(3):5221–5230. With permission.)

putation, or both. For a particular choice of methodology, the reader is referred to several references: 2, 6, 8 to 10, and 12 to 15.

Which method best computes what properties?

Most methods will yield orbitals with appropriate orbital energies and with appropriate nodal behavior and symmetries to support frontier molecular orbital arguments when applied to molecules of first and second main group elements. Most methods will also generate charge distributions that can be used to discuss electrophilic and nucleophilic reactions and Wiberg indices dealing with bonding nature. See References 2, 8, 10, and 14 and Appendix I for a description of the different methods cited.

The extended Hückel (EHT) model supports ligand or crystal field theory, since it will yield metal d and f orbitals as HOMOs and LUMOs systematically. Charge distributions in an EHT are examined through a Mulliken population calculation,[16,17] as will ZDO-type calculations (such as CNDO) but with potentially different interpretations and accuracies for metals. Geometries are well predicted by MNDO, AM1, PM3, and SINDO, with PM3 the most accurate. SINDO1 and PM3 are best for extended valency systems. INDO/1 will offer reasonable geometries, especially for transition metal complexes, and will often favor small rings. Heats of formation are successfully predicted by AM1, PM3, and SINDO1. Simple CNDO and INDO are not generally useful for total energy comparisons. Care must be taken that the molecular structures have been optimized. EHT, IEHT, MNDO, AM1, PM3, SINDO1, and INDO/S predict the first few ionization potentials of non-metal-containing systems well. INDO/S seems best for higher ionization processes. Only EHT models yield systematically useful ionization potentials. Dipole moments are well reproduced by MNDO, AM1, PM3, and SINDO1.

Molecular mechanics programs MM2, MM3, and MMX, with appropriate approximation software included, will predict the properties mentioned.

Results and discussion

Physical and biological applications

The integration of molecular orbital calculations with QSAR analyses has resulted in many "theoretical" indices (see Parameters section) and a small sampling of their applications is presented in this section (adapted from Reference 18). Using the results obtained from various levels of approximation of molecular mechanical[2,15,19] or quantum mechanical calculations, investigators have applied atomic charge density values, ε,[21-24] total atomic net charges, q and Q^T,[26-30] atomic σ-electron net charges, q^σ and Q^σ,[26,30] and atomic π-electron net charges q^π and Q^π,[26,30,31] to equations relating structure to function or activity. Hermann et al.[23] have applied the energy of the lowest unoccupied molecular orbital, E_{LUMO}, occupation numbers, δc, eigenvalue difference, Δ_E, incipient-transition-state differences, δ_E, and the total interaction term, Θ, in linear free energy relationship (LFER) equations. Further examples by Wohl included the use of E_{HOMO} (energy of the highest occupied molecular orbital), $\pi_{N,N'}$, the frontier self-atom polarizability of an atom N, $\pi_{N,N+1}$, the frontier atom–atom polarizability between atoms N and N+1, F[A], the intermolecular coulombic interaction energy for a molecule and point [A], and E[A], the electric field created at point [A] by a set of charges on a molecule, in structure–activity relationships.[30] The total electronic energy, energies of the frontier orbitals E_{HOMO} and E_{LUMO}, their difference ΔE and calculated dipole moments are often used as global electronic characteristics of molecules.[32]

Many of these same indices have been used to explain the quantum mechanical relationship of LFER to:

- **Chromatography** — topological index (T^E), total energy (E_T), and polarity parameter (D), for GC retention indices of amines[33-35]; $DH_{formation}$, ionization potentials (I), dipole moments (m), and LUMO for retention indices of PCDDs and PCDFs[36]; ionization potential (I), electron affinity (E_A), and dipole moments (μ^2), for retention indices of PCBs.[37]
- **Solvents** — HOMO and LUMO energies for hydrogen-bonding tendencies and dipole moments (μ), for polarity estimation;[38] use of superdelocalizabilities (S_E), for molecular polarizabilities (α), and use of LSER π^*, β, for dipole moments, μ, and μ/(HOMO-LUMO), respectively.[39]
- **Partitioning properties** — correlation of electrophilic superdelocalizability (S_E), as lipophilicity, and total charge density (Q^T) for hydrophilicity, with log P (E_{HOMO}, E_{LUMO}), total π energy also tried.[27,28]

- **Charge density** — q_i[40] and changes in localization and delocalization energies, ΔE[41] to determine Log P, and general solute-solvent interaction-based properties.

For the relatively new but powerful predictive LSER theory,[42] Famini,[43-46] an LSER pioneer, derived a quantum theoretical alternative (TLSER), using a polarizability index (π_I), molecular orbital basicity, ε_b, electrostatic basicity (q_-), molecular orbital acidity (ε_a), and electrostatic acidity (qH_+), where

$$\pi_I = \pi^*, \; (\varepsilon_b + q_-) = \beta, \text{ and } (\varepsilon_a + q_+) = \alpha$$

Quantum indices, such as π-electron charge density, ρ, free valence, and F_r, have been used to predict spectroscopic shifts.[47,48] Superdelocalizabilities, — electrophilic, $S^{(E)}$, free radical, $S^{(R)}$, and nucleophilic, $S^{(N)}$ — of substituted naphthalenes,[47] substituted pyridines, quinolines, and quinoxalines,[48] have been correlated with aspects of their ^{1}H-NMR spectra.

Quantum mechanics have been applied to biodegradability regarding active site carbon electronic superdelocalizability, S_E, correlation with biological oxygen demand (BOD).[49] Quantum mechanics have also been applied to the estimation of the environmental and metabolic fate of pesticides utilizing frontier electron density, f, a reactivity index for reactivity and regioselectivity, and photolysis index, ΔM, as bond electron population change with excitation.[11] The use of quantum parameters to examine the physical interpretation of LFER in biological systems and the relationship between chemical reactivity, structure and toxicity (including carcinogenicity) is ongoing. For example, net atomic charge, q, and electrophilic and nucleophilic superdelocalizabilities, $S^{(E)}$ and $S^{(N)}$, are used to compute linear free energy change in chemical reactions.[50] Charge density, q, frontier electron density, f, dipole moment, μ, total electron density, E^T, and superdelocalizability, $S^{(E)}$, are correlated with indole biological activity such as enzyme inhibition and hallucinogenicity.[51] Frontier π electron density, f, is used as an index for PCB, PCDF metabolism.[52] Frontier electron densities, f, and superdelocalizabilities help explain condensed PAH site carcinogenicity.[53] A wide range of indices are used to characterize the role of the bay and L-regions in carcinogenicity,[54-56] including electrostatic potentials.[57] E_{HOMO}, E_{LUMO}, ($E_{HOMO} - E_{LUMO}$), total electron densities, q_r, are used to explain pyrimidine derivative analgesic activity.[58] E_{HOMO}, E_{LUMO}, ($E_{HOMO} - E_{LUMO}$), total electron densities, q, superdelocalizabilities, $S^{(E)}$ and $S^{(N)}$ are used to explain biological potencies of benzylamines, tetracyclines, and 1-4-benzodiazapines.[59] Use of E_{HOMO} and E_{LUMO} help explain PCDD and PCDF metabolism and toxicity.[60] Methylnaphthalene bond reactivities, R_{ij}, superdelocalizabilities, S, and E_{HOMO} are correlated with toxicity to sea urchin and fish eggs.[61] There is a general review of quantum indices as applied to chemical carcinogenesis SAR, with multiple examples.[62] Use of charge density, q, and E_{HOMO}, E_{LUMO} help explain nitrobenzene and aniline derivative toxicity to Photobacterium phosphoreum.[63] Dipole moments, μ, ($E_{HOMO} - E_{LUMO}$), and atomic charges, q, explain mutagenic activity of dialdehydes in the Ames Salmonella assay.[64] Use of charge density, q, and E_{LUMO} help explain toxicity to fathead minnow, *Pimephales promelas*.[65] Superdelocalizability, $S^{(N)}$, and electronegativity, EN, are used for acute fish toxicity QSAR.[66] Acute fish toxicity QSAR of organophosphorothionates is determined using electronegativity, EN, absolute hardness, η, (super)delocalizability, D, ($E_{HOMO} - E_{LUMO}$), and atomic charges, q.[67] Ionization potential, I, and hardness, η, are used to rationalize phenylurea herbicide toxicity to numerous organisms.[68] The polarizability index, π_I, molecular orbital basicity, ε_b, electrostatic basicity, q_-, molecular orbital acidity, ε_a, and electrostatic acidity, qH_+, are applied to determine *Photobacterium phosphoreum* (Microtox test) and golden orfe fish (*Leuciscus idus melanotus*) toxicity, tadpole and guppy (*Poecilla reticulata*) narcosis, and frog (*Rana pipens*) muscle activity inhibition.[69] One ever-current topic for modeling is the interaction of compounds with a receptor site, for example:

- correlation of atom net total charge, q, atom π-electron density, Q^π, and atom frontier electron density with tryptamine derivative potency at a rat LSD receptor site;[70]
- use of E_{HOMO}, E_{LUMO} to indicate degree of alcohol charge–transfer interaction in aniline hydroxylation inhibition;[71]
- use of molecular electrostatic potentials (MEP) to explain chemical interactions (see References 72–74), and properties[75] such as toxicity;[76]
- elucidation of the charge–transfer nature of the PCDD and PCDF binding affinities with rat hepatic cytosol Ah receptor, using orbital symmetry with charge density, q, and E_{HOMO}, E_{LUMO},[77] and electrostatic potentials;[78]
- use of PCDD QSAR for biological activity (Ah site interaction) using ($E_{HOMO} - E_{LUMO}$) and orbital symmetry.[79-81]

Nucleophilic, electrophilic and free radical superdelocalizability indices, S_r^N, S_r^E, and S_r^R, respectively, have also been used to explain electronic drug-receptor interactions[47,48,82] and to guide drug design.[83] Molecular electrostatic potentials, E_A,[84] dipole moments, μ,[85] charge distribution, q,[86] frontier electron density values, f,[25] free valences, F_r,[47-48] valence force constants, Δf_r,[87-89] and atomic orbital coefficients, c,[25,29,50] have also been used successfully. Quantum chemistry has been applied to the broad area of chemical reactivity and reaction paths,[90,91] specifically using generalized perturbation theory.[92]

Comprehensive lists of MO indices used in QSAR studies have been compiled periodically (see References 18 and 93). Appendix II contains a large compilation of indices, categorized[1] as those related to charge and charge density, those related to energy, and those combining both charge density and energy. Formulae and application citations are included.

Mechanistic interpretation

Mekenyan and Bonchev[32] outlined three quantum chemical approaches to the electronic factors influencing the interaction of organic molecules:[26,41,94-97]

Static method.[98] In the static approach, the chemical reactivity at a position in a molecule is correlated with the density of all π-electrons at that position in the isolated molecule. One can use as static indices the well-known charge densities, free valency, bond orders, and atom polarizabilities. These reactivity indices are appropriate for systems where the transition state resembles the initial state, i.e., for very reactive species.

Localization Method.[99-104] In the localization approach, based on the Wheland intermediate,[105] the rate of reaction at a certain position in a molecule is determined by the "localization energy" which is required in localizing a requisite number (2,1, or 0) of π-electrons at that position. The assumption is that a direct correlation exists between the localization energy and the chemical reactivity. The method is applicable when the transition state is structurally dissimilar to the initial state, or to systems where the Brown non-crossing rule[106] holds.

Delocalization method.[107-110] In the delocalization approach, the electronic interaction between the molecule and the reagent has been divided into σ- and π-electron parts. The transition state is formed by the interaction of a pseudo-π orbital of the electrophile with the π-system and it resembles the charge–transfer complex. The important concept here is the frontier electron theory.[107-110] The electronic characteristics proposed by the delocalization approach are the "frontier" indices[107,109,110] such as frontier charges and superdelocalizability indices widely used in QSAR.[53,94,111] The idea of superdelocalizability indices is extended to all MO within the Klopman and Hudson[90,112-115] perturbation theory of chemical reactivity. The indices S_r^N and S_r^E characterize the electron acceptor and donor properties of the attacking reagent at the site r, respectively. When the charge transfer is considerable and the reaction is orbitally controlled, S_r correlates well with the activity. Such an effect specifies a covalent interaction. When the superdelocalizability indices do not correlate with activity but some atomic charges do, this is an indication for a small charge transfer and for a charge-controlled reaction as well as an ionic interaction. Figure 3 outlines what parameters indicate which situation in the molecule.

Errors

Many good correlations between chemical and biological activity and quantum chemical indices have been found, and these can be used to predict the activity of other untested molecules.[1] The electronic indices can successfully describe the influence of the electronic factors on activity in congeneric series, even without absolute data, since the errors in QSAR caused by the approx-

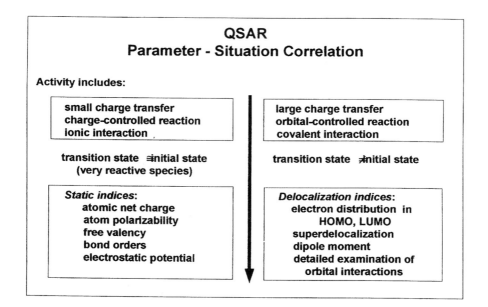

Figure 3 Mechanistic applications of quantum parameters.

imate nature of the MO method, inaccurate experimental geometry of the calculated compounds, neglect of solvation effects, etc., are consistent within structurally related series.[32] Thus, trends and differences among the indices might be meaningful even though the absolute values are not. However, limited by the nature of the approximations used in their calculations, these indices must be used with caution in inferring mechanisms of action.[1] Inappropriate interpretations of the mechanistic meaning of these correlations and of correlations with parameters describing electronic effects[116] have led to criticism of the methods of quantum mechanics rather than of the abuse of these methods.[117] Also, many of the MO indices are calculated for individual atoms, creating a large parameter pool for correlation analysis.[83] Parameter selection should be based on sound chemical (or biochemical) reasoning since the likelihood of obtaining chance correlations increases considerably with the number of parameters tried.

Acknowledgments

The author wishes to thank both Ann Zimmerman and Tracy Myers Janevic for their creative, relentless, and invaluable literature search and retrieval efforts (and, always with a smile). Use of trade names does not constitute government endorsement of commercial products. Contribution #914, Great Lakes Science Center, Ann Arbor, Michigan 48105.

Appendix I Methods glossary

See References 7–9; most available through QCPE.

MM *Molecular Mechanics* — uses classical spring mechanics operating through the chemical bond on the connected masses of the atoms making up the molecule. Hooke's Law, torsional forces, van der Waals attractions and repulsions replace the spring analog. Electrons are assumed to be at optimum distribution, potential energy functions describe interactions between nuclei, and the electronic system accounted for *de facto*. Any deviation from an "ideal" molecular geometry will correspond to an increase in energy. Parameterized for ground states only.

MM2, MM3 *Molecular Mechanics programs* — iterative energy minimization of molecular structure.

MMX *Molecular Mechanics program* — includes pi-electron SCF calculations, with option to drive dihedral angles.

PPP *Semi-empirical method* — self-consistent field method (SCF) developed both by Pariser and Parr, and by Pople.

SCF-MO *Self-Consistent-Field Molecular Orbitals* — use of repetitive energy minimization and solution of orbital equation matrix.

LCAO *Linear Combination of Atomic Orbitals* — assumes that the electron distribution in a molecule can be approximately represented as a sum of electronic distributions. The wave function of a molecule can thus be approximated as a linear combination of atomic orbitals having the individual wave functions χ_n. Thus $\psi_{molecule} \sim c_1\chi_1 + c_2\chi_2$, and the atomic coefficients c_1 and c_2 are chosen to minimize the wavefunction energy, and will be used to determine electron population densities and orbital energies.[118]

NNDO *Neglect of Negligible Diatomic Overlap (series of methods)* — all approximation schemes differ in the point, during the logical development, at which approximations are made and in the infusion of empirical data that takes place. All entail neglect of interactions that are not in the valence shell. The molecular core is defined as the nucleus in the *ab initio* calculations, but in NDO computations the core is defined as the nucleus and the non-valence electrons. The core influences valence electrons through its potential energy field. Coulomb and exchange energies arising from orbitals that are not neglected contribute to the total electronic energy. In semiempirical MO methods these integrals are not evaluated but approximated by empirical parameters.

ZDO *Zero Differential Overlap* — (general type of method).

HMO *Hückel Molecular Orbital Theory* — treats only π electron orbitals (noninteracting).

EHT   *Extended Hückel Theory* — includes all bonding orbitals in the secular matrix rather than just all π orbitals. Diagonal matrix elements considered measures of electron attraction and replaced by valence state ionization potentials. Complexity of computation increased to point where computer computation a necessity.

IEHT   *Iterative Extended Hückel Theory* — matrix elements are weighted to account for unequal electron distribution in molecule; results of simple EHT used to estimate charge distribution in molecule, which then adjusts matrix valence state ionization potentials. Iteration used to produce a self-consistent solution.

PCILO   *Perturbation Configuration Interaction using Localized Orbitals* — based on the expansion of molecular orbitals in terms of localized bond orbitals, which are constructed from pairs of hybrid orbitals. Pairs of localized antibonding orbitals form excited configurations for limited configuration interaction. ε and ψ are produced by perturbation theory; energy contributions come from simple term summation.

CNDO   *Complete Neglect of Differential Overlap* — eliminated most multi-center integrals; ignored most of the integrals used in the *ab initio* calculations and approximated, by simple expressions, all retained integrals. Many terms, such as the two-electron, one-center integral, were derived from experimental data. Atomic ionization potentials frequently replace integrals approximately representing energy with which an electron is held by an atom.

CNDO/1   *Modified CNDO* — uses two-center, two electron integral.

CNDO/2   *Modified CNDO/1* — includes electron affinities.

INDO   *Intermediate Neglect of Differential Overlap* — includes all one-center, two-electron terms (exchange integrals); close adherence to *ab initio* principles; could not represent lone-pair lone-pair interactions.

NDDO   *Neglect of Diatomic Differential Overlap (next series of methods)* — includes all two-electron two-center integrals involving charge clouds arising from pairs of orbitals on an atom; all interactions except those arising from diatomic differential overlaps are considered.

MINDO   *Modified Intermediate Neglect of Differential Overlap* — a more empirical approach to INDO.

MINDO/3   *(Succeeded MNDO, MNDO/2, and MNDO/2´)* — first practical general semiempirical program, extended to many elements (with some difficulty) and applicable to a wide range of studies. Several parameters were made adjustable to fit experimental data.

SINDO1   *Slater Intermediate Neglect of Differential Overlap*

MNDO   *Modified Neglect of Differential Overlap* — used monatomic (replaced diatomic) parameters in resonance integrals and core-repulsion terms, and improved parameters for a wide variety of elements. Could not reproduce hydrogen-bonding in biological systems.

AM1 *Austin Model 1 (modified MNDO)* — series of spherical Gaussian functions assigned for each atom to mimic correlation effects, and correct for excessive repulsions at van der Waals' distances. Increased parameters for atoms from 6–7 to 13–16. Reparameterization of elements extraordinarily difficult.

PM3 *Parametric Method Number 3 (modified AM1)* — all one-center two-electron integrals were optimized, as well as seven MNDO parameters and two AM1-type Gaussians; atom parameters optimized using automatic optimization routine that used a large set of reference molecular data.

ab initio Solution of Schrödenger's equation without assumptions or approximations.

MOPAC *Molecular Orbital PACkage* — general purpose, semiempirical molecular orbital program for the study of chemical reactions involving molecules, ions, and linear polymers. It implements the semiempirical Hamiltonians MNDO, AM1, MNDO/3, and MNDO-PM3, and combines the calculations of vibrational spectra, thermodynamic quantities, isotopic substitution effects, and force constants in a fully integrated program. Elements parameterized at the MNDO level include H, Li, Be, B, C, N, O, F, Al, Si, P, S, Cl, Ge, Br, Sn, Hg, Pb, and I; at the PM3 level the elements H, C, N, O, F, Al, Si, P, S, Cl, Br, and I are available. Within the electronic part of the calculation, molecular and localized orbitals, excited states up to sextets, chemical bond indices, charges, etc. are computed. Both intrinsic and dynamic reaction coordinates can be calculated. A transition-state location routine and two transition-state optimizing routines are available for studying chemical reactions. (Available through QCPE.)

Appendix II Parameters

Quantum chemical calculations generate a wide range of electronic data and indices that are of value as molecular descriptors. The indices most frequently used include[93]

1. atomic charges, electron and orbital densities
2. energies of the highest occupied MO (HOMO) and lowest unoccupied MO (LUMO)
3. nucleophilic and electrophilic superdelocalizabilities
4. molecular electrostatic potentials (MEP) and fields (MEF)
5. dipole moments

It is important to realize that the accuracy and reliability of these parameters strongly depend on the level of sophistication of the theoretical method, increasing from empirical to quantum mechanical semi-empirical to nonempirical (*ab initio*) methods.[119]

The quantum mechanical parameters used in QSAR can be divided into three categories: those related to charge and charge density, those related to energy, and those combining both charge density and energy. (Modified from Reference 1.)

Charge and charge density

1. *Atomic orbital coefficients*, c: orbital coefficients for linear combination of atomic orbitals (LCAO) molecular orbitals.[12,25,29,50]
2. *Bond order*, P_{ij}: measure of the probability of finding a π electron between atoms i and j,

$$P_{rs} = \sum_{\substack{all \\ electrons}} p_{rs}^j = \sum_{j} n_j c_{jr} c_{js}$$

where p_{rs} is the bond order for the r–s bond, n_j is the number of electrons, and c_{jr} and c_{js} are the coefficients for atoms r and s, all in the jth MO.[6,12,120]

3. *Free valence index*, F_r: a measure of reactivity

$$F_r = N_{max} - N_r$$

where N_r is the sum of the orders of all bonds joining atom r and N_{max} is the maximum value, usually taken as $\sqrt{3}$.[47,48]

4. *Atomic charge*, δ_A:

$$\delta_A = \sum_i n_i \left[\sum_m \left(c_{im}^2 + \sum_{B,m} c_{im} c_{in} S_{mn} \right) \right]$$

where δ_A is the charge on atom A, n_i is the number of electrons on molecular orbital i, $c_{im}(c_{in})$ are the coefficients of atomic orbital m(n) on atom B(A) in molecular orbital i, and S_{mn} is the overlap between orbitals m and n.[16,17,121]

5. *Occupation number*, δc: orbital charge at a molecular reactive center.[23]
6. *Atomic charge/electron densities*, q, ε, Q_b, ρ. (See References 21–24, 30, 47, 55, 58, 59, 65, 67, and 86.)
7. *Total electron density*, q_r: sum of electron densities contributed by each electron in each MO.

$$q_r = \sum_j n_j c_{jr}^2$$

where c_{jr} is the coefficient of atom r in the jth MO, which is occupied by n_j electrons. The sum is taken over all of the MOs.[12]

8. *Atomic net charges*, q, q_i: electron probability density relative to the neutral atom.

$$q_i = 1.0 - \sum Nc_i^2$$

where the sum is the electron probability density at atom *i* over all MOs and q_i is the total charge density (positive or negative) relative to the neutral situation.[6,11,26,30,31,40,71,122]

9. *Total (σ + π) net charge*, Q^N, Q^T: net total (σ + π) charge on atom N.[25-30,70,123-125]

10. *Absolute value of total (σ + π) net charge*, $\sum |Q^T|$: net total (σ + π) charge.[27-29,124-127]

For other charge density parameter references, see References 10, 49, 62, 63, 65, 69, and 128.

11. *Electrostatic basicity*, q_o: absolute value of the most negative formal charge in the solute (contaminant). Replaces solvatochromic (LSER) β as sum with molecular orbital basicity ε_b. Represents the charge-charge interaction between solvent and solute.[45,46,69]

12. *Electrostatic acidity*, qH_+: absolute value of the most positive formal charge of any hydrogen atom in the solute (contaminant). Replaces solvatochromic (LSER) α as sum with molecular orbital basicity ε_a. Represents the hydrogen or Brønstead acidity in the molecule.[45,46,69]

13. *Charge transfer constant*, C_T: direct measure of the capacity of a compound to act as a donor in π-electron complexes; analogous to the hydrophobicity constant π, when considered as the product of ρ and σ, where ρ is a reaction constant which can differ with different acceptors, and with different classes of donors, and where σ is a true constant for the substituted aromatic residue. Not strictly a quantum-mechanically-derived constant.

$$C_r = \log K_X - \log K_H$$

where K_X is the association constant for charge transfer complex formation with tetracyanoethylene of the substituted and K_H is that for the unsubstituted compound.[129]

14. *Dipole moment*, μ: vector sum of all charge densities q_i.[24,36,37,39,59,70,85,130]

15. *Polarity parameter*, Δ: largest difference in atomic charges; reflects the highest local intermolecular dipole moment; characterize ability of solutes to participate in polar interactions.[33,35]

16. *Atomic sigma, pi net charge*, q^σ, Q^σ, q^π, Q^π: See References 26, 30, 31, and 70.

17. *Frontier electron density*, f: electron density for the highest occupied molecular orbital (HOMO).[11,25,51,52,70,126,127,131,132]

18. *Topological electronic index*, T^E: descriptor of molecular polarity; reflects differences in solute size, shape and constitution.

$$T^E = \sum_{i,j} \frac{|q_i - q_j|}{r_{ij}^2}; \quad i \neq j$$

where q is the electronic excess charge on an atom and r_{ij} is the distance between each pair of atoms.[34,93]

19. *Valence force constants*, Δf_r: in units of dynes/cm; force acting to restrain the relative displacement of the nuclei in a molecule; can also be quantum-mechanically derived.

$$\Delta f_r = \frac{\pi^2 C^2 (v_s + v_{as})^2}{1/M_1 + 1/M_2}$$

where v_s and v_{as} are frequencies of symmetric and asymmetric stretching of a chemical bond, M_1 and M_2 are the atomic weights of the two atoms, and C is the velocity of light.[87-89]

Energy

1. *Highest occupied/lowest unoccupied MO energy*, E_{HOMO}, E_{LUMO}/E_{LEMO} E_{HOMO}: is a relative measure of the ability of an electron in the highest occupied orbital of a molecule to be transferred to an acceptor molecule. E_{LUMO} or E_{LEMO} can be a relative measure of the activation energy for a reaction, or the ease with which a molecule accepts an electron. (HOMO and LUMO are the reaction-determining orbitals in frontier-controlled reactions.) See References 23, 30, 31, 36, 38, 45, 46, 56, 59, 60, 62, 63, 65, 66, 69–71, 77, 79–81, 96, 132, and 133.
2. *Energy (HOMO-LUMO) difference*, GAP: reflects reaction transition energy.[54,58]
3. *MO eigenvalue differences*, Δ_E: substituent parameter reflecting frontier orbital (lowest unoccupied molecular orbitals) energy differences between molecules.[23,79]
4. *Ionization potential or energy*, I: The energy per unit charge needed to remove an electron from a given kind of atom or molecule to an infinite distance.[36,37,68,134,135]
5. *Electron affinity*, E_A: The work needed in removing an electron from a negative ion, thus restoring the neutrality of an atom or molecule.[37,134]
6. *Absolute electronegativity*, EN: a measure of the electron attraction tendency; a characteristic of a specific valence state of an atom.

$$EN = \frac{(IP + EA)}{2}$$

(a) average of the ionization potential IP and the electron affinity EA.[134]

$$EN = -\frac{(E_{HOMO} + E_{LUMO})}{2}$$

(b) average of the energies for the highest occupied and lowest unoccupied molecular orbitals, using MNDO calculations.[67]

7. *Absolute hardness,* η: resistance of chemical potential to change if number of electrons is changed.

$$\eta = \frac{(IP - EA)}{2}$$

(a) average of the difference between ionization potential IP and the electron affinity EA.[136]

$$\eta = -\frac{(E_{HOMO} - E_{LUMO})}{2}$$

(b) average of the difference of the energies for the highest occupied and lowest unoccupied molecular orbitals, using MNDO calculations.[67,68]

8. *Energy of lowest acceptor orbital,* $\langle \varepsilon^* \rangle$: energy of the lowest unoccupied molecular orbital (LUMO) on an acceptor molecule.[137]

9. *Incipient-transition-state differences,* δ_E: substituent constant showing degree of difference between energy differences between ground and transition states for parent molecule and derivative; examination of energy differences between ground state and proposed transition-state configurations for a reaction.

$$\delta_E = [-(E_\pi - E_G)_{deriv} + (E_\pi - E_G)_{parent}] \times 100$$

where E_{IT} = total energy of the incipient transition state and E_G = total energy of ground state.[23]

10. *Conformational energy,* E: total energy of a structure's particular conformation.[10,137]

11. *Total energy,* E_T: total energy for a molecule; a measure of ability of molecule to participate in nonspecific, nonpolar interactions.[33,35]

12. *Total interaction energies,* Θ: representations of trends toward a transition state between molecules; summations of Coulomb integral matrix elements representing repulsion integrals for interacting atoms between molecules multiplied by the occupation numbers ($\simeq$charge densities) of the respective atoms.[23]

13. *Coulombic interaction energies,* $F^{[A]}$: intermolecular Coulombic interaction energy for an atom/molecule and point [A].

$$F[A] = \sum_d \frac{q_d}{R_d}[A]$$

where q_d is a charge on atom d and R_d is the separation between atom and point $[A]$.[30,123]

Parameters incorporating both charge densities and energies

1. *Bond reactivity, R_{ij}:* weighted average of π-bond electron densities.

$$R_{ij} = \sum_k^{occ} \frac{p_{ij}^k(\pi)}{-E_k}$$

where p_{ij} is the electron density between atoms i and j in the *k*th MO K and E_k is the energy of the *k*th MO K.[55,56,61]

2. *Frontier polarizabilities:*
   - self-atom, $\pi'_{N,N}$: frontier self-atom polarizability of atom N; roughly measures the electric moment induced at an atom by itself.

$$\pi_{r,r} = 4 \sum_{j=1}^m \sum_{k=m+1}^n \frac{c_{rj}^2 c_{rk}^2}{\varepsilon_j - \varepsilon_k}$$

where c is the coefficient for the molecular orbital *j* or *k* on atom r, and ε is the corresponding orbital energy.
   - atom-atom, $\pi'_{N,N+1}$: frontier atom-atom polarizability between atom N and atom N+1; roughly measures the electric moment induced at an atom by an adjacent atom.

$$\pi_{s,r} = 4 \sum_{j=1}^m \sum_{k=m+1}^n \frac{c_{rj} c_{sj} c_{rk} c_{sk}}{\varepsilon_i - \varepsilon_k}$$

where c is the coefficient for the molecular orbital *j* or *k* on atom r or s, and ε is the corresponding energy for orbitals *j* or *k*.[30,123,128]

3. *Superdelocalizabilities, S_r:* energy-weighted averages of the π-electron densities on each atom; a measure of the energy stabilization due to an electron transfer to or from a reagent at center *r*. A measure of the tendency for interacting electron-rich or electron-poor centers to be localized at an atom; a measure of the ability of atoms in a molecule to form a weak π bond with an appropriate group in an acceptor site.

electrophilic   nucleophilic   radical

$$S_r^{(E)} = 2 \sum_{j=1}^m \frac{c_{rj}^2}{\lambda_j} \quad S_r^{(N)} = 2 \sum_{j=m+1}^N \frac{c_{rj}^2}{\lambda_j} \quad S_r^{(R)} = 2 \sum_{j=1}^m \frac{c_{rj}^2}{\lambda_j} + 2 \sum_{j=m+1}^N \frac{c_{rj}^2}{-\lambda_j}$$

where c_{rj} is the coefficient of the rth atomic orbital in the jth molecular orbital, and λ_j is the coefficient in the orbital energy, which is given as $\varepsilon_j = \alpha + \lambda_j \beta$ (α is an ionization potential and β an empirical energy parameter used in simple π-electron Hückel theory). See References 10, 26, 27, 29–31, 47, 48, 50, 51, 54–56, 58, 59, 62, 82, 83, 111, 123–125, and 138–140.

4. *Sum of electrophilic delocalizabilities for all atoms,*

$$S_E = 2\sum_i^{occ} \frac{C_i^2}{E_i}$$

where C_i is the orbital coefficient for occupied molecular orbital i and E_i is the occupied orbital energy.[39,108]

5. *Atomic reactivity, S_i:* energy-weighted average of π-electron densities on each atom (atomic superdelocalizability).

$$S_i = \sum_k^{occ} \frac{q_i^k(\pi)}{-E_k}$$

where q_i is the electron density on atom i in the molecular orbital k and E_k is the energy of the kth MO K.[61,108,141]

6. *Molecular Polarizability, α:*

$$\alpha_A = 4\sum_i^{occ}\sum_j^{unocc} \frac{c_j^2 c_i^2}{(E_i - E_j)}$$

where C_j is the orbital coefficient for the unoccupied molecular orbital j and E_i is the unoccupied orbital energy. Also,

$$\alpha(ahc) = \frac{4}{N}\left[\sum_A \tau_A\right]^2 (\mathring{A})^3$$

where N is the number of electrons in the molecule, and ahc refers to the atomic hybrid components τ_A of α for each atom in the molecule.[39,128,142-146]

7. *Molecular Orbital Acidity, ε_a:* interaction energy for covalent solute-to-solvent electron transfer. Replaces LSER solvatochromic α as a sum with qH_+, the electrostatic acidity.

$$\varepsilon_a = E_{LUMO_{solute}} - E_{HOMO_{solvent}}$$

where E_{LUMO} is the energy of the LUMO of the solute, and E_{HOMO} is for the HOMO of the solvent.[45,46,69]

8. *Molecular Orbital Basicity, ε_b:* interaction energy for covalent solvent-to-solute electron transfer. Replaces LSER solvatochromic β as a sum with q_-, the electrostatic basicity.

$$\varepsilon_b = E_{LUMO_{solvent}} - E_{HOMO_{solute}}$$

where E_{LUMO} is the energy of the LUMO of the solvent, and E_{HOMO} is for the HOMO of the solute.[45,46,69]

9. *Polarizability Index, π_I:* ability of the solute valence shell electrons to be polarized. (LSER Solvatochromic π^*)

$$\pi_I = \frac{\text{Polarization Volume}}{\text{Molecular Volume}}$$

where polarization volume (from MNDO calculations) is the change in the van der Waal's molecular volume due to the introduction of an electrical field.[44,46,69]

10. *Molar Electronic Polarization, P_E:* a measure of volume; alternative name: molar refraction; not strictly a quantum-mechanically-derived parameter.

$$P_E = \frac{n^2 - 1}{n^2 + 2} \frac{M}{D} = \frac{4}{3} \pi N \alpha_E$$

where n is the refractive index of the substance, M is its molecular weight, D is its density, N is Avogadro's number, and α_E is the electronic polarizability.[130,147]

11. *Molecular electrostatic field potentials,*
   - from point charges, E[A]: electric field created by a series of charges upon the midpoint of a bond at a given distance (point [A]) away.

$$E[A] = \sum_d \frac{q_d}{R_{d[A]}^3} r_{d[A]}$$

where q_d is a charge on atom d, $R_{d[A]}$ is the separation between atom d and point [A], and $r_{d[A]}$ is the vector distance.[10,30,72-74,84,123,127,148,149]
   - from electronic charge distribution, E[A] (or V(r)): electric field created at point [A] by a set of charges on a molecule, and described by molecular wavefunction; the energy of interaction of a point charge, with the unperturbed charge distribution of the molecule.

$$V(\vec{r}) = \sum_A \frac{Z_A}{|\vec{r} - \vec{R}_A|} - \sum_i N_i \int \frac{\psi_i^*(\vec{r}')\psi_i(\vec{r}')d\vec{r}'}{|\vec{r} - \vec{r}'|}$$

where, for a potential at point r, ψ_i is a molecular orbital containing N_i electrons, and Z_A is the positive charge on nucleus A, situated at R_A.[10,57,75,76,78,148,150-157]

References

1. Osman, R., Weinstein, H., Green, J.P. 1979. Parameters and methods in quantitative structure-activity relationships. In *Computer-Assisted Drug Design*, Eds. E.C. Olson, R.E. Christofferson, ACS Symposium Series 112, 2:21–77. Washington, DC: American Chemical Society.
2. Bowen, J.P., Allinger, N.L. 1991. Molecular mechanics: The art and science of parameterization. In *Reviews in Computational Chemistry*, Eds. K.B. Lipkowitz, D.B. Boyd, 3:81–97. New York, NY: VCH Publishers, Inc.
3. *Directory of Chemistry Software 1992*. Eds. W. Warr, P. Willett, G. Downs, Oxford, U.K.: Cherwell Scientific Publishing, and Washington, D.C.: American Chemical Society.
4. Boyd, D.R. 1990. Appendix: Compendium of software for molecular modelling. In *Reviews in Computational Chemistry*, Eds. K.B. Lipkowitz, D.B. Boyd, pp. 383–392. New York, NY: VCH Publishers, Inc.
5. Boyd, D.R. 1991. Appendix: Compendium of software for molecular modelling. In *Reviews in Computational Chemistry II*, Eds. K.B. Lipkowitz, D.B. Boyd, pp. 481–498. New York, NY: VCH Publishers, Inc.
6. Rogers, D.W. 1990. *Computational Chemistry Using the PC*, New York, NY.:VCH Publishers, Inc.
7. Stewart, J.J.P. 1990. MOPAC: a semiempirical molecular orbital program. *J. Computer-Aided Design*. 4:1–105.
8. Zerner, M.C. 1991. Semiempirical Molecular Orbital Methods. In *Reviews in Computational Chemistry II*, Eds. K.B. Lipkowitz, D.B. Boyd. 8:313–365. New York, NY.: VCH Publishers, Inc.
9. Stewart, J.J.P. 1990. Semiempirical Molecular Orbital Methods. In *Reviews in Computational Chemistry*, Eds. K.B. Lipkowitz, D.B. Boyd. 2:45–82. New York, NY: VCH Publishers, Inc.
10. Richards, W.G. 1977. *Quantum Pharmacology*. Chap. 10, p. 115. Woburn, MA.: Butterworth Publishers Inc.
11. Katagi, T. 1992. Quantum mechanical estimation of environmental and metabolic fates of pesticides. *J. Pesticide Sci.* 17:5221–5330.
12. Streitwieser, Jr., A. 1961. *Molecular Orbital Theory for Organic Chemists*, New York, NY.:John Wiley & Sons, Inc.
13. Hinchliffe, A. 1988. *Computational Quantum Chemistry*, New York, NY: John Wiley & Sons, Inc.
14. Boyd, D.R. 1990. Aspects of Molecular Modelling. In *Reviews in Computational Chemistry*. Eds. K.B. Lipkowitz, D.B. Boyd, pp. 321–354. New York, NY: VCH Publishers, Inc.
15. Rogers, D.W., McLafferty, F.J. 1991. *Ab Initio* molecular orbital calculations using the PC. *American Laboratory*, September, pp. 20–33.
16. Mulliken, R.S. 1955. Electronic population analysis on LCAO-MO molecular wave functions. I. *J. Chem. Phys.* 23:1833–1839.
17. Mulliken, R.S. 1955. Electronic population analysis on LCAO-MO molecular wave functions. II. Overlap populations, bond orders, and covalent bond energies. *J. Chem. Phys.* 23:1841–1846.

18. Purcell, W.P., Bass, G.E., Clayton, J.M. 1973. *Strategy of Drug Design: A Guide to Biological Activity*, pp. 1–65. New York, NY: John Wiley & Sons, Inc.
19. Duchamp, D.J. 1979. Molecular mechanics and crystal structure analysis in drug design. In *Computer-Assisted Drug Design*, Eds. E.C. Olson, R.E. Christofferson, ACS Symposium Series 112, 3:79–102. Washington, DC, American Chemical Society.
20. Jurs, P.C., Chou, J.T., Yuan, M. 1979. Studies of chemical structure-biological activity relations using pattern recognition. 1979. In *Computer-Assisted Drug Design*, Eds. E.C. Olson, R.E. Christofferson, ACS Symposium Series 112, 4:103–129. Washington, DC: American Chemical Society.
21. Hansch, C., Deutsch, E.W., Smith, R.N. 1965. The use of substituent constants and regression analysis in the study of enzymatic reaction mechanisms. *J. Am. Chem. Soc.* 87:2738–2742.
22. Hansch, C. 1968. The use of substituent constants in structure-activity studies. In *Physico-Chemical Aspects of Drug Action*.(also: *Proc. Int. Pharmacol. Meet.*, 3rd., Vol. 7) Ed. E.J. Ariens. pp. 141–167. New York, NY:Pergamon Press.
23. Hermann, R.B., Culp, H.W., McMahon, R.E., Marsh M.M. 1969. Structure–activity relationships among substrates for a rabbit kidney reductase. Quantum chemical calculation of substituent parameters. *J. Med. Chem.* 12:749–754.
24. Lien, E.J.C. 1970. Structure-activity correlations for anticonvulsant drugs. *J. Med. Chem.* 13:1189–1191.
25. Martin, Y.C. 1970. Electronic factors in drug-receptor interactions. *J. Med. Chem.* 13:145–147.
26. Cammarata, A., Stein, R.L. 1968. Molecular orbital methods in the study of cholinesterase inhibitors. *J. Med Chem.* 11:829–833.
27. Rogers, K.S., Cammarata, A. 1969. Superdelocalizability and charge density. A correlation with partition coefficients. *J. Med. Chem.* 12:692–693.
28. Rogers, K.S., Cammarata, A. 1969. A molecular orbital description of the partitioning of aromatic compounds between polar and nonpolar phases. *Biochim. Biophys. Acta.* 193:22–29.
29. Cammarata, A. 1968. Some electronic factors in drug-receptor interactions. *J. Med. Chem.* 11:1111–1115.
30. Wohl, A.J. 1970. Electronic molecular pharmacology: the benzothiadiazine antihypertensive agents. II. Multiple regression analyses relating biological potency and electronic structure. *Mol. Pharmacol.* 6:195–205.
31. Allen, R.C., Carlson, G.L., Cavallito, C.J. 1970. Choline acetyltransferase inhibitors: physicochemical properties in relation to inhibitory activity of styrylpyridine analogs. *J. Med Chem.* 13:909–912.
32. Mekenyan, O., Bonchev, D. 1986. OASIS method for predicting biological activity of chemical compounds. *Acta Pharm. Jugosl.* 36:225–237.
33. Osmialowski, K., Halkiewicz, K., Radecki, A., Kaliszan, R. 1985. Quantum chemical parameters in correlation analysis of gas-liquid chromatographic retention indices of amines. *J. Chromatography.* 346:53–60.
34. Osmialowski, K., Halkiewicz, K., Kaliszan, R. 1986. Quantum chemical parameters in correlation analysis of gas–liquid chromatographic retention indices of amines. II. Topological Electronic Index. *J. Chromatography.* 361:63–69.
35. Kaliszan, R. 1987. Quantitative structure–retention relationships in chromatography. *Chromatography.* June, pp 19–29.
36. Koester, C.J., Hites, R.A. 1988. Calculated physical properties of polychlorinated dibenzo-*p*-dioxins and dibenzofurans. *Chemosphere.* 17:2355–2362.

37. Makino, M., Kamiya, M., Matsushita, H. 1992. Computer-assisted prediction of gas chromatographic retention times of polychlorinated biphenyls by use of quantum chemical molecular properties. *Chemosphere*. 25:1839–1849.

38. Chastrette, M., Rajzmann, M., Chanon, M., Purcell, K. 1985. Approach to a general classification of solvents using a multivariate statistical treatment of quantitative solvent parameters. *J. Am. Chem. Soc.* 107:1–11.

39. Lewis, D.F.V. 1987. Molecular orbital calculations on solvents and other small molecules: correlation between electronic and molecular properties μ, π*, and β. *J. Comput. Chem.* 8:1084–1089.

40. Klopman, G., Iroff, L.D. 1981. Calculation of partition coefficients by the charge density method. *J. Comput. Chem.* 2:157–160.

41. Peradejordi, F., Martin, A.N., Chalvet, O., Daudel, R. 1972. Molecular orbital calculations on some nitrogen derivatives of conjugated hydrocarbons: Base strength of benzacridines and their amino derivatives. *J. Pharm. Sci.* 61:909–913.

42. Hickey, J.P., Passino-Reader, D.R. 1991. Linear solvation energy relationships: "Rules of thumb" for estimation of variable values. *Environ. Sci. Technol.* 25:1753–1760.

43. Famini, G.R. 1988. *Using Theoretical Descriptors In Structural Activity Relationships I. Molecular Volume*; CRDEC-TR-88031; U.S. Army Chemical Research Development and Engineering Center: Aberdeen Proving Ground, MD, January.

44. Famini, G.R. 1988. *Using Theoretical Descriptors In Structural Activity Relationships II. Polarizability Index*. CRDEC-TR-88137; U.S. Army Chemical Research Development and Engineering Center: Aberdeen Proving Ground, MD, September.

45. Famini, G.R. 1988. *Using Theoretical Descriptors In Structural Activity Relationships IV. Molecular Orbital Basicity and Electrostatic Basicity*; CRDEC-TR-085; U.S. Army Chemical Research Development and Engineering Center: Aberdeen Proving Ground, MD, November.

46. Famini, G.R. 1989. *Using Theoretical Descriptors In Structural Activity Relationships V. A Review of the Theoretical Parameters*; CRDEC-TR-085; U.S. Army Chemical Research Development and Engineering Center: Aberdeen Proving Ground, MD, July.

47. Sasaki, Y., Suzuki, M. 1970. Studies on the proton magnetic resonance spectra in aromatic systems. XVI. On the electronic indices of substituted benzene and naphthalene derivatives. *Chem. Pharm. Bull.* 18:1759–1773.

48. Sasaki, Y., Suzuki, M. 1970. Studies on the proton magnetic resonance spectra in aromatic systems. XVII. On the electronic indices of 3- and 4-substituted pyridine, 6-substituted quinoline, and 6-substituted quinoxaline derivatives. *Chem. Pharm. Bull.* 18:1774–1787.

49. Dearden, J.C., Nicholson, R.N. 1987. Correlation of biodegradability with atomic charge difference and superdelocalizability. In *QSAR in Environmental Toxicology — II*, Ed. K.L.E. Kaiser, pp. 83–89. Amsterdam, The Netherlands: Reidel Publishing Co.

50. Cammarata, A. 1969. An analysis of biological linear free-energy relations. *J. Med. Chem.* 12:314–317.

51. Green, J.P., Kang, S. 1970. The correlation of electronic structures of indole derivatives with their biological activities. In *Molecular Orbital Studies in Chemical Pharmacology*. Ed. L.B. Kier. pp. 105–120. New York, NY: Springer Verlag.

52. Kannan, N., Wakimoto, T., Tatsukawa, R. 1989. Possible involvement of frontier (π) electrons in the metabolism of polychlorinated biphenyls (PCBs). *Chemosphere.* 18:1955–1963.

53. Nagata, C., Fukui, K., Yonezawa, T., Tagashica, Y. 1955. Electronic structure and carcinogenic activity of aromatic compounds I. Condensed aromatic hydrocarbons. *Cancer Res.* 15:233–239.

54. Miyashita, Y., Takahashi, Y., Daiba, S., Abe, H., Sasaki, S. 1982. Computer-assisted structure–carcinogenicity studies on polynuclear aromatic hydrocarbons by pattern recognition methods. The role of the bay and L-regions. *Anal. Chim. Acta.* 143:35–44.

55. Loew, G.H., Phillips, J., Wong, J., Hjelmeland, L,. Pack, G. 1978. Quantum chemical studies of the metabolism of polycyclic aromatic hydrocarbons: bay region reactivity as a criterion for carcinogenic potency. *Cancer Biochem. Biophys.* 2:113–122.

56. Loew, G.H., Sudhindra, B.S., Ferrel, J.E. Jr. 1979. Quantum chemical studies of polycyclic aromatic hydrocarbons and their metabolites: correlation to carcinogenicity. *Chem. Biol. Interact.* 26:75–89.

57. Politzer, P., Daiker, K.C., Donnelly, R.A. 1976. Molecular electrostatic potentials: a new approach to the study of the metabolic and carcinogenic activities of hydrocarbons. *Cancer Lett.* 2(1):17–23.

58. Miyashita, Y., Seki, T., Yotsui, Y., Yamazaki, K., Sano, M., Abe, H., Sasaki, S. 1982. Quantitative structure-activity relations in pyrazolylpyrimidine derivatives for their analgesic activities. *Bull. Chem. Soc. Jpn.* 55:1489–1492.

59. Lukovits, I. 1983. Quantitative structure-activity relationships employing independent quantum chemical indexes. *J. Med. Chem.* 26:1104–1109.

60. Veerkamp, V., Serné, P., Hutzinger, O. 1983. Application of molecular orbital calculations to PCDD and PCDF metabolism and toxicity. *Chemosphere.* 12:581–584.

61. Sæthre, L.J., Falk-Petersen, I.-B., Sydnes, L.K., Lønning, S., Naley, A.M. 1984. Toxicity and chemical reactivity of naphthalene and methylnaphthalenes. *Aquat. Toxicol.* 5:291–306.

62. Loew, G.H., Poulsen, M., Kirkjian, E., Ferrell, J., Sudhindra, B.S., Rebagliati, M. 1985. Computer-assisted mechanistic structure-activity studies: application to diverse classes of chemical carcinogens. *Environ. Health. Perspect.* 61:69–96.

63. Gough, K.M., Kaiser, K.L. 1988. QSAR of the acute toxicity of para-substituted nitrobenzene and aniline derivatives to Photobacterium phosphoreum. In *Proceedings of QSAR '88: The Third International Workshop on Quantitative Structure-Activity Relationships In Environmental Toxicology,* Eds. J.E. Turner, M.W. England, T.W. Schultz, N.J. Kwaak. pp. 111–121. Knoxville, Tennessee. May 22–26, 1988.

64. Nilsson, L.M., Carter, R.E., Sterner, O., Liljefors, T. 1988. Structure-activity relationships for unsaturated dialdehydes. 2. A PLS correlation of theoretical descriptors for six compounds with mutagenic activity in the Ames salmonella assay. *Quant. Struct.-Act. Relat.* 7:84–91.

65. Purdy, R. 1988. Quantitative structure-activity relationships for predicting toxicity of nitrobenzenes, phenols, anilines, and alkylamines to fathead minnows. In *Proceedings of QSAR '88: The Third International Workshop on Quantitative Structure–Activity Relationships In Environmental Toxicology,* Eds. J.E. Turner, M.W. England, T.W. Schultz, N.J. Kwaak. pp. 99–110. Knoxville, Tennessee. May 22–26, 1988.

66. Schüürman, G., Klein, W. 1988. Quantum chemical parameters for the QSAR description of the acute fish toxicity. In *Proceedings of QSAR '88: The Third International Workshop on Quantitative Structure–Activity Relationships In Environmental Toxicology*, Eds. J.E. Turner, M.W. England, T.W. Schultz, and N.J. Kwaak. p. 205. Knoxville, Tennessee. May 22–26, 1988.

67. Schüürman, G. 1990. QSAR analysis of the acute fish toxicity of organic phosphorothionates using theoretically derived molecular descriptors. *Environ. Tox. Chem.* 9:417–428.

68. Nendza, M., Dittrich, B., Wenzel, A., Klein, W. 1991. Predictive QSAR models for estimating ecotoxic hazard of plant-protecting agents: target and non-target toxicity. *Sci. Total Environ.* 109–110:527–535. (special issue: QSAR in environmental toxicology).

69. Wilson, L.Y., Famini, G.R. 1991. Using descriptors in quantitative structure-activity relationships: some toxicological indices. *J. Med. Chem.* 34:1668–1674.

70. Johnson, C.L. and Green, J.P. 1974. Molecular orbital studies on tryptamines active on the LSD receptor of the rat fundus strip. *Int. J. Quantum Chem., Quantum Biol. Symp.* 1:159–167.

71. Testa, B. 1981. Structural and electronic factors influencing the inhibition of aniline hydroxylation by alcohols and their binding to cytochrome P-450. *Chem. Biol. Interact.* 34:287–300.

72. Hayes, D.M., Kollman, P.A. 1976. Electrostatic potentials of proteins. 1. Carboxypeptidase A. *J. Am. Chem. Soc.* 98:3335–3345.

73. Hayes, D.M., Kollman, P.A. 1976. Electrostatic potentials of proteins. 2. Role of electrostatics in a possible catalytic mechanism for carboxypeptidase A. *J. Am. Chem. Soc.* 98:7811–7816.

74. Šolmajer, T., Hodošček, M., Hadži, D., Lukovits, I. 1984. Correlation between affinity towards β-adrenergic receptors and electrostatic potentials of phenylethylamine derivatives. *Quant. Struct. Act. Relat.* 3:51–55.

75. Scrocco, E., Tomasi, J. 1973. The electrostatic molecular potential as a tool for the interpretation of molecular properties. *Fortschr. Chem. Forsch.* 42:95–170.

76. Weinstein, H., Rabinowitz, J. Liebman, M.N., Osman, R. 1985. Determinants of molecular reactivity as criteria for predicting toxicity: problems and approaches. *Environ. Health. Perspec.* 61:147–162.

77. Cheney, B.V., Tolly, T. 1979. Electronic factors affecting receptor binding of dibenzo-p-dioxins and dibenzofurans. *Int. J. Quant. Chem.* 16:87–110, and references therein.

78. Waller, C.L., McKinney, J.D. 1992. Comparative molecular field analysis of polyhalogenated dibenzo-p-dioxins, dibenzofurans and biphenyls. *J. Med Chem.* 35:3660–3666.

79. Kobayashi, S., Saito, A., Ishii, Y., Tanaka, A., Tobinaga, S. 1991. Relationship between the biological potency of polychlorinated dibenzo-p-dioxins and their electronic states. *Chem. Pharm. Bull.* 39:2100–2105.

80. Kafafi, S.A., Afeefy, H.Y., Said, H.K., Hakimi, J.M. 1992. A new structure-activity model for Ah receptor binding. Polychlorinated dibenzo-p-dioxins and dibenzofurans. *Chem. Res. Toxicol.* 5:856–862.

81. Kafafi, S.A., Said, H.K., Mahmoud, M.I., Afeefy, H.Y. 1992. The electronic and thermodynamic aspects of Ah receptor binding. A new structure–activity model: I. The polychlorinated dibenzo-p-dioxins. *Carcinogenesis.* 13:1599–1605.

82. Cammarata, A. 1970. Quantum perturbation theory and linear free energy relationships in the study of drug action. In *Molecular Orbital Studies in Chemical Pharmacology*, Ed. L.B. Kier, pp. 156–190. New York, NY: Springer-Verlag.

83. Redl, G., Cramer, R.D. III, Berkoff, C.E. 1974. Quantitative drug design. *Chem. Soc. Rev.* 3:273–292.

84. Hadži, D., Koller, J., Hodǒšcek, M., Kocjan, D. 1987. Molecular electrostatic potential — a critical assessment of its role in drug-receptor recognition. In *Modelling of Structure and Properties of Molecules,* Ed. Maksíc, Z.B. New York, NY: Halsead Press, John Wiley & Sons,

85. Lien, E.J., Guo, Z.R., Li, R.L., Su, C.T. 1982. Use of dipole moment as a parameter in drug–receptor interaction and quantitative structure-activity relationship studies. *J. Pharm. Sci.* 71:641–655.

86. Donné-Op den Kelder, G.M., Haaksma, E.E.J., van der Schaar, M.G.W., Veenstra, D.M.J., Timmerman, H. 1988. QSAR studies on mifentidine and related compounds. *Quant. Struct. Act. Relat.* 7:60–71.

87. Kakeya, N., Yata, N., Kamada, A., Aoki, M. 1969. Biological activities of drugs. VIII. Structure–activity relation of sulfonamide carbonic anhydrase inhibitors. (3). *Chem. Pharm. Bull.* 17:2558–2564.

88. Kakeya, N., Aoki, M., Kamada, A., Yata, N. 1969. Biological activities of drugs. Part VI. Structure activity relationship of sulfonamide carbonic anhydrase inhibitors (1). *Chem. Pharm. Bull.* 17:1010–1018.

89. Kakeya, N., Yata, N., Kamada, A., Aoki, M. 1970. Biological activities of drugs. IX. Structure–activity relation of sulfonamide carbonic anhydrase inhibitors (4). *Chem. Pharm. Bull.* 18:191–194.

90. Klopman, G. 1974. Chemical reactivity and reaction paths: general introduction. In *Chemical Reactivity and Reaction Paths,* Ed. G. Klopman. pp. 1–11. New York, NY: John Wiley & Sons.

91. Simonetta, M. 1974. Empirical and semi-empirical calculations of chemical reactivity and reaction paths. In *Chemical Reactivity and Reaction Paths*, Ed. G. Klopman. pp. 13–21. New York, NY.:John Wiley & Sons.

92. Klopman, G. 1974. The generalized perturbation theory of chemical reactivity and its applications. In *Chemical Reactivity and Reaction Paths*, Ed. G. Klopman. pp. 55–165. New York, NY: John Wiley & Sons.

93. van de Waterbeemd, H., Testa, B. 1987. The Parameterization of Lipophilicity and Other Structural Properties in Drug Design. In *Advances in Drug Research,* Vol. 16, pp. 85–225 New York, NY: Academic Press.

94. Fukui, K., Nagata, C., Yonezawa, T. 1958. Electronic structure and auxin activity of benzoic acid derivatives. *J. Am. Chem. Soc.* 80:2267–2270.

95. Fujita, T., Komazawa, T., Koshimizu, K., Mitsui, T. 1961. Studies on plant growth substances. Part XVI. Plant growth activity of substituted 1-naphthoic acids (III). *Agr. Biol. Chem.* 25:719–725.

96. Crow, J., Wassermann, O., Holland, W.C. 1969. Molecular orbital calculations on a new series of substituted-phenyl choline ethers. *J. Med. Chem.* 12:764–766.

97. Peradejordi, F., Martin, A.N., Cammarata, A. 1971. Quantum chemical approach to structure–activity relations of tetracycline antibiotics. *J. Pharm. Sci.* 60:576–582.

98. Dewar, M.J.S. 1965. Chemical Reactivity. *Adv. Chem. Phys.* 8:65–131.

99. Dewar, M.J.S. 1952. A molecular orbital theory of organic chemistry. I. General principles. *J. Am. Chem. Soc.* 74:3341–3345.

100. Dewar, M.J.S. 1952. A molecular orbital theory of organic chemistry. II. The structure of mesomeric systems. *J. Am. Chem. Soc.* 74:3345–3350.

101. Dewar, M.J.S. 1952. A molecular orbital theory of organic chemistry. III. Charge displacement and electromeric substituents. *J. Am. Chem. Soc.* 74:3350–3353.

102. Dewar, M.J.S. 1952. A molecular orbital theory of organic chemistry. IV. Free radicals. *J. Am. Chem. Soc.* 74:3353–3354.
103. Dewar, M.J.S. 1952. A molecular orbital theory of organic chemistry. V. *J. Am. Chem. Soc.* 74:3355–3356.
104. Dewar, M.J.S. 1952. A molecular orbital theory of organic chemistry. VI. Aromatic substitution and addition. *J. Am. Chem. Soc.* 74:3357–3363.
105. Wheland, G.W. 1942. A quantum mechanical investigation of the orientation of substituents in aromatic molecules. *J. Am. Chem. Soc.* 64:900–908.
106. Brown, R.D. 1952. Molecular orbitals and organic reactions. *Quart. Rev.* 6:63–99.
107. Fukui, K., Yonezawa, T., Shingu, H. 1952. A molecular orbital theory of reactivity in aromatic hydrocarbons. *J. Chem. Phys.* 20:722–725.
108. Fukui, K., Yonezawa, T., Nagata, C. 1954. Theory of Substitution in Conjugated Molecules. *Bull. Chem. Soc. Jpn.* 27:423–427.
109. Fukui, K., Yonezawa, T., Nagata, C. 1957. MO-Theoretical approach to the mechanism of charge transfer in the process of aromatic substitutions. *J. Chem. Phys.* 27:1247–1259.
110. Fukui, K., Nagata, C. 1959. Reply to the comments on the "Frontier Electron Theory". *J. Chem. Phys.* 31:550–551.
111. Fukui, K., Kato, H., Yonezawa, T. 1961. A new quantum-mechanical reactivity index for saturated compounds. *Bull. Chem. Soc. Jpn.* 34:1111–1115.
112. Klopman, G. 1964. A semiempirical treatment of molecular structures. II. Molecular terms and application to diatomic molecules. *J. Am. Chem. Soc.* 86:4550–4557.
113. Klopman, G. 1967. Solvations: A semiempirical procedure for including solvation in quantum mechanical calculations of large molecules. *Chem. Phys. Lett.* 1:200–202.
114. Klopman, G. 1968. Chemical reactivity and the concept of charge- and frontier-controlled reactions. *J. Am. Chem. Soc.* 90:223–234.
115. Klopman, G., Hudson, R.F. 1967. Polyelectronic perturbation treatment of chemical reactivity. *Theoret. Chim. Acta.* 8:165–174.
116. Nichols, D.E., Shulgin, A.T., Dyer, D.C. 1977. Directional lipophilic character in a series of psychotomimetic phenethylamine derivatives. *Life Sci.* 21:569–576.
117. Hansch, C. 1977. On the predictive value of QSAR. In *Biological Activity and Chemical Structure*, Ed. J.A.K. Buisman. Pharmacochemistry Library, Vol. 2. pp 47–61; 279–306. Amsterdam, The Netherlands:Elsevier Press.
118. Hanna, M.W. 1969. *Quantum Mechanics in Chemistry*, Second edition. New York, NY: W.A. Benjamin, Inc.
119. Testa, B. 1979. *Principles of Organic Stereochemistry*, pp. 201–208. New York, NY:Marcel Dekker.
120. Coulson, C.A. 1939. The electronic structure of some polyenes and aromatic molecules. VII. Bonds of fractional order by the molecular orbital method. *Proc. R. Soc.* 169A:413–428.
121. Mullay, J. 1986. A simple method for calculating atomic charges in molecules. *J. Am. Chem. Soc.* 108:1770–1775 (and references therein).
122. Andrews, P.R. 1972. Are calculated electron populations suitable parameters for multiple regression analyses of biological activity? *J. Med Chem.* 15:1069–1072.
123. Wohl, A.J. 1971. A Molecular Orbital Approach to Quantitative Drug Design. In *Drug Design*, Vol. 1, pp. 381–404. New York, NY: Academic Press.

124. Cammarata, A., Rogers., K.S. 1971. Electronic representation of the lipophilic parameter π. *J. Med. Chem.* 14:269–274.

125. Zavoruev, S.M., Bolotin, V.A. 1982. Relation between values for the distribution coefficients of substituted 1-phenyl-3,3-dialkyltriazenes and quantum chemical parameters. *Khim.-Farm. Zh.* 16:1361–1364.

126. Abdul-Ahad, P.G., Webb, G.A. 1982. Trends in dehydrogenase inhibitory potencies of some quinolones, using quantum chemical indices. *Eur. J. Med. Chem. Chim. Ther.* 17:301–306.

127. Abdul-Ahad, P.G., Webb, G.A. 1982. The use of molecular orbital calculations in the interpretation of the variation in enzyme inhibitory potency of a series of substituted pteridines and triazines. *J. Mol. Struct. THEOCHEM* 6:25–33.

128. Coulson, C.A., Longuet-Higgins, H.C. 1947. The electronic structure of conjugated systems. I. General theory. *Proc. R. Soc. (Lond.). Ser. A.* 191:39–60.

129. Hetnarski, B., O'Brien, R.D. 1975. The charge-transfer constant. A new substituent constant for structure–activity relationships. *J. Med. Chem.* 18:29–33.

130. Lien, E.J. 1969. The use of substituent constants and regression analysis in the study of structure-activity relationship. *Am. J. Pharm. Educ.* 33:368–375.

131. Weinstein, H., Chou, D., Kang, S., Johnson, C.L., Green, J.P. 1976. Reactivity characteristics of large molecules and their biological activity: indolealkylamines on the LSD/serotonin receptor. *Int. J. Quantum Chem., Quantum Biol. Symp.* 3:135–150.

132. Green, J.P., Johnson, C.L., Weinstein, H. 1978. Histamine as a neurotransmitter. In *Psychopharmacology: A Generation of Progress.* Eds. M.A. Lipton, A. DiMascio, K.F. Killam, pp. 319–332. New York, NY: Raven Press.

133. Green, J.P., Johnson, C.L., Weinstein, H., Kang, S., Chou, D. 1978. Molecular determinants for interaction with the LSD receptor: biological studies and quantum chemical analysis. In *The Psychopharmacology of Hallucinogens,* Eds. R.C. Stillman, R.E. Willette. pp. 28–60. New York, NY: Pergamon Press.

134. Mulliken, R.S. 1934. A new electronegativity scale; together with data on valence states and on valence ionization potentials and electron affinities. *J. Chem. Phys.* 2:782–793.

135. Agin, D., Hersch, L., Holtzman, D. 1965. The action of anesthetics on excitable membranes: a quantum-chemical analysis. *Proc. Natl. Acad. Sci. USA.* 53:952–958.

136. Parr, R.G., Pearson, R.G. 1983. Absolute hardness: companion parameter to absolute electronegativity. *J. Am. Chem. Soc.* 105:7512–7516.

137. Cheney, B.V., Christoffersen, R.E. 1983. Structure-activity correlations for a series of antiallergy agents. 3. Development of quantitative model. *J. Med. Chem.* 26:726–737.

138. Green, J.P., Malrieu, J.P. 1965. Quantum chemical studies of charge–transfer complexes of indoles. *Proc. Nat. Acad. Sci. USA.* 54:659–664.

139. Hansch, C. 1969. A quantitative approach to biochemical structure-activity relationships. *Accounts Chem. Res.* 2:232–239.

140. Miertus, S., Filipovic, P. 1982. Relationship between electronic structure and biological activity of 2-formylpyridine thiosemicarbazones. *Eur. J. Med. Chem.-Chim. Ther.* 17:145–148

141. Fujimoto, H., Fukui, K. 1974. Intermolecular interactions and chemical reactivity. In *Chemical Reactivity and Reaction Paths.* Ed. G. Klopman, pp. 23–54. New York, NY: John Wiley & Sons.

142. Leo, A., Hansch, C., Church, C. 1969. Comparison of parameters currently used in the study of structure-activity relationships. *J. Med. Chem.* 12:766–771.

143. Miller, K.J., Savchik, J. 1979. A new empirical method to calculate average molecular polarizabilities. *J. Am. Chem. Soc.* 101:7206–7213.
144. Koch, R. 1982. Ökotoxikologische stoffbewertung und strukturchemische parameter *Chemosphere.* 11:497–509.
145. Nadasdi, L., Medzihradszky, K. 1983. The use of a steric parameter (Y-γ) in QSAR calculations for peptide-hormones. *Peptides.* 4:137–144.
146. Dewar, M.J.S, Stewart, J.J.P. 1984. A new procedure for calculating molecular polarizabilities; applications using MNDO. *Chem. Phys. Lett.* 111:416–420.
147. Cammarata, A. 1967. An apparent correlation between the *in vitro* activity of chloramphenicol analogs and electronic polarizability. *J. Med Chem.* 10:525–527.
148. Weinstein, H. 1975. Some new quantum chemical procedures for the analysis of drug-receptor interactions. *Int. J. Quantum Chem., Quantum Biol. Symp.* 2:59–69.
149. Abdul-Ahad, P.G., Webb, G.A. 1982. Quantitative structure–activity relationships for some antitumor platinum(II) complexes. *Int. J. Quantum Chem.* 21:1105–1115.
150. Weinstein, H., Srebrenik, S., Paunscz, R., Maayani, S., Cohen, S., Sokolovsky, M. 1974. Characterization of drug reactivity in cholinergic systems by molecular interaction potentials. In *Chemical and Biochemical Reactivity*, Eds. E.D. Bergmann, B. Pullman, pp. 493–512. Dordrecht, Holland: D. Reidel Publ. Co.
151. Petrongolo, C., Macchia, B., Macchia, F., Martinelli, A. 1977. Molecular orbital studies on the mechanism of drug-receptor interaction. 2. β-Adrenergic drugs. An approach to explain the role of the aromatic moiety. *J. Med. Chem.* 20:1645–1653.
152. Weinstein, H., Srebrenik, S., Maayani, S., Sokolovsky, M. 1977. A theoretical model study of the comparative effectiveness of atropine and scopolamine action in the central nervous system. *J. Theor. Biol.* 64:295–309.
153. Weinstein, H., Osman, R. 1977. Models for molecular mechanisms in drug-receptor interactions. Serotonin and 5-hydroxyindole complexes with imidazolium cation. *Int. J. Quantum Chem., Quantum Biol. Symp.* 4:253–268.
154. Petrongolo, C. 1978. Quantum chemical study of isolated and interacting molecules with biological activity. *Gazz. Chim. Ital.* 108:445–478.
155. Weinstein, H., Osman, R., Edwards, W.D., Green, J.P. 1978. Theoretical models for molecular mechanisms in biological systems: tryptamine congeners acting on an LSD-serotonin receptor. *Int. J. Quantum Chem., Quantum Biol. Symp.* 5:449–461.
156. Weinstein, H., Osman, R., Green, J.P. 1979. The molecular basis of structure-activity relationships: quantum chemical recognition mechanisms in drug-receptor interactions. In *Computer-Assisted Drug Design*, Eds. E.C. Olson, R.E. Christoffersen. ACS Symposium Series 112, pp.161–187. Washington, D.C.: American Chemical Society.
157. Cammarata, A. 1972. Interrelation of the regression models used for structure–activity analyses. *J. Med Chem.* 15:573–577.

chapter twenty-three

Linear solvation energy relationships (LSER): "rules of thumb" for $V_i/100$, π^*, β_m, α_m estimation and use in aquatic toxicology

James P. Hickey

Introduction

Researchers, manufacturers, and regulating agencies must evaluate properties for chemicals that are either present in or could be released into the environment. The routine use of over 70,000 synthetic chemicals stresses the need for this information. However, minimal physical data and no toxicity data are available for about 80% of these compounds. The cost of testing these myriad present or potential chemicals is prohibitive, so researchers and managers increasingly rely on predictive models, i.e., quantitative structure–activity relationships (QSARs) for chemical property estimation, hazard evaluation and information to direct research and set priorities.

The Linear Solvation Energy Relationship (LSER) is recognized as a powerful predictive model, quite suitable for estimations of aquatic toxicity.[1] In the LSER model[2-8] many solubility and solvent-dependent properties of a substance (e.g., toxicity) are related to its chemical structure by a linear combination of three simple and conceptually explicit types of energy terms:

$$\text{Property} = \underset{\text{Term}}{\text{Cavity}} + \underset{\text{Term}}{\text{Dipolar}} + \underset{\text{Terms}}{\text{Hydrogen Bonding}} \tag{1}$$

Each fragment of the compound contributes to the energy required both to organize solvent (water or biosystem medium) molecules into a suitably-

1-56670-149-X/96/$0.00+$.50
© 1996 by CRC Press, Inc.

configured cavity around the solute molecule (cavity term, $mV_i/100$) and to contain the molecule in that cavity through formation of electrostatic (dipolar term, $s\pi^*$) and hydrogen bonds (hydrogen bonding terms, $b\beta_m$ and $a\alpha_m$) between the solute molecule and the medium. In practice, all four energy terms are used without units.

The general form of the equation presented here deals with the properties of a set of different solutes in a single solvent, or with distributions between pairs of solvents, and relates specifically to the solute variables.

$$\text{Property} = mV_i/100 + s\pi^* + b\beta_m + a\alpha_m \tag{2}$$

where: V_i is the intrinsic (van der Waals) molecular volume; π^* scales solute ability to stabilize a neighboring charge or dipole by non-specific dielectric interactions; β_m and α_m scale solute ability to accept or donate a hydrogen (or donate or accept an electron pair) in a hydrogen bond; and the coefficients m, s, b, and a are constants for a particular set of conditions, determined by multiple linear regression of the LSER variable values for a series of chemicals with the measured value for a particular chemical property.

This chapter provides a listing of the increasing variety of organic moieties and heteroatom groups for which LSER values are available, and the LSER variable estimation rules.[2] The listings include values for typical nitrogen-, sulfur- and phosphorus-containing moieties, and general organosilicon and organotin groups. The contributions by an ion pair situation to the LSER values are also offered in Table 1, allowing estimation of parameters for salts and zwitterions. The guidelines permit quick estimation of values for the four primary LSER variables $V_i/100$, π^*, β, and α by summing the contributions from its components. The use of guidelines and Table 1 significantly simplifies computation of values for the LSER variables for most possible organic compounds in the environment, including the larger compounds of environmental and biological interest.

Materials required

The high-tech equipment required for this technique is primarily a pad of paper, pencils with erasers, a pocket calculator (optional), and the values and guidelines presented in this chapter. (Fluency in organic chemistry and a creative imagination are also quite useful.) Access to a desktop PC is almost a necessity, in order to store all of your hard-won sets of LSER parameters, and to perform the regression analyses described later on. A working knowledge of statistical methods (or access to a statistician) will also help.

Procedure

The general computation rules[2] presented here and in Appendix I for the estimation of the LSER variables are used with the current LSER variable contributions for typical molecular structures and functional groups in Table

Table 1 Basic Values for LSER Variables ($V_i/100$, π^*, β, α)

Compound	(a)	$V_i/100$	π^*	β	α
Aliphatic basis structures					
n-Butane		0.455	0.00	0.00	0.00
n-Pentane		0.553	0.00	0.00	0.00
2-Methylbutane		0.543	0.00	0.00	0.00
n-Hexane		0.648	0.00	0.00	0.00
2-Methylpentane		0.638	0.00	0.00	0.00
n-Heptane		0.745	0.00	0.00	0.00
2-Methylhexane		0.735	0.00	0.00	0.00
n-Octane		0.842	0.00	0.00	0.00
2-Methylheptane		0.832	0.00	0.00	0.00
2,2,4-Trimethylpentane		0.812	0.00	0.00	0.00
Cyclopropane		0.310	−0.02	0.00	0.00
Cyclobutane		0.450	−0.01	0.00	0.00
Cyclopentane		0.500	0.00	0.00	0.00
Cyclohexane		0.598	0.00	0.00	0.00
Cycloheptane		0.690	0.00	0.00	0.00
Cyclooctane		0.815	0.00	0.00	0.00
t-Octahydro-1H-indene		0.884	0.02	0.00	0.00
Decahydronaphthalene		0.982	0.02	0.00	0.00
Ethylene oxide		0.249	0.56	0.50	0.00
Tetrahydrofuran		0.455	0.58	0.51	0.00
Tetrahydropyran		0.553	0.51	0.50	0.00
1,4-Dioxane		0.508	0.55	0.41	0.00
Tetrahydrothiophene		0.509	0.44	0.27	0.00
Pyrrolidine		0.460	0.14	0.70	0.00
Imidazolidine		0.431	0.27	0.54	0.00
Piperidine		0.556	0.17	0.70	0.00
Aromatic basis structures					
Benzene		0.491	0.59	0.14	0.00
Indan		0.784	0.52	0.14	0.00
Tetrahydronaphthalene		0.883	0.50	0.14	0.00
Biphenyl		0.920	1.20	0.28	0.00
9H-Fluorene		0.960	1.18	0.25	0.00
Naphthalene		0.753	0.70	0.20	0.00
Azulene		0.753	0.90	0.35	0.00
Anthracene		1.015	0.81	0.20	0.00
Phenanthrene		1.015	0.81	0.20	0.00
Furan		0.370	0.40	0.35	0.00
Dibenzofuran		1.581	0.60	0.30	0.00
Dibenzo-*p*-dioxin		1.616	0.45	0.60	0.00
Thiophene		0.445	0.70	0.25	0.00
Pyrrole		0.428	0.74	0.69	0.41
Imidazole		0.401	0.87	0.64	0.41
Pyridine		0.472	0.87	0.43	0.00
Pyridazine		0.451	0.35	0.54	0.00
Pyrimidine		0.451	0.87	0.64	0.00

Techniques in aquatic toxicology

Table 1 (continued)   Basic Values for LSER Variables ($V_i/100$, π^*, β, α)

Compound	(a)	$V_i/100$	π^*	β	α
Pyrazine		0.451	0.92	0.69	0.00
Triazine		0.430	1.15	0.72	0.00
Hydrocarbons					
CH_3-, $-CH_2-$, $>CH-$ and $>C<$					
(prim, sec)	al	0.098	0.00	0.00	0.00
(tert, quart)	al	0.088	0.00	0.00	0.00
1–3 $-CH_3$	ar	0.098	−0.04	0.01	0.00
4–6 $-CH_3$	ar	0.098	−0.04	0.02	0.00
1–3 $-CH_2-$	ar	0.098	−0.02	0.01	0.00
4–6 $-CH_2-$	ar	0.098	−0.02	0.02	0.00
$-C_6H_5$	al/ar	0.491	0.59	0.14	0.00
$-C(C_6H_5)_3$	al/ar	1.485	1.45	0.40	0.00
Unsaturation and salts					
Olefin	al/ar	−0.026	0.10	0.10	0.05
Alkyne	al/ar	−0.036	0.20	0.20	0.13
$-C\#CCH_3$	al/ar	0.315	0.20	0.17	0.13
Ion Pair {+,−}	al/ar	0.000	0.50	0.50	0.00
Halogens					
$-F$	al	0.030	0.08	0.19	0.06
	ar	0.030	0.03	−0.05	0.08
	py	0.030	0.04	0.09	0.00
$-CF_3$	al/ar	0.188	0.25	−0.25	0.15
$-Cl$	al	0.090	0.35	0.15	0.06
	ar	0.090	0.12	−0.04	0.00
	ar,wog	0.090	0.05	−0.05	0.00
	py	0.090	0.04	0.09	0.00
$-CCl_3$	al	0.368	0.35	−0.15	0.15
	ar	0.368	0.35	−0.10	0.15
	py	0.368	0.35	−0.10	0.10
$-Br$	al	0.131	0.43	0.17	0.05
	ar	0.131	0.20	−0.08	0.10
	ar,wog	0.131	0.04	−0.04	0.00
	py	0.131	0.04	0.07	0.07
$-CH_2Br$	al	0.257	0.38	0.05	0.00
	ar	0.257	0.05	−0.05	0.00
$-CBr_3$	al	0.491	0.40	−0.10	0.12
	ar	0.491	0.40	−0.10	0.15
	py	0.491	0.40	−0.10	0.10
$-I$	al	0.181	0.45	0.18	0.04
	ar	0.181	0.22	0.02	0.10
	ar,wog	0.181	0.04	−0.04	0.00
	py	0.181	0.04	0.05	0.05

Table 1 (continued)   Basic Values for LSER Variables ($V_i/100$, π^*, β, α)

Compound	(a)	Variables			
		$V_i/100$	π^*	β	α
Carboxy derivatives					
–OH	al	0.045	0.40	0.47	0.33
	al,or	0.045	0.45	0.51	0.31
	ar	0.045	0.13	0.23	0.60
–OOH	al/ar	0.080	0.41	0.36	0.40
–O–	al	0.045	0.27	0.45	0.00
	al,ir	0.045	0.54	0.51	0.00
	ar	0.045	0.10	0.22	0.06
–OO–	al/ar	0.080	0.28	0.34	0.00
:C=O		0.098	0.81	0.65	0.00
–C(=O)–	al	0.098	0.67	0.48	0.00
	al,wog	0.098	0.30	0.35	0.00
	al,ir	0.098	0.76	0.52	0.00
	ar	0.098	0.39	0.39	0.06
HC(=O)H		0.140	0.69	0.43	0.00
–C(=O)H	al	0.115	0.65	0.41	0.00
	al,wog	0.115	0.33	0.33	0.00
	ar	0.115	0.33	0.42	0.00
HC(=O)OH		0.224	0.65	0.38	0.65
HC(=O)SH		0.294	0.55	0.25	0.05
–OC(=O)H	al/ar	0.225	0.62	0.37	0.00
–SC(=O)H	al/ar	0.294	0.55	0.30	0.00
–C(=O)OH	al	0.139	0.60	0.45	0.55
	ar	0.149	0.15	0.30	0.59
–C(=O)SH	al/ar	0.294	0.55	0.25	0.05
–C(=O)OOH	al/ar	0.174	0.61	0.34	0.62
–C(=O)O–	al	0.139	0.55	0.45	0.12
	al,or	0.139	0.55	0.49	0.12
	lactone, ir	0.139	0.68	0.51	0.12
	ar	0.139	0.17	0.29	0.12
	o-phthalate	0.683	0.68	0.76	0.12
–C(=O)S–	al/ar	0.294	0.50	0.30	0.00
–C(=O)OO–	al/ar	0.174	0.56	0.34	0.12
–C(=O)OC(=O)–	al/ar	0.395	0.65	0.55	0.00
–C(=O)OOC(=O)–	al/ar	0.430	0.66	0.44	0.00
–OC(=O)OOC(=O)O–	al/ar	0.522	0.46	0.30	0.12
R_zX–C(=O)–$X'R_y$					
–OC(=O)O–	al/ar	0.185	0.45	0.38	0.12
–OC(=O)OH	al/ar	0.185	0.55	0.48	0.55
HOC(=O)OH		0.185	0.45	0.60	0.65
–OC(=O)SH	al/ar	0.339	0.45	0.35	0.15
–OC(=O)S–	al/ar	0.339	0.42	0.30	0.14
–SC(=O)OH	al/ar	0.339	0.35	0.38	0.50
HSC(=O)SH		0.374	0.35	0.45	0.10

Table 1 (continued)   Basic Values for LSER Variables ($V_i/100$, π^*, β, α)

Compound	(a)	Variables			
		$V_i/100$	π^*	β	α
–SC(=O)SH	al/ar	0.374	0.45	0.38	0.05
–SC(=O)S–	al/ar	0.374	0.35	0.33	0.00
HSC(=O)SH		0.339	0.40	0.52	0.65
–OC(=O)NH$_2$	al/ar	0.202	0.48	0.78	0.55
–SC(=O)NH$_2$	al/ar	0.312	0.52	0.65	0.38
–HNC(=O)OH	al/ar	0.202	0.78	0.82	0.70
–HNC(=O)SH	al/ar	0.312	0.50	0.63	0.23
–OC(=O)NH–	al	0.202	0.76	0.62	0.36
	ar	0.202	0.76	0.57	0.36
–OC(=S)NH–	al/ar	0.270	0.56	0.42	0.36
–HNC(=O)S–	al/ar	0.312	0.50	0.63	0.19
–SC(=S)NH–	al/ar	0.312	0.53	0.40	0.39
>NC(=O)OH	al/ar	0.202	0.76	0.84	0.60
>NC(=O)SH	al/ar	0.312	0.48	0.62	0.05
>NC(=O)O–	al/ar	0.228	0.75	0.65	0.00
>NC(=O)S–	al/ar	0.312	0.48	0.60	0.00
H$_2$NC(=O)NH$_2$		0.265	0.90	0.74	0.76
–HNC(=O)NH$_2$	al/ar	0.265	0.89	0.75	0.65
>NC(=O)NH$_2$	al/ar	0.265	0.88	0.77	0.38
–HNC(=O)NH–	al/ar	0.265	0.87	0.77	0.38
–HNC(=S)NH–	al/ar	0.307	0.67	0.55	0.38
>NC(=O)NH–	al/ar	0.265	0.85	0.78	0.19
>NC(=O)N<	al/ar	0.265	0.83	0.74	0.00
Sulfur derivatives					
–SH	al	0.117	0.35	0.16	0.03
	ar	0.117	0.35	0.02	0.23
–S–	al/ar	0.117	0.36	0.28	0.00
–SS–	al/ar	0.234	0.58	0.10	0.00
–S(=O)–	al	0.150	1.00	0.78	0.00
–S(=O)$_2$O–	ar	0.154	1.00	0.62	0.00
	al/ar	0.221	0.85	0.55	0.00
–OS(=O)O–	al/ar	0.250	0.70	1.02	0.00
–S(=O)$_2$–	al	0.170	1.00	0.48	0.00
	ar	0.174	1.00	0.42	0.00
–S(=O)$_2$O–	al/ar	0.221	0.85	0.55	0.00
–OS(=O)$_2$O–	al/ar	0.270	0.70	0.72	0.00
–SO$_3$H	al/ar	0.266	1.00	0.76	0.75
Amides, Imines, Nitriles					
–C(=O)NH$_2$	al	0.185	0.95	0.74	0.56
	ar	0.185	0.35	0.65	0.49
–C(=O)NH–	al	0.183	0.85	0.74	0.25
	lactam, ir	0.183	0.72	0.70	0.28
	ar	0.183	0.30	0.65	0.30
–C(=S)NH–	al	0.225	0.65	0.52	0.31
	ar	0.225	0.65	0.48	0.44
–C(=O)N<	al	0.183	0.76	0.66	0.00

Table 1 (continued) Basic Values for LSER Variables ($V_i/100$, π^*, β, α)

Compound	(a)	$V_i/100$	π^*	β	α
	lactam, ir	0.183	0.74	0.80	0.00
	ar	0.185	0.35	0.65	0.00
HC(=O)NH$_2$		0.185	0.95	0.65	0.49
–HNC(=O)H	al/ar	0.185	0.91	0.67	0.25
>NC(=O)H	al/ar	0.185	0.80	0.66	0.00
–C(=O)NHC(=O)–	al/ar	0.430	0.70	0.60	0.33
–C(=O)N(R)C(=O)–	al/ar	0.430	0.65	0.70	0.00
–N=CH$_2$	al/ar	0.152	0.45	0.80	0.15
–N=CH–	al/ar	0.152	0.35	0.78	0.10
–N=C<	al/ar	0.152	0.30	0.75	0.00
>C=NOH	al/ar	0.197	0.55	0.45	0.32
–N=C=O	al/ar	0.206	0.75	0.35	0.00
–N=C=S	al/ar	0.278	0.63	0.22	0.00
–C≡N	al	0.100	0.65	0.44	0.22
	ar	0.099	0.20	0.37	0.22
	py	0.099	0.20	0.37	0.20
Amines, Hydrazines					
–NH$_2$	al	0.080	0.32	0.69	0.00
	ar	0.080	0.13	0.38	0.26
–NH–	al	0.080	0.25	0.70	0.00
	ar	0.080	0.13	0.30	0.17
–N<	al	0.080	0.15	0.65	0.00
	ar	0.080	0.13	0.73	0.00
–NH–NH$_2$	al	0.150	0.75	0.90	0.15
	ar	0.138	0.55	0.90	0.45
–NH–NH–	al	0.150	0.60	0.85	0.05
	ar	0.138	0.45	0.85	0.35
>N–NH$_2$	al	0.150	0.65	0.90	0.15
	ar	0.138	0.65	0.90	0.20
>N–NH–	al	0.150	0.50	0.85	0.00
	ar	0.138	0.33	0.85	0.17
>N–N<	al	0.150	0.30	0.80	0.00
	ar	0.138	0.30	0.75	0.00
–N=N–	al/ar	0.125	0.15	0.15	0.00
–NHOH	al	0.130	0.56	0.93	0.26
	ar	0.130	0.26	0.83	0.46
–NHO–	al	0.130	0.52	0.90	0.05
	ar	0.120	0.35	0.80	0.22
>NO–	al	0.128	0.35	0.90	0.05
	ar	0.116	0.35	0.70	0.05
>NOH	al	0.130	0.35	0.90	0.14
	ar	0.116	0.35	0.70	0.14
–NHSH	al	0.200	0.65	0.80	0.10
	ar	0.188	0.55	0.60	0.25
–NHS–	al	0.200	0.60	0.75	0.00
	ar	0.188	0.45	0.55	0.17

Table 1 (continued)   Basic Values for LSER Variables ($V_i/100$, π^*, β, α)

Compound	(a)	$V_i/100$	π^*	β	α
>NS–	al	0.198	0.45	0.85	0.00
	ar	0.178	0.40	0.45	0.00
>NSH	al	0.198	0.45	0.70	0.05
	ar	0.178	0.45	0.50	0.05
–N=O	al/ar	0.100	0.50	0.15	0.00
–NO$_2$	al	0.140	0.79	0.25	0.12
	al,wog	0.140	0.35	0.20	0.00
	ar	0.140	0.42	0.20	0.16
	ar,wog	0.140	0.10	0.25	0.16
Inorganics					
R$_3$P	al	0.160	0.30	0.65	0.00
	ar	0.160	0.30	0.75	0.00
R$_3$P=O	al	0.195	0.90	1.05	0.00
	ar	0.195	0.90	0.92	0.00
R$_3$P=S	al/ar	0.237	0.75	0.47	0.00
(–O)$_3$P	al	0.295	0.45	0.72	0.00
	ar	0.295	0.45	0.50	0.00
(–O)$_2$(R)P(=O)	al	0.270	0.75	0.75	0.00
	ar	0.270	0.75	0.55	0.00
(–O)$_3$P=O	al	0.315	0.65	0.77	0.00
	ar	0.315	0.65	0.62	0.00
(–O)$_3$P=S	al	0.387	0.60	0.38	0.00
	ar	0.387	0.60	0.92	0.00
(–O)(–S)$_2$P(=O)	al/ar	0.459	0.55	0.90	0.00
(–O)$_2$(–S)P(=S)	al/ar	0.459	0.55	1.02	0.00
(–O)(–S)$_2$P(=O)	al/ar	0.459	0.55	0.90	0.00
(–O)(–S)$_2$P(=S)	al/ar	0.531	0.50	1.07	0.00
–O(R)(R′)P(=O)	al/ar	0.235	0.85	0.70	0.00
(–O)$_x$(HO)$_y$P(=O)	al/ar	0.270	0.85	0.60	0.75
(N–)$_3$P(=O)	al/ar	0.420	0.95	1.87	0.00
(N–)$_3$P(=S)	al/ar	0.462	1.40	2.55	0.00
Misc. Inorganics					
R$_4$Si	al	0.208	0.00	0.00	0.00
	ar	0.188	0.00	0.00	0.00
R$_4$Sn	al	0.240	0.10	0.05	0.00
	ar	0.220	0.10	0.05	0.00

Note: In column (a), aliphatic (al), aromatic (ar) and aliphatic/aromatic (al/ar) indicate environments where values apply; py, on pyridine; wog, with other groups present; or, on a ring; ir, in a ring.

1. The values for substituents reflect their use in an aliphatic, aromatic, or nonspecific system.

The general LSER computation method has five steps:

1. Determine the composition and nature of molecule's skeleton. Which segments of the molecule are aliphatic (chains, rings); which are aromatic?
2. Inventory functional groups on each molecular segment. How many of what functional groups are present on what segment?
3. Add up LSER contributions/substituent for each segment. Use Table 1 for most of the values needed, and observe certain rules for specific situations (see Appendix I).
4. Add up LSER contributions for all segments to get one set of four values for $V_i/100$, π^*, β, α. Each molecule is described by a set of four LSER values.
5. Modify the individual π^*, β, α values if indicated. A simple sum will usually do, but a chemist's intuition is helpful, too.

There is currently no provision to differentiate between possible geometric (cis/trans) or optical isomers of compounds. The different forms have different properties, but no applicable weighting scheme has been developed as yet for the present system.

Parameter information

Volume: (V_i or V_x)/100

Intrinsic volume, V_i: The solute intrinsic (van der Waals) molar volume, V_i, is based on Leahy's computer-calculated intrinsic volume.[9,10] With the present system, V_i can be estimated by simple additivity methods. The intrinsic volume contributions in Table 1 for structures and functional groups are strictly additive for the volume of the whole molecule and have been scaled by 1/100 to give comparable weight to all four LSER variables.

Characteristic volume V_x: The rules for computation of V_x are also included here and also must be scaled by 1/100. An acceptable alternative, Abraham and McGowan's[11] characteristic volume V_x is easily calculated with atomic volumes and accounting for the number of bonds.

The characteristic atomic volumes V_x in cm^3/mol are:

C	16.35	N	14.39	O	12.43	F	10.48	H	8.71
Si	26.83	P	24.87	S	22.91	Cl	20.95	B	18.32
Ge	31.02	As	29.42	Se	27.81	Br	26.21		
Sn	39.35	Sb	37.74	Te	36.14	I	34.53		

The total atomic volume is calculated, and 6.56 cm^3/mol is subtracted for each bond between atoms. A multiple bond (olefin or alkyne) is treated as one bond; the unsaturation is accounted for by subtracting the number of bonds from the total volume.

V_x and V_i are entirely equivalent, and related by the equation:

$$V_i = 0.597 + 0.6823 \, V_x \qquad r^2 = 0.998$$

For an example, see benzoic acid in Appendix II.

Dipolarity π^*

The π^* variable[3-8] is an indicator of solute "polarizability," scaling a solute chemical's ability to stabilize a charge or dipole on a neighboring solvent molecule through van der Waals and dipole–dipole electrostatic interactions. The contributions to π^* for each segment of a molecule given in Table 1 are considered additive for a molecule to a first approximation. See the coming section on Corrections to π^*, β, and α for recommended modifications.

π^* values are very nearly equivalent to molecular dipole moments for compounds with a single dominant bond dipole moment.[6,12] For compounds with known dipole moment (μ), π^* can also be estimated[12,13] by:

$\pi^* = 0.03 + 0.23\ \mu$ for non-polyhalogenated aliphatics
$\pi^* = 0.27 + 0.35\ \mu$ for polyhalogenated aliphatics
$\pi^* = 0.56 + 0.11\ \mu$ for aromatic compounds

π^* can be modified[4,6,14] using a "polarizability correction" variable δ in the term ($s\pi^* + d\delta$). Certain classes of compounds have specific δ values: for aromatic systems, $\delta = 1.00$ for phenyl and polynuclear aromatic hydrocarbon systems. For nonpolyhalogenated aliphatic systems, $\delta = 0.00$; for polyhalogenated systems, $\delta = 0.25$ for F, 0.50 for Cl, Br, I substitution. The value for the coefficient d is determined during the multiple linear regression analysis. The value for d is approximately 0.00 for highly polar compounds (π^* maximum polarizability), and around –0.40 for nonpolar compounds (π^* nearly zero).

Hydrogen bond acceptor basicity β and donor acidity α

The β_m and α_m variables scale a solute's ability to accept or donate a proton (or donate or accept an electron pair), respectively, in a solute–solvent hydrogen bond.[2-8]

The subscript m indicates that, for compounds that are capable of self-association (amphihydrogen bonding compounds), the parameter applies to the monomer solute, rather than the self-associated oligomer solvent. For compounds that are not capable of self-association, $\beta_m = \beta$ and $\alpha_m = \alpha$. The contributions to β and α for each segment of a molecule given in Table 1 are considered additive for a molecule to a first approximation. See the section ahead on Corrections to π^*, β and α for recommended modifications.

The correction factor ξ[4,6,14,15] is infrequently used in ($b\beta + e\xi$) as a coordinate covalency functional group parameter and is dependent on electronegativity and other factors that act on an HBA base electron lone pair(s). The variable has specific values for certain functional groups: ξ (eta) = 0.00 for aldehydes, ketones, acids, acid halides, anhydrides, amides, carbamates, and ureas (C=O); sulfoxides, sulfones, and sulfates (S=O); and amine N-oxides, nitrosos, and nitros (N=O). Amines are sensitive to the degree of N orbital hybridization, where ξ (eta) = 1.00 when sp^3 0.60 when sp^2 (including pyridines), and 0.10 when sp. The value for the coefficient e is set during the multiple linear regression analysis.

Structural factors

For acyclic units, Table 1 has the values for the basic chain structures. All substituent contributions are summed for the segment or molecule. Aliphatic ring systems with simple double or triple bonds are treated as saturated systems, and each multiple bond unit is treated as a separate substituent (see Table 1). Contributions for all substituent groups are summed with the values for the aliphatic rings.

An aromatic ring is treated as a complete unit, and benzene is the basic structure (see Table 1).

Multiple and Condensed Ring Systems: To form multiple ring systems (such as bicyclohexyl or biphenyl), add contributions from both ring systems for all parameters unless otherwise dictated by the specific system and subtract $V_i/100 = 0.128$ for each ring–ring junction. For aromatic ring systems, subtract $\pi^* = 0.08$ from total for each ring–ring link.

To extend a saturated fused ring system, add contributions for each ring system and subtract $V_i/100 = 0.107$ and $\pi^* = 0.01$ for each shared carbon atom. For shared carbon atoms in an aromatic-aliphatic fusion, subtract $V_i/100 = 0.103$ for each shared carbon. For an example, see octahydropentalene in Appendix II.

To extend a polynuclear aromatic ring system, sum the variable values for each ring and account for duplicated volume contributions by subtracting $V_i/100 = 0.1145$/redundant carbon and add for $\pi^* = 0.59$(1st) then 0.11/ring, and for $\beta = 0.14$(1st ring), 0.06(2nd ring), 0.00(3rd ring) and 0.05/all subsequent rings.

Alternatively, add to the values for phenanthrene or anthracene in Table 1 the variable contributions $V_i/100 = 0.262$, $\pi^* = 0.11$, and $\beta = 0.05$ for each additional ring. Sum any substituent contributions from Table 1 for each ring system. For an example of each alternative, see benzo(a)anthracene in Appendix II.

Substituent addition

For substituent groups attached to linear or branched chains and aliphatic rings, sum the values for each segment, including substituents, using the values in Table 1.

For alkyl and aryl groups on aromatic ring systems, calculate the values for the LSER variables (Table 1) for the ring system and the substituents/side chains separately. For multiple ring and polynuclear aromatic derivatives, determine the substituent variable values for each ring separately and then sum the results. The hydrocarbon skeletal values are in Table 1.

A convention exists for computation of halogen contributions to π^* and β for polyhalogenated aromatic ring systems (see Appendix I). Values for $V_i/100$ and α are determined by summation using Table 1. The convention for π^* and β depends on the ring system. The halide contributions are summed over successive substitutions and depend on the position on the ring. The same contributions to π^* and β apply to analogous systems where a ring oxygen (where applicable) was replaced by other heteroatom(s) in the ring system. The values for the heteroatom ring systems have to be derived or determined by some means if not found in Table 1. Values for substituents other than halides are determined as outlined elsewhere in this chapter.

Corrections to π^*, β, and α

In some instances, such as for aromatic systems with multiple substituents and for the π^* variable in aliphatic systems, the sum of contributions may yield an intuitively unrealistic value. At the discretion of the investigator, these variable values may be adjusted by several approximations, and each variable can be considered separately. While computations of π^*, β, and α values are presently designed to be simple sums of component group contributions, some users may feel that a vector sum or a sum with a component group hierarchy of importance (use/not use, and to what degree) would give a better value. At present, that must be left to the discretion of the user. Also, the values given for aromatic substituents in Table 1 do not reflect the effects of ring position or hydrogen bonding, and the investigator may want to weigh contributions for resonance and induction effects using tables of Hammet sigma constants.[16] The values used by the investigator are valid for comparison, providing the method of computation is consistent throughout. These corrections are minor in most cases and tend to fall well within the experimental error of most measured datasets. The examples should clarify the computation process and guide the user in these subjective adjustments. The different suggested modifications are not expected to give the same results for a certain compound because they can reflect adjustments for different factors.

Approximate vector sums

Discounting a minor component group contribution from the sum or multiplying the total parameter value(s) by 0.8 to 0.9 can simulate a vector sum or reflect diminishing contributions from multiple similar substituents. See Appendix II for the example 3-trichloromethyl-5-cyanophenol.

Dominant substituent group(s)

When a dominant substituent (e.g., nitro, nitroso, cyano, etc.) is present on an aromatic ring, for each subsequent substituent, use π^* = (ortho) 0.10, (meta) 0.05 or (para) 0.00 instead of full contribution.

Also, for β values for the following groups, use: $-NMe_xH_{2-x}$, 0.10; $-OH$ or $-OR$, 0.10; $-F$, $-0.02/-0.05$; $-Cl$, $-Br$, $-I$, $-0.04/-0.10$. The halide contributions to β depend on the presence of additional electron-withdrawing groups (halogens, alkyls, etc.) or electron-donating groups (amines, sulfides, etc.), respectively. The values for α remain unchanged (see Table 1). For example, see 3-ethoxy-4-iodonitrobenzene in Appendix II.

Resonance and hydrogen bonding

The variable values for aromatic systems can be modified to account for induction and resonance at an aromatic ring position. Accounting for hydrogen bonding between two neighboring groups is at present highly subjective.

Tables of Hammet σ^+ constants[16] for the various functional groups have proven very helpful. Presently, β and α values for the participating groups are multiplied by 0.1 to 0.3, inversely proportional to the suspected strength of the hydrogen bond. For example, see the nitrophenol congeners explanation in Appendix II.

LSER equation formation

The predictive LSER equation results from a multiple linear regression correlation of the four LSER parameters for each compound in a series with the corresponding physical property value (e.g., toxicity, solubility, etc.). The correlation works best with a series of compounds that has the widest variety of structures and functional groups, giving the widest possible numerical range for each variable. (With too narrow a range, the coefficient/variable is rendered statistically insignificant.)

The correlation coefficient, r^2, describes how much of the data variation is accurately described by the equation. The r^2 value can range from 0.00 to 1.00; values for LSER equations can range 0.90 to 0.98.

Any data set used was gathered preferably under high quality assurance/quality control conditions, assuring comparability of data. The coefficients m, s, b, and a are thus unique to a particular set of conditions and organism (if applicable).

As an example of the finished product, we developed an LSER equation for acute general narcosis for the fathead minnow (*Pimephales promelas*):

$$Log(LC_{50}) = -0.34 - 5.26 \ V_i/100 - 0.80 \ \pi^* + 3.98 \ \beta_m - 0.80 \ \alpha_m$$

for N-phenyldiethanolamine,

$$
\begin{aligned}
Log(LC_{50}) &= -0.38 - 5.24(1.038) - 0.76(1.24) + 3.93(1.32) - 0.83(0.66) \\
&= -0.38 - 5.44 - 0.94 + 5.19 - 0.55 \\
&= -2.12 \ (\text{actual value}, -2.39)
\end{aligned}
$$

for menthol,

$$
\begin{aligned}
Log(LC_{50}) &= -0.38 - 5.24(1.06) - 0.76(0.80) + 3.93(0.96) - 0.83(0.65) \\
&= -0.38 - 5.55 - 0.61 + 3.77 - 0.54 \\
&= -3.31 \ (\text{actual value}, -3.92)
\end{aligned}
$$

Results and discussion

This chapter is an inventory of the LSER variable values for molecular functional groups and variable estimation rules. Except for certain cases, the values for the whole molecule are simply the sums of the contributions for each component group. The computation guidelines and list of functional group values should facilitate widespread application of the LSER model of QSAR by a broad spectrum of investigators, including environmental toxicologists, hazard assessment personnel, and analytical chemists.

Why use LSER? The availability of the rules and the functional group values in Table 1 should encourage extensive application of this accurate QSAR model by investigators with widely varying objectives. Practice makes perfect; Appendix III lists interesting compounds with suggested parameter values. The LSER method applies best to situations where the widest variety of chemical functional groups are encountered, and where the interaction of the compound with the medium dominates the process (e.g., solubility, toxicity, etc.).

Many types of chemical properties that depend on solute–solvent interactions have been addressed by LSER. Aqueous solubility,[13,17] octanol/water partition,[14,18,19] solubility in and partition between blood and body organs,[9,20-22] bioaccumulation,[23] soil adsorption coefficients,[23] HPLC capacity factors with several mobile and stationary phases,[9,22,24] and toxicity to a variety of species[1,12,25-28] have been predicted by equations of LSER relationships. We have successfully applied the LSER model in developing predictive equations for toxicity of a wide range of environmental chemicals to *Photobacterium phosphoreum* (the Microtox test), *Tetrahymena pyriformis*, *Daphnia pulex*, *Daphnia magna*, the fathead minnow (*Pimephales promelas*), rainbow trout, and the rat Ah bioassay (enzyme induction, binding affinity). We have differentiated between modes of action, and successfully modeled a reactive mode. Where comparisons were possible, these regression equations were consistently more accurate in their predictions than other widely used QSAR models such as log K_{ow}.

Blum and Speece[1] compared the accuracy of three QSAR methods for prediction of toxicity of varied types of chemicals to heterotrophic bacteria. Using the same data set, they found the correlation coefficient (r^2) to be 0.82 with log P, the most widely used QSAR. The correlation coefficient was 0.78 for molecular connectivity, a QSAR parameter for which the parameters may be readily calculated with little knowledge of chemical properties. For the LSER method, their correlation coefficient was 0.92 and demonstrated the greater accuracy of LSER. They recommended extensive use of LSER contingent upon the further development of methods for calculating the LSER variables.

The National Biological Service/Great Lakes Science Center has been developing Predictox expert system software[29,30] that determines the LSER parameters for a molecule and then uses these variable values to predict the acute toxicities for *Photobacterium phosphoreum* (the Microtox test), *Daphnia pulex*, *Daphnia magna*, and the fathead minnow (*Pimephales promelas*). This software can translate SMILES string formulations into compounds with ring and/or noncyclic skeletons, double and triple bonds, and all common functional groups and then assign the LSER values. In its present form, the software has been routinely used at NBS/GLSC to predict acute toxicities of contaminants prior to laboratory bioassays using *Daphnia pulex*, thus saving both time and expense in range-finding tests. Predicted toxicity from the LSER model is also useful in directing priorities in hazard assessment. The presented methods and this software have been used to evaluate sites of concern in the Great Lakes by screening contaminants for toxicity.

Acknowledgments

The author wishes to thank Dr. Dora Passino-Reader for giving impetus to the development of this manuscript and for her endless encouragement. Use of trade names does not constitute government endorsement of commercial products. Contribution #915 from the Great Lakes Science Center, Ann Arbor, MI 48105.

Appendix I Aromatic ring substitution values

Benzenes, biphenyls and PAHs

Substitution pattern	π^* F,Cl/Br,I	β F,Cl/Br,I
First X/ring (any position)	0.10/0.15	−0.03/−0.04
Next X/ring		
2,3 or 4,5	0.10/0.15	−0.04/−0.05
2,(4 or 6) or 3,5	0.05/0.10	
2,5	−0.05/−0.10	
Third X/ring		
2,3,4 or 3,4,5	0.05/0.10	−0.03/−0.04
2,(3 or 4),5	−0.05/−0.15	
2,4,6	0.00/0.10	
Fourth X/ring 2,3,4,5 or 2,3,5,6	−0.05/−0.15	0.00
Fifth X/ring 2,3,4,5,6	−0.05/−0.15	0.00
Sixth X/ring (1,2,3,4,5,6)	−0.05/−0.15	0.00

The numbering scheme assumes substitution by some other group in the "1" position (e.g., the other ring in biphenyl). In addition, to calculate π^* and β for biphenyl and fluorene, sum both ring totals and subtract 0.8 from the result for π^*. For naphthalene, biphenylene, and higher PAHs, use analogous benzene positions and substitution guidelines and sum contribution totals for each ring. A good rule of thumb is to add 0.05 to π^* if the next substitution increases the dipole moment and to subtract 0.05 if the dipole would decrease. See pentachlorophenol and 3,4,5,3′,4′-pentachlorobiphenyl in Appendix II.

Dibenzofurans

The halogen substitution hierarchy on each ring proceeds from the most reactive to the least reactive position. The variable contributions are determined by the following.

Rings	Ring Positions	π^* F,Cl/Br,I	β F,Cl/Br,I
First X/ring	3 or 7	+0.15/+0.20	−0.03/−0.04
	2 or 8	+0.10/+0.20	
	4 or 6	+0.05/+0.10	
	1 or 9	−0.05/−0.10	
Next X/ring	2 or 8	+0.10/+0.20	−0.03/−0.04
	4 or 6	+0.05/+0.10	
	1 or 9	−0.05/−0.15	
Third X/ring	4 or 6	+0.05/+0.10	−0.03/−0.04
	1 or 9	−0.05/−0.15	
Fourth X/ring	1 or 9	−0.00/−0.15	0.00

For an example, see 1,3-difluoro-6,8-dibromodibenzofuran in Appendix II.

Dibenzo-p-dioxins

The halogen substitution hierarchy on each ring proceeds from the most reactive to the least reactive position. The variable contributions are determined by the following.

Rings	Ring Position	π^* F,Cl/Br,I	β F,Cl/Br,I
First X/ring	2,3,7,or 8	+0.15/+0.20	−0.03/−0.04
	4 or 6	+0.05/+0.10	
	1 or 9	−0.05/−0.10	
Next X/ring	2 or 8	+0.15/+0.20	−0.03/−0.04
	4 or 6	+0.05/+0.10	
	1 or 9	−0.10/−0.15	
Third X/ring	4 or 6	+0.05/+0.10	−0.03/−0.04
	1 or 9	−0.10/−0.15	
Fourth X/ring	1 or 9	−0.10/−0.15	0.00

For an example, see 2-nitro-1,3,6,9-tetrachlorodibenzo-*p*-dioxin in Appendix II.

Appendix II Examples

The following are worked-out examples referred to in the text and are listed alphabetically.

Benzo(a)anthracene

	$V_i/100$	π^*	β	α
4 Benzene rings	.491	.59	.14	.00
	.491	.11	.06	.00
	.491	.11	.00	.00
	.491	.11	.05	.00
	1.964	.92	.25	.00
8 Shared carbons	−.687	.00	.00	.00
	1.277	.92	.25	.00
Alternatively,				
Anthracene	1.015	.81	.20	.00
Next ring	.262	.11	.05	.00
	1.277	.92	.25	.00

Benzoic acid

for V_x,	c	$6 \times 16.35 = 98.10$	for $V_i/100$, (from Table 1)		
	C	$1 \times 16.35 = 16.35$	phenyl	C_6H_5-	0.491
	O	$2 \times 12.43 = 24.86$	aromatic	$-C(=O)OH$	0.149
	H	$6 \times 8.71 = 52.26$			0.640
		191.57			

−bonds $15 \times 6.56 = −98.40$

$ 93.17$

And, $V_i = 0.597 + 0.6823\ V_x$,

$ = 0.597 + 0.6823(93.17)$

$ = 64.17$

$V_i = 0.640\ (100)$

$ = 64.00$

1,3-Difluoro-6,8-dibromodibenzofuran

		$V_i/100$	π^*	β	α
Dibenzofuran		1.581	.60	.30	.00
Fluorines	1	.030	−.05	−.03	.00
	3	.030	+.15	−.03	.00
Bromines	6	.131	+.10	−.04	.00
	8	.131	+.10	−.04	.00
		1.843	.90	.16	.00

3-Ethoxy-4-iodonitrobenzene

	$V_i/100$	π^*	β	α
Benzene	.491	.59	.14	.00
Nitro	.140	.10	.25	.16
Ethoxy	.241	.05	.10	.06
Iodide	.181	.04	−.04	.00
	1.053	.74	.45	.22
Table 1:	1.053	.83	.58	.22

Nitrophenol congeners

	$V_i/100$	π^*	b	a
Phenol	.536	.72	.33	.60
Nitro	.140	.42	.20	.16
	.676	1.14	.53	.76

ortho

meta

para

Hammet constants[16] indicate that the nitro group (σ_m = .71; σ_p = .78) is a very strong electron withdrawing unit in both meta and para positions. A nitro substituent increases the phenolic acidity in the meta and para positions but is a very strong hydrogen bond acceptor in the ortho position. Beta is largely unaffected, and the polarizability increases from minimum with ortho substitution to maximum with para substitution. So,

	$V_i/100$	π^*	β	α
Ortho	.676	1.11	.50	.12
Meta	.676	1.16	.50	.80
Para	.676	1.21	.53	.90

2-Nitro-1,3,6,9-tetrachlorodibenzo-p-dioxin

		$V_i/100$	π^*	β	α
	Dibenzo-p-dioxin	1.616	.45	.60	.00
	Chlorines				
	1	0.090	−.05	−.03	.00
	3	0.090	+.05	−.03	.00
	6	0.090	+.05	−.03	.00
	9	0.090	−.10	−.03	.00
	Nitro group	0.140	.10	.25	.16
		2.116	.50	.73	.16

Octahydropentalene

		$V_i/100$	π^*	β	α
	2 cyclopentane rings	.500	.00	.00	.00
		.500	.00	.00	.00
		1.000	.00	.00	.00
	−2 "Shared" C	−.107	−.01	.00	.00
		−.107	−.01	.00	.00
		.786	−.02	.00	.00

3,4,5,3',4'-Pentachlorobiphenyl

		$V_i/100$	π^*	β	α
	Biphenyl	.920	1.20	.28	.00
	Chlorines				
	3	.090	.10	−.03	.00
	4	.090	.10	−.04	.00
	5	.090	.05	−.03	.00
	3'	.090	.10	−.03	.00
	4'	.090	.10	−.04	.00
	From π^*		−.80		
		1.370	.85	.11	.00

Pentachlorophenol

	$V_i/100$	π^*	β	α
Phenol	.536	.72	.33	.60
Chlorines				
2	.090	.10	−.03	.00
3	.090	.10	−.04	.00
4	.090	.05	−.03	.00
5	.090	−.05	.00	.00
6	.090	−.05	.00	.00
	.986	.87	.23	.60

3-Trichloromethyl-5-cyanophenol

	$V_i/100$	π^*	β	α
Benzene	.491	.59	.14	.00
Hydroxyl	.045	.13	.23	.60
Cyano	.099	.20	.37	.22
$-CCl_3$	.368	.35	−.10	.10
	1.003	1.27	.64	.92

A leveling effect for both the π^* and α may be more likely. The approximation for π^* could be $(1.27 \times .8) = 1.02$. The effect could be less for acidity, like $(0.92 \times .9) = 0.83$. Or, the α contribution from $-CCl_3$ could be considered redundant and α becomes $(.92 − .10) = 0.82$.

Appendix III Problem set

Using Table 1 and any other tools such as imagination and cunning (within professional guidelines), try your hand at determining values for the four LSER variables for the following chemicals. Suggested solutions (author's) have been included. Some of the solutions are simple sums, and others reflect application of one or more of the modifications discussed above. Patience and peace! Remember, there may be more than one "correct" answer, depending on your instincts, day of the week, phase of the moon, etc.

Compound	CAS#	$V_i/100$	π	β	α
2,3,7,8-Tetrachlorodibenzo-*p*-dioxin	1746016	1.821	1.05	0.48	0.00
5,5-Dimethyl-1,3-cyclohexanedione	126818	0.837	0.85	0.85	0.00
Acetophenone	98862	0.690	0.90	0.49	0.06
4-Chloroacetophenone	99912	0.780	0.90	0.45	0.03
Benzophenone	119619	1.040	1.20	0.70	0.00
Benzaldehyde	100527	0.606	0.92	0.44	0.00
Trifluoroacetic acid	62748	0.351	0.80	0.40	0.70
p-Aminobenzoic acid (PABA)	62237	0.790	0.81	0.27	0.90
4-Aminobiphenyl	92671	0.991	1.32	0.60	0.16
4,4'-Diaminobiphenyl	92875	1.062	1.46	1.00	0.32
Benzonitrile	100470	0.590	0.90	0.42	0.20
Urethane (ethyl carbamate)	51796	0.501	0.48	0.78	0.55
Carbaryl (sevin; 1-naphthyl-N-methylcarbamate)	63257	1.183	1.17	0.90	0.30
Carbazole	86748	0.980	0.71	0.49	0.20
Nicotine	54115	0.975	1.01	1.17	0.00
Phenylacetylene	536743	0.636	0.65	0.25	0.11
4-Chloromethylstyrene	15922007	1.480	0.78	0.52	0.00
3,4,5,3',4'-Pentachlorobiphenyl	57465288	1.370	1.65	0.03	0.00
Bis(chloromethyl)sulfide "mustard"	505602	0.752	0.80	0.20	0.00
p,p'-DDT	50293	0.632	1.90	0.24	0.00
Parathion	56382	1.410	1.28	1.16	0.16
Heptachlor epoxide	1024573	1.603	1.30	0.75	0.51
Bis(tri-*n*-butyltin) oxide	56359	2.877	0.47	0.55	0.00
Permethrin	52645531	1.771	1.45	0.80	0.10

References

1. Blum, D.J.W., Speece, R.E. 1990. Determining chemical toxicity to aquatic species. *Environ. Sci. Technol.* 24:284–293.
2. Hickey, J.P., Passino-Reader., D.R. 1991. Linear Solvation Energy Relationships: "Rules of Thumb" for Estimation of Variable Values. *Environ. Sci. Technol.* 25:1753–1760.
3. Abraham, M.H., Doherty, R.M., Kamlet, M.J., Taft, R.W. 1986. A New Look at Acids and Bases. *Chem. Britain.* 22:551–554.
4. Kamlet, M.J., Doherty, R.M., Abboud, J.-L.M., Taft, R.W. 1986. Solubility — A New Look. *Chemtech.* 16:566–576.
5. Kamlet, M.J., Abboud, J.-L.M., Taft, R.W. 1981. An examination of linear solvation energy relationships. *Prog. Phys. Org. Chem.* 13:485–630.
6. Kamlet, M.J., Abboud, J.-L., Abraham, M.H., Taft, R.W. 1983. Linear Solvation Energy Relationships. 23. A comprehensive collection of the solvatochromic parameters, π^*, α and β, and some methods for simplifying the generalized solvatochromic equation. *J. Org. Chem.* 48: 2877–2887.
7. Kamlet, M.J., Taft, R.W. 1985. Linear solvation energy relationships. Local empirical rules — or fundamental laws of chemistry? A reply to the chemometricians. *Acta Chem.Scand.* B39:611–628.
8. Taft, R.W., Abboud, J.-L.M., Kamlet, M.J., Abraham, M.H. 1985. Linear solvation energy relations. *J. Solution Chem.* 14:153–175.

9. Leahy, D.E., Carr, P.W., Pearlman, R.S., Taft, R.W., Kamlet, M.J. 1986. Linear solvation energy relationships. 42. A comparison of molar volume and intrinsic molar volume as measures of the cavity term in reversed-phase liquid chromatography. *Chromatographia* 21:473–478.

10. Leahy, D.E. 1986. Intrinsic molecular volume as a measure of the cavity term in linear solvation energy relationships: octanol-water partition coefficients and aqueous solubilities. *J. Pharm. Sci.* 75:629–636.

11. Abraham, M.H., McGowan, J.C. 1987. The use of characteristic volumes to measure cavity terms in reversed phase liquid chromatography. *Chromatographia*. 23:243–246.

12. Kamlet, M.J., Doherty, R.M., Abraham, D.J., Taft, R.W. 1988. Solubility properties in biological media. 12. Regarding the mechanism of nonspecific toxicity or narcosis by organic nonelectrolytes. *Quant. Struct. Activ. Relat.* 7:71–78.

13. Kamlet, M.J., Doherty, R.M., Abboud, J.-L.M., Abraham, M.H., Taft, R.W. 1986. Linear solvation energy relationships. 36. Molecular properties governing solubilities of organic nonelectrolytes in solution. *J. Pharm. Sci.* 75:338–349.

14. Taft, R.W., Abraham, M.H., Famini, G.R., Doherty, R.M., Kamlet, M.J. 1985. Solubility properties in polymers and biological media. 5. An analysis of the physicochemical properties which influence octanol-water partition coefficients of aliphatic and aromatic solutes. *J. Pharm. Sci.* 74:807–814.

15. Kamlet, M.J., Gal, J.-F., Maria, P.-C., Taft, R.W. 1985. Linear solvation energy relationships. Part 32. A co-ordinate covalency parameter, ξ which, in combination with the hydrogen bond acceptor basicity parameter, β, permits correlation of many properties of neutral oxygen and nitrogen bases (including aqueous pK_a). *J. Chem. Soc. Perkin Trans. II.* 1583–1589.

16. Hansch, C., Leo, A. 1979. *Substituent constants for correlation analysis in chemistry and biology.* pp 49–54. New York: J. Wiley & Sons.

17. Kamlet, M.J., Doherty, R.M., Abraham, M.H., Carr, P.W., Doherty, R.F., Taft, R.W. 1987. Linear solvation energy relationships. 41. Important differences between aqueous solubility relationships for aliphatic and aromatic solutes. *J. Phys. Chem.* 91:1996–2004.

18. Kamlet, M.J., Doherty, R.M., Carr, P.W., Mackay, D., Abraham, M.H., Taft, R.W. 1988. Linear solvation energy relationships. 44. Parameter estimation rules that allow accurate prediction of octanol/water partition coefficients and other solubility and toxicity properties of polychlorinated biphenyls and polycyclic aromatic hydrocarbons. *Environ. Sci. Technol.* 22:503–509.

19. Kamlet, M.J., Doherty, R.M., Abraham, M.H., Taft, R.W. 1988. Linear solvation energy relationships. 46. An improved equation for correlation and prediction of octanol/water partition coefficients of organic nonelectrolytes (including strong hydrogen bond donor solutes). *J. Phys. Chem.* 92:5244–5255.

20. Kamlet, M.J., Doherty, R.M., Abraham, D.J., Taft, R.W. Abraham, M.H. 1986. Solubility properties in polymers and biological media. 6. An equation for correlation and prediction of solubilities of liquid organic nonelectrolytes in blood. *J. Pharm. Sci.* 75:350–355.

21. Kamlet, M.J., Doherty, R.M., Fiserova-Bergerova, V., Carr, P.W., Abraham, M.H., Taft, R.W. 1987. Solubility properties in biological media. 9. Prediction of solubility and partition of organic nonelectrolytes in blood and tissues from solvatochromic parameters. *J. Pharm. Sci.* 76:14–17.

22. Sadek, P.C., Carr, P.W.,Doherty, R.M., Kamlet, M.J., Taft, R.W., Abraham, M.H. 1985. Study of retention processes in reversed-phase high-performance liquid chromatography by the use of the solvatochromic comparison method. *Anal. Chem.* 57:2971–2978.

23. Park, J.H., Lee, H.J. 1993. Estimation of bioconcentration factor in fish, adsorption coefficient for soils and sediments and interfacial tension with water for organic nonelectrolytes based on the linear solvation energy relationships. *Chemosphere.* 26:1905–1916.

24. Carr, P.W., Doherty, R.M., Kamlet, M.J., Taft, R.W., Melander, M., Horvath, C. 1986. Study of temperature and mobile-phase effects in reversed-phase high-performance liquid chromatography by the use of the solvatochromic comparison method. *Anal. Chem.* 58:2674–2680.

25. Kamlet, M.J., Doherty, R.M., Taft, R.W., Abraham, M.H., Veith, G., Abraham, D.J. 1987. Solubility properties in polymers and biological media. 8. An analysis of the factors that influence toxicities of organic nonelectrolytes to the golden orfe fish (*Leuciscus idus melanotus*). *Environ. Sci. Technol.* 21:149–155.

26. Kamlet, M.J., Doherty, R.M., Taft, R.W., Abraham, M.H. 1986. Solubility properties in polymers and biological media. 7. An analysis of toxicant properties that influence inhibition of bioluminescence in *Photobacterium phosphoreum* (the Microtox Test). *Environ. Sci. Technol.* 20:690–695.

27. Passino, D.R.M., Hickey, J.P., Frank, A.M. 1988. Linear solvation energy relationships for toxicity of selected organic chemicals to *Daphnia pulex* and *Daphnia magna*. In *QSAR 88: Proceedings of the Third International Workshop on Quantitative Structure-Activity Relationships (QSAR) in Environmental Toxicology*; Eds. Turner ,J.E., Williams, M.W., Schultz, T.W., Kwaak, N.J.CONF-880520- (DE88013180), pp 131–146. U.S. Department of Energy: Oak Ridge, TN.

28. Passino, D.R.M. 1986. Predictive models in hazard assessment of Great Lakes contaminants for fish. In *Proceedings of the Technology Transfer Conference*. ISSN 0-825-4591; Part B, pp. 1–26, Toronto, Ontario: Ontario Ministry of the Environment.

29. Hickey, J.P., Aldridge, A.J., Passino, D.R.M., Frank, A.M. 1990. An Expert System for Prediction of Aquatic Toxicity of Contaminants. In *Environmental Expert Systems*. Ed. Hushon, J., American Chemical Society Symposium Series 431, 7:90–107, Washington DC: American Chemical Society.

30. Hickey, J.P., Aldridge, A.J., Passino-Reader, D.R.M., Frank, A.M. 1992. Expert System Predicts Toxicity from Contaminant Chemical Structure. *Drug Inform. J.* 24:487–495.

chapter twenty-four

Solid phase microextraction

**G. R. Barrie Webster, Leonard P. Sarna,
and Kristina N. Graham**

Introduction

Contamination of surface and groundwater with pesticide residues and hydrocarbon fuel is a problem of increasing importance to society. Whether the focus is from the point of view of agriculture, the industrial sector, or environmental enforcement agencies, the need to monitor such contamination is compelling and is of intense interest. The protection of drinking water, surface waters for aquatic life, and the fishery must be addressed. Further, actions to prevent contamination of the environment through runoff and leaching from contaminated sites is strengthened by the use of reliable and convenient analytical methods.

Research into the interaction of organic environmental contaminants with environmental substrates is limited by time requirements for the procedures involved in the quantification and identification of analytes. Extraction, cleanup, solvent evaporation, and solvent disposal all require expenditures of time and money beyond the resources of most laboratories without substantial support of granting agencies or other interested agencies.

Simple, inexpensive analytical methods are required. Currently used technologies for water samples include liquid–liquid extraction or solid phase cartridge extraction. Solvent is required in volumes which range from liters to several tens of milliliters for each sample. Solid phase cartridge extraction has shortened the extraction step and reduced the amount of solvent required, but the procedure is still often relatively long and laborious. Solvents of high purity are required and are expensive to purchase or prepurify. Following their use, disposal of these solvents is also expensive and inconvenient. Sample extract manipulation also leaves open the danger of the introduction of contaminants or loss of sample through multiple-step manipulation.

Supercritical fluid extraction promises substantially to shorten extraction procedures for solid matrices found in aquatic systems. The use of solvents such as carbon dioxide is an attractive alternative to conventional solvent

1-56670-149-X/96/$0.00+$.50
© 1996 by CRC Press, Inc.

use; however, high pressure pumps and apparatus and tanks of compressed CO_2 must be used in such procedures.

A still simpler procedure is solid phase microextraction (SPME). The advent of the innovative Canadian SPME technology by Pawliszyn and co-workers at the University of Waterloo[1] opened the door to the development of new analytical methods for trace organic contaminants such as pesticide residues, polychlorinated dibenzo-*p*-dioxins (PCDDs), polychlorinated dibenzofurans (PCDFs), polychlorinated biphenyls (PCBs), and petroleum hydrocarbon residues in soils, sediments, and water.

Solid phase microextraction — preview

The SPME technique uses a polymer-coated fused silica fiber which is attached to a syringe-like apparatus in place of the normal syringe needle (Figure 1). Effectively, the analyst uses a silicone resin "solvent-on-a-stick." The fiber can be retracted into a sheath which has the function of both protecting the fiber and enabling it to be inserted through septa into sample vials. The fiber is exposed to a sample contained in a sample vial in the extraction/sorption step. During this step, a portion of the organic compounds in the sample which have appropriate partition coefficients migrate to the fiber and are sorbed by the polymer coating. After a fixed time (or alternatively, a time which allows full equilibrium to be attained), the fiber is retracted into the sheath, and the needle is removed from the vial. Immediately, it is taken and inserted into the injection port of a gas chromatograph (GC) where the analytes are thermally desorbed and focused at the head of the capillary column (held at cryogenic temperatures for volatile organic contaminants, or the GC program initial temperature, e.g., 40 to 80° for most analytes). The thermal desorption step takes 1 to 5 min, and normally leaves the fiber cleaned and ready to be used for the next run. The GC run is begun when the desorption is complete, and analysis proceeds in the usual way.

SPME theory

Theoretical aspects of the operation of SPME have been described by Pawliszyn and co-workers.[2] The principle behind SPME is the partitioning of the analyte between the sample matrix and the extraction medium. If a viscous liquid polymeric coating on a fine silica fiber is used as the extraction medium, the amount of analyte sorbed by the fiber coating at equilibrium is directly related to its concentration in the sample:

$$K = \frac{C_s}{C_{aq}} \tag{1}$$

where C_s = the concentration of analyte in the fiber coating (stationary phase), and C_{aq} = the concentration of analyte in the aqueous phase.

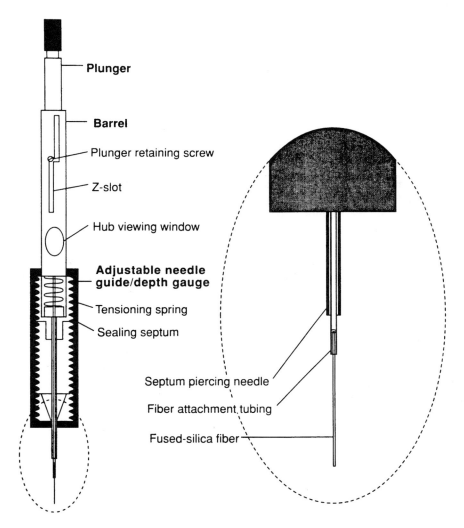

Figure 1 Solid phase microextraction (SPME) apparatus: size comparable to that of a normal GC syringe. (Reprinted with permission from Zhang, Z. et al. 1994. *Anal. Chem.* 66:844A–853A. Copyright 1994 American Chemical Society.)

The partition ratio

$$k' = \frac{C_s V_s}{C_{aq} V_{aq}} = \frac{n_s}{n_{aq}} = K \frac{V_s}{V_{aq}} \qquad (2)$$

where n_s = the number of moles of analyte in the fiber coating (stationary phase); n_{aq} = the number of moles of analyte in the aqueous phase; V_s = the volume of the stationary phase; and V_{aq} = the volume of the aqueous phase.

Rearranging Equation 2 yields

$$n_s = \frac{KV_s n_{aq}}{V_{aq}} \tag{3}$$

Substituting $C_{aq}V_{aq}$ for n_{aq}

$$n_s = KV_s C_{aq} \tag{4}$$

Assuming KV_s to be a constant, n_s becomes directly proportional to C_{aq}.
Louch et al.[3] subsequently have shown that in order to describe systems where the analyte aqueous concentration at equilibrium is changed during the extraction process (a situation occurring when distribution constants for analytes are very high and the volume of aqueous sample small), Equation 1 does not apply and a mass balance equation which relates analyte concentrations before and after extraction is considered to take into account analyte depletion in the aqueous phase

$$n_s + C_{aq}^{\infty} V_{aq} = C_{aq}^{o} V_{aq} \tag{5}$$

where C_{aq}^{∞} = the concentration of the analyte in the aqueous phase at equilibrium; C_{aq}^{o} = the initial concentration of the analyte in the aqueous phase; and V_{aq} = the volume of the aqueous phase.
Substituting

$$k = \frac{C_s^{\infty}}{C_{aq}^{\infty}} \tag{6}$$

into the above Equation 5 and simplifying, the following relationship is obtained:

$$n_s = \frac{KV_s V_{aq} C_{aq}^{o}}{KV_s + V_{aq}} \tag{7}$$

The amount of analyte absorbed by the fiber coating is proportional to the initial analyte concentration for both the infinite volume case and the finite volume case. The presence of the KV_s term in the denominator will only have an effect when the distribution coefficient is extremely large, since V_s for the SPME fiber is very small.
Zhang and Pawliszyn[4] have derived a relationship for fiber performance in headspace analysis based on the equilibrium of the analyte in a three-phase (fiber coating, aqueous solution, and headspace) system. Essentially, the total amount of the analyte in such a system can be expressed as

$$C_{aq}^o V_{aq} = C_s^\infty V_s + C_{aq}^\infty V_{aq} + C_h^\infty V_h \tag{8}$$

where C_{aq}^o = the initial concentration of the analyte in the aqueous phase; C_s^∞ , C_{aq}^∞ , C_h^∞ = the concentration of analyte at equilibrium in the fiber, the aqueous phase, and the headspace, respectively; and V_s , V_{aq} , V_h = the volume of the fiber coating, the aqueous phase, and the headspace, respectively.

Defining the coating/gas partition coefficient as K_1 = C_s^∞ / C_h^∞ and the gas/water partition coefficient as K_2 = $C_h^\infty / C_{aq}^\infty$, the amount of analyte absorbed by the coating (i.e., the capacity of the coating) can be expressed as

$$n = \frac{K V_s V_{aq} C_{aq}^o}{K V_s + K_2 V_h + V_{aq}} \tag{9}$$

Except for the term $K_2 V_h$, which is related to the capacity of the headspace ($C_h^\infty V_h$), the expression for headspace analysis is exactly the same as the expression defining analysis of the aqueous system. Both describe the mass absorbed by the polymeric coating at equilibrium, and detection limits for headspace analysis are expected to be similar to those of direct aqueous analysis.

Essentially then, any analyte having a greater affinity for the polymeric fiber coating than for the sample matrix (i.e., air, water, or soil) will partition into the fiber coating and concentrate there until an equilibrium is reached. Thus the sensitivity of the method is dependent on the type of coating and its volume. Extraction and concentration occur in the same step, and all analytes extracted are introduced into the gas chromatograph via thermal desorption of the fiber in the heated injection port.

SPME of contaminants in water

Pesticide residues

Reports in the literature on the use of SPME for a wide variety of analytes are growing. Work in Pawliszyn's laboratory has demonstrated the use of prototype SPME apparatus in the extraction of a wide range of analytes including chlorinated organics and phenols.[5,6] Recent work in the authors' laboratories has demonstrated the feasibility of the analysis of water and soil for several pesticides such as atrazine, metolachlor, diclofop-methyl, or lindane.[7,8] Such residues may be present in surface- or groundwater or in soils following agricultural application or accidental spills. Usual analyses commence with liquid–liquid extraction or solid phase cartridge extraction. GC analysis can be carried out using flame ionization (FID) or electron capture (EC) detection in the case of compounds which contain halogen constituents as was the case with the pesticides mentioned above. In the case of lindane, both direct SPME of the water or SPME of the headspace over the water can be performed successfully.[8] Lindane in soil solution has also been successfully analyzed.[9]

BTEX hydrocarbons and related analytes

Benzene, toluene, ethyl benzene, and the three isomers of xylene (*o-*, *m-*, and *p*-xylene) (BTEX), common industrial solvents and fuel components, are important as common contaminants of groundwater[10] and landfill leachates.[11] Their presence may be due to incomplete combustion of gasoline, leaking underground fuel storage tanks, or accidental spills of gasoline or other types of hydrocarbon fuels, or industrial accidents.

Analytical methodology for these compounds usually involves liquid–liquid extraction,[12,13] purge-and-trap,[14,15] or headspace sampling[16] followed by capillary column gas chromatography using an appropriate detector, such as flame ionization detection (FID), photoionization detection (PID), or mass spectrometry (MS). The poly(dimethylsiloxane)-coated extraction fiber technique has been successfully used to extract a range of analytes including hydrocarbons and substituted aromatics from aqueous solution and further described in a series of recent papers.[17-22] Thermal desorption of the sorbed BTEX within the GC injector coupled with cryofocusing was used in conjunction with conventional capillary GC and FID or ion-trap detection (ITD). The technique has also demonstrated potential application to the extraction of chlorobenzenes and PCBs,[17] caffeine and fragrances,[18] and U.S. EPA priority pollutants.[20]

Conventional capillary GC of BTEX has usually been carried out using nonpolar columns, e.g., DB-1 and DB-5 columns (Chromatographic Specialties, Brockville, ON, Canada).[23-25] As mentioned above, use of these columns for SPME analysis has required cryofocusing following thermal desorption in the injector of the GC. Recently, GC analysis of the BTEX contaminants split-injected in the conventional manner has been reported using the recently developed carbon layer open tubular (CLOT) capillary column[26,27] from Supelco (Canada) Ltd. (Ltée.), Mississauga, ON, Canada). This technology has been incorporated into the SPME-GC-FID analysis of BTEX in which all six components can be resolved.[28] Combination of SPME with Raman spectroscopic detection for BTEX analysis in water has also been described.[29]

Materials required

The SPME procedure is extremely simple and the materials required are surprisingly inexpensive. The technology is simple but "smart." The SPME fibers are available with several different coating types and thicknesses on the fused silica fiber from Supelco Inc. Two formats are available: the manual one (# 5-7300) which requires a special fiber holder resembling a normal syringe (Figure 1), and the one (# 5-7301) which fits the SPME-modified Varian 8200 autosampler (Varian Canada Ltd., Mississauga, ON, Canada) which allows autoinjection into a gas chromatograph. Sample vials for the

SPME technique can be of several sizes: the normal 2-ml autosampler septum-capped vial (screw-caps are the most convenient) or larger vials which can be equipped with a septum of varying capacities up to 40 ml. In our experience, the somewhat larger volumes of water enable more sensitive analyses to be performed for most hydrophobic analytes.[7,8,28]

There is an increasing choice of types of fiber coating available commercially to suit the variety of analytes of different polarities and volatilities. The more volatile non-polar analytes are best dealt with using the 100 μm polydimethylsiloxane coated fibers. Less volatile non-polar compounds may be more easily analyzed using the 20 or 30 μm coating. More polar analytes can be extracted using the polyacrylate fiber coating. Each of these is available commercially from Supelco.

Procedures

Initial considerations

SPME has a number of procedural features which require comment. Considerations such as the length for time to the attainment of equilibrium, the desorption time in the GC injection port, the initial temperature of the GC column oven, the effect of stirring the sample during the sorption step, the effect of heating the sample during the sorption step, and the effect of the size of the sample are important to the optimization of the SPME technique.

Length of time to equilibration.   The amount of analyte taken up by the coating on the fiber at equilibrium is proportional to the partition coefficient of the analyte between the water and the coating. The speed with which this proportion of the analyte is sorbed also depends on the rate of diffusion of the analyte to the fiber and the ease with which the analyte crosses the boundary layer of water surrounding the fiber. Equilibrium times of tens of seconds to tens of minutes are typical; whenever analyses are to be conducted using SPME, the time to equilibrium should be determined. Extractions which involve sorption for periods near to or in excess of the equilibration time ought to be reproducible. Those which are carried out for less than this period may also be utilized in reproducible analytical techniques providing that extraction conditions and sorption period for samples are held constant.

Injection setup.   The optimum depth of penetration for the exposed fiber in the injection port of the GC was similar to that for a 10-ml syringe needle. For manual injection, the SPME fiber assembly was clamped upright over the GC injection port. The GC septum was pierced using the septum piercing needle with the barrel of the fiber assembly resting on the GC injection port. The fiber was lowered into the injection port with the purge off. After the desorption time (e.g., 2 min; purge off), the fiber was retracted into the septum piercing needle, the needle was withdrawn from the injection port, the purge was turned on, and the chromatography was begun.

Desorption time.   Desorption of the analyte from the fiber coating occurs within the heated inlet of the GC. The temperature of the inlet must be sufficiently high that the desorption proceeds completely and rapidly without thermal degradation of the fiber coating or loss of the resin on the fiber. Characteristically, 100-μm fiber coatings are desorbed at 200 to 220° (all temperatures are in degrees Celsius). Above 220°, loss of integrity of the fiber takes place. Fibers with thinner coatings can, however, be taken to higher temperatures, e.g., up to 320° for the 7-μm coating. Typical desorption times are of the order of 2 to 10 min. The completeness of the thermal desorption should be confirmed by carrying out "blank" runs of the previously desorbed fiber on the GC using the same thermal desorption time and temperature.

Initial GC column temperature.   During the thermal desorption of the analyte from the fiber, the GC column must be held at a temperature low enough that the analyte will condense at the head of the column. Thus if the analyte is particularly volatile (e.g., VOCs), the column oven may have to be cooled below ambient temperature to retain the analyte. Loss of the volatile end of a series of analytes will easily be seen as a disproportionate lowering of the recovery of these analytes in a series of compounds of equal concentration but varying volatility.

Stirring (water and suspended material).   SPME will function well in water or headspace sampling; however, the time to equilibrium can be significantly shortened through agitation of the solution being sampled. The simple action of magnetic stirring (Figure 2) of the sample during SPME is perhaps the simplest and cheapest approach, but any other agitation technique may also be used. Not only is the agitated solution being presented to the fiber in such a manner that diffusion of the analyte is hastened, but the water layer surrounding the fiber in the water being sampled is also being disrupted, allowing the partition process to proceed. The agitation option is just now available as part of the autosampler available from Varian, the only company licensed to sell SPME modified autosamplers for GC work. Manual SPME, on the other hand, allows the analyst to stir the sample vial contents during extraction.

Heating (solid samples).   Heating is a useful technique to increase the transfer of analytes having a sufficient Henry's law constant from liquid samples or solid samples (e.g., soil) to the headspace. The sensitivity of headspace analysis by SPME can therefore be enhanced. As the temperature increases, however, the ability of the fiber coating to sorb the analyte is diminished. (Heating of the fiber is the means by which desorption of the analyte is effected.) Temperatures in the range of 40 to 60° typically provide the enhanced sensitivity sought without interfering with the sorption of the analyte by the fiber coating.

Effect of increasing the vial size.   The detection limits of the procedure in the case of hydrophobic analytes are often limited by the solubility of

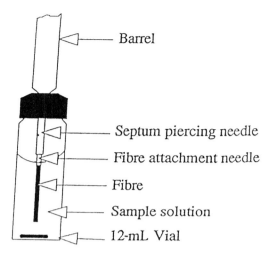

Figure 2 Septum-cap vial with SPME assembly illustrating position of the fiber and septum-piercing needle during direct solid phase microextraction. Small magnetic stir bar is shown. (Reprinted with permission from Sarna, L.P. et al. 1994. *J. Chromatogr.* 677:201. Copyright 1994 Elsevier Science B.V.)

these analytes in water. Stirring enhances the speed with which sorption can occur, but the quantity of analyte which has been sorbed at equilibrium is limited by the total amount of analyte in the sample. At the same low concentration, there is more analyte available to be sorbed in a somewhat larger sample than in a smaller one. Hence, SPME of analytes from samples in 40-ml vials gives lower detection limits than SPME of the same analytes at the same concentration in the usual 2-ml autosampler vials.

For headspace analysis, the same principle holds regarding the size of the headspace over the sample being analyzed. Larger headspace allows more of the same analyte to partition into the headspace. Larger headspace volume at the same concentration allows the fiber to sorb more analyte from the headspace, providing the analyte can diffuse to and sorb into the fiber coating. Similarly, a larger headspace over a smaller sample of soil containing a known concentration of analyte will provide a greater GC detector response than a smaller headspace over a larger sample of the same soil in the same sized vial.

SPME of pesticide residues in water

Standards. Conventional standard solutions of atrazine, metolachlor, diclofop-methyl, and lindane isomers were made up in toluene (pg/μl to ng/μl). Injections of these solutions were used to plot standard curves over at least two orders of magnitude. Standard solutions in methanol were prepared for use in the spiking of water samples.

Extraction. SPME extraction was performed using the 100-μm poly-dimethylsiloxane fiber assembly; in the case of lindane and isomers, the

Table 1   GC Conditions Used for SPME-GC of the Herbicides Metolachlor, Atrazine, and Diclofop-Methyl

	Varian 3400 (FID)	HP 5890 (EC)	HP 5890 (ITD)
Column	30 m × 0.32 mm 0.1 µm DB 5	30 m × 0.32 mm 0.25 µm DB 5	60 m × 0.32 mm 0.1 µm DB 5
Injector temperature	200°	220°	200°
Temperature program	100° 5 min, 5°/min, 250°, 3 min, 2°/min, 280° 6 min	100° 2 min, 10°/min, 280° 5 min	100° 2 min, 15°/min, 280° 5 min
Detector temperature	280°	350°	280° (transfer line)

Note: The Varian 3400 GC was equipped with an 8200 autoinjector modified for SPME and a flame ionization detector (FID). Manual SPME work used the same GC or Hewlett-Packard 5890 GCs equipped with either electron capture (EC) or ion trap detection (ITD).

20-µm fiber was also used. Automated SPME was performed on atrazine, metolachlor, and diclofop-methyl using 2-ml vials containing 1.5 ml of solution. Sorption times of 10 to 30 min were used. Manual SPME was used for lindane and isomers; 10-ml aliquots of spiked water samples were placed in 12-ml screw-cap vials equipped with stir bars and fitted with silicone/PTFE septa. After the sorption time, the fiber was retracted into the septum piercing needle, and the needle was withdrawn from the vial.

Analyses.   A Varian 3400 GC operating in the splitless mode and equipped with an FID and an 8200 autoinjector modified for SPME were used for automated analyses. Separations were conducted using a DB-5 30 m × 0.32 mm id column (Table 1). The 100-µm polydimethylsiloxane-coated SPME fibers were used. In general, a sorption time of 2 to 10 min and a desorption time of 2 min in the GC injector were used; other adsorption times were also investigated (e.g., 5, 10, 30 min). On the autosampler, sample solutions were limited to 1.5 ml.

Manual SPME work used the same GC or a Hewlett-Packard 5890A GC equipped with either electron capture (EC) or ion trap detection (ITD). Conditions are given in Table 1. Both 100- and 20-µm SPME fibers were used. Manual SPME was performed on samples from 1.5 to 25 ml for atrazine, 1.5 ml for metolachlor, 1.5 to 25 ml for the diclofop-methyl, and 1.5 to 25 ml for lindane.

SPME of the BTEX hydrocarbons in water

Standards.   Conventional standard solutions of benzene, toluene, ethyl benzene, and the xylenes (0.002% v/v) were made up in dichloromethane. Injections of 1.7 to 170 ng/µl of each component were used to plot a standard curve.

A standard solution in methanol of benzene, toluene, ethyl benzene, and *o-*, *m-*, and *p*-xylene (0.002% v/v) was prepared for use in spiking water samples at levels from 35 to 850 ng component per ml.

Extraction.   SPME extraction was performed using the manual 100-μm polydimethylsiloxane fiber assembly. Ten-ml aliquots of spiked water samples were placed in 12-ml, screw-cap vials equipped with stir bars and fitted with silicone/PTFE septa. The vial septum was pierced with the septum piercing needle, and the fiber was lowered into the solution so that the stainless steel needle attachment was just below the meniscus (Figure 2). After an extraction time of 2.0 to 20 min, the fiber was retracted into the septum piercing needle, and the needle was withdrawn from the vial.

Analysis.   A Hewlett Packard 5890 GC equipped with an FID and operating in the splitless mode was used for the analysis of the BTEX. Separations were conducted using a Supelco 30 m × 0.32 mm id CLOT column. The chromatographic conditions were as follows: injector, 220°; column, 40° for 2 min, 15°/min to 180°, hold 1 min; detector, 250°; flow rates: He carrier, 1.8 ml/min; He makeup gas, 30 ml/min; H_2, 30 ml/min; air, 150 ml/min. SPME samples from BTEX solutions containing 35 to 850 ng/ml were analyzed and the FID responses related to the concentrations sampled.

Results and discussion

Pesticide residues

The detection limits of the method were influenced by the partition coefficient of the pesticide; e.g., the specific responses for metolachlor and atrazine differ (Figure 3). From solutions at the same concentrations, the quantities of these analytes sorbed parallel their K_{ow} (octanol–water partition coefficient) values. The K_{ow} values for the pesticides are as follows: atrazine 219, metolachlor 2,820, and diclofop-methyl 37,800.[30] Linear response was obtained for metolachlor and atrazine over three orders of magnitude by

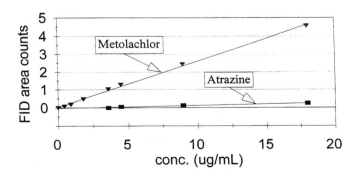

Figure 3   Comparative responses for metolachlor and atrazine at the same concentrations in water.

SPME-GC-FID (e.g., 0.45 to 180 ng/μl). Sensitivity was limited by the sample size accommodated by the autosampler. Linear response by SPME-GC-EC was obtained for the diclofop-methyl from 0.2 ng/μl for stirred 1.5 ml samples. Stirred 25 ml samples gave usable results down to 0.01 ng/μl with a 10 min sorption time. Extended sorption times ensured that equilibrium had been attained for the pesticides being analyzed.

These results indicate that SPME-GC can be used for residues of metolachlor, atrazine, and diclofop-methyl in water. SPME-GC gives linear response for analytes from water over several orders of magnitude of concentration with only a very slight deviation at the low end of the response curve. The 100-μm fiber provides good analytical extraction for these compounds. Recent work which is being prepared for publication has shown that the SPME technique is robust in the presence of real world water samples of these pesticides. The addition of salt to the water can also enhance the extraction.[31] Automated SPME in conjunction with GC can be improved by the addition of agitation of the aqueous samples.

The chromatographic response for lindane and three isomers in water is shown in Figure 4. The responses for one μl injection of the same concentrations of these analytes in liquid solution are given in Figure 5. Not only is the chromatographic separation excellent, the response through the SPME technique is significantly enhanced over the response through a normal injection of the same concentration of the analytes (in an organic solvent).

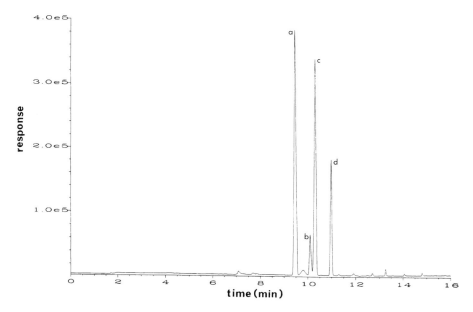

Figure 4 Direct SPME-GC responses for lindane and three isomers with identities and concentrations: (a) α-HCH 5.45 ng/ml, (b) β-HCH 13 ng/ml, (c) γ-HCH (lindane) 5.8 ng/ml, and (d) δ-HCH 5.5 ng/ml.

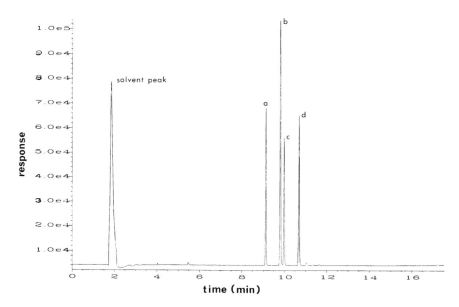

Figure 5 GC responses of liquid sample injection of lindane and isomers with identities and concentrations: (a) α-HCH 99 ng/ml, (b) β-HCH 436 ng/ml, (c) γ-HCH (lindane) 113 ng/ml, and (d) δ-HCH 135 ng/ml. Note that these concentrations are ca. 20 × those used for the SPME in Figure 4.

SPME promises to be a useful analytical tool for pesticide residues in water, both by direct sorption and by headspace analysis.

SPME of the BTEX hydrocarbons in water

New analytical methodology based on a combination of the previously separately described techniques of SPME and CLOT column capillary GC with FID has been successfully demonstrated to be feasible for the analysis of BTEX hydrocarbon contaminants in water. The apparatus for each of these techniques is now available "off the shelf."

Little difference in BTEX levels recovered was shown between the exposure times chosen, and was consistent with earlier reports.[19,20] While a 2-min extraction time was shown to be sufficient for BTEX from water, followed by a 2-min thermal desorption time in the injection port of the GC, we chose a 3.5-min exposure time for this work. Complete desorption after 2 min at 220° was demonstrated. Total run time for a desorption and GC analysis was 10 min.

Figure 6 demonstrates the chromatography and recovery of BTEX components from a 10-ml aliquot spiked at 170 ng/ml per component.[28] Two points are noteworthy. The first is that the chromatogram was generated without the need of the cryotrapping required for all previous work involving the use of SPME. Cryotrapping was previously required to minimize peak broadening associated with the relatively long desorption times

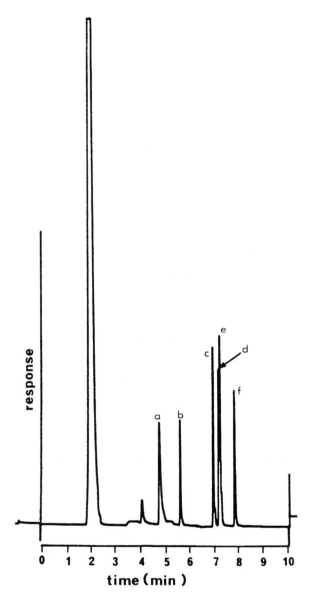

Figure 6   Direct SPME-GC-FID of BTEX showing the separation of each component (170 ng/ml) on the CLOT column: (a) benzene, (b) toluene, (c) ethylbenzene, (d) *p*-xylene, (e) *m*-xylene, and (f) *o*-xylene. (Reprinted with permission from Sarna, L.P. et al. 1994. *J. Chromatogr.* 677:201. Copyright 1994 Elsevier Science B.V.)

involved in sample introduction by SPME. The second point is that the chromatogram clearly illustrates resolution of the *m*- and *p*-xylene isomers, not seen in earlier papers using either conventional or SPME analysis.

Since SPME relies on the partition of the BTEX from water into the polymer coating of the extraction fiber, the response seen for each component

Table 2   Retention Relative to Benzene, Minimum Detectable Limit (mdl) by
Direct Fiber Introduction of Individual BTEX Components from Fortified Water,
and Linear Regression Results (Coefficient of Determination) for the BTEX
Components Extracted from Water by SPME and Desorbed and Analyzed by GC

Compound	Relative retention	MDL (ng)	Range of fortification (ng/ml)	r^2
Benzene	1.00	0.4	35–850	0.9507
Toluene	1.34	0.4	35–850	0.9701
Ethylbenzene	1.63	0.05	35–850	0.9930
p-Xylene	1.68	0.2	35–850	0.9923
m-Xylene	1.69	0.2	35–850	0.9904
o-Xylene	1.81	0.1	35–850	0.9973

Note: Benzene retention time was 4.18 min.[28]

will vary according to its partition coefficient. Table 2 presents data on the retention time and minimum detectable limit for each of the BTEX components. The linearity of the response for each of the component compounds compared to its concentration in the spiked water sample was determined. The r^2 values in Table 2 reflect the goodness of fit over the range of 35 to 850 ng/ml of BTEX components in water. The differences in minimum detectable limits for the BTEX components more closely reflect the trend in K_{ow} values reported by Arthur et al.[19] than the distribution coefficients in their original study and support their premise that the K_{ow} values of analytes can be used to predict the linear range and limit of quantification in new method development using these fibers before any experimental work is begun. The current study used the commercially supplied SPME fiber; Arthur et al. used fibers prepared in their own laboratory.

Analysis of BTEX in water by SPME, thermal desorption in the injection port, and GC-FID was directly possible using the commercially available CLOT column with linear response demonstrated for water samples containing 35 to 850 ng/µl. The previously prescribed cryofocusing of the desorbed BTEX is now unnecessary, making the SPME/CLOT column technique for BTEX readily usable by most analytical laboratories equipped with normal FID-GC instrumentation.

SPME-GC of other related analytes

SPME-GC can also be used to detect other analytes in water such as chlorinated dioxins and diesel fuel hydrocarbons (Figures 7 and 8). The dioxins were sorbed for 15 min at ambient temperature on a 7-µm fiber and desorbed for 5 min at 250°. GC conditions: 30 m × 0.32 mm id DB-5 capillary column; 100° for 5 min, 5°/min to 250° (3 min), 2°/min to 280° (3 min); purge 5 min; detector 350°; makeup 60 ml/min. The water soluble fraction of diesel was sorbed for 3 min with stirring at ambient temperature and desorbed for 5 min at 220°. Analysis was performed on a 60 m DB-5 column. Individual peaks were identified by SPME-GC-ITD (ion trap detector) and are identified in Figure 8.

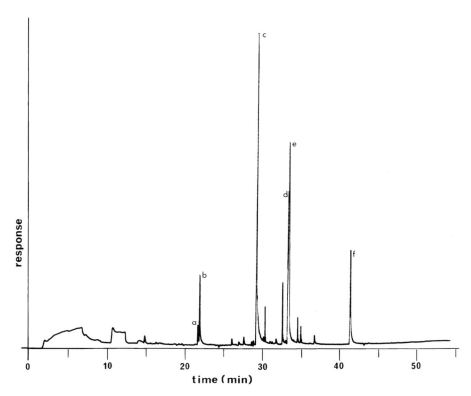

Figure 7 Direct SPME-GC-EC of several chlorinated dibenzo-*p*-dioxins from water:
(a) 2-MCDD 26.6 ng/ml, (b) 1-MCDD 25 ng/ml, (c) 1,2,4-TrCDD 30.4 ng/ml, (d)
1,2,3,4-TCDD 25.8 ng/ml, (e) 1,2,3,7-TCDD 12.9 ng/ml, and (f) 1,2,3,4,7,8-HxCDD
26.6 ng/ml.

Conclusions

SPME-GC has been used for residues of metolachlor, atrazine, diclofop-
methyl, and lindane isomers in water and has the potential to be applied to
a wide range of compound classes. SPME gives linear extraction of analytes
from water over several orders of magnitude of concentration. The use of
the 100-µm fiber is practicably limited to compounds of K_{ow} greater than ca.
1000. Other polymer coatings may be suitable for more polar analytes. Automated SPME in conjunction with GC can be improved by the addition of
agitation of the aqueous samples and the accommodation of somewhat larger
sample vials. SPME promises to be a useful analytical tool for pesticide
residues in water.

BTEX, related compounds, and the methylnaphthalenes in water can be
analyzed conveniently by SPME-GC-FID and PID using commercially available SPME apparatus and the autosampler modified by Varian. Samples up
to 30 ml can be handled conveniently by manual SPME. Volumes on the
autosampler are limited to 1.5 ml. Detection limits possible with the autosam-

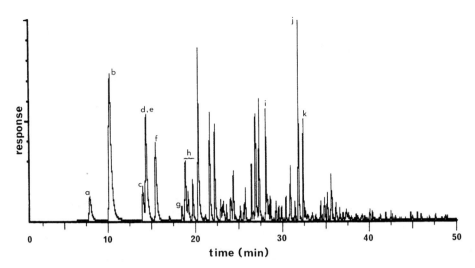

Figure 8 Direct SPME-GC-ITD of the water soluble fraction of diesel in water showing (a) benzene, (b) toluene, (c) ethylbenzene, (d) *p*-xylene, (e) *m*-xylene, (f) *o*-xylene, (g) *n*-propylbenzene, (h) trimethylbenzenes, (i) naphthalene, (j) 2-methyl-naphthalene, and (k) 1-methylnaphthalene. Identities were confirmed by use of authentic standards.

pler in its current configuration could be improved with the use of larger sample volumes. Stirring of the sample hastens extraction and reduces the reliance on diffusion for the analyte to reach the SPME fiber. Linearity of responses for the hydrocarbons analyzed extended over several orders of magnitude.

Other analytes including chlorinated dibenzodioxins can also be analyzed effectively by SPME-GC.

References

1. Belardi, R.P., Pawliszyn, J. 1989. The application of chemically modified fused silica fibres in the extraction of organics from water matrix samples and their rapid transfer to capillary columns. *Water Pollut. Res. J. Canada* 24:179–191.
2. Zhang, Z., Wang, M.J., Pawliszyn, J. 1994. Solid phase microextraction. *Anal. Chem.* 66:844A–853A.
3. Louch, D., Motlagh, S., Pawliszyn, J. 1992. Dynamics of organic compound extraction from water using solid phase microextraction with fused silica fibers. *Anal. Chem.* 64:1187–1199.
4. Zhang, Z., Pawliszyn, J. 1993. Headspace solid phase microextraction. *Anal. Chem.* 65:1843–1852.
5. Buchholz, K.D., Pawliszyn, J. 1993. Determination of phenols by solid phase microextraction and gas chromatographic analysis. *Environ. Sci. Technol.* 27:2844–2848.
6. Buchholz, K.D., Pawliszyn, J. 1994. Opitmization of solid phase microextraction conditions for determination of phenols. *Anal. Chem.* 66:160–167.

7. Webster, G.R.B., Sarna, L.P., Graham, K.N. 1994. Innovative Analysis of the Pesticides Metolachlor, Atrazine, and Diclofopmethyl in Water by SPME, 29th Annual Pesticide and Trace Analysis Seminar for Estern Canada, Vancouver, B.C., Canada, May 3–5.

8. Webster, G.R.B., Sarna, L.P., Anderson, K.A., Graham, K.N. 1994. Application of solid phase microextraction to the analysis of chlorinated contaminants in environmental matrices. 15th Annual Meeting, Society of Environmental Toxicology and Chemistry, Denver, CO, U.S.A., October 30–November 3 (Abstr.).

9. Popp, P., Kalbitz, K., Oppermann, G. 1994. Application of solid-phase microextraction and gas chromatography with electron-capture and mass spectrometry detection for the determination of hexachlorocyclohexanes in soil solutions. *J. Chromatog. A* 687:133–140.

10. Mackay, D.M., Roberts, P.V., Cherry, J.A. 1995. Transport of organic contaminants in groundwater: distribution and fate of chemicals in sand and gravel aquifers. *Environ. Sci. Technol.* 19:384–392.

11. Barker, J.F. 1987. Volatile aromatic and chlorinated organic contaminants in groundwater at six Ontario landfills. *Water Pollut. Res. J. Canada* 22:33–48.

12. Eiceman, G.A., McConnon, J.T., Zaman, M., Shuey, C., Earp, D. 1986. Hydrocarbons and aromatic hydrocarbons in groundwater surrounding an earthen water disposal pit for produced water in the Duncan Oil Field of New Mexico. *Int. J. Environ. Anal. Chem.* 24:143–162.

13. U.S. EPA. 1992. Test Methods for Evaluating Solid Waste, Physical/Chemical Methods, SW-846, 3rd Edition, Methods 3520A, July.

14. Hawthorne, S.B., Sievers, R.E. 1984. Emission of organic air pollutants from shale oil wastewaters. *Environ. Sci. Technol.* 18:483–490.

15. U.S. EPA. 1992. Test Methods for Evaluating Solid Waste, Physical/Chemical Methods, SW-846, 3rd Edition, Methods 5030A, July.

16. Pankow, J.F. 1991. Technique for removing water from moist headspace and purge gases containing volatile organic compounds: application in the purge with whole column cryotrapping (P/WCC) method. *Environ. Sci. Technol.* 25:123–126.

17. Arthur, C.L. Pawliszyn, J. 1990. Solid phase microextraction with thermal desorption using fused silica optical fibers. *Anal. Chem.* 62:2145–2148.

18. Hawthorne, S.B., Miller, D.J., Pawliszyn, J., Arthur, C.L. 1992. Solventless determination of caffeine in beverages using solid phase microextraction with fused silica fibers. *J. Chromatogr.* 603:185–191.

19. Arthur, C.L., Killam, L.M., Motlagh, S., Lin, M., Potter, D.W., Pawliszyn, J. 1992. Analysis of substituted benzene compounds in groundwater using solid-phase microextraction. *Environ. Sci. Technol.* 26:979–983.

20. Arthur, C.L., Potter, D.W., Bucholtz, K.D., Motlagh, S., Pawliszyn, J. 1992. Solid-phase microextraction for the direct analysis of water: theory and practice. *LC/GC* 10:656–661.

21. Potter, D.W., Pawliszyn, J. 1992. Detection of substituted benzenes in water at the pg/M1 level using solid-phase microextraction and gas chromatography-ion trap mass spectrometry. *J. Chromatogr.* 625:247–255.

22. Arthur, C.L., Killam, L.M., Bucholtz, K.D., Pawliszyn, J., Berg, J.R. 1992. Automation and optimization of solid phase microextraction. *Anal. Chem.* 64:1960–1966.

23. Cline, P.V., Delfino, J.J., Rao, P.C.S. 1991. Partitioning of aromatic constituents into water from gasoline and other complex solvent mixtures. *Environ. Sci. Technol.* 25:914–920.

24. Hutchins, S.R. 1991. Optimizing BTEX biodegradation under denitrifying conditions. *Environ. Toxicol. Chem.* 10:1437–1448.

25. Hutchins, S.R., Sewell, G.W., Kovacs, D.A., Smith, G.A. 1991. Biodegradation of aromatic hydrocarbons by aquifer microorganisms under denitrifying conditions. *Environ. Sci. Technol.* 25:68–76.

26. Spock, P.S., Long, R.E., Sidisky, L. 1992. Rapid screening of petroleum/chemical samples. *Supelco Reporter* XI(4):23–25.

27. Sidisky, L., Robillard, M.V. 1993. Carbon layer open rubular capillary columns. *J. High Res. Chromatogr.* 16:116–119.

28. Sarna, L.P., Webster, G.R.B., Friesen-Fischer, M.R., Sri Ranjan, R. 1994. Analysis of the petroleum components benzene, toluene, ethyl benzene, and the zylenes in water by commercially available solid-phase microextraction and carbon-layer open tubular capillary gas chromatography. *J. Chromatogr.* 677:201–205.

29. Wittkamp, B.L., Tilotta, D.C. 1995. Determination of BTEX compounds in water by solid-phase microextraction and raman spectroscopy. *Anal. Chem.* 67:600–605.

30. Worthing, C.R., Hance, R.J. 1991. *The Pesticide Manual*, 9th Edition, British Crop Protection Council, Farnham, Surrey, U.K. 1141 pp.

31. Zhang, Z., Pawliszyn, J. 1993. Analysis of organic compounds in environmental samples by headspace solid phase microextraction. *J. High Res. Chromatogr.* 16:689–692.

chapter twenty-five

Detection of herbicide residues in lipid-rish tissue using tandem mass spectrometry

John V. Headley, Kerry M. Peru, and Michael T. Arts

Introduction

A method of selecting lipid-rich tissue with detection of analytes by a tandem mass spectrometry technique is illustrated for the identification of herbicide residues in amphipods. The method was applied to specimens of wild populations of amphipods collected from microcosms spiked with a single pulse of either triallate (Avadex-BW®) or diclofop-methyl (Hoe-Grass®) at three different concentrations (1.0, 10.0, and 100.0 µg · l⁻¹). The technique provides a complementary tool to conventional procedures which require solvent extractions with derivitization steps. Likewise, the technique can be used for confirmation of residues of herbicides in lipid-rich tissue, 10 to 30 days following an initial single spike of 10 µg · l⁻¹ triallate or diclofop-methyl. The primary advantages of the technique with respects to aquatic toxicology are (a) the requirement for small quantities of tissue (<1 mg wet weight), (b) the ability to detect picogram levels of target analytes and, (c) user flexibility in the choice of tissue to be analyzed (e.g., lipid-rich storage tissue, muscle, nervous tissue). The technique can be readily adapted for detection of other contaminants in specific tissues in a diversity of aquatic and terrestrial species.

The methodology can be used for assessment of the potential effects a given contaminant may have on an aquatic ecosystem. Likewise, the procedure facilitates the study of the fate of compounds in the environment and their relative sensitivity and susceptibility within target organisms. For the examples illustrated in this manuscript, successful application of the technique is dependent on the ability to select lipid-rich tissue.

1-56670-149-X/96/$0.00+$.50
© 1996 by CRC Press, Inc.

lipid rich areas

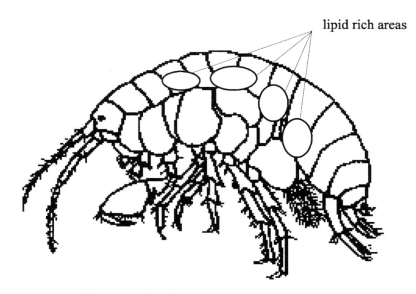

Figure 1   Schematic of amphipod.

Lipid-rich tissue

Many of the compounds displaying aquatic toxicity are lipophilic.[1] The lipid-rich tissues in which these contaminants are dissolved provide a matrix with a higher concentration of contaminants compared to the entire organism. For example, autoradiography has revealed that the herbicide triallate was preferentially associated with lipid-rich tissue alongside the gut where energy-rich triacylglycerols were stored[2] (see Figure 1).

The incorporation of lipophilic contaminants into discrete lipid droplets within storage tissues of invertebrates has been linked to the persistence of pesticides.[2] For confirmation of residues of pesticides in this application, neutral lipid-specific stain Nile red[3] was used to aid in the identification of lipid-rich storage tissues. Amphipods were selected as the study organisms because of their role as detritivores and their proximity to littoral regions receiving direct runoff. These factors make them especially susceptible to herbicide residues in aquatic ecosystems.

Materials and methods

Authentic standards can be obtained from Monsanto and Hoechst Canada for triallate and diclofop-methyl, respectively. These standards are used to obtain library spectra for confirmation of the identity of the herbicides in amphipod tissue. The main materials required are

1. Triallate, research grade.

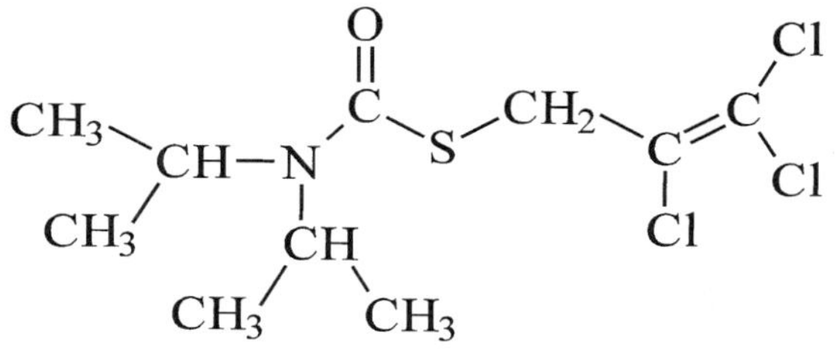

Figure 2 Structure of triallate, $C_6H_{16}Cl_3NOS$, molecular weight = 303.00.

(CAS) and IUPAC name S-2,3,3-trichloroallyl diisopropyl (thiocarbamate).[4] It is an amber oil with a molecular formula of $C_{10}H_{16}Cl_3NOS$. Triallate was introduced into Canada in 1962 by Monsanto Agricultural Co. and is currently marketed under the trade names Avadex®-BW and Fortress®. Triallate is a popular, moderately lipophilic (log octanol/water partition coefficient K_{ow} = 4.6), pre-emergence, herbicide recommended for control of wild oats in barley, lentils, durum wheat, spring wheat, and dry peas.[5] It is also recommended for wild oat control on rapeseed, flax, sugar beets, and mustard.[6] Triallate can persist from several weeks to several months in soils depending on the organic matter content.[7] Triallate entering these aquatic systems may be acutely toxic to some aquatic invertebrates[4,8,9] which in turn are critical for ducks, especially during their reproductive period.[10]

2. Diclofop-methyl, research grade.

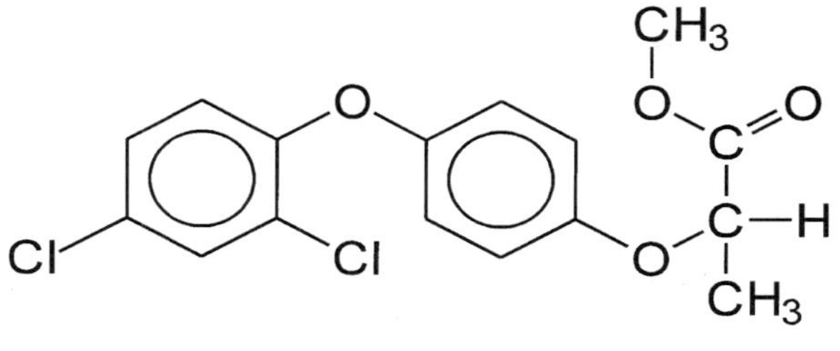

Figure 3 Structure of diclofop-methyl, $C_{16}H_{14}O_4Cl_2$, molecular weight = 340.02.

(IUPAC name 2-[4-(2,4-dichlorophenoxy)phenoxy]propionate, K_{ow} = 4.57). Diclofop-methyl is used to control graminaceous annual weeds such as wild oats, green foxtail, and witch grass primarily in wheat, barley, and soybeans.[11] At application rates ranging from 67 to 132 kg · km^{-2}, of active ingredient there is strong adsorption of the herbicide to soil and low potential for volatilization.[12,13]

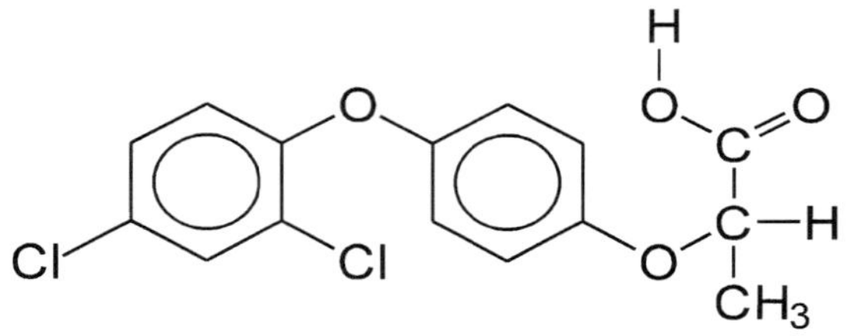

Figure 4   Structure of diclofop, $C_{15}H_{12}O_4Cl_2$, molecular weight = 326.01.

3. Diclofop, research grade.
   Under field conditions, the herbicide ester, diclofop-methyl is known to undergo hydrolysis to its corresponding acid, diclofop, 2-(4(2,4-dichlorophenoxy)-phenoxy)propionic acid.[14-16]
4. Perfluorokerosene, MS calibration standard.
5. Dry-ice.
6. Sterile razor-blade.
7. Microscope.
8. Hypodermic syringe, 10 µl.
9. Eppendorf f pipette, 200 µl.
10. Shallow quartz specimen holders (Fisons Instruments, Type C #7021901).
11. Water-cooled direct insertion probe (DIP).
12. Propane torch for flame-cleaning the quartz cups.
13. Petri dishes for preparation of lipid-rich tissue.
14. Amber lamp to illuminate the stained lipid-rich tissue. The ideal lamp should have an excitation frequency of 450 to 500 nm and the specimen should be viewed through a barrier filter of 520 nm. Tissues containing neutral lipids and stained with the Nile red will fluoresce brightly.
15. Nile red (0.25 mg in 10 ml acetone).

Procedures

Tissue preparation

The preparation of tissue samples is illustrated below for procedures applied to amphipods (*Gammarus lacustris* Sars) in our laboratory.[2]

   Before preparing the tissue for MS analysis, it is necessary to first condition a batch of new shallow quartz specimen holders, using a hot flame (propane torch will suffice). It is important to allow sufficient time (10 to 20 min) for the holders to cool to room temperature to minimize possible losses from volatilization of the analytes. The stepwise preparation of tissues for MS analysis is as follows:

1. Upon collection of amphipods, store the specimen immediately, frozen (–75°C) in the dark.

2. Just prior to instrumental analysis (within approximately 2 h to ensure sample integrity is not compromised), cut the frozen amphipods in half longitudinally with a sterile razor-blade. (Dry-ice can be used to ensure that the specimens do not thaw during this step. The frozen tissue is much easier to handle if thawing is kept to a minimum).

3. Stain the tissue with approximately 200 μl of Nile red (0.25 mg in 10 ml acetone) using an Eppendorff pipette to reveal lipid-rich tissues. Nile red has been shown to bind specifically to tissues rich in neutral-lipid such as triacylglycerol.[3] A microscope is required for seeing the stained lipid-rich droplets.

4. Remove a small piece of lipid-rich tissue (approximately 2 to 4 μl) via the needle of a hypodermic syringe, and place directly into a shallow quartz specimen holder (Fisons Instruments, Type C #7021901).

MS analysis

After preparation of the tissues, samples are subjected directly to instrumental analysis with no further sample preparation, cleanup or pre-concentration steps. For instrumental analyses, advance training in the operation of a tandem mass spectrometer is required. Procedures are illustrated in this section for a Fisons AutospecQ mass spectrometer with EBEQ geometry (Figure 5), equipped with a 4000-60 VAX data system, Digital Equipment

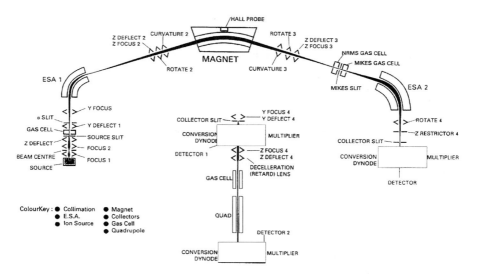

ION OPTICS - AutoSpec Q

Figure 5   Schematic of AutospecQ (EBEQ geometry). (Reprinted with permission from Chris Woodward on the *VG* Organic poster.)

Co., and Opus 3.1X software. These procedures may be adapted to other less expensive instrumentation employing quadrupole or ion-trap technology. The stepwise procedure is as follows:

1. Tune the mass spectrometer to a minimum of 1300 resolution, operating under electron impact conditions in the positive-ion mode, with perflourokerosene as the calibration reference standard.
2. Typical conditions and mass spectrometer parameters are as follows: ion source operated under electron impact conditions at 70 electron volts, 250°C, trap current 250 μA, with the MS operated at a scan speed 1 sec/decade, and mass range 50 to 600 Da.
3. Place the samples in shallow cups of the direct insertion probe (DIP) for 30 min at room temperature (approximately 23°C), prior to introduction to the ion source. This is necessary to reduce the moisture content of the samples and avoid tripping the vacuum protection system (set at 5×10^{-5} torr) of the mass spectrometer.
4. For best results, the direct insertion probe should be water-cooled, and heating limited to the radiant heat from the ion source with no additional heat supplied by the probe heaters. The DIP can also be used without being water-cooled. This may be more convenient where multiple inlets requiring water cooling are already interfaced to the MS. However, the run time of experiments will increase significantly, since sufficient time must be allowed for the probe to cool to room temperature between runs.

Detection of herbicides

The detection of analytes is best illustrated using the examples of the determination of the herbicides, triallate, diclofop-methyl and the hydrolysis product diclofop. The procedure for the specific identification of these analytes is as follows:

5. Examine the electron impact (EI) mass spectra of the authentic standards, examples of which are given in Figure 6a–c, for the purpose of selecting diagnostic ions suitable for evaluating the presence of the target analytes in the full scan spectra of amphipod tissue.
6. The ions at m/z 268 (triallate), m/z 340 (diclofop-methyl) and m/z 326 (diclofop) are suitable for tandem MS experiments.[17,18] The criteria for the selection of diagnostic ions for MS identification, or precursor-ions for subsequent tandem MS experiments are based on (a) their occurrence at relatively high mass, (b) their relative abundances, and (c) low probability of formation from fragmentation or re-arrangement in the ion source.
7. In general, positive electron impact MS spectra alone without the use of (a) chemical ionization, (b) negative ion electron capture, or (c) tandem MS/MS will not suffice for the confirmation of the presence of herbicide residues in lipid-rich tissue.[17,18] This is largely due to

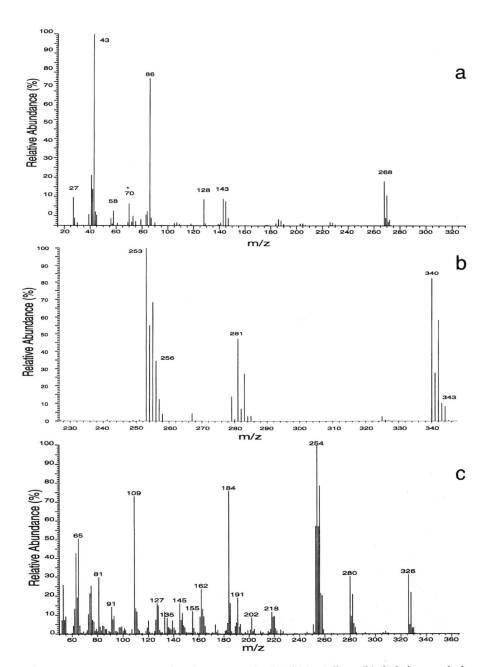

Figure 6   EI mass spectra of authentic standards of (a) triallate, (b) diclofop-methyl, and (c) diclofop.

interference from ions produced from the matrix. This is illustrated in the examples of the total-ion chromatogram and corresponding mass spectra of amphipod tissue containing diclofop, given in Figure 7.

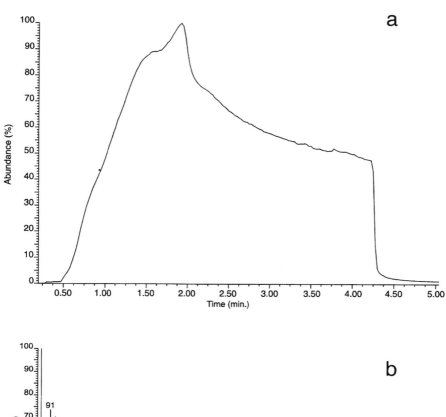

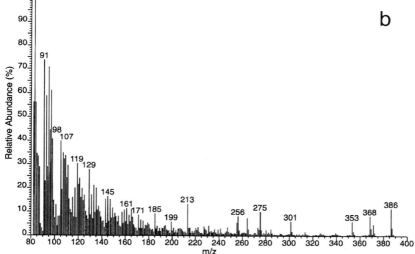

Figure 7 (a) Total-ion chromatogram and (b) corresponding mass spectrum of amphipod tissue containing diclofop. The presence of diclofop is masked by high background chemical noise created by interfering ions from the matrix. (Reprinted with permission from Arts, M.T. et al. 1996. *Environ. Toxicol. Chem.* 15(4):481–488. Copyright 1996 SETAC.)

MS/MS analysis

For the MS/MS experiments,[17,18] precursor-ions are selected manually and the collision cell was located in the fourth field free region (see Figure 5). In brief, selection of the precursor-ion is performed using the first MS, and subsequent fragmentation is induced by collisions with an inert gas in the collision cell. The MS/MS product-ions are detected using the second MS.[19] In this application, the product-ion spectra serve to verify that the selected extracted-ions are diagnostic for the respective analytes and are used to enhance the sensitivity of the method for identification of the herbicide residues relative to the MS full scan procedure.[17,18] The stepwise procedure is as follows:

1. Reduce the ion-beam (m/z 331 ion of perfluorokerosene) to 50% transmission using xenon as the collision gas.
2. Using neat standards of the herbicides optimize the abundance of the precursor-ions m/z 340, 326, and 286 by adjusting the eV of the ion source and the pressure of the collision gas for diclofop-methyl, diclofop, and triallate, respectively. Typical values observed fall in the range 20 to 30 eV.
3. Likewise optimize the low energy collisions in the collision cell. This is achieved while monitoring for example, the abundances of product-ions at m/z 253, 281, and 226, 184, for diclofop-methyl and triallate, respectively. Typical values observed fall in the range 12 to 20 eV and 20 to 35 eV (laboratory frame of reference) for precursor ions m/z 326 and 268, respectively.
4. Collect the product-ions by scanning the quadrupole in the mass range 30 to 350 Da at unit resolution.[20]
5. Representative results[17,18] of tandem MS product-ion scans for spiked tissue and field-derived amphipod tissue samples are given in Table 1. The data are representative of scans from amphipod tissues spiked with 50 pg of analytical standard and from field specimens obtained from microcosms treated with 100 µg/l of the herbicides. The observed differences in the abundances of the product ions for the field tissue and the laboratory spiked tissue (Table 1) can be attributed to matrix interference contributing to the abundance of the precursor-ion.

Table 1   Relative Abundances of Product-Ions of Triallate and Diclofop

m/z	Triallate			Diclofop		
	184	226	268	253	281	326
Spiked tissue	2.5	3.8	100.0	16.0	30.9	100.0
Field tissue	5.1	10.1	100.0	30.9	61.7	100.0

Note: m/z = mass to charge ratio.

Reprinted with permission from Arts, M.T. et al. 1996. *Environ. Toxicol. Chem.* 15(4):481–488. Copyright 1996 SETAC.

6. Under optimum instrumental conditions, the detection limit of the procedure is suitable for the confirmation of 20 pg of analyte in lipid-rich tissue.[17,18] Representative product-ion spectra are given in Figure 8a, b. The spectrum shown were (a) obtained using spiked amphipod tissue (2 mg), prepared by injecting lipid-rich tissue with 50 pg of triallate (2 μl of 25 pg/μl triallate in methanol), and (b) obtained for field amphipod tissue containing triallate.[17,18]

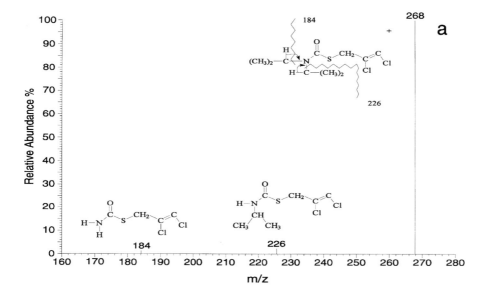

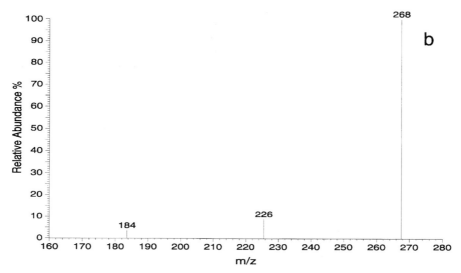

Figure 8 Product-ion spectra (a) spiked amphipod tissue and (b) field amphipod tissue. (Reprinted with permission from Headley, J.V. et al. 1995. *Anal. Chem.* 67:43–52. Copyright 1995 American Chemical Society.)

Table 2 Tandem MS Confirmation of Triallate
in Lipid-Rich Amphipod Tissues Collected
Over a 30-Day Period from *In Situ*
Microcosms Spiked with Triallate

Day	Control	Initial triallate concentration ($\mu g \cdot l^{-1}$)		
		1	10	100
1	N	Y	Y	Y
3	N	T	Y	Y
15	N	N	Y	Y
30	N	N	Y	Y

Note: Y, positive detection, both product-ions (m/z, 226 and 184); T, trace detection, one product ion; N, no detection.

Reprinted with permission from Headley, J.V. et al. 1995. *Anal. Chem.* 67:43–52. Copyright 1995 American Chemical Society.

Results and discussion

A practical advantage of the tandem MS methodology is the direct application to small amounts of tissue. This has been shown to facilitate the sampling of small-bodied invertebrates[18] and is complementary to conventional analytical techniques.[21-23]

For full assessment of the sensitivity of the procedure, a rigorous examination of the factors effecting the quantification of the analytes is required. As reported for earlier work with biofilm samples,[20,23] good product-ion MS/MS spectra can be obtained for tissue spiked with 20 pg of the herbicides. The identification detection limit of diclofop acid (based on signal:noise ratio of 3:1 for m/z 326) is approximately 300 pg in the MS experiments compared to 20 pg (based on observation of product-ions at m/z 281 and 253) for the MS/MS procedure. This level of performance is subject to cleaning the inner ion source following the introduction of approximately 30 to 50 samples.[17,18]

For some herbicides, the performance of the technique is adversely affected by the chemical-physical properties of the analyte. For example, poor performance was observed during a preliminary study of lipid-rich tissue spiked with bromoxynil (Headley, J.V., K.M. Peru, and M.T. Arts, unpublished work). For this herbicide, alternate methods of ionization, employing negative-ion electron capture are under investigation. It is anticipated that although the sensitivity of some chlorinated contaminants will be enhanced for MS experiments, in general, negative-ions will be more resistant to collision induced dissociations.[24] The stability of the negative-ions will likely limit their diagnostic value for the tandem MS experiments.

Representative results for amphipods collected from microcosms exposed to triallate are given in Tables 2 and 3.[17,18] These results demonstrate that triallate could be detected after 30 days in the 10 and 100 $\mu g \cdot l^{-1}$ treatments and up to day 3 in the 1 $\mu g \cdot l^{-1}$ treatment but was not detected

Table 3 Tandem MS Confirmation of the Hydrolysis
Product, Diclofop in Lipid-Rich Amphipod Tissues
Collected Over a 30-Day Period from *In Situ*
Microcosms Spiked with Diclofop-Methyl

Day	Control	Initial diclofop-methyl concentration ($\mu g \cdot l^{-1}$)		
		1	10	100
1	N	Y	Y	Y
3	N	Y	Y	Y
6	N	N	Y	Y
10	T	Y	Y	Y
15	N	N	N	Y
20	N	N	T	T
25	N	N	T	T

Note: Y, positive detection, both product-ions (m/z, 281 and
253); T, trace detection, one product ion; N, no detection.

in the control microcosm (Table 2). In general diclofop could be detected in
amphipod tissues up to and including day 10 in the 1, 10, and 100 $\mu g \cdot l^{-1}$
treatments,[18] and in general was not detected in tissues from the control
microcosm (Table 3). The significance of the levels of herbicide residues in
the lipid-rich tissue compared to levels observed for the water and sediment
compartments are discussed elsewhere.[18]

Conclusion

The examples illustrated demonstrate the utility of the tandem MS technique
for the detection of herbicides in lipid-rich tissue. The technique is well suited
for the identification of herbicide residues. For some contaminants, such as
bromoxynil, some refinement of the tandem MS application will be required.
However, in general, it is anticipated that the technique can be adapted for
study of a wide range of contaminants of environmental significance.

Acknowledgments

Technical assistance was provided by Brenda Headley and Mary Ferguson.
Funding from Pestplan, Environment Canada is gratefully acknowledged.

References

1. Barron, M. G. 1990. Bioconcentration. *Environ. Sci. Tech.* 24:1612–1618.
2. Arts, M.T., M.E. Ferguson, N.E. Glozier, R.D. Robarts, and D.B. Donald. 1995.
   Spatial and temporal variability in lipid dynamics of common amphipods:
   Assessing the potential for uptake of lipophilic contaminants. *Ecotoxicology*
   4:91–113.

3. Greenspan, P., E.P. Mayer, and S.D. Fowler. 1985. Nile Red: A selective fluorescent stain for intracellular lipid droplets. *J. Cell Biol.* 100: 965–973.

4. Kent, R. A., M. Taché, P.-Y. Caux, S. De Silva, and K. Lemky. 1992. Canadian water quality guidelines for triallate. Ecosystem Sciences and Evaluation Directorate, Eco-Health Branch, Ottawa, Ontario. Scientific Series No. 195, 47 pgs.

5. Worthing, C.R., and S.B. Walker (eds.) 1987. The pesticide manual: a world compendium. 8th ed. British Crop Protection Council, Thorton Heath, UK. 1081 pp.

6. Agriculture Canada. 1982. Guide to the chemicals used in crop protection. 7th ed. Research Branch, Ottawa, Canada. Publication #1093.

7. Jury, W.A., R. Grover, W.F. Spenser, and W.J. Farmer. 1980. Modelling vapour losses of soil-incorporated triallate. *Soil Sci. Soc. Am. J.* 44:445–450.

8. Johnson, B. T. 1986. Potential impact of selected agricultural chemical contaminants on a northern prairie wetland: A microcosm evaluation. *Environ. Tox. and Chem.* 5:473–485.

9. Buhl, K.J., and N.L. Faber. 1989. Acute toxicity of selected herbicides and surfactants to larvae of the midge *Chironomus riparius. Arch. Environ. Contam. Toxicol.* 18:530–536.

10. Krapu, G.L., and K.J. Reinecke. 1992. Foraging ecology and nutrition. In *Ecology and management of breeding waterfowl.* Batt, B.D.J., A.D. Afton, M.G. Anderson, C.D. Ankney, D.H. Johnson, J.A. Kadlec, and G.L. Krapu (editors). University of Minnesota Press, London. pp 1–29.

11. Herbicide Handbook. 1989. Weed Science Society of America. Herbicide Handbook Committee. N. E. Humburg (chairman). Champaign, IL. 301 pp.

12. Matthiessen, P., G. F. Whale, R. J. Rycroft, and D. A. Sheahan. 1988. The joint toxicity of pesticide tank-mixes to rainbow trout. *Aquat. Toxicol.* 13:61–76.

13. Smith, A.E., R. Grover, A.J. Cessna, S.R. Schewchuk, and J.H. Hunter. 1986. Fate of diclofop-methyl after application to a wheat field. *J. Environ. Qual.* 15:234–238.

14. Martens, R. 1978. Degradation of the herbicide [^{14}C]-diclofop-methyl in soil under different conditions. *Pestic. Sci.* 9:127–134.

15. Smith, A.E. 1979. Transformation of [^{14}C]diclofop-methyl in small field plots. *J. Agric. Food Chem.* 27:1145–1148.

16. Gaynor, J.D. 1984. Diclofop-methyl persistence in southwestern Ontario soils and effects of pH on hydrolysis and persistence. *Can. J. Soil Sci.* 64: 283–291.

17. Headley, J.V., K.M. Peru, and M.T. Arts. 1995. Tandem mass spectrometry of herbicide residues in lipid-rich tissues. *Anal. Chem.* 67(23):43–52.

18. Arts, M.T., J.V. Headley, and K.M. Peru. 1996. Detection and persistence of herbicide residues in *Gammarus lacustris* (Crustacea:Amphipoda) in Prairie Wetlands. *Environ. Toxicol. Chem.* 15(4):481–488.

19. Headley, J.V., and A.G. Harrison. 1985. Structure and Fragmentation of $C_5H_{11}O^+$ Ions Formed by Chemical Ionization. *Can. J. Chem.* (63):609–618.

20. Headley, J.V., K.M. Peru, J.L. Lawrence, and G.M. Wolfaardt. 1995. MS/MS identification of transformation products in degradative biofilms. *Anal. Chem.* 67:1831–1837.

21. Bruns, G.W., S. Nelson, and D.G. Erickson. 1991. Determination of MCPA, bromoxynil, 2,4-D, trifluralin, triallate, picloram, and diclofop-methyl in soil by GC-MS using selected ion monitoring. *J. Assoc. Off. Anal. Chem.* 74:550–553.

22. Headley, J.V., J.R. Lawrence, B.N. Zanyk, and P.W. Brooks. 1994. Transformation of the herbicide diclofop-methyl in a large scale physical model. *Water Poll. Res. J. Canada* 29:557–569.

23. Wolfaardt, G.M., J.R. Lawrence, J.V. Headley, R.D. Robarts, and D.E. Caldwell. 1994. Microbial exopolymers provide a mechanism for bioaccumulation and transfer of contaminants. *Microbial Ecol.* 27:279–291.

24. Headley, J.V., and K.M. Peru. 1994. Collision-induced dissociation mass spectrometry of the herbicide diclofop-methyl. *J. Rapid Commun. Mass Spectrom.* 8:484–486.

chapter twenty-six

Measuring metals and metalloids in water, sediment, and biological tissues

Michael C. Newman

Introduction

Outbreaks of cadmium (Itai-Itai disease) and mercury (Minamata disease) poisoning during the 1950s made us acutely aware of the adverse consequences of high concentrations of metals in our environment. Quickly, measurement of metals and metalloids became an integral component of our efforts to monitor and correct effects of anthropogenic emissions. The widespread introduction of commercial atomic absorption spectrophotometers (AAS) in the early 1960s contributed enormously to the rapid increase in essential data.

The first quantitative AAS was developed in the 1940s.[1] The number of commercial AAS units was increasing exponentially by 1963.[2] The introduction of flameless atomization methods lowered limits of detection several orders of magnitude by allowing the ground state metal to stay in the analytical light path longer than with flame atomization. Today flame and flameless capabilities are incorporated together in AAS units, allowing convenient measurement of elements present in mg/g to μg/kg concentrations. Well-established, preconcentration procedures are used to remove analytes from interfering matrices as well as to concentrate them in small volumes. Flame and furnace chemistries are now sufficiently well understood to allow effective matrix modification for most elements.

Although measurement of metal concentrations has become routine and convenient, considerable work remains to be done relative to assessing metal bioactivity and speciation. Numerous sample pretreatments exist that imperfectly reflect bioavailable metal. Discussed in detail, such methods and associated assumptions would easily fill an entire book. Further, the rapid changes in this area of metal ecotoxicology would make such a volume obsolete within a few years. Consequently, only the most fundamental techniques for measuring metals dissolved in waters, or present in solid samples are described here. Hopefully, these methods will have the most general utility. Space limitations also exclude adequate description of the cold vapor methods for mercury and hydride-generation methods for arsenic or selenium.

Materials required

General

Chemical safety

The analyst should not prepare or use any of the following reagents without fully understanding steps necessary for their safe use. Eye protection, protective clothing, and plastic gloves should be worn when handling strong acids. Always handle strong acids in a fume hood. The methods described here avoid the use of perchloric acid as it can react violently with organic materials. Nitric acid is a safer and spectroscopically cleaner acid; however, digestion of some samples may be unacceptable with this nitric acid digestion.

High purity acids

High purity nitric and hydrochloric acids may be purchased or produced. Commercially available acids include Ultrex® (Baker, Phillipsburg, NJ), Suprapur® (Merck, Darmstadt, Germany) and Aristar® (BDH Chemical Ltd., Poole, England). Sub-boiling distillation with a Teflon® still such as that sold by Berghof/America, Inc. (Raymond, NH) may also be used to generate these acids. Store distilled acids in acid-cleaned Teflon containers.

Acid cleaning of laboratory ware

No one procedure ensures both optimal allocation of effort and contamination-free analyses for all situations. Initially, the analyst should use procedural blanks to gain an understanding of the effort required for efficient, contamination-free analyses. Contamination control is one of the most crucial aspects in analysis of many trace metals.

Although expensive, it is preferable to use Teflon® containers and laboratory ware. Linear polyethylene, polycarbonate plasticware and, perhaps, Pyrex® are less costly alternatives. For example, some analysts may find a Pyrex® filtration apparatus more convenient than the suggested polycarbonate unit in the dissolved metals procedure below. Pyrex® volumetric flasks are a necessary compromise for careful volumetric measurement, although liquids may be weighed to avoid potential contamination from volumetric glassware. Rubber stoppers or seals, soft glass, Bakelite® caps, and caps made with contaminating seals such as paper or aluminum should not be used.

Clean all plasticware and glassware with a noncationic detergent (e.g., Acationox®, American Scientific Products, Stone Mountain, GA) before acid cleaning. Acid soak all materials that will touch the sample at least 24 h in 50% (v/v) concentrated nitric acid in deionized water. Rinse items seven to ten times with deionized water and place them in a contamination-free area to dry (e.g., under a Class 100 laminar-flow clean hood). Adjust the duration of soaking, rinsing, and level of contamination control during drying in balance with the concentration of analyte expected. Items for the analysis of lead in water may require extreme attention to restricting contact with dust particles during drying yet those used for the analysis of calcium in sediment may require much less attention. Use of procedural blanks aids in adjusting procedures to the appropriate level of rigor. The results from procedural blanks should dictate the specific details of washing and preparation, not rote adherence to any prescribed cleaning steps.

Many problems are eliminated if laboratory ware used for trace element measurement are used only for those analyses. Such dedicated laboratory ware should be stored in sealed plastic bags, e.g., Ziploc® bags. Acid-cleaned pipette tips can be conveniently stored in an acid-cleaned, Teflon® or linear polyethylene bottle with a polyethylene cap. Use wares as soon as reasonable after acid cleaning to minimize contamination during storage.

Soak membrane filters for separation of dissolved metals in dilute, high purity hydrochloric acid at least 4 h. A 0.5 N HCl soaking solution can be made for this purpose by carefully adding 4 ml of concentrated, high purity hydrochloric acid to 92 ml of deionized water. Soak filters for longer if blanks indicate contamination. Soaking can be done in a covered, acid-cleaned Teflon® beaker (100 ml). Rinse filters thoroughly with metal-free deionized water prior to use. Acid-washed plastic or Teflon®-coated tweezers should be used carefully to handle the filters.

If dissection is required in preparation of biological samples, metal cutting tools can easily introduce metals such as iron, zinc, and chromium. The author uses plastic, disposable utensils such as those sold in grocery stores for picnics as dissecting utensils. Plastic knives may be sharpened with nonmetallic abrasives but they should be carefully cleaned and checked for contamination before use. These plastic utensils may be destroyed by prolonged soaking in strong acid. A dilute (10% (v/v)) concentrated nitric acid soak for 12 h is probably adequate. However, the researcher should test one utensil in any cleaning procedure prior to general implementation.

Standard materials

Standard reference materials are the most effective means of troubleshooting analytical problems during method development and documenting accuracy during sample analysis. Sample spikes are useful in the absence of a standard material or to augment method troubleshooting. The U.S. National Institute of Standards and Technology (NIST) has standard materials for water, sediment, and animal tissue. The author also uses materials supplied by the National Resource Council of Canada. Veillon[3] and Van Loon[4] list other sources of standard materials for biological tissues. At this time, convenient suppliers include:

NIST
Standard Reference Material
 Program
Room 204, Building 202
National Inst. of Standards and
 Technology
Gaithersburg, MD 20899 USA

U.S. Geological Survey
National Center
Reston, VA 22092 USA

Brammer Standard Co., Inc.
5607 Fountainbridge Lane
Houston, TX 77069 USA

National Resource Council of
  Canada
Chemistry Division
Marine Analytical Chemistry
  Standards Program
Montreal Road
Ottawa, Canada K1A OR9

British Chemical Standards
Bureau of Analyzed Samples
Newham Hall
Middlesbrough, England

Deutsche Vertretung
Dipl-Met. G. Winopal
Universal-Forschungsbedarf
Echternfeld 25, Postfach 40
3000 Hannover 51
Federal Republic of Germany

International Atomic Energy Agency
Analytical Quality Control Services
Laboratory Selberdorf, P.O. Box 100
A-1400 Vienna, Austria

Community Bureau of Reference
Commission of the European
  Communities
200 Rue de la Lol
B-1049 Brussels, Belgium

Standard metal solutions

Stock solutions of 1000 µg/ml for preparing AAS standards can be purchased from several sources including Fisher Scientific (Pittsburgh, PA), Sigma (St. Louis, MO) and Perkin-Elmer (Norwalk, CT). Standards have a finite shelf life and should not be used beyond their expiration dates.

Matrix modifiers

Atomic absorption spectrophotometry measures the absorption of light at a specific wavelength (λ) by the ground state atom (M^o) during excitation: $M^o + E_\lambda \rightarrow M^*$. The absorption is proportional to the number of groundstate atoms in the light path. In turn, the number of ground state atoms is a function of the concentration of the element in the sample and physicochemical reactions occurring in the flame or graphite furnace (Figure 1). The reactions occurring in the flame or furnace may be manipulated with matrix modifiers to optimize production of ground state atoms and, thus, enhance absorption (sensitivity).

For flame AAS, the sample is aspirated to form small droplets in a premix chamber. These droplets are swept into the flame and form dry "clotlets" of salts and other solids. The clotlets melt and vaporize in the flame. Simple metal compounds such as the metal monoxides illustrated here undergo a sequence of transitions that, at equilibrium, determine the measured population of ground state atoms (M^o).

$$M^o + E_{Heat} \rightleftharpoons M^o + O^o \tag{1}$$

$$M^o + E_\lambda \rightleftharpoons M^* \tag{2}$$

$$M^o + E_{Heat} \rightleftharpoons M^+ + e^- \tag{3}$$

Selection of a hot flame (nitrous oxide-acetylene, 2900°C) instead of a cool flame (air-acetylene, 2300°C) for elements with metal monoxide disso-

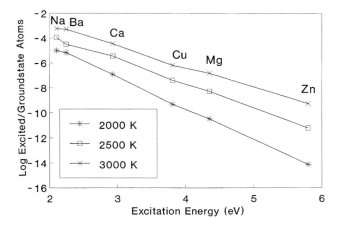

Figure 1 The relative amount of atoms excited and at groundstate for flames with three different temperatures (Na, Ba, Ca, Cu, Mg, and Zn). Both temperatures of the flame and the excitation energy determines the proportion of atoms that are excited. Regardless of excitation energy, the log (number of excited atoms/number of ground-state atoms) increases with flame temperature. For any given temperature, the proportion of atoms excited decreases with increasing excitation energies. (Data from Table VIII-5 in Reynolds et al.[50])

ciation constants greater than approximately 5 eV assures reaction (1) will produce ample amounts of M^O (Price[2]). (Pertinent dissociation constants are listed in the Appendix.) A reagent (e.g., La) may also be added that reacts with components in a droplet that would otherwise combine with the analyte to make it resistant to dissociation. If the flame is too hot, ionization (reaction 3) shifts the equilibrium to decrease the amount of M^O present (Figure 2).

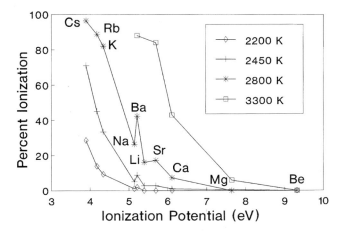

Figure 2 Percent of the total number of atoms that are ionized for ten elements aspirated into different temperature flames. Note that some (e.g., Cs) have high proportions of atoms ionized even in a lower temperature (2450°K, oxygen-hydrogen) flame. Others (e.g., Mg) have minimal ionization even in the hottest (3300°K, nitrous oxide-acetylene) flame. (Data from Table 12-3 in Willard et al.[51])

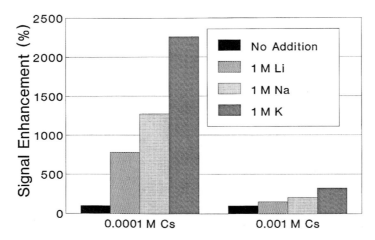

Figure 3 Signal enhancement (percentage increase relative to the signal for the solution with no addition of Li, Na, or K) for two different concentrations of Cs. Cesium, an easily ionized element (Figure 2), has a greatly enhanced signal if an excess of another easily ionized element is added. Elements with the lowest ionization potentials (see Figure 2) produce the best enhancement. The enhancement increases as more ionization buffer is added relative to the amount of Cs, e.g., 1 M ion buffer to 0.0001 M Cs versus 1 M ionic buffer to 0.001 M Cs. (Air-acetylene flame data from Table 12-4 in Willard et al.[51])

Elements with low ionization energies (see Appendix) such as the alkali and alkaline-earth metals are most prone to ionization.[5] Rare earth elements can also have ionization problems with a hot (nitrous oxide-acetylene) flame.[5] Adding excess amounts of any element that ionizes readily (i.e., an ionization buffer) can shift the equilibrium back to favor M^O (Figure 3). As is evident from their ionization potentials (E_i), cesium ($E_i = 3.89$), potassium ($E_i = 4.34$), sodium ($E_i = 5.14$) or, even, lanthanum ($E_i = 5.58$) can be used as ionization buffers.

For flameless AAS, a matrix modifier may be added for several reasons. In the furnace, a sample is dried onto a graphite surface and then heated to drive off any compounds that would interfere during subsequent measurement of analyte. After such pyrolysis, the sample is finally heated to a temperature sufficient to dissociate compounds containing the analyte. The amount of ground state atoms is measured at this atomization stage. Obviously, any material remaining after the pyrolysis stage can interfere with measurement during atomization. The presence of several salts of the analyte with their different dissociation constants may also change the shape (broaden or skew) of the absorbance curve during atomization and compromise quantification. Matrix modifiers (e.g., the Pd–Mg modifier described below or thiourea addition for cadmium analysis[6]) can be added to the sample that increase the temperature at which the analyte dissociates. In this way, more potentially interfering compounds can be driven off with higher temperatures during the pyrolysis stage without significant loss of analyte.

This is especially helpful for elements with low boiling temperatures (see Appendix) or those that form volatile compounds.

A matrix modifier may also be added to make components of the sample more volatile. Ediger[7] recommended additions of ammonium nitrate to samples containing large amounts of sodium chloride because the resulting sodium nitrate and ammonium chloride salts could be driven off at low pyrolysis temperatures.

The excess of matrix modifier may also result in the preponderance of only one salt to dissociate during atomization. For example, the addition of excess H_3PO_4 to lead samples greatly improves the shape of the atomization curve by favoring formation of a single lead phosphate salt prior to atomization.[8] Similarly, Ediger[7] suggested that cadmium can be stabilized during furnace AAS by adding ammonium phosphate.

In general, the reagents recommended by Perkin-Elmer[9,10] and Schlemmer and Welz[11] are used here. Additional discussion of matrix modifiers can be obtained from Varma[1] and Carnrick et al.[8] The reagents described in the Appendix for analysis of each element are produced and used as described below. Preparation of $Pd(NO_3)_2$ and $Mg(NO_3)_2$ solutions are taken from Schlemmer and Welz.[11] The $Pd(NO_3)_2$ and $Mg(NO_3)_2$ reagents may also be purchased from Perkin-Elmer (Norwalk, CT).

$Pd(NO_3)_2$ stock solution (1% (w/v) Pd). A minimum volume of Ultrex® or equivalent concentrated nitric acid is used to dissolve 1.0 g of palladium metal powder (Aldrich Chemical Co., Milwaukee, WI) and the dissolved metal is brought to 100 ml with deionized water. A small amount of metal may be dissolved first with mild heating and additional small amounts added until all is dissolved. $Pd(NO_3)_2$ can also be purchased as a salt (Sigma, St. Louis, MO or Aldrich, Milwaukee, WI) to make this solution.

$Mg(NO_3)_2$ stock solution (1% (w/v) $Mg(NO_3)_2$). Dissolve 1.729 g of $Mg(NO_3)_2 \cdot 6H_2O$ in 100 ml of deionized water.

$NH_4H_2PO_4$ stock solution (1% (w/v) $NH_4H_2PO_4$). Dissolve 1.000 g of ammonium phosphate dibasic in 100 ml of deionized water.

5 μg Pd + 3 μg $Mg(NO_3)_2$ reagent. Use Eppendorf® or similar calibrated pipettes with acid-cleaned tips to add the following volumes of stock solutions to a 25-ml, acid-cleaned volumetric flask: 1.250 ml of 1% Pd stock solution and 0.750 ml of the 1% $Mg(NO_3)_2$ stock solution. Bring to volume with deionized water. Add 10 μl per furnace injection for matrix modification as indicated in the Appendix.

5 μg $Mg(NO_3)_2$ reagent. Use Eppendorf or similar calibrated pipettes with acid-cleaned tips to add 1.250 ml of the 1% $Mg(NO_3)_2$ stock solution to a 25 ml, acid-cleaned volumetric flask. Bring to volume with deionized water. Add 10 μl per furnace injection for matrix modification as indicated in the Appendix.

15 µg Mg(NO₃)₂ reagent. Use Eppendorf or similar calibrated pipettes with acid-cleaned tips to add 3.750 ml of the 1% $Mg(NO_3)_2$ stock solution to a 25-ml, acid-cleaned volumetric flask. Bring to volume with deionized water. Add 10 µl per furnace injection for matrix modification as indicated in the Appendix.

50 µg NH₄H₂PO₄ + 3 µg Mg(NO₃)₂ reagent. Use Eppendorf or similar calibrated pipettes with acid-cleaned tips and acid-cleaned, volumetric pipettes to add the following volumes of solutions to a 25 ml, acid-cleaned volumetric flask: 12.500 ml of 1% $NH_4H_2PO_4$ stock solution and 0.750 ml of 1% $Mg(NO_3)_2$ stock solution. Bring to volume with deionized water. Add 10 µl per furnace injection for matrix modification as indicated in the Appendix.

LaCl₃ reagent (50 g/l La). CAUTION: The following solution reacts vigorously during preparation. Dissolve very small amounts of lanthanum oxide (La_2O_3) at a time and, after complete dissolution of each small amount, add a small amount more. Cautiously continue this process until all is dissolved. Carefully and slowly dissolve a total of 29.3 g of La_2O_3 in 250 ml of concentrated hydrochloric acid. Then cautiously dilute to 500 ml with deionized water. Add 1 ml of this $LaCL_3$ solution to 10 ml of each sample, blank and standard prior to analysis. Alternatively, 133.6750 g of lanthanum chloride heptahydrate ($LaCl_3 \cdot 7H_2O$) can be added to 1 l of deionized water to make this 50 g/l La solution.

KCl reagent (50 g/l K). Dissolve 95 g of KCl in 1 l of deionized water. Add 200 µl of this reagent to 10 ml of each sample, blank and standard prior to analysis.

Dissolved metals in water

> Acid-cleaned 500-ml Teflon® or linear polyethylene sample bottles
> Acid-cleaned 60-ml, wide-mouth, Teflon or linear polyethylene bottles for filtered samples
> Acid-cleaned 47-mm, polycarbonate filtration funnel and receiving flask
> Acid-soaked 0.45-µm membrane filter (e.g., Millipore® HAWP 047 or AAWP 047 filters)
> Acid-cleaned plastic or Teflon-coated tweezers for handling the filters
> Acid-cleaned 500-ml squirt bottle containing deionized water
> Source of ultrapure, deionized water with consistently low concentrations of the elements of interest (e.g., ASTM Type I reagent grade water)
> Vacuum pump with a liquid trap
> Class 100 laminar flow hood if required
> Plastic gloves (without talc powder)
> Calibrated fixed-volume pipettes (e.g., Eppendorf pipettes) of appropriate volumes
> Acid-cleaned pipette tips of appropriate sizes

An appropriate standard material
Insulated chest containing ice for field chilling of samples
Plastic Ziploc bags

Metals in biological tissue or sediments

Acid-cleaned Teflon beakers with a spout to allow effective pouring of the digest (25, 50 or 100 ml volume depending on the weight of sample to be digested, one per sample including blanks and standard materials)

Acid-cleaned Teflon watchglasses adequate to completely cover the Teflon beakers or loose-fitting beaker lids such as those sold by Berghof/American, Inc. (Raymond, NH)

Acid-cleaned Teflon funnels for digest transfer from beakers to volumetric flasks

Acid-cleaned 60-ml, Teflon or linear polyethylene bottles for sample digests

Acid-cleaned, plastic dissecting utensils if required

Acid-cleaned, Class A Pyrex 25-ml volumetric flasks with stoppers

Acid-cleaned, Class A Pyrex 5-ml volumetric pipette

Acid-resistant pipette bulb

Calibrated fixed-volume pipettes (e.g., Eppendorf pipettes) of appropriate volumes

Acid-cleaned pipette tips of appropriate sizes

Acid-cleaned 500-ml squirt bottle of deionized water

Source of ultrapure deionized water with consistently low concentrations of the elements of interest (e.g., ASTM Type I reagent grade water)

Plastic gloves (without talc powder)

Zero to 100°C thermometer (prefereably traceable to an NIST thermometer)

Hot plate capable of holding a constant temperature of 80°C.

Pyrex pan to serve as a water bath into which the Teflon beakers may be placed

Appropriate standard materials

Procedures

Dissolved* metals in water

Sampling

Sufficient numbers of acid-cleaned sample bottles should be carried to the field inside sealed plastic bags. Bring enough to take replicate samples at all or a subset of sample sites. Travel blanks may be produced by pouring a deionized water "sample" into acid-cleaned containers in the field. Travel blanks should be handled like all other samples. Field spikes or field pro-

---

* Dissolved metals are procedurally-defined as those passing through a 0.45-μm filter.

cessing of solutions of known metal concentrations are also helpful in tracking accuracy and precision of the entire sampling and measurement process.

Rinse each sample bottle (including the cap) with sample at least four times prior to filling with the final sample. Do not allow the sample to touch your hand or any other contaminating materials prior to entering the sample bottle. Do not agitate sediments or dislodge materials from vegetation during sampling. Do not take a sample near the water surface as some metals are concentrated in the surface microlayer. Seal the sample bottle in a Ziploc plastic bag and place it into a ice-filled cooler chest. Process the sample as soon as reasonable, usually within 24 h of sampling. Ideally, the sample should be filtered immediately. Adjust the above sampling instructions based on results from spiked samples and travel blanks.

Filtration

Ideally, one acid-cleaned filtration apparatus should be used for each sample. If this is unreasonable, samples can be filtered in sequence through one or a few filtration apparatus. If information is available about the relative concentrations expected in the samples, samples suspected to have low concentration should be filtered first so as to minimize potential contamination among samples. If there are extreme differences in metal concentrations among samples, samples can be grouped and filtered through separate apparatus. Regardless, results from filtration blanks should be used in any final judgment of the method adequacy.

With deionized water, thoroughly rinse all surfaces of the filtration apparatus that will contact the sample. Rinse the filter with deionized water and place it into the apparatus. Filter 50 to 100 ml of deionized water through the unit and discard the filtrate. Repeat this process at least three times. Use caution when disconnecting the vacuum tubing from the filtration flask as contaminating particulates can be sucked into your sample during the abrupt equalization of pressure. Initially use blanks and known metal solutions to assess the adequacy of these procedures for your specific needs and adjust as required.

Place 50 ml of water sample into this thoroughly-rinsed apparatus. Allow at least 25 ml to filter into the receiving flask. Rinse the receiving flask with this filtered sample and discard the rinse. Repeat this sample rinsing process. Place more sample into the filtration funnel and collect approximately 60 ml of filtered sample. Transfer approximately 10 ml of this filtered sample into an acid-cleaned, Teflon or polyethylene bottle. Use this sample aliquot to completely rinse the bottle and then discard the rinse. Fill the bottle with the remaining 50 ml of sample. Add 100 μl of concentrated, ultraclean nitric acid to the sample using a pipette with an acid-cleaned tip. This 0.2% (v/v) nitric acid preserved sample should have a pH less than 2.* Add more acid if required for adequate adjustment of your sample type.

Other methods exist for sampling dissolved metals. Windom et al.[12] pumped samples to cartridge filters in the field. The filtrate flowed directly

---

* Do not place the pH probe into the sample to check the pH as this could result in gross contamination. Instead, use another aliquot of filtered sample for assessing the adequacy of acidification.

into a polypropylene sample bottle and was acidified under a portable clean bench. This procedure requires considerable field effort and equipment but, if feasible, eliminates many ambiguities regarding changes in unfiltered water during transport and potential contamination upon contact with various laboratory ware.

Often, dissolved element concentrations are below the detection limit or, in the case of marine waters, present in a high salt matrix. The metals must then be removed from their original matrix and concentrated before analysis using one of several methods. They may be removed with solid resins (e.g., Colella et al.,[13] Koide et al.[14]), chelated and extracted with a solvent (e.g., Brooks et al.,[15] Bruland et al.,[16] Danielsson et al.,[17] Jan and Young,[18] Kinrade and Van Loon,[19] Pakalns and Farrar,[20] Koirtyohann and Wen[21]), or chelated and co-precipitated (e.g., Boyle and Edmond[22]). If the chelation-extraction procedure (e.g., APHA[23]) is used, it is important to keep in mind that pH of the sample is critical to effective extraction,[21] that humic acids can interfere,[20] and that the potential for contamination increases with preconcentration.

Storage and analysis

Minimal storage time of dissolved metal samples is recommended. Although regulatory guidelines for maximum storage times range from 28 days to 6 months, spiked samples and sample blanks should be used to assess the best storage time for your particular needs. Store samples in the cold (4°C) away from any potential source of contamination. They may be stored in the dark to avoid precipitation if silver concentrations are to be measured. Evaporation during storage can increase sample concentrations and should be avoided. A mark indicating the sample volume on the outside of each container can help in monitoring any potential volume changes.

Allow the sample to come to room temperature before analysis. To avoid contamination, never sample directly from the sample bottle. Instead, pour an aliquot from the bottle into an acid-cleaned container and discard the aliquot after use. Analyze the sample according to the Appendix and the AAS manufacturer's instructions.

Metals in biological tissue or sediments

Sampling, dissection, and storage

It is especially important with solid samples such as sediments or tissues that a large enough sample be taken so as to be representative. Too often treated as homogeneous materials, these materials range from poorly- to well-mixed, heterogeneous samples. Methods for calculating representative weights are provided in Ingamells and Switzer,[24] Ingamells,[25] Wallace and Kratochvil,[26] and Newman.[27]

Sediments should be taken from a specified depth as metal concentrations and species can change drastically with depth. If possible, nonmetallic or samplers with nonmetallic coatings (e.g., a Teflon-coated dredge) should be used to minimize contamination. The sediment may be mixed, put into an acid-cleaned, Teflon or plastic bottle, and stored on ice until return to the laboratory. It should be stored cold (4°C) away from any potential source of contamination until processed.

Biological samples may be tissues, organs or entire organisms depending on the goals of the study. If the entire animal or the gut alone will be analyzed, the organism must be allowed ample time to clear its gut of contaminating food. Tissues and organs may be carefully dissected from the animal using acid-cleaned, plastic utensils. Whole animals or dissected tissues should be placed into acid-cleaned, wide mouth containers (Teflon or plastic) and stored frozen until freeze drying can be performed. Twenty ml polypropylene scintillation vials with linerless caps are ideal for small samples. Placing sample vials in a Ziploc bag along with several ice cubes can reduce sublimation from samples during prolonged storage. Appropriate standard reference materials should be processed in tandem with samples from this step onward. They will reflect the quality of the process from storage through measurement.

Freeze drying

Loosening or removing caps from storage vials allows freeze drying of samples in their storage containers. Freeze dry until no more weight (water) is lost from the sample. Tighten caps onto vials prior to storage under desiccation. (Do not use desiccant that sheds small particulates that may contaminate the sample.) Digest the samples including the standard reference materials as soon reasonable after freeze drying. Note that volatile metals including those forming volatile organic compounds such as mercury can be lost during freeze drying.[28] Wet tissue can be analyzed for mercury and expressed later as dry weight using a wet:dry weight ratio. Relative to volatile elements, it is important to realize that most standard reference materials are freeze-dried materials that may have lost volatile compounds of analytes before they were analyzed for certified concentrations. Consequently, the lack of any apparent loss during freeze drying of a standard material in tandem with samples may not accurately reflect loss from sample. Splits of materials analyzed before and after freeze drying would better reflect such loss.

Digestion

Wear proper eye protection, protective clothing and plastic gloves while working with strong acids. Digestion should be done in a hood capable of efficiently removing the generated acid fumes.

Weigh 0.1000 to 0.5000 g of freeze-dried tissue or sediment and record the exact weight for later calculations. (The optimum weight needed to be representative can be estimated as described in Newman.[27]) It is best to use the same weight for all samples. Place the sample into an acid-cleaned Teflon beaker and cover immediately with an acid-cleaned Teflon watchglass or lid. Approximately 5 to 10% of all samples to be digested should be reference materials and an additional 5 to 10% should be procedural blanks (beakers without sediment or tissue added). Whether 1 in 5 or 10 samples is a blank or standard material will depend in the element of interest and its concentration in the materials being analyzed. Low concentrations of easily contaminated analytes require numerous blanks and standard materials.

After all samples have been placed into beakers, slowly add 5 ml of concentrated, ultraclean nitric acid to each. More acid may be required to completely wet and cover some types of samples, e.g., light powders. Make certain that all of the material contacts acid and no internal pockets or edges remain dry. Allow the covered beakers containing sample and acid to sit for 1 h at room temperature. This predigestion reduces the risk of vigorous foaming and loss of sample upon heating. Next, carefully place the covered beakers onto the hot plate and allow them to reflux at 80°C for 4 h. (A beaker filled with water and placed among the beakers containing samples can be used to monitor digestion temperature.) If there are wide deviations in temperature on the surface of the hot plate, a Pyrex pan partially filled with water can be placed onto the hot plate to provide more uniform heating of samples. Rotation of samples on the hot plate may also ensure that heating is less variable among samples. Periodically check to ensure that the entire bottom of each beaker remains wet and that the partially digested samples remain covered with acid. Add more acid if required to keep the entire sample wet with hot acid.

After 4 h, carefully remove the watchglass or lid from each beaker. (Some samples may require additional time to digest adequately. Use the digests of standard materials to assess the completeness of digestion or loss of analyte by excessive digestion.) Gently and carefully rinse the droplets of acid from the watchglass or lid back into the beaker using deionized water in a squirt bottle. Place the samples back onto the hot plate and allow the digest volume to evaporate to approximately 2.5 ml. After cooling, quantitatively transfer the digest to a 25 ml, acid-cleaned Class A volumetric flask using successive rinses with deionized water. Allow each digest to cool in the volumetric flask and bring the volume to nearly 25 ml with deionized water. Gently mix the digest in the flask and allow it to cool again. Carefully bring the digest to 25 ml volume. This digest has a 10% (v/v) nitric acid matrix. If concentrations allow further dilution, a 50 ml volumetric can be used instead to produce a 5% (v/v) nitric acid matrix. Carefully transfer the digest to an acid-cleaned Teflon, polypropylene, or polyethylene storage container.

This procedure has been used successfully by the author[29,30] for Pb and Zn analysis in biological tissues. Standard materials were used to indicate the adequacy of the digestion for Pb[29] (National Research Council Canada TORT-1 standard) and Zn[30] (US EPA Quality Control Samples, Metals in Fish). However, it may not be adequate for samples with high concentrations of lipids. Incompletely-digested lipids produce a turbid digest with a surface film. An inadequate digestion will be indicated by spurious values upon repeated analysis of digest aliquots or unacceptably low concentrations for standard reference materials. More rigorous digestion is required in this case. Clegg et al.[31] and Sinex et al.[32] compare alternative digestion procedures for biological tissue and sediments, respectively. Siemer and Brinkley[33] and

Chung and Tsai[34] describe simple refluxing systems for acid digests, and Uhrberg[35] describes an acid digestion bomb. Very effective methods are described that use conventional microwave ovens[36,37] or microwave ovens specifically designed for sample digestion.[38-40] CEM Corporation (Matthews, NC) produces microwave digestors and provides specific details for their application.

The nitric acid digest of sediments will not result in a good estimate of bioavailable metal in many cases. Procedures have been developed for determining metal concentrations in various sediment fractions. Concentrations in the various fractions may be related to solid phase species and bioavailability. Tessier and co-workers[41,42] describe one such technique in wide use today. If applied, it is important to acknowledge that the resulting fractions are procedurally-defined and artifacts can be present.[43,44] Pai et al.[45] provide specific recommendations for graphite furnace AAS analysis of the complex sediment extract matrices generated with this procedure.

Storage and analysis

Digests should be analyzed as soon as reasonable. Storage under refrigeration will reduce change in volume due to evaporation. Store the digests away from any contaminating materials.

Allow the sample to come to room temperature before analysis. To avoid contamination, never sample directly from the sample bottle. Instead, pour an aliquot from the bottle into an acid-cleaned container. Discard the aliquot after use: do not return it to the sample bottle. Analyze the sample according to the Appendix and the AAS manufacturer's instructions. Any dilutions of sample digests should be performed in acid-cleaned containers with special attention given to changes in the matrix, e.g., a significant change in viscosity associated with dilution. Matrix matching is often required for many elements and sample dilutions. The method of standard addition can be used to assess the amount of matrix matching required.

Results and discussion

Concentrations are estimated using aqueous standards or standards with matrices matching the sample. The method of standard addition can be used if there are matrix effects. Particularly with graphite furnace AAS and samples requiring various dilutions, it is prudent to assume that a matrix effect is present unless shown to be otherwise.

Although method detection limit is more commonly reported, measurement relative uncertainty also provides the end user with sufficient information to assess data quality.[27] If the data includes "below detection limit" observations, they must be treated as censored data in subsequent statistical analyses.[46,47] It is also essential that results of standard reference materials or, minimally, spiked samples recoveries, be reported. Expected reference material values and acceptable limits should also be reported. Comment must also be made regarding results from procedural blank analysis. Without this information, the end user has no means of determining the quality of measurement.

Acknowledgments

This work was supported by a contract DE-AC09-76SROO819 between the U.S. Department of Energy and the University of Georgia's Savannah River Ecology Laboratory.

Appendix of analytical details for specific elements

Information was extracted from Dean,[48] Lide,[49] Van Loon,[4] Varma,[1] Perkin-Elmer,[9,10] and Price.[2] Sensitivities are expressed in mg/l to produce a 0.0044 Abs unit (1% absorption) for flame AAS. For furnace AAS, characteristic masses are pg to produce a 0.0044 Abs unit for a Perkin-Elmer transverse heated graphite atomizer.[10] Because values for sensitivity and characteristic mass vary among instruments, those given reflect only general ranges and relative values among wavelengths. AW = atomic weight, D_o = dissociation energy of the metal monoxide in eV, BP = boiling point in °C, and E_i = ionization energy in eV.

Ag (silver, AW 107.868, D_o 2.2, E_i 7.57, BP 2163)

General: Silver solutions should be stored in an amber or opaque container as light could cause silver to precipitate.[9] Hydrochloric acid should not be used because silver will precipitate with chloride.

Flame technique: Use a lean air-acetylene flame to obtain sensitivities of 0.1 mg/l (λ = 328.1 nm) and 0.2 mg/l (λ = 338.3 nm).[2] Large amounts of aluminum can cause signal suppression.[2] Bromide, chromate, iodate, iodide, permanganate, tungstate, and chloride may precipitate silver from solution.[9]

Flameless technique: Perkin-Elmer[10] recommends addition of 5 µg Pd and 3 µg of $Mg(NO_3)_2$ to the sample. (Reagent described above.) The pyrolysis and atomization temperatures should be in the ranges of 800° and 1500°C, respectively. The characteristic mass is approximately 4.5 pg at 328.1 nm.

Al (aluminum, AW 26.98154, D_o 5.3, E_i 5.99, BP 2520)

General: During preparation of working standards, aluminum can quickly adsorb to glass volumetric flasks. Always add aliquots of the stock solution to acidified (nitric acid) deionized water at the bottom of each volumetric flask to avoid adsorption.

Flame technique: Respectively, the 308.2, 309.3 and 396.2 nm wavelengths have sensitivities of approximately 1.5, 1.1 and 1.1 mg/l[9] with the use of a rich nitrous oxide-acetylene flame and the presence of excess La or K to suppress ionization. ($LaCl_3$ and KCl solutions described above.) Price[2] reported slight signal suppression by calcium, silicon, and perchloric and hydrochloric acids. Acetic acid, titanium and iron may enhance signal.[9]

Flameless technique: Perkin-Elmer[10] recommends addition of 15 µg of $Mg(NO_3)_2$ to the sample. (Reagent described above.) The pyrolysis and atomization temperatures should be in the ranges of 1200° and 2300°C, respectively. The characteristic mass is approximately 31 pg at 309.3 nm.

As (arsenic, AW 74.9216, D_o 5.0, E_i 9.81, BP 615 and of hydride AsH_3–55)

Flame technique: Generally, a hydride generation method is used. However, a rich air-acetylene flame will produce sensitivities of 0.78, 1.0 and 2.0 mg/l at the 189.0, 193.7 and 197.2 nm wavelengths. Background absorption is high at these wavelengths.

Flameless technique: Perkin-Elmer[10] recommends addition of 5 µg Pd and 3 µg of $Mg(NO_3)_2$ to the sample. (Reagent described above.) The pyrolysis and atomization temperatures should be in the ranges of 1200° and 2000°C, respectively. The characteristic mass is approximately 40.0 pg at 193.7 nm.

Ba (barium, AW 137.33, D_o 5.8, E_i 5.21, BP 1805)

Flame technique: Price[2] and Varma[1] suggest a rich nitrous oxide-acetylene flame with an ionization buffer (KCl solution described above) at 553.6 nm to obtain a sensitivity of approximately 0.4 mg/l. Price[2] recommends analysis without an ionization buffer if the less sensitive 455.4 nm wavelength is used (sensitivity = 2.0 mg/l).

Flameless technique: The pyrolysis and atomization temperatures should be in the ranges of 1200° and 2300°C, respectively. The characteristic mass is approximately 15.0 pg at 553.6 nm.

Be (beryllium, AW 9.01218, D_o 4.6, E_i 9.32, BP 2472)

General: This toxic element should be handled with appropriate caution.

Flame technique: A rich nitrous oxide-acetylene flame is recommended. Price[2] reports a sensitivity of 0.05 mg/l at 234.9 nm. Absorption may be enhanced by nitric and sulfuric acids.[2] High concentrations of aluminum, sodium, silicon and magnesium can depress sensitivity.[1,2,9]

Flameless technique: Perkin-Elmer[10] recommends addition of 15 µg of $Mg(NO_3)_2$ to the sample. (Reagent described above.) The pyrolysis and atomization temperatures should be in the ranges of 1500° and 2300°C, respectively. The characteristic mass is approximately 2.5 pg at 234.9 nm.

Bi (bismuth, AW 208.9804, D_o 3.6, E_i 7.29, BP 1564)

Flame technique: Price[2] reports the following approximate sensitivities: 0.8 (223.1 nm), 1.5 (222.8 nm), 10 (227.7 nm), and 2.2 (306.8 nm) mg/l with a lean air-acetylene flame.

Flameless technique: Perkin-Elmer[10] recommends addition of 5 µg Pd and 3 µg of $Mg(NO_3)_2$ to the sample. (Reagent described above.) The pyrolysis and atomization temperatures should be in the ranges of 1100° and 1700°C, respectively. The characteristic mass is approximately 60.0 pg at 223.1 nm.

Ca (calcium, AW 40.08, D_o 4.8, E_i 6.11, BP 1494)

Flame technique: With a lean nitrous oxide-acetylene flame and La (or alkali salt) additions sensitivity is approximately 0.03 mg/l but increases twofold with use of a stoichiometric or lean air-acetylene flame at 422.7 nm.[2] Sensitivity is 20 mg/l at 239.9 nm (air-acetylene flame). (KCl and $LaCl_3$ solutions described above can be used to reduce ionization (KCl) or chemical interferences ($LaCl_3$).) With the lean air-acetylene flame, high concentrations of aluminum, beryllium, phosphorus, silicon, titanium, vanadium or zirconium can interfere with analysis.[1,2,9]

Flameless technique: The pyrolysis and atomization temperatures should be in the ranges of 1100° and 2500°C, respectively. The characteristic mass is approximately 1.0 pg at 422.7 nm.

Cd (cadmium, AW 112.41, D_o 1.5, E_i 8.99, BP 767)

General: With graphite furnace AAS, contamination can be a significant problem. Use special care in acid-cleaning and sample preparation. This element is relatively toxic and should be handled appropriately.

Flame technique: With a lean air-acetylene flame, sensitivity is 0.015 mg/l at 228.8 nm and 20 mg/l at 326 nm.[2] High concentrations of silicate interfere with analyses.[9]

Flameless technique: Perkin-Elmer[10] recommends addition of 50 µg $NH_4H_2PO_4$ and 3 µg of $Mg(NO_3)_2$ to the sample. (Reagent described above.) The pyrolysis and atomization temperatures should be in the ranges of 700° and 1400°C, respectively. The characteristic mass is approximately 1.3 pg at 228.8 nm.

Co (cobalt, AW 58.9332, D_o 3.8, E_i 7.88, BP 2928)

Flame technique: Price[2] reports the following sensitivities (mg/l): 240.7 nm – 0.08, 242.5 nm – 0.2, 252.1 nm – 0.5, and 341.3 nm – 4.0 with a lean air-acetylene flame. Perkin-Elmer[9] and Varma[1] note that large amounts of transition and heavy metals depress signal.

Flameless technique: Perkin-Elmer[10] recommends addition of 15 µg of $Mg(NO_3)_2$ to the sample. (Reagent described above.) The pyrolysis and atomization temperatures should be in the ranges of 1400° and 2400°C, respectively. The characteristic mass is approximately 17.0 pg at 242.5 nm.

Cr (chromium, AW 51.996, D_o 4.4, E_i 6.77, BP 2672)

Flame technique: Sensitivities using a rich air-acetylene flame for the 357.9, 359.3, 360.5, and 425.4 nm wavelengths are approximately 0.05, 0.15, 0.2, and 0.5 mg/l, respectively.[2] Phosphate, nickel and iron may interfere with an air-acetylene flame.[1,2,9] Perkin-Elmer[9] and Varma[1] indicate that the addition of calcium can eliminate the effect of phosphates and the addition of 2% (w/v) ammonium chloride can reduce the effect of iron.

Flameless technique: Perkin-Elmer[10] recommends addition of 15 μg of $Mg(NO_3)_2$ to the sample. (Reagent described above.) The pyrolysis and atomization temperatures should be in the ranges of 1500° and 2300°C, respectively. The characteristic mass is approximately 7.0 pg at 357.9 nm.

Cu (copper, AW 63.546, D_o 3.6, E_i 7.73, BP 2563)

Flame technique: Using a lean air-acetylene flame, Price[2] reports the following sensitivities for 217.9, 222.6, 244.2, 249.2, 324.8, and 327.4 nm: 0.6, 2.0, 40, 10, 0.04, and 0.1 mg/l, respectively.

Flameless technique: Perkin-Elmer[10] recommends addition of 5 μg Pd and 3 μg of $Mg(NO_3)_2$ to the sample. (Reagent described above.) The pyrolysis and atomization temperatures should be in the ranges of 1200° and 1900°C, respectively. The characteristic mass is approximately 17.0 pg at 324.8 nm.

Fe (iron, AW 55.847, D_o 4.2, E_i 7.90, BP 2862)

Flame technique: Price[2] lists the following sensitivities for the 248.3-, 252.3-, 271.9-, 296.7-, 302.1-, 344.1-, 372.0-, 382.4-, and 386.0-nm wavelengths: 0.08, 0.3, 0.5, 1.0, 0.7, 5.0, 1.0, 30, and 2.0 mg/l for a lean air-acetylene flame. Cobalt, copper, nickel or nitric acid can depress sensitivity, but their effects are much reduced by adjusting the flame to a very lean flame.[1,9] Organic acid such as citric acid may depress signal but this effect may be minimized by using phosphoric acid.[1] Silicon effects can be reduced by the addition of 0.2% (w/v) calcium chloride.[1,9]

Flameless technique: Perkin-Elmer[10] recommends addition of 15 μg of $Mg(NO_3)_2$ to the sample. (Reagent described above.) The pyrolysis and atomization temperatures should be in the ranges of 1400° and 2100°C, respectively. The characteristic mass is approximately 12.0 pg at 248.3 nm.

K (potassium, AW 39.0983, D_o 2.5, E_i 4.34, BP 759)

Flame technique: Price[2] lists the following sensitivities for 404.4-, 766.5-, and 769.9-nm wavelengths with a lean air-acetylene flame: 10.0, 0.2, 0.1 mg/l, respectively. Ionization is overcome by the addition of La, Na, or Cs.[1,2,9] The $LaCl_3$ solution described above may be used for this purpose.

Flameless technique: The pyrolysis and atomization temperatures should be in the ranges of 900° and 1500°C, respectively. The characteristic mass is approximately 2.0 pg at 766.5 nm.

Mg (magnesium, AW 24.305, D_o 4.1, E_i 7.65, BP 1090)

Flame technique: Use a lean nitrous oxide-acetylene flame to get approximate sensitivities of 25, 0.2, and 0.005 mg/l for the 202.5, 279.6 and 285.2 nm wavelengths.[2] La ($LaCl_3$ solution described above) or K (KCl solution described above) additions may be required as well as the hot nitrous oxide-acetylene flame. Perkin-Elmer[9] recommends a lean air-acetylene flame. The

LaCl$_3$ solution can reduce chemical interference from aluminum, silicon, titanium, zirconium and phosphorus.[1] The KCL solution reduces ionization.

Flameless technique: The pyrolysis and atomization temperatures should be in the ranges of 800° and 1900°C, respectively. The characteristic mass is approximately 0.4 pg at 285.2 nm.

Mn (manganese, AW 54.9380, D$_o$ 4.2, E$_i$ 7.43, BP 2062)

Flame technique: Price[2] reports sensitivities of 1, 0.025, 100, and 0.5 mg/l using a stoichiometric to lean air-acetylene flame for the 222.2, 279.5, 321.7, and 403.1 nm wavelengths. Perkin-Elmer[9] suggests that interference by silicon may be overcome by the addition of 0.2% (w/v) calcium chloride. Varma[1] notes interferences from phosphate, perchlorate, iron, nickel and cobalt that are reduced or eliminated using a lean air-acetylene or nitrous oxide-acetylene flame. Varma[1] suggests addition of K (KCl solution described above) to eliminate any ionization.

Flameless technique: Perkin-Elmer[10] recommends addition of 5 µg Pd and 3 µg of Mg(NO$_3$)$_2$ to the sample. (Reagent described above.) The pyrolysis and atomization temperatures should be in the ranges of 1300° and 1900°C, respectively. The characteristic mass is approximately 6.3 pg at 279.5 nm.

Mo (molybdenum, AW 95.94, D$_o$ 6.3, E$_i$ 7.09, BP 4639)

Flame technique: Price[2] reports sensitivities of 0.5, 1.5, 10.0, and 2.0 mg/l for the 313.3-, 317.0-, 320.8-, and 379.8-nm wavelengths using a rich air-acetylene flame. A rich nitrous oxide-acetylene flame has a sensitivity of 0.2 mg/l for the 313.3 nm wavelength. Interference from calcium, chromium, manganese, nickel, strontium, sulfate, and iron can be reduced by using a nitrous oxide-acetylene flame.[1,2,9] Varma[1] suggests the addition of either 0.5% (w/v) aluminum, 2% (w/v) ammonium chloride or 0.1% (w/v) sodium sulfate to reduce interferences.

Flameless technique: Perkin-Elmer[10] recommends addition of 5 µg Pd and 3 µg of Mg(NO$_3$)$_2$ to the sample. (Reagent described above.) The pyrolysis and atomization temperatures should be in the ranges of 1500° and 2400°C, respectively. The characteristic mass is approximately 12.0 pg at 313.3 nm.

Na (sodium, AW 22.98977, D$_o$ 2.7, E$_i$ 5.14, BP 883)

Flame technique: Sensitivities of 5.0 and 0.02 mg/l are reported by Price[2] for 330.2 and 589.0 nm wavelengths using a lean air-acetylene flame. The addition of an alkali salt (KCl solution described above) is recommended.[1] Noise produced by calcium can interfere at the 589.0 nm wavelength.[1]

Flameless technique: The pyrolysis and atomization temperatures should be in the ranges of 900° and 1500°C, respectively. The characteristic mass is approximately 1.2 pg at 589.0 nm.

Ni (nickel, AW 58.69, D_o 4.0, E_i 7.64, BP 2914)

Flame technique: For the wavelengths 232.0, 234.6, 247.7, 339.1, 341.5, 346.2, and 352.5 nm, Price[2] reports sensitivities of 0.1, 0.5, 50, 5, 0.2, 1.0, and 2.5 mg/l, respectively with a lean air-acetylene flame. Perkin-Elmer[9] and Varma[1] report that high concentrations of iron, cobalt, and chromium can depress the signal. A nitrous oxide-acetylene flame can be used to reduce this interference.

Flameless technique: The pyrolysis and atomization temperatures should be in the ranges of 1100° and 2300°C, respectively. The characteristic mass is approximately 20.0 pg at 232.0 nm.

Pb (lead, AW 207.2, D_o 3.9, E_i 7.42, BP 1750)

General: This element is toxic and should be handled appropriately. Low concentration work requires extreme care to avoid contamination.

Flame technique: Price[2] reports sensitivities of 0.12, 6.0, and 0.2 mg/l for the 217.0-, 261.4-, and 283.3-nm wavelengths with a lean air-acetylene flame. Note that, although the 217.0-nm wavelength is more sensitive than the 283.3-nm wavelength, the associated noise is higher than that of the 283.3-nm wavelength. The lower noise and only marginally lower sensitivity of the 283.3-nm wavelength make it slightly more useful.

Flameless technique: Perkin-Elmer[10] recommends addition of 50 µg $NH_4H_2PO_4$ and 3 µg of $Mg(NO_3)_2$ to the sample. (Reagent described above.) The pyrolysis and atomization temperatures should be in the ranges of 850° and 1500°C, respectively. The characteristic mass is approximately 30.0 pg at 283.3 nm.

Pt (platinum, AW 195.08, D_o 3.6, E_i 9.0, BP 3827)

Flame technique: Using a lean air-acetylene flame, Price[2] reports respective sensitivities for the 217.5-, 265.9-, 299.8-, and 306.5-nm wavelengths of 10, 2.5, 15, and 5 mg/l. Interferences from high concentrations of many elements, ammonium ion, sulfuric acid, perchloric acid and phosphoric acid can be reduced by adding 0.2% (w/v) La in 1% (v/v) hydrochloric acid.[1,9] Perkin-Elmer[9] also note that a nitrous oxide-acetylene flame will reduce these interferences.

Flameless technique: The pyrolysis and atomization temperatures should be in the ranges of 1300° and 2200°C, respectively. The characteristic mass is approximately 220.0 pg at 265.9 nm.

Sb (antimony, AW 121.75, D_o 3.9, E_i 8.64, BP 1587 and hydride SbH_3–17)

Flame technique: For the 206.8-, 217.6-, and 231.2-nm wavelengths and a lean to stoichiometric air-acetylene flame, Price[2] reports sensitivities of 0.5, 1.1, and 1.9 mg/l, respectively. Although the 206.8 nm wavelength is more sensitive, it is noisier than the 217.6 nm wavelength.

Flameless technique: Perkin-Elmer[10] recommends addition of 5 μg Pd and 3 μg of $Mg(NO_3)_2$ to the sample. (Reagent described above.) The pyrolysis and atomization temperatures should be in the ranges of 1300° and 1900°C, respectively. The characteristic mass is approximately 55.0 pg at 217.6 nm.

Se (selenium, AW 78.96, D_o 4.4, E_i 9.75, BP 685)

Flame technique: Price[2] reports a sensitivity of 0.8 mg/l for a "just luminous" lean air-acetylene flame with the 196.1-nm wavelength. At this wavelength, there is much light scatter from the flame and background correction is essential.[9]

Flameless technique: Perkin-Elmer[10] recommends addition of 5 μg Pd and 3 μg of $Mg(NO_3)_2$ to the sample. (Reagent described above.) The pyrolysis and atomization temperatures should be in the ranges of 1300° and 1900°C, respectively. The characteristic mass is approximately 45.0 pg at 196.1 nm.

Sn (tin, AW 118.69, D_o 5.7, E_i 7.34, BP 2603 and hydride SnH_4–52)

Flame technique: Sensitivities at 224.6 and 286.3 nm using a rich nitrous oxide-acetylene flame are 0.8 and 2.5 mg/l, respectively.[2] Price[2] reports sensitivities for the 270.6-, 286.3-, and 303.4-nm wavelengths to be 25, 10, and 50 mg/l with a rich air-acetylene flame, respectively. A rich nitrous oxide-acetylene flame and 328.1-nm wavelength seem the best combination because of the slightly lower noise at this wavelength relative to the slightly more sensitive 224.6-nm wavelength.

Flameless technique: Perkin-Elmer[10] recommends addition of 5 μg Pd and 3 μg of $Mg(NO_3)_2$ to the sample. (Reagent described above.) The pyrolysis and atomization temperatures should be in the ranges of 1400° and 2200°C, respectively. The characteristic mass is approximately 90.0 pg at 286.3 nm.

Ti (titanium, AW 47.88, D_o 6.9, E_i 6.83, BP 3289)

Flame technique: Price[2] recommends a rich nitrous oxide-acetylene flame for all three wavelengths (364.3, 365.4, and 337.2 nm). At 364.3, he gives a sensitivity of 1.5 mg/l. Addition of an excess of alkali salt (KCl solution described above) is recommended to reduce ionization.[1,2] The presence of hydrofluoric acid, iron and many other elements will enhance the signal for this element.[1,2] Sulfuric acid will greatly reduce sensitivity.[1]

Flameless technique: The pyrolysis and atomization temperatures should be in the ranges of 1500° and 2500°C, respectively. The characteristic mass is approximately 70.0 pg at 364.3 nm.

V (vanadium, AW 50.9415, D_o 6.7, E_i 6.75, BP 3409)

Flame technique: Price[2] gives a sensitivity of 1.0 mg/l for the doublet (318.39 and 318.34) using a stoichiometric nitrous oxide-acetylene flame. High concentrations of aluminum, titanium, iron and phosphoric acid can enhance the signal.[1,2] An ionization buffer (KCl solution described above) is required.

Flameless technique: The pyrolysis and atomization temperatures should be in the ranges of 1200° and 2400°C, respectively. The characteristic mass is approximately 42.0 pg at 318.4 nm.

Zn (zinc, AW 65.38, D_o 2.9, E_i 9.39, BP 907)

General: Zinc contamination can be a problem when working with low concentrations.

Flame technique: Using a stoichiometric to lean air-acetylene flame, Price[2] reports sensitivities of 0.012 and 150 mg/l for the 213.9- and 307.6-nm wavelengths, respectively.

Flameless technique: Perkin-Elmer[10] recommends addition of 5 µg of $Mg(NO_3)_2$ to the sample. (Reagent described above.) The pyrolysis and atomization temperatures should be in the ranges of 700° and 1800°C, respectively. The characteristic mass is approximately 1.0 pg at 213.9 nm.

References

1. Varma, A. 1984. *CRC Handbook of Atomic Absorption Analysis. Vol. 1.* Boca Raton, FL: CRC Press, 510 pp.
2. Price, W.J. 1972. *Analytical Atomic Absorption Spectrometry.* London: Heyden & Son, Ltd. 239 pp.
3. Veillon, C. 1986. Trace element analysis of biological samples. *Anal. Chem.* 58(8): 851A–66A.
4. Van Loon, J.C. 1985. *Selected Methods of Trace Metal Analysis: Biological and Environmental Samples.* New York: John Wiley & Sons. 357 pp.
5. Pinta, M. 1978. *Modern Methods for Trace Element Analysis.* Ann Arbor, MI: Ann Arbor Science Publishers. 492 pp.
6. Suzuki, M., Ohta, K. 1982. Reduction of interferences with thiourea in the determination of cadmium by electrothermal atomic absorption spectrometry. *Anal. Chem.* 54: 1686–89.
7. Ediger, R.D. 1975. Atomic absorption analysis with the graphite furnace using matrix modification. *Atom. Abs. Newslett.* 14: 127–30.
8. Carnrick, G., Schlemmer, G., Slavin, W. 1991. Matrix modifiers: their role and history for furnace AAS. *Am. Lab. (FAIRFIELD CONN)* Feb. 1991: 118–31.
9. Perkin-Elmer, 1982. *Analytical Methods for Atomic Absorption Spectrophotometry.* Norwalk, CT: Perkin-Elmer Corp.
10. Perkin-Elmer, 1992. *The THGA Graphite Furnace: Techniques and Recommended Conditions.* Norwalk, CT: Perkin-Elmer Corp.
11. Schlemmer, G., Welz, B. 1986. Palladium and magnesium nitrates, a more universal modifier for graphite furnace atomic absorption spectrometry. *Spectrochim. Acta* 41B(11): 1157–65.
12. Windom, H.L., Byrd, J.T., Smith, Jr., R.G., Huan, F. 1991. Inadequacy of NASQAN data for assessing metal trends in the nation's rivers. *Environ. Sci. & Technol.* 25(6): 1137–42.
13. Colella, M.B., Siggia, S., Barnes, R.M. 1980. Poly(acrylamidoxime) resin for determination of trace metals in natural waters. *Anal. Chem.* 52: 2347–50.

14. Koide, M., Lee, D.S., Stallard, M.O. 1984. Concentration and separation of trace metals from seawater using a single anion exchange bead. *Anal. Chem.* 56: 1956–59.

15. Brooks, R.R., Presley, B.J., Kaplan, I.R. 1967. APDC-MIBK extraction system for the determination of trace elements in saline waters by atomic-absorption spectrophotometry. *Talanta* 14: 809–16.

16. Bruland, K.W., Franks, R.P., Knauer, G.A., Martin, J.H. 1979. Sampling and analytical methods for the determination of copper, cadmium, zinc and nickel at the nanogram per liter level in sea water. *Analytica Chimica Acta* 105: 233–45.

17. Danielsson, L.-G., Magnusson, B., Westerlund, S. 1978. An improved metal extraction procedure for the determination of trace metals in sea water by atomic absorption spectrometry with electrothermal atomization. *Analytica Chimica Acta* 98: 47–57.

18. Jan, T.K., Young, D.R. 1978. Determination of microgram amounts of some transition metals in seawater by methyl isobutyl ketone-nitric acid successive extraction and flameless atomic absorption spectrophotometry. *Anal. Chem.* 50: 1250–53.

19. Kinrade, J.D., Van Loon, J.C. 1974. Solvent extraction for use with flame atomic absorption spectrometry. *Anal. Chem.* 46: 1894–98.

20. Pakalns, P., Farrar, Y.J. 1977. The effect of surfactant on the extraction-atomic absorption spectrophotometric determination of copper, iron, manganese, lead, nickel, zinc, cadmium and cobalt. *Water Res.* 11: 145–51.

21. Koirtyohann, S.R., Wen, J.W. 1973. Critical study of the APDC-MIBK extraction system for atomic absorption. *Anal. Chem.* 45: 1986–89.

22. Boyle, E.A., Edmond, J.M. 1975. Determination of trace elements in aqueous solution by APDC chelate co-precipitation. In: *Analytical Methods in Oceanography*, Ed. T.R.P. Gibb, Jr. 6: 44–55. Washington, DC: American Chemical Society. 238 pp.

23. APHA, AWWA, WPCF. 1989. *Standard Methods for the Examination of Water and Wastewater*, Washington, DC: American Public Health Association. 3:1–163.

24. Ingamells, C.O., Switzer, P. 1973. A proposed sampling constant for use in geochemical analysis. *Talanta* 20: 547–68.

25. Ingamells, C.O. 1974. New approaches to geochemical analysis and sampling. *Talanta* 21: 141–55.

26. Wallace, D., Kratochvil, B. 1987. Visman equations in the design of sampling plans for chemical analysis of segregated bulk materials. *Anal. Chem.* 59: 226–32.

27. Newman, M.C. 1995. *Quantitative Methods in Aquatic Ecotoxicology.* Boca Raton, FL: CRC/Lewis Publishers. 426 pp.

28. Sivasankara Pillay, K.K., Thomas, Jr., C.C., Sondel, J.A., Hyche, C.M. 1971. Determination of mercury in biological and environmental samples by neutron activation analysis. *Anal. Chem.* 43: 1419–25.

29. Newman, M.C., Mitz, S.V. 1988. Size dependence of zinc elimination and uptake from water by mosquitofish *Gambusia affinis* (Baird and Girard). *Aquat. Toxicol.* 12: 17–32.

30. Newman, M.C., Mulvey, M., Beeby, A., Hurst, R.W., Richmond, L. 1994. Snail (*Helix aspersa*) exposure history and possible adaptation to lead as reflected in shell composition. *Arch. Environ. Contam. Toxicol.* 27: 346–351.

31. Clegg, M.S., Keen, C.L., Lönnerdal, B., Hurley, L.S. 1981. Influence of ashing techniques on the analysis of trace elements in animal tissue. I. Wet ashing. *Biol. Trace Element Res.* 3: 107–15.

32. Sinex, S.A., Cantillo, A.Y., Helz, G.R. 1980. Accuracy of acid extraction methods for trace metals in sediments. *Anal. Chem.* 52: 2342–46.
33. Siemer, D.D., Brinkley, H.G. 1981. Erlenmeyer flask-reflux cap for acid sample decomposition. *Anal. Chem.* 53: 750–51.
34. Chung, S.-W., Tsai, W.-C. 1992. Atomic absorption spectrometric determination of heavy metals in foodstuffs using a simple digester. *Atomic Spectroscopy* 13(5): 185–89.
35. Uhrberg, R. 1982. Acid digestion bomb for biological samples. *Anal. Chem.* 54: 1906–8.
36. Hewitt, A.D., Reynolds, C.M. 1990. Dissolution of metals from soils and sediments with a microwave-nitric acid digestion technique. *Atomic Spectroscopy* 11(5): 187–92.
37. Kojima, I., Kato, A., Iida, C. 1992. Microwave digestion of biological samples with acid mixture in a closed double PTFE vessel for metal determination by "one-drop" flame atomic absorption spectrometry. *Analytica Chimica Acta* 264: 101–6.
38. Lajunen, L.H.J., Piispanen, J., Saari, E. 1992. Microwave dissolution of plant samples for AAS analysis. *Atomic Spectroscopy* 13(4): 127–31.
39. Mincey, D.W., Williams, R.C., Giglio, J.J., Graves, G.A., Pacella, A.J. 1992. Temperature controlled microwave oven digestion system. *Analytica Chimica Acta* 264: 97–100.
40. Lan, W.G., Wong, M.K., Sin, Y.M. 1994. Comparison of four microwave digestion methods for the determination of selenium in fish tissue by using hydride generation atomic absorption spectrometry. *Talanta* 41(2): 195–200.
41. Tessier, A., Campbell, P.G.C., Bisson, M. 1979. Sequential extraction procedure for the speciation of particulate trace metals. *Anal. Chem.* 51(7): 844–51.
42. Tessier, A., Campbell, P.G.C. 1988. Partitioning of trace metals in sediments. In: *Metal Speciation. Theory, Analysis and Application*, eds. J.R. Kramer, Allen, H.E., 9: 183–217. Chelsea, MI: Lewis Publishers. 357 pp.
43. Rendell, P.S., Batley, G.E., Cameron, A.J. 1980. Adsorption as a control of metal concentrations in sediment extracts. *Environ. Sci. Technol.* 14(3): 314–18.
44. Tipping, E., Hetherington, N.B., Hilton, J., Thompson, D.W., Bowles, E., Hamilton-Taylor, J. 1985. Artifacts in the use of selective chemical extraction to distributions of metals between oxides of manganese and iron. *Anal. Chem.* 57: 1944–46.
45. Pai, S., Lin, F., Tseng, C., Sheu, D. 1993. Optimization of heating programs of GFAAS for the determination of Cd, Cu, Ni and Pb in sediments using sequential extraction technique. *Int. J. Environ. Anal. Chem.* 50: 193–205.
46. Newman, M.C., Dixon, P.M., Looney, B.B., Pinder, III, J.E. 1989. Estimating mean and variance for environmental samples with below detection limit observations. *Water Res. Bull.* 25: 905–16.
47. Newman, M.C., Dixon, P.M. 1990. UNCENSOR: a program to estimate means and standard deviations for data sets with below detection limit observations. *Am. Env. Lab. (FAIRFIELD CONN)*: April 1990: 27–30.
48. Dean, J.A. 1992. *Lange's Handbook of Chemistry.* New York: McGraw-Hill, Inc.. 4: 23–35.
49. Lide, D.R. (Ed.) 1992. *CRC Handbook of Chemistry and Physics.* 73rd Edition. Boca Raton, FL: CRC Press.
50. Reynolds, R.J., Aldous, K., Thompson, K.C. 1970. *Atomic Absorption Spectroscopy. A Practical Guide.* New York, NY: Barnes & Noble, Inc.
51. Willard, H.H., Merritt, Jr., L.L., Dean, J.A. 1974. *Instrumental Methods of Analysis.* New York: D. Van Nostrand Co.

chapter twenty-seven

Analysis of non-ortho-PCBs in fish, bird eggs, sediments, soils, and SPMD samples by gas chromatography/high resolution mass spectrometry

Paul H. Peterman, Robert W. Gale, Donald E. Tillitt, and Kevin P. Feltz

Introduction

Polychlorinated biphenyls (PCBs) have been analyzed extensively in various environmental samples since they were first identified and found to be very persistent.[1] PCBs have usually been analyzed and reported as the sum of all PCB peaks, the sum of peaks for each chlorine homolog (Cl_{1-10}), or as PCB commercial mixtures, which were widely used from the 1940s until the 1970s, when their production was banned in the U.S. and phase-out began.[2] In the environment, the lipophilic PCBs[3] bioconcentrate in fish and other aquatic biota and bioaccumulate in predators, especially in piscivorous birds and higher mammals.[4] As many as 150 PCB congeners may be found in environmental samples. All 209 possible congeners are now commercially available as standards, but not all congeners can be simultaneously separated and individually determined.

Capillary gas chromatographic (GC) analysis of PCBs is essential for separating as many PCB congeners as possible. About 80 individual PCB congeners and 40 additional co-eluting peaks can be detected.[5-7] Some chromatographic peaks are comprised of two or three co-eluting congeners, usually varying in both toxicity and concentration. More congeners can be resolved using two complementary capillary columns joined in series,[8] or in parallel in separate GC ovens,[9] and/or using mass spectrometry (GC/MS) instead of electron capture detection (GC/ECD).[10,11] GC/MS cannot definitively mass resolve co-eluting PCB isomers (having same number of chlorines), and fragment ions of higher chlorinated PCBs at much higher concentrations can interfere with quantitation of lower chlorinated PCBs.

Structure–activity relationships for Cl_{4-8}-PCBs show that *meta* and *para*-chlorine substitution combined with one or no *ortho*-chlorine substitution (*o*-chlorine) affects planarity and toxicity.[12] Many PCBs have at least one *o*-chlorine substituted atom and some of the most abundant PCBs have two or more *o*-chlorine substitutions. Only 20 PCB congeners have no *o*-substitution; eight of these are Cl_{4-6} PCBs, of which four are specifically targeted for analysis because they are both prevalent and toxic. These nearly planar PCBs have the same mode of toxicity as 2,3,7,8-TCDD (dioxin),[12] inducing the biosynthesis of cytochrome P450-1A1 enzyme (aryl hydrocarbon hydroxylase) because they are stereochemically similar to dioxin. Dioxin-like effects in piscivorous birds have been attributed to these PCBs.[13-15] PCB congener 3,3′,4,4′,5-pentachlorobiphenyl (IUPAC No. 126) or simply PCB 126[16] is the most potent congener, with a dioxin equivalency of approximately 0.1,[12,17] compared with 1 by definition for 2,3,7,8-TCDD. Approximate dioxin equivalencies for three other non-*o*-PCBs are 0.005 for PCB 81, 0.0005 for PCB 77, and 0.01 for PCB 169.[17,18]

Any analytical method for non-*o*-PCBs, as well as any other potent dioxin-like compound of interest, must be sensitive and selective for a wide range of concentrations in various sample matrices.[19-21] Very low concentrations (pg/g or ppt) of non-*o*-PCBs, especially PCB 126, are toxicologically significant and their detection requires injection of nearly a gram-equivalent of a sample extract into a GC/MS to detect below 1 pg/g. Before GC/MS analysis, sample extracts must be extensively cleaned up to remove percent levels of lipids or other biogenic material. Also, as part of the enrichment, interfering PCBs must be removed. Non-*o*-PCBs and other dioxin-like compounds are separated based on congener planarity.[22-29] We routinely analyze four discrete fractions collected from a high performance liquid chromatography-carbon (HPLC-C) system.[22] In order of planarity the fractions contain: bulk PCBs (di- to tetra-*o*-PCBs) and other nonplanar compounds, e.g., polychlorinated terphenyls (PCTs); mono-*o*-PCBs, halogenated diphenyl ethers, and other partially planar compounds; non-*o*-PCBs and most polychlorinated naphthalenes (PCNs); and polychlorinated dibenzofurans (PCDFs) and dioxins (PCDDs).

Detection methods for non-*o*-PCBs at very low concentrations (pg/g) must be more selective than is possible with GC/ECD because sample extracts also contain ECD-responsive compounds, e.g., halogenated diphenyl ethers, PCNs, PCDFs, PCDDs, PCTs, as well as much higher concentrations (ng/g to µg/g) of DDE, chlordane, and potentially interfering *o*-substituted PCB congeners. Non-*o*-PCBs are present at parts per million levels in Aroclor® mixtures, span a wide range of concentrations, yet are at much lower levels (5 to 5000 times) than the close-eluting *o*-PCBs (Table 1).[9,20,30] For example, a sample containing 1 µg/g of total PCBs from Aroclor 1248 would typically have non-*o*-PCBs ranging from <2 pg/g for PCB 169 to about 74 pg/g for PCB 126 and to 3500 pg/g for PCB 77. Non-*o*-PCB concentrations vary between Aroclor lots and with the environmental fate (e.g., uptake,

Table 1 Concentrations (µg/g Aroclor) and Ratios of Closely Eluting o-PCBs Compared with Non-o-PCBs in Aroclor Mixtures

PCB Structure[a]	GC Ret. time[b] min:sec	Congener number	Number o-Cls	Aroclor 1242 µg/g[e,f]	Aroclor 1242 Ratio o:non-o	Aroclor 1248 µg/g[e,g]	Aroclor 1248 Ratio o:non-o	Aroclor 1254 µg/g[e,f]	Aroclor 1254 Ratio o:non-o	Aroclor 1260 µg/g[e,f]	Aroclor 1260 Ratio o:non-o
3,4,4',5-Cl$_4$	36:20	81[c]	0	140		260		10		<5	
2,2',3,4,5'-Cl$_5$	36:27	87	2	7,700	55	6,600	25	37,800	3,800	7,700	>1,500
2,2',3,3',6,6'-Cl$_6$	36:55	136	4	700	0.3	800	0.2	11,200	50	22,300	370
3,3',4,4'-Cl$_4$	37:04	77[c]	0	2,200		3,500		220		60	
2,3,3',4',6-Cl$_5$	37:14	110	2	15,300	7	19,000	5	58,500	270	19,000	320
2,2',3,3',4,5-Cl$_6$	43:59	129[d]	2	<500	<25	500	5	2,300	70	11,100	>3,700
3,3',4,4',5-Cl$_5$	44:01	126[c]	0	20		74		33		<3	
2,2',3,3',5,5',6-Cl$_7$	44:31	178	3	<500	<25	800	7	13,500	410	16,200	>5,400
3,3',4,4',5,5'-Cl$_6$	50:31	169[c]	0	<1		<2		<1		<1	

[a] Non-o-PCB and closely-eluting PCB(s) are grouped together in elution order from 50 m Ultra-1 column.

[b] Retention times of PCBs from 50 m Ultra-1 column programmed 120°C (1 min hold) to 240°C at 2.2°C/min, then to 305°C at 5°C/min (9 min hold).

[c] Denotes AHH-active non-o-PCB congener.

[d] Eluting 2 sec earlier than PCB 126, PCB 129 is the only o-PCB listed that is not resolved from non-o-PCBs with a 50 m Ultra-1 column.

[e] Values of AHH-active PCBs in Aroclor Mixtures by GC/HRMS, see Reference 20.

[f] Values of closely-eluting o-PCBs in Aroclors 1242, 1254, and 1260,[9] whose detection limit was 500 ppm.

[g] Values of closely-eluting o-PCBs in Aroclor 1248.[30]

biotransformation, or degradation) of each PCB in a sample. GC/MS is more selective than GC/ECD, and GC/high resolution MS (GC/HRMS) is even more selective, limited only by interfering ions approximately within 0.05 Da of targeted masses. GC/HRMS is also very sensitive with a detection limit (<0.1 pg) that rivals any GC/MS. Its calibration range can extend four orders of magnitude, e.g., 0.25 to 2500 pg, using a quadratic curve fit. With a highly efficient capillary GC column, a 50-m Ultra-1 (dimethyl polysiloxane) column, Cl_5-PCB 126 is the only non-*o*-PCB that is not resolved from another PCB (Cl_6-PCB 129) as listed in Table 1.[31] In contrast, all closely-eluting *o*-PCBs in Table 1 co-elute with the corresponding non-*o*-PCBs on a 5% phenyl-dimethyl polysiloxane capillary column (e.g., DB-5), which many analysts use for non-*o*-PCB analysis.[19]

Selective detection of multiple compounds with ions of the same nominal mass, but with sufficiently different accurate masses, is a unique capability of GC/HRMS. This paper shows definitive results for non-*o*-PCBs by GC/HRMS analysis and selective detection of two to three compound groups with the same nominal mass: ^{13}C-labeled PCB internal standards, and closely-eluting PCNs, and PCTs. Cl_{4-8} PCNs are present in some environmental samples at ng/g levels and are recovered in the same HPLC-C fraction as non-*o*-PCBs.[22] Some PCNs have dioxin equivalence, may be toxicologically important,[32] and with only nominal mass resolution some PCN congeners interfere both chromatographically and mass spectrally with ^{13}C-labeled PCB internal standards. PCT congeners that are recovered in the non-*o*-PCB fraction do not interfere with non-*o*-PCBs; however, their substitution patterns and potential toxicities are unknown.

GC/HRMS cannot fully resolve all fragment ions of co-eluting higher chlorinated PCBs from molecular ions of lower chlorinated non-*o*-PCBs, as shown by three low resolution electron ionization mass spectra for Cl_5-PCB 87, Cl_6-PCB 136, and non-*o*-Cl_4-PCB 77 (NIST Mass Spectral Library, PC Version 4.0). Ions in low resolution MS spectra, resolved to their unit mass, may consist of multiple accurate mass ions contributing to the same unit mass. For non-*o*-Cl_4-PCB 77 (or 81), its isotopic molecular ion ($C_{12}H_4{}^{35}Cl_3{}^{37}Cl$) at m/z 291.9194 (Figure 1A) is compared with the fragment ion, m/z 291.9008, for the closely-eluting tetra-*o*-Cl_6-PCB 136 (Figure 1C). Separation of PCB 77 from PCB 136 can be achieved in three ways. To spectrally resolve PCB 77 (m/z 291.9194) from PCB 136 (m/z 291.9008) requires a Resolving Power (R.P.) above 16,000, however, sensitivity drops as R.P. increases. Operating at 10,000 R.P. reduces the interfering ion of PCB 136 by 50% while increasing sensitivity from 16,000 R.P.; a methyl silicone GC bonded phase chromatographically resolves it from PCB 77. Also during enrichment, PCB 136 can be removed (>99.5%) by HPLC-C to minimize potential interference with PCB 77.

Di-*o*-Cl_5-PCB 87 and non-*o*-Cl_4-PCB 81 cannot be resolved by HRMS alone, either. GC separation is necessary and achievable with a long (50 m), methyl silicone thin phase (Ultra-1, 0.11 μm), narrow diameter (0.20 mm) capillary column. The m/z 292 fragment ion of PCB 87 (Figure 1B) consists

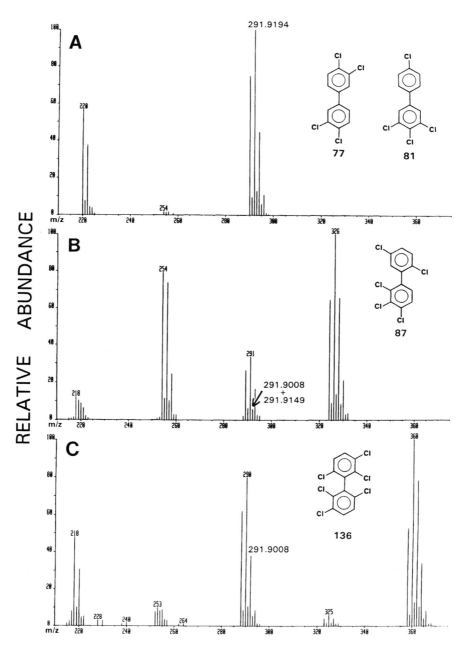

Figure 1   Mass spectra from NIST Library of non-*o*-Cl$_4$-PCB 77 (A), di-*o*-Cl$_5$-PCB 87 (B), and tetra-*o*-Cl$_6$-PCB 136 (C). Accurate masses constituting isotopic molecular ion of PCB 77 and of fragment ions of PCBs 87 and 136 are listed.

of m/z 291.9008 (loss of HCl) and 291.9149 ($C_{11}{}^{13}C$ $H_5{}^{35}C_{13}{}^{37}Cl$). The former ion can be partially resolved from PCB 81 (m/z 291.9194) as discussed above, but is relatively unimportant because its low ion abundance contributes <20% to m/z 292. The overlap of the latter ion can only be slightly reduced (<15%) by HRMS at 10,000 R.P. Most HRMS systems cannot achieve the 65,000 R.P. necessary for complete resolution.

Fragment ion abundances can be reduced by using an electron energy of about 35 eV. Potential interferences from *o*-PCB fragment ions are decreased while increasing abundances of isotopic molecular ions including those of the non-*o*-PCBs. Consequently, fragment ion abundances of *o*-PCBs 87 and 129 are 3%, while PCB 110 is only 0.7% of its molecular ion response. More abundant fragment ions of 2,2'-chloro-substituted PCBs 87 and 129 are present compared with those from 2,6-substituted PCB 110. These abundances can be used to estimate the response necessary to contribute 10% bias to non-*o*-PCB determined concentrations. PCB 110, measured as molecular ion response, can be 12 times higher than PCB 77, but PCBs 87 and 129 can be only 3 times higher than PCBs 81 and 126, respectively.

PCB 126 cannot be resolved from di-*o*-Cl$_6$-PCB 129 by GC/HRMS, but PCB 129 can be removed (>99.5%) by HPLC-C. Because residual amounts of PCB 129 may still exceed those of PCB 126, and PCB 129 fragment ions may interfere, it is necessary to monitor the molecular ion for PCB 129. As shown for PCB 81 and 87 above, HRMS cannot achieve the 75,000 R.P. needed to resolve the fragment ion of PCB 129 (m/z 325.8760) from PCB 126 (m/z 325.8804). Therefore, to assure accurate determinations of non-*o*-PCBs, it is important to combine HPLC-C, high resolution GC with a selective capillary column (e.g., bonded methyl silicone), HRMS, and interference monitoring.

Materials required

Reagents

Only distilled-in-glass, pesticide, or nanograde solvents are used. All reagents are stored in clean, glass containers with Teflon-lined caps. Sodium sulfate (anhydrous reagent grade, Fisher) is solvent washed and baked at 475°C for 8 h, then cooled and stored. Two sizes of silica gel (SG): Silica Gel-60 (70–230 mesh, EM Science) and a coarser Davisil® grade 635 (60–100 mesh, Fisher) (coarse SG), are washed with dichloromethane, dried in a stream of purified nitrogen, activated at 130°C for 24 h, passively cooled, and stored or used to prepare related reagents. Sulfuric acid/silica gel (SASG), is prepared by slowly adding sulfuric acid to activated coarse SG in a 3/7 gravimetric ratio (for coarse SASG), or in a 4/6 gravimetric ratio (for standard SASG). The mixtures are tumbled until evenly-coated, free-flowing powders are obtained. Potassium silicate (KS) is prepared by adding 300 g of SG to a solution of 168 g potassium hydroxide (Aldrich high purity) dissolved in 750 ml of methanol. The mixture is stirred for 90 min at 55°C, allowed to cool, filtered using a Buchner funnel to collect the KS on filter paper, washed three times with 100 ml of methanol, dried under nitrogen, and dried and stored at 130°C. Copper (EM Science, light-fine turnings) is activated with

0.75 *M* nitric acid for 20 s, flushed with tap water for 5 min, and rinsed three times with acetone, twice with methylene chloride, and then used immediately. The HPLC-C column packing is prepared by dispersing 300 mg activated carbon (Amoco PX-21, 2 to 10 μm) on 9.6 g Lichroprep RP-18 (15 to 25μm, EM Science).[22]

Specialized glassware

For sample extraction, a glass column 45 cm × 1, 2, or 4 cm with stopcock and 500-ml reservoir is used, depending on sample size. A 30 cm × 2 or 4 cm glass column with a 300-ml reservoir is used for the first reactive cleanup column for tissue samples. A 30 cm × 1 cm glass column with a 100-ml reservoir is used for reactive cleanup of sediment samples and for the second cleanup of tissue samples. Chromacol® semi-conical 0.9-ml Teflon-lined crimp-top disposable vials are used to hold final extracts for GC/HRMS and Chromacol semi-conical 1.1-ml snap-top vials to hold extracts for HPLC. All reused glassware is washed with detergent, rinsed with water (hot tap and deionized) and solvents, and baked at 475°C for 8 h in a muffle furnace. Prior to use, the baked glassware is rinsed with dichloromethane.

Spiking and standard solutions used for isotope dilution method

Compositions of the procedural internal standard (PIS), instrumental internal standard (IIS), and calibration standard solutions for non-*o*-PCBs are shown in Table 2. All ^{13}C-labeled PCBs were acquired from Cambridge Isotopes, Woburn, MA; individual unlabeled PCB solutions were obtained from Ultra Scientific, N. Kensingston, R.I. Also listed in Table 2 is a standard containing potentially-interfering PCBs 87, 110, and 129. Five other non-*o*-PCB congeners (3,4,4'-PCB 37; 3,3',4,5-PCB 78; 3,3',4,5'-PCB 79; 3,3',5,5'-PCB 80; and 3,3',4,5,5'-PCB 127) were acquired from Ultra Scientific and Accustandard, New Haven, CT, to confirm their identity. A mixed Aroclor PCB standard (A1111) containing equal amounts (by weight) of Aroclors 1242, 1248, 1254, and 1260 technical mixtures (Monsanto, St. Louis, MO) was used for spiking matrix QC samples. A Halowax 1014 PCN standard (Ultra Scientific) was used in some analyses to verify adequate MS resolution for the ^{13}C-labeled PCBs. Perfluorodecalin (PCR Inc., Gainesville, FL) was used to determine high resolution and to provide consistent accurate mass calibration during one or more days of GC/HRMS analysis. A mixed calibration standard for high performance-gel permeation chromatography (HP-GPC) contains di-(ethylhexyl)-phthalate, biphenyl, benzo(*g,h,i*)perylene, and sulfur at 100, 6, 10, and 20 μg, respectively.

Compressed gases

Ultra-high purity helium is the carrier gas for the GC/HRMS. Ultra-high purity hydrogen is the carrier gas and high purity nitrogen the make-up gas

Table 2 Spiking and Calibration Solutions (pg/µl) for Non-o-PCB Analysis

Compound (PCB congener no.)	Procedural internal standard[a]	Instrumental internal standard[b]	Interfering PCBs standard[c]	Calibration standards in nonane in autosampler vial (pg/µl, 100 µl total)						
				93 W	94 W	95 W	96 W	97 W	98 W	99 W
3,3',4,4'-TCB (77)	—	—	—	0.25	1	5	25	100	500	2500
3,4,4',5-TCB (81)	—	—	—	0.25	1	5	25	100	500	2500
3,3',4,4',5-PeCB (126)	—	—	—	0.25	1	5	25	100	500	2500
3,3',4,4',5,5'-HxCB (169)	—	—	—	0.25	1	5	25	100	500	2500
$^{13}C_{12}$-3,3',4,4'-TCB (77)	100	—	50	50	50	50	50	50	50	50
$^{13}C_{12}$-3,3',4,4',5-PeCB (126)	100	—	50	50	50	50	50	50	50	50
$^{13}C_{12}$-3,3',4,4',5,5'-HxCB (169)	100	—	50	50	50	50	50	50	50	50
$^{13}C_{12}$-2,2',4,4',5-PeCB (101)	—	100	50	50	50	50	50	50	50	50
2,2',3,4,5'-PeCB (87)			250							
2,3,3',4,6-PeCB (110)			250							
2,2',3,3',4,5-HxCB (129)			250							

[a] The procedural internal standard is spiked at 5000 pg into all samples before extraction.

[b] Just before GC/MS analysis each extract is spiked with 5000 pg of the instrumental internal standard.

[c] 50-m Ultra-1 column practically resolves (10% overlap) PCB 87 from PCB 81, resolves PCB 110 from PCB 77, and does not resolve PCB 129 from PCB 126.

for GC/ECD. High purity nitrogen with a carbon filter to trap residual hydrocarbons is used for evaporating small volumes of sample extracts.

Equipment

Several pieces of equipment are used for sample preparation, cleanup, and fractionation. A rotary evaporator (Büchi-Brinkman, Switzerland) efficiently concentrates sample extracts containing toluene. The evaporator with a glass vapor trap and greaseless ground glass fittings must be kept very clean by periodic washings with solvents and using a carbon filter on the air vent line. The vapor trap separates the main evaporator body from each sample extract. The isocratic HP-GPC system consists of an ISCO 2350 HPLC pump, a Perkin Elmer ISS 100 autosampler (1-ml injector loop) using a Phenogel column (25 cm × 22.5 mm i.d., 10-μm particle size, 100-Å pore size, Phenomenex, Inc., Torrance, CA), with an ISCO Foxy 200 fraction collector and ISCO V[4] UV absorbance detector at 254 nm. The dichloromethane mobile phase is pumped at 4.0 ml/min. For gradient HPLC-C, the same models of autosampler and fraction collector are used in conjunction with a Perkin Elmer Series 410 quaternary pump. The HPLC-C system typically uses a single 300-mg PX-21 carbon dispersed on C_{18} stainless steel column made in-house.[22] A 6-port air-actuated switching valve (Vici VALCO, Houston, TX) in the HPLC-C system reverses the column flow for collection of the final fraction containing PCDF/PCDDs.

GC/ECD and GC/MS equipment are used for screening or quantitating PCBs. GC/ECD systems consist of Hewlett Packard 5890 Series II GCs with [63]Ni electron capture detectors and HP 7673 GC autosamplers injecting 1-μl via cool on-column injectors. A 1 m × 530 μm retention gap is joined by a Restek® tapered glass press-tight connector to a 30 m × 0.25 mm × 0.25 μm DB-1 (J&W Scientific, Folsom, CA) column with hydrogen carrier gas at 12 psig (82 kPa) and nitrogen make-up gas at 60 ml/min. The ECD signal is processed by a Perkin Elmer GC chromatographic data system. The GC/HRMS consists of a Hewlett Packard 5890A GC/Fisons-VG 70-250S HRMS with HP 7673 GC autosampler injecting 2 μl via a cool on-column injector and VG PDP 11/73-based data system. A 2.5 m × 530 μm deactivated fused silica retention gap is connected to the injector and is joined via a Restek connector to a capillary column of either a 30 m × 0.25 mm × 0.25 μm DB-1 or a 50 m × 0.20 mm × 0.11 μm Ultra-1 (Hewlett Packard) directly interfaced into the MS source via a heated transfer line at 310°C. Helium carrier gas is maintained at either 12.5 psig (85 kPa) for the 30 m column with an initial linear velocity 30 cm/s or 45 psig (315 kPa) for the 50 m column with an initial linear velocity 27 cm/s. The low resolution GC/MS system consists of an HP 5890 Series II GC interfaced to a Finnigan 4023 quadrupole MS. A 30 m × 0.25 mm × 0.25 μm DB-5 column (J&W Scientific) is used. The oven is temperature programmed in a manner similar to PCDF/PCDD analyses: 120°C (1 min hold) to 210°C at 20°C/min, then to 300°C at 4°C/min, with a 5-min final hold time. The Finnigan 4023 GC/MS was also used for a full-scan EI analysis in an attempt to identify contaminants co-eluting with PCB 126.

Procedures

Sample preparation, spiking, and extraction

Each tissue sample is ground while partially frozen and homogenized by passing through a stainless steel power grinder (Hobart) three times. Each sediment sample is spread evenly in a glass pan, covered loosely with sol-vent-rinsed aluminum foil, and air-dried in a laboratory hood (with occasional stirring). The air-dried sediment is then milled to a finer texture using a Waring® blender. For sample extraction, a known amount (≤50 g) of tissue or sediment is removed and blended with 3 to 4× its weight of anhydrous sodium sulfate.

With every set of ten environmental samples, two replicates of one sample, plus a procedural blank, matrix blank, spiked matrix blank, and positive control matrix sample for quality assurance (QA) are prepared. QA samples are used to evaluate PCB background contamination, accuracy, and precision at both low and higher concentrations of analytes. Because PCBs are ubiquitous, efforts must be made to minimize laboratory sources of PCBs, e.g., PCB-containing ballasts in fluorescent light fixtures made before the mid-1970s.[33] We select matrix blank, spike, and positive control samples to match the sample matrices being prepared and analyzed. A matrix blank is chosen to represent a sample ideally of the same matrix (e.g., fish species or bird egg) which contains analytes (PCBs) not detectable or as low as possible. Matrix blanks consisted of pond-raised bluegill and grass carp, grocery-store chicken eggs, and pond sediment. A spiked matrix blank provides method QA of targeted analytes (PCBs) at various concentrations, representative of the concentrations in environmental samples. Low-level spiked QA samples test an analytical method's performance at low but environmentally signif-icant levels. A positive control QA sample is typically collected from an area impacted by targeted pollutants and usually also contains other pollutants which may interfere with the analytes. Our positive control carp and sedi-ment samples are composites (collected from Saginaw Bay, MI, an area known to have both PCB and PCDF/PCDD contamination and also repre-senting Lake Huron, an important aquatic resource, the Great Lakes) and Standard Reference Material 1941A marine sediment (NIST, Gaithersburg, MD). QA samples are analyzed with all sample sets to evaluate background contamination, accuracy, and precision at both low and higher concentra-tions of analytes.

While in the mixing container, each sample is spiked with ^{13}C-procedural internal standards (Table 2) and allowed to equilibrate (1 h). Matrix spike samples are also spiked with the A1111 mixed Aroclor standard, which contains both non-*o*-PCBs and potentially interfering PCB congeners in con-centrations similar to those found in environmental samples. Each sample is packed dry into an extraction column and extracted with 700 ml of dichlo-romethane by first adding a portion of the solvent to the column, just to the top of the homogenate, and allowing this solvent to permeate for 30 min; then the rest of the solvent is added, the flow rate adjusted to about 3 ml/min, and the eluate is collected in a flask. Solvent exchange to hexane is done by rotoevaporating to approximately 50 ml, adding 40 ml of hexane, and roto-evaporating to about 10 ml. Each extract now contains analytes, co-contam-inants, and mostly biogenic (naturally occurring) compounds. Percent lipid for each tissue extract is determined gravimetrically using a small known aliquot evaporated at 105°C to constant weight.

Cleanup

For tissue extracts, two SASG reactive cleanup columns are needed to sequentially remove biogenic materials, but for sediments, only the second column is necessary. The first reactive clean-up column is packed, from bottom to top with a 1-cm segment of sodium sulfate, 25 ml of KS, 25 ml of SASG, a 1-cm segment of sodium sulfate, and finally, 50 ml of coarse SASG. After rinsing the packed columns with 100 ml of dichloromethane, the tissue extracts are applied and 150 ml of dichloromethane is added to elute the analytes. The eluates collected from the columns are concentrated by rotary evaporation and solvent exchanged to isooctane. Tissue extracts containing more than 5 g of lipid are divided between multiple columns and the eluates later combined.

The second reactive column is packed, from bottom to top with a 1-cm segment of sodium sulfate, 10 ml of SG, 3 ml of KS, 5 ml of SASG, and finally, a 1-cm segment of sodium sulfate. After rinsing the packed columns with 25 ml of 3% (v/v) dichloromethane in hexane, the extracts are applied and 70 ml of 3% dichloromethane in hexane is added to elute the analytes. A small amount of isooctane is added as a keeper before the eluates are con-centrated to about 1 ml by rotary evaporation and then a gentle stream of nitrogen. The eluates are transferred into 1.1-ml vials for autosampling by HP-GPC.

HP-GPC cleanup using the Phenogel column is important to remove residual biogenic compounds which would otherwise impair GC perfor-mance. The HP-GPC elution is calibrated using the four-compound calibra-tion mixture. Fraction collection begins at about 14 min, just after the first compound di-(2-ethylhexyl)-phthalate elutes and extends to about 18:30 min, just before the elution of sulfur. Sample extracts are concentrated to 0.5 ml for HPLC-C fractionation.

HPLC-C fractionation

Because a single 300-mg HPLC-C column can be overloaded by >600 µg of total PCBs, a small portion of the HP-GPC eluate (2 to 5%) is removed for GC/ECD screening to estimate a total PCB concentration.[22] For the 300-mg carbon column, a flow rate of 2.5 ml/min is maintained during a multi-step solvent program that contains both isocratic and linear gradient steps for solvent A (5% methanol:10% dichloromethane:85% hexane) and solvent B (toluene). The program steps are (1) 45 min of solvent A; (2) 40 min gradient from 100% solvent A to 75% solvent A; (3) 25 min gradient from 25% solvent B to 100% solvent B; (4) 100% solvent B for 18 min; (5) reverse flow through the column (128 min total time) with 100% solvent B for an additional 90 min, and continued for another 200 min to wash the column; (6) column re-equilibration for 25 min with 100% solvent A. The column is returned to forward flow at 442 min total time. The collection times for the four fractions are: 0 to 60 min (containing bulk PCBs); 60 to 90 min (mono-*o*-Cl-PCBs); 90 to 128 min (non-*o*-Cl-PCBs); and 128 to 218 min (PCDF/PCDDs). The fractionation times are based on GC/ECD analysis of the elution of early and late-eluting PCB congeners within each fraction: PCB 170, at end of fraction 1, PCBs 118 and 189 for fraction 2, PCBs 37/81 and 169 for fraction 3, and 2,3,7,8-TCDF and OCDD for fraction 4.[22]

The non-*o*-PCB eluates are concentrated, transferred to 15-ml test tubes, evaporated to about 200 µl using a nitrogen stream, and transferred to 0.9 ml vials by glass transfer pipette. The extract in the vial is carefully evaporated under nitrogen to about 50 µl, then spiked with IIS (Table 2).

GC/HRMS analysis

The 50 m Ultra-1 column (preferred) is temperature programmed with an initial 1-min hold at 120°C followed by a ramp to 240°C at 2.2°C/min, another ramp to 305°C at 5°C/min, and a final hold of 9 min. The final hold time chosen at 305°C is sufficient for most known late-eluting residual compounds, e.g., polychlorinated terphenyls, to elute. The 30-m DB-1 column is held 1 min at 120°C followed by a ramp to 250°C at 3°C/min, another ramp to 300°C at 6°C/min, and a final hold of 2 min.

The HRMS is tuned in normal electron ionization (EI) mode at m/z 293 using the calibrant perfluorodecalin (<1 µl) bled in from a heated reservoir. Tuning for high resolution (10,000 R.P. with a 10% peak overlap) requires optimum settings on the slits and lenses to narrow the ion beam peak width sufficiently but still have ion beam transmission as high as possible (4 to 8% of full low resolution). Response is optimized by adjusting the electron energy near 35 eV to minimize the space-charge effect caused by ionizing helium carrier gas atoms.[34] With a filament trap current of 200 µA, or 500 µA for more sensitivity, the ion repeller is optimized (slightly negative, e.g., –2.5 V) for the VG HRMS using responses of higher mass ions at lower accelerating voltages (e.g., 6000 V) during selected ion monitoring (SIM). HRMS source pressure is about 5×10^6 mbar at initial GC oven temperature.

SIM EI GC/HRMS is highly sensitive because the HRMS measures a few highly-resolved accurate ions selected with more dwell time (>35 ms) during 1-sec scans than it would for each ion of a full-scan mass range (<1 ms). Two groups of ions are used (Table 3). The first group of 22 ions are scanned each second from 22 to 47 min elapsed time, with 35 ms dwell times and interscan times (allowed for switching voltage to the next ion) of 10 ms. After a 5-sec delay for switching, the second group of 13 ions are scanned from 47:05 to 60 min with 60 ms dwell times and interscan times of 10 ms. The switching time from Group 1 to 2 is about midway between the elution of PCBs 126 and 169. The two most abundant ions each (Table 3) are used for non-*o*-PCBs and the ¹³C-labeled PISs and IIS, to assure their positive identification and quantitation by matching both their ion ratio (within ± 15%) and their GC retention times (within 3 sec). The IIS normalizes differ-

Table 3 Selected Ions Monitored for QA and Analysis of Non-o-PCBs and PCNs by GC/High Resolution MS

Analyte(s)	Reason for monitoring[a]	More abundant ion (m/z)	Less abundant ion (m/z)	Ion ratio (less abundant)/(more abundant)
Ion group 1:[b]				
Cl$_3$-PCB 37	Non-o-PCB	255.9614	257.9584	0.98
Cl$_4$-PCNs	Dioxin-like compounds	265.9038	263.9067	0.78
Cl$_4$-PCBs 77, 81	AHH-active non-o-PCBs	291.9194	289.9224	0.78
Lock mass/check	HRMS SIM Calibrant	292.9825		
Cl$_5$-PCNs	Dioxin-like compounds	299.8648	301.8618	0.66
^{13}C$_{12}$-PCB 77	Procedural Internal Standard (PIS)	303.9597	301.9626	0.78
Cl$_4$-PCDFs	HPLC-carbon breakthrough	305.8987		
Cl$_6$-PCNs	Dioxin-like compounds	333.8258	335.8229	0.82
Cl$_5$-PCBs 126; 87,110	Non-o-PCB; Close-eluting	325.8804	323.8834	0.62
^{13}C$_{12}$-PCB 126; 101	PIS; Instrumental Internal Standard	337.9207	335.9236	0.62
Cl$_6$-PCBs 129; 136	Interference; close-eluting	359.8415		
Cl$_6$-Diphenyl ethers	HPLC-carbon carryover	375.8364		
Cl$_7$-PCBs	Potential Interference	393.8025[d]		
Ion group 2:[c]				
Cl$_3$-Terphenyls	Late-eluting relative to PCB	333.9897		
Cl$_6$-PCB 169	AHH-active non-o-PCB	359.8415	361.8385	0.80
Cl$_7$-PCNs	Dioxin-like compounds	367.7868		
Cl$_4$-Terphenyls	Late-eluting relative to PCB	367.9507		
^{13}C$_{12}$-PCB 169	PIS	371.8817	373.8788	0.80
Lock mass/check	HRMS SIM Calibrant	392.9760		
Cl$_7$-PCB 189	HPLC-carbon carryover	393.8025		
Cl$_8$-PCN	Dioxin-like compound	401.7479		
Cl$_5$-Terphenyls	Late-eluting relative of PCB	401.9117		
Cl$_8$-PCBs	Potential Interference	429.7606[d]		

[a] Ions for non-o-PCBs (left); Ions for other compounds and for QA reasons (right).

[b] Each ion (except lock mass check for 5 ms) was sequentially monitored for 35 ms and 10 ms interscan to next ion.

[c] Each ion (except lock mass check for 5 ms) was sequentially monitored for 60 ms and 10 ms interscan to next ion.

[d] Interscan time of 20 ms after last ion was longer for settling time after MS reset to full accelerating voltage.

ences in final volume and amount of sample extract injected, to determine the percent recoveries of the PISs and GC/HRMS performance during each sample analysis. Cl_{4-6}-PCNs are monitored to ensure that mass resolution is sufficiently above 4000 R.P. to avoid potential PCN interference. Cl_5- and Cl_6-PCNs respectively elute close to the ^{13}C-PCBs 77 and 126 and differ in mass by only a 0.1 Da. Additional ions (Table 3) monitor Cl_{7-8} PCNs and Cl_{3-5}-PCTs, o-PCBs (potential interferences), mono-o-PCB 189, Cl_4-PCDFs, and Cl_6-diphenyl ethers (HPLC-C checks), and as lock mass and lock mass check ions for the HRMS.

Just before data acquisition, a slow 20-sec scan of the perfluorodecalin calibrant is used to calibrate the HRMS within 25 ppm ($\pm$ 0.01 Da). During acquisition, the calibrant is constantly bled into the HRMS source, enabling the data system to keep the magnet on the lock mass ion while the voltage is scanned for all other ions in the group. The lock mass check ion in each group monitors stability during the analysis and over the entire sample set (up to several days). Calibrant typically lasts 18 to 24 h before needing to be replenished. Large amounts (e.g., 10 μg) of co-eluting contaminants can significantly reduce sensitivity while eluting, by either having an intense ion close enough to the lock mass to interfere or by reducing the ionization efficiency of the source. The lock mass check ion chromatogram will show a corresponding negative peak.

The GC/HRMS data system detects analyte peaks, integrates using signal/noise threshold criteria, identifies within narrow retention time windows that are adjusted by normalizing slight shifts in retention times to retention times of ^{13}C-labeled PCBs, quantitates using calibration curves, and finally reports as concentrations to a printer, disk, or exported into PC spreadsheet programs. For quantitation, a list of the PCB analytes and procedural internal standards with appropriate ions, retention times, and allowable time windows is prepared. Generally, the better a GC column's resolution is, the easier, faster, and more accurate peak identification and quantitation will be. By using ^{13}C-labeled PISs, the concentration of each PCB congener, C, is inherently self-corrected to account for losses throughout the entire method (extraction, isolation of analytes, and GC/HRMS analysis), as shown in the equation: $C = A_a S / R_{ap} A_p M$ where A_a = Peak area of native PCB analyte, A_p = Peak area of the corresponding ^{13}C-PCB PIS, M = Spiked-sample mass (g), R_{ap} = Response factor derived from regression curve fitting, $A_a S/A_p^*$ (Varying amounts of analyte), and S = Amount of ^{13}C-PCB PISs spiked into sample.

For each native congener ion monitored in each sample set, a quadratic calibration curve best describes the ratio of native ion peak area responses from the wide range of amounts to those of the ^{13}C-labeled PIS amount in the calibration standards (Table 2). PCB 81 is calibrated and quantitated using ^{13}C-PCB 77, until ^{13}C-PCB 81 is acquired. Each calibration curve is specifically matched to the range of analyte responses in the sample set. Calibration standards are analyzed at the beginning of each set and are interposed after each four analyses. Samples are diluted whenever the native PCB/PIS area ratio in the sample exceeds 50, the ratio in the highest standard. Samples should be diluted by a factor of at least 20 by transferring ≤5% to a different vial containing the same amount (5000 pg) of ^{13}C-PIS (plus ≤5%). PCB concentrations in samples are computed and the values are averaged from both ions for each analyte, provided that the ion ratios are within ±15%.

Before computed PCB concentrations are reported, several QC criteria must be met:

1. Peak area for both ions must exceed three times the background noise, otherwise "ND" (Not Detected) is reported.
2. Peaks must occur at retention times from −1 to +3 sec within the corresponding ^{13}C-labeled PIS peaks, which elute about 1 sec earlier than the native peaks, or from −0.2 to 0.5% relative retention times of PCB 81, which lacks the corresponding ^{13}C-PIS.
3. The ion ratio of an analyte must be within ±15% of the theoretical isotopic abundance, otherwise the lesser concentration value determined from the two ions is reported along with "NQ" (Not Quantifiable).
4. Response of the isotopic molecular ion of interfering Cl_6-PCB 129 (Table 3) should be less than three times that of PCB 126 to ensure that PCB 129's fragment ions (3% abundant) do not contribute >10% to non-*o*-PCB 126. If the fragment ions contribute >10% but <30%, the interfering amount must be subtracted from PCB 126's computed value, and if >30%, the sample extract must be rerun through HPLC-C.
5. The acceptable range of ^{13}C-PCB PIS percent recoveries is 25 to 125%. Native concentrations based on lower PIS recoveries are still valid, but may not be as accurate at PIS-levels approaching the detection limit.

Recoveries of ^{13}C-PCB PISs in the final sample extract before any dilution is made are calculated based on a known amount of IIS (Table 2) spiked into each final extract and into each calibration standard. Overall method efficiency is measured by comparing the amount of the PISs detected in the *final* extract with the amount originally spiked into the sample. Linear calibration curves for the ^{13}C-PCB PISs average the PIS/IIS response factors of constant amounts of ^{13}C-PCB PISs in each sample and standard. Selective adsorption of the more planar PISs onto the vials, the GC retention gap, or the GC column may decrease the response factors. If chromatographic resolution degrades significantly (<50% resolution of PCB 81 from 87), a new retention gap should be installed and/or sample extracts cleaned up further. Whenever percent recoveries are outside the acceptable range, those samples in the sample set are reextracted and analyzed, if possible.

Results and discussion

Quality assurance

During the past two years, method and instrument performance were improved by consistently following the QC criteria listed above and carefully evaluating the results from the QC samples. Initially, GC/HRMS ion ratios (criterion 3) were consistently low (–10 to –25%) for analytes in both samples and standards, especially for higher mass ions within an SIM group, i.e., those ions detected at reduced voltages. An improperly positioned conversion dynode inside the MS was responsible for the low ion ratios, by producing ion responses that were highest at a reduced voltage (6500 V) instead of at full operating voltage (8000 V).

Occasional low recoveries of ^{13}C-PCB 126 (criterion 5) and the stability of the lock mass check ion indicated an unknown co-eluting contaminant in a few environmental and QC samples. The responses of ^{13}C-PCB 126 and native PCB 126 decreased at the same time and to the same extent that the lock mass check ion chromatogram showed a corresponding negative chromatographic peak. Full-scan GC/MS analysis of a QC chicken egg blank using the Finnigan 4023 quadrupole MS detected a very large contaminant (~10 µg) whose mass spectrum did not match any spectrum in the NIST Library. The molecular ion is apparently absent and the most abundant fragment ions are: m/z 105 (base peak or 100% abundance), 343, 43, 75, and 91 (each 35%), 115 and 131 (5%), 311 (2%), 326 (1%), 463 (0.5%), 358 (loss of 105 from 463, 0.2%), and 293 (0.2%). Ions m/z 91 and 105 suggest alkyl-phenyl substituted fragment ions such as benzyl and xylyl constituents, the loss of 105 from m/z 463 to 358 indicates that the molecular mass may be just above 463, and low abundance high mass ions suggest that much of the compound is not aromatic. This co-eluting peak with a very small m/z 293 ion may have interfered with the lock mass calibration ion and/or suppressed the MS ionization process. Quantitation of native PCB 126 was not significantly affected because both native and ^{13}C-PCB 126 ions were reduced by the same factor (about 3×). Relative to the IIS, the ^{13}C-PCB 126 ion responses had to be adjusted higher, based on the relative drop of the lock mass ion to properly quantitate its recovery. Although HRMS at 10,000 R.P. should resolve a non-halogen containing ion at m/z 326.1 or .2 from m/z 325.8804 for PCB 126, the ion ratio for PCB 126 must be checked for any interference.

Residual amounts of interfering o-PCBs, primarily PCB 129, in the non-o-PCB fraction after HPLC-C removal (criterion 4), are checked in each sample. Selected ion chromatograms in Figure 2, show just 1 min of the analysis using a 30 m DB-1 column. Analysis of the nonfractionated A1111 standard (30 µg) is shown on the left, and, on the right, is a chicken egg QC sample spiked with A1111 (20 µg) that has been cleaned up, and fractionated by HPLC-C. The top chromatograms are for Cl_5-PCB 126, the middle chromatograms for Cl_6-PCBs 158, 129, and 166, and the bottom chromatograms for Cl_7-PCB 178. Each chromatogram is shown normalized to its largest peak. For example, the PCB 178 peak height (Figure 2C) is 476 and a PCB 178 fragment ion (Figure 2A) peak height of 13 indicates 3% abundance (13/476). A fragment ion of Cl_6-PCB 129 is a significant interference in non-o-PCB analysis because PCB 129 is much higher in Aroclors than that of PCB 126 (Table 1) and is not chromatographically resolved. PCB 129's molecular ion is about 230 units (Figure 2B). Separate GC/HRMS analysis of PCB 129 found its fragment ion to be 3% abundant, which represents about 7 units or most of the combined peak response for both PCB 126 and 129's fragment ion (Figure 2A). In the QC chicken egg spike with HPLC-C, PCB 126 (Figure 2D) can be accurately determined because PCB 129 (Figure 2E) is less than three times background noise, a removal of >99.5% of PCB 129 (and 178) by HPLC-C.

Residual amounts of o-PCBs 87, 110, and 136, remaining after HPLC-C fractionation of the same chicken egg spike, are compared with initial amounts in the Aroclor mixture (Figure 3). HRMS only partially resolves fragment ions of the o-PCBs as discussed earlier, requiring chromatographic resolution and removal by HPLC-C. Although PCB 110 is about 20 times higher and PCB 136 is five times higher than PCB 77 in the Aroclor mixture, the DB-1 column (Figure 3A) provides adequate resolution to measure PCB 77. If not resolved, the fragment ion from PCB 110 would have contributed just 14% to PCB 77, because it is only 0.7% abundant at 10,000 R.P. PCB 81, however, cannot be accurately determined in the Aroclor mixture with the 30 m DB-1 column because the fragment ion from PCB 87 interferes (Figure 3A) and because the PCB 87 concentration is about 140 times higher than PCB 81 in the Aroclor mixture. The fragment ion from PCB 87 is about 3.5% abundant and in the analysis of the Aroclor mixture would contribute about 500% to the response of PCB 81, without adequate chromatographic resolution or removal by HPLC-C. HPLC-C removed >99.5% of the closely-eluting o-PCBs including 87, as shown by the large dynamic range of the GC/HRMS. The ion response for PCB 110 dropped from about 13,000 to 3 (Figure 3B, E), and the ion response for PCB 87 drops proportionally in Figure 3D, allowing PCB 81 to be determined without interference.

Analyses of positive, negative, and A1111-spiked negative control materials accompanied each analytical sample set and are summarized in Table 4. Whenever possible, the control materials are of the same or similar matrices as environment samples. Procedural blanks normalized to 25-g sample sizes and matrix blanks provide periodic QC checks for detection limits, background levels of PCBs, and for higher PCB levels from possible carryover from HPLC-C. PCBs 77 and 126 are generally low but ubiquitous in all negative control materials. The high percent relative standard deviation (% RSD) values in the negative control materials are typical results of analyses near the detection limits and variability in maintaining low-level laboratory background. Good indicators of carryover from HPLC-C of biota samples are PCBs 126 and 169, which are very low in Aroclor mixtures and abiotic samples, and relatively high in biota. Sample carryover from HPLC-C was minimized by optimizing the fractionation and more effectively rinsing the HPLC's autosampler injection loop following injections.

Analyzing positive control samples requires the method to selectively and consistently determine the non-o-PCBs in the presence of various co-contaminants such as PCNs, pesticides, hydrocarbons, PAHs, sulfur, and bulk PCBs. The non-o-PCBs in the positive control matrices span a wide range of concentrations (Table 4), which necessitates the full calibration range by the GC/HRMS. Percent RSD values for the non-o-PCBs present above 10 pg/g in the positive control samples are all less than 20%, and compare favorably with an interlaboratory study.[19] The high precision resulted from the use of ^{13}C-PCBs to correct for variable method recoveries. Without [13]C-

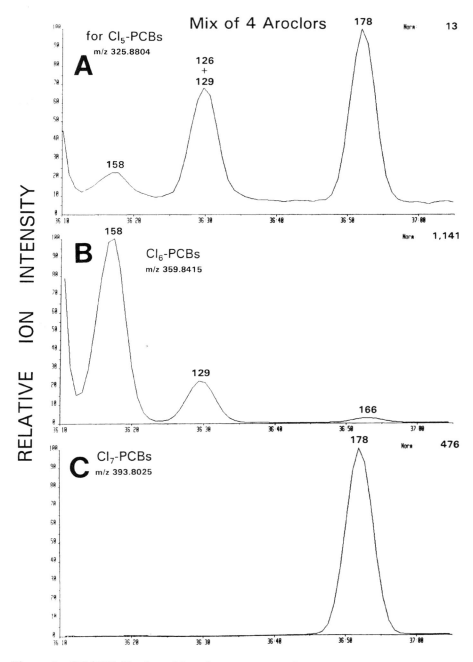

Figure 2 GC/HRMS selected ion chromatograms (36–37 min segment) for non-*o*-Cl$_5$-PCB 126 (A,D), co-eluting *o*-Cl$_6$-PCBs 129 (B,E), and *o*-Cl$_7$-PCB 178 (C,F). Chromatograms (left half) show GC/HRMS of an equal mix of Aroclors 1242, 1248, 1254, and 1260 (A1111); chromatograms (right half) show GC/HRMS of the A1111 spiked into a chicken egg, cleaned up, and >99.5% removal of *o*-Cl$_6$ and Cl$_7$-PCBs by HPLC-C.

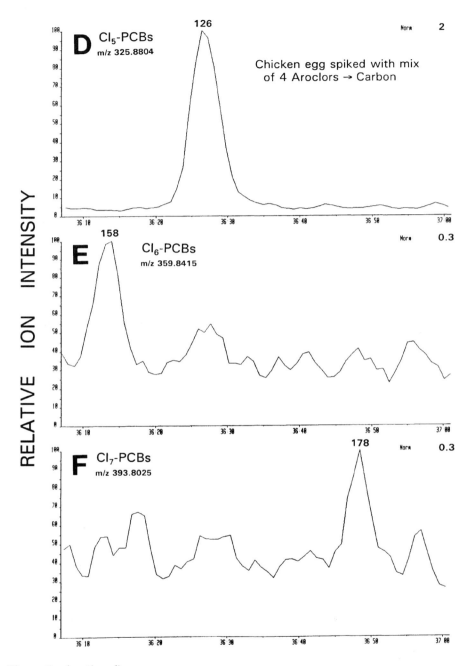

Figure 2 *(continued)*

PCB 81, however, several positive control replicates had much lower reported concentrations of PCB 81 and, as outliers, are not included with the data set. HPLC-C fractionation was adjusted, during method development, to more completely recover PCB 81. Accuracy could be estimated but not

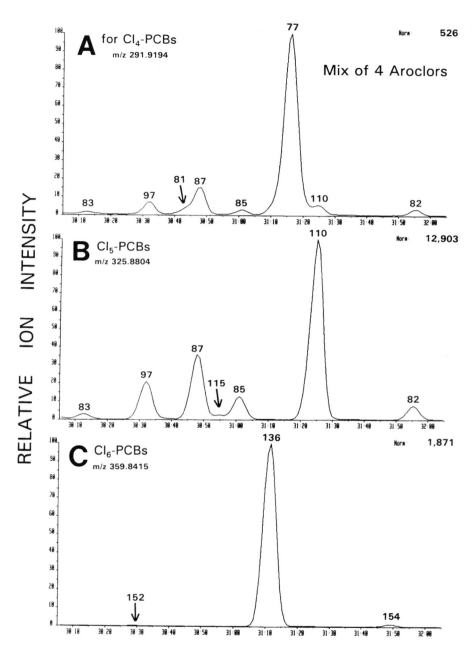

Figure 3 GC/HRMS selected ion chromatograms (30–32 min segment) for non-*o*-Cl$_4$-PCBs 81 and 77 (A,D), closely-eluting *o*-Cl$_5$-PCBs 87 and 110 (B,E), and closely-eluting *o*-Cl$_6$-PCB 136 (C,F). Chromatograms (left half) show GC/HRMS of the A1111 standard; chromatograms (right half) show GC/HRMS of the A1111 spiked into a chicken egg, cleaned up, and >99.5% removal of *o*-Cl$_5$ and Cl$_6$-PCBs by HPLC-C.

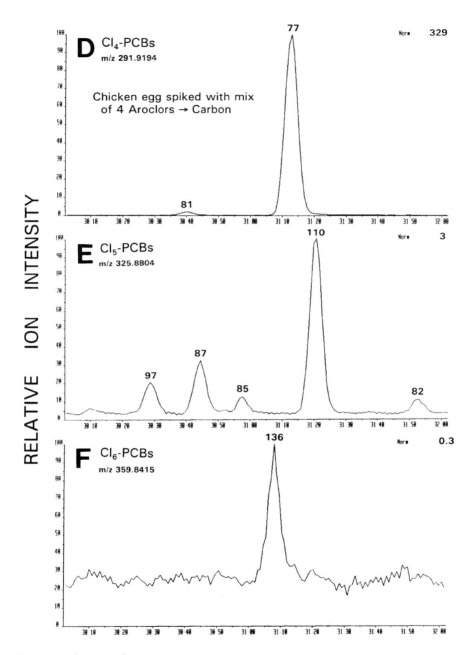

Figure 3   *(continued)*

fully evaluated because no certified values for non-*o*-PCBs exist for any of these samples.

Results of negative control materials spiked with Aroclors are reported as pg of congener per µg Aroclor (Table 4). A1111 spikes vary from 20 to 100 µg, which covers the full calibration range of the GC/HRMS for all non-*o*-

Table 4 Mean Concentrations and (% Relative Standard Deviations) of Non-o-Polychlorinated Biphenyls (pg/g) in Quality Control Samples and (pg/μg Aroclor) in Aroclor-Spiked Quality Control Samples

QC Material — collection site	Number of replicates[a]	Non-o-polychlorinated biphenyls			
		Tetra: 3,4,4',5-TCB congener 81	Tetra: 3,3',4,4'-TCB congener 77	Penta: 3,3',4,4',5-PeCB congener 126	Hexa: 3,3',4,4',5,5'-HxCB congener 169
Positive control materials					
Carp — Saginaw Bay, MI	26[b]	430 (16)	3,150 (11)	1,100 (12)	69 (19)
Sediment — Saginaw Bay, MI	15[c]	425 (16)	17,900 (18)	400 (19)	7 (26)
NIST SRM 1941A Sediment	3	30 (3)	610 (3)	52 (8)	8 (44)
Negative control materials					
Grass carp — NBS/MSC	7	3 (31)	22 (45)	5 (32)	2 (–) 6 ND (1–2)[d]
Bluegill — NBS/MSC	4	6 (78)	55 (42)	9 (48)	2 (63)
Channel catfish — NBS/MSC	3	1 (–) 2 ND (1)[e]	8 (10)	5 (16)	2 (28)
Chicken eggs — local grocery	16	5 (90)	17 (63)	8 (120)	11 (140) 8 ND (1–3)
Pond sediments — NBS/MSC	5	15 (31) 3 ND (8)	105 (75)	22 (90)	15 (71) 2 ND (5)

	[a]				
SPMD blanks — NBS/MSC	2	— (-)	8 (10)	2 (-)	— (-)
Procedural blanks (25-g basis)	39	2 ND (1)	9 (120)	1 ND (2)	2 ND (1)
		2 (65)		4 (80)	3 (56)
		17 ND (1-2)		6 ND (1-2)	23 ND (0.5-2)
Matrix spikes with 1:1:1:1 Aroclor mix[f,g] (pg/μg Aroclor)					
Chicken egg spikes w/Aroclors	11	97 (15)	1,890 (5)	72 (35)	4 (59)
Bluegill spikes w/Aroclors	4	96 (3)	1,875 (2)	86 (4)	6 (40)
Sediment spikes w/Aroclors	2	95 (1)	1,825 (1)	80 (1)	3 (2)

[a] Number of sample replicates for each congener.

[b] Concentrations of PCB 81 from 5 sample replicates are not included because HPLC-carbon recovered $<33\%$ of PCB 81.

[c] Concentrations of PCB 81 from 2 sample replicates are not included because HPLC-carbon recovered $<33\%$ of PCB 81.

[d] Number of replicates that were Not Detected and (Range of Detection Limits).

[e] Number of replicates that were Not Detected at (Detection Limit).

[f] Spiked with equal amounts of Aroclors 1242, 1248, 1254, and 1260 totaling 20 to 100 μg.

[g] Non-o-PCB values of 104, 1500, 33, and < 2 (pg/μg Aroclor) from Schwartz et al.[20] without correction for recovery of ^{13}C-PCB PISs.

PCBs and requires effective HPLC-C fractionation. PCB 126 is present in the A1111 mixture at a much lower concentration than in biota and lower in concentration relative to the interfering PCB 129. HPLC-C removed sufficient PCB 129 from all spikes, avoiding the need to correct for PCB 129 interference with PCB 126. Values of non-o-PCBs in the spikes are reported as is, without subtracting any concentrations inherent in the negative control materials. Percent RSD values are <10% for many of the spiked negative control matrices (Table 4). When used with positive control materials, spiked samples provide additional data about method accuracy and precision in instances where other PCB congeners are present at much higher levels than the non-o-PCBs.

Non-o-PCBs in environmental samples

The method was used to analyze over 200 samples in about 15 environmental sample sets, including fish, bird eggs, soils, sediments, and SPMDs (Table 5). (SPMDs are semipermeable plastic membrane devices containing thin films of triolein, that concentrate compounds via passive partitioning[35] similar to respiratory uptake of contaminants in fish.[36]) Samples were collected from point-source small and regional problem areas (Crab Orchard National Wildlife Area, IL, New York City harbor, and Saginaw Bay, MI) and nationwide from selected sites of the National Contaminant Biomonitoring Program.[37] Concentrations of non-o-PCBs in Table 5 range from about 1 pg/g to thousands of pg/g and were determined with little need for dilution because of the wide calibration range of the GC/HRMS. Calibration with weighted quadratic regression curve fitting allowed the lower response data to better fit the curve near the detection limit. Bird eggs had the widest range of non-o-PCB concentrations. PCBs 126 and 169 best exhibit bioconcentration in fish and biomagnification by birds. PCB 126 contributes most of the non-o-PCB dioxin-equivalents.

Other non-o-PCBs were detected using this method, but their concentrations in environmental and QC samples (Tables 6 and 7) were only estimated, since they were not specifically targeted for analysis and a full set of calibration standards was not run. PCB 37 is the only prevalent congener of the additional non-o-PCBs in the Aroclor mixture (Table 7) and is much higher in abiotic samples than in biota (Table 6). GC retention times (min:sec) of these PCB congeners using the 50-m Ultra-1 column (same conditions as Table 1) are 29:05 (PCB 37), 34:57 (PCB 79), 35:45 (PCB 78), and 41:48 (PCB 127). The column resolves Cl_3-PCB 37 from Cl_4-PCB 44 (which is 7 sec earlier) and Cl_4-PCBs 42 and 59 (8 sec later). Also resolved are Cl_4-PCB 78 from Cl_5-PCB 83 and Cl_5-PCB 127 from Cl_6-PCB 153 (12 sec earlier). PCB 127 elutes 34 sec later than the mono-o-Cl_5-PCB 105, but might be masked by the ~5000 times higher level of PCB 105 were it not removed by HPLC-C. PCB 79 is not resolved from the abundant di-o-Cl_5-PCB 99, which also is removed by HPLC-C. Lastly, non-o-Cl_4-PCB 80 is present at trace levels in the Aroclor mixture but is collected in the mono-o-PCB fraction from HPLC-C.

Table 5 Concentration Ranges of Target Non-o-Polychlorinated Biphenyls (pg/g) in Environmental Matrices

Sample matrix description — collection location	No. of samples	Non-o-polychlorinated biphenyls				
		Tetra: 3,4,4′,5-TCB congener 81	Tetra: 3,3′,4,4′-TCB congener 77	Penta: 3,3′,4,4′,5-PeCB congener 126	Hexa: 3,3′,4,4′,5,5′-HxCB congener 169	
Fish						
Variety,[a] 28 sites from U.S. National Contaminant Biomonitoring Program	30	4–150	27–3,600	15–1,500	2–340	
Variety[b] — Saginaw Bay, MI	24	9–610	43–6,800	32–1,400	3–94	
Caged channel catfish — Mississippi R. and Saginaw Bay, MI	32	1–14	8–130	5–20	1–4	
Bird eggs						
Eagle eggs — Great Lakes, Washington, Alaska	58	12–4,200	83–12,000	85–16,000	24–2,600	
Falcon eggs — New Jersey	11	19–240	180–6,100	270–4,300	60–670	
Variety[c] — Saginaw Bay, MI	15	16–1,500	110–5,100	72–9,500	7–1,100	
Starling eggs — Crab Orchard NWR, IL	10	5–200	13–2,300	7–1,800	5–110	
Soils, sediments, and SPMDs						
Soils — Crab Orchard NWR, IL	2	750–290,000	14,000–4.6 million	3,600–570,000	120–21,000	
Sediments — New Jersey — New York City Harbor	20	13–320	260–9,500	9–240	5–40	
Semipermeable membrane devices — Saginaw Bay, MI	11	7–36	160–1,100	5–17	0.6–1	

[a] Fish samples were white sucker, channel catfish, brown bullhead, carp, lake trout, and northern pike.

[b] Fish samples were white sucker, rainbow smelt, yellow perch, quillback carpsucker, walleye, gizzard shad, white bass, freshwater drum, logperch, spottail shiner, alewife, and channel catfish.

[c] Bird eggs were from Canada goose, mallard, black-crowned night heron, ring-necked gull, double-crested cormorant, herring gull, great egret, Forster's tern, and Caspian tern.

Table 6　Concentration Ranges of Other Non-o-Polychlorinated Biphenyls (pg/g) in Environmental Matrices

Sample matrix description — collection location	No. of samples	Non-o-polychlorinated biphenyls			
		Tri: 3,4,4'-TCB congener 37	Tetra: 3,3',4,5'-TCB congener 79	Tetra: 3,3',4,5-TeCB congener 78	Penta: 3,3',4,4',5'-HxCB congener 127
Fish					
Variety[a] — 28 sites from U.S. National Contaminant Biomonitoring Program	30	<2–50	ND <2–8[d]	ND <2–4	ND <2–10
Variety[b] — Saginaw Bay, MI	16	30–900	2–200	ND <0.6–1	<2–50
Channel catfish caged (1 month) — Saginaw Bay, MI	16	ND <2–60	ND <0.3–6	ND <0.3–<3	ND <0.3[e]
Bird eggs					
Falcon eggs — New Jersey	11	5–40	NA[f]	NA	NA
Variety[c] — Saginaw Bay, MI	15	20–2,000	1–10	ND <0.6–0.8	ND <1–10
Starling eggs — Crab Orchard NWR, IL	10	20–400	ND <1–10	ND <1–7	4–30
Soils, sediments, and SPMDs					
Soils — Crab Orchard NWR, IL	2	8,000–6 million	400–40,000	30–20,000	40–3,000
Sediments — New Jersey-New York City Harbor	20	700–10,000	ND <10–30	ND <10	ND <10
Semipermeable membrane devices — Saginaw Bay, MI	11	600–6,000	6–20	ND <0.5–3	ND <0.5–10

a Fish samples were white sucker, channel catfish, brown bullhead, carp, lake trout, and northern pike.

b Fish samples were white sucker, rainbow smelt, yellow perch, quillback carpsucker, walleye, gizzard shad, white bass, freshwater drum, logperch, spottail shiner, alewife, and channel catfish.

c Bird eggs of Canada goose, mallard, black-crowned night heron, ring-necked gull, double-crested cormorant, herring gull, great egret, Forster's tern, and Caspian tern.

d ND < Value 1–Value 2: Not detected at estimated detection limit (Value 1) and detected as high as estimated value (Value 2).

e ND < Value: Not detected at estimated detection limit.

f Not analyzed.

Table 7 Estimated Mean Concentrations and (% Relative Standard Deviations) of Other Non-o-Polychlorinated Biphenyls (pg/g) in QC Samples and (pg/μg Aroclor) in Aroclor-Spiked QC Samples

Quality control material — collection site	No. of replicates[a]	Non-o-polychlorinated biphenyls			
		Tri: 3,4,4'-TCB congener 37	Tetra: 3,3',4,5'-TCB congener 79	Tetra: 3,3',4,5-TCB congener 78	Penta: 3,3',4,4',5'-PeCB congener 127
Positive control materials					
Carp — Saginaw Bay, MI	11[b]	450 (38)	20 (43)	3 (65)	9 (23)
Sediment — Saginaw Bay, MI	3	24,000 (47)	100 (35)	20 (240) 1 ND (10)[c]	40 (−)
NIST SRM 1941A Sediment	3	1,200 (4)	30 (17)	10 (−) 2 ND (5)	2 ND (10) 10 (25)
Negative control materials					
Bluegill — NBS/MSC	4	20 (99)	4 (59)	— (−) 4 ND (1–5)[d]	1 ND (5) 8 (−)
Chicken eggs — local grocery	7[e]	20 (79)	2 (10) 5 ND (1–3)	— (−) 7 ND (1–3)	3 ND (1–6) 7 (87)
Pond sediments — NBS/MSC	5	60 (40)	— (−) 5 ND (5–10)	20 (160) 3 ND (5–10)	40 (10) 3 ND (5–10)
SPMD blanks — NBS/MSC	2	6 (9)	— (−) 2 ND (0.5)	— (−) 2 ND (0.5)	— (−) 2 ND (0.5)
Matrix spikes with 1:1:1 Aroclor mix[f] (pg/μg Aroclor)					
Chicken egg spikes	7	4,200 (24)	5 (48)	3 (31)	3 (62)
Bluegill spikes	4	4,300 (14)	8 (58)	6 (64)	— (−) 4 ND (1)
Sediment spikes	2	5,200 (8)	3 (10)	2 (10)	1 (−) 1 ND (1)

a Number of replicates for each congener unless otherwise stated.

b 13 Replicates for PCB 37.

c Number of replicates that were not detected at (Detection Limit).

d Number of replicates that were not detected and (Range of Detection Limits).

e 11 Replicates for PCB 37.

f Spiked with equal amounts of Aroclors 1242, 1248, 1254, and 1260 totaling 5 to 100 μg.

Selective detection of PCNs and PCTs

Interference with ^{13}C-PCBs 77 and 126 from Cl_5- and Cl_6-PCNs, respectively, can occur because both classes of compounds have some isotopic molecular ions at the same unit masses. Cl_6-PCNs can interfere with ^{13}C-PCB 126 at unit mass resolution as shown in Figure 4 by the analysis of a positive control carp (using a 30-m DB-5 column). Four selected ion chromatograms show an isotopic molecular ion (m/z 326) of PCB 126 (Figure 4A), ions m/z 332 and 334 from Cl_6-PCNs (Figure 4B,C), and m/z 336 (Figure 4D). The broadened peak at scan 1046 (13:05, m/z 336, Figure 4D), is the combined response of both ^{13}C-PCB 126 and a Cl_6-PCN. Also shown in the middle chromatograms is a Cl_6-PCN co-eluting with ^{13}C-1,2,3,4-TCDD, used as an instrumental internal standard for PCDF/PCDD analyses. Cl_6-PCNs can also interfere with ^{13}C-2,3,7,8-TCDD at unit mass resolution, but >90% of Cl_{4-8} PCNs are fractionated into the non-o-PCB fraction using HPLC-C. Alumina can be used to remove PCNs from the PCDF/PCDD fraction.

With HRMS at 10,000 R.P., the ^{13}C-PCBs and the PCNs are effectively differentiated using separate accurate masses. Using the same carp extract of Figure 4, HRMS analysis (Figure 5) shows native PCB 126 at 1100 pg/g (Figure 5A), Cl_6-PCNs (m/z 335.8229, Figure 5B), ^{13}C-PCB 126 (m/z 335.9236, Figure 5C), and finally Cl_6-PCBs, especially PCB 129 (Figure 5D). Using a 30-m DB-1 column with a slower temperature program than used for the DB-5 column in Figure 4, Cl_6-PCNs are also chromatographically resolved from ^{13}C-PCB 126. Ten Cl_6-PCNs are possible, and it is not known whether other Cl_6-PCN congeners, which were not detected, exist and might therefore co-elute with ^{13}C-PCB 126 in other samples. Figure 6 shows the PCB 81 and 77 segment of the GC/HRMS analysis (Figure 6A), Cl_5-PCNs (Figure 6B) at 0.1 Da less than ^{13}C-PCB 77 (Figure 6C), and finally residual o-Cl_5PCBs including 87 and 110 (Figure 6D). ^{13}C-PCB 77 elutes between small Cl_5-PCN peaks on the DB-1 column, but ^{13}C-PCB 81 (when available) would elute sufficiently close to another small Cl_5-PCN peak to require HRMS determinations using accurate masses.

Cl_{4-6}-PCN congener patterns detected in sediments more closely resemble the pattern in the PCN mixture Halowax 1014 or as a byproduct in PCB mixtures, e.g., Aroclor 1242,[38] than the pattern found in biota. Cl_6-PCNs detected in three matrices and Halowax 1014 are compared in Figure 7: a falcon egg from the Kola Peninsula, Russia (A), Saginaw Bay carp (B), Saginaw Bay sediment (C), and Halowax 1014 (D). The top chromatogram is representative of the pattern of Cl_6-PCNs found in bird eggs collected in the U.S. and that reported for Swedish fish and other biota.[39,40] The earliest eluting Cl_6-PCN peak, at 38:07, is reported to be a combination of 1,2,3,4,6,7- and 1,2,3,5,6,7-PCN congeners.[41] It appears as a very small peak in Halowax 1014 (Figure 7D), but is much more persistent in biota. Dioxin equivalency of the congener pair is 0.002;[32] other PCNs were also found to be AHH-active, but much less persistent. Williams et al.[42] found the Cl_6-PCN congener pair in human adipose tissue, but were unable to resolve the pair using any of eight different GC phases.

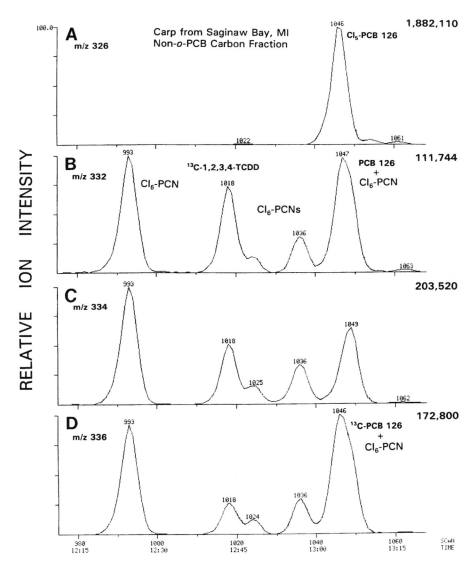

Figure 4 Co-elution of ^{13}C-PCB 126 and Cl_6-PCN at scan 1046 (D) shown in GC/MS (quadrupole) selected ion chromatogram (12:10–13:20 segment) of m/z 336 using a 30 m DB-5 column and the non-*o*-PCB HPLC-C eluate from the Saginaw Bay carp QA sample. Other chromatograms of m/z 326 show native Cl_5-PCB 126, and m/z 332 and 334 show both Cl_6-PCNs and ^{13}C-TCDD.

Chlorinated terphenyls (PCTs) are only 0.07 Da higher than ^{13}C-PCBs, but they can be selectively detected from both ^{13}C-PCBs and PCNs by HRMS above 5000 R.P. PCTs elute later and do not interfere with any targeted analytes (Figure 8). This figure shows a 16-min portion of the analysis of the Saginaw Bay positive control sediment. The two most abundant Cl_6-PCN ions (Figure 8A, C) are mass resolved from tenfold higher amounts of Cl_3-

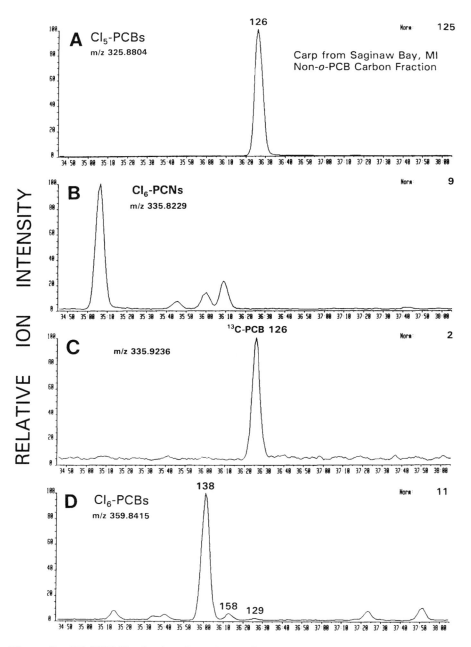

Figure 5   GC/HRMS selective detection and DB-1 column resolution of Cl$_6$-PCNs (B) from ^{13}C-PCB 126 (C) in analysis (34:45–38:10 segment) of the same Saginaw Bay carp QA sample eluate as in Figure 4. Ion chromatogram for native PCB 126 (A) has minimal (< 1%) co-eluting interference from PCB 129 chromatogram (D).

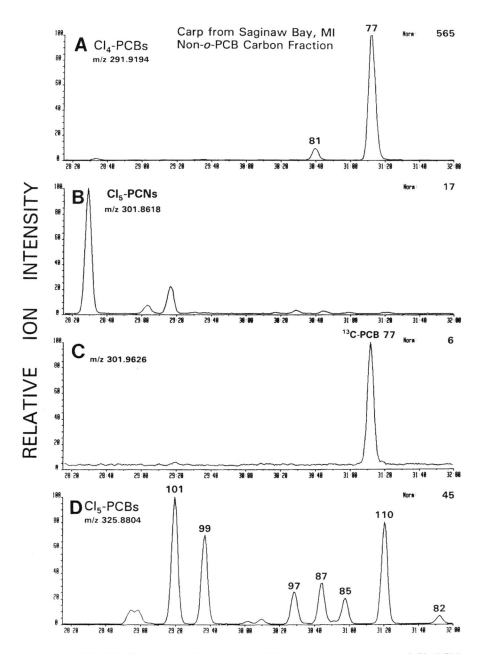

Figure 6 GC/HRMS selective detection and DB-1 column resolution of Cl_5-PCNs (B) from ^{13}C-PCB 77 (C) in analysis (28:15–32:00 segment) of the same Saginaw Bay carp QA sample eluate as in Figure 4. Ion chromatogram for native PCBs 81 and 77 (A) shows no interference from closely-eluting but resolved PCBs 87 and 110, respectively (D).

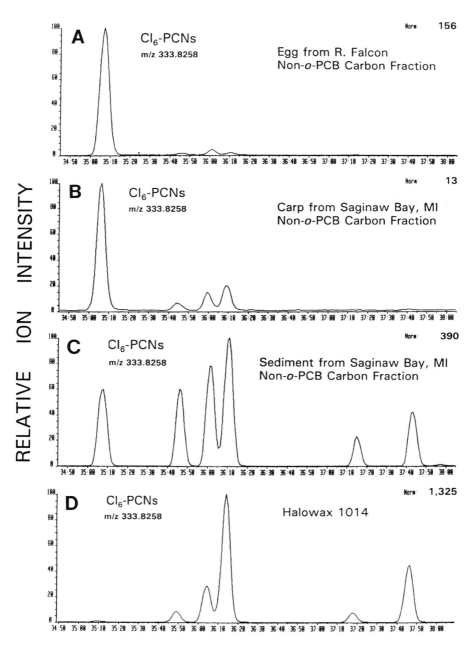

Figure 7 Fate of Cl_6-PCN congeners shown comparing GC/HRMS analysis (34:45–38:00 segment) of Halowax 1014 mixture (D), and three environmental matrices: QA sediment sample from Saginaw Bay, MI (C), QA carp from Saginaw Bay, MI (B), and an egg from a falcon collected in Russia (A).

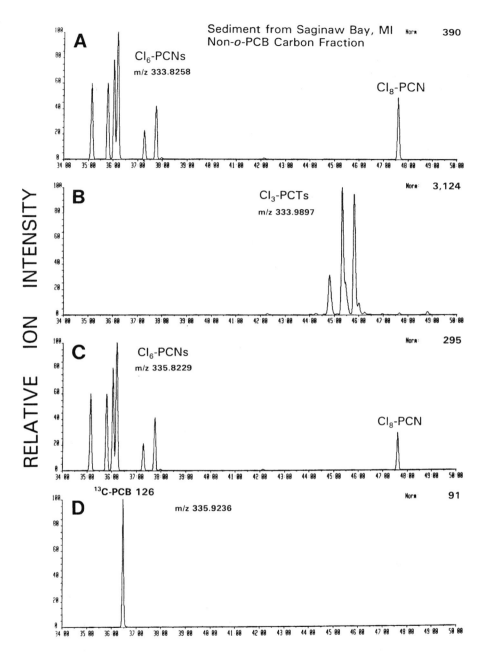

Figure 8 GC/HRMS selective detection of three groups of chlorinated aromatic compounds with similar unit mass ions but sufficiently different accurate ions in the non-*o*-PCB eluate from HPLC-C of QA sediment sample from Saginaw Bay, MI. Selected accurate ion chromatograms (34–50 min segment) of Cl$_6$-PCNs and fragment ion from Cl$_8$-PCN (A,C), higher amount of Cl$_3$-PCTs (B), and ^{13}C-PCB 126 (D).

PCTs (Figure 8B) and from ^{13}C-PCB 126 (Figure 8D). (Another ion for Cl$_3$-PCTs could be measured at m/z 335.9867 to also show its resolution from the other compounds.) Octachloronaphthalene at 47:35 was also detected using accurate mass ions for Cl$_6$-PCNs because its fragment ions from the loss of two chlorines were sufficiently close in mass to be detected.

Summary

This method, combining reactive adsorbent clean-up, HPLC-C fractionation, selective high resolution GC, and high resolution MS, was used to determine AHH-active non-o-PCBs without significant interferences in many different environmental matrices at concentrations ranging from <1 ppt to >10 ppb. HPLC-C removed greater than 99.5% of potentially interfering o-PCBs, high resolution capillary gas chromatography with a methyl silicone column resolved all non-o-PCBs from o-PCBs except PCB 126 from 129. From a single low resolution mass, HRMS selectively detected accurate mass ions of ^{13}C-PCBs, PCNs, and PCTs separated by <0.1 Da. Quantitation of analytes with weighted quadratic regression curve fitting of response data provided calibration over four orders of magnitude. Four other non-o-PCBs were detected using the method; with the exception of Cl$_3$-PCB 37 in abiotic samples, they were found at low concentrations. Cl$_{4-8}$ PCNs were also recovered in the non-o-PCB fraction. PCN patterns in biota differ significantly from that in Halowax 1014 or in PCB mixtures as by-products. Persistent AHH-active PCNs were found in biota.

Acknowledgments

We thank Carl Orazio, Kathy Echols, and Colleen Rostad for manuscript review, Ted Schwartz, John Meadows, and Dennis Schroeder for technical assistance.

References

1. Jensen, S., 1972. The PCB Story. Ambio 1(4):123–131.
2. Erickson, M.D., 1986. Analytical Chemistry of PCBs. Boston: Butterworth Publishers. 508 pp.
3. Mackay, D., Paterson, S., Eisenreich, S.J., Simmons, M.S., 1983. Physical Behavior of PCBs in the Great Lakes. Ann Arbor, MI: Ann Arbor Science Publishers. 442 pp.
4. Waid, J.S., 1986. PCBs and the Environment. Vol. I–III. Boca Raton, FL: CRC Press, 691 pp.
5. Larsen, B., Bøwadt, S., Tilio, R., 1992. Congener Specific Analysis of 140 Chlorobiphenyls in Technical Mixtures on Five Narrow-Bore GC Columns. Int. J. Environ. Anal. Chem. 47:47–68.
6. Brown, J.F. Jr., 1994. Determination of PCB Metabolic, Excretion, and Accumulation Rates for Use as Indicators of Biological Response and Relative Risk. Environ. Sci. Tech. 28:2295–2305.

7. Mullin, M.D., Pochini, C.M., McCrindle, S., Romkes, M., Safe, S.H., and Safe, L.M., 1984. High Resolution PCB Analysis: Synthesis and Chromatographic Properties of All 209 PCB Congeners. *Environ. Sci. Tech.* 18:468–476.

8. Bøwadt, S. and Larsen, B. 1992. Improved Congener-Specific GC Analysis of Chlorobiphenyls on Coupled CPSil-8 and HT-5 Columns. *J. High Res. Chromatog.* 15: 377–380.

9. Schulz, D.E., Petrick, G., Duinker, J.C., 1989. Complete Characterization of Polychlorinated Biphenyl Congeners in Commercial Aroclor and Clophen Mixtures by Multidimensional Gas Chromatography-Electron Capture Detection. *Environ. Sci. Tech.* 23:852–859.

10. Gebhart, J.E., Hayes, T.L., Alford-Stevens, A.L., Budde, W.L., 1985. Mass Spectrometric Determination of Polychlorinated Biphenyls as Isomer Groups. *Anal. Chem.* 57:2458–2463.

11. Onuska, F.I., Terry K.A., 1986. Characterization and Determination of PCB Isomers by High Resolution Gas Chromatography and HRGC/Mass Spectrometry. *J. High Res. Chrom. Commun.* 9:671–675.

12. Safe, S., 1990. Polychlorinated Biphenyls (PCBs), Dibenzo-p-dioxins (PCDDs), Dibenzofurans (PCDFs), and Related Compounds: Environmental and Mechanistic Considerations Which Support the Development of Toxic Equivalency Factors (TEFs). *Crit. Rev. Toxicol.* 21(1):51–88.

13. Kubiak, T.J., Harris, H.J., Smith, L.M., Schwartz, T.R., Stalling, D.L., Trick, J.A., Sileo, L., Docherty, D., Erdman, T.C., 1989. Microcontaminants and Reproductive Impairment of the Forster's Tern in Green Bay, Lake Michigan-1983. *Arch. Environ. Contam. Toxicol.* 18:706–727.

14. Tillitt, D.E., Ankley, G.T., Giesy, J.P., Ludwig, J.P., Kurita-Matsuba, H., Weseloh, D.V., Ross, P.S., Bishop, C.H., Sileo, L., Stromberg, K.L., Larson, J., Kubiak, T.J., 1992. Polychlorinated Biphenyl Residues and Egg Mortality in Double-Crested Cormorants from the Great Lakes. *Environ. Toxicol. Chem.* 11: 1281–1288.

15. Giesy, J.P., Ludwig, J.P., Tillitt, D.E., 1994. Deformities in Birds of the Great Lakes Region. *Environ. Sci. Tech.* 28:128A–135A.

16. Ballschmiter, K. and Zell, M. 1980. Analysis of Polychlorinated Biphenyls (PCB) by Glass Capillary Gas Chromatography. *Fresenius Z. Anal. Chem.* 302: 20–31.

17. Ahlborg, U.G., Becking G.C., Birnbaum, L.S., Brouwer, A., Derks, H.J.G.M., Feeley, M., Golor, G., Hanberg, A., Larsen, J.C., Liem, A.K.D., Safe, S.H., Schlatter, C., Waern, F., Younes, M., Yrjänheikki, E., 1994. Toxic Equivalency Factors for Dioxin-like PCBs. *Chemosphere* 28:1049–1067.

18. Harris, G.E., Kiparissis, Y., Metcalfe, C.D., 1994. Assessment of the Toxic Potential of PCB Congener 81 (3,4,4′,5-Tetrachlorobiphenyl) to Fish in Relation to Other Non-ortho-substituted PCB Congeners. *Environ. Toxicol. Chem.* 13(9):1405–1413.

19. De Voogt, P., Haglund, P., Reutergårdh, L.B., De Wit C., Waern, F., 1994. Fishing for Quality in Environmental Analysis. Interlaboratory Study on Non- and Mono-Ortho Chlorinated Biphenyls. *Anal. Chem.* 66(5):305A–311A.

20. Schwartz, T.R., Tillitt, D.E., Feltz, K.P., Peterman, P.H., 1993. Determination of Mono- and Non-*o,o′*-Chlorine Substituted Polychlorinated Biphenyls in Aroclors and Environmental Samples. *Chemosphere* 26(8):1443–1460.

21. Creaser, C.S., Krokos, F., Startin, J.R., 1992. Analytical Methods for the Determination of Non-ortho Substituted Chlorobiphenyls: A Review. *Chemosphere* 25(12):1981–2008.

22. Feltz, K.P., Tillitt, D.E., Gale, R.W., Peterman, P.H., 1995. Automated HPLC Fractionation of PCDDs and PCDFs, and Planar and Non-planar PCBs on C_{18}-Dispersed PX-21 Carbon. *Environ. Sci. Tech.* 29:709–718.

23. Stalling, D.L., Huckins, J.N., Petty, J.D., Johnson, J.L., Sanders, H.O., 1979. An Expanded Approach to the Study and Measurement of PCBs and Selected Planar Halogenated Aromatic Environmental Pollutants. *Ann. N.Y. Acad. Sci.* 320:48–59.

24. Kuehl, D.W., Butterworth, B.C., Libal, J., Marquis, P., 1991. An Isotope Dulution High Resolution Gas Chromatographic-High Resolution Mass Spectrometric Method for the Determination of Coplanar Polychlorinated Biphenyls: Application to Fish and Marine Mammals. *Chemosphere* 22(9–10): 849–858.

25. Ford, C.A., Muir, D.C.G., Norstom, R.J., Simon, M., Mulvihill, M.J., 1993. Development of a Semi-Automated Method for Non-ortho PCBs: Application to Canadian Arctic Marine Mammal Tissues. *Chemosphere* 26(11):1981–1991.

26. Tanabe, S., Kannan, N., Wakimoto, T., Tatsukawa, R. 1987. Method for the Determination of Three Toxic Non-orthochlorine Substituted Coplanar PCBs in Environmental Samples at Part-per-trillion Levels. *Int. J. Environ. Anal. Chem.* 29:199–213.

27. De Boer, J., Stronck, C.J.N., Van der Valk, F., Wester, P.G., Daudt, M.J.M., 1992. Method for the Analysis of Non-ortho Substituted Chlorobiphenyls in Fish and Marine Mammals. *Chemosphere* 25(7–10):1277–1283.

28. Storr-Hansen, E., Cederberg, T., 1992. Determination of Coplanar Polychlorinated Biphenyl (CB) Congeners in Seal Tissues by Chromatography on Active Carbon, Dual-Column High Resolution GC/ECD and High Resolution GC/High Resolution MS. *Chemosphere* 24(9):1181–1196.

29. Haglund, P., Asplund, L., Järnberg, U., Jansson, B., 1990. Isolation of Mono- and Non-ortho-Polychlorinated Biphenyls from Biological Samples by Electron-Donor Acceptor High Performance Liquid Chromatography Using a 2-(1-Pyrenyl)ethyldimethylsilylated Silica Column. *Chemosphere* 20(7–9):887–894.

30. Mullin, M.D., 1985. Congener Specific, PCB Analysis Techniques Workshop (Chairman), EPA Large Lakes Research Station, Gross Ile, MI, June 11–14. 38 pp.

31. Larsen, B., Bøwadt, S., Tilio, R., Facchetti, S., 1992. Choice of GC Column Phase for Analysis of Toxic PCBs. *Chemosphere* 25(7–10):1343–1348.

32. Hanberg, A., Wærn, F., Asplund, L., Haglund, E., Safe, S., 1990. Swedish Dioxin Survey: Determination of 2,3,7,8-TCDD Toxic Equivalent Factors for Some Polychlorinated Biphenyls and Naphthalenes Using Biological Tests. *Chemosphere* 20: 1161–1164.

33. MacLeod, K.E., 1979. Sources of Emissions of Polychlorinated Biphenyls into the Ambient Atmosphere and Indoor Air. U.S. EPA Report No. 600/4-79-022 as cited in Reference 2.

34. Green, B.N., Gray, B.W., Guyan, S.A., Krolik, S.T., 1986. An EI Source Optimised for Low Level Pollutant Analysis by GC-HRSIR. Presented at the 34th Annual Conference on Mass Spectrometry and Allied Topics, Cincinnati, OH, June 8–13.

35. Huckins, J.N., Manuweera, G.K., Petty, J.D., Mackay, D., Lebo, J.A., 1993. Lipid-Containing Semipermeable Membrane Devices for Monitoring Organic Contaminants in Water. *Environ. Sci. Technol.* 27:2489–2496.

36. Huckins, J.N., Petty, J.D., Lebo, J.A., Orazio, C.E., Prest, H.F., Tillitt, D.E., Ellis, G.S., Johnson, B.T., Manuweera, G.K. 1995. Semipermeable Membrane Devices (SPMDs) for the Concentration and Assessment of Bioavailable Organic Contaminants in Aquatic Environments. In *Techniques in Aquatic Toxicology*, Ed. G.K. Ostrander. Boca Raton, FL: CRC/Lewis Publishers. 1996, Chapter 34.

37. Schmitt, C.J., Zajicek, J.L., Peterman, P.H., 1990. National Contaminant Biomonitoring Program: Residues of Organochlorine Chemicals in U.S. Freshwater Fish, 1976–1984. *Arch. Environ. Contam. Toxicol.* 19:748–781.

38. Haglund, P., Jakobsson, E., Asplund, L., Athanasiadou, M., Bergman, Å., 1993. Determination of Polychlorinated Naphthalenes in Polychlorinated Biphenyl Products Via Capillary Gas Chromatography-Mass Spectrometry after Separation by Gel Permeation Chromatography. *J. Chromatogr.* 634:79–86.

39. Jansson, B., Asplund, L., Olsson, M., 1984. Analysis of Polychlorinated Naphthalenes in Environmental Samples. *Chemosphere* 13(1): 33–41.

40. Järnberg, U., Asplund, L., de Wit, C., Grafström, A.-K., Haglund, P., Jansson, B., Lexén, K., Strandell, M., Olsson, M., Jonsson, B., 1993. Polychlorinated Biphenyls and Polychlorinated Naphthalenes in Swedish Sediment and Biota: Levels, Patterns, and Time Trends. *Environ. Sci. Technol.* 27:1364–1374.

41. Jakobsson, E., Eriksson, L., Bergman, Å., 1992. Synthesis and Crystallography of 123467-Hexachloronaphthalene and 123567-Hexachloronaphthalene. *Acta Chem. Scand.* 46: 527–532.

42. Williams, D.T., Kennedy, B., LeBel, G.L., 1993. Chlorinated Naphthalenes in Human Adipose Tissue from Ontario Municipalities. *Chemosphere* 27(5): 795–806.

chapter twenty-eight

Analysis of tributyltin and congeners in fish and shrimp tissue

J. M. O'Neal, W. H. Benson, and J. C. Allgood

Introduction

This is a method for the extraction, derivatization, and internal standard gas chromatographic (GC) quantitation of mono-, di-, tri-, and tetrabutyltin in fish and shrimp tissue. The analytes are triple extracted from the tissue using hexane containing 0.20% tropolone and sonication. The extract is dried over anhydrous sodium sulfate and then reduced to a 3 to 5 ml volume. The extract is quantitatively transferred to a 125-ml bottle and derivitized using *n*-pentylmagnesium bromide. Following derivitization the extract is reduced in volume to 1 to 2 ml and cleaned up using a Florisil®-packed liquid chromatographic column. After elution of the derivitized analytes has been accomplished using hexane, the volume is finally reduced to a volume of 1.0 ml for gas chromatographic analysis. The gas chromatograph is equipped with a flame photometric detector (FPD) using a 610-nm filter. The method of internal standards is used for quantitation. The internal standards are tetrapropyl tin and tripropyl tin. Detection limits for mono-, di-, tri-, and tetrabutyl tin are 7.0, 3.5, 2.0, and 1.0 ug/kg dry weight as determined from the Code of Federal Regulations 40 CFR Part 136. The percentage recovery for all analytes is nearly quantitative. This method permits the high precision and accurate determination of the butyltin analytes which are of importance to aquatic toxicology due to their known toxicological effects.[1]

Materials required

Capillary GC with autosampler and flame photometric detector — H-P 5890 Series II

Bandpass interference filter, 610 nm — Ealing Electro-Optics, 35-3847

Data station and GC software — H-P — Vectra 386/25 and Chemstation software

Capillary GC column, DB-5 (registered trademark by J&W Scientific), 0.25-mm id, 0.25-µ film thickness, 30-m — Baxter, C4587-69

Sonicator, 600 watt — Tekmar Model TSD-600, 10-0458-000

Centrifuge — IEC Clinical Centrifuge with 4 place rotor (4 × 50 ml), trunions, and shields; centrifuge — Fisher 05-101-5; rotor — Fisher 05-103B; trunions — Fisher 05-207; shields — Fisher 05-180

Nitrogen evaporator, 24 position, N-EVAP, Organomation Associates, Inc., 11250-RT

Column with reservoir, 200 × 11 mm OD, 280 mm overall — Supelco, 6-4747

Centrifuge tubes, 40 ml — Fisher, 05-556

Glass square wide-mouth bottle w/TFE-lined closure, 4 oz — Fisher, 03-326-4D

Beakers, 150 ml — Fisher, 02-539J

Flasks, 125 ml — Fisher, 10-040D

Florisil (registered trademark of Floridin Co.), pesticide grade, 60/100 mesh — Supelco, 2-0280

Glass wool, silane treated — Supelco, 2-0411

Concentrator tubes, 25 ml — Fisher, K570050-2525

Tropolone, 98% purity — Aldrich, T8,970-2

Hexane, pesticide grade — Fisher, H303-4

Hydrochloric acid, Baker Instra-Analyzed — Baxter, 9530-33 BC

Sulfuric acid, Baker Instra-Analyzed — Baxter, 9673-33 BC

Pentylmagnesium bromide, 1.5–2.5 M in ether — Alfa, 87296

Tetra-n-propyltin — Alfa, 71137

Tri-n-propyltin chloride, 99% — Alfa, 71122

n-Butyltin trichloride, 95% — Alfa, 71125

Di-n-butyltin dichloride, 97% — Alfa, 14116

Tri-n-butyltin chloride, 95% — Alfa, 14119

Tetra-n-butyltin, 96% — Alfa, 14115

Sodium sulfate, anhydrous, Certified ACS, tested for Pesticide Residue Analysis, 10–60 mesh — Fisher, S415-500

Procedures

Standards preparation

Following this procedure will result in the detection and quantitation of mono-, di-, tri-, and tetrabutyl tin compounds at the method detection limits of 7.0, 3.5, 2.0, and 1.0 µg/kg , dry weight basis. This method will perform well for fish tissue and shrimp. Over 1000 samples of fish and shrimp tissue have been analyzed by our laboratory using this method. The fish included gafftopsail catfish, hard head catfish, blue catfish, Atlantic croaker, and grouper. The shrimp included brown shrimp and white shrimp. It is strongly suggested that standards and spiking solutions be prepared first before beginning sample workup.

The internal standards of tetra-*n*-propyltin and tri-*n*-propyltin, the four *n*-butyltin standards, and the spiking solutions of the four *n*-butyltin analytes used with QA/QC and laboratory control material (LCM) samples all individually have a concentration of 60 ng of tin/ml in the final 1 ml extract to be analyzed by GC. Note that tri-*n*-propyltin, mono-*n*-butyltin, di-*n*-butyltin, and tri-*n*-butyltin must be derivitized with the Grignard reagent pentylmagnesium bromide. Tetra-*n*-propyltin and tetra-*n*-butyltin do not react with the Grignard reagent. For efficiency and convenience it is suggested that mono-, di-, and tributyltin standards be derivitized together for use as one of the standards and part of the spiking solution for QA/QC samples. The tri-*n*-propyltin internal standard is derivitized separately. In summary, before beginning the analysis of samples, a total of four standard solutions should be available: (1) a standard of tetra-*n*-propyltin, (2) a standard of tetra-*n*-butyltin, (3) a derivitized standard of tri-*n*-propyltin, and (4) a mixed derivitized standard containing mono-, di-, and tri-*n*-butyltins. As an aid in preparing the derivitized standards a separate solution of each standard to be derivitized is prepared in hexane. To prepare the mixed mono-, di-, tri-*n*-butyl standard a volume of each individual standard containing 60 μg of tin of that standard is taken, placed in a 4-oz bottle, and mixed. Add 2.0 ml of approximately 2.0 *M* pentylmagnesium bromide in ethyl ether to the mixed butyltin solution. Allow this solution to react for 15 min. Shake every 5 min. Place the bottle in an ice bath. Add 15 ml of deionized water to the solution and swirl. Slowly add 10 ml of 5 *M* sulfuric acid to the solution. Swirl during the addition. After the addition of the acid allow the reaction solution to rest for 15 min. Remove the hexane layer from this solution and place it in a 125-ml Erlenmeyer flask. Add 5 ml of hexane to the reaction solution and swirl. Remove the hexane layer and combine it in the Erlenmeyer flask with the first hexane layer. Prepare a 200 × 11 mm OD liquid chromatography column by placing a small plug of silane-treated glass wool in the botton of the column. Pack the column with 16 g of Florisil (activated by heating at 130°C for 24 h) and top the Florisil with 4 g of anhydrous sodium sulfate. Rinse the column with 50 ml of hexane and discard the rinse. Quantitatively transfer the hexane solution of derivitized mono-, di-, tri-*n*-butyltin compounds from the Erlenmeyer flask to the chromatography column. Use a 100-ml volumetric flask to collect the eluant from the column. As the hexane solution containing the derivitized compounds reaches the top of the sodium sulfate layer add approximately 40 to 50 ml of hexane. After elution is complete use hexane to bring the volume of the 100 ml flask up to the calibration mark. The mixed standard is now prepared. Use 100 μl of this solution to spike samples, or dilute 100 μl of this standard to a volume of 1.0 ml to achieve a concentration of 60 ng of tin/ml of each analyte in the mixture. The tri-*n*-propyltin standard is derivitized in the same manner. Tri-*n*-propyltin and tetra-*n*-propyltin were chosen as internal standards for the following reasons: (1) to permit monitoring of extraction efficiency (tetra-*n*-propyltin), (2) to permit monitoring of derivitization efficiency and extraction efficiency (tri-*n*-propyltin), and (3) to permit a choice of internal standards for quantitation purposes.

Sample workup

In the procedure below all solution transfers are made by quantitative techniques. Approximately 2 g of fish or shrimp tissue (wet weight) are weighed and transferred to a 40-ml centrifuge tube. The internal standards of tetra-*n*-propyltin and tri-*n*-propyltin (dervitized) are added to the centrifuge tube at a level of 60 ng of tin for each internal standard. Add 1.0 ml of concentrated hydrochloric acid. Add 30 ml of hexane which is 0.20% in tropolone. Sonicate for 90 sec using the pulse mode set at 80%. Following sonication, centrifuge for 4 min. Remove the hexane extract and transfer to a 150-ml beaker containing 10 to 15 g of anhydrous sodium sulfate. Add another 30 ml of hexane/tropolone and repeat the procedure, adding the hexane extract to the 150 ml beaker. Repeat the hexane extraction procedure once more. This results in a total of 90 ml of hexane extract in the 150 ml beaker. Allow the hexane extract to dry over the sodium sulfate for 15 min. Occasionally swirl the mixture during the 15 min drying time. Transfer the extract to a 125-ml Erylenmeyer flask. Rinse the beaker and sodium sulfate a minimum of five times with hexane and add to the flask. Concentrate the sample to 3 to 5 ml using the N-Evap with the bath temperature set at 45°C. Transfer the extract to a 4-oz bottle using a minimum of 5 hexane rinses. Add 2.0 ml of approximately 2.0 M pentylmagnesium bromide in ethyl ether. Allow to react for 15 min. Shake every 5 min. Place the bottle in an ice bath. Add 15 ml of deionized water and swirl. Slowly add 10 ml of 5.0 M sulfuric acid. Swirl during the addition and allow to rest for 15 min. Transfer the hexane layer to a 125-ml Erlenmeyer flask and reduce the volume to 1 to 2 ml using the N-Evap. Pack a 200 × 11 mm liquid chromatography column with 16 g of Florisil (activated at 130°C for 24 h) and top with 4 g of anhydrous sodium sulfate. Rinse the column with 50-ml of hexane and discard the rinse. Transfer the extract to the column. Elute the column with 100 ml of hexane and collect in a 125-ml Erlenmeyer flask. Concentrate the extract to 5 to 10 ml using the N-Evap with the bath temperature set at 45°C. Transfer the extract to a 25 ml concentrator tube using a minimum of 5 hexane rinses. Concentrate to 1.0 ml. Transfer the sample to a GC autosampler vial, store under nitrogen, and store in a refrigerator until analyzed.

Gas chromatography analysis

Splitless injection is used. A sample volume of 1.0 μl is injected by the autosampler. Ultra high purity helium is used as the carrier gas and the carrier flow is 1.2 ml/min. Nitrogen is used as the flame photometric detector makeup gas. The flow rate is 65 ml/min. The capillary GC column being used is a J&W 30 m 0.25 mm ID 0.25 μm film thickness DB-5 column. The injector and detector temperatures are both 250°C. The following temperature program is used: initial temperature of 75°C and hold for 1 min; ramp at 30°C/min to 150°C with no hold time; ramp at 10°C/min to 210°C and hold for 6 min. This results in a GC run time of 15.5 min. The chromatographic data were collected by the data station and quantitative analysis reports were generated.

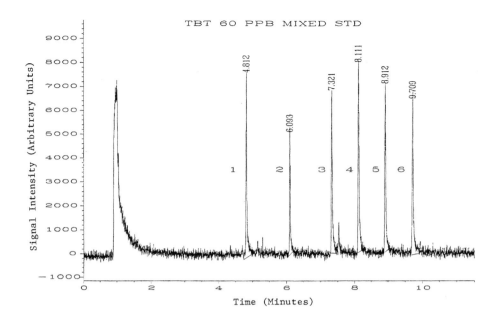

Figure 1   GC chromatogram of standards. The identities of the analytes are 1-tetra-*n*-propyltin (ISTD), 2-tri-*n*-propyltin (ISTD), 3-tetra-*n*-butyltin, 4-tri-*n*-butyltin, 5-di-*n*-butyltin, and 6-mono-*n*-butyltin.

Results and discussion

This method produces an excellent extraction, cleanup, GC separation, and quantitation for the *n*-butyltin analytes. The separation of the analytes is baseline. A typical chromatogram for a 60-ppb standard for each analyte and internal standard is shown in Figure 1. A typical spiked sample of fish tissue produces a similar chromatogram. Based upon our experience of using this method for the analysis of over 1000 samples we find the method to be very reliable and accurate and the percent recovery is nearly quantitative (>95%). The percent recovery is based upon the recovery of the analytes from the Laboratory Control Material which is spiked grouper. While the preparation of the derivitized standards may seem laborious we have found that the standards thus prepared are stable for a period of at least 3 months provided that they are kept sealed and refrigerated when not in use. The use of the flame photometric detector with the 610-nm filter eliminates many of the interferences that may be encountered with other filters or detectors. It is rather specific for tin-containing compounds.

References

1. Huggett, R.J., Unger, M.A., Seligman, P.F., Valkirs, A.O. 1992. The marine biocide tributyltin. *Environ. Sci. Technol.*, Vol. 26, No. 2.

chapter twenty-nine

Analysis of organochlorine pesticides and PCBs in fish and shrimp tissue

J. M. O'Neal, W. H. Benson, and J. C. Allgood

Introduction

This is a method for the extraction and internal standard gas chromatographic (GC) quantitation of 38 organochlorine pesticides and PCBs (DDT and its metabolites, 11 other selected organochlorine pesticides, and 21 individual PCB congeners) from fish tissue. The method is unique in that it avoids fractionation of the analytes in the extraction/cleanup steps as would be encountered in traditional methods, and separation and quantitation are achieved in a single gas chromatographic run. The analytes are triple extracted from the fish tissue using acetonitrile and sonication. The combined extract is added to de-ionized water and triple extracted with pentane. The pentane extract is reduced in volume and passed through a cleanup chromatographic column packed with Florisil. This cleanup procedure effectively removes biogenic material from the extract. The analytes are eluted from the column using a mixture of ethyl ether/hexane. The eluant is reduced to a volume of 1 ml and it is now ready for capillary gas chromatographic analysis using electron capture detectors (ECD). The method detection limits for the analytes are in the 1 to 2 ppb range as determined by the procedures outlined in the Code of Federal Regulations 40 CFR Part 136. The average percentage recovery for all analytes is 95.5%. This method permits the high precision and accurate determination of 38 analytes which are of importance to aquatic toxicology due to their known toxicological effects.

Materials required

Capillary GC with autosampler and electron capture detectors — Hewlett-Packard (H-P) 5890 Series II

Data station and GC software — H-P, Vectra 386/25 and Chemstation
  software
Capillary GC column, DB-5 (registered trademark by J&W Scientific),
  0.25-mm ID, 0.25 µm film thickness, 60-m — Baxter, C4587-71
Sonicator, 600 watt — Tekmar Model TSD-600, 10-0458-000
Centrifuge — IEC Clinical Centrifuge with 4 place rotor (4 × 50 ml),
  trunions, and shields; centrifuge — Fisher 05-101-5; rotor — Fisher
  05-103B; trunions — Fisher 05-207; shields — Fisher 05-180
Nitrogen evaporator, 24-position, N-EVAP, Organomation Associates,
  Inc., 11250-RT
Column with reservoir, 200 × 11 mm OD, 280 mm overall — Supelco, 6-
  4747
Centrifuge tubes, 40 ml — Fisher, 05-556
Separatory funnel with PTFE stopcock, 250 ml — Fisher, 10-437-10C
Beakers, 150 ml — Fisher, 02-539J
Flasks, 125 ml — Fisher, 10-040D
Florisil (registered trademark of Floridin Co.), pesticide grade, 60/100
  mesh — Supelco, 2-0280
Glass wool, silane treated — Supelco, 2-0411
Concentrator tubes, 25 ml — Fisher, K570050-2525
Acetonitrile, pesticide grade — Fisher, A999-4
Ethyl ether for organic residue analysis — J. T. Baker, 9259-02
Pentane, pesticide grade — Fisher, P400-4
Hexane, pesticide grade — Fisher, H303-4
Chlorinated pesticides in hexane — NIST, SRM 2261
Chlorinated biphenyls in iso-octane — NIST, SRM 2262
Endrin, 100 µg/ml — AccuStandard, Inc., P-045S
Endosulfan , 100 µg/ml — AccuStandard, P-091S
PCB 77, 35 µg/ml — AccuStandard, C-077S
PCB 99, 35 µg/ml — AccuStandard, C-099S
PCB 103, 35 µg/ml — AccuStandard, C-103S
PCB 126, 35 µg/ml — AccuStandard, C-126S
PCB 198, 35 µg/ml — AccuStandard, C-198S
Sodium sulfate, anhydrous, Certified ACS, tested for Pesticide Residue
  Analysis, 10–60 mesh — Fisher, S415-500

Procedures

Sample workup

Following this procedure will result in the detection and quantitation of 38
analytes, if present, which include DDT and its metabolites, 11 other selected
organochlorine pesticides, and 21 individual PCB congeners at the 1 to 2
ppb detection level. This method will perform well for fish tissue and shrimp.
Over 1000 samples of fish tissue and shrimp have been analyzed using this
method. The fish included gafftopsail catfish, hard head catfish, blue catfish,
Atlantic croaker, and grouper. The shrimp included white and brown shrimp.
The cleanup portion of this method does not perform well for oysters. It is

Table 1 Analytes and GC Elution Order

Compounds

1	C-8	21	O,P'-DDD
2	HCB	22	C-77
3	Lindane	23	Endrin
4	C-18	24	C-118
5	Endosulfan I	25	O,P'-DDT
6	C-28	26	P,P'-DDD
7	Heptachlor	27	C-153
8	C-52	28	C-105
9	Aldrin	29	P,P'-DDT
10	C-44	30	C-138
11	C-103 (ISTD)	31	C-126
12	Heptachlor epoxide	32	C-187
13	C-66	33	C-128
14	O,P'-DDE	34	C-180
15	Alpha-Chlordane	35	Mirex
16	C-101	36	C-170
17	trans-Nonachlor	37	C-198 (ISTD)
18	C-99	38	C-195
19	Dieldrin	39	C-206
20	P,P'-DDE	40	C-209

strongly suggested that standards and spiking solutions be prepared first before beginning sample workup. Standards and spiking solutions have the following concentrations in the final 1 ml extract to be analyzed by GC: DDT and its metabolites and organochlorine pesticides except for endrin and endosulfan, 49.3 ng/ml; endrin and endosulfan, 50 ng/ml; and PCB congeners, 52.5 ng/ml. Take note that PCBs 103 and 198 are used as internal standards (ISTDs) for GC quantitation. The complete list of the 38 analytes and the two internal standards (PCBs 103 and 198) are listed in Table 1. Please note that each PCB congener is identified in Table 1 by its Ballschmiter and Zell (BZ) number and is preceded by "C" (for example, PCB congener 103 is listed as C-103). In the procedure below all solution transfers are made by quantitative techniques. Approximately 10 g of fish or shrimp tissue (wet weight) are weighed and transferred to a 40-ml centrifuge tube. The internal standards of PCB 103 and PCB 198 were each added to the centrifuge tube at a level of 52.5 ng. Ten ml of acetonitrile were added to the centrifuge tube and the mixture was sonicated for 90 sec using an 80% duty cycle. Following sonication the mixture was centrifuged for 4 min. After the centrifugation was complete the acetonitrile extract was removed from the centrifuge tube and reserved. The acetonitrile extraction procedure was repeated twice more which resulted in a total of 30 ml of extract. The 30 ml of acetonitrile extract were transferred to a 250-ml separatory funnel which contained 70 ml of deionized water which had been saturated with pentane and contained 2%

sodium sulfate. This solution was extracted with 10 ml of pentane and the pentane layer was removed and placed in a 150-ml beaker which contained 10 g of anhydrous sodium sulfate for drying purposes. The pentane extraction was repeated twice more which resulted in a total of 30 ml of pentane extract. After completion of the drying step (about 5 min with swirling) the pentane extract was transferred to a 125-ml flask and the volume was reduced to approximately 3 to 5 ml by nitrogen evaporation using the N-EVAP. The 200 × 11 mm OD liquid chromatography column was prepared by first placing a small plug of glass wool in the end of the column. A total of 3.5 g of Florisil, which had been activated by heating at 130°C for 24 h, was then added. The Florisil was then topped by 1.5 g of anhydrous sodium sulfate. A volume of 20 ml of hexane was added to the column for rinse purposes. The hexane rinse eluting from the column was discarded. As the last of the 20 ml of hexane reached the top of the sodium sulfate layer at the top of the column the pentane extract was added to the column and collection of the eluant began. The eluant was collected in the same 125 ml flask that originally contained the 3 to 5 ml pentane extract. As the pentane extract reached the top of the sodium sulfate layer, 40 ml of a hexane solution containing 10% ethyl ether was added to elute the analytes from the column. The eluant was reduced to a volume of approximately 10 ml using the N-EVAP.

The remaining 10 ml of eluant was transferred to a concentrator tube and the volume of the eluant was reduced to one ml using the N-EVAP. In practice the volume was reduced to about 0.8 ml and the walls of the concentrator tube were rinsed to bring the volume to 1.2 ml. This procedure was repeated twice to ensure that very little, if any, analytes remained on the exposed walls of the concentrator tube. In the final repetition of this procedure the volume was reduced to 1.0 ml. This final 1.0 ml extract was placed in a GC autosampler vial and then refrigerated until analysis by gas chromatography.

Gas chromatography analysis

Splitless injection was used. A sample volume of 1.0 µl was injected by the autosampler. Ultra high purity helium was used as the carrier gas and the flow rate was 1.8 ml/min. Nitrogen was used as the electron capture detector makeup gas and the flow rate was 60 ml/min. The injector temperature was 280°C and the detector temperature was 310°C. The following oven temperature program was used in the analysis: initial temperature of 150°C and hold for 40 min; ramp temperature at 1°C/min to 195°C and hold for 20 min; ramp at 1°C/min to 220°C and hold for 2 min; ramp at 3°C/min to 280°C and hold for 17 min. The total instrument runtime was 169 min. The column used in the analysis was a J&W Scientific DB-5 capillary column, 0.25-mm ID, 0.25-µm film thickness, and 60 m in length. An external calibration curve was generated for backup or emergency use only. For the external calibration curve, nominal analyte concentrations of the standards were 20, 50, and 100 ng/ml. For normal operation the method of internal standards was used. As noted earlier the internal standards of PCB 103 and PCB 198 were each spiked into every sample at a level of 52.5 ng. Chromato-

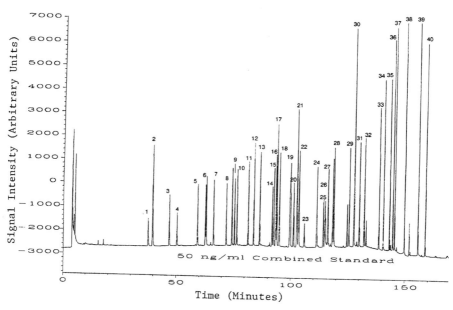

Figure 1   GC chromatogram of standards. See Table 1 for number key. (Reprinted with permission from *Chrom. Connection,* July, 1994. p. 3. Copyright 1994 Baxter [Scientific Products].)

graphic data was collected by the data station that produced chromatograms and quantitative analyses reports for each sample.

Results and discussion

This method produces an excellent cleanup of the sample and achieves virtually baseline resolution of 38 analytes and 2 internal standards. A typical chromatogram for a nominal 50 ng/ml standard is shown in Figure 1. Refer to Table 1 for identification of each analyte remembering that each PCB congener is identified as "C-BZ number." A typical spiked sample of fish or shrimp tissue produces a similar chromatogram. The average percent recovery of the analytes based upon recovery data from the laboratory control material sample (spiked grouper) that was analyzed with each batch of 10 samples (N = 33 batches) was 95.5%. Based upon our experience this method saves time over the traditional methods involving fractionation particularly when an extensive QA/QC program is involved. Further, it avoids the potential problems of one or more analytes or internal standards being split between fractions thereby creating difficulties and possible errors in quantitation. This method also offers high extraction efficiency which leads to improved accuracy and high percent recoveries.

Section IV

Techniques for aquatic toxicologists

chapter thirty

Extremely RAPD fingerprinting: a versatile tool for population genetics

David L. Lattier, Denise A. Gordon, Richard N. Silbiger, Frank McCormick, and M. Kate Smith

Introduction

In recent years there have been burgeoning advances in uses of the polymerase chain reaction (PCR) to detect polymorphisms and genetic markers.[1,2] Short oligonucleotide PCR primers, with lengths varying between 5 and 10 nucleotides, produce banding patterns that are resolved by agarose or polyacrylamide electrophoresis.[3] This technique, commonly referred to as Random Amplified Polymorphic DNA (RAPD), or DNA Amplification Fingerprinting (DAF), is based upon the amplification of genomic DNA with individual primers of arbitrary nucleotide sequence, and provides a practical means to construct genetic linkage maps of plant and animal species. The RAPD/DAF procedure also makes possible lineage evaluation[4] since a proportion of amplification fragment length polymorphisms (AFLPs) are inherited by offspring in a Mendelian fashion. This method provides data comparable to results of standard M13 DNA fingerprinting and restriction fragment length polymorphisms (RFLP) analysis, but with greater sensitivity and less labor.

We present a procedure for use in DAF analyses, which eliminates the task of isolating DNA from field collected animals. This technique is based on direct use of cleared tissue digests in place of purified genomic template in the PCR amplification scheme. Reproducible genetic results have been accomplished on tissue digests prepared from several species of teleosts. Amplification patterns of DNA present in tissue digests, when electrophoresed, are identical to those observed when purified genomic template is used in thermal amplification (PCR) reactions.

We are currently using this method to test our hypothesis that there is a statistically significant link between reduction in genetic diversity and environmental insult. Therefore, we are correlating the overall allelic variance of three populations of brown bullhead catfish, taken from three inland waterways in Ohio. The integrity of the streams has been extensively characterized, and each exhibits a varying degree of xenobiotic impact.

Materials required

- Standard rodent ear punch, National Band and Tag Co., Newport, KY
- DNA thermal cycler (PCR machine)
- Amplitaq™ Stoffel Fragment (thermal stable DNA polymerase), Perkin Elmer/Cetus, Norwalk, CT
- Oligonucleotide 10-base primers, Operon, Inc. Alameda, CA
- Proteinase K, Boehringer-Mannheim, Inc., Indianapolis, IN (20 mg ml^{-1} stock)
- SeaKem LE agarose. FMC, Inc., Rockland, ME
- Agarose gel electrophoresis unit, including power source
- Tabletop microcentrifuge

Procedures

Species of fish were taken from Nine Mile Creek, located in Hamilton County, Southwestern Ohio. The fish were gathered by electroshocking, and following non-invasive, caudal fin tissue acquisition, all individuals were released at the site of capture. Laboratory maintained yellow perch (*Perca flavescens*) and fathead minnow (*Pimephales promelas*) were used as control animals, from which high molecular weight DNA and caudal fin lysate was prepared concurrently.

Prior to electroshocking, 100 µl lysis/digestion buffer (50 mM KCl, 1.5 mM MgCl$_2$, 10 mM Tris-Cl (pH 8.5, 25°C), 0.01% gelatin, 0.45% Nonidet P-40, 0.45% polyoxyethylene-sorbitan monolaurate [Tween 20]) was aliquoted into several 0.5 ml Eppendorf microcentrifuge tubes. To the aliquoted lysis buffer was added 1 mg ml^{-1} proteinase K from a stock solution of 20 mg ml^{-1}. Addition of the proteinase K, prior to sample collection, eliminated the requirement for refrigerating samples during transport to or from sites of field collection.

Fish were identified at the collection site, and a standard rodent identification ear punch was used to remove a 2-mm circular "disk" from caudal fins, proximal to the body. Micro-dissection forceps were used to place the excised tissue in a marked tube containing the lysis buffer, after which the sample tubes were flicked repeatedly until the tissue was immersed in buffer. Since the technique of PCR is exquisitely sensitive, care was taken to avoid DNA cross-contamination from individual fish; therefore, the ear punch was thoroughly cleansed with alcohol between tissue sampling. At this point, tubes containing tissue samples can be transported to the laboratory at ambient temperature, and transportation time has no consequence on the integrity of the samples or subsequent results.

Tissue samples were placed in a dry air incubator at 55°C, overnight, to ensure complete digestion by proteinase K. Following proteinase K digestion, tubes containing tissue samples were centrifuged at 12,000 rpm, room temperature, for 3 min to remove particulate and undigested material. A volume of 80 μl of each lysed sample was removed to a fresh 0.5-ml Eppendorf microcentrifuge tube, and lysates were heated at 91°C, 10 min, to inactivate proteinase K. Following these manipulations, tissue lysates, if stored at –20°C, could be used recurrently in PCR reactions and displayed explicitly reproducible results, with no variation in PCR amplified genetic data.

We attempted to determine DNA concentrations in tissue lysates prepared from caudal fin punctures, using a Shimadzu RF5000U Spectrofluorophotometer. Following a standard protocol,[5] this sensitive fluorometric method functioned at the limit of nucleic acid detection, and all lysate samples were determined to be ≤37 ng μl^{-1} DNA.

Prior to analyzing the tissue lysates by PCR, the lyophilized 10-base primers used in DNA amplification reactions, were suspended in 3.0 mM Tris, 0.2 mM EDTA (3.0:0.2 TE buffer, pH 8.0) at a final concentration of 5×10^{-6} M (5 pmoles μl^{-1}).

PCR reactions comprised the following reagents, all of which, except for the DNA template/lysate, were combined in a master mix.

Per reaction:
2.5 μl 10× Amplitaq™ Stoffel buffer
3.0 μl 25 mM MgCl$_2$
2.5 μl 2 mM d[A,T,C,G]TP mix (prepared from 10 mM stocks)
5.0 μl (5 pmoles μl^{-1}) single 10-base oligonucleotide primer
0.4 μl (4.0 units) Amplitaq™ Stoffel Fragment DNA polymerase
9.6 μl sterile deionized water
23 μl

Tissue lysate was added to the individual reactions as the final step, prior to thermal cycle DNA amplification. Reaction tubes containing 23-μl aliquots from the master mix were chilled on ice, or at 3°C in the DNA thermal cycler. A volume of 2 μl tissue lysate was added to each tube, and components were manually mixed with a Gilson Pipetman. To avoid aerosol DNA cross-contamination of reaction mixes, all reagent manipulations were performed with aerosol resistant pipette tips (ART). Completed mixes were overlaid with 2 drops of mineral oil, and amplified in a DNA Thermal Cycler 480 (Perkin Elmer/Cetus) for 45 cycles with the following profile: 45 sec at 94°C, 45 sec at 39°C, and 1 min at 72°C.

In order to verify that the fidelity and banding pattern of PCR reactions using lysate was indistinguishable from reactions in which purified genomic template was used, genomic DNA was isolated from laboratory maintained yellow perch and fathead minnow. Tail muscle from respective fish was placed in 0.5 ml digestion buffer (50 mM Na$_2$EDTA, 1% sodium dodecyl sulfate, 50 mM Tris-Cl (pH 7.4, 25°C), 100 mM NaCl) containing 1 mg ml^{-1} proteinase K, and incubated overnight at 65°C. Following digestion, 200 µl 5 M potassium acetate (3 M potassium/5 M acetate) was added to the 1.5-ml Eppendorf tubes which were vortexed vigorously, and placed on ice for 1.5 h. The digests were cleared by centrifugation at 15,000 rpm, 4°C, for 15 min. The supernatant was immediately removed to new tubes on ice. Cold absolute ethanol was added to the top of 1.5-ml tubes, and tubes were gently inverted once. High molecular weight DNA was isolated by spooling on to sterile 50 µl, heat sealed, capillary tubes. DNA was washed once with 70% ethanol and dried briefly in air. DNA was dissolved in 250 µl 10.0 mM Tris, 1.0 mM EDTA (10.0:1.0 TE buffer, pH 8.0).

DNA was extracted once with Tris-equilibrated phenol, once with phenol-chloroform-isoamyl alcohol (25:24:1 vol/vol/vol), and once with chloroform-isoamyl alcohol (24:1). Following organic extractions, the aqueous phase was precipitated by addition of 0.33 volume of 10.0 M ammonium acetate and 2.5 volumes of absolute ethanol. Genomic DNA was dissolved in 100 µl 3.0 mM Tris, 0.2 mM EDTA (3.0:0.2 TE buffer, pH 8.0). DNA purity and concentration were determined by A_{260} and A_{280} on a Beckman DU-8 spectrophotometer. For thermal cycle amplification reactions, genomic DNA was prepared to a concentration of 12.5×10^{-9} g µl^{-1}. Amplification reactions using purified, high molecular weight genomic DNA template were prepared as described earlier, except that 2 µl (12.5×10^{-9} g µl^{-1}) per tube of purified DNA was added at the start of the reaction.

Amplified DNA products were electrophoretically separated in 1.65% SeaKem LE agarose (FMC, Inc., Rockland, ME). Gels and electrophoresis buffer (1 × TBE) were prepared with 10× TBE [1× = 89 mM Tris base, 89 mM boric acid, 2 mM Na$_2$EDTA (pH 8.0), 0.5 µg ml^{-1} ethidium bromide]. From each 25 µl amplification reaction, a volume of 12 µl was removed and combined with 1.5 µl 10 × SuRE/Cut Buffer B (Boehringer Mannheim), and 1.5 µl of gel loading dye (0.025% bromophenol blue, 0.025% xylene cyanol FF and 50% glycerol). Adjusting the salt concentration with restriction enzyme Buffer B, prior to electrophoresis, provided sharper DNA bands and greater resolution. The 15 µl volume of each reaction was loaded into the agarose gel and electrophoresed at 300 V, 8°C, using an IBI (Kodak) HRH gel box equipped with a cooling plate.

Following electrophoresis, agarose gels were visualized on a UV light box and photographed with type 55 positive-negative film (Polaroid®, Cambridge, MA). The negatives were processed per manufacturer's instructions and scanned with a Personal Densitometer® (Molecular Dynamics, Sunnyvale, CA). Densitometric results were analyzed with the ImageQuant® computer program (Molecular Dynamics, Sunnyvale, CA).

Results and discussion

The method described was developed to support correlation indices of genetic diversity with environmental impact in wild aquatic and terrestrial species. The number of individual field samples required for this type of analysis renders DNA isolation, purification, and quantitation prohibitive with respect to time and labor. The capability of detecting DNA amplified polymorphisms by using unrefined tissue digests offered an exciting alternative; however, this method required confirmation that PCR reaction efficiency was not subject to inhibition by components present in the crude lysate. To address this question, purified genomic template DNA was prepared from a single yellow perch concurrently with unprocessed tissue lysate. Additionally, genomic DNA was prepared from a fathead minnow. DNA amplification reactions, containing either 25×10^{-9} g of genomic template DNA or 2 µl tissue lysate, were performed using oligonucleotide primer OPJ-18 (5'-TGGTCGCAGA-3'; Operon, Inc.). Identical electrophoretic banding patterns of amplified DNA products were observed in yellow perch (lanes A and B). Lane C represents the amplified banding pattern generated from purified genomic template prepared from a laboratory maintained fathead minnow. The banding pattern in lane D is the result of a PCR reaction using caudal fin lysate prepared from a separate, wild caught fathead minnow. When lanes C and D are compared, numerous identical species-specific bands are observed in the amplified patterns of both individuals; however, distinct polymorphisms are prominent. This demonstrates that the parallel amplification patterns observed in yellow perch lysate, and genomic DNA, are not a spurious result. The amplified DNA banding patterns, using caudal fin lysates prepared from several indigenous species of teleosts, are shown in Figure 1 (Lanes E through K).

To optimize this method of DNA amplification fingerprinting, we tested a range of primer annealing temperatures, various thermostable DNA polymerases, and a series of $MgCl_2$ concentrations in PCR reactions. Oligomer primer annealing temperatures were examined between a range of 35°C and 41°C. Throughout the temperature range, there were no observable differences in the electrophoretic DNA banding patterns in agarose. Three thermal stable DNA polymerases were tested — native *Taq*®, Amplitaq®, and Amplitaq Stoffel Fragment® (Perkin Elmer/Cetus), all of which displayed acceptable results. Fish tissue lysate was amplified with Amplitaq Stoffel Fragment DNA polymerase, using the manufacturer's recommended $MgCl_2$ concentration of 3.0 mM. Magnesium titration curves, for reactions including either tissue lysate or purified genomic template, indicated that when using native *Taq* or Amplitaq DNA polymerase, a $MgCl_2$ concentration of 2.5 mM is required for short oligonucleotide directed thermal amplification.

The method presented combines the powerful RAPD/DAF tools of genetic analysis with a non-lethal, relatively non-invasive method of tissue acquisition. We have also applied this method to analyze genetic polymorphisms in terrestrial animals, such as the Gray-Tailed voles (*Microtus canicaudus*). By treating, as described, a minute quantity of tissue obtained from an ear punch, or a tail section, we were able to generate reproducible amplified DNA banding patterns. This simple and rapid technique obviates the need for analytical radionuclides and the isolation, purification, and quantitation of genomic template DNA for thermal cycle amplification reactions. The strategy is particularly useful when collecting field specimens for population or forensic analysis, or species verification. The method may also be used with locus-specific oligo primers, and as genomic sequence information

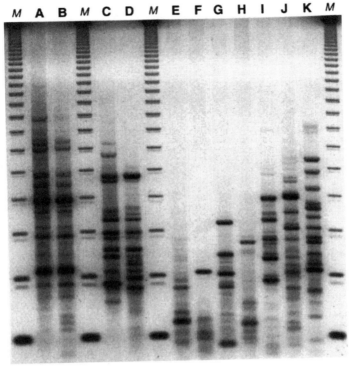

Figure 1 DNA amplification fingerprint patterns generated from purified, high molecular weight genomic template DNA, or DNA contained in caudal fin tissue lysates. The amplification reactions were prepared as described in the Procedures section, using oligo primer OPJ-18 and 4.0 units of Amplitaq Stoffel Fragment DNA polymerase. (A) Banding pattern generated from a PCR reaction using 25×10^{-9} g DNA from yellow perch (*Perca flavescens*). (B) Banding pattern generated from a PCR reaction from DNA contained in 2 μl of tissue lysate from the same individual in A. (C–D) The identical experimental scheme as A and B, except that the genomic DNA (Lane C) and the caudal fin lysate (Lane D) were prepared from separate fathead minnows (*Pimephales promelas*). Lanes E–K depict the PCR amplification patterns resulting from use of fin tissue lysates acquired from the following diverse species; common white sucker (*Catostomus commersoni*), northern creek chub (*Semotilus atromaculatus*), northern hog sucker (*Hypentelium nigricans*), yellow bullhead catfish (*Ictalurus natalis*), johnny darter (*Etheostoma nigrum*), rainbow darter (*Etheostoma caeruleum*), and smallmouth bass (*Micropterus dolomieu*). (M) 123-base pair DNA ladder (GIBCO/BRL Gaithersburg, MD).

becomes available, it provides an exciting application for polymorphically significant microsatellite probes.

Acknowledgment

This research was supported in part by an appointment to the Research Program at the National Exposure Research Laboratory/U.S. Environmental Protection Agency administered by the Oak Ridge Institute for Science and

Education through an interagency agreement between the U.S. Department of Energy and NERL/EPA.

References

1. Williams, J.G.K., Kubelik, A.R., Livak, K.J., Rafalski, J.A., Tingey, S.V. 1990. DNA polymorphisms amplified by arbitrary primers are useful as genetic markers. *Nucl. Acids Res.* 18:6531–6535.
2. Woodward, S.R., Sudweeks, J., Teuscher, C. 1992. Random sequence oligonucleotide primers detect polymorphic DNA products which segregate in inbred strains of mice. *Mammalian Genome.* 3:73–78.
3. Caetano-Anolles, G., Bassam, B.J., Gresshoff, P.M. 1991. DNA amplification fingerprinting using very short oligonucleotide primers. *BIO/Technol.* 9:553–557.
4. Scott, M.P., Haymes, K.M., Williams, S.M. 1992. Parentage analysis using RAPD PCR. *Nucl. Acids Res.* 20:5493.
5. Daxhelet, G.A., Coene, M.M., Hoet, P.P., Cocito, C.G. 1989. Spectrofluorometry of Dyes with DNAs of Different base composition and conformation. In *Current Protocols in Molecular Biology,* Eds. F.M. Ausubel, R. Brent, R.E. Kingston, A.3.10 (Appendix 3). Greene Publishing Associates and John Wiley & Sons, New York.

chapter thirty-one

Preparation of whole small fish for histological evaluation

John W. Fournie, William E. Hawkins, Rena M. Krol, and Marilyn J. Wolfe

Introduction

Toxicologic pathology, which is primarily concerned with chemically-induced structural changes in cells or tissues, depends on the proper histological processing of test specimens. In fishes, histopathological examination is widely recognized as a reliable method for disease diagnosis[1] and for assessing acute and chronic effects of exposure to toxicants at the cellular level in both marine and freshwater species.[2] Published histological methods are available for fishes in general,[3] for particular species such as the channel catfish, *Ictalurus punctatus*,[4] the striped bass, *Morone saxatilis*,[5] and salmonids.[6] In some cases, custom methods have been developed for specific organs or tissues that are of particular interest. For example, a recent study evaluated several fixation protocols on the liver of the Atlantic salmon, *Salmo salar*.[7]

The studies cited above deal mainly with large fish species. The increased use of small fishes in aquatic toxicology, carcinogenicity testing, and biomedical research (see Reference 8), also requires that samples are properly processed so that reliable histopathological evaluations can be performed. One of the acknowledged advantages of using small fish in toxicological research is that histological processing and evaluation are facilitated because whole specimens, rather than dissected tissues, can be processed, sectioned, and studied. Whole small fish, particularly the medaka (*Oryzias latipes*) and the guppy (*Poecilia reticulata*), have been used in several recent studies to examine histological effects of environmental toxicants, including β-hexachlorocyclohexane,[9] dimethylbenz[a]anthracene,[10] benzo[a]pyrene,[11] acetylaminofluorene,[12] and organotins.[13]

For histological data from the small fish models to be useful for regulatory purposes, they must meet quality assurance criteria with regards to accuracy and consistency of pathology data similar to the criteria that are currently being applied to the rodent models.[14] The quality of pathology data depends in large part on the histological methods used to obtain the data.[15] Whether dealing with tissues of rodents or fish, critical elements such as choice of fixative, processing protocols, sectioning strategies, and avoiding postmortem autolysis are the same.

In this paper, procedures and methods covering all aspects of the preparation of small fish for histologic examination are presented. The methods described result in properly prepared, slide-mounted tissue sections of whole small fish. Details regarding fixation, processing, sectioning, and staining of fish tissues are given. The specific protocols presented are from those used at the U.S. Environmental Protection Agency, Gulf Ecology Division (EPA-GB), Gulf Breeze, FL, the Gulf Coast Research Laboratory (GCRL), Ocean Springs, MS, and Experimental Pathology Laboratories (EPL), Herndon, VA.

Materials required

The following two lists contain equipment, supplies, and reagents used in many modern histology laboratories. All the equipment items are not required to properly prepare fish tissues for histological examination because many steps in the process such as embedding, staining, and coverslipping can be performed manually. However, the use of automated equipment makes tasks easier and provides consistency in tissue preparation. Mention of commercial products does not constitute endorsement by the U.S. Environmental Protection Agency.

Equipment

SHUR-MARK™ cassette marking system
Shandon Hypercenter® XP enclosed tissue processor
Tissue-Tek® tissue embedding console system
Reichert-Jung 2030 rotary microtome or an automated microtome
Lighted tissue flotation bath
Surgipath® VSP slide printer
Shandon Varistain™ XY robotic slide stainer
Hacker-Meisei RCM-3655 robot coverslipping machine

Supplies and reagents

Iridectomy scissors, forceps, scalpel, and other dissecting tools
Tissue-Tek® Uni-cassettes
Formaldehyde
Graded ethanols (70%, 80%, 95%, 100%)
Xylene
Paraplast® X-TRA
Surgipath® disposable molds
Disposable knife blades

Microscope slides
Acid alcohol
Ammonia water
Harris' hematoxylin
Eosin Y
Glass coverslips
Permount®

Procedures

To produce high quality histologic sections of whole small fish, attention must be given to all facets of preparation. The quality of pathology data can be compromised by inconsistency in histological procedures. Consistency is improved by working from written standard operating procedures (SOPs). Development and application of SOPs is required for laboratories conducting studies under Good Laboratory Practices (GLP) guidelines but the use of SOPs is advisable even for non-GLP laboratories. Below, the approaches for fixing, processing, sectioning, and staining procedures for whole small fish are detailed.

Fixation

After collection, the fish are euthanized by immersion in an overdose (at least 0.4%) solution of tricaine methanesulfonate (TMS; MS-222). Weight, standard length (distance in mm from the tip of the snout to the base of the caudal fin), and gender are recorded for each specimen. All unusual external morphological features (e.g., lordosis, kyphosis, scoliosis), and signs of neo-plastic changes, infectious disease, or other abnormalities are also recorded. The caudal fin and posterior portion of each fish are examined for gross lesions and if none are present, the tail is removed by an angular cut begin-ning near the anal vent and proceeding anteriorly to the dorsal side of the fish. This is done so that the carcass can fit into the tissue cassette and it also improves fixative penetration. The body cavity is then opened. Using iridec-tomy scissors, a mid-ventral slit is carefully made from the anal vent to the operculum. The plane of sectioning must be determined before gross trim-ming so that each specimen in a particular study is sectioned similarly. Although small aquarium fish in toxicologic studies are usually sectioned in a longitudinal plane, exceptions may be made for individual fish that have eye or skin lesions on the lateral body wall that would be cut away by longitudinal sectioning. Fish with such lesions may be trimmed in a trans-verse plane prior to processing. Alternatively, fish with such lesions may be embedded, after processing, for sectioning in a horizontal plane.

After gross trimming, each fish is placed in a prelabeled cassette and the cassette placed in a container of Bouin's fixative for 24 to 72 h at a 1:20 v:v tissue:fixative ratio. The fixation process is terminated by pouring off fixative and rinsing the cassettes two to three times with tap water. At this point, processing can be suspended indefinitely by immersing the cassettes in 10% neutral buffered formalin. If fish are to be processed immediately, they are

Table 1   Processing Schedule for Fixed Fish Tissues

Fluid	Time (min)	Temperature (°C)
70% ethanol	60	A[1]
80% ethanol	60	A
95% ethanol	60	A
95% ethanol	60	A
100% ethanol	50	A
100% ethanol	50	A
100% ethanol	60	A
Xylene	60	A
Xylene[2]	60	A
Xylene[2]	60	A
Paraffin[2]	60	60°
Paraffin[2]	60	60°

[1] Ambient.

[2] Stations under vacuum.

rinsed in flowing tap water for 24 h after fixation and then processed as described below.

For some studies, specific fixation protocols may be necessary. For example, if time or sample size allows, the blunt end of a small syringe containing Bouin's solution is inserted into the mouth to run fixative over the gills and into the abdominal cavity to flush it with fixative. This procedure ensures optimum fixation of all tissues and reduces fixation artifacts. Because formaldehyde, a component of Bouin's, is considered carcinogenic, it should be handled with appropriate care in adequately ventilated areas.

Processing

Fixed tissue samples in cassettes are dehydrated, cleared, and infiltrated with paraffin either by hand or using an automatic tissue processor. Older automatic processors move the tissues through a graded series of ethyl alcohol (70 to 100%), two or three changes of xylene or a xylene substitute, and two changes of melted paraffin. With newer, totally enclosed tissue processors (e.g., Shandon Hypercenter® XP), the tissue samples remain in a reaction chamber and solutions are pumped into and out of the chamber making the procedure more reliable and safer. A processing protocol is described in Table 1. After processing, whole fish or tissues are properly oriented for microtomy and embedded in stainless steel or plastic molds containing clean, melted paraffin. After the paraffin hardens on a cold plate, the molds are removed, and the paraffin blocks are trimmed for sectioning.

Sectioning

The trimmed blocks are placed into cassette clamps on a precision rotary microtome (e.g., Reichert-Jung 2030) or an automated microtome. Thin ribbons of paraffin (4 to 6 μm) containing sections of tissue are cut from the blocks using disposable microtome blades. The ribbons are carefully placed on a heated waterbath (38°C) containing gelatin, albumin, or a commercially available product such as Surgipath® Sta-on as an adhesive. The stretched sections are cut to the desired length using a heated tissue separator and picked up on clean glass slides. Slides are then placed on a warming tray or in an oven at 58°C to promote adherence between the tissue and the slide.

In studies such as carcinogen bioassays, tissue accountability is important and specific sectioning protocols are employed. For example, medaka might be sectioned longitudinally at a thickness of 4 to 5 μm so that five step sections are taken through the fish: two right paramedian sections, one midsagittal section, and two left paramedian sections. This method allows for the histologic evaluation of approximately 30 tissues in over 85% of the specimens.[16] Regardless of the sectioning protocol employed, consistency is important. Each specimen from each test group should be treated as uniformly as possible.

Staining

The slide-mounted tissue sections are now ready for staining. For routine histopathological examination, the slides are usually stained with hematoxylin and eosin (H&E). A sample staining schedule is presented in Table 2. This is a "regressive" staining procedure in which the sections are overstained in a relatively neutral solution of Harris' hematoxylin. Excess stain is removed with an acid alcohol solution, then the sections are neutralized with an alkaline solution of ammonia water. This results in a blue staining of nuclei. The most widely used counterstain is eosin which stains the cytoplasm a deep pink.

Table 2   Slide Staining Schedule for Tissue Sections

Reagent	Number of stations	Time in each station
Xylene	3	3 min
100% Ethyl alcohol	2	1 min
95%	1	1 min
70%	1	1 min
Water	1	1 min
Harris' hematoxylin	1	9 min
Water	1	1 min
Acid alcohol	1	1–2 sec
Water	1	1 min
Ammonia water	1	1 min
Water	1	1 min
70% Ethyl alcohol	1	1 min
Eosin	1	1 min
95% Ethyl alcohol	2	1 min
100% Ethyl alcohol	3	1 min
Xylene	3	1 min

A "progressive" staining procedure can also be employed. This type of staining is accomplished by using a hematoxylin solution which contains an excess of aluminum salts or acid. This increases the selectivity of the stain for nuclei. After staining with hematoxylin, the sections are washed well with water and then the secondary or counterstain is applied.

Test of histology protocols on medaka

A test using 840 one-year-old medaka was conducted to compare some fixation and decalcification parameters. Several histological procedures were tested for their effects on processing time and section quality. High quality sections can be difficult to produce because small fish are composed of tissues that range in consistency from soft visceral organs to hard bones and scales. Scaling by gently scraping the body with an angled scalpel blade promotes good fixative penetration and ease of sectioning. However, scaling also removes the skin, which is often a key target organ, from the fish.

Figure 1 shows that in this study, one group of medaka was scaled, a second group was not scaled, and a third group was not scaled but later decalcified (after fixation and storage) in TBD-1 (Shandon Inc.), a 14% hydrochloric acid solution, for 4, 8, or 12 h. Specimens from each of these five groups of fish were exposed to three types of fixative (Lillie's, Bouin's, or 10% neutral buffered formalin) for varying time periods (2, 3, 4, or 5 days). After fixation, specimens were placed in 10% neutral buffered formalin (NBF) for storage. Times in NBF storage varied from 3 to 12 days. Each treatment group consisted of 15 fish except for the groups of unscaled fish where 10 fish were used for each test. Fewer fish were used in those tests because historically that treatment produced the poorest sections. Two sections, a mid-lateral and mid-sagittal section, were taken from each fish. All organs in the sections were examined, with the liver expected to be the most important indicator of fixation quality. The results of this test of fixation procedures are discussed below.

Results and discussion

Good fixation is necessary to produce histologic sections that allow accurate interpretation of cellular and tissue morphology. Autolysis, the natural degenerative changes that occur in tissues after death, is halted by fixation. The widely used aldehyde-containing fixatives preserve cell constituents and tissue morphology in as life-like a condition as possible. Most fixative mixtures contain coagulants that facilitate embedding together with anticoagulants that control both the texture and the staining reactions of the tissue. The amount of fixative relative to the mass of tissue should be at least 20:1 and the time of fixation should be 24 to 48 h for most tissues. Fixation is terminated by washing the tissues in running water. The most commonly used fixative for light microscopy is 10% NBF because it is inexpensive, easy to prepare, maintains cell morphology and antigenicity well, and is compatible with most stains. Furthermore, tissues can be stored in 10% NBF indefinitely. A recent study on Atlantic salmon liver morphology showed that

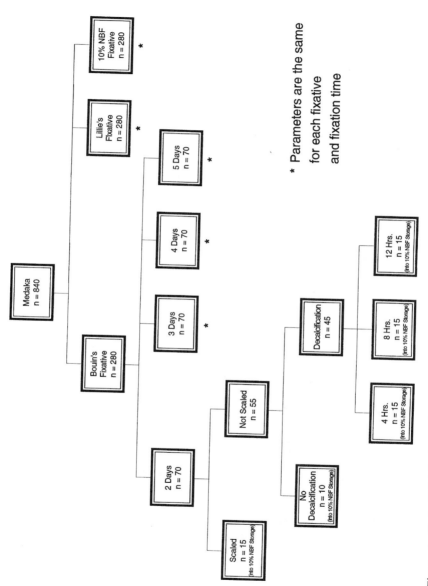

Figure 1 Test of histologic methods for adult medaka.

10% NBF was superior to several other routinely used fixatives.[7] Hepatocyte cytoplasm and nuclei were homogeneous, well outlined, and exhibited only moderate shrinkage.

Compound fixatives such as Bouin's solution,[17] Davidson's solution,[18] Lillie's,[19] and Dietrich's solution[20] are also frequently used. Bouin's solution which is composed of formalin, acetic acid, and picric acid, is widely used in fish pathology. It is especially useful for fixation of certain tissues (e.g., eyes, gills, skin), and its demineralizing properties minimize distortion and tearing artifacts from bony structures, thus allowing whole-body sectioning of small fishes.[1] Bouin's fixative also promotes excellent tissue staining. Additionally, the shrinkage properties commonly associated with Bouin's are minimal. In the Atlantic salmon liver study, hepatocyte shrinkage with Bouin's fixative was minimal.

Preparing small whole fish from studies with large sample sizes presents some problems common to fixation of all animals and some problems unique to the study size. It is always necessary to find the best fixative and decalcification routine to produce good sections. If sample sizes are large, it is important to store tissues in a fluid where they may have to stay for extended periods of time until processing can be done. Ten percent NBF is an excellent fixative for long term fixation and storage because tissues can be left in it indefinitely. Fixatives such as Bouin's and Lillie's are both good mixtures because they contain properties that counteract each other. For example, in Bouin's solution the swelling action of acetic acid is balanced by the shrinking action of picric acid. In Bouin's and Lillie's solutions the hardening and minor shrinkage properties of formaldehyde are balanced by the softening and shrinking actions of picric acid.[21]

Because fish scales cause sectioning knife marks and tissue tears when small whole fish are microtomed, decalcification is also an important consideration in preparation. Some fixative mixtures contain decalcifying agents, such as the formic acid in Lillie's fixative. Sometimes this type of fixative can be used without additional decalcifying agents. In other cases, a preparation process combining fixation followed later by decalcification may be used.

The liver is the major organ of interest in many carcinogenicity studies using small fish. In the test described in "test of histology protocols on medaka," the best overall protocols were Bouin's and Lillie's fixatives for 2 to 4 days, followed by storage in 10% NBF, then an 8- or 12-h decalcification step in TBD-1. There was more gill epithelial lifting and shrinkage of tissue around kidney tubules with Lillie's fixative than with Bouin's. Since there seems to be no definitive combination that produced clear results above all others, a fixation time of 3 days in Bouin's with an 8-h decalcification in TBD-1 is recommended for adult medaka. The decalcification time was chosen because it fits more easily into the typical work schedule in most laboratories. Figure 2 is a micrograph showing a medaka specimen processed according to the above protocol with acceptable results.

Histopathologic evaluation of whole small fish is an important endpoint in toxicologic pathology. Careful planning for the sectioning of whole fish with a view to capturing all gross lesions in the sections as well as main-

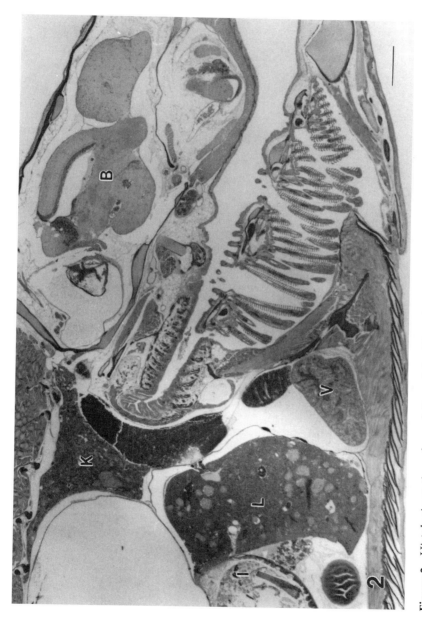

Figure 2   Histologic section of an adult medaka. Liver (L); kidney (K); ventricle of heart (V); brain (B); intestine (I). Bar = 50 mm.

taining consistency in sections from one fish to another is vital to the success of the evaluation. Fixation, gross trimming, and sectioning methods may be customized to the objectives of the study to ensure meaningful results.

Acknowledgment

This study was supported in part by U.S. Environmental Protection Agency, Washington, D.C., Contract No. CR-816007-01-0. This paper is contribution number 936 of the U.S. EPA Gulf Ecology Division.

References

1. Ferguson, H. W. 1989. *Systemic Pathology of Fish.* Ames, IA: Iowa State University Press, Ames. 263 pp.
2. Hinton, D.E., Baumann, P.C., Gardner, G., Hawkins, W.E., Hendricks, J.D., Murchelano, R., Okihiro, M.S. 1992. Histopathologic biomarkers. In *Biomarkers: Biochemical, Physiological, and Histological Markers of Anthropogenic Stress,* Eds. R.J. Huggett, R.A. Kimerle, P.M. Mehrle, H.L. Bergman. Boca Raton: CRC/Lewis Publishers. pp. 155–209.
3. Hinton, D.E. 1990. Histological techniques. In *Methods for Fish Biology,* Eds. C.B. Schreck, P.B. Moyle. Bethesda, MD: American Fisheries Society. pp. 191–209.
4. Grizzle, J.M., Rogers, W.A. 1976. *Anatomy and Histology of the Channel Catfish.* Auburn AL: Auburn University Agricultural Experiment Station. 94 pp.
5. Groman, D.B. 1982. *Histology of the Striped Bass.* Bethesda, MD: American Fisheries Society Monograph 3. 116 pp.
6. Yasutake, W.T., Wales, J.H. 1983. *Microscopic Anatomy of Salmonids: An Atlas.* Washington, DC: United States Department of the Interior. Resource Publication 150. 189 pp.
7. Speilberg, L., Evensen, O., Bratberg, B., Skjerve, E. 1993. Evaluation of five different immersion fixatives for light microscopic studies of liver tissue in Atlantic salmon *Salmo salar. Dis. Aquat. Organ.* 17: 47–55.
8. Hawkins, W.E., Walker, W.W., Overstreet, R.M. 1995. Practical carcinogenicity tests with small fish species. In *Fundamentals of Aquatic Toxicology: Effects, Environmental Fate, and Risk Assessment,* Ed. G.M. Rand, 14:421–446. Washington, DC: Taylor and Francis.
9. Wester, P.W., Canton, J.H. 1986. Histopathological study of *Oryzias latipes* (medaka) after long-term β-hexachlorocyclohexane exposure. *Aquat. Toxicol.* 9:21–45.
10. Hawkins, W.E., Walker, W.W., Lytle, J.S., Lytle, T.F., Overstreet, R.M. 1989. Carcinogenic effects of 7,12-dimethylbenz[a]anthracene in the guppy (*Poecilia reticulata*). *Aquat. Toxicol.* 15:63–82.
11. Hawkins, W.E., Walker, W.W., Overstreet, R.M., Lytle, T.F., Lytle, J.S. 1988. Dose-related carcinogenic effects of waterborne benzo[a]pyrene on livers of two small fish species. *J. Ecotox. Environ. Safety* 16:219–231.
12. James, M.O., Hawkins, W.E., Walker, W.W. 1994. Phase 1 and phase 2 biotransformation and carcinogenicity of 2- acetylaminofluorene in medaka and guppy. *Aquat. Toxicol.* 28: 79–95.
13. Wester, P.W., Canton, J.H. 1987. Histopathological study of *Poecilia reticulata* after long term exposure to bis(tri-*n*-butyltin)oxide (TBTO) and di-*n*-butyltindichloride (DBTC). *Aquat. Toxicol.* 10: 143–165.

14. Boorman, G.A., Montgomery, C.A., Eustis, S.L., Wolfe, M.J., McConnell, E.E., Hardisty, J.F. 1985. Quality assurance in pathology for rodent carcinogenicity studies. In: *Handbook of Carcinogenesis Testing*, Eds. H. Milman, E. Weisberger. Park Ridge, NJ: Noyes Publications. pp. 345–357.

15. Bucci, T.J. 1991. Evaluation of altered morphology. In *Handbook of Toxicologic Pathology*, Eds. W.M. Haschek, C.G. Rousseaux, New York: Academic Press. pp. 23–35.

16. Wolfe, M.J., Punglia, R. 1989. Analysis of the five step section method for histology of the Japanese Medaka (*Oryzias latipes*). Compendium of the FY1988 and FY1989 Research Reviews for the Research Methods Branch. U.S. Army Biomedical Research and Development Laboratory, Fort Detrick, MD, pp. 173–183.

17. Luna, L.G. 1968. *Manual of Histologic Staining Methods of the Armed Forces Institute of Pathology.* New York: McGraw-Hill. 258 pp.

18. Shaw, B.L., Battle, H.I. 1957. The gross and microscopic anatomy of the digestive tract of the oyster, *Crassostrea virginica* (Gmelin). *Can. J. Zool.* 35:325–347.

19. Humason, G.L. 1972. *Animal Tissue Techniques, 3rd ed.* San Francisco: W.H. Freeman and Co. pp. 31, 463.

20. Yevich, P.P., Yevich, C.A. 1994. Use of histopathology in biomonitoring marine invertebrates. In *Biomonitoring of Coastal Waters and Estuaries*, Eds. Kees I., M. Kramer. Boca Raton: CRC/Lewis Publishers. pp. 179–204.

21. Carson, F.L. 1990. *Histotechnology.* Chicago: ASCP Press. 294 pp.

chapter thirty-two

Construction and use of a large-scale dosing system to expose fish to waterborne contaminants

Andrew S. Kane, Suzanne V. Jacobson, and Renate Reimschuessel

Introduction

Aquatic animals are important models for research and monitoring the effects of environmental pollution. Comparison of contaminant effects, however, is often difficult due to intra- and inter-species variation. These variables or other stressors may be further confounded by differences between laboratories, variance in animal husbandry, and fluctations in contaminant concentration and water quality. There exists a need for an exposure system design which (1) is simple to construct and use for both general fish maintenance and experimental dosing of non-volatile waterborne agents, (2) can accommodate larger fish (20 to 30 cm), and (3) will provide consistent results. We have designed and constructed a large-scale dosing system to conduct bioassays with fish. The flow-through design offers more accurate assessment of toxicant effects as compared with static or static renewal systems which tend to be more stressful to test animals due to diminished water quality (and associated increased stress) throughout exposures.

Utility of the exposure/maintenance (E/M) system described stems from the ability to non-chemically treat large volumes of water from a city water source and deliver consistent volumes to multiple aquaria. This E/M system should be regarded as a provocative template from which to custom design a system for an investigator's specific purposes. We will detail the apparatus built at the University of Maryland Aquatic Pathobiology Center (APC). However, the University of Maryland does not endorse any particular manufacturer and encourages system designers and builders to investigate local suppliers who may offer helpful product support and technical assistance.

Applications of this E/M system include early life stage, acute, and chronic toxicity tests. The system can be used to investigate median effect concentrations (e.g., EC_{50}), i.e., the concentration of a substance which produces a specified sublethal effect in the mean of a given test population for a specific amount of time (e.g., 24, 72, or 96 h). Examples of sublethal endpoints include alterations in physiology, histopathology, behavior, bioaccumulation, immunotoxicity, and neurotoxicity. In addition, the system lends itself for conducting lethality tests such as median lethal concentration (LC_{50}) assays. In contrast to an LD_{50}, where the administered dose is known (because the xenobiotic is introduced by injection, gavage, etc.), the LC_{50} is the concentration of an aqueous substance to which an animal is exposed. Since the actual dose internalized by test animals in an LC_{50} assay is not accounted for in aqueous exposure, measurement of xenobiotic concentrations in plasma or target tissues is important when interpreting dose-response. Applications of LC_{50} data include determination of species-specific sensitivities and appropriate sublethal doses for various treatments and experimental protocols.

System use with copper and brown bullheads

Use of the E/M system for lethality studies is exemplified in this paper using research conducted in our laboratory on acute copper exposure of the brown bullhead, *Ameiurus nebulosis*. Copper was selected as a test metal because it is ubiquitous in the aquatic environment, has application in aquaculture, and its effects on fish have been well studied.[1-5]

Brown bullhead were selected for use in our study because of their regional environmental relevance and sensitivity to metals *in vivo*.[6-8] Brown bullhead are also sensitive to a variety of infectious/antigenic agents *in vitro*.[9-12] Data from this acute copper exposure study includes 24 and 96 h LC_{50s}, hematocrit and serum protein values, behavioral observations, and alterations in skin color adaptation.

Materials

Animals

Fish should be acquired from reliable sources which can provide records on water quality, exposure to chemicals (including therapeutics), and any history of illness or parasite infections.

Chemicals and buffers

All reagents should be of the highest purity commercially available.

Apparatus and hardware

15-ml Polypropylene tubes (Corning Inc., Corning, NY)

Activated granular charcoal, SECA 30 or Filtersorb 200 (Encotech, Inc., Donora, PA or Calgon Carbon Corporation, Bridgewater, NJ, respectively)

Adsorber, 900-pound carbon capacity (Encotech, Inc. model W-3600, Donora, PA)

Borosilacate (Pyrex) glass tubing, 4-mm O.D. (2-mm I.D.), and 10-mm O.D. (8-mm I.D.)

Flameless atomic absorption spectrophotometer (Perkin Elmer model 5100, Norwalk, CT)

Ionanalyzer (Orion model EA 920, Boston, MA)

Microhematocrit tubes, heparinized

Peristaltic pump, low-flow, multichannel (Ismatec model G-78001, Cole-Parmer, Chicago, IL)

Pleated fiber filter housing, Micro Star filter C-120-Z3 (Haywood Pool Products, Inc., Elizabeth, NJ)

Polypropylene float valve, $\frac{1}{2}$-in (Special Plastic Systems model FVL-50, Alhambra, CA)

Polypropylene tank test vessels, 200-l rectangular, natural colored (Chem-Tainer Industries, Inc., Babylon, NJ)

Prefilter, Haywood model S-240 high rate pool filter (Haywood Pool Products, Inc., Elizabeth, NJ)

PVC automatic pressure control valve (Plast-O-Matic, Inc., Totowa, NJ)

Reichert refractometer (Cambridge Instruments, Inc., Buffalo, NY)

Silastic tubing, medical grade, non-sterile, $\frac{1}{8} \times \frac{1}{4}$ inch, and $\frac{5}{16} \times \frac{1}{2}$ inch

Syringe filters, 0.45 μm filter, sterile (Gelman Sciences, Ann Arbor, MI)

Syringes, 1 cc tuberculin with $25\frac{5}{8}$ gauge needles

Syringes, 60 cc

Procedures

Fish acquisition and acclimation

Pond-reared brown bullhead, 22 to 30 cm (total length), were obtained from a reputable supplier and transported via same-day air shipment. Upon arrival, fish were acclimated gradually to laboratory test conditions over a 6-h period. Holding and test water conditions were pH 7.2 to 7.4, hardness 80 mg/l, temperature 20° ± 2°C, with a photoperiod of 16 h light:8 h dark. Fish were held for a minimum of 4 weeks prior to testing to insure good health. During this holding period, several animals were sampled and thoroughly examined by gross necropsy and histopathology. Holding tanks consisted of 500-l aquaria, in which loading of fish did not exceed 5 g/l. Flow-through water (100% exchange/d) with coarse and biological filtration provided adequate water quality (with respect to pH, ammonia, nitrite and dissolved oxygen).[13-15] Fish were fed 38–480 trout grower diet (Zeigler Brothers, Inc., Gardners, PA) daily. Test animals were not fed for 2 d prior to or during testing.

LC$_{50}$ design and protocol

Definitive 96-h LC$_{50}$ tests consisted of flow-through bioassays.[3] Bioassays began by introducing test animals randomly into the exposure vessels. In preliminary range-finding tests, 5 animals were exposed, in replicate, to each exposure concentration, 0, 0.01, 0.1, 1.0, and 10 mg Cu/l. Mortality data from these exposures were also used to determine a 24-h LC$_{50}$. For definitive 96-h LC$_{50}$ tests, 6 to 10 animals were exposed, in replicate, to nominal concentrations of 0, 25, 50, 75, 100, and 150 µg Cu/l. Actual measured concentrations of filtered copper were 3.0, 23.7, 50.7, 75.3, 96.7 and 154.0 µg Cu/l. Test animals were observed several times daily throughout the exposure period for morbidity, mortality, and abnormal behavior. An animal was considered dead when there was no obvious respiratory movement and it did not respond to gentle prodding. Statistical LC$_{50}$ concentrations with 95% confidence intervals were calculated by probit analysis[3] using computer software.

As an alternative to traditional methodology, an acute LC$_{50}$ may also be derived using the up and down method.[16] Compared to conventional assays, this method reduces the number of animals used by up to 5 times. In brief, animals are dosed one at a time, starting with an exposure concentration approximating the LC$_{50}$. If the animal survives the exposure, then the next animal is exposed to a log dose higher. Alternatively, if the first animal succumbs following exposure, then the next animal is exposed to a log dose lower. This procedure is continued until doses bracketing survival and death are obtained. Survival data is then entered into a standard equation and values (from a table) are applied to derive LC$_{50}$ values and 95% confidence intervals. Disadvantages to this method include larger confidence intervals and the necessity for endpoints to be derived within a short, i.e., 24-h period. In our study the up and down method was not used because the endpoint we selected was 96 h and we anticipated a significant difference in the LC$_{50}$ values obtained at 24 and 96 h.

Copper treatment and sampling

Toxicant stock solutions were made using copper sulfate (CuSO$_4$ · 5H$_2$O) dissolved in distilled deionized water. Water samples for copper analysis were collected using a 60-cc syringe which had been acid washed with 10% HCl. Two syringe volumes of sample water were used to rinse the syringe prior to taking each sample, starting with the lowest concentration. Samples were taken from the center of each tank, 10 cm below the surface of the water, and filtered through a sterile 0.45-µm filter. Sample filtration in this manner is useful in estimating aqueous metal concentrations which are bioavailable. Factors which may cause discrepancy between filterable and non-filterable (presumably non-bioavailable) copper include water quality constituents which contribute to alkalinity and hardness, presence of organics, and excess mucus production from test animals. Filters were washed with 20 ml of the sample prior to delivering 10 ml sample into 15 ml acid-washed polypropylene tubes. To stabilize samples until analysis, 100 µl of ultrapure nitric acid was added to each sample and then capped tightly and stored. The samples were analyzed for copper concentration on a flameless atomic absorption spectrophotometer. Means of concentration values obtained at 0, 48 and 96 h were used as exposure values to calculate the LC_{50}s.

Hematocrit and total protein determinations

Blood samples were drawn from the caudal vein of control and exposed fish with a 1 cc tuberculin syringe fitted with a $25^{5}/_{8}$ gauge needle. Blood samples were taken from moribund animals at 48 and 72 h, as well as surviving animals at the end of the 96-h exposure. Hematocrit, as packed cell volume (PCV), was determined in heparinized microhematocrit tubes after centrifugation for five minutes at $10,000 \times$ g. Total plasma protein values were determined using a Reichert refractometer.

Statistical analysis

Statistical LC_{50} concentrations with 95% confidence intervals were calculated using probit analysis[3] on computer-based software (Pharmacologic Calculation System). Blood parameters of copper exposed animals were compared with controls using analysis of variance (SAS, general linear models procedure).

E/M system design

Water supply and filtration

Water used for animal maintenance at the University of Maryland Aquatic Pathobiology Center (APC) is nonchemically treated for organics and chlorine using a series of mechanical and carbon filters (Figure 1). Water enters the APC from a direct, uninterrupted polypropylene line from the building's city main. Polypropylene piping is used to connect the building water supply with that of the APC. Polypropylene is a very durable plastic which is difficult for "unknowing" maintenance workers to tap into, and it requires a specialized heat sealing tool to join pipe lengths and joints. Two backflow preventers are placed in-line; one downstream of the junction where the facility draws water from the building main, and one downstream of the building main from where the facility taps. This prevents potential cross contamination between the APC and the rest of the building in the event of a water shut down. All piping materials downstream of the APC entry point are schedule 40 PVC.

Downstream of the entry point into the facility, water passes through a coarse, in-line stainless steel basket filter (Figure 1A). Since water delivery requirements for carbon filtration and animal maintenance and dosing are generally low pressure, water pressure is reduced from the "city main" pressure of approximately 60 psi down to 20 psi using an in-line pressure reducing valve (Figure 1B). Water is then directed through two pleated fiber swimming pool filters arranged in parallel (Figure 1C). These fiber filters

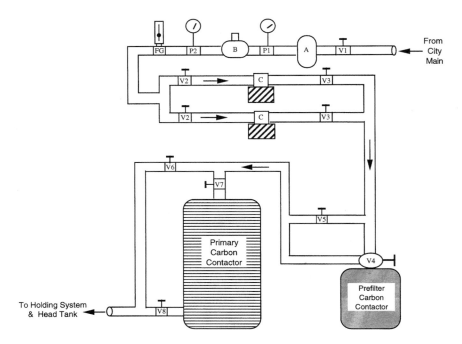

Figure 1 Schematic design of water supply and filtration system inside the Aquatic Pathobiology Center. From top right (arrows indicate direction of flow): (V1) emergency facility shut-off valve; (A) stainless steel basket filter; (P1) upstream pressure gauge; (B) automatic pressure-control valve; (P2) downstream pressure gauge; (FG) in-line flow gauge; (V2) respective upstream shut-off valves for pleated fiber filter cartridges; (C) pleated fiber filter cartridges; (V3) respective downstream shut-off valves for pleated fiber filter cartridges; (V4) Hayward prefilter multiport valve; (V5) prefilter contactor bypass valve; (V6) primary contactor bypass valve; (V7) primary contactor inlet valve, closed if bypassing primary contactor; (V8) primary contactor outlet valve, closed if bypassing primary contactor. Valve arrangement allows bypassing one of the pleated fiber filters and/or one of the contactors to perform maintenance without entirely stopping flow to the holding system and head tank.

remove suspended materials; the parallel arrangement allows for maintenance of one of the filters without stopping flow through the other (see valving arrangement immediately up- and downstream of these filters). Downstream of the fiber filters are two activated carbon contactors for dechlorination and organics removal. The first contactor consists of a swimming pool filter (150-lb carbon capacity) and serves principally as a prefilter for fine suspended and flocculent materials. This prefilter is easily maintained and backwashed. The second contactor serves as the primary filter (900-lb carbon capacity) for removal of organics and chlorine.

The prefilter contactor consists of a modified Haywood model S-240 high rate pool filter which comes factory-fitted with a multiport valve assembly enabling flow, backwash, rinse, and waste flow. The Hayward filter was modified to hold a greater quantity of charcoal by extending the length of the internal central pipe with a custom PVC extension sleeve. This alteration allowed for an increase in charcoal capacity from 100 to 150 lb. Inside the filter, water upwells from the central pipe. Water then flows downward through the charcoal bed to a set of radial arms with 0.2-mm slits. This slit size prevents 0.5-mm charcoal from passing through.

The prefilter delivers water to the primary carbon contactor (Adsorber W-3600). Piping up- and downstream of both contactors is fitted with appropriate ball valves to permit continued water flow to fish systems when bypassing one or the other filters for backwashing or maintenance. Both filters utilize activated granular charcoal.

The modified prefilter alone should be sufficient to deliver dechlorinated water to six, 140-l capacity, test chambers with five exchanges/day, plus an additional 500-l holding tank with two exchanges/day. With both carbon contactors on-line and in proper maintenance, the filtration system can deliver up to 15,000 l/day. This is, of course, dependent on the free chlorine concentration of the incoming water (on average for most areas 1 ppm) and an acceptable chlorine breakthrough concentration (0.1 ppm at APC). Free residual chlorine can be measured using an ion-selective electrode;[17] some individuals can detect chlorine in solution (as low as 0.3 ppm) by rapidly filling a clean bucket and immediately smelling the water .

Dosing system

We custom designed and built a large scale dosing system for the described experiments with brown bullhead (Figure 2). Flow of diluent and toxicant stocks through any dosing system should be initiated and calibrated for the specific exposure concentrations at least 1 week prior to testing (or until stable).

Non-chemically dechlorinated water from the APC carbon contactors was diverted to an all-glass head tank above the exposure vessels (Figure 3). This head tank evenly distributed water, by gravity feed, to exposure vessels via glass 2-mm I.D. (0.8-mm wall) flow tubes and 8-mm I.D. (1-mm wall) distribution lines. The I.D. and length of the flow tube determines the flow rate to the exposure vessels. Flow tubes were securely inserted into the holes in the bottom of the head tank by means of silastic tubing sleeves. Flow to each exposure vessel was maintained at 420 ml/min using 1.5-cm long 2-mm I.D. flow tubes with a head height of 25 cm. This allowed for approximately 5 water exchanges per exposure vessel per day. A low-flow multichannel peristaltic pump delivered a precise flow of different concentrations of copper stock solutions into T-connectors in the respective distribution lines leading to the exposure vessels.

Head tank water level was maintained constant by means of a ½-inch polypropylene float valve. Holes (¼-inch) were drilled in the bottom of the head tank to receive flow tubes which drain into the larger distribution lines leading to exposure vessels. Alternatively, for facilities using other (or unlimited) water sources, water level in the head tank can be kept constant by using a standpipe in lieu of a float valve. In this case, incoming water flow should be maintained in slight excess of drainage through the flow tubes and the standpipe.

Figure 2   Double-tier version of large scale dosing system. Exposure tanks are supported by a ¾-inch plywood support on a steel frame (all epoxy painted). Note the glass distribution lines going to the back of each tank. Tanks drain via 1-inch standpipes, through 1 inch polypropylene bulkhead fittings, and into a 1-½ inch drain manifold (seen under the upper tanks). Portion of head tank with its suspended ceiling support is seen at upper left. This larger number of tanks (12 vs. 6 as described in text) requires use of both prefilter contactor and primary contactor to provide sufficient flow and dechlorination.

Test vessels consisted of 200-l rectangular "natural" polypropylene tanks. Water drained from the exposure vessels via an internal (1 inch) standpipe. Standpipes were fitted with external sleeves (2-½ inches) to allow for upwelling bottom drainage, a feature that enhances organic debris removal. An airstone may be placed between the internal standpipe and the sleeve to favor discharge of any solids (without aerating the tank water). Standpipe height was adjusted to allow for desired exposure volume, 115 l in our case. The E/M system is setup in an isolated exposure room. Lighting in the exposure room for our experiment was set at 16-h light:8-h dark, and consisted of four 60-watt incandescent fixtures with reflection shields pointed toward the ceiling (indirect lighting) to reduce additional stress during testing. (When bright lights abruptly come on and off on timers, fish tend to be startled and may bang into sides of tanks. A light ramping system which allows for lights to slowly come up at "dawn" and down at "dusk" is an appropriate, but costly, investment.)

Water quality

Temperature of incoming water to our entire facility is tempered and held constant using a flow-through (steam) heat exchanger with stainless steel

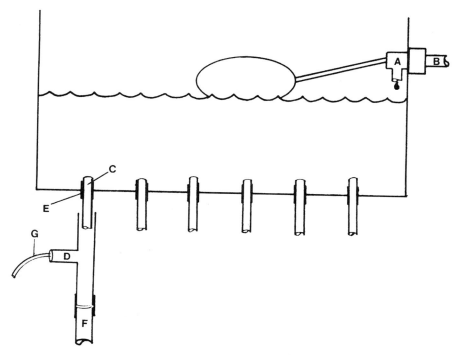

Figure 3 Head tank configuration. A, ½-inch PVC float valve; B, ½-inch water inlet line from carbon contactors; C, 2-mm flow tube; D, ⅜-inch polypropylene tee; E, silastic tubing connector sleeve; F, 8-mm I.D. distribution line leading to exposure vessels (not shown); G, ⅛-inch toxicant stock flow line from peristaltic pump (connection to tee is via polypropylene reducer, not shown).

exchange elements. However, a much less expensive alternative for a small number of aquaria (such as the 6 exposure vessels described in this paper) might utilize a smaller device such as a 1.5-kw stainless steel spa heater with an incoly element (as made by C&R Specialties, Fontana, CA). This heater may be wired for 120 or 240 V. With a ΔT of 10°C this heater should provide the E/M system with constant 20°C (±2°) water temperature during exposures. Dissolved oxygen (D.O.) was measured using a membrane probe on an Orion EA 920 ionanalyzer. D.O. remained within 80% of saturation in all test vessels throughout the exposures. The pH was measured potentiometrically. Hardness was measured by EDTA titration as $CaCO_3$ equivalents.[17] Bioassay water had a pH of 7.2 to 7.4 and a hardness of 80 mg/l. Residual chlorine (OCl⁻) levels were below 0.1 mg/l, reduced by the charcoal contactor from 0.8 to 1.5 mg/l, the normal concentration of Baltimore City water. Residual chlorine was measured with an ion specific probe.

System maintenance

Periodic monthly backwashing of the prefilter (and less often the primary contactor) charcoal bed is necessary to remove surface deposits of flocculent organic material which may settle on the bed and impede flow. Flow impedance will raise the filter operating pressure which in turn may cause channelization of the bed. Backwashing should be done as routine maintenance and before (not during) use of the dosing system for toxicity exposures. Coarse fiber filters upstream of the contactors (Figure 1C) reduce the amount of suspended flocculants and solids passing through the system. These fiber filters also require routine washing and eventual replacement. Use of two fiber filters allows for maintenance of one without the need to stop flow through the other.

Tank maintenance

Prior to (and between) experiments it is necessary to remove any organic material, microorganisms, or sorbed metals from the tank interior. To disinfect tanks, scrub the walls and bottom of drained tank, standpipe, any filter materials, siphon tubes and nets with a 1:10 dilution of 5.25% sodium hypochlorite (bleach). After disinfection, rinse tanks and associated articles with regular tap water until the smell of sodium hypochlorite is no longer detectable, then rinse a final time with filtered water. It is best to leave tanks dry until used again. If metals were utilized in a previous experiment, the process described above should be followed with a 10% HCl acid wash followed by appropriate rinsing. In either case, any previously used air tubing and airstones should be replaced or sterilized.

In protocols in which animals are maintained in experimental tanks for extended periods of time (i.e., greater than 4 days), it is necessary to physically clean the tank interior, including the standpipe and filter, using filter floss and then siphoning out the debris. Failure to remove organic films can significantly affect the concentration of bioavailable copper or other toxicants.

Results

LC_{50}

The 24-h LC_{50} with 95% confidence intervals for copper to brown bullhead was 880 (650 to 1,420) µg Cu/l; the 96-h LC_{50} was 40 (32 to 48) µg Cu/l. No mortality occurred in control groups. Tables 1 and 2 show probit analyses of 24- and 96-h dose responses, respectively. Acute dose–response (mortality) data are graphically presented in Figure 4, and cumulative mortality (time-to-death) for bullhead throughout the 96-h exposure period is shown in Figure 5.

Hematocrit and serum protein

Hematocrits for fish exposed to 23.7, 50.7, 75.3, and 96.7 µg Cu/l were significantly higher ($p \leq 0.05$) than control fish values (Table 3). Plasma total protein from fish exposed to 75.3 and 96.7 µg Cu/l was significantly higher

Table 1   24-h LC$_{50}$ Response Data for Brown Bullhead Exposed to Copper (A) with Estimated LC Values and Confidence Limits (B)

A. Copper exposure

Conc. (µg Cu/l)	Number exposed	Number responding	Observed proportion responding	Adjusted proportion responding	Predicted proportion responding
3	10	0	0.0000	0.0000	0.0000
100	10	0	0.0000	0.0000	0.0000
150	10	0	0.0000	0.0000	0.0004
370	10	1	0.1000	0.1000	0.0500
720	10	2	0.2000	0.2000	0.3514
1,190	10	8	0.8000	0.8000	0.7165

B. Estimated LC values and confidence limits

Point	Conc. (µg/l)	95% confidence limits	
		Lower	Upper
LC 01	258.5	28.1	423.7
LC 05	370.1	79.0	537.9
LC 10	448.1	135.9	617.1
LC 15	509.9	194.2	682.8
LC 50	880.2	649.3	416.7
LC 85	519.5	1,076.5	5,927.9
LC 90	1,729.0	1,178.4	8,563.5
LC 95	2,093.7	1,339.8	14,852.0
LC 99	2,997.8	1,688.3	42,118.6

Table 2   96-h LC$_{50}$ Response Data for Brown Bullhead Exposed to Copper (A) with Estimated LC Values and Confidence Limits (B)

A. Copper exposure

Conc. (µg Cu/l)	Number exposed	Number responding	Observed proportion responding	Adjusted proportion responding	Predicted proportion responding
3.0	20	0	0.0000	0.0000	0.0000
23.7	20	3	0.1500	0.1500	0.1417
50.7	20	13	0.6500	0.6500	0.6957
75.3	13	13	1.0000	1.0000	0.9093
96.7	15	14	0.9333	0.9333	0.9684
154.0	12	12	1.0000	1.0000	0.9977

B. Estimated LC values and confidence limits

Point	Conc. (µg/l)	95% confidence limits	
		Lower	Upper
LC 01	12.9879	6.0288	18.8651
LC 05	18.0108	10.0013	24.2239
LC 10	21.4408	13.0580	27.7667
LC 15	24.1178	15.5998	30.5111
LC 50	39.6562	31.5968	47.5819
LC 85	65.2055	53.8571	88.1758
LC 90	73.3468	59.6831	104.4523
LC 95	87.3149	68.9619	135.2876
LC 99	121.0833	89.2280	222.7289

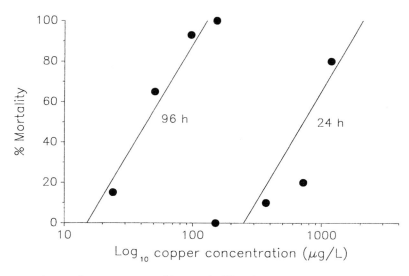

Figure 4 Acute dose–response of brown bullheads to copper.

($p \leq 0.05$) than control fish. The hematocrit:total protein ratio for 23.7, 50.7, and 96.7 μg Cu/l exposed fish was also higher ($p \leq 0.05$) than control fish.

Behavioral changes

Behavioral observations of copper-intoxicated fish included exaggerated respiratory movements and increased rate of respiration (gilling), change of pitch attitude (from horizontal to a 45° head up angle), lethargy, and eventually loss of equilibrium.

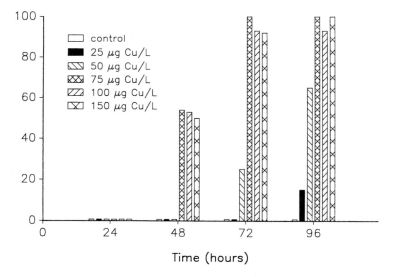

Figure 5 Cumulative acute mortality of brown bullheads exposed to copper showing response over the 96-h exposure period.

Table 3 Effect of Copper Exposure on Brown Bullhead
Packed Cell Volume (PCV), Total Plasma Protein (TP),
and PCV:TP Ratio

	Nominal exposure concentration (µg Cu/l)				
	0 n = 17	25 n = 14	50 n = 8	75 n = 3	100 n = 12
PCV	34.0 (±8.0)	43.0 (±4.4)*	45.6 (±7.5)*	52.0 (±4.0)*	46.6 (±7.2)*
TP	4.9 (±0.7)	5.3 (±0.7)	5.3 (±0.6)	7.0 (±0.9)*	5.8 (±1.2)*
PCV:TP	6.9 (±1.5)	8.2 (±1.3)*	8.6 (±1.0)*	7.5 (±0.8)	8.2 (±1.5)*

Note: Values are means (± SD). Asterisk denotes exposure values significantly different from control values ($p \leq 0.05$).

Color adaptation

A distinct color change was noted in the brown bullhead tested in the dosing
system. Brown bullhead, normally black to dusky dorsally, turned light gray
when moved from the dark green holding tanks to the off-white, natural
color, polypropylene exposure vessels (Figure 6). This color change was
gradual over the 4 d exposure period. Control (3 µg Cu/l) animals and those
exposed to 23.7 µg Cu/l exhibited a more rapid and extreme dark-to-light
color transition than did animals exposed to 50.7 and 96.7 µg Cu/l.

Figure 6 Light and dark color adapted brown bullheads 72 h after being transferred
from a dark holding tank to relatively light colored exposure tanks (refer to text for
additional details).

Discussion

Dosing system

The dosing system used in this study is straightforward in its design, relatively inexpensive to construct and consistent in its performance. The system and the activated charcoal contactor(s) may be plumbed for a variety of laboratory culture or dosing needs.

A major consideration for system construction includes adequate planning and appropriate use of materials. All fittings, valves and connections should be made of PVC or other relatively inert products (copper, zinc-plated, and other types of materials may serve as a source of contamination). Further, a newly-constructed system needs to be "seasoned." New PVC systems should be put into operation without animals several weeks to months prior to use, depending on the amount of piping and number of fittings (glue joints) in line. This is due to leaching of solvents from the cement and primers used to join PVC pipe fittings. PVC pipes and fittings may also have contaminated surfaces prior to installation. To evaluate whether a newly-constructed (glued) PVC piping system is ready for use it is recommended that system water be analyzed for organics (particularly for methyl ethyl ketone, acetone and tetrahydrofuran) by GC/MS. Samples for organic analysis should be taken conservatively, i.e., after water is allowed to sit static in the pipes for several hours or overnight. Care should be taken to follow appropriate sampling procedures for organic solvents since solvent residues are volatile. Use of smell to detect organic residues from water which was static in the system pipes (and then immediately discharged) is effective when solvent concentrations are initially high; this may help reduce the cost of multiple, repeated GC/MS analyses when monitoring the decline of solvent concentrations over time.

When designing holding and exposure systems for aquatic animals one should take into account local resources and materials. These include experts from regional laboratories, universities, and aquaria, as well as reputable, knowledgeable plumbing and manufacturing professionals and distributors. It may be helpful to browse through aquaculture supply catalogs (such as those distributed by Aquatic Eco-Systems, Apopka, FL, 800-422-3939 and Aquanetics Systems, San Diego, CA, 619-291-8444) to become familiar with available materials. Also, consider taking advantage of parts (or all) of pre-fabricated systems which are commercially available. Ingenuity is often the key to successful and cost-effective system design.

Data derived using dosing system

The acute 96-h LC_{50}, 40 µg/l (pH = 7.2 to 7.4; hardness = 80 mg/l) reported in this study for the brown bullhead is the lowest (indicating highest toxicity) yet determined for this species. For comparison in other flow-through bioassays, Brungs et al.[18] derived an LC_{50} range for brown bullhead of 179 to 190 µg Cu/l using carbon filtered tap water as diluent (pH = 7.2 to 8.2; hardness = 190 to 215 mg/l). Christensen et al.[19] reported an LC_{50} of 180 µg

Cu/l (pH = 7.6; hardness = 202 mg/l). Additional flow-through data using "standard water" as diluent yielded a 96-h LC_{50} to juvenile brown bullhead of 540 (95% CI = 430 to 770) µg/1.[20] This "standard water" consisted of a mixture of demineralized spring water with carbon-filtered tap water, and had a hardness of 196 to 205 mg/l and a pH of 7.9 to 8.1. The highest 96-h LC_{50} we found in the literature for copper was calculated by Benedetti et al.,[21] 2,500 µg/l, for adult brown bullhead. However, these authors did not define their diluent water.

Water hardness has a substantial effect on the toxicity of copper (and other metals) in aqueous exposures. An increase in hardness causes reduction in copper toxicity.[22-24] For example, fathead minnows, *Pimephales promelas*, exposed to copper in soft vs. hard diluent water (20 and 360 mg/l hardness, respectively) showed differences of two orders of magnitude in LC_{50} values, 22 and 1,760 µg/l, respectively.[25] Reduced metal toxicity in harder water is, in part, a result of binding with hydroxyl, carbonate, phosphate, and other dissolved anions.[26-28] Considering the level of hardness of the diluent water used in the present study (80 mg/l), the LC_{50} value for brown bullhead appears reasonable. The effect of pH on copper toxicity was investigated by Howarth and Sprague.[23] According to these authors, Cu^{2+} and $CuOH^+$ are the primary toxic copper species. The effects of pH on copper toxicity appear greatest at increased hardness and may be dependent on copper concentration.

Copper-exposed brown bullhead in this study showed an increase in hematocrit relative to the controls. Other authors have also observed increased hematocrit in copper-exposed fish. Christensen et al.[19] found that brown bullhead exposed to 107 µg Cu/l for 6 d had a 41.9% (±4.6) hematocrit relative to unexposed control fish (37.1% ± 3.6 hematocrit). This difference, although found to be statistically significant ($p \leq 0.05$), may not be functionally significant. Waiwood[29] exposed rainbow trout (*Oncorhynchus mykiss*) to 30 µg Cu/l for 48 h and showed significantly higher hematocrit (38.97%) in copper-exposed fish compared with controls (31.78%). This author also measured erythrocyte counts; statistical analysis indicated that hematocrit and red cell count were not significantly different between exposure and control groups.

Christensen et al.[19] suggested that increased hematocrit was due to mortality of small immature erythrocytes or erythrocyte swelling, a symptom demonstrated for chromium- and zinc-treated rainbow trout.[30,31] However, Waiwood[29] suggested that increased hematocrit was not due to red cell swelling, but due to hemoconcentration accompanied by a shift in water from blood to muscle. Increased hematocrit in the present study also appeared due to hemoconcentration based on increased serum protein values in copper-exposed (75 and 100 µg/l) groups. Hemoconcentration (dehydration), based on increased hematocrit:total protein ratios, was significant in all copper-exposed groups except the 75 µg/l exposure group. Lack of significance in the 75 µg Cu/l exposure group was due to small sample size; only three fish were bled because others had succumbed prior to obtaining blood samples. Data from this group, however, shows a similar trend in hemoconcentration as seen in the other exposure groups.

In this study brown bullhead showed the capacity to change color in response to their surroundings. Fish exposed at low (3 and 23.7 µg/l) copper concentrations color-adapted (from a very dark gray or black to a light gray) faster than animals exposed at higher concentrations (50.7 to 154.0 µg/l). The color adaptation response was observed after animals were moved from the dark green holding tanks to the lighter, natural colored polypropylene tanks. In fishes, this is generally a physiological response associated with camouflage. Many fish, however, will not color adapt. The mechanism by which copper alters this response is unclear. It is known that sensory organs, including both chemo- and mechanoreceptors, are very sensitive to copper.[32] However, color adaptation to the surrounding environment is most likely based on visual cues.[33] A speculative explanation is that threshold levels of copper have an effect on some part of the hypophyseal-hypothalamic system or on melanophores directly.

Use of brown bullhead or any fish model, in research or for controlled biomonitoring, requires that animals be healthy and in stable condition prior to use. A holistic approach, incorporating veterinary, toxicology, and husbandry perspectives, is key to the successful collection of meaningful data.

In summary, the E/M system and water filtration scheme described in this paper represent cumulative design concepts from previous efforts in our facility[34] and from other laboratories.[35-44] The described systems are comparatively easy to construct and have been adapted for use with relatively larger exposure tanks. When planning to construct a new system, the design should be able to accommodate alternate toxicants or test species as well as potential system "add-ons" or upgrades. Careful planning, including review of the design with experienced colleagues, will help to minimize the total effort and the need for potential redesign.

Acknowledgments

The authors thank Richard Bennett, Eric May, Ann Muhvich, and Tom Bell for their support throughout these studies. We also thank Mark Meyers for reviewing this manuscript. Additional support for this project came from Benjamin Trump and the University of Maryland Department of Pathology, and The Maryland Department of Natural Resources, contract # F180-89-008.

References

1. Nriagu, J.O. 1979. *Copper In The Environment Part II: Health Effects*. New York:John Wiley & Sons. 489 pp.
2. EPA (U.S. Environmental Protection Agency). 1989. Short-term methods for measuring the chronic toxicity of effluents and receiving waters to freshwater organisms, 2nd edition. Prepared by C.I. Weber, W.H. Peltier, T.J. Norberg-King, W.B. Horning, II, F.A. Kessler, J.R. Menkedick, T.W. Neiheisel, P.A. Lewis, D.L. Klemm, Q.H. Pickering, W.L. Robinson, J.M. Lazorchak, L.J. Wymer and R.W. Freyberg. EPA/600/4-89/001.

3. EPA (U.S. Environmental Protection Agency). 1991. Methods for measuring the acute toxicity of effluents and receiving waters to freshwater organisms, 4th edition. Edited by C.I. Weber. EPA/600/4-90/027.

4. Sorensen, E.M. 1991. *Metal Poisoning In Fish*. Boca Raton:CRC Press. 374 pp.

5. Di Giulio, R.T., Benson, W.H., Sanders, B.M. and Van Veld, P.A. 1995. Biochemical mechanisms: metabolism, adaptation and toxicity. In *Fundamentals of Aquatic Toxicology*, 2nd edition, Ed. G.M. Rand, 17:523–561. Washington, D.C.: Taylor and Francis. 1,125 pp.

6. Garofano, J.S. and Hirshfield, H.I. 1982. Peripheral effects of cadmium on the blood and head kidney in the brown bullhead (*Ameiurus nebulosis*). *Bull. Environm. Contam. Toxicol.* 28:552–556.

7. Hargis, W.J. and Zwerner, D.E. 1988. *Proc. Understanding the Estuary: Advances in Chesapeake Bay Research Conf. Some histologic gill lesions of several estuarine finfishes related to exposure to contaminated sentiments: a preliminary report, Baltimore, 1988, Chesapeake Research Consortium publication 129. CBP/TRS 24/88.*

8. Reimschuessel, R., Kane, A.S., Muhvich, A.G. and Lipsky, M.M. 1993. *Recovery of Histological Lesions in the Gills of Brown Bullheads Exposed to Copper*. Mar. Env. Res. 35(1–2):198–99 (Abstr.).

9. Bowser, P.R. and Plumb, J.A. 1980. Channel catfish virus: comparative replication and sensitivity of cell lines from channel catfish ovary and the brown bullhead. *J. Wildl. Dis.* 16(3):451–454.

10. Isbell, G.L. and Pauley, G.B. 1983. Characterization of immunoglobulins from the brown bullhead (*Ameiurus nebulosis*) produced against a naturally occurring bacterial pathogen, *Aeromonas hydrophila*. *Dev. Comp. Immunol.* 7(3):473–482.

11. Buck, C.C. and Loh, P.C. 1985. Growth of brown bullhead (BB) and other cell lines on microcarriers and the production of channel catfish virus (CCV). *J. Virol. Methods.* 10(2):171–184.

12. Martin-Alguacil, N., Babich, H., Rosenberg, D.W. and Borenfreund, E. 1991. *In vitro* response of the brown bullhead catfish cell line BB to aquatic pollutants. *Arch. Environ. Contam. Toxicol.* 20(1):113–117.

13. Munro, A.L.S. and Roberts, R.J. 1989. The aquatic environment. In *Fish Pathology*, 2nd edition, Ed. R.J. Roberts, 1:1–12. Philadelphia:Baillere Tindale. 467 pp.

14. Moe, M.A. 1992. *The Marine Aquarium Reference: Systems And Invertebrates*, Plantation, FL:Green Turtle Publications. 512 pp.

15. Spotte, S. 1979. *Fish and Invertebrate Culture: Water Management in Closed Systems*, 2nd edition. New York:John Wiley & Sons. 179 pp.

16. Dixon, W.J. and Massey, F.J. 1983. *Introduction to Statistical Analysis*, 4th edition, 19:426–441. New York:McGraw Hill. 678 pp.

17. APHA (American Public Health Association), American Water Works Association, and the Water Pollution Control Federation. 1985. *Standard methods for the examination of water and wastewater*, 16th edition. APHA, Washington, D.C.

18. Brungs, W.A., Leonard, E.N. and McKim, J.M. 1973. Acute and long-term accumulation of copper by the brown bullhead, (*Ameiurus nebulosus*). *J. Fish. Res. Bd. Can.* 30:583–586.

19. Christensen, G.M., McKim, J.M., Brungs, W.A. and Hunt, E.P. 1972. Changes in the blood of the brown bullhead (*Ameiurus nebulosis*) following short and long term exposure to copper (II). *Toxicol. Appl. Pharmacol.* 23:417–427.

20. EPA (U.S. Environmental Protection Agency). 1976. Validity of laboratory tests for predicting copper toxicity in streams. EPA-600/3-76-116.

21. Benedetti, I., Albano, A.G. and Mola, L. 1989. Histomorphological changes in some organs of the brown bullhead, (*Ameiurus nebulosus*), following short- and long-term exposure to copper. *J. Fish Biol.* 34:273–280.

22. Miller, T.G. and Mackay, W.C. 1980. The effects of hardness, alkalinity and pH of test water on the toxicity of copper to rainbow trout (*Salmo gairdneri*). *Wat. Res.* 14:129–133.

23. Howarth, R.S. and Sprague, J.B. 1978. Copper lethality to rainbow trout in waters of various hardness and pH. *Wat. Res.* 12:455–462.

24. Lauren, D.J. and McDonald, D.G. 1986. Influence of water hardness, pH, and alkalinity on the mechanisms of copper toxicity in juvenile rainbow trout (*Salmo gairdneri*). *Can J. Fish. Aquat. Sci.* 43:1488–1496.

25. Pickering, Q.H. and Henderson, C. 1966. The acute toxicity of some heavy metals to different species of warmwater fishes. *Air Wat. Pollut. Int. J.* 10:453–463.

26. Hartung, R. 1973. Biological effects of heavy metal pollutants in water. In *Metal Ions In Biological Systems: Studies Of Some Biochemical And Environmental Problems*, Ed. S.K. Dhar, 161–172. New York:Plenum Press. 306 pp.

27. Chapman, G.A. and McCrady, J.K. 1977. Copper toxicity: a question of form. In *Recent Advances in Fish Toxicology, A Symposium*, Ed. R.A. Tubb, 132–151. Corvallis Environmental Research Laboratory, U.S. Environmental Protection Agency.

28. Leland, H.V. and Kuwabara, J.S. 1985. Trace Metals. In *Fundamentals of Aquatic Toxicology*, Ed. G.M. Rand and S.R. Petrocelli, 13:374–415. New York:Hemisphere Publishing. 666 pp.

29. Waiwood, K.G. 1980. Changes in hematocrit of rainbow trout exposed to various combinations of water hardness, pH, and copper. *Trans. Am. Fish. Soc.* 109:461–463.

30. Schiffman, R.H. and Fromm, P.O. 1959. Chromium-induced changes in the blood of rainbow trout (*Salmo gairdneri*). *Sewage and Industrial Wastes.* 31:205–211.

31. Hodson, P.H. 1974. *The effect of temperature on the toxicity of zinc to fish of the genus Salmo.* PhD thesis. University of Guelph, Guelph, Canada.

32. Kleerekoper, H., Waxman, J.B. and Matis, J.H. 1973. Interaction of temperature and copper ions as orienting stimuli in the locomotor behavior of goldfish (*Carassius Auratus*). *J. Fish. Res. Board Can.* 30:725–728.

33. Fugii, R. 1993. Coloration and chromatophores. In *The Physiology of Fishes*, Ed. D.H. Evans, 17:535–562. Ann Arbor: CRC Press. 592 pp.

34. Kane, A.S., Bennett, R.O. and May, E.B. 1988. A dosing system to vary pH, salinity and temperature. *Wat. Res.* 22(10):1339–1344.

35. Benoit, D.A., Mattson, V.R. and Olson, D.L. 1982. A continuous-flow mini-diluter system for toxicity testing. *Wat. Res.* 16:457–464.

36. Benoit, D.A., Phipps, G.L. and Ankley, G.T. 1993. A sediment testing intermittent renewal system for the automated renewal of overlying water in toxicity tests with contaminated sediments. *Wat. Res.* 27:1403–1412.

37. Carlson, R.W., Lien, G.J. and Holman, B.A. 1989. An automated monitoring system for fish physiology and toxicology. EPA/600/S3-89/011. Available through U.S. EPA, Duluth, or through NTIS (order #PB89-155212/AS).

38. Gingerich, W.H., Seim, W.K. and Schonbrod, R.D. 1979. An apparatus for the continuous generation of stock solutions of hydrophobic chemicals. *Bull. Environm. Contam. Toxicol.* 23:265–269.

39. Lemke, A.E., Brungs, W.A. and Halligan, B.J. 1978. Manual for construction and operation to toxicity testing proportional diluters. EPA/600/3-78-072.

40. McKim, J.M. and Goeden, H.M. 1982. A direct measure of the uptake efficiency of a xenobiotic chemical across the gills of brook trout (*Salvelinus fontinalis*) under normoxic and hypoxic conditions. *Comp. Biochem. Physiol.* 72C:65–74.

41. McKim, J.M., Olson, G.F., Holcombe, G.W. and Hunt, E.P. 1976. Long term effects of methyl mercuric chloride on three generations of brook trout (*Salvelinus fontinalis*): toxicity, accumulation, distribution and elimination. *J. Fish. Res. Board Can.* 33:2726–2739.

42. Mount, D.I. and Brungs, W.A. 1967. A simplified dosing apparatus for fish toxicology studies. *Wat. Res.* 1:21–29.

43. Pinkey, A.E., Klauda, R.J. and Wright, D.A. 1987. Manual for design and operation of a solenoid-based delivery system for aquatic toxicity testing. *Environ. Technol. Lett.* 8:153–158.

44. Riley, C.W. 1975. Proportional diluter for effluent bioassays. *J. Fed. Wat. Poll. Control Fed.* 47(11):2620–2626.

chapter thirty-three

Assessment of sediment toxicity at the sediment–water interface

Brian S. Anderson, John W. Hunt, Michelle Hester, and Bryn M. Phillips

Introduction

There is growing recognition that sediment contamination may have a significant deleterious effect on aquatic ecosystems through impacts on benthic community structure and associated effects on overlying water bodies.[1-3] State and federal government regulatory agencies currently rely on several tools to assess the extent of sediment pollution and to determine effects of contaminants on marine communities; these include measures of sediment chemical concentrations, bioaccumulation, biomarkers, assessments of benthic community structure, and biological toxicity testing. Together, these measures provide for an integrative approach to defining sediment quality.

Toxicity tests are considered to be a powerful assessment tool because they can be used to describe the effects of complex mixtures of chemicals on ecologically important organisms.[4] Sediment protocols generally involve testing some combination of three different exposure matrices to evaluate toxicity: solid phase (whole sediment), interstitial water, and elutriates. While all exposure matrices have specific applications, it has been recognized that there is a clear need for more toxicity test protocols in which sublethal endpoints are used in tests of solid phase samples. One approach is to apply toxicity test protocols traditionally used for water column toxicity testing to assessing toxicity of solid phase samples at the sediment–water interface (SWI). Variations on this approach have been reported by Chapman and Morgan[5] and Burgess et al.[6]

Sediment toxicity test procedures have largely ignored the SWI, although it has been the subject of numerous geochemical studies. This is an ecologically important habitat where significant densities of benthic and epibenthic species occur. In addition to being a likely contaminant exposure location for strictly epibenthic species, the gametes and embryonic stages of many infaunal, epibenthic, and water column species may spend critical phases of their early development associated with this environment. Due to the flux of contaminants out of sediments, the constant deposition of new material from the water column, and remineralization of organic matter within the sediments, this habitat has potential for containing toxic concentrations of contaminants.[7-10]

This paper describes an exposure system designed to assess toxicity at the SWI. In this procedure, sediment samples are placed into test chambers which are then filled with uncontaminated overlying seawater. Screen tubes are then placed into the test chambers so that the screen is almost in contact with the sediment. After a 24-h equilibration period, test organisms are inoculated into the screen tubes, where they develop in proximity to the sediment. The size of the screen is small enough to retain embryos but large enough to allow for passive diffusion of chemicals into the screen tube. At the termination of the test the screen tube is removed and the animals are washed into a separate container for microscopic evaluation. In this test system most of the animals are retained, allowing accurate quantification of developmental abnormalities.

This exposure system has several advantages. First, it complements the current suite of exposure matrices by allowing sediment toxicity assessment in an ecologically important habitat in which toxic effects have not been adequately investigated. The exposure system facilitates the use of relatively sensitive embryo/larval toxicity test protocols in solid phase exposures, addressing the current lack of sensitive solid phase tests which incorporate sublethal endpoints. Although the method has been developed for marine and estuarine applications using sea urchin and fish embryos, it has not yet been tested using protocols developed for freshwater species. The toxicity testing procedures are general enough, however, to be adaptable for most applications.

An additional advantage of this exposure system is that it may be used to assess sediment toxicity in intact, unhomogenized cores, thereby eliminating artifacts that result from the manipulation of sediment and pore water samples. Thus, this procedure may better approximate *in situ* exposure conditions. The screen employed in this procedure minimizes any interaction between test organisms and resident infauna which may be present in intact samples. The screen also reduces grain size effects, eliminating a possible false-positive artifact for which some solid phase protocols have been criticized.[11] The exposure system has also been combined with analyses of toxicant flux measures to provide biological information on the effects of contaminated sediments on the water column.[12]

Materials required

Polycarbonate tubing for sediment cores (7.5-cm id, AIN Plastics, Santa Clara, CA)

Polyethylene plastic caps for cores (7.5 cm, AIN Plastics, Santa Clara, CA)

Polycarbonate tubing for exposure screen tubes (e.g., 5-cm ID, AIN Plastics, Santa Clara, CA)

Plastic cement for screen construction (Craftics® Brand, Chicago, IL)

Polyethylene (PECAP®) screen (AREA, Homestead, FL)

Parafilm® for sealing cores

Refractometer — for determining salinity

Thermometers, glass or electronic, laboratory grade — for measuring water temperatures

Thermometer, National Bureau of Standards Certified (see U.S. EPA METHOD 170.1, U.S. EPA, 1979) — to calibrate laboratory thermometers

pH and DO meters — for routine physical and chemical measurements. Unless the test is being conducted to specifically measure the effect of one of these two parameters, portable, field-grade instruments are acceptable

Standard or micro-Winkler apparatus — for determining DO (optional)

Electrodes and reagents for measuring hydrogen sulfide and ammonia in seawater

Benchtop centrifuge for extraction of interstitial water for measurement of NH_3 and H_2S

Beakers, 1,000-ml borosilicate glass

Wash bottles — for dilution water; for topping off graduated cylinders

Wash bottles — for deionized water; for rinsing small glassware and instrument electrodes and probes

Constant temperature chambers or water baths — for keeping dilution water supply, gametes, and embryo stock suspensions at test temperature (15°C) prior to the test

Water purification system — Millipore Super-Q, deionized water (DI) or equivalent

Pipettes, automatic — adjustable, to cover a range of delivery volumes from 0.010 to 1.000 ml

Inverted or compound microscope — for inspecting gametes and making counts of embryos and larvae

Hemacytometer

Sedgwicke-Rafter counting cell

Mixing Plunger (for mixing gametes)

Counter, two-unit, 0 to 999 — for recording counts of embryos and larvae

Water bath, incubator, or room with temperature control — for maintaining test solution temperature (18 or 20°C) during the 96-h test

Graduated cylinders — Class A, borosilicate glass or non-toxic plastic labware, 50 to 1000 ml for making test solutions. (*Note*: not to be used interchangeably for gametes or embryos and test solutions).

Pipette bulbs and fillers — PROPIPET®, or equivalent

Fume hood — to protect the analyst from effluent or formaldehyde fume

Tape, colored — for labeling tubes and other containers

Markers, waterproof — for marking containers, etc.

Teflon spoons

Gloves, disposable — for personal protection from contamination

Data sheets (one set per test) — for data recording

Formaldehyde, 37% (Concentrated Formalin) — for preserving embryos and larvae

Hexane, HCl, and deionized water for cleaning glass and plasticware (reagent grade)

Tanks, trays, or aquaria for holding adult sea urchins, e.g., standard salt-water aquarium (capable of maintaining sea water at 15°C), with appropriate filtration and aeration system

Air pump, air lines, and air stones — for aerating water containing adult urchins (for static systems and emergency aeration for flow-through systems)

0.5 *M* KCl

Syringe and 24-gauge needle for injecting urchins

Procedures

The following procedures are divided into sections for testing homogenized sediment and intact (non-homogenized) surficial sediment samples. The SWI method has been used primarily for testing marine and estuarine sediment samples and the screen tubes described here were designed for exposing invertebrate and vertebrate embryos larger than 60 µm in diameter. Modifications may be necessary for smaller animals or life stages.

Screen tube construction

Screen tubes have been constructed from a variety of materials including PVC and polycarbonate. Polycarbonate tubing has several characteristics which better suit this application. It is a durable and relatively inert plastic which holds up well under continued use and is easily cleaned with conventional solvents. Because it is clear, it facilitates observation of test organisms and conditions. Screen tube size will vary with species and protocols; for using the sea urchin embryo/larval protocol we have constructed screen tubes using 4-cm (ID) diameter stock (AIN Plastics, Santa Clara, CA) which is cut into 15-cm high sections on a conventional bandsaw. The wall thickness is 3 mm. A 1-cm section is cut from the bottom of the tube and this serves as the pedestal which sits on the sediment surface (Figure 1). Polyethylene screen (AREA, Homestead, FL) is glued to the tube using clear-thickened acrylic plastic glue and the pedestal is then glued back on the tube to sandwich the screen. A small hole is drilled into the side of the pedestal which is used to purge any air trapped under the screen during immersion. Screen size will very depending upon the application; we use 37-µm mesh screen for the sea urchin embryo development protocol. This size screen works well for life-stages greater than approximately 60 µm. Polyethylene mesh is stronger than conventional nylon mesh and better withstands repeated solvent rinses.

Polycarbonate core tubes are used for collecting intact, unhomogenized samples (see below). Cores are constructed by cutting 7.3-cm (id) stock into 20-cm high sections. Polytheylene caps may be purchased for sealing samples inside the tubes.

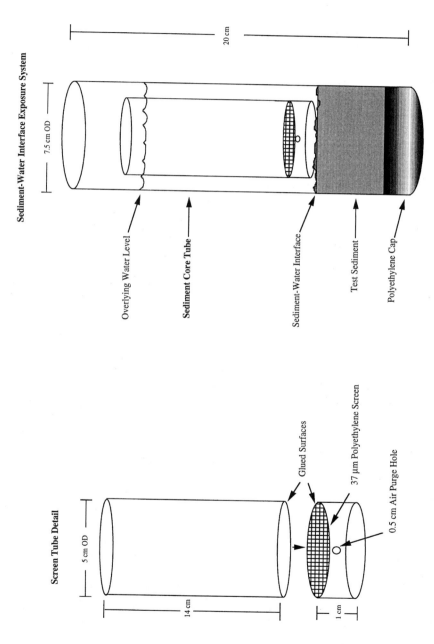

Figure 1 Schematic diagram of sediment–water interface exposure system.

Screen tubes, cores and caps are cleaned using the following procedure:

Rinse in warm tap water (soap and scrub if necessary)
Rinse three times with reagent-grade hexane
Rinse three times in deionized water
Rinse three times with 3 N HCl acid
Rinse three times in deionized water
Soak 48 h in deionized water
Soak 48 h in seawater
Rinse three times in milli Q water
Place in drying oven (55°C)

Homogenized sample handling

The following procedures are for conducting solid-phase SWI tests using the 96-h sea urchin (*Strongylocentrotus purpuratus*) embryo development proto-col. Sediment collection and processing procedures follow guidelines described in ASTM.[13] It is assumed that test sediments are homogenized according to this method prior to loading into the test containers. Sediment and overlying water are added to the cores 24-h prior to inoculation of the test organisms to allow samples to equilibrate.

Prior to loading homogenized sediment into the cores, polyethylene caps are placed on one end of each core and this end is sealed with Parafilm. After labeling, the cores are arranged in groups of five replicates per sample with one extra core per sample to be used for interstitial sulfide and ammonia measurements at the termination of the test.

Samples to be tested are identified and the interstitial water salinity of each sample is checked. Interstitial salinity must be greater than 30‰ to be tested with purple urchin embryos. Samples with interstitial salinity below 30‰ are adjusted following methods described in ASTM.

Using a Teflon spoon, 200 ml of sediment is placed into each tube, forming a layer 5 cm deep. After placing a clean plastic disk on the surface of the sediment using forceps, 500 ml of "clean" overlying seawater (33 ± 2‰; 15°C) is added carefully with a siphon to minimize disturbance of the sediment; the disc is removed after the water is added. The cores are arranged in a constant temperature room and covered with acrylic sheets. The sediment and overlying water are allowed to equilibrate overnight before introduction of the screens and test organisms. A screen tube is added to each core container the following morning. The screen tube is gently placed in the core so that the bottom collar rests on the substrate, this leaves the screen itself 1 cm above the substrate. The water in the outer core tube should be 12 cm deep, resulting in 150 ml of water in the screen tube itself.

Intact sediment sample handling

Methods for handling intact core samples are essentially the same as those described above, with the following modifications designed to minimize disturbance of the sample. The core may be taken directly from the sampling device or, using divers, directly from the sediment surface. Depth of the sample will depend on the study goals. We generally take 5-cm deep cores because this is the practical sampling depth of the Van Veen grab sampler used at our laboratory. The core is pressed into the sediment and a pre-cleaned acrylic plate is inserted under the bottom of the core to prevent leakage of sample or interstitial water as the sample is removed. It is convenient to mark the 5-cm height for reference using a plastic cable tie wrapped around the outside of the core. After the core is removed from the sediment, the bottom is capped quickly as the acrylic plate is removed; then the top is capped. A small pinhole in the top cap relieves positive pressure on the sample and minimizes leakage as the cap is attached. Sample integrity is verified by the presence of sediment overlying water; if an inordinate volume of interstitial water or sediment leaks out, the sample is discarded and a new one taken. The outside of the tube is dried and the bottom wrapped tightly in Parafilm to prevent leakage, and the core is stored upright on ice.

Approximately 24 h prior to initiation of the toxicity test, the overlying water in the core (from the sample site) is gently siphoned off of the top of the sample leaving about 0.5-cm of water remaining to minimize disturbance of the sediment surface. Five hundred milliliters of fresh 33 ± 2‰ overlying water is then introduced into the cores using acrylic disks to minimize sample disturbance, as described above. Unlike homogenized samples, salinity of interstitial water may not be adjusted in the intact samples. As described above for homogenized samples, the cores are usually arranged in groups of five replicates per sample with one extra core per sample to be used for interstitial sulfide and ammonia measurements at the termination of the test. Screen tubes are gently added to the cores at the beginning of the day of the test. The test is run under ambient laboratory lighting conditions.

Toxicity testing

The following procedures are patterned after those described by ASTM.[14] The purpose of the embryo/larval development test with the sea urchin, *Strongylocentrotus purpuratus*, is to determine if sediment samples cause abnormal development of exposed embryos relative to embryos exposed to control or reference samples.

Test animals

Newly fertilized embryos of the sea urchin, *Strongylocentrotus purpuratus* are used in the embryo development toxicity test. Adult broodstock animals may be collected in the field or obtained from a commercial supplier. *S. purpuratus* is distinguished from *S. franciscanus*, a less common congener, by its purple to occasionally green color. Urchins can be maintained easily in aquaria or other tanks provided with running seawater. Urchins will eat a wide variety of marine algae, but prefer giant brown kelp, *Macrocystis pyrifera*. To ensure year-round spawning, culture animals at 12 to 15°C, in complete darkness. Water quality parameters should be monitored weekly and salinity maintained at $33 \pm 2‰$. It is preferable to separate urchin brood-stock by sex to prevent accidental spawning. The sex of the urchins may be determined at the time of spawning. After animals are spawned for toxicity tests, males and females should be placed in separate culture tanks. Once these animals are re-conditioned they may be spawned for use in later toxicity tests.

Dilution water

Dilution water for toxicity tests may be either filtered natural seawater or artificial seawater ($33 \pm 2‰$). Although there have been no reports demonstrating the utility of artificial seawater for long-term cultures of *Strongylocentrotus*, artificial sea salts (Forty Fathoms® brand, Marine Enterprises, Inc. Baltimore, MD) have been used successfully to culture sexually mature *Arabacia punctulata* sea urchins and to perform fertilization tests with this species. An alternative artificial seawater is made from GP2 seawater formulation.[15] This water is used to prepare eggs and sperm for toxicity tests, and for overlying water in sediment exposures. Tests are conducted at 15°C, so all water must be adjusted to this temperature prior to use.

Collection and preparation of gametes

To perform this test, quality gametes must first be collected and the eggs fertilized; the embryos are then diluted to the appropriate density for addition to the test containers.

Spawning of urchins

Eight to ten urchins are placed on several layers of paper towels on a clean surface and each urchin is injected with 0.5 cc of 0.5 M KCl in the soft tissue surrounding the Aristotle's lantern. The animals are gently shaken once or twice to stimulate gamete release. Gametes are released aborally with females releasing orange-colored eggs, and males releasing cream-colored sperm. Any urchins that have not spawned after 10 min are re-injected with 0.5 cc of 0.5 M KCl.

For tests conducted near the beginning or end of the purple sea urchin spawning season (approximately September to April) eggs must be inspected for uniformity and roundness. Only females with uniformly round eggs lacking follicles are selected. Eggs are not used if a high proportion of germinal vesicles are present or if eggs are released from females in large clumps. Eggs and sperm are washed off the urchins with a squirt bottle filled with 1-μm filtered seawater. Gametes are rinsed into separate beakers using special care to avoid cross contamination between sperm and eggs. To assess gamete viability, one drop of eggs are placed onto a well-slide and a small amount of sperm is added to test for fertilization. The presence of fertilization membranes is determined after ~5 min. If fertilization membranes are not present, new eggs are isolated. The eggs are placed in a constant temperature room (15°C) until ready for counting.

Preparation of appropriate sperm dilution

Sperm density must be 500 sperm per egg. Sperm density is determined by pipetting one ml of sperm into a designated graduated cylinder containing 75 ml of dilution water and 5 ml of buffered formalin or acetic acid to kill the sperm. The graduated cylinder is filled up to the 100 ml mark with dilution water. This is the sperm solution used to determine sperm density. While counting the sperm, the covered container holding the sperm solution to be used for fertilizing the test eggs is held at 15°C to maintain viability.

Sperm density is measured using a hemacytometer, and egg density is determined using a Sedgewick-Rafter Slide. Once the number of eggs per ml is determined, appropriate volumes of diluted sperm and eggs are determined to give a fertilized dilution of 500 sperm per egg. The sperm and eggs are then mixed in a 1-liter glass beaker and after 10 min, fertilization is checked. At least 95% of the eggs must be fertilized. If this percentage is not reached, wait an additional 10 min. If 95% fertilization is still not achieved, start over with new sperm and eggs.

The screen tubes are then inoculated with 1000 embryos per tube. It is important that the volume necessary to deliver 1000 embryos is small to minimize dilution of the overlying water. We generally use ≤5% dilution as our target, this gives an inoculation volume maximum of 7.5 ml in 150 ml of overlying water.

The negatively buoyant embryos will sink to the bottom and develop on the screen surface. Within approximately 24 h, larvae will have hatched and developed into the gastrula stage. These are visible swimming in the water column (assuming no toxicants inhibit development). By 48 h, the larvae progress to the pluteus stage. The test is terminated between 72 and 96 h.

Test termination

To terminate the test, the screen is taken from the sediment vessel and gently rinsed in a bucket of clean seawater to remove extraneous sediment on the outside of the tube. The contents of each screen tube are washed through a funnel into an appropriate container (such as a 20-ml scintillation vial) using a squirt bottle filled with seawater. After first rinsing down the inside of the tube itself so that all of the animals are on the screen surface, water from the squirt bottle is applied from the outside of the screen to the inside, washing the contents of the tube into the funnel. The contents are fixed with buffered formalin (5% final concentration) and capped and stored for later evaluation.

Water quality

The following water quality parameters are measured at the beginning and end of the test:

Overlying water dissolved oxygen (DO), pH, salinity
Overlying water ammonia
Overlying water hydrogen sulfide
Interstitial water ammonia
Interstitial water hydrogen sulfide

Overlying water samples are taken approximately 1 cm from the sediment surface (outside the screen tube). Interstitial water samples are extracted using a benchtop centrifuge. The ASTM sea urchin development protocol requires that dissolved oxygen levels must be between 60 and 100% of saturation. Test sediments which are high in organic material may cause low D.O. levels. If this is the case, it may be necessary to slowly aerate the overlying water in the screen tubes. This is best accomplished by slow bubbling through a disposable Pasteur pipette. If one treatment is aerated due to low D.O., all treatments should be aerated.

Data collection and tabulation

Using normal brightfield microscopy (100×), count a minimum of 100 individuals (e.g., embryos, plutei, etc.) per sample using a hand counter with multiple keys (such as a blood cell counter). The animals are categorized as either "normal" or "abnormal" and percent abnormal is calculated from the total count. For more descriptions of toxicity test endpoints using sea urchins, see Anderson et al.,[16] Hose,[17] and Bay et al.[18] The latter paper is a thorough review of the use of echinoids in toxicity testing.

Normally developed larvae have the following characteristics:

- Four skeletal rods that extend at least half the distance from the base to the apex.
- A gut which demonstrates signs of differentiating into three chambers. In some cases it is not possible to distinguish all three chambers of the gut. If the apex of the gut appears lobed and constricts distally, then normal gut development may be inferred.

Abnormal or inhibited larvae may fit into any one of the following categories:

- Pathological prehatched: unfertilized eggs and fertilized eggs with the membrane still visible.
- Pathological hatched: larvae that have no fertilization membrane and demonstrate an extensive degree of malformation or necrosis. Most larvae appear as dark balls of cells or dissociated blobs.
- Retarded: larvae at blastula and gastrula stages that have no gut differentiation or have underdeveloped spicules.
- Gut abnormalities: larvae that are otherwise normal with exogastrulated guts, lacking differentiation or lacking guts.

- Skeletal abnormalities: misshapen larvae as a result of missing or underdeveloped skeletal rods, rods growing in abnormal directions or extraneous rods.

Note: Skeletal rods that are slightly separated at the apex might be caused by formalization and should not be considered abnormal. Calculate Percent Normal Development for each control replicate test. Test acceptability is 70% normal development in the seawater control.

Data analysis

Data are recorded on standardized Urchin Development Test Data Sheets. Normally, percent development in each treatment is compared to an appropriate reference treatment (seawater control or reference sediment from an uncontaminated environment). Statistical comparisons are made using analysis of variance (ANOVA) or *t*-tests.[19] Since both assume that responses are independently and normally distributed with a common variance within treatment levels, a test of the validity of these assumptions is recommended. The raw data are transformed using an arcsine/square root transformation for normalization. Bartlett's test or Levin's test may be used to test for homogeneity of variances. Non-parametric statistical tests may be used as an alternative.

Example data set

The following data were generated from preliminary SWI experiments assessing the toxicity of surficial sediments (top 2 cm) and interstitial water collected from 13 sites in San Diego and Los Angeles Harbors. Sediment from individual Van Veen grab samples were composited to provide a homogenate sample for each site. Interstitial water was extracted from the homogenized sediment in order to compare interstitial water toxicity to that of solid phase samples. Interstitial water was extracted using a piston squeezer[20] with some modifications based on the work of Carr et al.[21] In addition, intact (not homogenized) core samples were taken from two separate grab samples at each site using polycarbonate core tubes. Integrity of the intact samples was determined by verifying the presence of overlying site water. For solid phase SWI exposures, uncontaminated 1-µm filtered natural seawater was added to the core tubes containing either sediment homogenate or intact sample to provide 500 ml of clean overlying water; this resulted in 150 ml in each screen tube exposure chamber. Overlying site water was siphoned off the top of the intact samples prior to introduction of clean overlying water. Screen tubes were placed into each core tube and all samples were allowed to equilibrate for 24 h at 15°C prior to initiation of the toxicity tests. All samples were then tested with the 96-h purple sea urchin (*Strongylocentrotus purpuratus*) embryo-larval development test. For additional toxicity comparisons, sediment homogenates from all 13 sites were also tested with the 10-day amphipod protocol using the phoxocephalid amphipod *Rhep-*

oxynius abronius.[22] Bulk phase metal and organic chemicals were measured on each sample for correlation with toxicity.

Results and discussion

Results of these experiments indicate a range of toxicity in these samples, with more toxicity in interstitial water relative to solid-phase SWI exposures using homogenized sediment (Figure 2). Eleven out of 13 sites (85%) were significantly toxic relative to the seawater control using interstitial water as the test medium; 9 out of 13 (69%) were toxic relative to the control using SWI exposures. Because these experiments were initiated with uncontaminated overlying water, toxicity in the SWI exposures is presumably caused by toxicants diffusing into the overlying water from the sediment. Because of dilution effects by the overlying water, one would expect less toxicity at the SWI relative to the interstitial water.

There was greater toxicity to sea urchin embryos exposed to sediment homogenate at the SWI relative to amphipods (*Rhepoxynius abronius*) exposed concurrently to the same samples (Figure 2). While 69% were significantly toxic to the urchin embryos using SWI exposures, 46% were significantly toxic to amphipods relative to the controls (Yaquina Bay, OR home sediment).

Results of Spearman Rank correlations of bulk phase metal and organic contaminants with toxicity indicated that both the amphipods and urchins were responding to similar contaminants in the sediment homogenates. There were statistically significant negative correlations between amphipod survival and normal sea urchin development with copper and total PAH concentrations in these samples (Table 1). There was no significant correlation between sea urchin development and sediment grain size. There were no significant correlations between urchin development in interstitial water with bulk-phase chemicals measured in the homogenates from which the interstitial water was extracted.

Toxicity to sea urchin embryos was consistently greater in intact (unhomogenized) sediment samples relative to homogenized samples using SWI exposures. All 13 sites demonstrated significant toxicity to sea urchin embryos in the intact samples (Figure 3). Although it is not clear what caused the difference in toxicity, there was a clear difference in toxic response between homogenized and intact samples. While it is obvious that these manipulations may have significant effects on the concentration and bioavailability of contaminants in a given sediment sample, only a few studies have addressed the magnitude of these effects on toxicity.[23,24] It has been suggested that homogenization alters sediment integrity and may result in changes in speciation through oxidation-reduction or other chemical processes, as well as volatilization, sorption, and desorption. Previous studies have demonstrated that metal flux into the overlying water is considerably less in homogenized samples relative to intact samples.[12]

In conclusion, this exposure system facilitates the use of relatively sensitive embryo/larval toxicity test protocols in solid phase exposures, addressing the current lack of sensitive solid phase tests which incorporate sublethal endpoints. The screen employed in this procedure serves to retain

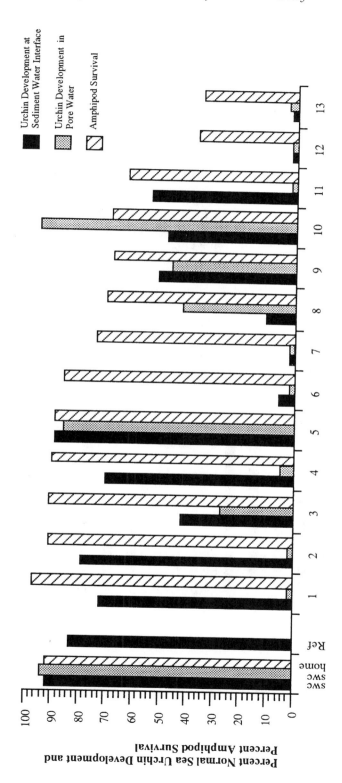

Figure 2 Amphipod survival in sediment and percent normal development of sea urchin larvae in interstitial water and sediment–water interface (SWI) exposures using sediment homogenate from 13 samples collected in San Diego and Los Angeles Harbors. SWC = sea water control; Ref = Reference site (note: no urchin development at SWI was tested from this site).

Table 1   Significant Spearman Rank Correlations between Toxicity
and Selected Contaminants

Contaminants	Urchin development at SWI	Amphipod survival	Urchin development in interstitial water
Copper	−.501 ($p < 0.05$)	−.793 ($p < 0.01$)	NS
Total PAHs	−.536 ($p < 0.05$)	−.613 ($p < 0.05$)	NS
Grain size	NS	−.487 ($p < 0.05$)	NS
TOC	NS	NS	NS

Note: N = 13 samples; NS = not significant; SWI = Sediment–water interface. All samples
used were homogenized.

all of the exposed organisms, allowing for a more accurate assessment of
toxic effects. An additional advantage is that this system may be used to
assess sediment toxicity in intact, unhomogenized cores, thereby eliminating
artifacts that result from the manipulation of sediment and pore water samples, and thus more closely approximating *in situ* exposure conditions. Preliminary results indicate that artifacts resulting from sediment homogenization may in some cases significantly reduce sediment toxicity. The screen
also minimizes any interaction between test organisms and resident infauna
which may be present in intact samples. Future work will investigate the
effects of sediment homogenization on metal and organic compound flux
into the overlying water column and the interaction between contaminant
flux and toxicity at the sediment–water interface.

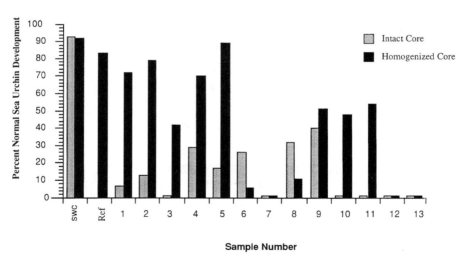

Figure 3   Percent normal development of sea urchin larvae in sidement–water interface (SWI) exposures using intact (undisturbed cores) and sediment homogenate
from 13 samples collected in San Diego and Los Angeles Harbors. SWC = sea water
control; Ref = Reference site (note: no intact core was tested from this site).

Acknowledgments

The authors thank John Newman (organics) and John Goetzl (metals) for conducting chemical analyses on these samples. Dr. Russel Fairey was responsible for sample collection.

References

1. Giesy, J.P., C.J. Rosiu, R.L. Graney, and M.G. Henry. 1990. Benthic invertebrate bioassays with toxic sediment and pore water. *Environ. Toxicol. Chem.* 9(2):233–248.
2. Power, E.A. and P.M Chapman. 1992. Assessing sediment quality. In *Sediment Toxicity Assessment*, Ed. G.A. Burton, Jr. CRC/Lewis, Boca Raton, FL. 457 pp.
3. Long, E.R., M.F. Buchman, S.M. Bay, R.J. Breteler, R.S. Carr, P.M. Chapman, J.E. Hose, A.L. Lissner, J. Scott, and D.A. Wolfe. 1990. Comparative evaluation if five toxicity tests with sediments from San Francisco Bay and Tomales Bay, California. *Environ. Toxicol. Chem.* 9 (9): 1193–1214.
4. Chapman, P.M., E.R. Long, R.C. Swartz, T.H. DeWitt, and R. Pastorok. 1991. Sediment toxicity tests, sediment chemistry and benthic ecology do provide new insights into the significance and management of contaminated sediments — a reply to Robert Spies. *Environ. Toxicol. Chem.* 10(1):1–4.
5. Chapman. P.M. and Morgan, J.D. 1983. Sediment bioassays with oyster larvae. *Bull. Environ. Contam. Toxicol.* 31, 438–444.
6. Burgess, R.M., K.A. Schweitzer, R.A. McKinney, and D.K. Phelps. 1993. Contaminated marine sediments: Water column and interstitial toxic effects. *Environ. Toxicol. Chem.* 12(1): 127–138.
7. Cross, F.A. and W.G. Sunda. 1978. Relationship between bioavailability of trace metals and geochemical processes in estuaries. *Estuarine Interactions*: 429–442.
8. Duursma, E.K. and M. Smies. 1982. Sediments and transfer at and in the bottom interfacial layer. *Poll. Transfer Transport Sea* 2:101–139.
9. Officer, C.B. and D.R. Lynch. 1989. Bioturbation, sedimentation and sediment–water exchanges. *Estuar. Coast. Shelf. Sci.* 28: 1–12.
10. Reidel, G.F., J.G. Sanders, and R.W. Osman. 1989. The role of three species of benthic marine invertebrates in the transport of arsenic from contaminated estuarine sediment. *J. Exp. Mar. Biol. Ecol.* 13: 143–155.
11. Spies, R.B. 1989. Sediment bioassays, chemical contaminants and benthic ecology: New insights or just muddy water? *Mar. Environ. Res.* 27: 73–75.
12. Anderson, B.S., J.W. Hunt, R. Fairey, M.D. Stephenson, and M. Puckett. 1994. Assessment of sediment toxicity at the sediment–water interface. Society of Environmental Toxicology and Chemistry (SETAC), Denver, CO. MC09 (Abstr.)
13. American Society for Testing and Materials (ASTM). 1993. *Standard Guide for Collection, Storage, Characterization, and Manipulation of Sediments for Toxicological Testing*. ASTM, Philadelphia, PA. pp. 1200–1214.
14. American Society for Testing and Materials (ASTM). 1995. *Standard Guide for Conducting Static Acute Toxicity Tests with Echinoid Embryos*. ASTM, Philadelphia, PA. pp. 1029–1047.

15. Environmental Protection Agency. 1988. Short-Term Methods for Estimating the Chronic Toxicity of Effluents and Receiving Waters to Marine and Estuarine Organisms. Eds. Weber, C.I., W.B. Horning, II, D.J. Klemm, T.W. Neiheisel, P.A. Lewis, E.L. Robinson, J. Menkedick and F. Kessler. EPA-600/4-87/028. Environmental Protection Agency, Cincinnati, OH. 417 pp.

16. Anderson, S.L., J.E. Hose, and J.P. Knezovich. 1994. Genotoxic and developmental effects in sea urchins are sensitive indicators of effects of genotoxic chemicals. *Environ. Toxicol. Chem.* 13(7): 1033–1041.

17. Hose, J.E. 1985. Potential uses of sea urchin embryos for identifying toxic chemicals: Description of a bioassay incorporating cytologic, cytogenetic, and embryologic endpoints. *J. Appl. Toxicol.* 5: 245–254.

18. Bay, S.M., R. Burgess, and D. Nacci. 1993. Status and applications of echinoid (phylum Echinodermata) toxicity test methods. In *Environmental Toxicology and Risk Assessment*. Eds. Landis, W.G., Hughes, J.S., and Lewis, M.A., American Society for Testing and Materials, Philadelphia, PA, pp. 281–302.

19. Sokal, R.R. and J.F. Rohlf. 1981. *Biometry.* Second edition, W.H. Freeman and Co., New York. 859 pp.

20. Bender, M., W. Martin, J. Hess, F. Sayles, L. Ball, and C. Lambert. 1987. A whole-core squeezer for interfacial pore-water sampling. *Limnol. Oceanogr.* 32(6): 1214–1225.

21. Carr, R.S., J.W. Williams, and C.T.B. Fragata. 1989. Development and evaluation of a novel marine sediment pore water toxicity test with the polychaete *Dinophilus gyrociliatus. Environ. Toxicol. Chem.* 8:533–543.

22. American Society for Testing and Materials (ASTM). 1993. *Standard Guide for Conducting 10-Day Static Sediment Toxicity Tests with Marine and Estuarine Amphipods.* ASTM, Philadelphia, PA. pp. 1138–1163.

23. Burton, G.A., Jr. 1993. Sediment toxicity assessments using *in situ* assays. Society of Environmental Toxicology and Chemistry (SETAC), Lisbon, Portugal. 191 pp. (Abstr).

24. Shea, D. 1993. Do traditional toxicity tests adequately represent *in situ* sediment conditions? Society of Environmental Toxicology and Chemistry (SETAC), Houston, TX . 122 pp. (Abstr.).

chapter thirty-four

Semipermeable membrane devices (SPMDs) for the concentration and assessment of bioavailable organic contaminants in aquatic environments

J.N. Huckins, J.D. Petty, J.A. Lebo, C.E. Orazio, H.F. Prest,
D.E. Tillitt, G.S. Ellis, B.T. Johnson, and G.K. Manuweera

Introduction

The well-known tendency of fish and other organisms to concentrate trace environmental contaminants to potentially harmful levels in their tissues is a major ecosystem and human health concern. Chemicals are concentrated from environmental media via respiration, ingestion, and in some cases direct dermal contact. Bioconcentration is the thermodynamically driven partitioning process by which a chemical is concentrated in lipoidal tissues of an organism directly from water.[1]

Huckins, et al.[2] designed a lipid-containing semipermeable membrane device (SPMD) to mimic the key aspects of the bioconcentration process and to passively sample (*in situ*) bioavailable aqueous organic contaminants. The SPMD membrane is generally made of low density polyethylene (LDPE) layflat tubing. This type of polymer is often described as "nonporous," even though free volume exists as transient cavities generally <10 Å in diameter.[3] Other nonporous polymers such as silicone or Silastic®, polypropylene, polyvinyl chloride, etc., and plasma-treated microporous tubing are also suitable as SPMD membranes for some applications.

Triolein (1,2,3-tri[cis-9-octadecenoyl]glycerol [a neutral triglyceride]), a mixture of lipids characteristic of the organism of interest, or high-molecular weight (>600 Da) lipid surrogates (e.g., certain silicone fluids) can be used as SPMD sequestration phases. Although the compositions of neutral lipids in organisms are known to vary considerably with species, sex, age, health, diet, and season, triolein represents a major portion of the neutral lipids in most fishes, and is often the major component of the triglycerides.[4]

Smith et al.[5] suggested that a one-to-one relationship between a contaminant's log K_{tw} (equilibrium triolein-water partition coefficient) and its log bioconcentration factor (BCF) should be observed. Chiou[6] demonstrated that log K_{tw} and log K_{ow} (equilibrium octanol-water partition coefficient) were closely correlated for many hydrophobic organic compounds. This finding suggests that, for first approximations, the K_{tw}s of hydrophobic contaminants can be estimated from their K_{ow} values, which are generally available in the literature. These considerations, the availability of synthetic, contaminant-free triolein, and the relatively low melting point of triolein suggest that, for many applications, triolein is the best lipid for use in SPMDs.

Several aspects of SPMDs mimic the bioconcentration process. The permeant size limitation of nonporous polymers is conveniently similar to that estimated for biomembranes.[3,7] Also, nonmediated or diffusional transport of nonelectrolytes through biomembranes has been shown to be phenomenologically similar to diffusion across nonporous polymeric membranes such as LDPE.[8] It is generally assumed that a hydrophobic compound must be in solution (each molecule surrounded by a hydration shell) to permit its uptake by the absorbing epithelium of an organism. The same constraint likely applies to nonporous LDPE SPMDs, as the SPMD appears to sample only dissolved organic contaminants[9] or that fraction of chemical not associated with particulates and dissolved organic carbon such as humic acids. Also, ions of organic and inorganic compounds are not sampled because of the very high resistance to mass transfer of charged species through nonpolar, nonporous membranes and the low neutral lipid solubility of ionic compounds.

A key element of SPMD design is that the LDPE membrane is the rate-limiting barrier in the contaminant uptake process.[10] This feature permits sampling rates independent of environmental turbulence (water and air velocity), but limits analyte uptake fluxes. To circumvent this constraint, the SPMD employs a high membrane surface-area-to-lipid-volume design (i.e., a layflat tube with a thin lipid film). Using this configuration, water sampling rates of 1 to 10 l per day can be expected for most organic contaminants with a 1-g triolein SPMD.[11] However, equilibrium is generally not approached during SPMD exposures of less than 28 d because of the high K_{tw} values of most hydrophobic contaminants.

In this work we describe the utility, performance characteristics, and the precautions and limitations related to the use of SPMDs. Similarities and differences between SPMDs and biomonitoring organisms are discussed. The potential use of SPMDs as tools for concentrating bioavailable contaminants for evaluation by *in vitro* bioassay is demonstrated. Mathematical modeling of SPMD uptake is reviewed with respect to the estimation of ambient environmental concentrations of dissolved organic contaminants. Also, the possible effects of temperature and biofouling on analyte uptake are examined, and results from studies demonstrating the efficacy of the device for concentrating contaminants from water and sediment are presented.

The studies highlighted in this work are (1) three laboratory flow-through exposures for SPMD aqueous calibration; (2) several laboratory static exposures (single application of chemical) to determine equilibrium triolein-water (K_{tw}) partition coefficients of SPMDs and the uptake of chemicals from sediments; (3) comparison of concentrations of selected pesticides in ultrafilter permeates of Mississippi River water to those derived from SPMDs deployed at the same sites; (4) exposure of SPMDs to a small urban freshwater stream to locate a possible polycyclic aromatic hydrocarbon (PAH) point source; (5) side-by-side marine exposures of SPMDs and bivalves to examine the potential differences in the types of chemicals concentrated in the two sampling matrices; and (6) demonstrations of the amenability of SPMD concentrates to chosen *in vitro* bioassay techniques. These studies were chosen to help illustrate the importance of SPMD technology to the field of aquatic toxicology. Through the use of SPMDs, it now appears to be feasible to estimate the time-integrated dose of waterborne organic contaminants to aquatic organisms.

Materials required

SPMD availability

The SPMD technology is the subject of two government patents (Huckins, et al., U.S. Patents, #5,098,573 and #5,395,426) which were licensed to Environmental Sampling Technologies (EST), a division of Custom Industrial Analysis Labs, 1717 Commercial Drive, St. Joseph, MO 64503. The patents cover both assembly of SPMDs and dialytic recovery of analytes from SPMDs. Some of the work presented here represents the original research used in patenting SPMDs, and thus was not a patent infringement. Currently, SPMDs are available only from EST or sublicensees.

Chemicals

Most of the LDPE layflat tubing used in this research was purchased from Brentwood Plastics, Inc., Brentwood, MO. Lot no. 12184, a general purpose, clear tubing was used in most single-application static tests. This lot was considered untreated by the manufacturer but did contain 0.1% w/w of erucamide, a slip additive. The wall thicknesses ranged from 72 to 76 μm. To reduce the possibility of analytical interferences, Lot no. 18387, a No. 940 (untreated, no erucamide or other additives) clear tubing, was used for all other studies. The wall thicknesses of this lot ranged from 84 to 89 μm. The intended thickness for both of these LDPE tubing lots was 76 μm. Both lots of layflat tubings were 2.5 cm wide. LDPE used in the Corio Bay exposure was a general purpose clear tubing and was purchased from Cope Plastics, 4441 Industrial Drive, Godfrey, IL. The width of this tubing was 5 cm and the wall thickness was about 50 μm. Removal of potential analytical interferences from LDPE membranes used in SPMDs has been described in detail and this step is required for the assembly of contaminant free SPMDs.[2,9,10]

Two grades (95 and 99%) of triolein were obtained from Sigma Chemical Co., St. Louis, MO. Most laboratory aqueous exposures were performed with 99% triolein, whereas sediment exposures, one laboratory flow-through aqueous exposure, and field exposures employed less expensive 95% triolein. Samples of the two grades of triolein were analyzed by high resolution gas chromatography (HRGC) or HRGC-mass spectrometry (MS) for background substances that might interfere with environmental contaminant analysis. This step was required to ensure that SPMDs were free of contaminants when assembled.

Both UL-^{14}C-2,2′,5,5′-TCB (tetrachlorobiphenyl, ≥99% purity, 14.2 mCi/mM), a model polychlorinated biphenyl (PCB), and ^{14}C-9-phenanthrene (≥98% purity, 19.4 mCi/mM), a model PAH, were purchased from Sigma Chemical Co., St. Louis, MO. UL-^{14}C-Mirex (≥ 98% purity, 7.25 mCi/mM); UL-^{14}C-2,2′,4,4′,5,5′-HCBi (hexachlorobiphenyl, ≥98% purity, 14.06 mCi/mM); and UL-^{14}C-HCB (hexachlorobenzene, ≥98% purity, 13.49 mCi/mM) were obtained from Pathfinder Laboratories Inc., St. Louis, MO. ^{14}C-dibenz[a,h]anthracene (≥95% purity, 8.8 mCi/mM) and ^{14}C-dieldrin (≥95% purity, 22 mCi/mM) were purchased from Amersham Life Sciences, Arlington Heights, IL. ^{14}C-1-naphthalene (≥95% purity, 0.88 mCi/mM) was obtained from ICN, Costa Mesa, CA. Autoradiography was used to confirm that the purity of these test chemicals remained ≥95%.

Neat, high-purity standards of organochlorine pesticides (OC) were obtained from EPA's Pesticide Repository, Research Triangle Park, NC or Ultra Scientific, 250 Smith Street, North Kingstown, RI. These OCs were used in flow-through SPMD exposures and were hexachlorobenzene, pentachloroanisole, alpha-benzenehexachloride, beta-benzenehexachloride, lindane, Dacthal®, heptachlor, heptachlorepoxide, oxychlordane, trans-chlordane, cis-chlordane, trans-nonachlor, cis-nonachlor, o,p′-DDT, o,p′-DDD, o,p′-DDE, p,p′-DDT, p,p′-DDD, p,p′-DDE, dieldrin, endrin, mirex, and methoxychlor.

Organic solvents used were Optima® grade from Fisher Scientific Company, Pittsburgh, PA. Net-Dip or Sanaqua® didecyl dimethyl ammonium chloride solution (7.5%) was purchased from Aqua Science Research Group, North Kansas City, MO.

Apparatus

The Clamco Corp. Model 251B heat-sealer used to make SPMDs was obtained from Consolidated Plastics Co., Inc., Twinsburg, OH. Static water exposures were conducted in 1-l glass microcosm chambers or 0.9-l glass jars with aluminum foil-lined lids.[2] Two flow-through diluters were used for SPMD aqueous calibration exposures, and were modified Mount and Brungs systems (personal communication, David Zumwalt, Midwest Science Center, Columbia, MO). Dual micromedic automatic pipetters from Micromedic Systems, Horsham, PA, were interfaced with each diluter to deliver the appropriate amounts of test chemicals for SPMD uptake studies. Glass diluter aquaria used for flow-through uptake and dissipation studies were 30-l volume, and the water was exchanged every 0.5 or 5 h (OC pesticide exposure). Contaminant-free deep-well water was used in SPMD laboratory exposures. All chemical stock bottles were amber glass, and light was minimized during the conduct and processing phases of studies to minimize the potential for photodegradation.

SPMD deployment structures used in field exposures were of several designs. For the exposure to assess freshwater PAH contamination in a shallow stream, each SPMD was enclosed in a 46-cm by 5.1-cm diameter galvanized conduit.[12] The SPMD loops were suspended between wire rods and the ends of the conduits were covered with copper screens. For the Corio Bay deployment, SPMDs were woven on Teflon spindles which were mounted on protective stainless steel (type 316) frames.[13] The specifics of deployment structure materials and configurations can vary, but the following must be considered: (1) metal containment structures should be free of cutting oils or other potential interferences; (2) use of plastic components should be minimized because of the potential for leachable residues; (3) the design of the structures should minimize abrasion of the membranes, even in turbulent environments; (4) screening or other protective materials should be of large enough mesh size to allow adequate exchange of ambient water at the SPMD surface, even after biofouling of the structure; (5) shading structures may be required for analytes that undergo photolysis; and (6) the entire structure should be securely tethered to prevent loss during minor flood events and be amenable to hiding due to possible vandalism.

Instrumentation

Radioactivity measurements were performed with a model 3801 liquid scintillation counter from Beckman Instruments, Fullerton, CA. The GC analyses were performed using model HP5890A systems from Hewlett Packard, Palo Alto, CA. An electron capture detector (ECD) and a photoionization detector (PID) were used for sample analyses. Capillary HRGC columns used were DB-5 (30 m and 60 m) and DB-17 (60 m) from J & W Scientific, Folsom, CA.

The gel permeation chromatography (GPC) system used in some of the described studies included the following modules: Perkin-Elmer series 410 solvent delivery system with a Perkin-Elmer ISS-200 Autoinjector system, Perkin-Elmer Co., Norwalk, CT; an ISCO Foxy 200 fraction collector, ISCO, Inc., Lincoln, NE; and a 22.5-mm i.d. × 250 mm (10 μm particle size, 10 nm pore size) Phenogel® GPC column, Phenomenex, Inc., Torrance, CA.

Organisms

Mussels (*Mytilus edulis*) of uniform age and size were obtained from a commercial grower at Portarlington, Victoria, Australia, for the Corio Bay exposure. The farm site was within 200 m of the outermost experimental site. The mussels, which were grown on ropes, were from a single spatfall. Approximately 100 organisms were transferred into open containers (originally designed for oyster outgrowth) for deployment. Following exposure, the mussels were retrieved and returned at 4°C to the laboratory. The soft tissues were removed and stored at –25°C until analysis.

Procedures

SPMD assembly

The construction of SPMDs used in these studies has been described in detail.[2] However, key aspects are delineated herein. The clean LDPE tubing was cut into segments 10.2, 46, 176, and 229 cm long, and these segments were loaded with 0.1, 0.5, 2, and 5 ml, respectively, of triolein (density 0.91). Since triolein, at room temperature, is slightly viscous and can cling to surfaces, a pipetter equipped with a total displacement plunger was necessary for accurate volumetric delivery. The triolein was formed into thin films throughout the tubing segments as described earlier,[2] and the SPMDs were heat-sealed at both ends, and in some cases, were heat-sealed into loops. Each SPMD thus prepared had a membrane surface area-to-lipid-volume ratio of about 400 cm^2/ml. Because organic vapors are readily sequestered by triolein, LDPE tubing, and assembled SPMDs (e.g., PCB vapors in 1 to 10 m^3 air/day are extracted by a 1-g triolein SPMD[11]), care must be taken to minimize exposure time to laboratory air. In most cases, SPMD assembly was performed in an environmentally controlled room having charcoal-filtered air.

SPMDs should be placed inside clean gas-tight metal paint cans for transport on ice to the field. In some cases, Teflon®-lidded amber glass jars were used for transport of SPMDs. However, even Teflon is porous;[14] thus gas tight seals are not achieved, leading to possible SPMD contamination. Following field exposures, the SPMDs (including trip blanks) were recovered, placed in the same metal shipping cans (on ice), and immediately transported back to the analytical laboratory for sample processing and residue enrichment.

Laboratory exposures

Details of the experimental designs of flow-through laboratory exposures to calibrate SPMDs, i.e., to determine analyte sampling rate, have been reported.[10] Briefly, test chemicals were dissolved in acetone and the appropriate amounts of stock solutions (0.1 ml acetone per liter of water) were delivered into 30-l aquaria water by Micromedic pumps. The target or nominal concentrations of test chemicals in diluter water were 1 ng/l, 10 ng/l, and/or 100 ng/l. Exchange rates of aquaria water were sufficient to maintain relatively constant water concentrations. Water temperature was held constant at 18°C or at 26°C. Replicate SPMDs were suspended vertically in diluter aquaria from 30.5-cm diameter stainless steel rings whose centrums were the input points of exposure water. Replicated (n = 3 or n = 4) SPMD samples were removed for analysis at least weekly during these exposures.

The dissipation of chemicals from SPMDs into contaminant-free water was examined using phenanthrene. The ^{14}C-phenanthrene dissipation experiments were conducted at multiple temperatures and concentrations using a flow-through diluter (without Micromedic pipetters). The phenanthrene was spiked into SPMD triolein to obtain concentrations of 20, 200, and 2000 ng/g. Two aquaria were used for each phenanthrene concentration and each aquarium had 12 SPMDs at the beginning of the 14-d experiment. SPMDs were sampled at days 3, 5, 7, 9, 11, and 14. The flow rate of water through each aquarium was 2 liters every 10 min. Water temperatures of experiments were 18°C, 24°C, and 30°C.

Static exposures were also conducted as described earlier[2] and consisted of single applications of test chemicals (generally 1 µg dissolved in ≤100 µl of acetone) into ≈1 liter of well water. Test containers used for static exposures were glass jars (8.9 cm diameter by 22.9 cm height) equipped with polypropylene lids. Replicates (n = 3) of SPMDs, water, and container rinsates from these tests were analyzed to obtain mass-balance data points through time.

Exposures of SPMDs (0.1 g triolein in 10.2-cm LDPE layflat tubing) to spiked sediment were conducted in glass jars identical to those used in static aqueous exposures. Radiolabeled phenanthrene, dieldrin, and 2,2′,5,5′-TCB were separately spiked into dry 300-g portions of a soil. This soil had been previously characterized[15] and contained 1% organic carbon. The spiked soil was thoroughly homogenized, then poured into exposure chambers containing 900 ml of well water and partly coiled and weighted SPMDs resting on their edges. The SPMDs were thereby submerged in sediment. Twelve exposure chambers were prepared in this manner for each test chemical and three replicates (SPMDs, sediment, and water) were sampled for each compound on days 1, 7, 14, and 28.

Field exposures

Because of the marked differences in the environmental conditions of the field studies presented in this work, only limited similarities exist between study designs. Lebo et al.[12] have described details of the urban freshwater stream deployment. Briefly, the SPMDs were deployed at several locations in a small midwestern creek (Flat Branch Creek, Columbia, MO) thought to be contaminated with PAHs and were left in place for 21 d during August 1991. The SPMD deployment structures (see Apparatus section) were inspected periodically and after rainfall. Leaves and other debris were removed from the copper screens, and sand was sluiced from inside the conduits. At the end of the exposure, the SPMDs were removed and immediately sealed in individual glass jars for transport to the lab.

Prest et al.[13] have described details of the side-by-side SPMD-bivalve study in Corio Bay, Victoria, Australia. Contamination problems in Corio Bay appear to stem from a variety of industrial and urban sources, which include a major refinery effluent (cooling water used is disinfected by chlorination). All sites were located at beacons marking major shipping lines.

SPMDs and bivalves were deployed by scuba divers and exposed for a period of 60 d, between June and August 1992.

Finally, Ellis, et al.[9] have described details of a field study involving the analysis of ultrafiltered water samples (partitioned three times with methylene chloride), SPMDs, and caged and feral fish at three sites on the Upper Mississippi River (Lakes Onalaska, Pepin and Baldwin). Caged fish and most SPMDs used in this study were exposed for 28 days. Also, as part of this study several SPMDs were exposed for 58 days and transported (overnight) back to the laboratory in river water to evaluate the effects of biofouling on SPMD uptake rates.

Sample storage

Exposed SPMD samples should be kept sealed in the same cans used for transport or in other clean metal paint cans (gas-tight seal is essential), and stored frozen at –20°C until processing. Samples should not be stored in a freezer with standards or other possible sources of contamination.

SPMD processing and residue enrichment

Processing of the SPMDs generally involved the following steps: (1) removal of exterior surficial periphyton and debris; (2) organic solvent dialysis; (3) gel permeation chromatography; and/or (4) Florisil® chromatography (Figure 1). For PAH samples, columns of potassium silicate and silver nitrate-treated benzenesulfonic acid were used in place of Florisil.[12] In the case of SPMDs exposed in Corio Bay,[13] the SPMD lipid was rinsed from the membranes, (circumventing the need for membrane cleaning), placed in clean LDPE tubes, and dialyzed with cyclohexane.

Cleaning of fouled SPMD membranes was accomplished by placing the devices (note: the SPMD membrane should not be touched with bare hands prior to the surficial cleaning) in a 500-ml, wide-mouth, amber glass jar equipped with a Teflon-lined cap. Hexane (100 ml) was added to the glass jar, the lid was tightly closed, the jar was shaken for 20 to 30 sec, and the hexane was discarded. Afterwards the SPMDs were placed in a stainless steel pan and rinsed with copious amounts of running water and scrubbed with a clean brush. At this point, the SPMDs were checked for small holes in the membrane. If a hole was found and other replicate samples were not available, the hole was isolated by heat-sealing and the mass of lipid remaining in the SPMD was determined. After the integrity of the SPMDs were ensured, they were submerged in a glass tank containing 1 N HCl (mineral removal) for approximately 30 sec. Following the HCl wash they were rinsed with running water to remove any acid. All water on the membrane surfaces was removed using rinses of acetone followed by isopropanol. The SPMDs were allowed to air dry for approximately 5 min on a piece of solvent-rinsed aluminum foil in a clean room free of organic contaminants (note: unless the atmosphere is organic contaminant-free, air drying may result in uptake of airborne contaminants).

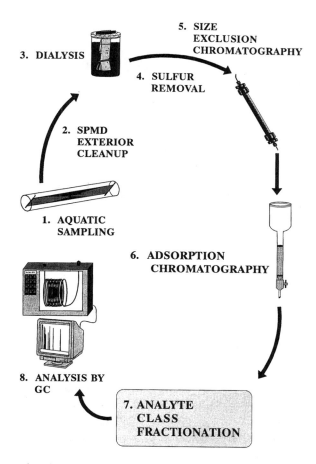

Figure 1   Generalized scheme used for the analysis of residues concentrated in exposed SPMDs. Fewer or additional steps may be required for analysis depending on the composition of concentrated residues. Also, bioassay of SPMD concentrates requires fewer steps (typically through dialysis or gel permeation chromatography) depending on the specific test.

Dialytic recovery of analytes from most SPMDs used in these studies was accomplished in glass canning jars (capacities of about 0.2, 0.4, 0.9, and 1.9 l) with solvent-rinsed aluminum foil under the screw-type lids to prevent contact with the rubber gaskets. For all studies SPMDs were dialyzed with at least 125 ml hexane per ml of SPMD triolein. Because SPMDs must be submersed during dialysis, a minimum solvent volume of 75 ml was required, even for 0.11-ml triolein SPMDs. In all cases, SPMDs were allowed to dialyze for 48 h (solvent was exchanged at 24 h for Corio Bay SPMDs[13]). Subsequently, SPMDs were removed from the jars (sulfur removal with copper wool can be performed at this point or later, if needed), and the dialysates were quantitatively transferred to round bottom flasks. Then the dialysate volumes were reduced to approximately 5 ml by a rotary evaporation system. The dialysates were quantitatively transferred to test tubes using several 2-ml hexane rinses and further reduced in volume ($\approx$0.5 ml) with streams of high-purity nitrogen.

SPMD dialysates from exposures conducted during early phases of SPMD research were further enriched using low-resolution GPC whereas those from more recent studies were enriched using high performance GPC. GPC enrichment was not used for Corio Bay samples. Dialysates enriched by low-resolution GPC were quantitatively transferred to the chromatography (GPC) columns containing 10 g of S-X3 Biobeads, as previously described.[9] Each GPC column was eluted with 100 ml of hexane/dichloromethane (80/20, v/v). Typically, small amounts, generally <5% (w/w) of the triolein mass, of codialyzed lipids and polyethylene waxes were found to elute in the first 27 ml of solvent. Analytes of interest eluted in the following 73 ml. Concentrated dialysates enriched by high performance GPC were loaded onto a 22.5-mm i.d. × 250-mm, Phenogel® column. The mobile phase consisted of hexane/dichloromethane (80/20, v/v) delivered at a flow rate of 4 ml/min. Prior to sample chromatography, the GPC was calibrated by injecting benzene and monitoring its retention. The collect cycle (typically 72 to 240 ml) was initiated at the beginning of the elution of benzene; all previous eluate (typically 0 to 72 ml) constituted the excluded (dump) fraction.

Following enrichment by GPC, the resulting solutions were subjected to Florisil chromatography. In most cases, activated Florisil (130°C) columns were prepared by inserting pledgets of glass wool into disposable Pasteur pipets and adding approximately 2 cm of Florisil. Each column was rinsed with 5 ml hexane followed by 5 ml dichloromethane prior to application of the analyte solution. Analytes were recovered using 20 ml of dichloromethane. Larger 1-cm i.d. columns with 27 cm of 0.5% deactivated Florisil were used for Corio Bay SPMDs and bivalve extracts. The analytes were eluted with 70 ml of hexane (nonpolar organochlorine analyte fraction) followed by 50 ml of 30% dichloromethane in hexane (v/v) to recover moderately polar pesticides such as benzenehexachloride isomers.

HRGC analyses were conducted with HP model 5890A GCs. Two different HRGC capillary columns (see Instrumentation section) or HRGC/MS were used to confirm compound identities. Operating conditions are given by Prest et al.,[13] Lebo et al.,[12] and Ellis et al.[9]

Quality control

Trip blank SPMDs (at least one for each field sample site) accompanied SPMD sampler arrays during deployment, recovery, and transportation. These trip blanks were processed and analyzed exactly as deployed samples. Similarly, laboratory SPMD controls (at least one for each treatment) were always included for sample sets from each treatment. Procedural blanks consisted of portions of hexane taken through the entire cleanup procedure. Often, low levels of a peak coincident with hexachlorobenzene and the presence of several early eluting hydrocarbons were observed in procedural blanks. While these low-level laboratory contaminants could not be completely eliminated (all lab procedures would have to be conducted in clean rooms) field-deployed and laboratory samples nearly always met detection and quantitation limit criteria.[16]

Analytical recoveries were determined by spiking (individual compounds or a mixture, in the case of OCs) the triolein of freshly prepared, control and/or field blank SPMDs, and processing these quality control samples along with exposed SPMDs. Recoveries through analytical procedures were generally >75% with good precision, i.e., <20% coefficient of variation. Generally, quality control samples represented 20 to 30% of a sample set.

The limit of detection (LOD) was typically measured by determining the values of coincident GC peaks for each analyte in the procedural blanks. The LOD was defined as the mean plus three standard deviations of values so determined. The limit of quantitation (LOQ) was defined as the mean plus ten standard deviations of these values. Also, when adequate numbers of field blank and laboratory control samples were available, they were used, as described above for procedural blanks, to determine overall limits of detection and quantitation.

Bioassay methods

Culture of H4IIE rat hepatoma cells followed standard methods reported by Tillitt et al.[17] The H4IIE bioassay procedures were modified for adaptation into a 96-well microtiter plate configuration (Tysklind et al.[18]). The H4IIE cells were seeded at 7000 cells/well in 250 µl of D-MEM culture media.[17] After a 24-h incubation, the cells were dosed for 72 h with SPMD dialysates (enriched through GPC) or reference standards in 5-µl volumes of isooctane. The potency of the SPMD dialysates were calibrated against tetrachlorodibenzo-*p*-dioxin (TCDD). After the 72-h incubation, the plates were washed three times with distilled water and the cells were lysed. Ethoxyresorufin O-deethylase (EROD) activity was measured in each well.[18] The amount of protein in each well was determined by the method of Bradford[19] as well as the values used to normalize dose to each well and EROD activity. The doses of each sample (g equivalents/mg cellular protein) or TCDD standard were plotted against EROD activity (pg TCDD/mg cellular protein) to develop dose–response curves and calculations of TCDD-equivalents (EQ), using the method of Ankley, et al.[20]

Several samples (concentrates from SPMDs exposed to Antarctica sediments or a small urban stream) were also assayed for genotoxicity and toxicity using two companion *in vitro* bioluminescent tests, Mutatox® and Microtox® (Microbics Corporation, Carlsbad, CA). Genotoxic (DNA-damaging) and toxic substances were detected by determining the ability of a sample extract or specific chemical (rat liver S9 fraction was used for exogenous metabolic activation of chemicals for Mutatox) to change the luminescent state of *Photobacterium phosphoreum*.[21,22] The degree of light increase (Mutatox) or loss (Microtox), when compared to controls, indicated the relative genotoxicity or the acute toxicity of the sample. The genotoxicity and acute toxicity of many chemicals, including PAHs, have been determined with these assays.[21-24]

In this study we first determined the sensitivity and specificity of Mutatox and Microtox using selected model PAHs of known genotoxic and toxic activity.[21-24] Then these tests were used to assay concentrates from SPMDs exposed to Antarctic sediments (known to contain hydrocarbon contaminants) in a 10-d microcosm study. Also, an SPMD concentrate from a 21-d exposure in Flat Branch Creek,[12] a small urban stream in Columbia, MO, suspected of PAH contamination, was examined for genotoxicity. The SPMD lipid from these exposures was diluted in a carrier solvent (acetone or dimethyl sulfoxide), or dialyzed with hexane,[2] and transferred to a carrier solvent. Note that use of an appropriate carrier solvent, such as acetone or dimethyl sulfoxide, for sample delivery to the test cells, is critical to the success of these assays.

Theory and ambient concentration estimation

Recently, Huckins, et al.[10] described much of the theory related to the uptake and dissipation of contaminants by SPMDs in aquatic systems and delineated several mathematical models for estimating water concentrations of analytes from their concentrations in SPMD triolein. The rise in analyte concentration during SPMD exposures (constant concentrations) was shown to follow first-order kinetics. Thus, depending on exposure duration, physicochemical properties of the analyte, temperature of the exposure medium, and extent of membrane biofouling, the concentration of residues sequestered in SPMDs may represent linear, curvelinear, or asymptotic regions of the overall uptake curve. The overall uptake curve can be described by

$$C_L = C_w K_{Lw}(1-\exp[-k_u t]) \tag{1}$$

where C_L and C_w are analyte concentrations in the triolein and water, K_{Lw} is the equilibrium lipid/water partition coefficient (approximated by the K_{ow}), and k_u is the overall uptake rate constant. The capacity of the SPMD to sequester an analyte is set by the magnitude of the K_{Lw} value (model amplitude), and k_u is $k_o A K_{mL}/V_L$ where k_o is the mass transfer coefficient of an analyte into the SPMD triolein (m/h), A is membrane surface area (m²), K_{mL} is the membrane/lipid partition coefficient, and V_L is the lipid volume (m³). Use of Eq. (1) to predict C_w requires multiple determinations of C_L through time or kinetics data. The number of estimated parameters (e.g., C_w and K_u), should generally be no more than half the number of C_L values measured. Also note that, because membrane fouling often occurs during environmental exposures, SPMD uptake may be biphasic in nature resulting in poor model fits to the data.

Assuming equilibrium between the SPMD lipid and exposure water is not approached for a particular analyte (i.e., strictly speaking $C_L/C_w<<K_{Lw}$ but $\leq 0.5\ K_{Lw}$ or less than one half-time), then for a constant temperature

$$C_L = C_w K_{mw} k_o At/V_L \tag{2}$$

where K_{mw} is the membrane/water partition coefficient. Here the group $K_{mw}K_oA$ is the effective daily sampling rate (R_s) of the SPMD for a particular analyte, i.e., m^3/d or l/d. The use of Eq. (2) to estimate analyte C_w requires that the SPMD is still in the linear sampling phase or that the average sampling rate during the interval of exposure is known. When C_L/C_w is approximately equal to K_{Lw}, the exponential term in Eq. (1) becomes negligible and then

$$C_L = C_wK_{Lw} \tag{3}$$

This is the region in which C_L reaches equilibrium with C_w. If K_{Lw} is known, the water concentration C_w can be determined from C_L.

If organic solvent dialysis is used to recover contaminant residues sequestered by SPMDs, analytes are extracted from both the lipid and membrane phases of the SPMD. In some cases the amount of analyte in the membrane can be 50% of the total SPMD residue. The following equation can be applied (note: required when calibration data [R_s] solely based on C_L.) to estimate C_L from the total amount of residue recovered by dialysis:[12]

$$C_L = A_{MD}/(M_L + K_{ML}M_M) \tag{4}$$

where A_{MD} is the total mass of the analyte in the dialysate, M_L is the mass of the triolein, K_{ML} (as defined earlier) is the membrane-triolein partition coefficient, and M_M is the mass of the membrane.

The data generally needed for generating SPMD estimates of the average analyte concentrations in ambient water include: (1) mean temperature during the exposure period; (2) visual observation of the extent of membrane biofouling; (3) R_s values of contaminants of interest for the exposure duration and temperature used; and (4) measured C_L values or total analyte mass sequestered by the SPMD.

Results and discussion

Observations on theory

These studies indicate that the uptake of organic compounds by SPMDs is in accordance with polymer permeability and equilibrium partition theory. The driving force for the uptake of dissolved nonpolar chemicals can be viewed as their SPMD K_{mw} and K_{Lw} partition coefficients, where $K_{Lw} > K_{mw}$. As K_{mw} and K_{Lw} (approximated by K_{ow}) increase with hydrophobicity and molecular size, R_s or the SPMD sampling rate of analytes is expected to rise proportionally. However, as molecular size approaches and exceeds the size of the average available membrane cavities, resistance to mass transfer should rise exponentially, i.e., SPMD sampling rates should steeply decline (see subsequent discussion on PAH R_s values for an illustration of this phenomenon). Thus, variable SPMD R_s values are expected for aqueous analytes based on their K_{mw}, K_{Lw}, or K_{ow} values and steric factors (molecular

Table 1   SPMD (Normalized to 1 g Triolein Configuration)
Sampling Rates (R_s, n = 3) for Organochlorine Pesticides (OCs) in
Flow-Through Aqueous Exposures (10 ng/l Nominal, 16 d, 26°C)

OCs	Log K_{ow}[a]	Water ng/l[b]	R_s l/d (c.v.)[c]	(SPMD water) concentration factor[d]
Hexachlorobenzene	6.2	1.9	8.2 (12)	24,100
Pentachloroanisole	>5.1	4.2	5.7 (11)	16,700
Alpha-BHC	3.8	2.9	1.4 (55)	4,200
Beta-BHC	3.8	11.5	0.2 (19)	700
Lindane	3.9	5.6	0.4 (73)	1,200
Dacthal		10.0	1.2 (15)	3,400
Heptachlor	4.4	2.7	7.1 (10)	20,900
Heptachlor epoxide	2.6	9.0	2.0 (4)	5,900
Oxychlordane		4.2	7.6 (8)	22,300
T-chlordane	4.1	5.1	6.8 (11)	20,000
C-chlordane	4.1	7.0	5.4 (13)	15,800
T-nonachlor	5.6	4.6	7.0 (14)	20,500
C-nonachlor		7.2	5.1 (18)	15,100
o,p'-DDT		6.5	6.0 (11)	17,600
o,p'-DDD		8.3	5.3 (9)	15,500
o,p'-DDE		4.5	7.6 (7)	22,400
p,p'-DDT	5.7	6.1	4.4 (6)	13,000
p,p'-DDD	6.1	8.2	5.0 (17)	14,600
p,p'-DDE	6.0	4.5	7.3 (29)	21,500
Dieldrin	3.5	9.5	3.8 (11)	11,100
Endrin	5.5	9.2	5.9 (27)	17,400
Mirex		3.2	6.2 (4)	18,100
Methoxychlor	4.2	12.3	3.2 (21)	9,400

[a] Values from various literature sources.

[b] Measured concentrations.

[c] Coefficient of variation in percent.

[d] Concentration factor in the whole device, i.e., lipid plus membrane.

size, shape, rotational and conformational freedom). No mention is made here about the aqueous diffusion layer at the membrane–water interface. Huckins et al.[10] have indicated that, for most contaminants, the SPMD membrane dominates overall resistance to the mass transfer of chemicals, thus controlling analyte R_s values. Without membrane control, R_s values would be expected to vary for individual analytes under different regimes of water flow and associated turbulence, because of variations in diffusion layer thickness.

Laboratory studies

Table 1 shows SPMD R_s values ($\bar{x}$, n = 3) derived from a 16-d flow-through exposure (10 ng/l nominal, 26°C) to 23 OC pesticides. These values ranged from a low of 0.2 l/d to a high of 8.2 l/d for a 1-g triolein SPMD (i.e., data normalized to an SPMD with 1 g triolein and about 4 g membrane). Analytical recoveries of OCs from SPMDs averaged about 80% and no recovery corrections were made. Based on molecular modeling,[11] only mirex and possibly methoxychlor should be large enough to evoke high steric imped-ance to membrane permeability or uptake. Generally, OCs with higher K_{ow} values were sampled at higher rates. However, it is clear from other work with PCBs[11] and this work as well that, in the case of isomers and homologs with about the same molecular dimensions, bond rotational freedom, con-formational freedom, and shape do affect SPMD R_s values. With the possible exception of mirex, HCB has the largest K_{ow} (Log K_{ow} = 6.2) value of the OCs, and it was sampled at the highest rate (8.2 l/d). Compounds with log K_{ow} values <4 were probably beyond the linear phase of SPMD uptake at the end of 16 d, resulting in reduced R_s values. This suggests that, during the initial phase of uptake, SPMD R_s values for analytes with log K_{ow} <4 were higher than the 16-d values given in Table 1. Although the data[10,11] are not presented in this work, SPMDs appear to have sampling rates indepen-dent of concentration, and the absolute amount of chemical sequestered is proportional to dissolved concentration, i.e., a first-order relationship.

In another study[11,26] SPMDs were exposed to 16 priority pollutant (PP) PAHs (1, 10, and 100 ng/l, nominal, at 10°C, 18°C, and 26°C) for 21 d in a flow-through diluter system. SPMD (1 g triolein configuration) R_s values ($\bar{x}$, n = 3) for PP PAHs at 14 d (26°C) ranged from 0.1 l/d for naphthalene to 4.2 l/d for pyrene. Note that analytical recoveries of PP PAHs averaged about 50% and that these R_s values were not adjusted for SPMD recovery. Eight of the 16 PP PAHs have higher K_{ow} values than pyrene (log K_{ow} = 5.3) yet their sampling rates were lower. Figure 2 shows the possible relationship between R_s and K_{ow} values for PP PAHs. This data suggests that, as molec-ular size increases with PP PAH K_{ow} values, a permeability threshold is reached resulting in a sharp reduction in SPMD sampling rates. Because PAHs are rigid, planar molecules with little conformational freedom, the effect of increasing molecular size on membrane permeability should be pronounced.

Both temperature and biofouling can affect the R_s of an SPMD. The reduction in analyte R_s by biofouling was assessed by recovering heavily fouled SPMDs (58-d exposure) from the Upper Mississippi River and expos-ing them along with fresh SPMDs to constant aqueous concentrations of phenanthrene.[9] The reduction in SPMD uptake due to fouling represented about 25 to 40% of the R_s value. Temperature affects molecular diffusion and polymer free volume, and thus higher temperatures should normally result in larger R_s values for analytes. Multiple temperature laboratory exposures (10°C to 26°C) of SPMDs to OC pesticides (unpublished data, Midwest Science Center, Columbia, MO) have indicated a 1.2- to 1.8-fold increase in SPMD R_s values for each 5°C increase in water temperatures. However, only small compound-specific changes in SPMD R_s values were observed for planar rigid PP PAHs under the same testing regime.[26] The use of a perme-ability reference compound (i.e., a noninterfering compound with moderate SPMD fugacity added to the triolein just prior to deployment) appears to offer promise as a method to correct SPMD R_s values for the effects of variable temperature and fouling.

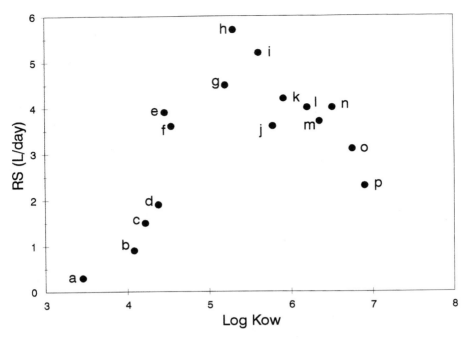

Figure 2 The potential relationship between SPMD sampling rates (R_s; 1-g triolein configuration) and PAH octanol–water partition coefficients (K_{ow}). Data used are from a 100 ng/l flow-through exposure (18°C) at 14 d. Each R_s value represents mean of three determinations (coefficients of variation <14% for all compounds except naphthalene [46%]). Analytes are designated by letters as follows: a, naphthalene; b, acenaphthylene; c, acenaphthene; d, fluorene; e, phenanthrene; f, anthracene; g, fluoranthene; h, pyrene; i, chrysene; j, benzo[*b*]fluoranthene; k, benz[*a*]anthracene; l, benzo[*k*]fluoranthene; m, benzo[*a*]pyrene; n, indeno[*1,2,3-cd*]pyrene; o, dibenz[*a,h*]-anthracene; and p, benzo[*g,h,i*]perylene.

The capacity of an SPMD to sequester a contaminant is largely controlled by the K_{Lw} value. The K_{Lw} value can be viewed as representative of the maximum volume of water that can be extracted by 1 ml of triolein in an SPMD, i.e., if the K_{Lw} or K_{ow} of a chemical is 10^4, then at equilibrium about 10^4 ml of water would have been extracted.

Figure 3 shows the rise to equilibrium of SPMD triolein with aqueous residues of four compounds. In these static exposures (single application of chemicals), water concentrations fall and lipid concentrations rise, resulting in a much more rapid approach to equilibrium than in flow-through, constant concentration exposures. Even with falling water concentrations, considerable time is often required for SPMDs to approach equilibrium with surrounding water. Table 2 shows the time required to reach ≥90% of equilibrium K_{Lw} values in SPMDs statically exposed to selected compounds. Also, literature values for K_{ow}, BCF, and conventional triolein–water partition coefficients (Table 1), show reasonable agreement with ≥90% K_{Lw} values determined in this work.

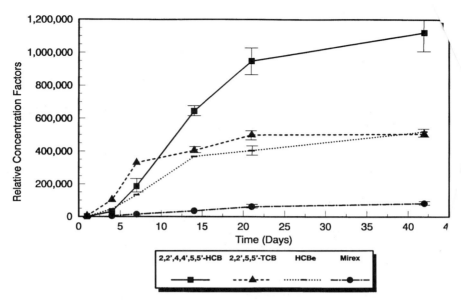

Figure 3 Comparisons of the triolein–water concentration factors (through tim SPMDs statically exposed to single applications of selected PCB congeners and pesticides. HCB, TCB, and HCBe are hexachlorobiphenyl, tetrachlorobiphenyl, hexachlorobenzene, respectively. Each point represents x̄ (n = 3) and error bar S.D.s. Some error bars are obscured by data point symbols.

Table 2   Comparison of Pseudoequilibrium ($\geq$90%) SPMD Triolein–Water
Partition Coefficients (K_{Lw}) to Data in the Literature

Compounds	Log SPMD K_{tw} (days)[a]	Log K_{ow}	Log BCF[b]	Log K_{tv}
2,2′,5,5′-TCB[d]	5.7 (21)	5.8	6.2	5.6
2,2′,4,4′,5,5′-HCBi[e]	6.1 (42)	6.7	6.2	6.2
Hexachlorobenzene	5.7 (42)	6.2	5.8	5.5
Naphthalene	3.6 (7)	3.4	3.6	
Phenanthrene	5.0 (21)	4.5	4.0[f]	

[a] Time required to reach $\geq$90% of steady-state value.

[b] Fish, values are normalized to an estimated 3% lipid content.

[c] Conventional triolein-water partition coefficients reported by Cary Chiou, R ence 6.

[d] Tetrachlorobiphenyl.

[e] Hexachlorobiphenyl.

[f] *Daphnia pulex.*

Exposure of SPMDs to episodic releases of chemicals in the environment is similar to our single application static exposures in that aqueous concentrations fall rapidly. However, in lotic environments advective transport generally results in a more rapid decline in water concentrations. Once aqueous contaminant levels fall below equilibrium concentrations (i.e., K_{Lw} and K_{mw}) with exposed SPMDs, desorption of sequestered residues is evoked. Unlike living organisms, which depurate chemicals by metabolism and active export of organic carbon (via waste removal and respiration), SPMDs are passive-thermodynamic systems. Thus, the aqueous dissipation of hydrophobic compounds from SPMDs is generally much slower than that observed for biota. In the case of recalcitrant compounds such as PCBs, where steady-state concentrations in organisms may approach or exceed thermodynamic equilibrium, the depuration rates of these compounds from aquatic organisms are expected to be closer in magnitude to SPMD dissipation rates for the same chemicals. For example, Huckins et al.[10] reported that the SPMD dissipation rate constant for 2,2′,5,5′-TCB was $\simeq 0.0003$ to 0.0004 h^{-1}, while the depuration rate constants (k_2) for the same congener in zebra fish and goldfish were 0.002 h^{-1} and 0.0006 h^{-1}, respectively.[27,28]

Since clearance of concentrated residues from fish and SPMDs follows first-order kinetics, half-lives can be determined from k_2 values (i.e., $t_{1/2} = 0.693/k_2$). The half-life of 2,2′,5,5′-TCB in SPMDs placed in clean flowing water is 82.5 d, whereas half-lives for the fish were 14.4 and 48.1 d. This data suggests that the use of SPMDs in monitoring applications would increase the likelihood of the detection of episodic releases of hydrophobic chemicals because of the greater retention times of analytes. This is particularly true for compounds that are readily metabolized, such as PAHs. The SPMD k_2 for phenanthrene is 0.0014 h^{-1} and its half-life is 20.5 d. Based on our phenanthrene dissipation studies, half-lives were independent of concentration for a 10^2 range in concentrations. It is apparent that compounds with K_{ow} values $>10^5$ will be sufficiently retained by SPMDs for detection even when an episodic release occurred during the first few days of a long-term exposure.

Interpretation of data from SPMDs exposed to water is relatively straightforward when compared to interpretation of similar data from SPMDs exposed to benthic sediments. This is because of the likelihood that the SPMD membrane is not the rate-limiting step in the uptake of most chemicals from the sediment phase. Based on theory of molecular diffusion attenuated by sediment sorption,[29] SPMDs should rapidly remove dissolved residues from adjacent pore water, distorting the local equilibrium and resulting in the desorption of chemicals from affected sediment particles. However, movement of additional residues from more remote locations to SPMD surfaces will be highly retarded by sediment particle adsorption–desorption kinetics and the tortuosity or porosity of the sediment matrix.

Other investigators have shown that desorption hysteresis may be the rate-limiting step in the overall mass transfer of many chemicals in sediments.[30] For compounds with K_{oc} (organic carbon normalized sediment–water partition coefficient) values in the range of 10^4 to 10^5 (assuming 1% organic carbon), pore water diffusion coefficients are estimated to range from 10^{-4} to 10^{-3} cm^2/d, whereas nonturbated bulk water diffusion coeffi-

Table 3   Exposure of SPMDs[a] to Contaminated Sediment[b]

Analyte	Concentration factor in triolein			
	1 d	7 d	14 d	28 d
Phenanthrene	2,300	14,600	17,800	42,500
Dieldrin	300	4,700	6,500	10,900
2,2',5,5'-TCB[c]	2,600	13,000	26,600	38,800

[a] SPMDs (n = 3/test chemical) contained 0.1 g triolein and were covered by sediment.

[b] 300 g of sediment (organic carbon = 1%) and 950 ml well water.

[c] Tetrachlorobiphenyl.

cients should be about 0.4 to 0.8 cm^2/d.[29] The low diffusivity of high-K_{oc} compounds in undisturbed sediments suggests that the amount of these chemicals sequestered by the SPMD will be representative of a relatively small volume of contiguous pore water and sediment. However, sediment bioturbation or other natural mixing events will reduce the significance of the interparticle diffusional term in the overall mass transfer process, thus permitting the sampling of much larger volumes of water and sediment relative to truly stagnant systems.

In view of these observations, concentration factors (CFs, nonequilibrium lipid concentrations divided by water concentrations) of compounds sequestered from sediment pore waters should generally be less than those measured in SPMDs exposed to bulk waters (includes static and flow-through systems). Table 3 shows the results of SPMD laboratory exposures to sediments containing phenanthrene, dieldrin and 2,2',5,5'-TCB. The sediments were undisturbed throughout the exposure, yet lipid CFs at 28 days (Table 3) ranged from a low of ≈11,000 for dieldrin to a high of ≈42,000 for phenanthrene. The 28-d CF for 2,2',5,5'-TCB was 38,800 in these sediment exposures, whereas the 28-d CF for 2,2',5,5'-TCB in flow-through aqueous exposures was ≈100,000.[31] Assuming the 28-d sediment concentrations divided by corresponding water concentrations were representative of K_{oc} values, then the lipid CFs shown in Table 3 were 25%, 73%, and 89% of the K_{oc} values for 2,2',5,5'-TCB, dieldrin and phenanthrene, respectively. Lipid CFs at 28 d for 2,2',5,5'-TCB and phenanthrene were only 10% and 42% of their respective equilibrium K_{Lw} values.[10] Thus, much longer exposures (>>28 d) to undisturbed sediments are required to approach equilibrium concentrations of these analytes in SPMDs. However, a more rapid approach to SPMD equilibrium has been demonstrated by Lefkovitz et al.[32] when sediments are continuously turbated.

Field studies

Since only single molecules of low to moderate molecular-weight (<600 Da) compounds are small enough to fit in nonporous LDPE (SPMD membrane) cavities,[3] only truly dissolved (surrounded by hydration shell) compounds should be sampled by SPMDs. To test this hypothesis, lindane, dieldrin, and pentachloroanisole concentrations, measured in ultrafiltered water from

Upper Mississippi River,[9] were compared to concentrations estimated from SPMD data. Linear and equilibrium models given in the Procedures section were used to estimate water concentration. The ratio of water concentrations generated from dividing SPMD values by ultrafilter permeate values was 1.0 ± 0.4 ($\bar{x} \pm$ S.D., n = 9) for lindane, 1.6 ± 0.8 (n = 7) for dieldrin, and 1.9 ± 1.3 (n = 9) for pentachloroanisole. Overall, a ratio of 1.51 ± 0.97 (n = 25) was observed, strongly suggesting that SPMDs sample only the truly dissolved phases of nonionic chemicals in water. The increasing SPMD/permeate ratios, as K_{ow} increases (i.e., K_{ow} of pentachloroanisole > dieldrin > lindane), suggests greater sorptive losses to the carbon-based ultrafilter with increasing analyte K_{ow} (personal communication, J. Leenheer, USGS, Arvada, CO).

Chromatograms from the analysis of SPMDs deployed in a small urban creek[12] suspected of PAH contamination, are shown in Figure 4. Site II is just downstream from the suspected source of hydrocarbons and PP PAHs, and it is shown at three different attenuation levels because of the high concentrations (≈ 0.5 mg total aromatics/1.0 g triolein SPMD) and complexity of the sequestered aromatic residues. Site I was well above the possible point source and represented runoff from the downtown section of Columbia, MO, a small midwestern city. Thus, asphalt leachates and combustion products would be expected to dominate residues from Site I. The marked difference between SPMDs from Sites I and II confirm the presence of a point source that predominantly emitted alkylated naphthalenes.[12]

Semipermeable membrane devices (SPMDs) and mussels (*Mytilus edulis*) were deployed side-by-side at several sites in Corio Bay, Victoria, Australia for a period of 60 d to examine their relative abilities to monitor a known gradient of chlorinated contaminants, which included PCBs and OC pesticides.[13] Figure 5 shows chromatograms from SPMD and mussel samples exposed at the same sites. Overall, levels of PCB and OC pesticide contamination were about the same, however, the HRGC profiles of PCBs and halogenated unknowns from the two matrices differed. SPMD data (Figure 5A) suggested that lower-chlorinated PCBs and unknowns (early-eluting HRGC components) were present at high levels in the water column, while mussel data (Figure 5B) implied essentially the reverse. These results were expected for several reasons. Although the early-eluting (less chlorinated) PCBs and other organics have lower SPMD sampling rates than more chlorinated compounds, a major portion of these early eluting compounds would reside in the aqueous phase of aquatic systems. This is due to their higher water solubilities and lower K_{oc} values. With respect to aquatic organisms, an inverse relationship has been shown to exist between log K_{ow} and depuration rate constants (k_2) of OCs.[27,28] Also, investigators[33] have shown that as log K_{ow} exceeds about 5.5, bioaccumulation is largely mediated by food and particulate ingestion rather than uptake from water.

Both Figures 5A and 6A illustrates that SPMDs contained many intense peaks in the HRGC-ECD chromatograms that were absent or minor in the mussels; roughly 80% of the signal from SPMD concentrates was due to unidentified compounds. Also, background residues in SPMD blanks were low (Figure 6B). Extracts from SPMDs placed in a refinery effluent stream,

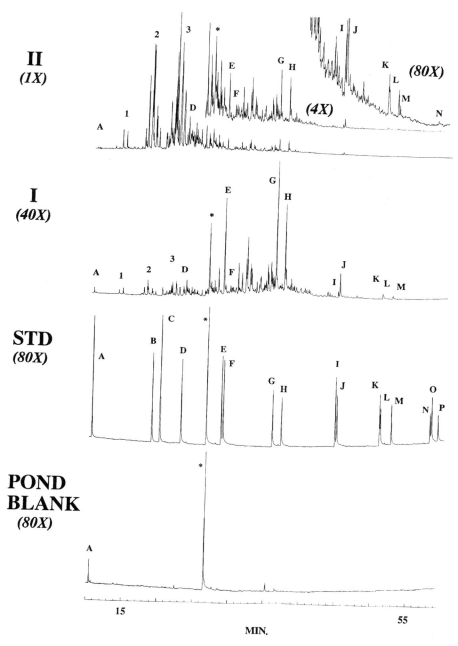

Figure 4 Attenuated GC/PID chromatograms of enriched dialysates from SPMDs exposed in an urban creek (sites I and II) as compared to chromatograms of a standard solution (2.0 μg/ml of each PP PAH) and of a pond blank SPMD. Priority Pollutant PAHs are labeled with upper case letters in the sequence of their retention on the DB-5 column. Additional labels denote: 1, monomethylnaphthalenes; 2, dimethyl-naphthalenes; 3, trimethylnaphthalenes; and *, 1-methylfluorene (the surrogate).

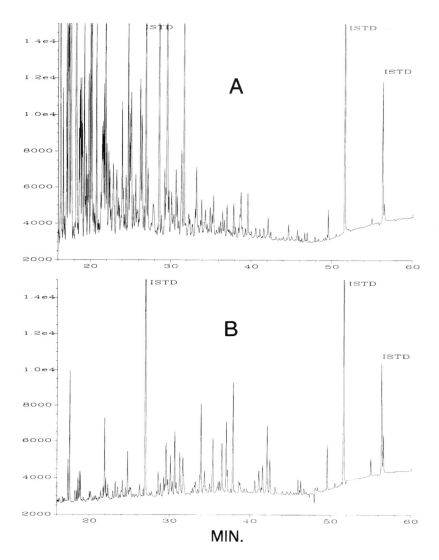

Figure 5 Comparison of the electron-capture GC detector response to an enriched SPMD lipid sample (A) and an enriched lipid extract from contiguously exposed bivalves (B). Peaks denoted by ISTD are instrumental internal standards. The 60-d exposure was conducted in Corio Bay, Victoria, Australia.

where bivalves were not viable (elevated temperature and turbidity), revealed extremely intense and complex HRGC-ECD chromatograms (Figure 6A) from halogenated hydrocarbon components (more than ten times as intense as those from the open-bay deployed SPMDs [Figure 5A]). Unlike aquatic organisms, SPMDs can be used in diverse environments and across a wide range of water qualities. However, because SPMDs do not sample ionized organic compounds and do not ingest particulates or food, the two

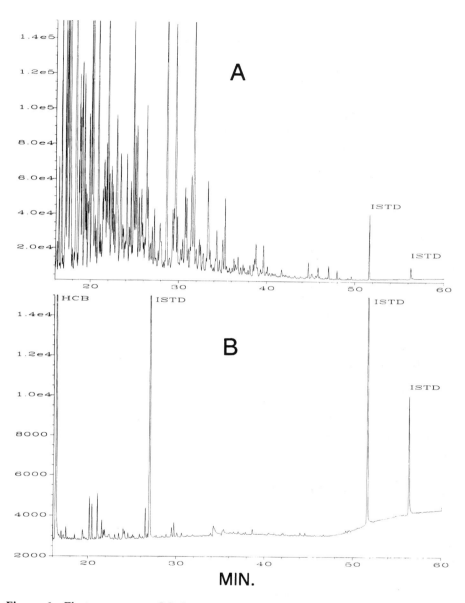

Figure 6 Electron-capture GC detector response to enriched lipid from an SPMD exposed in chlorinated intake water of a petroleum refinery (A), and enriched lipid from an SPMD blank (B). Peaks denoted by ISTD are instrumental internal standards. The 60-d SPMD exposure was conducted in Corio Bay, Victoria, Australia. This chromatogram represents a 10-fold greater detector response than those shown in Figure 5A.

approaches are viewed as complementary with respect to the types of information generated.

Appropriateness of lipid normalization

Comparisons of SPMD concentrations to those found in living organisms from the same site must be approached with caution. Typically, residue concentration data of multiple species are presented on a lipid-normalized basis. A key assumption in the normalization procedure is that organisms' lipids have reached steady state with ambient waters. This assumption is often valid for feral organisms, but it may not be valid for biomonitoring organisms, especially for high-K_{ow} contaminants and short-term (less than 56-d) exposures. The lipid percentages of aquatic organisms vary widely but generally lie in the 0.5 to 10% range, whereas an SPMD is usually >15% lipid, e.g., our SPMD configuration is $\simeq$20% triolein. Since the percent lipid content governs the relative masses of nonmetabolized chemicals accumulated by equal masses of organisms and SPMDs at equilibrium, SPMDs generally have much higher analyte capacities than aquatic organisms. However, their first-order uptake rates are about the same.[11,27,28] Thus, for short-term exposures, SPMDs, and in some cases biomonitoring organisms, may not reach equilibrium and lipid normalization of sequestered residue concentrations is inappropriate. If both SPMDs and organisms are in the linear phases of uptake, comparisons of relative sampling efficiencies can be made in terms of their first order rate constants, expressed as ml d^{-1}g^{-1}, where ml represents the volume of water extracted and g represents the weight of whole-body tissue or the whole SPMD (lipid plus membrane). Ultimately, comparisons of SPMDs and organisms should include considerations of the ratios of analyte signal (detector response to samples at quantitation windows of analytes), to background noise (detector response to control samples at quantitation windows of analytes), and how well the concentration in these matrices reflect analyte concentrations in the water sampled.

Bioassay of SPMD concentrates

Microtox and Mutatox bioassays have received considerable acceptance as screening tools to rapidly assess large numbers of sediment samples for toxic residues.[21-24] Because these *in vitro* tests use brief exposure periods (<30 min), they require high ng to µg amounts of toxicant to elicit measurable responses. Since most hydrophobic contaminants are present in water only at <µg/l concentrations (trace levels), direct use of Microtox and Mutatox assays to screen water for these contaminants is often not feasible. In this study we used a single 1 g triolein SPMD to concentrate trace contaminants from the water of a small urban stream for analysis with Mutatox. Also, we exposed SPMDs to Antarctica sediments in laboratory microcosms to concentrate bioavailable contaminants for analysis with both Microtox and Mutatox.

Table 4 shows Microtox readily identified samples with toxic compounds. Concentrates from SPMDs exposed to Winter Quarters Bay and

Table 4   Use of Microtox® and Mutatox® to Determine the
Toxicity of SPMD Concentrates[a]

Sample type	Microtox Toxicity (EC-50)[a]	Mutatox Genotoxicity (+/−)
SPMDs		
Winter Quarters Bay[b]	3.1 (2.9–3.3)	−
McMurdo Sound[b]	88 (28–275)	−
Flat Branch[c]	NA[d]	+
Quality control		
Procedural blank[e]	ND[f]	−
SPMD control[g]	ND	−
Microtox phenol reference toxicant (µg/ml H$_2$O)	19 (17–21)	NA
Mutatox benzo[a]pyrene reference toxicant (1.0 µg/vial)	NA	+

Note: Microtox values are 5-min EC-50s with 95% confidence intervals (in parentheses);
    Mutatox values show positive response as + and negative response as −.

[a] Assays were conducted on lipid diluent or dialysates and EC-50 values represent mg
    SPMD lipid/ml carrier solvent.

[b] SPMDs exposed to Antarctica sediments in microcosms.[25]

[c] SPMDs exposed to a small urban stream.[12]

[d] None analyzed.

[e] Solvents/reagents used in tests.

[f] None detected.

[g] Freshly prepared SPMD, carried through Microtox and Mutatox test.

McMurdo Sound sediments demonstrated greater than one order of magnitude difference in their Microtox EC-50 values.[25] Results from gas chromatographic analysis of these samples supported the Microtox data, as concentrations of contaminants (largely PAHs and PCBs) in Winter Quarters Bay SPMDs were about an order of magnitude higher than those in the McMurdo Sound SPMDs. The SPMDs exposed to Winter Quarters Bay sediment contained 2.4 mg PAHs/g triolein; the Microtox EC-50 value was 3.1 mg triolein/ml carrier solvent.

Table 4 also shows the results of Mutatox testing. Genotoxic compounds are presented as positive responses and compounds that have no genotoxic response are presented as negative. Thus, the mutatox is a yes-no test: the compound is either genotoxic or it is not genotoxic. Also, exogenous activation is required to make these genotoxic determinations. The Winter Quarters Bay SPMD concentrate was not genotoxic even though it was acutely toxic.[25] However, the SPMD concentrate from Flat Branch Creek, Site II (Figure 4), was genotoxic. This result is not unexpected because of the presence of significant amounts of 4- and 5-ring PAHs (known genotoxins) in this sample, in addition to high concentrations of 2- to 3-ring PAHs (Figure 4).

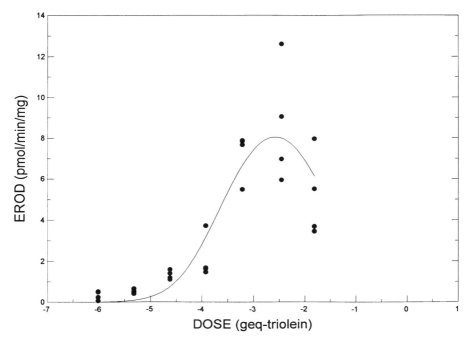

Figure 7 Ethoxyresorufin O-deethylase (EROD) induction in PLHC-1 fish hepatoma cells (*Poecilieopsis lucida*) exposed to SPMD dialysate. Ten SPMDs (1 ml triolein configuration) were deployed in bleach-kraft mill effluent (BKME) for seven days, with renewal of the BKME after days 2 and 4 of exposure. The SPMDs were separated into two five-SPMD samples, dialyzed in hexane and transferred to isooctane (150 µl). Doses were normalized to gram-equivalents of triolein.

Because of the diversity of modern bleach kraft mills (BKMs) and the inherent complexity of converting wood to bleached pulp, effluents from this industry contain complex and variable suites of organic compounds.[34] SPMDs were used to help characterize and identify bioavailable substances in a Canadian BKM effluent that are known to accumulate in fish and cause the induction of cytochrome P450 enzymes.[35,36] The dialysates of SPMDs exposed to the BKM effluent were tested for their ability to induce fish P450IA activity in a cell line (*Poeciliopsis lucida* hepatic carcinoma, PLHC-1). EROD activity was measured in the PLHC-1 cells to determine TCDD-EQs (D. Tillitt, unpublished data, Midwest Science Center). Indeed, SPMDs successfully concentrated EROD-inducing substances from the BKM effluent (Figure 7). The responses of the PLHC-1 cells indicates that the inducers are PAH-like compounds, due to the relatively small degrees of EROD induction (maximum approximately 8 pmol/min/mg) when compared to TCDD and other dioxin-like compounds, which have a maximum induction of 50 to 150 pmol EROD/min/mg. The other information that can be obtained from the dose–response curve shown in Figure 7 is the relative toxic potency of the material concentrated in the SPMD. The mass of SPMD lipid that contained enough inducing substances to cause half-maximal induction of EROD was

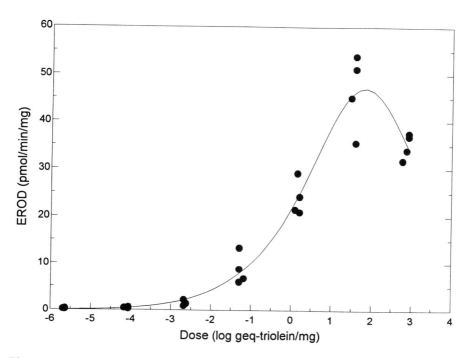

Figure 8 Ethoxyresorufin O-deethylase (EROD) induction in H4IIE rat hepatoma cells exposed to a SPMD dialysate. SPMDs (four each, 1-ml triolein configuration) were deployed for a 28-d period in Bayou Meto, Arkansas. Doses of dialysate were normalized to gram-equivalents of triolein per mg of cellular protein.

approximately 10 μg (equivalents). Comparison to the standard, TCDD, suggests a potency of approximately 2000 pg TCDD-EQ/g-triolein. Therefore, the bioavailable compounds in the SPMD concentrate from the BKM effluent were either very potent or by inference from SPMD sampling characteristics, present in the water in large amounts.

Bayou Meto, Arkansas is an area of known contamination from polychlorinated dibenzo-*p*-dioxins (PCDDs) and polychlorinated dibenzofurans (PCDFs). However, these compounds (TCDD in particular) are present in water (dissolved) only at ultra-trace levels.[37] Thus, SPMDs were used to determine the aqueous bioavailability of these chemicals at Bayou Meto. The dialysates of the SPMDs were subjected to *in vitro* bioassay (EROD) with H4IIE rat hepatoma cells to quantify the overall dioxin-like activity (Figure 8). The EROD activity in the cells produced by the SPMD dialysate had the same maximal response as TCDD (≈50 pmol/min/mg). Using the H4IIE bioassay, the estimated TCDD-EQ[17] in the SPMDs compared well with the analytical (HRGC/MS) results of the same samples.[38]

This work shows that screening sets of samples using the combination of SPMDs with the semi-quantitative H4IIE bioassay affords an expedient alternative to costly HRGC/MS analyses of all of the samples. The SPMDs concentrate bioavailable residues at a test site, while the H4IIE bioassay provides a semiquantitative estimate of the toxicity of the sequestered chemicals (P450 inducers) at that site, indicating whether HRGC/MS analysis is desirable.

Summary

In recent years, interest in SPMDs has grown and they have been the subject of three international workshops.[39-41] In this work SPMDs were shown to be an important new tool for aquatic toxicologists and other investigators. We demonstrated that the devices sequestered only dissolved nonionic compounds and that time-averaged aqueous concentrations of chemicals can be estimated from their SPMD concentrations. This SPMD-derived data provides an improved picture of the dose of waterborne chemicals experienced by aquatic organisms, and should be directly applicable to water quality criteria and aquatic toxicity data, which are based on dissolved residue concentrations. The similarities and differences in the uptake of contaminants by SPMDs and aquatic organisms were highlighted by experiments and theoretical discussions. Close similarity was shown to exist in the mass transfer of recalcitrant chemicals through an organism's respiratory membrane and through the LDPE SPMD membrane. Differences in the amounts and composition of residues concentrated in contiguously exposed aquatic organisms and SPMDs appear to largely stem from the ingestion of food and the active clearance of many chemicals by biota. The thermodynamic nature of SPMDs (i.e., lack of metabolism and active depuration) provides an advantage over biomonitoring organisms when monitoring the composition and quantities of labile waterborne contaminants, and should facilitate the identification of these toxicants. However, aquatic organisms such as bivalves provide complementary data, since SPMDs do not sample particulates and ionic compounds. SPMDs can be manufactured virtually free of contaminants and can be deployed in almost any environment, as they are not affected by water quality or other factors limiting organism uptake or survival. In many cases, quantities of waterborne residues can be sequestered by SPMDs sufficient for bioassay. This feature permits application of *in vitro* methods such as Microtox, Mutatox, and EROD for screening the toxicities of ultra-trace contaminants in water. Finally, it is hoped that this work will provide a sound experimental and theoretical basis for the further validation of SPMDs as an important new tool for the field of aquatic toxicology.

Acknowledgments

The authors wish to thank G. Gibson, R. Clark, L. Cleveland, B. Richardson, C. Rostad, J. Parrott, W. Gala, and D. MacKay for technical assistance and advice. Also, we gratefully acknowledge the support of the National Fish and Wildlife Foundation, the American Petroleum Institute, Chevron Oil Co., AMOCO Corp., the U.S. Fish and Wildlife Service, U.S. Department of Defense, and the National Science Foundation.

References

1. Neely, W.B. 1980. *Chemicals in the Environment: Distribution, Transport, Fate and Analysis.* Marcel Dekker. New York, N.Y. 245 pp.
2. Huckins, J.N., Tubergen, M.W., Manuweera, G.K. 1990. Semipermeable membrane devices containing model lipid: A new approach to monitoring the bioavailability of lipophilic contaminants and estimating their bioconcentration potential. *Chemosphere* 20:533–552.
3. Hwang, S.T., Kammermeyer, K., eds. 1975. *Membranes in Separations.* Robert E. Krieger Publishing. Malabar, Fl. 559 pp.
4. Henderson, J.R., Tocher, D.R. 1987. The lipid composition and biochemistry of freshwater fish. *Prog. Lipid Res.* 26:281–347.
5. Smith, J.A., Witkowski, P.J., Chiou, C.T. 1988. Partition of nonionic organic compounds in aquatic systems. In *Reviews of Environmental Contamination and Toxicology,* Ed. G.W. Ware, 103:127–151. Springer-Verlag: New York/London.
6. Chiou, C.T. 1985. Partition coefficients of organic compounds in lipid-water systems and correlations with fish bioconcentration factors. *Environ. Sci. Technol.* 19:57–62.
7. Opperhuizen, A., Velde, E.W.v.d., Gobas, F.A.P.C., Liem, D.A.K., Steen, J.M.D.v.d. 1985. Relationship between bioconcentration in fish and steric factors of hydrophobic chemicals. *Chemosphere* 14:1871–1896.
8. Lieb, W.R., Stein, W.D. 1969. Biological membranes behave as non-porous polymeric sheets with respect to the diffusion of non-electrolytes. *Nature* 224:240–243.
9. Ellis, G.S., Huckins, J.N., Rostad, C.E., Schmitt, C.J., Petty, J.D., MacCarthy, P. 1995. Evaluation of lipid-containing semipermeable membrane devices (SPMDs) and gas chromatography-negative chemical ionization-mass spectrometry for monitoring organochlorine contaminants in the Upper Mississippi River. *Environ. Toxicol. Chem.*, 14:1875–1884.
10. Huckins, J.N., Manuweera, G.K., Petty, J.D., MacKay, D., Lebo, J.A.. 1993. Lipid-containing semipermeable membrane devices for monitoring organic contaminants in water. *Environ. Sci. Technol.* 27:2489–2496.
11. Huckins, J.N., Petty, J.D., Orazio, C.E., Zajicek, J.L., Gibson, V.L., Clark, R.C., Echols, K.R. 1994. *15th Annual Meeting, Society of Environmental Toxicology and Chemistry.* Denver, CO. MB01 (Abstr.)
12. Lebo, J.A., Zajicek, J.L., Huckins, J.N., Petty, J.D., Peterman, P.H. 1992. Use of semipermeable membrane devices for in situ monitoring of polycyclic aromatic hydrocarbons in aquatic environments. *Chemosphere* 25:697–718.
13. Prest, H.F., Richardson, B.J., Jacobson, L.A., Vedder, J., Martin, M. 1995. Monitoring organochlorines with semipermeable membrane devices (SPMDs) and mussels (*Mytilus edulis*) in Corio Bay, Victoria, Australia. *Mar. Pollut. Bull.* 30:543–554.
14. Comyn, J., Ed. 1985. *Polymer Permeability.* Elsevier Applied Science Publishers: New York. 383 pp.
15. Ingersoll, C.G., Nelson, M.K. 1990. Testing sediment toxicity with *Hyalella azteca* (Amphipoda) and *Chironomus riparius* (Diptera) In *Aquatic Toxicology and Risk Assessment* ASTM STP 1096. ed. W.G. Landis and W.H. van der Schalie. 13:93–109. ASTM Philadelphia, PA.
16. Keith, L.H. 1991. *Environmental Sampling and Analysis: A Practical Guide.* CRC/Lewis Publishers: Boca Raton, FL. 143 pp.

chapter thirty-five

Considerations for the experimental design of aquatic mesocosm and microcosm experiments

Peter F. Chapman and Stephen J. Maund

Introduction

The physical design of mesocosm studies, appropriate sampling techniques, and choice of endpoints have all received a great deal of attention in recent years.[1-3] The selection of appropriate sizes of system, treatment types, and sampling strategies depends entirely on the objectives of the study. Study objectives are most commonly determined from the results of laboratory toxicity studies which enable the identification of those organisms which may be at potential risk in the field. Physical design criteria are briefly summarized in Table 1, and further details can be found elsewhere.[1,4]

From a chemical regulatory perspective, a common objective of mesocosm studies is to determine the effects of particular concentrations of a chemical on an aquatic ecosystem (often a predicted environmental concentration (PEC) and some factors or multiples of this) . PECs are derived from a combination of use pattern, and fate of the chemical concerned, often involving the use of standardized models.[5] The objective of such studies is to determine whether the selected concentration will elicit an effect on ecosystems at the concentration predicted to occur in the environment. Such experiments typically involve the comparison of the treated mesocosms to an untreated control and are most commonly analyzed using analysis of variance (so-called ANOVA design experiments). An alternative objective, which tends to be more research oriented, has the aim of determining the response of the ecosystem to the chemical across a wide range of concentrations (so-called dose–response or regression experiments).

A common criticism of mesocosm studies is their lack of power to detect differences between untreated control mesocosms and chemically-treated

Table 1 Design Criteria for Short-Term (up to 1 month) and Long-Term
(1–6 months) Aquatic Mesocosm Studies to Measure Pesticide Effects

Taxon	System	Endpoint(s)
Short-term studies[a]		
Fish	1–5 m³	LC_{50}/EC_{50}/limit value/% mortality
Zooplankton	1–5 m³	Abundance for major taxa
Macroinvertebrates	1–≥25 m³	Semi-quantitative abundance for major taxa
Phytoplankton	1–5 m³	Chlorophyll *a* concentration
Long-term studies[b]		
Fish	≥25 m³	Growth/condition
Zooplankton	≥25 m³	Population/diversity/recovery
Macroinvertebrates	≥25 m³	Population/diversity/recovery
Phytoplankton	1–5 m³	Chl *a*/diversity/recovery
Periphyton	1–5 m³	Chl *a*/biomass
Macrophytes	≥25 m³	Biomass/% cover

[a] Compounds whose effects persist for <1 month.

[b] Compounds whose effects persist for >1 month.

After SETAC-Europe, 1992.[1]

mesocosms. This lack of power is due in many cases to high variability between individual mesocosms and is only to be expected of complex ecological systems. A high degree of control is possible in laboratory studies with single species, but field mesocosms with large numbers of species are much more variable owing to the inability to sample the entire mesocosm, interactions between species, and a host of other factors. However, with careful attention to principles of experimental design it is possible to perform mesocosm experiments which can detect differences between control and treatment groups at reasonable levels of sensitivity.

The purpose of this chapter is therefore to highlight those issues that need to be considered when designing mesocosm experiments in order to provide the desired levels of sensitivity. The main emphasis is on ANOVA-type designs, although brief mention will be made of regression designs; the main emphasis is also on design rather than analysis. In a short chapter it is only possible to highlight issues that need to be considered. For the more detailed information necessary to design a mesocosm study, the interested experimenter will need to consult an appropriate statistical text[6] or consult a statistician.

Figure 1 is a flow diagram that shows the main statistical tasks and decision points that need to be tackled during the course of a typical mesocosm experiment. It follows the order of the chapter and is an aid to understanding the discussion.

Procedures

Definition of study objective

Before embarking on a field mesocosm experiment, it is essential to determine the likely impact of a chemical through laboratory experiments, and

Tasks   **Purpose**

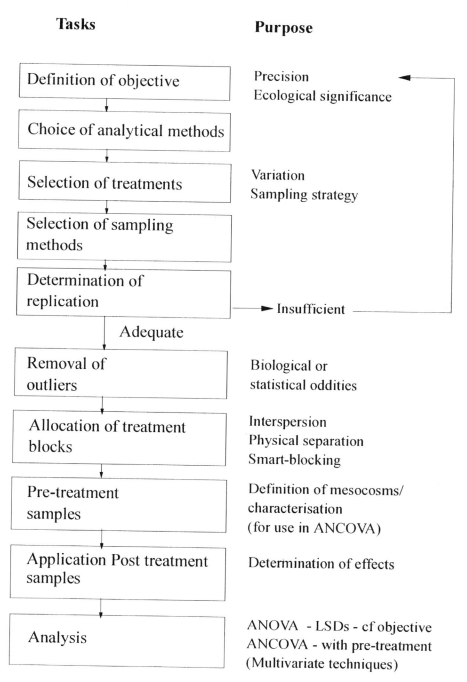

Figure 1   The main statistical tasks and decision points that need to be tackled during the course of a typical mesocosm experiment.

failure to do this may result in areas of effort being expended on endpoints that are of no practical importance. Given that it has been decided to proceed, it is essential that clear objectives are defined. Small differences in objective can lead to quite different designs and if the objective is precise a design can be selected that has a high probability of meeting its objective. The objective should include at least the following items, because without them the selection of the experimental design may be somewhat arbitrary.

- It should define the derived parameters of interest for each endpoint, such as an EC_{50} in regression or a no observed effect concentration (NOEC) in ANOVA experiments.
- The level of precision, i.e., width of confidence intervals, that are required for derived estimates. As an alternative, the desired power of a relevant hypothesis test can be stated.
- The size of effect to be detected, which should be of ecological significance. For example, to determine whether a chemical has an impact on a population with a high rate of intrinsic increase, differences of 100 to 200% may be appropriate (e.g., zooplankton). For organisms with low rates of increase, or where smaller decreases in biomass may have adverse economic impacts (for example, the production of fish biomass) differences of less that 50% may be more appropriate.

Armed with a clear, unambiguous objective an experimenter with a sound knowledge of the principles of experimental design and analysis will know what data are required from an experiment and how these should be analyzed. This in turn will enable the design of an experiment with a high probability of achieving its objectives or, alternatively, the estimation of derived parameters with the desired level of precision. Thus a clear objective leads directly to appropriate choices about the following:

- The number of treatments and, if the treatments represent increasing doses or concentrations of the test chemical, the choice of doses.
- How the treatments should be assigned to the mesocosms, for example, whether the treatments should be assigned completely at random or whether a constrained randomization should be employed, such as the arranging of treatments in replicate blocks.
- The total number of mesocosms, the number of replicate mesocosms per treatment, and whether or not the number of replicates needs to be the same for each treatment.
- The organisms to be sampled, the size of the sample, and how the sampling should be carried out within the mesocosm.

These considerations will be discussed in more detail in the remainder of the chapter.

Data analytical considerations

Although the statistical analysis comes at the end of the experiment, it is the form that the analysis takes that enables the objectives of the experiment to be met. Furthermore, the precise details of the design can only be worked out after the form of the analysis has been decided upon. Therefore, any discussion on experimental design should be preceded by a discussion on methods of analysis.

A mesocosm experiment will usually comprise an untreated control plus a number of treatments, each of which represents a specific dose or concentration of the test chemical. Often each treatment is replicated a number of times, although the degree of replication need not be the same for each treatment. An extreme form of mesocosm experiment has two replicated treatments, an untreated control, plus one treatment at a single dose or concentration. At the other extreme there will be an untreated control plus a number of treated mesocosms, each treated with a different dose and none of them replicated; this type of experiment can only be analyzed by use of regression methods which are discussed in a later section. These extreme forms of design address quite different objectives so it is not appropriate to consider either of them to be in any way superior. The replicated design with an untreated control and one treatment is the optimum design if determining the effect of a single chemical dose is the exclusive objective of the study. The unreplicated design with several different doses is capable of addressing a wide variety of objectives, such as the estimation of an EC_{50}.

The statistical analysis will usually take one of two forms, both of which are complementary, although one form is more informative than the other. For a full description of both methods see Sokal and Rohlf.[6] The first method involves calculating the mean response across replicates for each measured variable and presenting the means together with their standard errors or, preferably, their confidence intervals. If, as is likely, the purpose of the analysis is to compare each treatment with the control, the mean difference between control mean and treated mean should be calculated together with the confidence interval for the difference. The 95% confidence interval about the difference between the control mean and the mean of one of the treatments is expressed as:

$$\pm\, t_{0.05;df} s(1/T + 1/C) \tag{1}$$

where s is an estimate of the between mesocosm standard deviation, which is calculated as the square root of the residual mean square in an analysis of variance, df is the degrees of freedom for the estimate of the residual mean square, $t_{0.05;df}$ is the percentile of the distribution of t with df degrees of freedom, T is the number of replicates for the treatment, and C is the number of replicates for the untreated control. The between mesocosm sample standard deviation s is an estimate of the unknown population value σ which in turn is a function of σ_m and σ_s according to the equation:

$$\sigma^2 = \sigma_m^2 + \sigma_s^2 / N \tag{2}$$

where N is the size of the within mesocosm sample, σ_s^2 is the variance between sample values within a mesocosm, and σ_m^2 is the variance between mesocosms.

The second method of analysis is to carry out a hypotheses test under a null hypothesis that the mean of the treated mesocosms is equal to the mean of the untreated control. If a statistically significant result is obtained (confidence interval does not include zero) then it can be concluded that the two means are not equal. If the result is non-significant (confidence interval includes zero) then it cannot be concluded that the means are equal, but nor can it be concluded that the means are not equal.

Whichever approach to the analysis is used, estimation or hypothesis testing, analysis of covariance can be used to improve the precision of derived parameters provided that there is a relationship between endpoints sampled and the pre-treatment data. Effective use of covariance analysis calls for careful selection of appropriate pre-treatment variables, which are best provided by understanding the ecology of the system being studied. For a comprehensive description of the analysis of covariance see Cox.[7]

Endpoints to be sampled

The objective of the study will indicate which endpoints are to be sampled and will be one of two types.

- Some endpoints will be counts for which, theoretically, there is no upper limit. An example is the number of insects of a particular species found in emergence traps, and we can assume that such data are from a Poisson, negative binomial, or related distribution. Such data can be analyzed after a suitable transformation such as the log-arithmic transformation,[8] or by fitting an appropriate model with a non-normal error structure.[9,10] It needs to be emphasized that the purpose of data transformation is to stabilize the error variance, not to make the data appear as if it derives from a normal distribution. This is because both ANOVA and regression methods are relatively robust to deviations from normality, but homogeneity of variance is essential for an analysis to be valid.
- Some variables such as fish length or weight will be from a continuous distribution which is normal or can be easily transformed to normal-ity. Some physico-chemical variables will take the form of percentages but the range of values will be sufficiently small that we can assume they derive from a normal distribution.

It is unlikely that any of the variables measured will take the form of percentage mortality because this requires the experimenter to know both the total number of live organisms present and the number that died, and this is virtually impossible in a mesocosm experiment. An exception to this would be the use of *in situ* bioassays, such as caged fish, but these would not represent true ecosystem-level measurements.

In a complex ecosystem it is usually possible to measure a large number of different variables and it is a common criticism of mesocosm studies that too many endpoints are sampled. Sampling mesocosms is expensive and time consuming, and sampling large numbers of organisms often requires the resource that is available to be spread thinly, possibly resulting in data of poor quality. Furthermore, when large numbers of interrelated variables are measured, the task of summarizing and interpreting the results of the experiment becomes daunting. It is preferable therefore to measure a small number of key variables that enable the objectives of the experiment to be addressed and to ensure that the measurements are done to as high a standard as is possible.

Design issues

Precision

Experimental design enables the objectives of an experiment to be met in two senses:

- Objectives determine what derived parameters need to be estimated from an experiment and whether or not a particular parameter can be estimated depends entirely on choice of treatments. For example, if it is necessary to estimate the response at a particular concentration then mesocosms at that concentration must be included in the experiment. Alternatively, if it is necessary to estimate some feature of a dose–response curve, such as an EC_{20} then the chosen concentrations must be suitable to enable a curve to be fitted.
- In order to be useful, objectives must also specify the desired level of precision for estimates or the power required for hypothesis tests.

From Equations 1 and 2 it can be seen that the confidence interval for a treatment mean, or for the difference between two treatments, is affected by a number of features of the experimental material and the design imposed upon it:

- The variance between observations measured within a mesocosm, denoted by σ_s^2, e.g., the variance between counts taken from different emergence traps. Precision improves as this variance gets smaller. Different methods of sampling will affect this variance but, once chosen, there is little further scope for controlling it in an experiment.
- The variance between mesocosms, denoted by σ_m^2. Precision improves as this variance gets smaller and there is quite a lot of scope for controlling it. Use of pre-treatment information to reject outlier mesocosms and to impose a constrained randomization can both lead to a significant reduction in residual variance.
- The number of observations per mesocosm, denoted by N in Equation 2. If the within-mesocosm variation is large relative to the between mesocosm variation, then precision can be improved by increasing the size of the within-mesocosm sample. However, if the within-mesocosm variation is relatively small, there may be no benefit at all from increasing sample sizes. Either way this is completely under the

control of the experimenter and is a balance between additional costs incurred and benefits derived.

- The number of replicates per treatment, denoted by T and C in Equation (1). Precision improves as replication increases and this is again entirely under the control of the experimenter. A law of diminishing returns applies to replication in the sense that additional replication has large effects on precision if replication is low, but only small effects if replication is large. Thus, while in theory the standard error of a treatment mean can be reduced to very near zero, in practice economical and logistical considerations impose an upper limit on numbers of replicates.
- The number of degrees of freedom for estimating the residual mean square in the analysis of variance, denoted by df in Equation (2). Precision improves as this increases.

Thus, if the analysis and presentation of results is to take the form of an analysis of variance followed by the tabulation of treatment means and mean differences with associated confidence limits then, provided the experimenter has a good indication of the levels of variability likely to be experienced in an experiment, a good estimate can be made as to the level of replication required. For mesocosms there is a wealth of information in existence about the variability measured in completed experiments,[11,12] although this information is sometimes not freely available in collated forms.

Power

If hypotheses tests are to be performed to compare treated mesocosms with the untreated control, then an experimenter needs an appreciation of the concept of power in order to be able to design an experiment that is sufficiently powerful for his needs. And before power can be discussed, two types of errors that can arise out of a hypothesis test need to be defined:

Type I error rate: This is the probability of the test giving a significant result, when in reality there is no difference between treatments. This error rate is set by the person performing the test and by tradition is set at 5%. It is because the error rate is usually so low that obtaining a significant result does give one confidence that the treatments really are different.

Type II error rate: This is the probability of not obtaining a significant result when there is a real difference. It is not set by the person performing the test but is affected by various features of an experiment.

Power is equal to one minus the type II error rate and so is the probability of obtaining a significant result when there is a real difference. So it is the probability of concluding that there is a difference when there really is one. Increasing power is analogous to increasing the precision of an estimate and the power of a test is influenced by the same five features of an experiment or design that are outlined above. In addition, however, power is also influenced by the following:

- The true, but unknown, difference between the treatments. Thus, if the difference between the true means of two treatments is very small then the probability of a test giving a significant result is small and as the true difference increases the probability gets larger. This relationship is very important and implies that when we think about the power of a test we have to specify a difference that we wish to detect. Thus if we specify that for various reasons we wish to detect a difference of 20% between the control and a treatment, we can then speak meaningfully of the power of the test to detect this difference. If we are not prepared to specify a difference then the concept of power is of little value in helping to design an experiment.
- The choice of multiple comparison test. There are many different tests that can be used for comparing treatment means in the analysis of variance[13] with the *t*-test, Williams' test, and Dunnet's test being the most popular for use in ecotoxicology. Each test has well-defined and understood properties and they are not equally powerful.

Figure 2 shows a graph of a typical power curve. In this graph, the x-axis represents the true difference between the treatment means and the y-axis gives the probability of obtaining a significant result. In an ideal world, experiments would be designed to give high levels of power, so for example one could argue that all experiments should be designed so as to give powers of 95%, but in practice this is often impossible because the replication required is too large and leads to prohibitively expensive experiments. Often, a figure of 80% is quoted as a reasonable level of power to attempt. Familiarity with power curves is useful because it gives an insight into what is possible to achieve from a proposed experiment.

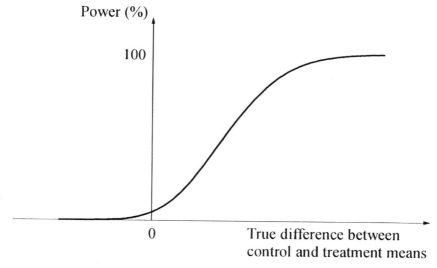

Figure 2 Power, the probability of obtaining a statistically significant result from a test, increases as the true difference between the treatments increases.

Choosing optimum numbers of replicates

While power curves give insight into the capabilities of experiments, it is easier to use a relatively straightforward formula, such as discussed in Sokal and Rohlf,[6] for determining the required number of true replicates:

$$T \geq 2(\sigma / \delta)^2 [t_\alpha + t_{2(1-P)}]^2 \tag{3}$$

where T is the number of replicates for each treatment, δ is the difference it is desired to detect, σ is the between mesocosm standard deviation (see Eq. 2) which is estimated by the sample value s, t_α is taken from tables of the t-distribution and can be either one or two tailed, α is the significance level of the test, $t_{2(1-P)}$ is also a two-tailed value from t-tables with a significance level of $2(1 - p)$, and finally P is the desired power of the test. Eq. (3) assumes that the number of replicates is identical for all treatments including the untreated control. In practice, if the untreated control is to be compared with several different treatments it will be more efficient to have a larger number of replicates for the control.

If the size of the standard deviation between mesocosms is known approximately, and this is often the case on the basis of previous experiments, the equations presented here, or similar equations, can be used to determine the replication required to give a desired level of power in an experiment. If it appears that the experiment is not capable of achieving its objective with available resources, then it may be necessary to consider either amending the objective or using alternative test systems. For example, in aquatic eco-systems, it has often been suggested that reductions in system size can reduce variability,[14] or alternatively, it may be possible to improve the sampling method to reduce measurement errors. However, the implications of this for the interpretation of the experiment must also be considered, e.g., whether reductions in scale will impact the relevance of the data. More realistic field studies are large and tend to be of high ecological relevance, but are usually more inherently variable; smaller systems tend to be less variable but their applicability to real ecosystems must be considered. The influence of these two factors needs to be considered in relation to the overall objective of the study.

If the size of the variability between mesocosms that is likely to be observed in an experiment is a complete unknown, steps should be taken to obtain information from external sources. If this fails then the mesocosms should be sampled prior to the pre-treatment sampling in order to provide relevant data so that the variability can be measured directly.

Assigning of treatments to experimental units

Differences between mesocosms at the beginning of an experiment, (such as community structure), or non-treatment related differences between them during the course of an experiment (such as operator effects, or environmental gradients such as exposure or shading), can have a large influence on the endpoints measured during the experiment. Such influences can lead to biased results if one treatment is affected to a greater extent than another. It is therefore essential that treatments are applied to mesocosms at random and, if there is a possibility that failure to do so will lead to biased results, it is advisable that all management and sampling activities are carried out in a random order.

Even if an experiment is properly randomized it is still possible for the randomization to lead to a biased experiment, especially if there is a small number of mesocosms or there is a large environmental gradient. The experimenter should therefore study the randomization produced and only implement it if he is satisfied that it will not lead to biased results. If the randomization is suspect it should be discarded and another generated.[15]

Mesocosm guidelines tend to recommend taking pre-treatment samples but do not give explicit guidance on how this information should be used. It is known that a post-treatment variable can be affected by the pre-treatment status of mesocosm; for example, unpublished research by the present authors has shown that the total numbers and weights of juvenile fish at the end of an experiment can be influenced by the amount of phyto-plankton present at the outset. In these circumstances, a constrained randomization should be considered and this can be achieved by grouping mesocosms into replicate blocks in such a way that within a block all mesocosms are similar with respect to the pre-treatment variable. Blocking can also be used to mitigate the effect of management practices, such as sampling on different days. The advantage of blocking is that some of the variation between mesocosms can be assigned to a block effect in the analysis of variance (or covariance) resulting in a smaller residual mean square and leading to more precise estimates or more powerful tests.

If possible, the number of mesocosms available should be more than is actually needed. Any individual mesocosms that appears to be an outlier[16] at the pre-treatment sampling stage can then be eliminated from the experiment completely. For example, any mesocosm that lies outside an interval of size three standard deviations either side of the mean could be excluded.

Results and discussion

A hypothetical example

It has been decided to carry out a study, the objective of which is to determine the effects of two treatment concentrations, the PEC and a multiple of three times this value, on a range of organisms, both phytoplankton and zooplankton. Special attention is to be given to two species of zooplankton, Cladocera and Copepoda. It is considered to be important to be able to detect a twofold difference between either of the chemical treatments and the untreated control using a test with a 5% significance level and 80% power.

The selected experimental set up comprises 18 glass fiber mesocosms, 1.25 m in diameter and depth, positioned within a concrete pond. The bottom of each mesocosm is then covered with 10 cm of thoroughly mixed sediment from a well-established pond and filled to a depth of 0.7 m with pond water from an established source. Populations of a wide variety of pond organisms are established through transfer in sediment, water, and other means and

Daphnia from laboratory cultures are also added. The mesocosms are then allowed to equilibrate for two months.

The objective calls for ANOVA as the method of analysis in order to estimate a common error variance. This will allow confidence intervals to be estimated for both treatment means and the mean difference between each treatment and the untreated control. Finally, a multiple comparison method, in particular a *t*-test, will be used to test the hypothesis that the treatment means are not different from the control. This set of requirements calls for three treatments, an untreated control, a chemical treatment equal to the PEC and finally a chemical treatment equal to three times the PEC.

A depth-integrating water column sample is used for sampling both phytoplankton and zooplankton, and ranked pre-treatment counts of Cladocera are shown in Table 2. Three of the mesocosms have very low initial counts and it is hoped to exclude them from the experiment. Therefore, calculations are performed on the remaining 15 counts. The coefficient of variation of the remaining 15 counts is 55% and this value is inserted as σ in Eq. (3). The remaining inputs to Eq. (3) are as follows: the size of effect to be detected (δ) is set at 100%; t_α is taken from a one-sided table of *t*-values with 5% significance level and 14 degrees of freedom for error so is set at 1.76; $t_{2(1-P)}$ is set at 0.868, the two-sided *t*-value with a significance level of 40%. The calculation suggests that the minimum number of replicates should be greater than 4.18 which is rounded up to 5.

Table 2 Allocation of Mesocosm to Replicate Block According to
Pre-Treatment Cladocera Abundance

Mesocosm	Cladocera count	Block allocation
4	25	Excluded from experiment
15	25	Excluded from experiment
18	30	Excluded from experiment
1	70	1
7	70	1
9	75	1
5	90	2
6	105	2
14	110	2
2	120	3
11	125	3
17	140	3
8	190	4
10	200	4
13	220	4
3	290	5
12	315	5
16	330	5
Mean count for included mesocosms		163.3
Standard deviation for included mesocosms		89.8
Coefficient of variation for included mesocosms		55%

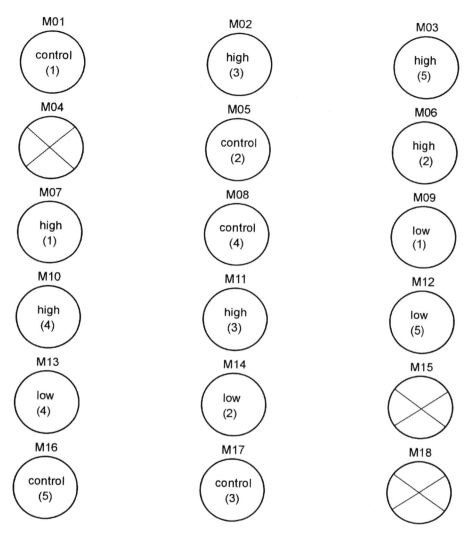

Figure 3 Hypothetical example: the arrangement of treatments (control, low, high), and blocks (numbers in brackets) to mesocosms (denoted by M numbers).

The power calculations therefore confirm that five replicates should be adequate in order to be able to detect a 2-fold difference with a 5% one-sided test. The three mesocosms with low counts can therefore be eliminated from the experiment. The remaining 15 ponds are then allocated at random to 5 replicate blocks as shown in Table 2 and Figure 3.

If there is a strong relationship between pre- and post-treatment Cladocera counts then the assigning of mesocosms into blocks should lead to a reduction in residual error variance in the analysis of variance and hence even greater power than is required. However, this cannot be guaranteed, so the conservative approach is to base the number of replicates on an assumption that the blocking may not improve the experiment.

A further refinement of the power calculation could be achieved by performing the calculation on the logarithm scale. This has not been done here but is probably a wise course to take in view of the fact that the variation between mesocosms within blocks increases as the mean count per block increases.

This is a very simple example of how the number of replicates can be chosen, and how mesocosms can be allocated to replicate blocks, by studying a single pre-treatment count. In practice there will be a number of different pre-treatment measurements made on different types of organisms, and power calculations based on different organisms will give different results. Therefore the practical application of the ideas presented here call for a greater amount of judgment and compromise than has been presented in this example. In particular, the use of pre-treatment information in allocating of mesocosms to blocks should always be based on sound ecological principles.

Additional consideration for dose–response experiments

Analysis of dose–response experiments is usually achieved by fitting regression models which can be considered under two headings.

- Linear models: These are a very wide class of regression models including straight line, polynomial and multiple regression models.[9,17] Furthermore, the error distribution can be any one of the exponential family of distributions which include the normal or Gaussian distribution.[9] So, for example, a logistic model with a binomial error structure is a linear model. Linear models with a normal error distribution have a common feature in that analytical solutions can be found for their parameters or, in other words, the parameter estimates can be expressed as functions of the data. In the ecotoxicological literature the phrase linear model or linear regression is often mistakenly used to describe the much more restricted case of straight line regression.[18] Use of linear models is an empirical approach to data analysis in that they cannot be used to model underlying scientific processes.
- Non-linear models: These are an equally wide class of models[19-21] that can take a large variety of shapes and, again, there is a wide choice of error distributions. Non-linear models can be derived from first principles, by solving differential equations, so they can be used to model underlying processes. All parameter estimates are obtained by an iterative process, including those for models with normal errors, and the existence of local optima can make it difficult to find solutions.

Regression models would appear to greatly expand the range of objectives that can be addressed in mesocosm experiments. So, for example, threshold concentrations that have no effect (NECs), EC_{50}s, and time to recovery can all be estimated with varying degrees of difficulty. However, there may be practical difficulties in using regression methods for analysis of mesocosm data owing to the interaction of species. A toxicant effect may manifest itself as a suppression of the population of one species and a complementary enhancement of the population of another. The resulting dose responses for the individual species may appear to be very strange or non-existent.[22] Not only will it be difficult to choose an appropriate model in these circumstances, but it will also be difficult to address a worthwhile objective. One way around these difficulties is to arbitrarily exclude data from the analysis, a practice usually frowned on by statisticians. An example of how this was achieved with some success is given in Liber et al.[18]

Most of the advice given on mesocosm design earlier in this chapter also applies to regression experiments, but regression does raise some additional design issues. For useful advice on design for regression experiments see Shaw et al.[23] The precision of parameter estimates is determined by many things already discussed, but also includes choice of model, number of doses, position of doses, and number of replicates at each dose. Furthermore, the ideal design for one derived parameter is not necessarily ideal for a second parameter. Often regression designs are proposed that have no replication[18] but this is not a necessary restriction. In fact, lack of replication requires much judgment in choice of model because it is impossible to perform a significance test to determine whether or not the chosen model is appropriate.

In summary, regression methods would appear to offer distinct advantages but there are many difficulties that need to be resolved and scope for a great deal of further research.

Future developments

There is an inherent problem in the measurement of ecological endpoints in a univariate approach because the response of an experimental unit, such as a mesocosm, to a treatment is essentially of a multi-variate nature. Methods for dealing with inherent variability of ecological systems are under development,[24-26] giving an indication of effects on the overall community, rather than one position of it. The advantages of these kinds of analysis are that they allow data on the whole community to be summarized into one or a small number of statistics.

Concluding remarks

Mesocosm studies are used in the assessment of chemicals to determine effects at population and community levels of organization. This can help reduce the uncertainty of risk assessments that are based on laboratory experiments alone, through more realistic exposure to the chemical and the inclusion of ecological processes. Although this increases the realism of the experiment, it also leads to difficulties in the interpretation of the studies because of the large number of interacting factors that control ecosystem processes.

This chapter has attempted to explain how many of these problems can be resolved through incorporating effective experimental design procedures into the study. In order to select an appropriate experimental design a number of factors should be carefully considered. These include the objectives of the study, selection of treatments, selection and allocation of experimental materials and units, sampling strategies, and the statistical analysis used.

Currently used routine methods of analyzing mesocosm data are mostly univariate and are hence inefficient because they only analyze single populations. In order to fully understand the effect of chemical treatments in mesocosm experiments, it is necessary to analyze measures of community structure. Whilst various researchers are looking at ways of achieving this by use of multivariate statistical measures, there is some way to go before these methods can be adopted routinely.

References

1. European Workshop on Freshwater Field Tests. 1992. Summary and recommendations of the European workshop on freshwater field tests (EWOFFT). Potsdam, Germany, June 25–26, 1992.
2. Society of Environmental Toxicology and Chemistry, Europe. 1992. Guidance document on testing procedures for pesticides in freshwater mesocosms. Report of the workshop "A meeting of experts on guidelines for static field mesocosm tests," Monks Wood Experimental Station, Huntingdon, U.K., July 1991. Society of Environmental Toxicology and Chemistry. 46 pp.
3. Society of Environmental Toxicology and Chemistry. 1991. Report of the workshop on aquatic microcosms for ecological assessment of pesticides. The SETAC Foundation for Environmental Education and RESOLVE (a program of the World Wildlife Fund). Wintergreen, VA, October 6–11, 1992.
4. AMEAC. 1993. Draft OECD guideline for testing of chemicals proposal for: freshwater lentic field testing xenbiotic chemicals. Aquatic Model Ecosystem Advisory Committee, Society of Environmental Toxicology and Chemistry — Foundation for Environmental Education, Pensacola, Florida, 27 pp.
5. European and Mediterranean Plant Protection Organisation (EPPO) and Council of Europe (1993). Decision-making scheme for the environmental risk assessment of plant protection products. *Bulletin OEPP/EPPO*, 23, Blackwell Scientific Publications.
6. Sokal, R.R., Rohlf, F.J. 1981. *Biometry*, 2nd edition. San Francisco: Freeman. 859 pp.
7. Cox, D.R. 1958. *Planning of Experiments*. New York: John Wiley & Sons. 308 pp.
8. Fisher, L.D., Belle, G.V. 1993. Biostatistics: *A Methodology for the Health Sciences*. New York: John Wiley & Sons. 956 pp.
9. McCullagh, P., Nelder, J.A. 1989. *Generalised Linear Models*. London: Chapman and Hall. 511 pp.
10. Dobson, A.J. 1983. *An Introduction to Statistical Modelling*. London: Chapman and Hall. 125 pp.
11. Graney, R.L., Kennedy, J.H., Rodgers, J.H. Jr., Eds. *Aquatic Mesocosm Studies in Ecological Risk Assessment*. 1994. Boca Raton: CRC/Lewis. 723 pp.
12. Hill, I.R., Heimbach, F., Leeuwangh, P., Matthiessen, P. Eds. 1994. *Freshwater Field Tests for Hazard Assessment of Chemicals*. 1994. Boca Raton: CRC/Lewis.
13. Day, R.W., Quinn, G.P. 1989. Comparisons of treatments after an analysis of variance in ecology. *Ecol. Monogr.* 59:433–463.
14. Heimbach, F. 1994. Methodologies of aquatic field tests: System design for field tests in still waters. In *Freshwater Field Tests for Hazard Assessment of Chemicals*, Eds. Hill, I.R., Heimbach, F., Leeuwangh, P., Matthiessen, P., 141–150. Boca Raton: CRC/Lewis.
15. Hurlbert, S. 1984. Pseudoreplication and the design of ecological field experiments. *Ecol. Momogr.* 54:187–211.

16. Barnett, V., Lewis, T. 1979. *Outliers in Statistical Data.* New York: John Wiley & Sons. 365 pp.
17. Draper, N.R., Smith, H. 1981. *Applied Regression Analysis.* New York: John Wiley & Sons. 709 pp.
18. Liber, K., Kaushik, N.K., Solomon, K.R., Carey, J.H. 1991. Experimental designs for aquatic mesocosm studies: a comparison of the "ANOVA" and regression design for assessing the impact of tetrachlorophenol on zooplankton populations in limnocorrals. *Environ. Toxicol. Chem.* 11:61–77.
19. Bates, D.M., Watts, D.G. 1988. *Nonlinear Regression Analysis and Its Applications.* New York: John Wiley & Sons. 365 pp.
20. Seber, G.A.F., Wild, C.J. 1989. *Nonlinear Regression.* New York: John Wiley & Sons. 768 pp.
21. Ross, G.J.S. *Nonlinear Estimation.* New York: Springer-Verlag.
22. Crane, M., Whitehouse, P. 1994. Biological effects of the pesticide SAN 527 I 240 EW in freshwater ponds. Proc. Brighton Crop Prot. Conf., Brighton, 1994, 1313–1318. Farnham, U.K.: The British Crop Protection Council.
23. Shaw, J.L., Moore, M., Kennedy, J.H., Hill, I.R. 1994. Design and statistical analysis of field aquatic mesocosm studies. In *Aquatic Mesocosm Studies in Ecological Risk Assessment,* Eds. Graney, R.L., Kennedy, J.H., Rodger, J.H., Jr. 85–103. Boca Raton: CRC/Lewis.
24. ter Braak, C.J.F. 1994. Canonical community ordination. Part 1: basic theory and linear models. *Ecoscience* 1:127–140.
25. Matthews, G.B., Matthews, R.A., Hachmoller, B. 1991. Mathematical analysis of temporal and spatial trends in the benthic macroinvertebrate communities of a small stream. *Can. J. Aquat. Sci.* 48:2184–2190.
26. Clarke, K.R. 1993. Non-parametric multivariate analyses of changes in community structure. *Austr. J. Ecol.* 18:117–143.

Index